POSSUMS and OPOSSUMS:
Studies in Evolution

POSSUMS and OPOSSUMS: Studies in Evolution

Volume 1

Edited by

Michael Archer

Published by

Surrey Beatty & Sons Pty Limited

In association with

The Royal Zoological Society of New South Wales

ISBN—0 949324 05 1
National Library Canberra

Published November, 1987

PRINTED AND PUBLISHED IN AUSTRALIA BY
SURREY BEATTY & SONS PTY LIMITED
43 Rickard Road, Chipping Norton, NSW 2170

The squirrel-sized Santacrucian (early Miocene) microbiotheriid *Microbiotherium tehuelchum* from Patagonia. Reconstructions by Peter Murray.

vi

LIST OF COLOUR PLATES

CONTENTS

VOLUME 1

INTRODUCTION

THE ORIGIN of "Possums and opossums: studies in evolution" is almost as obscured by time as the subject matter. It began in late 1984 as a symposium of the Royal Zoological Society of New South Wales in Sydney. At the time it was conceived that the subject matter would confine itself to Australian "possums" (a non-phylogenetic *pot pourri* generally interpreted as the more cuddly arboreal Australian marsupials) and American "opossums" (another muddly collective term for the 'didelphid-like' marsupials) with a focus on questions of intercontinential relationships. However, even during the Symposium it became evident that the scientists involved wanted the subject matter to extend to all of the Australian diprotodont marsupials (a larger but phylogenetically meaningful group) and to have at least a major focus on the fossil record of Australian diprotodont as well as early Cainozoic American marsupials.

In the months that followed, many papers covering a vast range of topics were submitted for the anticipated publication and it was decided that the best guide to eligibility (prior to refereeing) would be value in developing phylogenetic understanding. The intent was to compile a volume to compliment the phylogenetic studies previously presented in "Carnivorous marsupials" (Archer 1982) and the contributions on mainly modern 'possums' in "Possums and gliders" (Smith and Hume 1984). The only major group of marsupials then left unreviewed would be the bandicoots, an omission that would be redressed in the next few years.

To anticipate, the summary of this new data and its significance to phylogenetic understanding is presented in the chapter by Aplin and Archer (1987; this volume). The 'summary of this summary' is the syncretic marsupial classification presented in that chapter.

This focus on marsupial phylogenetics is of course only a beginning. Considering the now prodigious rate at which new ideas and data are being produced, the phylogenetic concepts presented in these volumes will no doubt become rapidly outdated — which is as it should be if systematics is to remain a healthy science. However, the primary morphological and palaeontological data presented here should serve as a more enduring brick in future reconstructions of phylogenetic understanding.

THE BALANCE OF SUBJECT MATTER

Although the chapters in these volumes range from immunology to parasitology, ultrastructure, embryology, gross anatomy, biochemistry, physiology and palaeontology, the majority involve biomolecular and palaeontological subjects. This reflects the fact that these two areas of relevance to marsupial phylogenetics are perhaps the fastest growing.

The relatively rapid growth in understanding of the immunological relationships of marsupials (which effectively begin with Kirsch's 1968 publication), combined with the complete lack of knowledge about Australia's early Tertiary fossil marsupial record, has meant that molecular systematics sometimes provides indications of interfamilial relationships otherwise unanticipated or ambiguously indicated by other studies. For example, the conclusion of Baverstock *et al.* (1984) that *Acrobates* (with *Distoechurus*) was not closely related to the other members of the family within which it was then classified (the Burramyidae), was the first clear rationalisation for a conclusion that we now all accept and for which there is now supporting evidence from many other independent studies (Aplin and Archer 1987; this Volume).

That is not to say, however, that the fossil record has failed to provide key information about marsupial phylogenetics. On the contrary, as many of the contributions in these volumes demonstrate, our sole knowledge about much of marsupial diversity comes from the American and Australian fossil records.

Much that was of significance in the Australian fossil record was collected from the Miocene deposits of central Australia over the last two decades but work commitments and other obstacles, as they often do, delayed immediate publication of these materials. Preliminary notifications of some of these were given in Stirton *et al.* (1968), Archer and Bartholomai (1978), Tedford *et al.* (1977), Archer (1984) and Woodburne *et al.* (1986). The decision to expand the phylogenetic sections of these volumes provided opportunity and impetus to formally describe many of these otherwise little-known but highly distinctive parts of the Australian marsupial radiation.

The result is not only description here of many new species and genera of Australian Cainozoic mammals, but of three entirely new families: the Ilariidae (Tedford and Woodburne 1987); the Miralinidae (Woodburne, Pledge and Archer 1987); and the Pilkipildridae (Archer, Tedford and Rich 1987).

The formal birthings in these volumes of the previously collected central Australian fossil marsupials also heralds a change in things to come. With the recent discoveries (since 1983) of new and rich Tertiary sequences in northwestern Queensland, it is apparent that this region will provide considerably more information about the diversity and evolution of Australia's Tertiary mammals than has so far been provided by all of the central Australian deposits.

These volumes have also provided an opportunity for formal publication by Marshall (1987) of many of the diverse and often bizarre marsupials previously collected from Brazil as part of the late Palaeocene Itaborian Local Fauna. Similarly, Reig et al. (1987) have reappraised current understanding about diversity and phylogenetic relationships of Neogene didelphimorphian marsupials.

These papers compliment the many systematic revisions of non-Australian marsupials by Marshall, Reig, Pascual, Crochet, Simpson, Sige and others (see Aplin and Archer 1987, this Volume).

But again, even here, while the chapters by Marshall (1987) and Reig et al. (1987) describe some of the most fascinating non-Australian marsupials so far discovered, more recent discoveries in South America of Cretaceous marsupials promise a great deal more of importance to marsupial phylogenetic studies.

One paper presented here, that by Rich et al. (1987), does *not* have its focus on marsupial phylogenetics. In fact, it deals with what is evidently a fossil rhinoceros from New Caledonia. It is included here because, as things stand in the literature, this fossil has been described as a new species of the diprotodontid genus *Zygomaturus*.

STANDARDISATION OF NOMENCLATURE AND STYLE

Inevitably with a compilation as large and diverse as this, involving authors from opposite ends of the earth, there will be conflicting terminologies. Further, with more than 60 contributions originally submitted for publication, it was simply not practical to exchange copies of all final contributions to all contributors in order that they appropriately revise their manuscripts. On the other hand, neither was it realistic to insist as an editor that all authors use one system of nomenclature when use of this system might well

conflict with a personal conviction that it was less appropriate than some other. In any case, in terms of marsupial taxonomy, this would have been impossible because the summary classification (Aplin and Archer 1987) deriving in large part from these papers did not in fact come together until late 1986. Consequently, the systematic nomenclature used in each contribution (and in particular in Marshall 1987; Reig et al. 1987; Harding 1987) should be interpreted within the context of that contribution alone. However, the taxonomy used in at least the following contributions does follow that of Aplin and Archer (1987): Archer, Tedford and Rich (1987); Tedford and Woodburne (1987); Woodburne, Tedford and Archer (1987); Woodburne, Tedford, Archer and Pledge (1987); Woodburne, Pledge and Archer (1987); Aplin (1987) and Hughes, Hall, Aplin and Archer (1987).

Taxonomic terms often used in these volumes in ways that conflict with usage in Aplin and Archer (1987) are: 'Metatheria' (used in effect as a synonym of Marsupialia); 'Eutheria' (used as a synonym of Placentalia); 'Vombatoidea' or 'Vombatimorphia' (used as a synonym of Vombatiformes); 'Phalangeroidea' (used as a synonym of Phalangerida). All we can do here is to encourage future authors to consider the recommendations and rationalisations provided in Aplin and Archer (1987) before perpetuating older and/or less justifiable junior synonyms.

In the interests of making all of the contributions more comprehensible to non-morphologists, authors were encouraged to use some standard dental nomenclature. In particular, most authors have used the P1-3 and M1-5 marsupial cheektooth nomenclature established on the basis of modern embryological and morphological studies (Archer 1978).

Duality in dental nomenclature occurs in regard to the name for the outer (cheek side) surface of the teeth. Most of the contributions herein use the term 'buccal' but some use 'labial'. They are one and the same although, in terms of etymology and structure, labial is more appropriate in humans than marsupials.

Of more importance is the fact that while most authors have used the traditional term 'hypocone' for the upper molar cusp in marsupials that occurs posterior to the protocone, Tedford and Woodburne (1987) have suggested that because this cusp appears to be homologous with the metaconule of tribosphenic marsupials, it should retain this name in quadritubercular marsupials. For this reason, they further recommend that the cusp commonly called the 'metaconule' in quadritubercular marsupial molars should be called the neometaconule. This new terminology has been used in all contributions in these volumes by Woodburne and in

Aplin and Archer (1987). Use of the older term hypocone in other contributions may only reflect the fact that the other authors have not yet had a chance to consider the alternative.

Finally, in all contributions except Reig *et al.* (1987) and Marshall (1987), a distinction is made between author/date statements for taxa and publications. Taxonomic names if accompanied by indications of original authorship have the author and date separated by a comma (e.g. Ameridelphia Szalay, 1982). In contrast, citations to publications do not have a comma between the author and the date (e.g., Szalay 1982).

ORGANISATION WITHIN THE VOLUMES

There are four sections in these volumes. The first contains the contribution by Aplin and Archer (1987) which represents as best we can provide a summary of the new data and an integration into a new synthetic interpretation of marsupial phylogenetics.

The second section (contributions 2-13) contains overviews of marsupial phylogenetics as a whole as well as contributions dealing specifically with non-Australian marsupials.

The third section (contributions 14-46) includes papers that focus specifically on living as well as extinct Australian diprotodont marsupials. These contributions were solicited in such a way that they could be grouped by family. thus the diprotodontans from the Hamilton Local Fauna are presented as four seperate papers.

The final section (contribution 47) deals with the New Caledonian *"Zygomaturus"* (which, as noted above, has been claimed to be a marsupial but evidently is not).

The superb full-page reconstructions used as lead pages are the work of Dr Peter Murray. Peter's skills in this regard are particularly notable because he is also a professional verte-brate palaeontologist (e.g. Murray *et al.* 1987). His reconstructions (in order of their appearance in these volumes) are: *Didelphis virginiana*, (with bird); *Microbiotherium tehuelchum; Wynyardia bassiana; Ilaria illumidens; Wakaleo vanderleuri; Neohelos* sp.; *Miralina doylei; Ektopodon stirtoni; Djilgaringa gillespiei;* undescribed Riversleigh pseudocheirid; *Burramys* sp. from Riversleigh; and *Bettongia moyesi.*

ACKNOWLEDGEMENTS

Many people and organisations have contributed to the production of these volumes. In terms of organisations, the Royal Zoological Society of New South Wales sponsored the original Symposium and the University of New South Wales provided the venue. Surrey Beatty & Sons were incredibly patient with all of the delays involved in bringing in the final versions of the manuscripts and provided all of the financial support for publication.

As sponsoring organisations that enabled collection of much of the material reported here, the Australian Reserch Grants Scheme, the National Estate (Queensland), the National Science Foundation and the National Geographic Society deserve particular thanks.

Many other organisations have helped support research at Riversleigh in northwestern Queensland but the results of most of this support will be reported elsewhere. The only Riversleigh materials reported here (Archer *et al.* 1987; Flannery *et al.* 1987 and Flannery and Archer 1987) were collected between 1983 and 1985 when this project was principally supported by the Australian Reserch Grants Scheme. However, additional vital support during this period came from the Royal Australian Air Force, the Queensland Museum, the Australian Museum, IBM Australia Limited, PROBE and an anonymous private donor. Other support given during this period is noted in the acknowledgements of the papers cited above. More recent vital support from the Department of Arts, Heritage and the Environment, from the National Estate Grants Programme (Queensland), Wang Computers Pty Ltd, Mount Isa Mines Pty Ltd, Australian Geographic and International Chemical Industries (ICI) ensures that Riversleigh will continue to produce important material.

Individuals who helped to run the Symposium, prepare photographs and figures and who have been generally supportive of many aspects of the work presented here are acknowledged in the individual contributions but several individuals deserve special thanks: Ms Jenny Taylor, Mr Henk Godthelp, Dr Mel Dickson, Dr Suzanne Hand, Ms Anna Gillespie, Mr Ken Aplin, Dr Tim Flannery, Mr Harry Parnaby, Mr Ross Arnett and Ms Lisa Jensen.

Sincere thanks are also extended to authors who contributed their papers on time and who then had to suffer the delays involved in gathering in the remainder of the volumes. Although there were many reasons for the delays, in the long run it was my responsibility and I extend sincere apologies to any authors who have been inconvenienced by these delays.

The photographs of the living Koala (*Phascolarctos cinereus*), Wombat (*Vombatus ursinus*), Cuscus (*Phalanger rufoniger*), Glider (*Petaurus breviceps*) and Feather-tailed Glider (*Acrobates pygmaeus*) were taken by (respectively): R. and D. Keller, E. Beaton, T. Flannery, E. Beaton and Debbie Andrews.

Thanks are also extended to the more than 60 referees who took on the daunting task of examining the manuscripts in their original form. Some of these were among the authors of other contributions but most had no involvement in the volumes.

My sincerest thanks, however, are reserved for Ivor Beatty and his miraculous production team at Surrey Beatty & Sons. The intermidable delays in gathering in the late contributions and in finalising these volumes were a tremendous burden to them. To add to the difficulties, in the middle of production, the Georges River decided to take a week-long and very disastrous short-cut through the middle of the factory. I sincerely doubt that any other publisher would have coped as well with these natural as well as unnatural disasters and for this the Royal Zoological Society of New South Wales and I are very grateful.

REFERENCES

APLIN, K. AND ARCHER, M., 1987. Recent advances in marsupial systematics with a new syncretic classification. Pp. xv-lxxii *in* "Possums and opossums: studies in evolution" ed by M. Archer. Surrey Beatty & Sons and the Royal Zoological Society of New South Wales: Sydney.

ARCHER, M., 1978. The nature of the molar-premolar boundary in marsupials and a reinterpretation of the homology of marsupial cheekteeth. *Mem. Qd Mus.* **18:** 157-64.

ARCHER, M. (ed.), 1982. "Carnivorous marsupials". Royal Zoological Society of New South Wales: Sydney.

ARCHER, M., 1984. The Australian marsupial radiation. Pp. 663-808 *in* "Vertebrate zoogeography and evolution in Australasia" ed by M. Archer and G. Clayton. Hesperian Press: Sydney.

ARCHER, M. AND BARTHOLOMAI, A., 1978. Tertiary mammals of Australia: a synoptic review. *Alcheringa* **2:** 1-19.

ARCHER, M., HAND S. J. AND GODTHELP, H., 1986. "Uncovering Australia's dreamtime". Surrey Beatty & Sons: Sydney.

BAVERSTOCK, P. R., 1984. The molecular relationships of Australasian possums and gliders. Pp. 1-8 *in* "Possums and gliders" ed by A. P. Smith and I. D. Hume. Surrey Beatty & Sons and the Australian Mammal Society: Sydney.

KIRSCH, J. A. W., 1968. Prodromus of the comparitive serology of Marsupialia. *Nature (Lond.)* **217:** 418-20.

SMITH, A. P. AND HUME, I. D. (eds.), 1984. "Possums and gliders". Surrey Beatty & Sons and the Australian Mammal Society: Sydney.

STIRTON, R. A., TEDFORD, R. H. AND WOODBURNE, M. O., 1968. Australian Tertiary deposits containing terrestrial mammals. *Univ. Calif. Publs Geol. Sci.* **77**.

SZALAY, F. S., 1982. A new appraisal of marsupial phylogeny and classification. Pp. 621-40 *in* "Carnivorous marsupials" ed by M. Archer. Royal Zoological Society of New South Wales: Sydney.

TEDFORD, R. H., ARCHER, M., BARTHOLOMAI, A., PLANE, M. D., PLEDGE, N. S., RICH, T. H. V., RICH, P. V. AND WELLS, R. T., 1977. The discovery of Miocene vertebrates, Lake Frome area, South Australia. *B.M.R. J. Aust. Geol. Geophys.* **2:** 53-57.

WOODBURNE, M. O., TEDFORD, R. H., ARCHER, M. TURNBULL, W. D., PLANE, M. D. AND LUNDELIUS, E. L., 1985. Biochronology of the continental mammal record of Australia and New Guinea. *Spec. Publ. S. Aust. Dept. Mines and Energy* **5:** 347-63.

Section 1

Synthesis and Overview

RECENT ADVANCES IN MARSUPIAL SYSTEMATICS WITH A NEW SYNCRETIC CLASSIFICATION

K. P. APLIN[1], and M. ARCHER[1]

All available data (including that presented in "Possums and opossums: studies in evolution") having apparent value in interpreting the phylogenetic relationships of marsupial families is reviewed. The marsupial classifications of Simpson (1945), Ride (1964), Kirsch (1977), Szalay (1982), Archer (1984b), Woodburne (1984a), Reig et al. (1987) and Marshall (1987) are reviewed and compared as a preliminary to presentation of a syncretic family-level classification. The new classification was constructed using phylogenetic systematic procedures but has avoided problems commonly inherrent in purely cladistic classifications. Only groups that are demonstrably monophyletic are built into the heirarchy although two possibly paraphyletic taxa (Pediomyidae and Peradectidae) are retained in an uncertain position outside of the marsupial heirarchy.

Higher level taxa within Theria that encompass marsupials are also considered. Eutheria Gill, 1872 is regarded to be the senior homonym of Eutheria Huxley, 1880. Placentalia Owen, 1837 is available as the appropriate name for the inappropriate Eutheria Huxley, 1880. Within Marsupialia, two Cohorts are recognised: Ameridelphia Szalay and Australidelphia Szalay. Within Ameridelphia, three orders are recognised: Didelphimorphia (Gill, 1872) for Didelphidae and Sparassocynidae; Paucituberculata (Ameghino), 1894 for four superfamilies and eight families; and Sparassodonta Ameghino, 1894 for Borhyaenidae and Thylacosmilidae. The interordinal relationships of the three orders are unclear. Within Australidelphia, five orders are recognised: Microbiotheria Ameghino, 1889 for Microbiotheriidae; Dasyuromorphia (Gill, 1872) for Thylacinidae, Dasyuridae and Myrmecobiidae; Peramelemorphia (Kirsch, 1968) for Peramelidae and Thylacomyidae; Notoryctemorphia (Kirsch, 1977) for Notoryctidae; and Diprotodontia Owen, 1866 for 18 families. The families of Diprotodontia are divided into two suborders: Vombatiformes Woodburne, 1984 for Phascolarctidae, Vombatidae, Thylacoleonidae, Wynyardiidae, Ilariidae, Diprotodontidae and Palorchestidae; and Phalangerida Aplin and Archer (new) for all other diprotodontian families.

The phylogenetic relationships of each family and the rationale for concluding that each is monophyletic are reviewed. As a consequence, many revisions in family content are proposed, particularly for non-Australian taxa. A new family, the Acrobatidae, is proposed to encompass the genera Acrobates and Distoechurus which are quite clearly not referable to any previously established family level taxon.

Key Words: Theria; Eutheria; Monodelphia; Placentalia; Didelphia; Protodelphia; Holoclemensiidae; Marsupialia; Ameridelphia; Didelphimorphia; Paucituberculata; Sparassodonta; Australidelphia; Microbiotheria; Dasyuromorphia; Peramelemorphia; Notoryctemorphia; Diprotodontia; Phascolarctimorphia; Vombatimorphia; Phalangerida; Acrobatidae; Evolution; Classification; Phylogenetic; Taxonomy.

Pages xv-lxxii in POSSUMS AND OPOSSUMS: STUDIES IN EVOLUTION ed by M. Archer. Surrey Beatty & Sons and the Royal Zoological Society of New South Wales: Sydney, 1987.

INTRODUCTION

THE rate of change in understanding about marsupial phylogenetics since the 1960s has been extraordinarily high. This has been for essentially three reasons. First, there has been a prodigious and steadily more successful effort to improve knowledge about the fossil record of Australia's mammals. Second, new studies of little-examined tissue systems (such as brain structure, tarsal bone morphology and sperm ultrastructure) have been aggressively pursued partly in response to what has been perceived as the inadequacies of the fossil record to provide satisfactory answers to the 'higher-level' phylogenetic questions. Third, advances in molecular systematics have provided a vast new data set useful for examining the interfamilial relationships of living forms.

On the palaeontological front, revised understanding about the late Cretaceous and early Tertiary marsupial record of North and South America has led several of the authors in these volumes (Fox, Reig, Kirsch and Marshall) to propose entirely new overviews of early marsupial interrelationships, some (but not all) of which have been brought forward here in our effort to synthesize often incongruous conclusions.

Palaeontological research being carried out now in Australia is not yet working on the same time scale as that being carried out in the America's. Because it has so far been largely constrained by an early Miocene lower age limit, its principal contribution has been to provide knowledge about marsupial diversity although it has provided some information about the interfamilial relationships of the living Australian groups.

Comparative anatomical studies of marsupials have been going on more or less continuously since the mid 1800s but often these early studies were carried out solely for the sake of understanding structure or function rather than

[1]School of Zoology, University of New South Wales, P.O. Box 1, Kensington, New South Wales, Australia 2033.

phylogenetic pattern. More recent studies, such as the seminal work by Szalay (1982) on marsupial tarsal bone morphology, have had a profound effect on phylogenetic theory. Szalay's study, for example, was the first to draw attention to the relationship between the South American microbiotheriids and the whole of the Australian marsupial radiation, a concept that effected a great deal of subsequent research some of which is presented in these volumes (Reig, Kirsch, Marshall, Temple-Smith, Woolley, Grant and the present chapter).

Comprehensive examination of the molecular relationships of Australian marsupials effectively began with the serological studies of Kirsch (1968, 1977). His results led to a major overhaul in understanding about marsupial family-level diversity and relationships. While in its basic construction it reflects many of the principal concepts of Ride's (1964) classification, it in turn has been the starting point for most subsequent changes in marsupial classification, including many of those proposed by us in this chapter.

Following Kirsch's 'pioneering' studies, Maxson *et al.* (1975), Baverstock *et al.* (1987; this Volume) and Sarich (pers. comm. to Archer) have each picked up and run with the molecular approach to marsupial phylogenetics principally through study of the antigenic properties of albumins. Baverstock's studies in particular are now providing a very provocative body of data to supplement as well as challenge current phylogenetic understanding.

The combination of newly available information has had a curious effect. Whereas on other continents discovery of the older fossil record frequently long preceded consideration of molecular systematics and hence in some cases mitigated against acceptance of its value in phylogenetic studies, in Australia molecular data has commonly been the only guide (apart from gross anatomy of living forms) for interpreting interfamilial relationships. As a result, during the last two decades molecular systematics has profoundly influenced our perceptions of marsupial phylogenetics.

On top of these developments has also come widespread use of a particular method of character analysis known as phylogenetic systematics (e.g., Wiley 1981). The main axiom of this method is that shared-derived character states constitute the only valid basis for recognising cladistic relationships (i.e., those based on common descent). Classifications arising from use of this method are often, though sometimes not exclusively, developed on the principle that they should contain only monophyletic taxa (groups that contain their common ancestor and all of its descendants).

Rigorous use of the phylogenetic systematic approach has characterised most recent attempts to come to grips with marsupial phylogenetics. In the present chapter, we have used this method when integrating the results of other similar studies as well as those arising from analyses that formally recognise paraphyletic as well as monophyletic taxa (e.g., Reig *et al.* this Volume).

In the sections that follow, we have provided a brief review of eight previous classifications, beginning with that of Simpson (1945) and an extended rationalisation for the syncretic classification proposed in this paper (Table 2). This classification has been constructed using the principles noted above to integrate what appear to us to be the most persuasive phylogenetic arguments previously presented or presented for the first time in these volumes.

In part because of the relatively recent convergence of research interests noted above, seven of the nine classifications discussed here have been developed in the last 10 years. For this reason we have no illusions about the probable lifespan of the new syncretic classification proposed herein. It is a measure of the excellent health of marsupial systematic studies that classifications which previously might have become the 'standard' for decades are now the immediate targets for the darts of new phylogenetic studies.

MARSUPIAL CLASSIFICATIONS SINCE 1945

Table 1 presents the marsupial classifications of Simpson (1945), Ride (1964), Kirsch (1977), Szalay (1982), Archer (1984b), Woodburne (1984a), Reig, Kirsch and Marshall (1987; this Volume) and Marshall (1987; this Volume). The last two were limited in their content to non-Australian didelphid-like marsupials.

In each case we have restricted the presentation to ranks at family level and above. If reasonably certain about a previous author's synonymy of taxa otherwise recognised here at family level, we have added these in parentheses.

Simpson (1945)

Simpson's (1945) classification recognises six superfamilies as the basic monophyletic units of marsupial diversity. One of the reasons this classification served for as long as it did was that it contained no statements about higher levels of relationship between the superfamilies and hence was rarely in the firing line of subsequent phylogenetic analysis. However, as more information about interrelationships between his superfamilies came to light, this same lack of specific phylogenetic statements eventually led to its demise as an appropriate summary statement of current phylogenetic understanding.

Table 1. Eight principal classifications of marsupials (down to family level) including and subsequent to that of Simpson (1945). In some cases, as discussed in the text, the classifications as shown here are amplifications of the originals insofar as we can be confident of the original author's intent. The classifications of Reig *et al.* (1987) and Marshall (1987) were originally and intentionally limited to subsets of non-Australian marsupials but are included here because they present significant departures from traditional concepts.

Simpson 1945

Infraclass Metatheria
Order Marsupialia
Superfamily Didelphoidea
Family Didelphidae
Family Caroloameghinidae
Superfamily Borhyaenoidea
Family Borhyaenidae
Superfamily Dasyuroidea
Family Dasyuridae
Family Notoryctidae
Superfamily Perameloidea
Family Peramelidae
Superfamily Caenolestoidea
Family Caenolestidae
Family Polydolopidae
Superfamily Phalangeroidea
Family Phalangeridae
Family Thylacoleonidae
Family Vombatidae
Family Macropodidae
Family Diprotodontidae

Ride 1964

Infraclass Metatheria
Superorder Marsupialia
Order Marsupicarnivora
Superfamily Didelphoidea
Family Didelphidae
Family Stagodontidae
Family Pediomyidae
Superfamily Borhyaenoidea
Family Borhyaenidae
Family Thylacosmilidae
Family Necrolestidae
Superfamily Dasyuroidea
Family Dasyuridae
(including Myrmecobiidae)
Family Thylacinidae
Order Paucituberculata
Family Caenolestidae
Family Polydolopidae
Family Groeberiidae
?Paucituberculata
Family Argyrolagidae
Order Perameline
Family Peramelidae
Order Diprotodonta
Family Phalangeridae
Family Wynyardiidae
Family Vombatidae
Family Diprotodontidae
Family Macropodidae
Marsupialia *incertae sedis*
Family Notoryctidae
?Marsupialia
Family for *Gashternia*

Kirsch 1977

Superorder Marsupialia
Order Polyprotodonta
Suborder Didelphimorphia
Superfamily Didelphoidea
Family Didelphidae
Family Microbiotheriidae
Family Pediomyidae
Superfamily Borhyaenoidea
Family Borhyaenidae
Family Thylacosmilidae
Family Necrolestidae
Superfamily Dasyuroidea
Family Dasyuridae
Family Myrmecobiidae
Suborder Peramelemorphia
Superfamily Perameloidea
Family Peramelidae
Family Thylacomyidae
Suborder Notoryctemorphia
Superfamily Notoryctoidea
Family Notoryctidae
Order Paucituberculata
Superfamily Caenolestoidea
Family Caenolestidae
Superfamily Polydolopoidea
Family Polydolopidae
?Paucituberculata
Superfamily Groeberioidea
Family Groeberiidae
Superfamily Argyrolagoidea
Family Argyrolagidae
Order Diprotodonta
Superfamily Vombatoidea
Family Vombatidae
Family Phascolarctidae
Superfamily Phalangeroidea
Family Phalangeridae
Family Ektopodontidae
Family Burramyidae
Family Thylacoleonidae
Family Macropodidae
Family Diprotodontidae
Superfamily Tarsipedoidea
Family Tarsipedidae
Superfamily Wynyardioidea
Family Wynyardiidae

Szalay 1982

Infraclass Metatheria
Cohort Ameridelphia
Order Didelphida
Suborder Didelphiformes
Family Didelphidae
Family Pediomyidae
(incl. Stagodontinae, Glasbinae, Peradectinae and Pediomyinae)
Suborder Borhyaeniformes
Superfamily Borhyaenoidea
Family Borhyaenidae
Family Thylacosmilidae
Family Necrolestidae
Superfamily Caenolestoidea
(?incl. Polydolopidae, Caenolestidae and Groeberiidae)
Superfamily Argyrolagoidea
Family Argyrolagidae
Cohort Australidelphia
Order Dromiciopsia
Family Microbiotheriidae
Order Dasyurida
Family Dasyuridae
Family Myrmecobiidae
Family Thylacinidae
Order Syndactyla
Suborder Syndactyliformes
Superfamily Notoryctoidea
Family Notoryctidae
Suborder Perameliformes
Suborder Phalangeriformes
Superfamily Phalangeroidea
Superfamily Vombatoidea
Superfamily Macropodoidea
Superfamily Diprotodontoidea

Archer 1984

Subclass Marsupialia
Cohort Didelphicia
(for *Holoclemensia*)
Cohort Metatheria
Superorder Stagodontia
Family Stagodontidae
Superorder Alphadontia
(*Alphadon* and *Albertatherium*)
Superorder Peradectadontia
Family Peradectidae
Superorder Pediomyidia
Family Pediomyidae
Superorder Caenolestoidea
Family Borhyaenidae
Family Thylacosmilidae
Family Caenolestidae
?Borhyaenimorphia
Family Groeberiidae
Family Argyrolagidae
Superorder Polydolopimorphia
Family Polydolopidae
Family Prepidolopidae
Family Glasbiidae
Family Carolo-ameghiniidae
Family Bonapartheriidae
Superorder Australidelphia
Order Dromiciopsia
Family Microbio-theriidae
Order Eometatheria
Suborder Dasyurida
Family Dasyuridae
Family Myrmecobiidae
Family Thylacinidae
Order Syndactyla
Infraorder Peramelina
Family Peramelidae
Family Thylacomyidae
Infraorder Diprotodonta
Superfamily Vomba-toidea
Family Vombatidae
Family Phascolarctidae
Family Thylacoleonidae
Family Wynyardiidae
Family Palorchestidae
Family Diprotodontidae
Superfamily Macro-podoidea
Family Potoroidae
Family Macropodidae
Superfamily Phalanger-oidea
Family Phalangeridae
Family Burramyidae
Family Ektopodontidae
Family Pseudocheiridae
Family Petauridae
Family Tarsipedidae
Family for *Acrobates* and *Distoechurus*

Woodburne 1984

Infraclass Metatheria
Cohort Ameridelphia
Order Didelphiformes
Suborder Didelphimorphia
Superfamily Didelphoidea
Family Didelphidae
Family Pediomyidae
Family Stagodontidae
Family Bonapartheriidae
Family Prepidolopidae
Superfamily Borhyaenoidea
Family Borhyaenidae
Family Thylacosmilidae
Family Necrolestidae
Superfamily Paucituberculata
Superfamily Caenolestidae
Family Caenolestidae
Family Polydolopidae
?Paucituberculata
Superfamily Groeberioidea
Family Groeberiidae
Superfamily Argyrolagoidea
Family Argyrolagidae
Cohort Australidelphia
Order Dromiciopsia
Superfamily Microbio-therioidea
Family Microbiotheriidae
Order Dasyurida
Suborder Dasyuromorphia
Superfamily Dasyuroidea
Family Dasyuridae
Family Myrmecobiidae
Family Thylacinidae
?Order
Suborder Notoryctemorphia
Superfamily Notoryctoidea
Family Notoryctidae
Order Syndactyla
Suborder Perameemorphia
Superfamily Perameloidea
Family Peramelidae
Family Thyla-comyidae
Order Diprotodonta
Suborder Vombatiformes
Superfamily Vombatoidea
Family Thylacoleonidae
Family Vombatidae
Family Diprotodontidae
Family Palorchestidae
Family Wynyardiidae
Superfamily Phasco-larctoidea
Family Phascolarctidae
Suborder Phalangeriformes
Superfamily Phalangeroidea
Family Phalangeridae
Family Ektopodontidae
Family Petauridae
(including Pseudocheirinae)
Family Burramyidae
Family Macropodidae
(including ?otoroinae)
Superfamily Tarsipedoidea
Family Tarsipedidae

Reig, Kirsch & Marshall 1987

Order Polyprotodonta
Suborder Didelphimorphia
Superfamily Microbio-therioidea
Family Peradectidae
Family Microbiotheriidae
(including Glasbinae, Pediomyinae and Microbiotheriinae)
Family Stagodontidae
Family Caroloameghinidae
Superfamily Didelphoidea
Family Didelphidae
Family Sparassocynidae
Didelphoidea *incertae sedis*
(Fourteen genera of Paleocene to Pliocene South American marsupials)

Marshall 1987

Suborder Didelphimorphia
Superfamily Microbiotherioidea
Family Peradectidae
Family Stagodontidae
Family Microbiotheriidae
(including Pediomyinae)
Superfamily Borhyaenoidea
Family Borhyaenidae
Family Thylacosmilidae
Superfamily Didelphoidea
Family Didelphidae
Family Sparassocynidae
Suborder Polydolopimorphia
Superfamily Carolo-ameghinioida
Family Caroloameghiniidae
(including Glasbinae)
Superfamily Caenolestoidea
Family Caenolestidae
Superfamily Argyrolagoidea
Family Argyrolagidae
Family Groeberiidae
Family Gashternidae
Superfamily Polydolopoidea
Family Prepidolopidae
Family Bonapartheriidae
Family Polydolopidae

Ride (1964)

While other classifications were produced in the 1950s and 1960s, the first one to significantly depart from Simpson's (1945) standard was that of Ride (1964; reproduced in Table 1 to family level with additions from Graham and Ride 1967). Ride's classification differs mainly in its recognition of the impropriety of trying to contain the whole of the marsupial radiation into a single order. Consequently, he divides the marsupials up into four orders, with three of these (Paucituberculata, Peramelina and Diprotodonta) more or less corresponding in family-level content to three of Simpson's superfamilies (Caenolestoidea, Perameloidea and Phalangeroidea). He departs more significantly in placing Simpson's remaining three superfamilies (Didelphoidea, Borhyaenoidea and Dasyuroidea) in a single paraphyletic ordinal taxon, the Marsupicarnivora.

An important phylogenetic aspect of Ride's analysis is that he regarded the bandicoots and the diprotodontians to be derived from a common syndactylous stock. *Notoryctes*, on the other hand, he considered to be of uncertain affinities, a view reflected in its position as *incertae sedis* within his classification.

In its expansion up to a multiordinal system, Ride's classification was strongly criticised by a number of leading authorities (e.g., Simpson 1970; Calaby *et al.* 1974 and Clemens and Marshall 1976). To others, however, it provided a new taxonomic freedom and the stimulus for major subsequent studies in marsupial systematics such as that of Kirsch (1968, 1977).

Kirsch (1977)

Kirsch's (1977) classification, a refined version of the classification he first presented in preliminary form in 1968, was the first to reflect understanding about the molecular relationships of living marsupials. His studies of serology (involving whole sera rather than single proteins) provide the first significantly different body of data for phylogenetic analysis.

Perhaps the most significant contribution of Kirsch's serological studies was the clarification of the previously under-estimated family-level diversity within the Order Diprotodontia. More specifically, he effectively demonstrated that within what had previously been regarded to be the single family Phalangeridae, there are at least four family-level taxa: Phalangeridae, Burramyidae (including the group we have recognised below as the family Acrobatidae), Tarsipedidae and Petauridae (including the group we have recognised below as the Pseudocheiridae).

Like Ride's classification, that of Kirsch is an amalgamation of monophyletic and paraphyletic taxa. Although Kirsch abandons Ride's paraphyletic taxon Marsupicarnivora, he revitalises an older and yet more complexly paraphyletic taxon, Polyprotodonta, as one of the three orders in his classification.

Equally significant in view of long-standing debates about specific inter-family phylogenetic relationships between South American and Australian marsupials is his conclusion that all of the Australian groups are more closely related to each other than any are to any of the American groups. However, rather than expressing this conclusion in his classification, he combines most of the American and Australian carnivorous forms in the paraphyletic taxon Polyprotodonta.

It is important to realise about Kirsch's classification that he does not use his serological results as the only guide to determining position in the phylogenetic heirarchy.

Szalay (1982)

Szalay's (1982) classification, based on his study of tarsal bone morphology, has had a profound effect on current understanding of marsupial systematics. The most striking hypothesis arising from this study is that the South American family Microbiotheriidae, as represented by its only living species *Dromiciops australis*, was part of a monophyletic group that otherwise contained all of the Australian but none of the other families of American marsupials.

Szalay formalised this concept by recognising two marsupial cohorts: the Ameridelphia for all of the American forms except the Microbiotheriidae; and the Australidelphia for all of the Australian forms plus the Microbiotheriidae. This classification was constructed in large part using the methodology of the phylogenetic systematic method (using shared-derived character states of tarsal morphology) and hence does not include concepts such as Polyprotodonta or Marsupicarnivora. However, implicit in Szalay's phylogenetic hypothesis (as evidenced in his Fig. 10) was the notion that the extinct American family Pediomyidae represented a paraphyletic 'stem group' from which the Ameridelphia and Australidelphia were both derived, the former having perhaps had as many as two separate derivations from within the Pediomyidae. Fox's suggestion (1987; this Volume) that the Pediomyidae as currently recognised probably represents a non-monophyletic taxon will be of particular interest here.

Another significant departure from common practise is Szalay's use of the concept Syndactyla as one of the four orders recognised in his classification. This highlights what has long been (and still is) a conflict in marsupial classification: the affinities of the bandicoots (peramelids and thylacomyids).

Archer (1984b)

Archer (1984b) presented a speculative classification following a wide-ranging review of marsupial phylogeny (Archer 1984a,b). This was an attempt to rigourously use phylogenetic systematics to express current understanding about marsupial phylogenetics.

One important feature of this classification is the formal recognition of the marsupial status of the early Cretaceous *Holoclemensia texana* (more commonly regarded as a 'eutherian-metatherian' grade mammal). This was distinguished as the plesiomorphic sister cohort (Protodelphia) of the rest of the Marsupialia (recognised formally as the Metadelphia). Within Metadelphia, Archer recognised a total of seven superorders, including one for the monophyletic australidelphian radiation (Australidelphia). This in turn contained two orders, the first for microbiotheriids (Dromiciopsia) and the second for the whole of the Australian radiation (Eometatheria; a term coined 'informally' by Simpson 1970). Szalay's Syndactyla was also used (though at subordinal level) on the grounds that syndactyly appeared to be a synapomorphic rather than convergent state.

Other departures from tradition included establishment of the concept Polydolopimorphia to include the polydolopid- like American groups including polydolopids, prepidolopids, bonapartheriids, glasbiids and caroloameghiniids and classification of Peradectidae, Stagodontidae, Pediomyidae and a group containing *Alphadon* and *Albertatherium* as coordinate superorders, a rank equal to that of the whole of the australidelphian radiation.

Within the australidelphian radiation, the Petauridae of Kirsch (1977) was split into two families, the other being the Pseudocheiridae. Similarly, within the Diprotodontia, the positions of the Diprotodontidae, Palorchestidae, Thylacoleonidae and Wynyardiidae, based on studies of basicranial morphology, differed from that of previous classifications. Most of these basic rearrangements were followed by Woodburne (1984a) and have been carried over, with additional changes, into our present classification.

Woodburne (1984a)

Woodburne's (1984a) classification, which was also intended to reflect a wide range of phylogenetically useful data, differs from that of Archer (1984b) in several major respects. First, it is more traditional in placing the whole of the American marsupial radiation (less the microbiotheriids) within the single order Didelphiformes.

On the other hand, it is more radical than all previous classifications in recognising two suborders (Vombatiformes, Phalangeriformes) within Diprotodontia, and in treating the Phascolarctidae as a superfamily within the suborder Vombatiformes.

Woodburne's use of the concept Syndactyla as an order is credited to Szalay (1982) yet departs significantly from all earlier usages of the term in not including the diprotodont marsupials; these are recognised by Woodburne as a coordinate order under their traditional label, Diprotodonta.

In its allocation of the diprotodont families to suborders and superfamilies Woodburne's classification in most regards conforms with the arrangements suggested by Archer (1984b); one point of distinction concerns the placement of the kangaroos and rat-kangaroos in a single family.

Further departures from Archer (1984b) and the classification presented herein include the lack of distinction between the pseudocheirids and petaurids and the inclusion within the Burramyidae of *Acrobates* and *Distoechurus* (which are now separated as the Acrobatidae).

Reig, Kirsch and Marshall (1987)

Reig, Kirsch and Marshall (1987; this Volume), present a broad review of phylogenetic data relevant to understanding the phylogenetics of the American marsupials and conclude with a significantly different classification of the American forms. They divide all didelphimorph marsupials into two basic categories: the Microbiotherioidea and the Didelphoidea. The Didelphoidea is presented as a monophyletic superfamily containing just didelphids and sparassocynids. Microbiotherioidea as used in this classification is a paraphyletic superfamily that contains four families of didelphimorphs. Most controversial of these included families is the Peradectidae which is interpreted by Reig *et al.* to be the probable stem group for all Cainozoic marsupials.

Establishment of the significantly paraphyletic taxon Microbiotherioidea is partly justified by Reig *et al.* on the grounds that it might serve to stimulate further research into the phylogeny and classification of the earliest marsupials. The present review and that of Marshall (1987; this Volume) on the Itaborian didelphid-like marsupials represent two immediate responses to this stimulus.

Other significant changes proposed by Reig *et al.* (1987) include their placement of Pediomyinae and Glasbiinae in Microbiotheriidae. Innovative allocations of various Cainozoic didelphine genera to particular subfamilies also represent significant advances in phylogenetic understanding.

Reig *et al.* (1987) give no support to the suggestion of Szalay (1982) that Microbiotheriidae is the sister taxon of the whole of the Australian marsupial radiation and conclude that the tarsal similarities are probably convergent.

Marshall (1987)

Marshall's (1987) classification focuses on the diverse assemblage of early Tertiary Itaborian marsupials from Brazil. As noted in Reig *et al.* (1987), the morphological diversity of Itaborian didelphimorph marsupials is greater than that for the whole of the rest of the American Cainozoic record and must inevitably affect any attempt at classification of didelphimorph marsupials.

As a result of his studies, Marshall (1987; this Volume) divides the American marsupials into two suborders, the Didelphimorphia and the Polydolopomorphia. His Didelphimorphia differs from that of Reig *et al.*(1987) in: 1, excluding the Caroloameghiniidae (which he places as a superfamily within the Suborder Polydolopimorphia); 2, including the Borhyaenoidea; and 3, in excluding the Glasbiinae (which he places as a subfamily within the Caroloameghiniidae).

Marshall's (1987) Polydolopimorphia is similar in content to Archer's (1984) Polydolopimorphia except that it contains the Caenolestidae and Argyrolagidae (groups which Archer tentatively placed in his Borhyaenimorphia following the interpretation of shared tarsal character states proposed by Szalay (1982)).

In his text, Marshall (1987) also departs from Reig *et al.* (1987) in suggesting that the results of his studies are consistent with Szalay's (1982) concept of 'special affinity' between *Dromiciops australis* and at least some Australian marsupials. Marshall went on to suggest that the Australian middle Miocene *Ankotarinja tirarensis*, described by Archer (1976d) as a probable dasyurid, actually displays "Microbiotheriinae-like" tooth characters. However, this intimation is not carried through into his classification, unless we are to infer (almost certainly incorrectly) that the Australian marsupials are intended by Marshall to be a subgroup of the Microbiotheriidae, having descended from early representatives of this family.

A NEW SYNCRETIC CLASSIFICATION OF THE MARSUPALIA

The classification which we offer here (Table 2) has been constructed to reflect as much as possible about current understanding of marsupial phylogenetics. In this classification, no single character system has determined hierarchical position if this indication has been contradicted by analyses of other character systems. Where contradictory phylogenetic interpretations occur, we have opted for the best-supported hypothesis. In this sense, this classification is a syncretic consensus of current phylogenetic understanding.

Systematic Methodology and Purpose

The notion that classification should reflect understanding of evolutionary history is widely accepted. As many biologists have pointed out, evolution is the only scientific model that makes sense of the diversity of living things. In this context, classification is the language for expressing understanding about evolutionary history.

On the other hand there is controversy about whether classifications should reflect just cladistic (i.e., evolutionary branching) relationships of organisms or other attributes such as morphological distinctness, taxonomic diversity or antiquity. The literature on this topic abounds with concepts such as relative objectivity, information content, preservation of stability, utility for general biologists and utility of grades versus clades.

In part because of these conflicting pressures, purely cladistic classifications have not fared all that well at least in so far as 'success' is judged by wide usage. For example, although McKenna's (1975) cladistic classification contains many fundamental and important phylogenetic concepts, these have generally only been used in a piecemeal fashion (e.g., by Prothero 1981 and Novacek 1986). His classification in its entirety appears to be less acceptable probably because of what is perceived to be its unwieldliness and its plethora of previously unfamiliar names.

Because it is our aim here to provide a syncretic classification that is fully cladistic and yet acceptable for general usage, we clearly need to carefully justify and explain the procedures employed in its construction.

As argued effectively by Ridley (1985), if objectivity is to be cited as a major criterion by which the worth of classifications can be judged, then cladistic classifications are the only ones that could theoretically attain this goal. This is because cladistic classifications seek only to reflect an intrinsic property of organisms: the pattern of evolutionary relationships that links all organisms. Understanding of this pattern is subject to examination and potential falsification which are among the primary goals of modern systematic studies.

In contrast, claims that other types of classification (e.g., those based on phenetic analysis) are more objective or have a greater 'information content' are doomed to flounder because the properties they are attempting to discover and reflect in classification are non-intrinsic (e.g., diversity) or relatively subjective (degree of

Table 2. The classification proposed in this work. Original authors and dates (separated by a comma) of taxa are cited immediately after the taxon unless it is used here for the first time. If the original application of the taxon was not at the hierarchical level used here, the original author and date are in parentheses and the author and date (not separated by a comma) of the first use of the taxon at the level employed here follows the name and date of the primary author. Any departures from this format are discussed in the text. Taxa in caps and small caps are those which are specifically superordinate to the marsupial orders.

SUBCLASS THERIA Parker and Haswell, 1897[1]
  Superlegion Kuhneotheria McKenna, 1975
       Family Kuhneotheriidae Kermack, Kermack and Musset, 1968
  SUPERLEGION TRECHNOTHERIA McKenna, 1975
   Legion Yinotheria Chow and Rich, 1982
       Family Shuotheriidae Chow and Rich, 1982
   LEGION YANGOTHERIA Chow and Rich, 1982
    Sublegion Symmetrodonta (Simpson, 1945) Chow and Rich, 1982[2]
    SUBLEGION CLADOTHERIA (McKenna, 1975) Chow and Rich, 1982
     Infralegion Dryolestoidea (Butler, 1939) new rank[3]
        Family Amphitheriidae Owen, 1846
        Family Paurodontidae Marsh, 1887
        Family Dryolestidae Marsh, 1879
    INFRAREGION ZATHERIA (McKenna, 1975) new rank
     Infraclass Peramura McKenna, 1975
        Family Peramuridae Kretzoi, 1960
    INFRACLASS TRIBOSPHENIDA McKenna, 1975
     Superdivision Aegialodontia (Butler 1978) new rank
        Family Aegialodontidae Kermack, Lees and Mussett, 1965[4]
    SUPERDIVISION EUTHERIA Gill, 1872
     Division Monodelphia De Blainville, 1816
      Supercohort Pappotherida (Butler, 1978) new rank
         Family Pappotheriidae Slaughter, 1965[5]
      Supercohort Placentalia Owen, 1837[6]
    DIVISION DIDELPHIA De Blainville, 1834
      Supercohort Protodelphia (Archer, 1984b) new rank
         Family Holoclemensiidae Aplin and Archer, 1987 new[7]
      SUPERCOHORT MARSUPIALIA (Illiger, 1811) Cuvier 1817
       COHORT AMERIDELPHIA Szalay, 1982
        Order Didelphimorphia (Gill, 1872) new rank s.m.[8]
          Family Didelphidae Gray, 1821
          Family Sparassocynidae (Reig, 1958) Archer 1984
        Didelphimorphia: family *incertae sedis*
           Subfamily Protodidelphinae Marshall, 1987[9]
        Order Paucituberculata (Ameghino, 1894) s.m.
          Superfamily Caroloameghinioidea (Ameghino, 1901) Marshall 1987 s.m.
          Family Caroloameghiniidae Ameghino, 1901
          (including Glasbiinae Clemens, 1966)
          Superfamily Caenolestoidea (Trouessart, 1898) Osborn 1910 s.m.
          Family Caenolestidae Trouessart, 1898
          Superfamily Argyrolagoidea (Ameghino, 1904) Simpson 1970 s.m.
          Family Gashterniidae Marshall, 1987
          Family Groeberiidae Patterson, 1952
          Family Argyrolagidae Ameghino, 1904
          Superfamily Polydolopoidea (Ameghino, 1897) Clemens and Marshall 1976 s.m.
          Family Prepidolopidae Pascual, 1981 s.m.
          Family Bonapartheriidae Pascual, 1981 s.m.
          Family Polydolopidae Ameghino, 1897 s.m.
        Order Sparassodonta Ameghino, 1894 s.m.
          Family Borhyaenidae Ameghino, 1894
          Family Thylacosmilidae (Riggs, 1933) Marshall 1976
       COHORT AUSTRALIDELPHIA Szalay, 1982
        Order Microbiotheria Ameghino, 1889 s.m.
          Family Microbiotheriidae Ameghino, 1887
        Order Dasyuromorphia (Gill, 1872) new rank s.m.
          Family Thylacinidae Bonaparte, 1838
          Family Myrmecobiidae Waterhouse, 1838
          Family Dasyuridae (Goldfuss, 1820) Waterhouse 1838
        Order Peramelemorphia (Kirsch, 1968) new rank s.m.
          Family Peramelidae (Gray, 1825) Archer and Kirsch 1977
          Family Thylacomyidae (Bensley, 1903) Archer and Kirsch 1977
        Order Notoryctemorphia (Kirsch, 1977) new rank s.m.
          Family Notoryctidae Ogilby, 1892
        Order Diprotodontia Owen, 1866 s.m.
         Suborder Vombatiformes Woodburne, 1984
         Infraorder Phascolarctomorphia Aplin and Archer, 1987 new
          Family Phascolarctidae Owen, 1839
         Infraorder Vombatomorphia Aplin and Archer, 1987 new
          Family Diprotodontidae Gill, 1872 s.m.
          Family Palorchestidae (Tate, 1948) Archer and Bartholomai 1978 s.m.

Table 2. — Continued.

Family Wynyardiidae Osgood, 1921 s.m.
Family Ilariidae Tedford and Woodburne, 1987 s.m.
Family Vombatidae Burnett, 1830 s.m.
Family Thylacoleonidae Gill, 1872 s.m.
Suborder Phalangerida Aplin and Archer, 1987 new
Superfamily Phalangeroidea (Thomas 1888) new concept s.m.
Family Phalangeridae Thomas, 1888
Family Miralinidae Woodburne, Pledge and Archer, 1987
Family Ektopodontidae Stirton, Tedford and Woodburne, 1967
Superfamily Macropodoidea (Gray, 1821) Archer and Bartholomai 1978 s.m.
Family Potoroidae (Gray, 1821) Pearson 1950
Family Macropodidae Gray, 1821
Superfamily Burramyoidea (Broom, 1898) new rank s.m.
Family Burramyidae (Broom, 1898) Kirsch 1968
Superfamily Petauroidea (Gill, 1872) new rank s.m.
Family Pseudocheiridae (Winge, 1893) Archer 1984
Family Petauridae (Gill, 1872) Archer 1984
Superfamily Tarsipedoidea (Gervais and Verreaux, 1842) Kirsch 1968 s.m.
Family Tarsipedidae Gervais and Verreaux, 1842
Family Acrobatidae Aplin, 1987 new
Phalangerida: superfamily *incertae sedis*
Family Pilkipildridae Archer, Tedford, Woodburne and Rich, 1987[10]
Marsupialia: cohort *incertae sedis*
Family Peradectidae Crochet, 1979 s.m.
Family Stagodontidae Marsh, 1889 s.m.
Family Pediomyidae (Simpson, 1927) Clemens 1966 s.m.
Didelphia: supercohort *incertae sedis*
Albertatherium, Bistius
Tribosphenida: superdivision *incertae sedis*[11]
Picopsis, Potamotelses, Deltatheridium, Deltatheroides, Beleutinus, Hyotheridium, Falepetrus, Kermackia, Kielantherium
and *Trinititherium*
?Theria: superlegion *incertae sedis*[12]
Order Monotremata Bonaparte, 1838
Family Ornithorhynchidae Burnett, 1830
Family Tachyglossidae Gill, 1872
Order Multituberculata Cope, 1884[13]

[1] The concept of Theria used here is a modification of McKenna (1975), Prothero (1981) and Chow and Rich (1982). Monotremata and Multituberculata may be therians as suggested by Archer *et al.* (1986) but if so their position within Theria is uncertain.

[2] Prothero's (1981) phylogenetic study of Jurassic trechnotherians does not provide a formal allocation of all forms to family level taxa. However, Symmetrodonta includes seven genera distributed in at least two major groups, Amphidontoidea and Spalacotherioidea.

[3] Prothero (1981) presents the three families of dryolestoids in a ranked sequence as shown here such that Paurodontidae is the sister group of Dryolestidae and these two together are the sister group of Amphitheriidae.

[4] Includes *Aegialodon* only although other taxa listed below under Tribosphenida *incertae sedis* may belong here.

[5] Includes *Pappotherium, Prokennalestes* and possibly *Slaughteria*.

[6] Included are the various cohorts of placental mammals none of which are listed here (see e.g., McKenna 1975).

[7] Includes only *Holoclemensia*.

[8] s.m. = *sedis mutabilis* (i.e., the phylogenetic position of this group within the next highest rank level is uncertain).

[9] Marshall (1987) established this subfamily within Caroloameghiniidae. Our doubts about this allocation are discussed in the text.

[10] This family, established in the present Volume, is either a petauroid or a phalangeroid. In the original work it is concluded that it is most probably a petauroid with greatest affinities to the Petauridae but we remain open-minded.

[11] Tribotheria Butler is available as a group name for most of the tribosphenid taxa here (plus *Aegialodon, Pappotherium* and *Holoclemensia*) but it is not used here. It was established as a paraphyletic concept and does not appear to contribute to an understanding of tribosphenid phylogenetic relationships.

[12] Inclusion of Monotremata and possibly Multituberculata in Theria follows Archer *et al.* (1985) and Jenkins *et al.* (in press); other interpretations (e.g., Kuhne 1977 and Clemens 1979) are noted in the text.

[13] There are at least 10 families of multituberculates (Clemens and Kielan-Jaworowska 1979).

morphological distinction) and these attributes depend signficantly on individual perception. They also fail insofar as they possess no theoretical justification equivalent to the theory of evolution that would explain why taxa should be arranged in a heirarchical system at all.

'Mixed classifications' (part cladistic; part something else) are no more defensible on objective grounds than those based on cladistic procedures. Because the objectivity of the parts based on cladistic analysis are despoiled by the attempts to include other kinds of relatively subjective information 'mixed classifications' might be more appropriately known as 'polluted classifications'.

We accept the arguments of Ridley (1985) and those of other cladistic systematists (e.g., Wiley 1981) that have paved the way for his argument. However, we recognise that there have been problems with acceptance of purely cladistic classifications.

There are three common objections to cladistic classifications:

1. Cladistic classifications result in a proliferation of categorical ranks and of new taxonomic names. Patterson and Rosen (1977) note that this common practice most significantly effects the more primitive taxa of any group because the relationships of these to other taxa in the group are often the least well understood.

2. Cladistic classifications encourage the practice of reclassification as a consequence of investigation of each new character system.

3. Cladistic classifications have no way of expressing patristic affinity or paraphyletic taxa which are common components in 'evolutionary' classifications.

Most of these points are adaquately addressed by Wiley (1981) and Ridley (1985). Suffice it to say that there have been a number of suggestions about procedures cladistic systematists could use to improve the acceptability of their classifications. We have incorporated some of these procedures in our present classification (e.g., ranking and the use of the indication *sedis mutabilis*). However, because we consider that our particular approach in developing this classification differs signficantly from what has gone before, it would be appropriate to briefly review this approach.

Present Procedure

The determination of understanding about phylogenetic relationships and the expression of this understanding are two quite distinct steps in systematic procedure (see also Kirsch 1977).

Estimation of Phylogenetic Relationships

Our methods for inferring phylogeny have been those of phylogenetic systematics ('cladistics'). Our only departure from a purely objective procedure in analysing data with potentially useful phylogenetic value has been, when faced with potential conflict in phylogenetic interpretation, to rely more heavily on indications from particular character systems. This clearly represents a 'weighting' procedure but one we regard to be theoretically justifiable. For whatever reasons, different character systems appear to demonstrate different evolutionary plasticities.

For example, basicranial morphology seems to be less subject to major change than tooth morphology. The contentious problem of weighting in character analysis has been considered elsewhere (e.g., by Archer 1982b and Kirsch 1982).

As a consequence, we have not accepted all apparent indications of affinity with the same degree of confidence. At the same time, however, we consider that the evidence from no single character system to be sacrosanct. The safest position is to assume that all character systems are liable to convergence and that all can on occasion undergo signficant changes in rates of change. In the present case, we have for example interpreted the often conflicting evidence about the relationships of *Tarsipes* and *Acrobates* to reflect differing rates of change within particular groups of diprotodontian marsupials.

The degree of certainty which we accord to phylogenetic hypotheses is as a result based on: 1, the degree of congruency of evidence from different character systems; and 2, the kinds of evidence involved.

Transition of Phylogenetic Understanding into Classification

Our procedure for converting phylogenetic hypothesis into classification has been relatively straightforward. The general principles of phylogenetic classification are discussed by Wiley (1981). The most important are that all taxa (insofar as can be determined) should be monophyletic and that sister taxa should be classified at equivalent rank levels. While the sister group concept might seem to some to exclude the value of recognising patristic relationships in a classification, Ridley (1985) demonstrates that patristic hyptheses are only a specialised type of cladistic statement and as such are concepts that can be incorporated into a cladistic classification.

However, in order to maximise the acceptability of our classification, we had to make two types of decisions about phylogenetic understanding: 1, which parts of the overall hypothesis should be formalised in taxonomy; 2, what are the appropriate rank levels for expressing this understanding.

Decisions about formalisation of taxomonic concepts: In the present work, we have formalised only those parts of the phylogenetic hypothesis that are well supported by independent bodies of data. This support was determined on the basis of congruency and the 'weight' accorded to evidence from particular character systems. An apparent exception here is our acceptance of the concept Ameridelphia which is in fact supported so far only by the feature of sperm-pairing. However, we have accepted it by presuming that the phenomenon of sperm-pairing is relatively significant.

In order to avoid creation of a plethora of names for each sister group relationship, we have wherever possible presented taxa in a 'ranked' manner as outlined by Wiley (1981). Thus, for example, under the Order Dasyuromorphia, the sequence of family names indicates our understanding of the sister group relationships such that Thylacinidae is indicated to be the sister group of a combined Dasyuridae-Myrmecobiidae sister group. This practice avoids otherwise having to recognise a formal rank level for the combined Dasyuridae-Myrmecobiidae taxon and a comparable superfamilial one for Thylacinidae.

Phylogenetic relationships that are poorly understood or significantly controversial we have expressed in the classification in one of two ways (both recommendations in Wiley 1981): polychotomies; and formalised statements of uncertainty.

In the first case. we have indicated unresolved polychotomies with use of the letters 's.m.' meaning *sedis mutabilis* (position changeable). For example, because we are uncertain of the interrelationships of the australidelphian orders one to another, each of the five names are followed by 's.m.'. This indicates that we have no confidence about which two (or for that matter three or four) of the five are more closely related to each other. Use of this indication of course only implies uncertainty about relative position within the next higher taxonomic rank in the classification. In other words, we are not in doubt that the five orders of australidelphian marsupials are correctly placed within Australidelphia.

In the second case, we have used the indication *'incertae sedis'* (uncertain position) after taxa whose position within a particular level of the heirarchy is either in significant doubt or threatened by clear indications that the taxon may not be monophyletic. This term is generally used in systematics to indicate an 'aberrant' taxon or one too poorly known to enable its relationships to be determined. This for example is our conclusion about Stagodontidae although we are in no doubt that it represents a monophyletic clade. On the other hand, other groups indicated to be *incertae sedis,* such as Peradectidae, contain mainly 'primitive' forms that have often been dumped into the taxon because of their lack of otherwise derived features. As a result, it is probable that among some of these relatively poorly understood and often paraphyletic taxa are the ancestors for the higher level groups within which they are placed.

Decisions about appropriate rank levels: Questions about determination of appropriate rank levels plague systematists of all procedural persuasions. Commonly choices are confined at the upper end by widely accepted rank levels for the group superordinate to the group being examined. In our case, all of the groups we have examined fall within Mammalia, a taxon widely accepted at the class level.

Within these limits, there are various ways to determine appropriate rank levels. One way would simply be to to count up the number of 'well-supported' branching points in the cladogram which one wished to formalise and then spread them out between the highest and the lowest categorical levels in use. This would be an objective approach to the problem but would result in non-equivalency of rank levels between groups because rank level would probably be determined primarily by diversity within the taxon.

A more subjective but clearly more acceptable method involves making an assessment of the relative degree of distinctness of the various taxa in order to achieve some degree of equivalency between groups. Although this sounds like a phenetic approach to classification, it only becomes so if an assessment of distinction was used during the determination of phylogenetic relationships that preceded the attempt to convert the phylogenetic conclusions into a classification. What we are talking about here is the 'jacking' up or down of complete already cladistically arranged, nested sets of taxa such that taxa falling at about the same heirarchical level display about the same degree of distinctness in morphology, ecological diversity and so forth.

Use of information of this kind in making decisions about appropriate rank levels does not threaten the integrity of the cladistic classification so long as the actual branching sequences are not affected and sister taxa are maintained at comparable rank level in the classification.

In building the present classification, we have taken the family and the order to be the two basic units of mammalian classification (see, e.g., Van Valen 1971; Lillegraven 1972 and references therein). Accordingly, we have attempted to recognise families and orders of marsupials that are roughly even in terms of distinctness.

But how is distinctness measured? To begin with, we should note that we have paid little attention to numerical diversity within the group. This diversity can be significantly effected by life style (i.e., there will be fewer carnivores than herbivores), historical factors (e.g., chance events) and geographic factors (groups on large or geographically diverse land masses are more likely to become diverse than otherwise equivalent groups on small or geographically homogeneous land masses). Hence, we are equally comfortable with a monotypic family or order as with one containing many times this diversity.

Instead, we have used three kinds of information to determine distinctness. First, we have considered morphological and ecological divergence (which of course are often correlated). With regard to morphological divergence, we have taken into consideration the degree of uniformity vs variability of the anatomical features used to determine degree of divergence. For example, we have placed more weight on variations displayed by spermatozoa, in the auditory regions and in the tarsus.

The second body of information used to determine relative distinctness is antiquity. We consider, all other things being equal, that groups with a long geological history have a greater 'distinction' than those with a short history. Again, it is important to note here that we have not used extrinsic features of this kind to determine the phylogenetic relationships of any group.

The third body of data for determining distinctness is biochemical divergence. It is, at least indirectly, an indicator of absolute age. However, this data needs to be treated with caution because major variations in the rate of change are to be expected particularly across very broad suites of taxa such as those dealt with here. Despite this caution, it is possible via the 'cladistic' method of outgroup comparison to assess the relative rate of molecular divergence for any given taxon relative to other members of a predefined monophyletic group. This data is of course limited to modern taxa but it can nevertheless give a rough idea about the antiquity of a group.

Our decision to rank marsupials at the supra-ordinal level: Relative here to the question of determining distinctness and appropriate rank level for a phylogenetically determined taxonomic concept, is our decision to classify marsupials into more than a single order. The traditional approach was the single order approach and it is still employed by some modern authors (e.g., Clemens 1979). We find this approach difficult to justify given the combined evidence for the antiquity of the major groups of marsupials based on: 1, the fossil evidence (e.g., the ameridelphian suborders represented in the late Cretaceous); 2, the degree of molecular divergence; 3, the breadth of morphological variation (often obscured in earlier works through reference only to 'typical' marsupials such as *Didelphis* and *Trichosurus*); and 4, the breadth of ecological adaptations.

Each of the eight major groups of marsupials recognised herein are at least equivalent in 'distinctness' to the placental orders. As a result, we can only conclude that the common reluctance to recognise the propriety of several orders of marsupials stems either from a deeply-rooted and virtually inexcusable anthropocentric chauvinism or a lack of willingness to change the names on museum cabinets or university teaching collections.

In view of the above, and in accordance with the principles and procedures set out above, we have taken what might be viewed as a radical action in boosting the number of marsupial orders from three (e.g., as recognised by Kirsch 1977) or four (as recognised by Ride 1964) to eight. However, in our view this is still conservative, even for the presently known taxa (e.g., ordinal status for the two subordinal taxa recognised herein within Diprotodontia could easily be defended) and must certainly be expected to rise further as the Australian fossil record becomes better known.

HIGHER LEVEL TAXA WITHIN MAMMALIA USED TO CLASSIFY MARSUPIALS

A point that commonly emerges when high-level classifications of therian mammals are compared is the lack of consistency in terms used for the highest ranks in marsupial and placental classifications. In particular, Metatheria and Marsupialia seem to be used interchangeably as are Eutheria and Placentalia. Still more confusing, Eutheria is used in two quite different ways, either as a synonym of Placentalia or as a name for the combined group of Placentalia plus Marsupialia. To add to the confusion there is an array of other taxonomic terms which appear to us to be senior although little used synonyms of taxa presently being used in a wide variety of different ways. Some of the background to these issues is discussed in Simpson (1945) and Marshall (1981). Our intention here is to provide a brief discussion of these and other taxonomic problems and to provide a justification for the classification proposed here. With regard to nomenclatural usage at higher taxonomic levels, we have followed the informal guidelines of Brothers (1983).

Subclass Theria Parker and Haswell, 1897 and Legion Rank Taxa

We have not considered the composition of subclasses other than Theria because Theria encompasses marsupials. We note recent suggestions that some groups traditionally regarded as non-therian (e.g., by Crompton and Jenkins 1979), such as Monotremata and Multituberculata, have been placed within Theria on the basis of new considerations of morphology (e.g., Kemp 1983) and the newly discovered remains of Mesozoic monotremes (Archer *et al.* 1986 and Jenkins *et al.* in press). Tentatively accepting these as therian mammals, because the precise placement of these groups within Theria is in doubt (Jenkins *et al.* ibid.), we have elected to use a phylogenetic hierarchy modified after McKenna (1975), Chow and Rich (1982) and Prothero (1981) and to leave Monotremata and Multituberculata as Theria incertae sedis.

Synapomorphies rationalising Trechnotheria, Yangotheria, Cladotheria, Zatheria and Tribosphenida are given in McKenna (1975) and Chow and Rich (1982). We have added Infralegion Dryolestoidea (whose synapomporphic states are those given by Prothero (1981) for Dryolestoidea) and Aegialodonta (whose synapomorphic states are those of the Aegialodontidae).

Infraclasses and Superdivisions

McKenna (1975) established Tribosphenida for all those mammals that have a protocone. This taxon thereby includes the Aegialodontidae. However, the appropriate name for the sister group taxon of Aegialodontia has been a matter of considerable controversy.

Gill (1872) divided living mammals into two groups: the Eutheria, containing the marsupials and placentals; and the Prototheria containing the monotremes. This is the first usage of the term Eutheria and clearly establishes its content. Although earlier authors recognised the monophyletic nature of a group containing the placentals plus marsupials (e.g., Haeckel 1866), Gill was the first to formalise the concept with a taxonomic name. This original concept of Gill's is unambiguous (see below in the discussion of supercohorts) and we have used it here.

Subsequent confusion arose from inappropriate use of the homonym Eutheria Huxley, 1880 as an objective synonym of Placentalia Owen (see below). While it is clear that Huxley (1880) did not intend his name Eutheria to become a formal taxon (as noted by Simpson 1945), it has been used in this way by a variety of subsequent authors (e.g., Gregory 1910).

As a result, in 1945 Simpson suggested that Theria Parker and Haswell, 1897 be used instead of its objective senior synonym Eutheria Gill, 1972 in order that Eutheria Huxley be available as an appropriate name for placental mammals and thereby an alternative to Placentalia Owen, 1837.

Theria in this sense, as a preferable although junior synonym of Eutheria Gill, has been used by, among others, Gregory (1910), Turnbull (1971), McKenna (1975), Lillegraven *et al.* (1979) and Prothero (1981).

Since its erection in 1897, Theria has been used in ways significantly different from its original intent. For example, Simpson (1945) redefined the concept to include, among other non-placental and non-marsupial groups, dryolestids, spalacotheriids, docodontids and paurodontids. Theria is now commonly applied to a vast array of mammalian subclasses even including (on the basis of recent work) the monotremes and possibly multituberculates (Archer *et al.* 1986; Jenkins *et al.* in press).

As recommended by Brothers (1983), as long as a redefined higher level taxon (above family group level) retains at least its original families, the name remains available despite addition of additional family level taxa.

As a result, while Theria is an available taxon, as now used it is no longer a junior synonym of Eutheria Gill, 1872 and the latter name is available as an appropriate sister taxon for Superdivision Aegialodontia. Eutheria Gill, 1872, used in this way, entirely preserves the concept of Gill (1872) that it be a taxon uniting marsupials and placentals and expressing understanding of their monophyly.

If, on the other hand, we were to accept Simpson's (1945) suggestion that Eutheria Gill be suppressed because it confounds use of its junior subjective homonym Eutheria Huxley, we would need to create a new superdivisional taxon to replace Eutheria Gill. For reasons noted below, we do not accept that Eutheria Huxley, 1880 should be treated as a senior synonym of Placentalia Owen, 1837 and, therefore, consider that Eutheria Gill is an available and appropriate taxon for this superdivisional level.

Divisions: Didelphia De Blainville, 1834 and Monodephia De Blainville, 1816

De Blainville (1816) was the first to recognise major subgroups of the modern kind within Mammalia. He named the placental mammals 'monodelphs' and a combination group, containing what we now call monotremes and marsupials, the 'didelphs'. Both of these he regarded to be subclasses of Mammalia. De Blainville subsequently revised 'didelphs' to exclude monotremes (which he called 'ornithodelphs'; De Blainville 1834).

We propose use of De Blainville's name Monodelphia (modified to satisfy nomenclatural convention) at the division level. Monodelphia has been used in this sense by, among others, Gill (1872), Gregory (1910), Jones (1923) and Gregory (1947). We have not used Placentalia Owen at the Division level because of the need to have a superordinate taxon that combines Placentalia (as the taxon containing all cohorts of what are commonly regarded to be placental mammals) with what we regard to be its plesiomorphic sister taxon containing Pappotheriidae (see below). Monodelphia De Blainville serves as this superordinate taxon.

Synapomorphic states (not autapomorphies) for Monodelphia would include marked reduction of stylar cusps, reduction to three adult molar teeth and molarisation of the last premolar.

We also propose use at the division level of De Blainville's (1934) concept and name Didelphia (similarly modified to satisfy convention).

Didelphia has been used in this sense by, among others, Gill (1872), Gregory (1910) and Jones (1923). Here also we have not used Marsupialia because of the need to have a superordinate taxon that contains Supercohort Marsupialia plus Supercohort Protodelphia, the latter being a new taxon (containing *Holoclemensia texana*) that we regard to be the plesiomorphic sister taxon of Marsupialia (see below).

Synapomorphic states of Didelphia would include twinning of the entoconid and hypoconulid, enlarged stylar cusp C, enlarged metacone, loss of dyphyodonty in the post-canine cheekteeth with functional replacement of M1 by P3, alisphenoid floor for the middle ear and pseudovaginal birth canal (the last two being unknown in *Holoclemensia*).

Supercohorts: Marsupialia (Illiger, 1811), Placentalia Owen, 1837, Protodelphia Archer, 1984 and Pappotherida (Butler, 1978) new rank

Huxley (1880) applied the names Prototheria (to the egg-laying monotremes), Metatheria (to the marsupials) and Eutheria (to the placentals) not as formal taxonomic names but as an indication of what he perceived to be a hypothetical sequence of evolutionary development. Thus he believed that ancestral monotremes gave rise to ancestral marsupials which in turn gave rise to ancestral placentals, the latter being in his concept the 'true' live-bearing mammals. In the same way Huxley (1880) conceived hypothetical categories for other groups like primates such that there were 'Protoprimates', 'Metaprimates' and 'Euprimates'.

These hypothetical terms were not intended by Huxley to become the formal taxonomic names for the groups they described (many of which do not even exist) although that is how some people have used them. Simpson, in particular, used Eutheria Huxley, 1880 in preference to Placentalia Owen, 1837 as the name for placental mammals citing as justification for ignoring nomenclatural priority and propriety, common usage of Huxley's term.

Use of Huxley's term Eutheria as a formal taxon of any kind is, however, inappropriate for one basic reason: it postdates Gill's (1872) original use of the same term for an entirely different concept (the one used in our present classification). Eutheria Huxley, 1880 is clearly a junior subjective homonym of Eutheria Gill, 1872.

Both concepts of Eutheria have been used since 1880 although, as Simpson (1945) notes, Eutheria Huxley has been more frequently used. Eutheria Gill has been used by, among others, Beddard (1902), Osborn (1910), Vandebroek (1961, 1964), Kermack (1964) and Turnbull (1971). As a result, lack of usage cannot be cited as justification for abandoning the senior homonym Eutheria Gill, 1872 in favour of Eutheria Huxley, 1880.

Placentalia Owen, 1837 is available as the senior synonym of Eutheria Huxley, 1880. It has been almost universally used in this sense (i.e., placental mammals) in the non-taxonomic scientific literature and is unambiguous in the strict taxonomic sense in contrast to the term Eutheria Huxley, 1880. Although it has not been used in the formal taxonomic sense as frequently as Eutheria Huxley, Placentalia has been used in the sense of Owen by, among others, Bonaparte (1837), Simpson (1945 p. 164: "It was De Blainville (in 1816) who first made subclasses on this basis, the monodelphs, our Placentalia, and the didelphs, our Marsupialia and Monotremata"), Gregory (1910), Winge (1923) and Gregory (1947).

The only common objection to use of the term Placentalia Owen, 1837 of which we are aware concerns the presence of a placenta in marsupial as well as placental mammals. An anatomical fact of this kind is not, however, a basis for rejecting the validity of a taxonomic name. Placentalia Owen is no less appropriate than thousands of other valid but etymologically inappropriate names like Monotremata (because most animals from protistans to pogonophorans also have just "one-opening"), Dasyuridae (because almost all other mammals have "hairy tails"), *Chaeropus* (because this bandicoot does not in fact have a "pig's foot") or *Dendrolagus* (because this kangaroo is not a "tree rabbit").

Even the conceptual and hence etymological basis for Huxley's term Eutheria is no longer appropriate, placental and marsupial mammals being regarded as independently specialised sister taxa of each other rather than an ancestor-descendant sequence as implied by Huxley's use of the prefixes 'proto', 'meta' and 'eu'.

The most appropriate term for the mammalian group informally called placentals is Placentalia Owen. Use of the term Eutheria Huxley for this concept runs counter to the general nomenclatural principle of priority and perpetuates nomenclatural ambiguity.

Related to use of the term Eutheria Huxley, 1880, is use of the term Metatheria Huxley, 1880. This has frequently been used as an objective synonym of Marsupialia Illiger, 1811. Since Simpson (1945), however, it has sometimes often been used at a hierarchical level superordinate to Marsupialia but universally to include nothing other then Marsupialia.

Its persistence seems to us to reflect nothing more than a common urge to have a euphonic term to balance Eutheria Huxley: Metatheria sounds better than Marsupialia in opposition to Eutheria. However, again Huxley's term, when used as nothing more than a euphonic synonym for Marsupialia, serves no purpose not already served by the senior synonym Marsupialia.

Authorship of Marsupialia in the sense used here is in some doubt. Although the term was coined as such by Illiger (1811), it was applied at the family level. Our use of the term at the super-cohort level more strictly follows Cuvier (1817) who used Marsupiaux at the ordinal level.

The two other supercohorts recognised here are Pappotherida and Protodelphia. The first contains *Pappotherium* and possibly *Slaughteria* (fide Fox 1980) and *Prokennalestes*.

Protodelphia includes *Holoclemensia*.

Although recognition here of the Supercohort Protodelphia for *Holoclemensia* does not necessitate establishment of a new family, we consider that because our classification is based primarily on understanding of the phylogenetic relationships of family level taxa, it would be appropriate to formalise a family level taxon for *Holoclemensia*. This family is defined as follows:

FAMILY HOLOCLEMENSIIDAE NEW

Type genus: Holoclemensia Slaughter, 1968.

Additional included genera: none.

Family diagnosis: didelphian mammals that differ from all others (i.e., from all marsupials *sens. strict.;* the present work) in having a hypertrophied stylar cusp C and in which stylar cusp A is greater in size than stylar cusp B.

Discussion of Holoclemensiidae: Holoclemensia texana, the only known species of the genus, was established as a marsupial by Slaughter (1968, 1971) based on features of upper and lower molar morphology interpreted by him to be synapomorphies shared with other marsupials. This determination has been supported by, among others, Sige (1972), Crompton and Kielan-Jaworowska (1978) and Fox (1975, 1980). Other authors, including Turnbull (1971), Tedford (1974) and Kielan-Jaworowska, Eaton and Bown (1979), have suggested that *Holoclemensia* would be better regarded as a therian of uncertain affinities. Most commonly, it has been classified as a therian of "metatherian-eutherian" grade.

We concur with Slaughter (1968) and Fox (1980) that *Holoclemensia* displays features otherwise unique to marsupials (e.g., the large size of stylar cusp C, the large metacone and an approximated entoconid and hypoconulid) and that this is sufficient reason for hypothesising the monophyly of a group containing marsupials plus *Holoclemensia*.

It is possible that *Albertatherium* should be included in Holoclemensiidae following the discussion of Fox (1980) although Fox did not specifically suggest a special relationship between this genus and *Holoclemensia*. Because of doubts about the significance of the range of morphology noted by Fox for species of this genus, we have not taken the step of including it in Holoclemensiidae.

It is also possible that the late Cretaceous *Bistius bondi* (Clemens and Lillegraven 1986) should be referred to Holoclemensiidae based on the text and illustrations given by Clemens and Lillegraven (1986). These show an enlarged metacone, very large stylar shelf that appears to share features with *Holoclemensia* such as hypertrophy of the parastylar region and the particularly large stylar cusp C (although this cusp is even larger in *Holoclemensia* where its extreme size is presumably an autapomorphic condition). Clemens and Lillegraven (1986) note particular resemblances to species of *Alphadon* (peradectid marsupials) and *Glasbius* (regarded here to be a caroloameghiniid marsupial) but nevertheless place it in the paraphyletic group "Tribotheria". While we consider that *Bistius*, like *Albertatherium*, should be placed in Division Didelphia, we are uncertain about its supercohort affinities.

Marsupial Cohorts: Ameridephia Szalay, 1982 and Australidelphia Szalay, 1982

The cohort level is the highest currently in use within placental and marsupial classifications. In general, this rank level is taken to represent the fundamental clusters of orders, the origins of which are regarded with some degree of confidence to be independent of each other.

Use of the cohort rank level has a long historical precedent within classifications of placental mammals (e.g., Gregory 1910, Simpson 1945, Szalay 1977 and Novacek 1986). Szalay's (1982) highly innovative recent classification represents the first attempt to recognise such fundamental divisions within Marsupialia and the first use of the cohort rank level in marsupial classification.

As noted previously, Szalay (1982; classification D in Table 1) established two cohorts, Australidelphia (Australian marsupials plus microbiotheriids) and Ameridelphia (all other marsupials).

Of these, Australidelphia appears to be strictly monophyletic on the basis of tarsal morphology (Szalay 1982) and preliminary results of MC'F using albumin (Sarich pers. comm. to Archer, 1986). Szalay (1982) postulated that australidelphians descended from a common ancestor whose tarsal morphology closely resembled that of the living *Dromiciops australis*. Derived features of the australidelphian tarsus include: 1, a lower ankle joint pattern (CLAJP) where a single continuous facet is present on the respective surfaces of the astragalus and the calcaneus; 2, the distal sustentacular facet (Sud) is absent; 3, the calcaneocuboid joint (CCJ) has been reshaped by a distolateral lengthening of the calcaneus to produce an angled articular surface; 4, the articulation between the ectocuneiform and the first metatarsal joint (EMIJ) is sellar-shaped which allows for greater mobility of the hypertrophied first digit.

In support of the evidence for Australidelphia from tarsal morphology, Szalay (1982) notes a number of similarities in dentition between *Dromiciops* and Australian marsupials as follows: 1, hypertrophied incisors; 2, well-developed heels on the lower incisors; 3, much less open trigonids on the lower molars; and 4, a small and transversely short mandibular condyle.

However, based on the results of wider studies of dental and mandibular morphology in marsupials as a whole (e.g., Archer 1976a; Kirsch and Archer 1982; Reig *et al.* 1987), we are cautious about the phylogenetic significance of these features. For example, although incisors for many of the important groups of Tertiary marsupials are unknown, they are relatively large (compared with modern didelphids) in at least some peradectids and North American and European Tertiary didelphids (Crochet 1980; Fox 1983). In accordance with this evidence, it therefore seems more likely that relatively large incisors are the plesiomorphic condition with the small incisors of living didelphid marsupials being the derived condition. In regard to the significance of the presence or absence of distal heels on the lower incisors, we consider that the degree of development of these structures is stongly correlated with size of the incisors, thereby contributing little if any independent phylogenetic information. The size of the mandibular condyle is also strongly correlated with body size. The condyle is small and narrow in all small marsupials and is larger and relatively broader in the larger species of nearly all marsupial groups.

Some additional support for Australidelphia might come from Marshall's (1987) suggestion of special dental similarities between the Palaeocene South American microbiotheriid *Mirandatherium* and the Miocene Australian *Ankotarinja*. According to Marshall, these similarities include: 1, a reduced M_5 relative to M_4; 2, M_{2-4} with talonids that are basined and wider than their trigonids; 3, a lack of compression on the M_2 trigonid with a low paraconid aligned with the metaconid and hypoconid; and 4, enlarged and subequal hypoconids and entoconids. Archer (1976d) noted similarities between *Ankotarinja* and other non-Australian forms but

considered these to be shared primitive features. For the present, we support the earlier interpretation, although we agree that *Ankotarinja* might well repay a fresh examination in the light of the many new phylogenetic hypotheses presented since 1976.

Other potential support for Australidelphia comes from aspects of reproductive anatomy. Temple-Smith (1987; this Volume) reports marked similarities between the spermatozoa of *Dromiciops* and several diprotodontian marsupials. Again, it is difficult to determine whether these similarities are more likely to be symplesiomorphic or synapomorphic. An important point, however, is that the material examined of *Dromiciops* (though not well preserved) gave no indication that sperm pairing occurs in this species which is otherwise a distinctive derived feature shared by all other living South American marsupials. More adequately preserved material of *Dromiciops* is urgently needed in order to determine whether or not sperm pairing occurs in this species.

Another potentially derived feature that *Dromiciops* shares with all Australian marsupials is the condition of having its teats restricted to the abdominal region within an at least incipiently developed pouch (Reig *et al.* 1987 and references noted therein). This contrasts with the seemingly more plesiomorphic condition found in certain didelphids (e.g., species of *Thylamys* and *Monodelphis*) wherein the teats are arranged as in placentals along latero-ventral lines from the pectoral to the inguinal regions. However, it is important to note that in certain other didelphids and caenolestids the teats also occur only in the abdominal region.

Cytology may provide further support for the monophyly of Australidelphia. Sharman (1982) suggests that the 2N=14 karyotype of *Dromiciops*, unlike that of didelphids and caenolestids, is more similar to that of certain Australian marsupials such as vombatids, peramelids and burramyids. On the other hand, results of G-banding studies carried out by Rofe and Hayman (1985) suggest that the 2N=14 of Australian marsupials and at least some didelphids (e.g., *Caluromys derbianus*) share a very high proportion of homologous, conserved autosomal sections. For this reason it is likely that any special similarity between the 2N=14 of *Dromiciops* and Australian marsupials may be the result of relatively minor rearrangements. More detailed banding studies of the *Dromiciops* karyotype are required before the significance of these similarities can be determined.

In terms of serology, Kirsch (1977) found *Dromiciops* to lie roughly equidistant from the main didelphid assemblage and the main cluster of Australian marsupials. While Kirsch has cited his results as evidence that contradicts the notion of australidelphian monophyly, we see no major conflict. It is clear from the fossil evidence that microbiotheriids were already distinct from other australidelphians by the late Palaeocene (Marshall 1987) and thus lack of a close serological resemblance between *Dromiciops* and any other australidelphian group is to be expected.

The serological results of Sarich (pers. comm. to Archer, 1986) provide an interesting contrast to those of Kirsch. Sarich used microcompliment fixation (MC'F) of albumin as a basis for examining serological relationships and found that *Dromiciops* clearly grouped with the Australian rather than the American marsupials. Even more specifically (as discussed below), it appeared to exhibit a particular affinity to peramelids and dasyurids.

Cohort Ameridelphia, as originally formulated by Szalay (1982), was clearly designed to be a convenient, geographically-based 'wastebasket' taxon in which were placed all marsupials that (a) either lacked the distinctly australidelphian tarsal specialisations or (b) were relatively ancient taxa unrepresented by tarsal elements. One group that he included within this cohort was the expanded family Pediomyidae which was itself explicitly stated to approximate the basal stock for the rest of the marsupials.

Our usage of Ameridelphia differs substantially from that of Szalay inasmuch as we regard our taxonomic concept to be strictly monophyletic. Accordingly, we have had to exclude from our concept of Ameridelphia (Table 2) three family level taxa regarded by Szalay to be part of this group: Stagodontidae, Pediomyidae and Peradectidae. The reasons for excluding each of these with details of their content and comments about their possible relationships to other groups are included in a later section of this paper.

In contrast to the variety of different tissue systems that indicate or at least suggest monophyly of Australidelphia (see above), the monophyletic status of Ameridelphia rests on a single derived feature: the presence of epididymal sperm-pairing.

Although obviously not knowable for any exclusively fossil taxa, sperm-pairing has been documented in members of two ameridelphian families (Didelphidae and Caenolestidae) each of which is placed in a separate order of Ameridelphia. The highly derived and decidedly singular feature of sperm-pairing thus serves to unite two of the three orders of ameridelphian marsupials recognised here. As shall be discussed further in a later section, the third order of entirely extinct ameridelphians (Sparassodonta) is united indirectly to these taxa via apparently synapomorphic tarsal morphology shared by members of this group and Paucituberculata.

Recognition of a higher taxon on the basis of a single derived feature clearly runs against the general principle, stated in our introductory remarks about systematic procedure, that any formal concept should be well-supported by a variety of congruent and independent lines of evidence.

In this case, however, we regard sperm-pairing to be an exceptionally 'weighty' indication of monophyly between those groups in which it occurs. By way of support for this conclusion, we note that sperm-pairing is not found in any of the australidelphians so far examined (only three families remain uninvestigated), in any placental mammal or monotreme, or indeed in any other lineage of vertebrates. Sperm-pairing of a somewhat different kind does occur, however, in several unrelated groups of invertebrates.

As indicated above, we are aware of no other biochemical or morphological evidence that would support the monophyly of Ameridelphia. Serologically, caenolestids fall at least as far from the didelphids as they do from the Australian marsupials (Hayman *et al.* 1971). As with the similar situation for *Dromiciops* and the Australian marsupials, however, such a great divergence between the two American families is entirely consistent with their full ordinal separation and their postulated divergence in the late Cretaceous or earliest Tertiary.

The question of whether any dental or skeletal features might prove to be diagnostic for ameridelphians is currently under study by Case, Marshall and Woodburne (pers. comm.). Preliminary results suggest that such a morphological diagnosis is possible although this will depend on the initial assumptions about the phylogenetic status of Peradectidae and Pediomyidae.

CLASSIFICATION WITHIN THE COHORT AMERIDELPHIA

As noted previously, the internal classification of Ameridelphia has been the subject of considerable conflict. The classifications of Szalay (1982), Reig *et al.* (1987) and Marshall (1987) each differ significantly in this regard.

One possible interpretation of this conflict might be that the phylogenetic relationships of the ameridelphian groups are too poorly understood to enable a concensus of opinion in classification. However, we suspect that this would be a misinterpretation of the basis for this conflict.

Comparisons of the strictly phylogenetic conclusions (as opposed to the taxonomic conclusions) of Szalay, Reig *et al.* and Marshall reveal substantial areas of basic agreement. The variation in classifications thus seems to mainly reflect differences in approach about how to express phylogenetic understanding in classification (see our introductory remarks about phylogenetic systematic procedure).

Based on our reading of the evidence, there are three major areas of consensus about ameridelphian phylogenetics. These are: 1, recognition that the modern didelphids represent a relatively close-knit clade which together with a restricted suite of close fossil relatives, are not at all closely related to the majority of late Cretaceous and Palaeogene marsupials with which they have in the past been commonly classified; 2, recognition of a major clade containing the key families Caenolestidae, Polydolopodidae and (in the case of Szalay 1982 and Marshall 1987) Argyrolagidae, along with other related taxa; and 3, rejection of didelphoid ancestry for borhyaenoids, which thus are seen as a major lineage in their own right.

The present consensus view would thus indicate an early differentiation of the Ameridelphia into three major clades: a 'didelphoid' clade leading to the modern didelphid radiation (our Didelphimorphia); a 'borhyaenoid' clade with no living representatives (our Sparassodonta); and a 'caenolestoid-polydolopoid' clade of which only caenolestids survive (our Paucituberculata). All three major groups are interpreted to have descended from a common 'basal' stock that displayed an essentially plesiomorphic marsupial dentition and tarsus. While there is some conflict about recognition of the taxonomic identity and composition of this ancestral group, it is generally agreed that it would have been similar in aspects of its dental morphology to the species of *Alphadon* (i.e., a peradectid in terms of our present classification), and would have had a tarsal complex similar to those described by Szalay (1982) as representatives of his basal 'Pediomyidae'.

One area of disagreement concerns interpretation of the interrelationships of these three major ameridelphian clades. Szalay (1982) recognised a sister group relationship between 'borhyaenoids' and 'caenolestoids-polydolopoids' because of what he interpreted to be shared-derived states in tarsal morphology. In contrast, the 'phylogram' of Reig *et al.* (1987) appears to imply a closer relationship between the 'didelphoids' and the 'borhyaenoids', with the caenolestids (plus their relatives) having been derived from within a basal 'microbiotherioid' stock. Marshall (1987) questions the validity of Szalay's (1982) concept, but seems to argue against recognition of any special 'didelphoid-borhyaenoid' relationship.

In arriving at the arrangement presented in our classification, we have accepted Szalay's (1982) conclusions about the phyletic distinction of the Didelphoidea and the sister group status of the 'borhyaenoids' and the 'polydolopoid-caenolestoid' clade. However, whereas Szalay (1982) included all of these groups within a single order (Didelphida), we recognise three separate orders, one for each of the three major

clades within Ameridelphia. In support of this arrangement, we note the striking degree of morphological differentiation between each of the three groups and their very long independent histories. This has been clearly demonstrated by the recent description of the caroloameghiniid *Roberthoffstetteria nationalgeographica* (a paucituberculatan) from the Maestrichtian (late Cretaceous) of Bolivia.

Order Didelphimorphia (Gill, 1872) new rank

Although the concept of Didelphimorphia has been used in a very broad sense in recent years (e.g., Kirsch 1977 to encompass all of the American 'marsupicarnivores'), we favour a more restricted use that more closely reflects the original concept of Gill (1872). We recognise two families of didelphimorphians: the Didelphidae (late Palaeocene? to Recent); and the Sparassocynidae (late Miocene to early Pleistocene).

As indicated above, recent morphological studies of Crochet (e.g., 1980), Szalay (1982), Reig *et al.* (1987) and Marshall (1987) have resulted in significant changes in the content of Didelphidae. Prior to Szalay's (1982) analysis, Didelphidae was treated as a taxonomic wastebasket for all American and European 'didelphid-like' marsupials. Hence, at one time or another, the following taxa (in the form used in the present classification) have been allocated to the Didelphidae: Peradectidae; Pediomyidae; Caroloameghiniidae; Glasbiinae; Stagodontidae; Microbiotheriidae; Prepidolopidae; Bonapartheriidae; and even the placental *Cimolestes* (e.g., by Simpson 1945). Even *Holoclemensia* has in the past been referred to the Didelphidae, an allocation that would have afforded to the Didelphidae the extraordinary duration (for any mammalian higher taxon) of over 100 million years.

Szalay's (1982) restricted family Didelphidae includes only those taxa exhibiting the following derived tarsal features: an upper ankle joint, UAJ, that is broad and curved due to the transverse size increase of the fibuloastragalar facet, AFi2; a calcaneocuboid joint, CCJ, that has a well-developed cuboid pivot (i.e., an odontoid process) that articulates with a distinct pit-like concavity at the distal end of the calcaneus; and an astragalonavicular joint, ANJ, that has a definite dorsal torsion resulting in the lateralmost area of the distal astragalonavicular contact being positioned anterodorsally. Based on Szalay's illustrations (Figs 6-7), these features are clearly represented in members of both didelphid subfamilies (i.e., the calauromyine species of *Glironia* and the didelphine species of *Didelphis* and *Chironectes*). The dental features commonly cited as a basis for a wider concept of Didelphidae were rejected by Szalay as either marsupial or even therian symplesiomorphies. However, restricting membership in the Didelphidae to marsupials that demonstrate the derived tarsal features noted above would mean exclusion of the vast majority of Cainozoic didelphid-like marsupials because very few of these are known from tarsal bones.

Based on his studies of the dentition of European 'didelphids', Crochet (e.g., 1980) divided the family into two groups, the Peradectini, including species of some of the late Cretaceous and early Tertiary genera such as *Alphadon* and *Peradectes,* and the Didelphini, including the living taxa and a variety of American and European fossil taxa. The primary basis for this division was the shape of the centrocrista (between the paracone and metacone): plesiomorphically straight in members of the Peradectini, but V-shaped in members of the Didelphini.

Reig *et al.* (1987) elevated Crochet's taxon to familial rank as the Peradectidae. Implicit in their action, though nowhere stated, is the converse assumption that didelphids *(sens. strict.)* are at least in part defined by their possession of a V-shaped centrocrista. Certainly this is the case for most of the taxa regarded by them to belong in Didelphidae; however, one Miocene species referred to the Didelphidae has a straight centrocrista *(Hondadelphys fieldsi;* see below).

Marshall (1987) used these and other dental criteria to determine the familial and subfamilial affinities of the diverse Itaborian marsupials. Although he recognised many of these Palaeocene taxa as didelphids on the basis of dental morphology, the extraordinary diversity of Itaborian didelphids led Marshall to erect several new subfamilies within this family.

We are cautious about accepting the conclusions of Marshall (1987) regarding the didelphid affinities of many of these Itaborian forms. Unhesitating inclusion of this vast range of morphological types within the family undermines the importance of Szalay's (1982) observations about the apparently monophyletic nature of his more resricted concept of Didelphidae. Didelphidae in the sense of Marshall (1987) would once again have an enormous range in time (60 ma) and morphology.

The problem is to determine which if any of the Itaborian taxa should be included within Didelphidae *(sens.* Szalay 1982), which would be more appropriately recognised as distinct families albeit allied to the Didelphidae and which, if any, should be excluded from any close relationship to the Didelphidae? From Marshall's (1987) comments, it is clear that he regards at least one of his subfamilies (Didelphinae) to be based entirely on symplesiomorphic features and another (Derorhynchinae) to be only tenuously referable to the Didelphidae (e.g., Derorhynchinae). Unfortunately, no associated tarsal bones

are available for any of the Itaborian 'didelphids'. However, Szalay (1982) does state that he examined a 'sample' of Itaborian tarsal elements which he lists as representing his paraphyletic stem group Pediomyidae *(sens. lat.).* For the present, we propose that the Itaborian 'didelphids' be listed collectively, though at the appropriate subfamilial or generic rank, under 'Didelphimorphia: Family *incertae sedis'*.

Similar uncertainty should probably focus on Marshall's (1987) new subfamily Protodidelphinae which was placed by Marshall in the Caroloameghiniidae because of the relatively brachyodont molars of its members. However, because Marshall (1987) notes that protodidelphines differ from other caroloameghiines in having V-shaped centrocristae, one should not rule out the possibility that they could not be part of a monophyletic group containing the didelphids *(sens. strict.).*

As a final consideration about the classification of didelphimorphians, we question the rationale for treating caluromyines as didelphids. We consider it worth noting that caluromyines and didelphines show consistent and significant differences (as thoroughly noted by Reig *et al.* 1987) in cheektooth morphology, palatal and auditory anatomy, teat number (reduced in caluromyines), penis and spermatozoan morphology and female reproductive anatomy. Further, Kirsch (1977) regarded didelphids *(sens. lat.)* to be more divergent serologically than other marsupial families, with *Caluromys* being the most divergent of the taxa examined. MC'F values between genera of didelphids have also been found to be higher than expected compared with MC'F values separating the genera of other marsupial families. Albumin immunological distances (AIDs) of up to 80 units are indicated, although values for *Caluromys* itself are somewhat lower (evidently because of a lower rate of molecular evolution for this form; Maxson *et al.* 1975).

While we have not taken the step of separating the caluromyines from the didelphids, we consider that the message is clear: diversity within even the undoubtedly monophyletic Neocainozoic Didelphidae probably exceeds that normally regarded to be 'acceptable' within a single family of mammals.

The species of *Sparassocynus* have been compared at different times with borhyaenoid and dasyuroid marsupials, but the studies of Reig (1958) and Reig and Simpson (1972) provide most support for affinity with the Didelphidae. As argued by Reig *et al.* (1987), however, presence of some unusual features in species of *Sparassocynus* suggest an early separation from other didelphimorphians. Following Reig *et al.* (1987) and others (e.g., Archer 1984b), we support the view that the species of *Sparassocynus* should be placed in a family distinct from the Didelphidae.

Order Sparassodonta Ameghino, 1894

There is no longer any significant doubt that the two families of borhyaenoids, the Borhyaenidae and the Thylacosmilidae, form a monophyletic clade. Borhyaenids and thylacosmilids dominated most of the predaceous niches in South America throughout the Tertiary, sharing these only with a few of the larger didelphids, flightless birds and later a few procyonids (Marshall 1977). The borhyaenids in particular demonstrate a wide variety of carnivorous types ranging from smaller types superficially resembling viverrids to larger almost bear-like forms. The thylacosmilids are strikingly convergent on the placental sabre-toothed cats although they achieved an even greater degree of sabre-tooth specialisation. The earliest known borhyaenoid is a late Palaeocene species of *Patene* from the Itaborian Local Fauna but it is almost certain from the occurrence of the only slightly younger (early Eocene) but highly specialised *Arminiheringia* that the family's origins must well predate the Palaeocene.

Until relatively recently, borhyaenoids were generally viewed as the carnivorous off-shoot of the otherwise mainly insectivorous to omnivorous didelphids. As noted above, this view is now giving way to acceptance of the fact that not only have borhyaenoids had a very long independent history, but that they were clearly not derived from any group of marsupials currently included within the Didelphidae.

Once again, Szalay's (1982) observations about tarsal morphology figure prominently in this changing perception of the phylogenetic relationships of the borhyaenoids. He concluded that borhyaenoids had been derived directly from a basal 'pediomyid' stock and that as a result they had nothing whatsoever to do with true Didelphidae *(sens. strict.)*.

On the other hand, Szalay (1982) does report a number of seemingly shared derived similarities between the tarsi of borhyaenoids and those of caenolestids (specifically *Caenolestes*), many of which he identifies as special adaptations to terrestrial/cursorial lifestyles in at least the more generalised members of both groups.

In accordance with this evidence, Szalay (1982) concluded that borhyaenids should be grouped with the caenolestoids in his suborder Borhyaeniformes. He also included in this group (but with less confidence) the Argyrolagidae, Groeberiidae and Polydolopoidea, the latter being unknown from tarsal elements but allied by previous authors (e.g., Simpson 1928) to the Caenolestoidea on the basis of dental morphology.

The overviews of dental evidence presented by Reig *et al.* (1987) and Marshall (1987) support phyletic separation of the borhyaenoids from Didelphidae but neither study specifically addresses the possibility of a special relationship between borhyaenoids and either all or some of the caenolestoids, argyrolagids, groeberiids or polydolopoids. The major feature of interest noted by them in the dentition of borhyaenoids is the plesiomorphically straight centrocrista on the upper molars, a feature that argues against derivation of borhyaenoids from didelphids *(sens. strict.)*. Other characteristic features of the borhyaenoid dentition include reduction of the stylar cusps and the stylar shelf, poor development of the protocone and of the associated conules, expansion of the metacone at the expense of the paracone, the related elongation of the postmetacrista and reduction of the metaconid and talonid on the lower molars. All of these features relate to carnivorous specialisations in the dentition.

The long-standing controversy (e.g., Bensley 1903; Sinclair 1905; Woods 1924; Simpson 1941; Archer 1976a; Marshall 1977; Szalay 1982; Sarich *et al.* 1982; Archer 1982b) about a possible close relationship between borhyaenoids and thylacinids is noted below in connection with dasyuromorphians. It was largely based on the striking similarity between the dentitions of members of each group as well as on a few postcranial features such as reduction of the epipubic bones in both groups. Suffice it to say here that the evidence most clearly supports the view that thylacinids are dasyuromorphians as argued by Simpson (1941), Marshall (1977), Szalay (1982), Sarich *et al.* (1982) and Archer (1982b).

Marshall (1977) noted similarities in the stylar shelf morphology of pediomyids and borhyaenoids and suggested that the latter might have originated from late Cretaceous pediomyids. This suggestion was challenged by Fox (1979) and subsequently abandoned by Marshall (1987).

Cranially, borhyaenoids are very distinctive. Features which taken together separate them from all other groups of ameridelphians include: 1, the broad naso-lachrymal contact; 2, the unfenestrated bony palate; and 3, the apparent absence of any mastoid contribution to the rear face of the cranium. In the postcranial skeleton, borhyaenids display a marked reduction of the epipubics, as well as a characteristic tarsal morphology.

As indicated in an earlier context, in the present classification we have elected not to formalise the possibly monophyly of borhyaenoids with caenolestoids and polydolopoids. There are three reasons for our decision. First, apart from the tarsal evidence, there is little evidence to support such a combined order and it would thus be considerably more vulnerable than any of the other orders recognised here (each of which is supported not only by tarsal morphology but by evidence from other independent tissue systems). Second, the diagnostic features of the 'borhyaeniform' morphotype as defined by Szalay are (as Szalay himself clearly recognises) adaptations to a terrestrial/cursorial life. While this in no way precludes the validity of the tarsal features noted by Szalay as potential synapomorphies, it seems to us that insufficient is currently known about the tarsal morphology of the earliest ameridelphians or even the members of the supposed 'Borhyaeniformes' to enable a decision about whether these features represent symplesiomorphic similarities, the product of convergence or, as Szalay (1982) presumes, indications of monophyly. Futher, without knowledge of the tarsal morphology of polydolopids or caroloameghiniids (which are placed by Szalay in his 'Borhyaeniformes'), there is not yet even tarsal morphology to support the whole of Szalay's concept. Third, for reasons noted above, we consider that there is more to recommend treatment of borhyaenoids as an order in their own right. Clearly they have had a very long, independent history (almost certainly stretching back into the Mesozoic) and display consistent morphological differences over this period. These points aside, if future study should support the monophyly of borhyaenoids with one or another of the other taxa in Szalay's Borhyaeniformes, our classification provides adequate scope for the formalisation of an appropriate supraordinal taxon to accommodate these groups.

Ameghino's (1984) term 'Sparassodonta', originally proposed at the ordinal level, is both formally available and appropriate for use as the ordinal name for 'borhyaenoids'. Unlike some of Ameghino's other ordinal names, Sparassodonta included no placental groups.

At the infrafamilial level, borhyaenid and thylacosmilid generic level diversity and phylogenetics have been thoroughly examined by Marshall (1976a, 1977, 1978, 1979 and 1981). As with the Didelphidae, however, we are of the opinion that the Borhyaenidae could be usefully subdivided into two or more family-level groups. In our view, this would simply better reflect the complex history and extraordinary adaptive breadth of the group. A relevant point here is that although Marshall's examinations (1976a, 1977, 1978, 1979) of all of the borhyaenid material enabled him to recognise four distinctive subgroups within the Borhyaenidae, it did not enable him to determine the origins or relationships of the groups one to another, three of which have independent histories at least as far

back as the early Eocene. Further, the present arrangement of just two families probably does not provide a suitable context within which to taxonomically represent the relationships of the thylacosmilids which surely must have had their origins from somewhere within the present limits of the Borhyaenidae rather than from outside of it. By way of comparison with placental classifications, we consider that the two-family system for sparassodonts would be roughly equivalent to placing all of the modern, terrestrial, placental carnivores into a single family, excepting the Giant Panda which might be singled out as a second family on account of its remarkable specialisations and (until recently) obscure origins.

As a footnote to this section we would like to draw attention to our doubts about the allocation by Marshall (1976b) of *Hondadelphys fieldsi* to the Didelphidae. The same features noted by Marshall as a demonstration of how this Miocene carnivorous marsupial had converged on borhyaenids suggest to us that it is in fact a sparassodont, albeit one with some atypically plesiomorphic features. This possibility was considered by Marshall but not pursued. In this regard we would draw attention to the presence in *H. fieldsi* of a straight centrocrista, a poorly developed alisphenoid tympanic process and no mastoid contribution on the hind surface of the skull. However, it lacks the naso-lachrymal contact otherwise ubiquitous in borhyaenids (but not thylacosmilids). One possible interpretation of these facts would be that *H. fieldsi* represents a persistant lineage that diverged from very early sparassodonts before they acquired the characteristic naso-lachrymal contact.

Order Paucituberculata (Ameghino, 1894)

The present concept Paucituberculata differs in only minor ways from Marshall's (1987) suborder Polydolopimorphia (e.g., in our exclusion of Protodelphinae). However, while we are thus in agreement with and applaud Marshall's major contribution in demarcating the monophyly of this major clade of ameridelphians, we have elected not to follow his formal concept. There are two basic reasons for this decision. In the first place, Marshall's concept differs only slightly in context (e.g., by inclusion of Caroloameghiniidae and *Gashternia*) from Paucituberculata Ameghino, 1894 in the sense that the latter name has been employed for more than twenty years (e.g., Ride 1964; Kirsch 1977 and Woodburne 1984). Continuity of the established and stable usage of Paucituberculata (Ameghino, 1894) seems preferable to adopting Marshall's new term Polydolopimorphia.

The second reason is that Marshall's term is a junior homonym of Polydolopimorphia Archer, 1984 and is thus not strictly available for use given our decision to follow the recommendations of Brothers (1983) regarding nomenclatural procedures involving higher level names. Archer's (1984b) taxon Polydolopimorphia is available nomenclaturally under these terms but differs significantly in concept from Paucituberculata as used here because it did not contain the families Caenolestidae or Argyrolagidae.

Our reasons for maintaining Paucituberculata as an independent order rather than subsuming it within some taxon equivalent to Borhyaeniformes Szalay, 1982 have been discussed above in relation to Sparassodonta.

Paucituberculata, as currently understood, encompasses marsupials of remarkable taxonomic and ecological diversity. With eight families, the family level diversity of this order is second only to that of the Australian Diprotodontia. Like Diprotodontia, it contains a variety of possum-like taxa (e.g., the caroloameghiniids, polydolopids and some of the caenolestids), more specialised ricochetal forms (the argyrolagids) and even rodent-like forms (the groeberiids). In some areas its diversity may even exceed that of Diprotodontia inasmuch as some of the shrew-like caenolestids are decidedly bandicoot-like. Further, it is not yet clear what the gashterniids, prepidolopids or bonapartheriids were like but they are certain to extend the ecological breadth of this order.

Considering its taxonomic diversity, it is not surprising that there are few common derived morphological features by which to characterise the whole order. Marshall (1987) suggested that in general members of the order display: 1, a general trend towards bunodont cheekteeth; 2, an enlargement of stylar cusps B and D to become the primary buccal cusps of the upper molars; and 3, a trend toward procumbent orientation of the lower canine and incisor dentition. These features are best seen in the early and least specialised representatives of the order, such as the caroloameghiniids and prepidolopids, but are also recognisable in the dentition of living caenolestids.

In agreement with Marshall (1987), we consider the late Cretaceous South American *Roberthoffstetteria nationalgeographica* (Marshall *et al.* 1983) and the late Cretaceous North American species of *Glasbius* (Clemens 1966) to be caroloameghiniids. In terms of known morphology, these taxa represent a suitable structurally ancestral stock for later carolameghiniids, but possibly one already too specialised by that time to have given rise to the other paucituberculatans. In either case, these Cretaceous records clearly demonstrate the great antiquity of the order as a whole.

As noted above in the discussion of didelphimorphians, we consider that the Protodidelphinae should be removed from the Caroloameghiniidae (where it was placed by Marshall 1987) on the grounds that protodidelphines appear to have V-shaped centrocristae. In this regard, they appear to be more similar to didelphids, although their bunodont specialisations suggest that they should perhaps be distinguished at the family level from Didelphidae. We have listed them here as 'Didelphimorphians: family *incertae sedis*'.

Of the remaining paucituberculate families recognised here, the Polydolopidae has the second longest and most complex fossil record. Polydolopids are known from the late Palaeocene through Oligocene of South America and from the Eocene of Antarctica (Seymour Island). The family encompasses considerable morphological diversity and may eventually be more appropriately divided into two or more families.

The Prepidolpidae and Bonapartheriidae are each so far monotypic and restricted to the Eocene of South America. Of these, prepidolopids may be structurally annectant to members of both Polydolopidae and Bonapartheriidae, however, *Prepidolops* itself, clearly occurs too late in time to have given rise to any of the other polydolopoids.

Inclusion of the Polydolopidae within Paucituberculata was justified in many earlier classifications (e.g., Simpson 1945; Ride 1964 and Kirsch 1977) primarily on the strength of a perceived close similarity in between caenolestids and polydolopids (e.g., Simpson 1928, 1948 and Paula Couto 1952). This view is now no longer accepted and specific similarities previously noted between polydolopids and caenolestids (e.g., diprotodonty) are now generally regarded to be the result of convergence. As noted above, the case for paucituberculate monophyly now rests primarily on Marshall's (1987) interpretation of the major buccal cusps in this group as being stylar cusps B and D, a view long ago advocated by Osgood (1921:121-26) for *Caenolestes*, but subsequently ignored.

The fossil record and intrafamilial taxonomy of the Caenolestidae has been reviewed by Marshall (1980). The earliest definite record of the group is from the lower Oligocene, although the diversity present at that time indicates a much earlier origin for the group (possibly even in the late Cretaceous). Known caenolestids have been divided into three groups recognised either as subfamilies (Caenolestinae, Palaeothentinae and Abderitinae; e.g., Marshall 1976; Pascual and Herrera 1973, 1975) or tribes (e.g., as employed by Marshall 1980). Some doubt must exist about the monophyly of the diverse forms placed within the Abderitinae (Marshall 1976). As Marshall (1980) recognised, at least two of these subfamilial groups are almost certainly paraphyletic with abderitines arising from within palaeothentines which in turn arose from basal caenolestines. The living caenolestids are placed in three genera within the Caenolestinae.

The other paucituberculates are the gashterniids, argyrolagids and groeberiids all of which were grouped by Marshall (1987) as argyrolagoids. Their relationships to each other and to other paucituberculates are less secure in large part because they are much less well known. In particular, very little is known about the gashterniids or groeberiids. However, based on the available evidence of dental morphology, each of these taxa appears to group with the Argyrolagidae.

Argyrolagus itself is known from reasonably complete skeletal material. This was monographed by Simpson (1970) but should perhaps be re-examined in the light of new morphological studies (e.g., Szalay 1982) and phylogenetic concepts expressed in these volumes. The earliest argyrolagid is described by Wolff (1984) from the early Oligocene of Brazil. *Gashternia* and *Groeberia* are both significantly older, being respectively at least upper Palaeocene (Itaborian) and early Eocene in age.

Considering hypotheses for higher-level phylogenetic relationships, Marshall (1987) suggests that the caroloameghiniids represent the most plesiomorphic family within Paucituberculata. However, it is unclear from this whether he regards the other paucituberculate superfamilies to represent a monophyletic clade. Certainly there are no obvious features that would support such a concept. For the present there seems to be little morphological basis or taxonomic necessity for formalising ranks above the level of superfamily within this order.

CLASSIFICATION WITHIN THE COHORT AUSTRALIDELPHIA

The rationalisation for the monophyly of Australidelphia is discussed above. It depends mainly on data from two character systems: tarsal bone morphology (Szalay 1982); and albumin serology (Sarich, pers. comm. to Archer 1986). Some additional supporting data comes from studies of cytology (Sharman 1982) and sperm structure (Temple-Smith 1987).

We have recognised five orders within Australidelphia: Microbiotheria, Dasyuromorphia, Peramelemorphia, Diprotodontia and Notorectemorphia. As best as we can determine from available and sometimes conflicting data, each of these orders appears to constitute a strictly monophyletic group and to be approximately equivalent in terms of morphological and ecological distinction (but not taxonomic diversity).

None of the orders are clustered here into specific superordinal groups because, although there are indications for such higher level clusters from particular character systems, such indications are commonly contradictory. A classic example would be interpretation of the relationships of the Peramelemorphia. If we accept that the syndactyl condition is monophyletic within Marsupialia (*fide* Szalay 1982), we would group bandicoots with Diprotodontia and Notoryctemorphia to the exclusion of Dasyuromorphia and Microbiotheria. On the other hand, if we accept serological interpretations (Kirsch 1977 and Sarich, pers. comm. to Archer), bandicoots would be grouped instead with Dasyuromorphia. As a result, we have chosen, for much the same reasons that Simpson (1945) chose not to cluster his five superfamilies, not to formalise any of the current hypotheses about the interordinal relationships of australidelphian marsupials.

Microbiotheria Ameghino, 1889

The practice of recognising the South American family Microbiotheriidae as a distinct order of australidelphian marsupials stems directly from Szalay's (1982) highly influential study of marsupial tarsal anatomy.

From the nomenclatural point of view, however, Szalay's concept Dromiciopsia (as followed in several subsequent classifications such as that of Archer 1984b and Woodburne 1984a) is strictly a junior synonym of the much older ordinal concept Microbiotheria Ameghino, 1889. In common with Szalay's Dromiciopsia, Microbiotheria as originally proposed contained only the single family Microbiotheriidae. Within this group, Ameghino placed *Microbiotherium* as well as a variety of other genera of South American fossil marsupials. Although many of these other forms have now been allocated to other groups (see Marshall 1982), this does not alter the familial name or its identity as specified by the type genus *Microbiotherium*. Accordingly, Microbiotheria Ameghino, 1889 and Dromiciopsia Szalay, 1982 are identical for nomenclatural purposes and we see no reason not to use the senior synonym.

In terms of the concept of Szalay (1982), Microbiotheria contained only *Dromiciops* and its undoubted fossil relative *Microbiotherium*, both included within the family Microbiotheriidae (following in part from the conclusions of Reig 1955).

More recently, the Microbiotheriidae (as used herein) was significantly revised by Marshall (1982, 1987) to include a number of other evidently closely related taxa of Palaeocene and Miocene age (*Mirandatherium*, *Eomicrobiotherium* and *Pachybiotherium*). We accept the microbiotheriid affinities of these taxa. Additionally, both Marshall

(1987) and Reig *et al.* (1987) have further expanded the concept of Microbiotheriidae to include, on the basis of dental morphology, the late Cretaceous to early Tertiary Pediomyinae. Reig *et al.* (1987) but not Marshall (1987) also included Glasbiinae in the Microbiotheriidae. Marshall (1987) placed Glasbiinae within Caroloameghiniidae.

We have followed the conclusions of Marshall (1987) regarding the position of Glasbiinae but not his or Reig *et al.'s* (1987) conclusions about Pediomyinae. In part our reluctance stems from the apparent absence of australidelphian-like tarsal bones in sites that yield pediomyines (see below in the section on Pediomyidae), something which we would expect if pediomyines are microbiotheriids and the Australidelphia is monophyletic.

Further, as discussed by Fox (1987), we consider that the present concept of Pediomyinae lacks evidence for monophyly and may well constitute a polyphyletic group containing a range of unrelated forms. Accordingly, we cannot justify to ourselves any conclusions about its relationships as a whole to other marsupials.

As a result, we follow the concept of Microbiotheriidae as revised by Marshall (1982) with the inclusion of *Mirandatherium travassosi* (*fide* Marshall 1987).

In terms of the interordinal affinities of Microbiotheria, it is clear that Szalay (1982) envisaged his Dromiciopsia to be a paraphyletic stem group from which all other australidelphians were derived. It was effectively defined by its members' possession of the derived australidelphian tarsal condition in combination with their otherwise 'primitive' features such as high incisor number and didactylous hind foot with a "large grasping hallux".

If Szalay's conclusions were carried over into a classification, Microbiotheria would be recognised as the plesiomorphic sister group (paraphyletic in intent if not content) of some other multiordinal taxon encompassing all other australidelphian taxa.

One reason why we have not followed this procedure here is that we now suspect that Microbiotheria may prove to be the sister group of just one or two of the other Australian orders. Such an indication comes, for example, from albumin serology (Sarich, pers. comm. to Archer) which suggests the possibility that *Dromiciops australis* may be more closely related to dasyurids and/or peramelids than to other australidelphian groups.

In this regard, it is interesting to recall the suggestion by Marshall (1987) that the mid Miocene Australian *Ankotarinja* exhibits derived microbiotheriid dental features of the kind seen in the Palaeocene South American *Mirandatherium*. This is provocative but we remain unconvinced that these similarities could not be symplesiomorphic. Otherwise, however, the dental morphology of known microbiotherians does not appear to exhibit any particularly close similarities to any other australidelphians and we presume that microbiotheriids are either plesiomorphic (e.g., incisor number) or autapomorphic (e.g., molar pattern) in dental morphology.

As Szalay (1982) notes, tarsal bone morphology of *Dromiciops australis* is at least superficially more similar to that of some diprotodontians such as acrobatids (e.g., *Distoechurus pennatus*) than it is to the members of the other australidelphian orders. With Szalay (1982) we view this similarity as symplesiomorphic retention in these groups of both a common ancestral arboreal lifestyle and an ancestral australidelphian tarsal pattern.

As noted previously by Archer (1984a), microbiotherians are uniquely primitive among australidelphians in having four lower incisors, all other australidelphians having either one, two or three. Archer (1982b) used this synapomorphic state as part of the rationalisation for distinguishing all Australian marsupials as the clade Eometatheria.

In terms of auditory structure, microbiotherians, although autapomorphically very specialised (as noted by Segall 1969 and Archer 1976), share at least some derived features (e.g., development of a large rostral tympanic process of the petrosal) with dasyuromorphians and possibly with *Notorcytes*.

A similar indication of affinity comes from the gross anatomy of the gut: *Dromiciops australis* shares with all of the members of two Australian orders (Dasyuromorphia and Notoryctemorphia) and with *Tarsipes* (alone within the Order Diprotodontia) the derived loss of an intestinal caecum (Hume 1982).

Spermatozoan morphology has been studied by Temple-Smith (1987) and its contribution to understanding of microbiotherian relationships has been discussed above (in the section on cohorts). On balance, the evidence from spermatozoan morphology supports placement of microbiotherians within Australidelphia. More specifically, Temple-Smith notes particular similarities between spermatozoan morphology of *Dromiciops* and that of phalangerids, burramyids and petaurids but concludes that these features are probably 'primitive' features of marsupial spermatozoa.

The evidence of serology is, as noted above, not clear. On the face of it, Kirsch's results (1977; and as reconsidered in Reig *et al.* 1987), based on study of the reactivity of polyvalent antisera, appear not to support the concept of Microbiotheria as advanced by Szalay (1982). On the other hand, Sarich's ongoing studies (pers. comm. to Archer), based on MC'F studies of albumin, *do* suggest a special serological relationship between *Dromiciops australis* and Australian marsupials.

On balance, there is clearly a growing body of evidence that ties microbiotheriids to the Australian marsupial radiation but none that ties it more closely to any of the other American groups. However, it is not yet clear whether the microbiotheriids represent the sister group of the whole of the rest of the Australian radiation or just a part of it. Our tentative solution to this phylogenetic uncertainty is to rank Microbiotheria as one of the five orders of Australidelphia.

Order Dasyuromorphia (Gill, 1872) new rank

The content of this taxon (established by Gill 1872 as a suborder and since used by Kirsch 1968, 1977; Archer 1984b and Woodburne 1984a), with its three contained families the Thylacinidae, Myrmecobiidae and Dasyuridae, is the same as Szalay's (1982) Order Dasyurida. The latter term is therefore a junior synonym of Gill's taxon.

Next to bandicoots (peramelemorphians), dasyurids display the most plesiomorphic australidelphian dentitions (Archer 1976a, 1984a). This fact, combined with their didactylous hindfeet (in contrast to the syndactylous hind feet of bandicoots), has led to the common view that dasyurids represent the group most similar to the stock that gave rise to the rest of the Australian marsupial radiation (e.g., Ride 1964). Although this is a justifiable view, we have doubts that it is the correct view (see comments below on peramelemorphians).

Living dasyurids are markedly homogeneous in terms of molecular biology (e.g., Baverstock *et al.* 1982), cytology (e.g., Young *et al.* 1982), spermatozoan morphology (Harding 1982) and general morphology (Archer 1976a, c, 1982a). Intrafamilial differences noted in other character systems such as dental morphology (e.g., Archer 1976a, 1982a), ecology (Lee *et al.* 1982 and Lee and Cockburn 1985) and reproductive structure (Woolley 1982) are not of the kind that lead to any doubts about the tight-knit monophyly of the family.

For this reason, we are convinced that living dasyurids represent a relatively recent radiation. This concept was suggested by Archer (1982b) in an effort to integrate otherwise noncongruent data from studies of the serology and basicranial morphology of dasyuroids.

Myrmecobius fasciatus, the only known myrmecobiid, was once thought to be a long-surviving representative of a mammal group otherwise restricted to the Mesozoic. As the Numbat became better known, this view gave way to the concept that it represented a specialised group within Dasyuridae (e.g., Leche 1891; Weber 1905; Ride 1964). More recently it has been classified as a distinct family (Archer and Kirsch 1977) and regarded to be the sister group of the Dasyuridae (e.g., Archer 1982b).

Thylacinids have been even more controversial. Most of the debate about their relationships has involved just the modern *Thylacinus cynocephalus* although there are now at least two additional older species (Archer 1982b). The principal controversy has involved the early suggestion (e.g., Sinclair 1906) that thylacines formed part of a monophyletic group with the South American borhyaenids. Opinions about this have bounced back and forth (e.g., in favour, Wood 1924; Archer 1976a; against, Simpson 1941; Marshall 1977). The most recent contributions of new data to the controversy were those of Szalay (1982; tarsal morphology) and Sarich *et al.* (1982; albumin serology), both studies concluding that thylacines were part of the Australian radiation and (from Sarich *et al.*'s data), close to (if not within) the Dasyuridae.

Archer (1982b) summarised all available phylogenetic data and concluded that Thylacinidae most probably represents the plesiomorphic sister group of Dasyuroidea (Dasyuridae plus Myrmecobiidae) and that thylacines had therefore converged dentally on borhyaenids. This view incorporates the basic conclusions of Szalay (1982) and Sarich *et al.* (1982) but not the view that Thylacinidae falls within Dasyuridae. The occurrence of at least three thylacines in the middle and late Miocene record suggested to Archer (1982b) that the relatively recent separation date for thylacines from dasyures interpreted by Sarich *et al.* (1982) on the basis of a molecular clock date must be in error and with it the conclusion that thylacines were a recently derived subgroup of the Dasyuridae. All that can be said with certainty at this point is that thylacines and dasyures are both present in middle Miocene assemblages and that all dasyurids and myrmecobiids share cranial and dental synapomorphies represented by plesiomorphic states in thylacinids.

The common ancestor of the three families would have had as derived features (within Australidelphia) a reduced incisor number (to I_{1-3}^{1-4}), an enlarged squamosal epitympanic sinus in an otherwise essentially plesiomorphic basicranium and loss of the intestinal caecum. It is probable that this ancestor was a terrestrial insectivore or carnivore.

The interrelationships of the Dasyuromorphia within Australidelphia are unclear although three principal hypotheses have at one time or another been advanced: that dasyuromorphians stand apart from all other groups of Australian marsupials; that they are the sister group of the Peramelemorphia; that they are somehow related to the Notoryctemorphia. Evidence for possible relationships to peramelemorphians and notoryctemorphians is considered below. To anticipate the conclusions, we consider that dasyuromorphians and peramelemorphians are probably more closely related to each other than either is to any other ordinal level group but are not as persuaded about a special relationship to *Notoryctes*.

Order Peramelemorphia (Kirsch, 1968) new rank

The monophyly of the members of this group have not been in any recent doubt although the number of families have. Until 1977, most authors regarded all bandicoots to be members of the single family Peramelidae. Archer and Kirsch (1977) suggest, after reviewing all potentially useful phylogenetic data, that the genera *Macrotis* and *Ischnodon* be set apart as the family Thylacomyidae. However, Archer (1984b) reviews the rationale for this conclusion and notes that it was made mainly on the basis of autapomorphic attributes. Evidence presented by Archer and Kirsch (1977) therefore would not discount the possibility that Thylacomyidae is a highly specialised sister taxon of a particular group of peramelids rather than the family as a whole. As a result, their act in excising Thylacomyidae may have rendered Peramelidae paraphyletic.

More recent studies of bandicoot morphology (Groves and Flannery in press) also suggest a bifamilial division of bandicoots but in this system *Macrotis* and *Ischnodon* are grouped with *Isoodon*, *Perameles* and *Chaeropus* as the Peramelidae while *Rhynchomeles*, *Echymipera*, *Microperoryctes* and *Peroryctes* are grouped as the second as yet unnamed family.

The broader questions of interordinal relationships of the bandicoots are far more complex and have had a long history of controversy. The common basic problem revolves about the significance of bandicoot syndactyly: is it or is it not a synapomorphic condition shared with diprotodontians? Researchers have in the past been divided (e.g., Jones 1923 'yes'; Abbie 1937 'no') and remain divided (e.g., Szalay 1982 'yes'; Kirsch 1977 'no') in their opinion about the best answer to this question. As noted above, recent morphological studies of the syndactylous condition (Marshall 1972; Szalay 1982 and Hall 1987) as yet have given no basis for suspecting that the condition is convergent (which is of course not necessarily the same thing as demonstrating that it is synapomorphic).

In many early classifications, single character systems were used to divide the Australian marsupials into two groups. Where dental morphology served as this guide, on the basis of polyprotodonty (and molar morphology) bandicoots were grouped with dasyuromorphians (e.g., Owen 1866; Thomas 1888; Gregory 1910); where the presence or absence of syndactyly served as the primary guide, bandicoots were grouped with diprotodontians (and sometimes *Notoryctes;* e.g., Bensley 1903).

In most recent classifications (e.g., Szalay 1982; Archer 1984a), the similarities between the dentitions of bandicoots and dasyuromorphians have been perceived to be plesiomorphic while their syndactylous condition has been identified as a synapomorphy shared with diprotodont marsupials. For these reasons, bandicoots have commonly been regarded as an annectant group linking the plesiomorphic dasyuromorphians to the diprotodont marsupials (e.g., Ride 1964; Szalay 1982; Archer 1984b).

In terms of phylogenetic systematics, the syndactylous condition would be perceived as a potential synapomorphy whereas the dental features that relate bandicoots to dasyuromorphians would be symplesiomorphies. Consequently, without additional contradictory data, Syndactyla would be a justifiable taxonomic concept.

Unfortunately for this appealing interpretation, biochemical studies have failed to support the monophyly of bandicoots with diprotodontians. Kirsch (1968, 1977), for example, presents evidence based on polyvalent antisera that bandicoots grouped with dasyuromorphians rather than diprotodontians. His conclusion was corroborated by Maxson *et al.'s* (1975) and Baverstock *et al.'s* (in press) microcomplement fixation (MC'F) studies of albumin. Sarich (pers. comm. to Archer, 1986) has also reiterated the conclusion based on studies of albumin that bandicoots and dasyuromophians are apparently more closely related to each other than to diprotodontian marsupials. The study of Baverstock *et al.* (1987) is particularly hard to fault because they also demonstrate that their conclusions could not have resulted from variations in the rate of albumin evolution.

Baverstock *et al.* (1987), in considering the phylogenetic significance of syndactyly, suggest either that it is (as has often been presumed) convergent or that it is in fact the symplesiomorphic condition for all Australian marsupials, the didactylous condition of dasyuromorphians being a derived state within the Australian marsupial radiation.

Consideration of other evidence from other character systems has not helped to resolve the conflict. Studies of bandicoot molar morphology (e.g., Bensley 1903; Archer 1976a, 1976b, 1984b and Woodburne 1984a) have produced a plethora of often contradictory suggestions. Bensley (1903), while noting the basic similarity of the bandicoot molar pattern to that of dasyurids and didelphids points out that they share with diprotodontian marsupials development of a hypocone (or metaconule in the sense of Tedford and Woodburne 1987; this Volume). Archer (1976a) notes that many of the dental features of bandicoots that make them stand out among Australian marsupials are also present in 'didelphid' marsupials and hence possibly indicative of the essentially plesiomorphic structure of bandicoots among Australian marsupials. Archer (1976b) also notes similarities between bandicoot dilambdodont upper molars and those of selenodont diprotodontians suggesting that the latter may have been derived from the former. Among the similarities noted were the enlarged stylar cusps B, C and D (accidentally misrepresented in that paper as stylar cusps A, B and D respectively) and their incorporation as the buccal ends of the dilambdodont crests.

However, because dilambdodont molar patterns have developed independently many times in mammal evolution, it is at least as likely that the selenodont pattern of diprotodontian marsupials developed independently from a non-peramelemorphian ancestor. Further, the common involvement of at least stylar cusps B and D may be a plesiomorphic predisposition within marsupials (Fox 1987; this Volume) and not in itself evidence for a special bandicoot-diprotodontian relationship.

Some support for the notion of a special phyletic relationship between bandicoots and diprotodontians might be drawn from studies of steroid-binding glycoproteins (SBG) by Sernia *et al.* (1979), Bradley (1982) and others (see references in these papers). In brief, these studies have revealed that whereas dasyurid and didelphid marsupials possess only a single SBG (CBG — corticosteroid binding globulin), bandicoots and diprotodontians tend to possess a second SBG similar to placental sex-hormone binding globulin (SHBG). For several reasons, we hesitate to attach too much phylogenetic data to these results. Firstly, it should be noted that SHBG was not found in all diprotodontians examined (for example, it was absent in some macropodids and in *Vombatus ursinus).* Moreover, the characterisations of the 'bandicoots' as having SHBG and of the 'didelphids' as lacking SHBG were based on tests involving only one species from each group. Clearly, while comparative endocrinological studies of this kind hold considerable potential for revealing phylogenetic relationships at all taxonomic levels, in this case we must conclude that there is as yet insufficient evidence to assess whether or not SBGs have anything of value to contribute to the debates about higher level relationships within marsupials.

Considering basicranial morphology, Archer (1976c) notes that bandicoots include among their rank forms that are remarkably didelphid-like. In relatively plesiomorphic species like *Microperoryctes murina,* bandicoots exhibit character states (e.g., the lack of a squamosal epitympanic sinus) more plesiomorphic than those of any other Australian group of marsupials. However, it is also true that some of the more 'derived' bandicoots are at least similar to dasyuromorphians in their pattern of middle-ear pneumatisation. Further, in contrast to all other Australian marsupials, *Notoryctes,* bandicoots and dasyuromorphians have imperforate stapes (Doran 1877; Segall 1970 and Archer 1976c).

Other features that at least suggest the possibility of a special dasyuromorphian relationship include aspects of spermatozoan morphology, seminiferous tubule structure and of host-parasite relationships. Among marsupial spermatozoa examined to date, those of bandicoots and dasyuromophians (and *Tarsipes)* are relatively very large (Harding 1987 and Temple-Smith 1987; both this Volume). Similarly, both bandicoots and dasyuromorphians have relatively large seminiferous tubules with few loops in the array (Woolley 1987). As noted by Harding *et al.* (1982), however, this feature may be strictly correlated with large sperm size and hence be the result of selective pressures that are obfuscating phylogenetic pattern. Humphrey-Smith (1987; this Volume) concludes from what appear to be host-specific parasites that bandicoots are more closely related to dasyuromorphians than they are to diprotodontian marsupials.

The fossil record for bandicoots has until recently been very poorly known. New discoveries being made at Riversleigh in north-western Queensland, however, are beginning to reveal a far greater breadth to the bandicoot radiation than is suggested by the living species. Of particular interest here is a diverse and extremely plesiomorphic group of tiny bandicoots which exhibit similarities to both didelphids and dasyurids. Exploration of the full phylogenetic and taxonomic significance of this group will have to await completion of the detailed study presently underway (J. Muirhead, in prep.). However, as far as we are aware, nothing here suggests phylogenetic ties in other directions than those already suggested above.

While we suspect that on balance the evidence favours a special bandicoot/dasyuromorphian phylogenetic relationship, we have not followed Kirsch (1968, 1977) in combining these taxa within a single order. The primary reason is that we remain cautious because not all of the data which suggests alliance of bandicoots with another order indicates that this alliance is necessarily with the Dasyuromorphia. Secondly, with Ride (1964), we consider that even were the monophyly eventually confirmed beyond reasonable doubt, the two groups would nevertheless warrant recognition at full ordinal rank. Relevant here is the growing palaeontological awareness of the former adaptive breadth of the bandicoot radiation and awareness of the considerable molecular distance of bandicoots from dasyuromorphians (e.g., Baverstock *et al.* in press) which indicates that bandicoots and dasyuromorphians may well have had an independent history at least as long as that which distinguishes the Diprotodontia (see below).

Considering the ordinal name for the bandicoots, Ride (1964) was the first to elevate the group to this rank and in so doing used Gray's name Peramelina. Gray (1825), however, clearly established his taxon at the tribal level. In terms of the recommendations of Brothers (1983), elevation of taxonomic names from one group level to another is a permissible action but one which should be avoided whenever possible. Since in this instance we have the option of using Kirsch's (1968) subordinal name Peramelemorphia, this has seemed to us to be the best alternative.

Order Notoryctemorphia (Kirsch, 1977) new rank

There is only one species in this order, the remarkably chrysochlorid-like marsupial mole, *Notoryctes typhlops*. Despite a considerable amount of intensive investigation, the phylogenetic relationships of *Notoryctes* remain so enigmatic that it cannot be placed with any confidence within any of the other four australidelphian orders.

Traditionally, *Notoryctes* was usually placed with dasyuroids (e.g., Stirling 1891; Gadow 1892; Carlsson 1904; Weber 1928; Winge 1923; Jones 1923 and Simpson 1945). Features often cited in support of this placement include its polyprotodont dentition (a plesiomorphic feature within Australidelphia), its lack of an intestinal caecum (a possible synapomorphic state shared with dasyuromorphians and microbiotherians), its pattern of distribution of Brunner's glands and the columelliform (imperforate) nature of its stapes (the last two features of uncertain character state polarity but also found in dasyuromorphians and bandicoots).

In contrast, it has frequently been suggested that the hind foot of *Notoryctes* betrays vestiges of syndactyly and that this, combined with its polyprotodont dentition, is evidence of affinity to bandicoots (e.g., Dollo 1899; Bensley 1903 and Szalay 1982). However, not all who have examined the foot of *Notoryctes* have agreed that it is syndactylous (e.g., Stirling 1891; Gadow 1892 and Jones 1923). As a result, the presumption of notoryctid syndactyly needs to be treated with care.

Cytologically, *Notoryctes* is highly specialised, with a unique diploid complement of 2N=20 (Calaby *et al.* 1974). Unfortunately, in the absence of evidence from G-banding, it is not clear to which of the various other marsupial karyotypes *Notoryctes* is most similar. At gross levels of comparison, however, its karyotype is most similar to that of certain diprotodontians (notably some petaurids).

Serology has also failed to provide unambiguous indications of relationship (Kirsch 1977). However, using tests on unabsorbed antisera, *Notoryctes* did appear to fall somewhat closer to dasyuromorphians and bandicoots than to other groups of Australian marsupials.

Microcomplement fixation of albumin has been undertaken using antigenic reactions only so that available results are as yet incomplete (Baverstock *et al.* in press). However, preliminary results suggest that *Notoryctes* is not closely related to any of the other major groups of Australian marsupials.

Studies of the auditory region have been attempted in the past (e.g., van Kampen 1905; Gregory 1910; Klaauw 1931; Segall 1970 and Archer 1976c) but have commonly been frustrated by the fact that the bones of the basicranial region fuse up very early in life thereby making phylogenetic deductions based on structure difficult. Klaauw (1931) suggests that the ectotympanic morphology of *Notoryctes* resembles that of diprotodontians while Gregory (1910) considers that its auditory region is more like that of dasyurids. Archer (1976c) notes the same similarities to diprotodontians pointed out by Klaauw (1931; in ectotympanic morphology) but also notes albeit distant similarities to bandicoots. More recent studies by Aplin (in prep.) have revealed no clear evidence for diprotodontian affinity.

As yet, no one has had a chance to examine either the spermatozoan morphology or the structure and development of the foetal membranes of *Notoryctes*. Both of these character systems have the potential to shed critical light on its interordinal relationships.

The fossil record for notoryctids has until now been a total blank. Recent discoveries in the middle Miocene deposits of Riversleigh in northwestern Queensland have produced three as yet unnamed taxa that, while bizarre in terms of dental morphology, may provide information useful for interpreting the interordinal relationships of notoryctids. Suffice it to say here they suggest an albeit very distant relationship between *Notoryctes* and peramelemorphians.

In summary, the affinities of *Notoryctes* within Australidelphia remain almost totally obscure. To some extent this apparent failure to resolve interordinal relationships may reflect a paucity of data available for phylogenetic analysis. On the other hand, *Notoryctes* itself is sufficiently well-known to make it obvious to us that it cannot be satisfactorily included within any of the other four orders of australidelphian marsupials.

This conclusion leaves us with two alternatives. Either we relegate *Notoryctes* once again to 'Australidelphia *incertae sedis*' (as in Ride 1964 and Kirsch 1968), or we create for it a fifth australidelphian order. Because this latter course of action better expresses current phylogenetic understanding (anticipated by Kirsch 1977), we herein propose elevation of Notoryctemorphia Kirsch, 1977 to full ordinal rank.

Order Diprotodontia Owen, 1866

Our concept of the order Diprotodontia conforms closely in content to the equivalently named taxon in most recent classifications including those of Kirsch (1968a, 1977a), Archer (1984b) and Woodburne (1984a). In regard to details of infraordinal classification, however, we propose a new arrangement. Before discussing details of the proposed changes, we shall deal briefly with three more general areas of interest: 1, evidence for monophyly; 2, possible interordinal relationships; and 3, comparative diversity of the group relative to other marsupial and placental orders.

The evidence for diprotodontian monophyly has been reviewed by Kirsch (1977a) and Archer (1984b). Features generally regarded to be diagnostic of this group are: syndactyly, diprotodonty, a superficial thymus gland and a fasciculus aberrans. The status of syndactyly in this group requires qualification; as argued earlier, syndactyly is generally regarded to be synapomorphic for some larger group made up of either bandicoots plus diprotodontians (e.g., Ride 1964), these taxa plus *Notoryctes* (e.g., Bensley 1903 and Szalay 1982), or all Australian marsupials (with loss of syndactyly in dasyuromorphians). Alternatively, as Kirsch (1977a) maintains, syndactyly might have arisen twice among Australian marsupials, once in an ancestral bandicoot and once in an ancestral diprotodontian. Only in this latter sense would syndactyly be synapomorphic for Diprotodontia; in all others it would be symplesiomorphic.

Each of the three remaining features does appear to be at least potentially synapomorphic for Diprotodontia. However, only incisor morphology (e.g., Bensely 1903) and the distribution of thymic tissue (Yadav 1973) have been determined in sufficiently large numbers of species to be confident that the derived conditions are indeed present in all diprotodontians. By contrast, the potentially very important feature of a fasciculus aberrans remains unconfirmed for many groups including a number of critically important taxa such as *Tarsipes rostratus* and the burramyids.

The phylogenetic significance of diprotodonty (hypertrophy of the medial lower incisors) was challenged by Hofer (1952), partly because of its documented multiple origins among placental mammals. Kirsch (1977a), however, returned that he knew of no such evidence for homoplaisy among Australian marsupials and therefore upheld diprotodonty as a diagnostic feature. More recently, however, we have learnt of just such a second major group (as yet undescribed) of fossil Australian marsupials which are fully diprotodont but which do not appear to belong within Diprotodontia.

While some doubt might thus be cast upon several of the traditional "diagnostic" features of Diprotodontia, the essential unity of the group remains unchallenged. Indeed, recent studies by Baverstock (1984 and Baverstock *et al.* 1987) on the evolution of marsupial albumin and by Aplin (1987) on the basicranial region of the skull, have each served to strengthen the concept of diprotodontian monophyly through the addition of new data and diagnostic characters.

As regards possible interordinal relationships, traditional comparative morphological studies (e.g., Bensley 1903 and Ride 1959, 1964) have tended to place Diprotodontia near to Peramelemorphia and Notoryctemorphia, based on the shared feature of syndactyly (although *Notoryctes* is not certainly syndactylous). As discussed in an earlier section, however, more recent studies challenge this view and suggest that Diprotodontia is probably the sister taxon of the Australian polyprotodonts (i.e., a combined Dasyuromorphia-Peramelemorphia-Notoryctemorphia clade).

The view that Diprotodontia could represent a fundamental and, by implication, very ancient section of the marsupial radiation, is supported by the great taxonomic and ecological diversity of the group, as well as by emerging details of its evolutionary history.

Comparisons of taxonomic diversity with other marsupial and placental orders are hindered by the lack of a pre-Miocene fossil record for Diprotodontia. However, some estimate of how the post-Miocene record might reflect the full, historical diversity of the group may be gained by extrapolation from the evolutionary histories of several of the better known placental orders (e.g., Lillegraven 1971 and Anderson and Jones 1984). Based on figures for some of the larger and more ancient orders (e.g., Primates, Artiodactyla, Lipotyphla, Perissodactyla) it appears that the average number of extinct families is usually well over one half of the total and not infrequently over two-thirds of the total. Given that there are 10 "modern" families of diprotodontians, the "actual" diversity of the group might thus reasonably be expected to lie somewhere around 25-30. Compared with the full gamut of placental and marsupial groups, Diprotodontia is thus potentially one of the largest mammalian orders, perhaps being on par with Artiodactyla (27 families) and Primates (25 families) and possibly exceeded only by Rodentia (*c.* 50 families; all diversity values abstracted from Anderson and Jones 1984).

Comparison of ecological diversity is clearly a highly subjective matter. As emphasised previously by Ride (1964) and Kirsch (1977a), however, Diprotodontia may be ecologically more diverse than any placental order. In terms of the variety of basic "lifestyles" (or "ways of life" sensu van Valen 1971), only Diprotodontia includes all of the following: large predaceous carnivores (e.g., thylacoleonids), a variety of generalised to highly-specialised browsers and grazers (e.g., macropodoids, vombatids, ilariids, diprotodontids), arboreal folivores and frugivores (various possum families), generalised insectivore-omnivores (e.g., *Cercartetus caudatus*; Smith 1985), as well as specialist nectivores (e.g., *Tarsipes*, *Acrobates*, some *Cercartetus* spp.; Turner 1982), gumnivores (e.g., *Petaurus* spp.), gramnivores (e.g., *Burramys parvus*) and entomophages (e.g., *Dactylopsila* spp; Smith 1981). In terms of locomotory diversity, diprotodontians include specialist bipedal saltators, ambulators and even volplaners. Also of note is the incredible size range, from the 15 g *Acrobates pygmaeus* to the rhinoceros-sized *Diprotodon optatum*. In comparative terms, this ratio of smallest to largest species is probably matched among placental orders only by Cetacea, wherein the blue-whale weighs approximately 4000 times that of the smallest dolphins.

The fossil record of Diprotodontia, like that of Australian marsupials in general, is presently confined to the post-Oligocene period. However, given that the earliest known Oligocene and Miocene faunas already contain a high proportion of essentially modern diprotodontians it thus seems reasonable to assume that the major radiation of the group had occurred well prior to this time. In this regard it is of interest to note that none of the known fossil diprotodontian families appear to be structurally annectant to any modern lineage. Patristic relationships are, of course, evident at the intrafamilial level.

Biochemical data coming to hand over the last 20 years are also consistent with the notion of an early origin of Diprotodontia. As noted by Kirsch (1977a:117), "Even without the inclusion of *Tarsipes*, the group of Australian diprotodont marsupials would be almost as internally divergent serologically as the rest of the Australian marsupials combined . . .". Other studies using the quantitative immunological MC'F technique (e.g., Maxson *et al.* 1975, Baverstock 1984 and Baverstock *et al.* 1987) give a comparable picture of high molecular diversity within Diprotodontia.

In overview, all available evidence points toward recognition of Diprotodontia as one of the largest and most diverse radiations of mammals.

Synonyms of DIPROTODONTIA Owen, 1866

"Diprotodontia" constitutes the original spelling of the term coined by Owen (1866) for the Australian "diprotodont" marsupials. Owen's term was followed by, among others, Thomas (1888), Weber (1928), Gregory (1910), Osborn (1910), Sonntag (1922), Abbie (1937), Simpson (1945), Boardman (1950a), van Valen (1971) and Marshall (1981).

The alternative spelling "Diprotodonta" was used (without comment over the last "i") by Thomas (1895) and Dederer (1909) among the earlier generation of systematists and again by Ride (1962, 1964). Most "modern" workers have followed Ride's usage. Notable examples include Kirsch (1968a, 1976, 1977a), Archer (e.g., 1981, 1984b) and Woodburne (1984a).

Because the change in ending from "ia" to "a" appears to represent an invalid emendation we have returned to Owen's original term Diprotodontia. We note that Marshall (1981) appears to have come to a similar decision.

Although most recent workers have followed Owen's Diprotodontia in spirit if not in letter, a number of workers have elected to depart from tradition and to recognise essentially the same group of animals under an entirely different name. These names constitute junior synonyms of Diprotodontia Owen, 1866. We are aware of three examples: Duplicicommissurala Abbie, 1937, proposed as a provisional label for those taxa (= Diprotodontia) which possess a fasciculus aberrans; Phalangeriformes Szalay, 1982, proposed at infraordinal level to emphasise the monophyletic status of the diprotodont marsupials within the order Syndactyla; and, "Syndactyla diprotodontia", used by Jones (1923) in the same sense and for the same group. Another taxonomic label used recently as a formal substitute for Diprotodontia is Phalangeroidea Weber, 1928 (e.g., Simpson 1945, 1970; Ride 1962 and Woodburne 1967). It should be noted, however, that this is not Weber's original usage;

he used Phalangeroidea in opposition to Vombatoidea and included both taxa in a suborder Diprotodontia Owen.

MAJOR GROUPINGS WITHIN DIPROTODONTIA

In contrast to the essential stability of the wider concept Diprotodontia, the taxonomic arrangement of its constituent genera and families has been subject to frequent change.

Prior to Kirsch's major overhaul of diprotodontian classification in the late 1960's, the dominant feature of the group was the broadly conceived familial taxon "Phalangeridae". To most workers (Winge is one of the few exceptions), this represented a basal group within Diprotodontia, such that all other families were derived ultimately from a "phalangerid" ancestry. As to exactly how many additional families should be recognised, opinion varied, but a total diversity of between five and seven families was usual.

Kirsch's (1968a) classification of Diprotodontia was radical inasmuch as he recognised a total of 11 diprotodont families divided between four superfamilies. Of these, the Phalangeroidea was in some ways equivalent to the earlier "Phalangeridae", since in addition to the various families of possums and kangaroos, it also included a variety of only tenuously related fossil taxa (e.g., Diprotodontidae, Thylacoleonidae).

In the decade since Kirsch's (1977a) extended discussion of his classification, the content and meaning of several of his groupings have been secondarily modified, leading in some cases to a taxonomic imbalance and conflict of ranks. Thus, while most recent workers would accept the phylogenetic placement of the kangaroos within the Phalangeroidea (*sensu* Kirsch), the wider grouping of kangaroos and rat kangaroos is now generally regarded as a superfamilial taxon, Macropodoidea (e.g., Archer and Bartholomai 1978; Flannery 1987 and refs. cited therein). In similar fashion, Diprotodontidae (*sensu* Kirsch), is now generally employed at superfamilial rank (Diprotodontoidea; e.g., Archer and Bartholomai 1978) and as such impinges taxonomically on the wider grouping (Vombatoidea *sensu* Kirsch) with which it is generally grouped phylogenetically.

In response to this problem of taxonomic imbalance, Szalay (1982) and Archer (1984b) each presented a new superfamilial arrangement of Diprotodontia. Szalay listed four superfamilies but did not define their content. Archer (1984b) reduced this to three: Macropodoidea, Phalangeroidea (all possums including *Tarsipes*) and Vombatoidea. This latter grouping was greatly expanded over Kirsch's earlier concept, including not only the two "diprotodontoid" families, but also Wynyardiidae and Thylacoleonidae. To

the extent that it accurately reflects the perceived phylogenetic affinities of these groups, and contains no taxa at conflicting ranks, this arrangement is clearly workable. On the other hand, each of Archer's Phalangeroidea and Vombatoidea contains an overwhelmingly diverse complex assemblage of families, yet allows little scope to formally express the detailed phyletic history of each group.

One obvious solution to this dilemma is to elevate some or all of the superfamilial taxa to a rank above that of superfamily but below that of order. This action was followed by Woodburne (1984a), following a general agreement as to its propriety among participants of the "Possums and Opossums" symposium. As reviewed earlier, Woodburne (1984a) recognised two sub-ordinal level taxa within Diprotodontia, Vombatiformes and Phalangeriformes, which correspond in content to Archer's (1984b) Vombatoidea, on the one hand, and to his Phalangeroidea plus Macropodoidea on the other.

In the present classification we follow the basic structure and composition of Woodburne's arrangement, but introduce an additional hierarchical rank (Infraorder) to further express the phylogenetic complexity of one group. In justification of this expanded intraordinal classification of Diprotodontia, we suggest that it more accurately portrays the level of ecological, morphological and taxonomic divergence within the order as a whole, as well as providing a greater degree of taxonomic freedom for the expression of phylogenetic concepts. As emphasised in earlier sections, however, this latter consideration is of secondary importance; quite emphatically, we do not consider numerical diversity at this or any other taxonomic level to constitute adequate grounds for elevation of taxonomic rank.

With regard to nomenclature, Woodburne's (1984b) term Vombatiformes is followed here for the major, monophyletic clade comprised of the living Phascolarctidae and Vombatidae, and the extinct relatives of the latter group. On the other hand, Phalangeriformes Woodburne, 1984 is a junior homonym of Phalangeriformes Szalay, 1982 (which is itself a junior synonym of Diprotodontia Owen, 1866). Accordingly, we suggest that Phalangeriformes Woodburne be considered unavailable and propose the term "Phalangerida" as a replacement name or nomen novum.

Suborder Vombatiformes Woodburne, 1984

The concept Vombatiformes has at its nub the recognition of special affinity between the living koala and wombats. As noted above, this concept is an old one, with its principal advocates among the late 19th and early 20th Century comparative anatomists (e.g., Weber 1928; Sonntag 1921a, 1921b and Pocock 1921). Later workers, however, often using a more phenetic yardstick of affinity, tended to emphasise a number of more superficial similarities (particularly dental) between the koala and the "possums" (e.g., Simpson 1945). Hughes (1965), however, championed the earlier view of koala-wombat affinity, based on his studies of spermatozoan morphology in Australian marsupials. Finally, Kirsch's (1968, 1977a) serological results provided clear confirmation of close affinity between the koala and wombats.

Acceptance of Kirsch's serological findings has prompted a reconsideration during recent years of much of the comparative anatomical evidence (e.g., Strahan 1978 and Archer 1984c). Morphological features that may be confidently cited in support of koala-wombat monophyly include: the highly characteristic sperm head morphology (Hughes 1965) and other features of the spermatozoan (Harding *et al.* 1987), the presence of a distinct cardiogastric gland (Krause and Leeson 1973) and reduction or absence of a thoracic thymus (Yadav 1973). The more extensive listings given by Strahan (1978) and Archer (1984c), as well as the results of a wider survey in progress by Aplin, suggest that the evidence will eventually prove even more extensive. Suffice it to say here that evidence of koala-wombat monophyly can be found in virtually all organ systems examined to date including external morphology, visceral anatomy, haematology, seminal and gastric biochemistry, cardiovascular arrangements, oro-nasal anatomy, palatal morphology, general myology and male reproductive anatomy.

As shall be discussed at greater length below, each of the five families of fossil vombatiforms appear to be related more closely to Vombatidae than they are to Phascolarctidae. As a result, the koala and wombats are clearly not "close relatives" in the usual sense of the term, since they are separated phyletically by a large number of fossil families.

Some indication of just how wide might be the phylogenetic gap between the koalas and wombats, can be obtained through a consideration of the nature of taxa concerned. As it happens, all known vombatiforms are highly specialised animals. Included among their ranks are rapacious carnivores such as the thylacoleonids and a great variety of specialised herbivores including the fossorial, rodent-like vombatids, chalicothere-like palorchestids, rhinoceros- or hippopotomas-like diprotodontids and possibly kangaroo-like (at least dentally) wynyardiids. Ilariids, too, were presumably herbivorous, although exactly what they might have looked like or consumed is not clear.

Another manifestation of this widespread specialisation is the fact that all known vombatiforms are moderately large to gigantic. The smallest known vombatiform is probably the mid-Pleistocene vombatid *Warendja wakefieldi* (Hope and Wilkinson 1982), which was the size of a large Brush-tailed Possum. Other taxa of roughly comparable size include the fossil koalas *Litokoala* spp. and the possible ilariids *Koobor* spp. Gigantism among vombatiforms was most pronounced during the late Tertiary and Pleistocene, culminating in taxa such as *Palorchestes azael*, *Diprotodon optatum* and *Phascolonus gigas*. On the other hand, certain Miocene taxa such as the species of *Ilaria* are also impressive for their size, particularly given the more densely forested environment in which they presumably lived.

Considering the fossil record of this group, vombatiforms as a group form a dominant component of the late Oligocene and earliest Miocene local faunas and all seven vombatiform families are present by this time (see below). Two families, Ilariidae and Wynyardiidae, are known only from deposits of this age.

In overview, the post-Oligocene vombatiform radiation appears to be largely relictual, presumably representing but the "tail-end" of a formerly more diverse and ecologically "balanced" radiation. In the absence of a fossil record, the time of origin and primary radiation of this group cannot be known for certain. However, given that each of the known families is clearly distinct by earliest Miocene times, we feel confident in postulating an early Tertiary origin for the group as a whole and a probable Palaeogene antiquity for the major vombatiform radiation.

Immunological evidence is divided over the probable age of the vombatiform radiation. On the one hand, Kirsch (1977:121) concluded from his results that ". . . the wombats and *Phascolarctos* are serologically distinct from each other at about the subfamily level". On the other hand, MC'F analysis of albumin suggests a more distant relationship in line with the present conclusions. Baverstock's results (1984; Baverstock *et al.* 1987), in fact, show no indication of special affinity between the koala and wombats, but fail also to place either family with any other group of diprotodontians. Given the limitations of this method for resolving very distant phyletic relationships, this result is consistent with the present interpretation of a great antiquity for the vombatiform clade.

Similarities in basic body proportions, as well as the common absence of a tail, have been used by some workers to suggest that the common ancestor of the koalas and wombats was a terrestrial animal, and the koala, secondarily arboreal. Although consideration of the wider assemblage of vombatiforms indicates that some of this evidence should be reassessed, in general the scenario appears to hold. "Diprotodontoids" were clearly terrestrial. Thylacoleonids, while often portrayed as arboreal, were most likely cursorial and scansorial, based on Finch's (1982) analysis of general postcranial proportions and Wells and Nichol's (1977) analysis of its manus and pes.

Groupings Within Vombatiformes

Woodburne (1984a) recognised two superfamilial taxa within Vombatiformes, one for the koalas and the other for the remaining vombatiform families. In keeping with the morphological, palaeontological and biochemical evidence reviewed above, we regard the split between the two groups to be a more fundamental one and suggest an equivalent division of Vombatiformes into two infraordinal taxa.

Infraorder Phascolarctomorphia new: This group contains a single family, Phascolarctidae. Apart from the living koala, *Phascolarctos cinereus,* and its immediate fossil relatives, this family includes the Miocene species of *Madokoala, Perikoala* and *Litokoala* (Archer 1978b and Woodburne *et al.* 1987a). All fossil koalas are currently known solely from tooth-bearing parts. Possible cladistic relationships among phascolarctids were reviewed by Archer (1978b) and more recently revised by Woodburne *et al.* (1987a).

The Pliocene species of *Koobor* (Archer and Wade 1976 and Archer 1977) are another group usually included within Phascolarctidae. However, as discussed by Pledge (1987b) and Tedford and Woodburne (1987), *Koobor* spp. are also similar to ilariids, particularly the diminutive *Kuterintja ngama* (Pledge 1987b). In accordance with this doubt about the familial affinities of *Koobor,* we herein refer the genus to Vombatiformes: incertae sedis.

Since all fossil koalas are currently referred to Phascolarctidae, the numerous peculiarities of the living koala (e.g., Martin 1836; Forbes 1881; Young 1887; Pocock 1921; Sonntag 1921; MacKenzie 1934; Werner 1964 and Strahan 1978) may be taken as provisionally diagnostic of the wider infraorder. More realistically, however, many of these features are almost certainly autapomorphies relative to the very restricted diet and remarkable physiological adaptations of the living species. Pending discovery of more complete material of the Miocene genera, however, little can be said regarding the true morphological and ecological diversity of Phascolarctomorphia.

Infraorder Vombatimorphia new: As noted earlier, the present concept Vombatomorphia corresponds closely to Vombatoidea as used recently by Archer (1984b) and Woodburne (1984a). It differs, however, in the addition of one extra fossil family (Ilariidae) and in its elevation to infraordinal rank.

Since five of the six families of Vombatomorphia are known exclusively from fossil remains, evolutionary questions such as the monophyly or otherwise of the group, and the pattern of interfamilial relationships, must be debated solely on the palaeontological evidence of bones and teeth. Fortunately for this task, many of the fossil vombatomorphians are known by relatively complete cranial and postcranial materials, a reflection, no doubt, of the generally large size and corresponding robusticity of their remains. On the other hand, much of this material is only now being analysed. For the present, therefore, discussion is necessarily limited to features of the dentition and some basic points of cranial morphology such as are noted in published accounts.

On the question of vombatomorphian monophyly, dental morphology provides few useful characters, simply because the dentitions of most vombatomorphian families are so highly autapomorphic. As first noted by Tedford *et al.* (1977), Miocene vombatomorphians (their "vombatoids") commonly show vestiges of a "selenodont" pattern in the upper molars. However, since phascolarctids are also markedly selenodont (Archer 1976b), this condition is presumably symplesiomorphic among vombatomorphians.

One potentially diagnostic feature of the vombatomorphian dentition is the gross reduction of the first two premolars in each toothrow. These teeth may either be retained as vestigial unicuspids (e.g., *Thylacoleo* spp.) or lost altogether (e.g., vombatids and "diprotodontoids"). Among other diprotodontians a similar pattern of reduction is found only in Burramyidae and in adult macropodoids (resulting from a different pattern of tooth replacement in this latter group; cf. Archer 1978a).

In terms of their known cranial morphology, vombatomorphians appear to share the significant derived condition of a "squamosal bulla", wherein the alisphenoid tympanic process which usually shields the anterior part of the tympanic cavity is replaced by an analagous process from the squamosal (Tedford *et al.* 1977; Archer 1984b; Woodburne 1984a and Aplin 1987). Importantly, the koala shows the more plesiomorphic condition of an "alisphenoid bulla", despite the otherwise highly apomorphic condition of the auditory region in this species (Aplin 1987).

Other cranial features which might ultimately prove diagnostic of Vombatomorphia (but for which relevant data are absent for several groups) include: the reduction or loss of the postglenoid emmissary vein (postglenoid canal transmits continuation of temporal veins; Aplin 1987); a characteristic modification of the infratemporal and pterygoid fossae, related to enlargement of the lateral pterygoid at the expense of the medial pterygoid musculature (Aplin in prep.); the expansion of the premaxillae at the expense of the maxilla, resulting in positioning of the infraorbital foramen close to the maxillo-premaxillary suture (Rich and Archer 1979); and, other detailed features of auditory anatomy (Aplin 1987 and in prep.).

In the postcranial skeleton, there is a strong superficial similarity between the major limb elements of the living vombatids, *Wynyardia bassiana,* the structurally generalised palorchestid *Ngapakaldia tedfordi* and *Thylacoleo carnifex,* yet the cladistic significance of this similarity is not known. Similarly, the proximal tarsal elements (calcaneum and astragalus) are at least phenetically very much alike in the vombatids, palorchestids and thylacoleonids. In this case, however, the similarity probably does hold phylogenetic significance, since the vombatomorphian tarsus differs considerably from that of all other diprotodontians including those taxa (e.g., *Distoechurus pennatus*) judged by Szalay (1982) to possess an essentially plesiomorphic australidelphian tarsus.

Overall, then, the case for vombatomorphian monophyly appears fairly strong, since it is supported by a variety of synapomorphic cranial features, as well as by a high degree of phenetic resemblance in the postcranium. On the other hand, we are quick to admit that much of the relevant morphological evidence has yet to be presented in sufficient detail to allow independent assessment. In addition, detailed studies of the abundant postcranial material, including that of *Wynyardia bassiana,* are urgently called for.

Given these conclusions, it is of interest to note that the unity of Vombatimorphia as recognised herein is already under attack. Murray *et al.* (1987), based on their analysis of a spectacularly well-preserved cranium of the medial Miocene thylacoleonid *Wakaleo vanderleueri,* have suggested that the marsupial lions evolved independent of the other vombatomorphians, both groups arising from an early, burramyid-like stock. In particular, they place considerable emphasis upon the presence in *Wakaleo* of a unique "bilaminar" bulla, formed jointly from alisphenoid and squamosal derivatives. This they interpret as a structurally intermediate condition between the exclusively "alisphenoid" bulla, such as occurs in polyprotodont marsupials (e.g., didelphids), and the fully "squamosal" bulla seen in *Thylacoleo* spp. and the remaining vombatomorphian families. Because *Wakaleo* spp., on other evidence, clearly belongs within Thylacoleonidae, they conclude that the fully "squamosal" bulla must have been achieved twice, once in *Thylacoleo* and once in the ancestor of the remaining vombatomorphians.

This point and others raised by Murray *et al.* (1987) will be discussed in full in a forthcoming paper on thylacoleonid relationships (Aplin in prep.). For the present, however, suffice it to say that we regard the bilaminar tympanic process of *Wakaleo* to be a curious but phylogenetically insignificant development, which might just as readily be viewed as a secondary modification upon the more typical condition in *Thylacoleo.* As a more general comment, however, we suggest that even were the squamosal bulla proven to be convergent between thylacoleonids and other vombatomorphians, this fact by itself would not outweigh the remaining cranial and postcranial features that together serve to hold Thylacoleonidae within Vombatomorphia.

As remarked in an earlier section, known vombatomorphians are remarkably varied in their "way of life" and include a variety of highly specialised browsing and grazing forms as well as the omnivorous (?*Wakaleo* spp.) to undeniably carnivorous (*Thylacoleo* spp.) marsupial lions. All known vombatomorphians are also moderately-large to gigantic and are the largest taxa in all adequately known local faunas save that of the present day.

Largely because of the strikingly specialised nature of virtually all known vombatomorphians, the pattern of interfamilial relationships remains poorly understood.

One area of general agreement in the past has been the phylogenetic unity of the various major groups of "diprotodontoids" (i.e., palorchestids, zygomaturines, diprotodontines). Until recently, these were placed in a single family, Diprotodontidae (e.g., Stirton *et al.* 1967). Archer and Bartholomai (1978), however, considered the known diversity to be better expressed by recognition of two families, but united these again at superfamilial level as Diprotodontoidea. More recently, Archer (1984b) has questioned the monophyly of each of the Palorchestidae and Diprotodontidae as currently recognised, as well as the basis for their higher-level association. As shall be discussed below, this uncertainty is further compounded by the possibility that some taxa currently included within the Wynyardiidae might actually be closer phyletically to one or other groups of "diprotodontoids". Clearly, the whole area of "diprotodontoid" systematics is urgently in need of review; pending such a study we prefer to recognise two separate families, Palorchestidae and Diprotodontidae, though with the qualification that Palorchestidae in particular may well be paraphyletic. No superfamilial taxon is recognised.

As indicated above, Wynyardiidae in its recently expanded sense (Tedford *et al.* 1977; Rich and Archer 1979 and Pledge 1987), also presents difficulties of interpretation. As shown by Aplin (1987), *Wynyardia bassiana* from the early Miocene of Tasmania is decidedly plesiomorphic in respect of its auditory and general basicranial anatomy. In contrast to other vombatomorphians, it retains the primitive features of an alisphenoid tympanic wing, a patent (albeit reduced) postglenoid emmisary vein and either an absence or poor development of a squamosal epitympanic sinus. Other, minor features of basicranial and wider cranial anatomy, however, hint at a closer relationship to Vombatomorphia than to any other major group of diprotodontians. Based on this evidence, Aplin (1987) tentatively recognises that *Wynyardia* as the plesiomorphic sister taxon of the remaining vombatomorphians.

By contrast, the recently described Central Australian "wynyardiids", *Namilamadeta snideri* (Rich and Archer 1979) and *Muramura williamsi* (Pledge 1987a) agree inasmuch as they are both decidedly vombatid — and "diprotodontoid"-like in details of dental and cranial morphology. As emphasised by Archer (1984b), dental resemblance to members of the latter group is particularly strong, involving such features as molar lophodonty and the detailed morphology of the upper premolar. Pledge (1987a), on the other hand, has stressed certain vombatid-like features of the *Muramura* cranium, in particular the presence of an "anterior masseteric fossa" on the front of the zygomatic arch.

Based on Pledge's (1987a) illustrations of *M. williamsi*, we are more impressed by its apparent cranial similarities to structurally plesiomorphic "diprotodontoids" such as *Ngapakaldia tedfordi* (Stirton 1967). In particular, we draw attention to the presence in both taxa of a well-developed masseteric process, a similarly proportioned rostrum with strongly projecting nasals and a marked expansion of the anterior interorbital region. The "anterior masseteric fossa" noted by Pledge, in our view, actually represents the origin of the maxillonasolabialis group of the fascialis musculature; the strong development of these muscles also represents a feature in common with *Ngapakaldia* spp. and other "diprotodontoids".

Whatever the final outcome of such comparisons, the important point in the present context is that the affinities of both *Namilamedata* and *Muramura* appear to be within Vombatomorphia, rather than peripheral to it as was concluded by Aplin for *Wynyardia*. One possible interpretation is that the Central Australian taxa are simply not wynyardiids, in which case they might be either referrable to some other family (one of the "diprotodontoid" taxa?) or, more likely, granted

familial status in their own right. Detailed studies of the *Muramura williamsi* skeleton currently underway by Pledge (in prep.) should provide a speedy solution to this problem. Pending completion of his study, however, we shall continue to use Wynyardiidae in its expanded sense and to include it within Vombatomorphia (though primarily on the characters of the referred genera).

The newly described Ilariidae from the early Miocene of Central Australia were initially thought to be giant koalas. More detailed study of the teeth and cranium, however, indicated a special resemblance to vombatids and their relatives (Tedford and Woodburne 1987). Of particular importance in this regard is Tedford and Woodburne's observation in one species of *Ilaria* of the characteristic vombatomorphian feature of a "squamosal bulla". Little else in the way of cranial structure is reported, pending further studies. In the dentition, the general trend towards molar lophodonty is reminiscent of the condition in "diprotodontoids", wynyardiids and in unworn teeth of vombatids. Whether this has any cladistic significance is unclear.

Thylacoleonids are more plesiomorphic than all other vombatomorphians inasmuch as they retain a full (albeit functionally reduced) complement of upper antemolar teeth. Otherwise, however, their dentition bears little resemblance to that of any other vombatomorphian family.

Interestingly, the molars in the more plesiomorphic members of this group (e.g., *Priscaleo pitikantensis* Rauscher 1987; *Wakaleo* spp., Clemens and Plane 1974 and Murray *et al.* 1987) appear to be at least superficially similar to those of certain "bunodont" possums such as petaurids and burramyids. One explanation for this dental similarity is that it represents the plesiomorphic diprotodontian condition; alternatively thylacoleonids and the "bunodont" possums could be convergent in dental morphology. Both possibilities warrant careful consideration.

The cranium of *Thylacoleo carnifex* was described in detail by Woods (1956). Although he noted some similarities to vombatids, his analysis led to no specific phylogenetic conclusions. A more recent study of this species by Aplin (in prep.) has likewise resulted in no clear indication of relationship to any other vombatomorphian family. Overall, the emerging picture thus appears to place thylacoleonids at some distance from the main cluster of "herbivorous" vombatomorphians. However, further studies, including a cladistic reanalysis of the manus and pes, should probably be attempted prior to formalisation of any such concept.

As indicated above, an alternative interpretation of thylacoleonid affinities, based largely on the cranial structure of *Wakaleo vanderleueri,* is given by Murray *et al.* (1987).

From the foregoing comments, any attempt to cluster the families of Vombatimorphia into superfamilies or even higher taxa would clearly be premature. In the present classification, therefore, we list the six families of vombatomorphians as coordinate and unranked taxa.

Suborder Phalangerida, new name: The Phalangerida as recognised herein is comprised of 11 families of which three are wholly extinct. In order of recognised generic diversity these are: Macropodidae, Pseudocheiridae, Potoroidae, Phalangeridae, Petauridae, Ektopodontidae, Pilkipildridae, Miralinidae, Acrobatidae, Burramyidae and Tarsipedidae. As shown in Table 2, these are arranged into five superfamilies; one family cannot be allocated at superfamilial level and is therefore listed as Phalangerida: incertae sedis. No infraordial taxa are recognised.

Compared with previous classifications, this grouping circumscribes two of Kirsch's (1968a, 1977a) superfamilies, namely Phalangeroidea (in part) and Tarsipedoidea, and two out of the three and four superfamilies recognised by Archer (1984b) and Szalay (1982) respectively. More distantly, yet as a point of considerable historical interest, this grouping corresponds almost precisely (only differing in knowledge of fossil groups) with Winge's (e.g., 1941) familial grouping Phalangistidae.

The monophyletic status of the Phalangerida is far less readily demonstrated than that of its sister taxon Vombatiformes. However, since in our view there is a reasonable degree of congruence between the results of serology and morphology over this point, we feel justified for the present in accepting Phalangerida as a "natural" clade, coordinate with the Vombatiformes.

Serological support for the Phalangerida comes primarily from Kirsch's (1968a, 1977a) studies using polyvalent antisera. These data indicate a relatively close affinity between three of the major groups of possums (i.e., Pseudocheiridae, Petauridae and Phalangeridae) and further demonstrate the affinity of the macropodoids to this group. Data for any of the smaller bodied taxa are far less complete, however, and serve only to demonstrate the general distinctness, from all other groups of possums, of the species of *Tarsipes, Acrobates, Cercartetus* and *Burramys*. Based on a combination of weak serological indications and other cytological and morphological evidence, Kirsch concluded that *Tarsipes* did not belong with the other possums, but that *Burramys, Cercartetus* and *Acrobates* did.

Results of MC'F on albumin presented by Maxson *et al.* (1975) and Baverstock (1984 and Baverstock *et al.* 1987) would appear to contradict the present concept of a monophyletic Phalangerida, inasmuch as the macropodoids

appear to be highly divergent from the other diprotodontians and each of the koala and the wombats appear to fall "within" the main cluster of "possum" families. At least with regard to the placement of the kangaroos, these results probably reflect major differences in the rate of albumin evolution between these groups. On the other hand, the failure of the MC'F data to effectively discriminate between our Vombatiformes and Phalangerida cannot be so explained. Even so it is not necessarily inconsistent with their existence: unless each of the postulated diprotodontian suborders were cladistically distinct for a substantial period of time prior to the origin of the known neogene lineages, the original monophyly of each group simply may not be recorded at molecular level or, more likely, it might not be detectable using the relatively coarse immunological techniques currently in use.

Morphological evidence that would support Phalangerida is likewise considerably less compelling than that marshalled behind Vombatiformes. In our view there are four reasons why this should be so. First, although several of the phalangeridan families are strikingly specialised (see below), the group as a whole appears to be more plesiomorphic in many aspects of gross anatomy than is Vombatiformes. One reason why this might be so is that the arboreal adaptation shown by the majority of phalangeridans represents a continuity of lifestyle from the original diprotodontian stock; intuitively, and in view of documented tarsal and pedal morphology, this concept has considerable merit (e.g., Huxley 1880; Dollo 1899; Bensley 1903 and Szalay 1982).

Secondly, in contrast to this general conservatism, several phalangeridans are strikingly specialised, the prime example of course being the remarkable Honey Possum, *Tarsipes rostratus* (Renfree *et al.* 1984). As argued below, macropodoids and acrobatids also fall into this category, although in these taxa the degree of specialisation is nowhere near as great.

Thirdly, anatomical knowledge of even the most basic kind is glaringly incomplete for a number of major groups of phalangeridans including burramyids, petaurids, pseudocheirids and *Tarsipes*. To a large degree this situation almost certainly reflects the former inclusion of all of these groups within a single family, as well as the still common misrepresentation of *Trichosurus vulpecula* as a "typical" possum. By contrast, both the koala and the wombats are the subject of a comprehensive anatomical literature, although much of it admittedly is in need of revision.

Finally, in postulating monophyly for Phalangerida we are struggling to overcome the overbearing conceptual legacy of more than 100 years of misrepresentation of the "Phalangeridae" as consistent "basal stock" within Diprotodontia as a whole. As a consequence, we (and probably others) have been reluctant to accept as synapomorphic for this group, features which, were they present instead in Vombatiformes, would be unhesitatingly granted the attention they deserve.

Turning back to the original question, what then do we consider to be synapomorphic features of the Phalangerida? As it turns out, there appears to be little or nothing which is both convincingly derived and shared by all phalangeridans; from the morphological standpoint Phalangerida may thus be virtually undiagnosable. However, since the problem of definition appears to derive at least in part from the presence of several highly autapomorphic taxa, we have taken the approach of basing an initial diagnosis upon only less dramatically specialised phalangeridans which do share numerous derived features of morphology. This restricted concept is then employed as a "standard" against which each of the more atypical taxa are individually assessed. Depending on this assessment, the "aberrant" taxa then are either accepted as specialised members of the group, or else excluded from Phalangerida altogether.

Features of "typical" phalangeridans: Four families are considered to be "typical" of "core" phalangeridans, namely Phalangeridae, Petauridae, Pseudochiridae and Burramyidae. Collectively these taxa display numerous seemingly shared derived features of basicranial, dental and external anatomy, as well as seemingly specialised features of spermatozoa and placentation.

Derived features of the basicranial region of these taxa include: completion of the bony floor of the tympanic cavity by posterior attenuation of the alisphenoid tympanic process and less marked anterior expansion of the caudal tympanic process of the petrosal; development of an extensive, deeply invasive and fully enclosed squamosal epitympanic sinus and (other than in burramyids) a corresponding mastoid epitympanic sinus; completion of a tubular bony outer ear canal, formed through precise articulation (sometimes involving fusion) of a laterally expanded ectotympanic (meatal process), a postglenoid process of the squamosal and the posttympanic process of the squamosal; development from the medioventral border of the ectotympanic of a "promontorial" process which rests directly upon and determines the form of the posteroventral surface of the promontory; and, location of the "postglenoid" foramen anterior to the postglenoid process, at the medial end of the mandibular fossa (posterior part of the glenoid fossa). Some or all of these features are noted by Kampen (1905), Klaauw (1931), Winge (1941), Segall (1969, 1971), Aplin (1987) and Flannery (1987).

In the dentition, the primary feature of interest is the trend towards marginal placement of the primary buccal cusps, with accompanying reduction in width or loss of the stylar shelf (also noted by Flannery 1987). As discussed by Woodburne *et al.* (1987), the identity of the buccal cusps in these taxa is not certainly known; they may represent the paracone and metacone but could also be enlarged stylar cusps (see earlier comments on dentally convergent paucituberculatans). Other possible dental synapomorphies include the presence of a "neometaconule" on the anterior molars (Woodburne *et al.* 1987) and the seemingly trivial but remarkably consistent presence of a small, vestigial lower incisor, located directly over the root of the main procumbent incisor (possibly functioning as a posterior "cingulum"). In two of the four "typical" families the anterior cheekteeth (P^3 and P_3-M_2) together form a precise sectorial shearing unit; in the M_2 this results in longitudinal "paralophid" shear (for further details see Woodburne *et al.* 1987). By contrast, modern pseudocheirids and petaurids both lack this feature, although in each case the anterior cheekteeth are clearly specialised in different ways. In at least one group of Miocene pseudocheirids, however, the P_3 appears to more closely resemble the "typical" sectorial type of the other phalangeridans (Archer *et al.* 1987).

Features of external morphology shared among "typical" phalangeridans include: the presence of a distal prehensile area on the ventral surface of the tail; fusion of the first interdigital and posthallucal plantar pads on the pes; presence of a distinct posterior antihelical process (or its hair-bearing rudiment) on the inner surface of the pinna; loss of the interramal group of vibrissae; presence of a strong post- and subauricular hair tract reversal, and associated postocular tractal convergence; and a typically oblong to lemon-shaped cross-sectional shape of the major pelage hairs (see Lynne 1959; Boardman 1950a, b, and earlier works; Brunner and Coman 1974; Valente and Woolley 1982; other features from Aplin in prep.).

The spermatozoa vary greatly in both size and shape and in ultrastructural detail among the various families of phalangeridans (Harding 1987a and earlier works cited therein). One feature common to the four "typical" phalangeridan families, however, is the development during final epididymal maturation of abundant midpiece plasma membrane invaginations. As far as is known, these unusual structures occur in no other group of marsupials.

Finally, placentation in the "core" phalangeridan families Pseudocheiridae and Phalangeridae is by yolk-sac placenta only, there being no chorio-allantoic fusion and no erosion of maternal tissues by foetal elements (Sharman

1961; Hughes and McNally 1968 and Hughes 1974). This contrasts with the more usual situation of combined yolk-sac and limited chorio-allantoic placentation, such as occurs in the living vombatiforms, dasyurids and at least some didelphids, as well as with the highly invasive chorio-allantoic placenta of the bandicoots (McCrady 1938; Pearson 1949 and Hughes 1974). On the question of which of these diverse modes of placentation might be the "primitive" one for marsupials, much has been written (see discussion in Sharman 1961 and Hughes 1974). Judged strictly on the criterion of outgroup analysis (e.g., Watrous and Wheeler 1981), however, the mode of placentation seen in the Koala and wombats may well approximate the plesiomorphic diprotodontian condition. Contrariwise, the yolk-sac placentation of the phalangeridans might thus be regarded as a potential synapomorphy of this group.

Assessment of "atypical" phalangeridans: Three taxa are herein regarded as "atypical" phalangeridans in the sense that they show at least some of the derived features of that group, yet differ in other significant respects. These are the macropodoids, acrobatids (diagnosed in later section) and *Tarsipes rostratus.*

Macropodoidae: Among the numerous species of macropodoids, the greatest number of derived phalangeridan features are found in the otherwise more plesiomorphic members of each family (i.e., *Hypsiprymnodon moschatus* within Potoroidae and *Dorcopsulus* spp. within Macropodidae). Characteristic phalangeridan features observed in these and some or all other macropodoids include: the same basic pattern of middle-ear enclosure (van Kampen 1905; Flannery 1987 and Aplin 1987); an essentially identical mode of placentation (Sharman 1961 and Hughes 1974); various of the external anatomical specialisations (e.g., fusion of plantar pads, pinnal morphology); and most of the characteristic dental specialisations (e.g., marginal buccal cusps, sectorial P_3^3, paralophid shear on M_2 and the retained I_2).

In certain other respects, however, macropodoids appear to be more plesiomorphic than the "typical" phalangeridans. Four points deserve special mention. First, in no macropodoid is there a specialised prehensile area on the tail, despite the fact that it may be of use during climbing in some taxa (e.g., *H. moschatus* and *Dendrolagus* spp.) and the well-known use of the tail to carry nesting material in various potoroids (e.g., *Bettongia* spp. and *H. moschatus;* Johnson and Strahan 1982). Secondly, midpiece plasma membrane invaginations are absent in all macropodoids investigated to date (Harding 1987a and refs. cited therein). Thirdly, as documented by Boardman (1950a, and earlier works), a post-auricular hair stream reversal does not occur in

any macropodoid, even though the overall hair tract pattern may be highly complex. Lastly, whereas all elements of the tympanic floor and external meatal tube are rigidly united in the possums, in macropodoids they retain a greater degree of independence. As noted by Flannery (1987), this is particularly noticeable in the case of the ectotympanic, which remains only loosely attached to the postglenoid process of the squamosal.

Based largely on the last-mentioned feature, Flannery (1987) postulated that macropodoids might represent the sister taxon of the remaining phalangeridans. With the greater weight of evidence now at hand, we are inclined to agree. However, with Flannery, we are also cautious not to reject outright other alternative hypotheses of relationship such as the possibility of a special relationship to either Phalangeroidea or to *Tarsipes.*

Acrobatidae: These previously little studied taxa are superficially very "possum"-like, but closer examination has shown them to possess numerous peculiarities of structure (see below and Aplin in prep.). In particular, acrobatids differ from "typical" phalangeridans in numerous features of basicranial and dental morphology, in the pattern of hair tracts, in the more complicated nature of their placentation, in the cross-sectional shape of the pelage hairs and in several other aspects of external anatomy.

Importantly, in several respects (e.g., vibrissal representation, aspects of dental morphology, hair tract pattern, pelage-hair structure), acrobatids appear to be more plesiomorphic than the "typical" phalangeridans.

On the other hand, acrobatids are decidedly phalangeridan-like in possessing a fully prehensile tail, a reduced number of plantar pads on the pes, a reduced vibrissal complement, a typical "phalangeridan" pinnal morphology, a number of shared features of basicranial anatomy specialisations (including presence of deeply invasive squamosal epitympanic sinus, similar external meatal and postglenoid relations, etc.) and the presence of midpiece plasma membrane invaginations on the spermatozoa.

As was the case with the previous group, acrobatids thus appear to possess a number of typically "phalangeridan" features, along with others which are seemingly more plesiomorphic. In our opinion, however, the combined phalangeridan-like features of the acrobatids are sufficiently compelling to locate this group either "within" the cluster of "typical" families, or else in close proximity to that group, i.e., as an immediate sister group to the "typical" phalangeridans. This and other possible interpretations of acrobatid affinities are discussed at length in a later section.

As noted earlier, Kirsch's (1968a; 1977a) serological data failed to indicate any special relationship between *Acrobates pygmaeus* and any other family of possums. On consideration of their wider features, however, Kirsch regarded both this species and the other pigmy possums (i.e., *Cercartetus* spp. and *Burramys parvus*) to be most closely related to the other possums and the macropodoids.

Baverstock's MC'F results are unambiguous inasmuch as they indicate a very close relationship between acrobatids and each of Petauridae and Tarsipedidae. In terms of the present discussion, a close affiliation with petaurids would draw the acrobatids well inside the limits of Phalangerida 'as recognised herein.

Tarsipedidae: Tarsipes rostratus is without doubt the paragon of autapomorphic specialisation within Diprotodontia. Kirsch (1977a:72), speaking generally, referred to *Tarsipes* as the kind of animal that is "so modified in all of their known characters that consideration of the whole complex of taxonomic features still leaves the species unplaced with respect to close relatives". Further, he emphasised the need for "the discovery of as yet undetected homologies" and for the examination of "new set(s) of characters". Unfortunately, Kirsch's own highly innovative serological investigations failed in respect of *Tarsipes*; he was able to confirm only its general diprotodontian affinities and its general distinctness from other diprotodont families.

As hinted at above, MC'F results presented by Baverstock (1984 and Baverstock *et al.* 1987) present a very different picture of affiinities. Based on these data, *Tarsipes* is placed approximately equidistant from each of Acrobatidae and Petauridae as recognised herein. More distantly, these taxa are linked then to Pseudocheiridae.

Morphological evidence is consistent with the MC'F result insofar as it also suggests a close relationship between *Tarsipes* and acrobatids (see below). However, there is substantially less indication of direct affinity between *Tarsipes* and the Petauridae.

Returning to the original question of whether *Tarsipes* can be linked with the "typical" phalangeridans, the answer must be a qualified "yes". Certainly, if acrobatids are to be included within this group, then so, too, must be *Tarsipes*, since the evidence of special affinity between these taxa is remarkably strong.

On the other hand, were *Tarsipes* assessed solely on its own merits, we suspect that it may have been rejected as probably not closely related to Phalangerida. In particular, it displays only a small number of the characteristic features of this group. Moreover, where phalangeridan features are present, the morphological correspondence is rarely compelling. For example, while *Tarsipes* agrees with the other phalangeridans in having a ventrally prehensile tail, the bare section appears to be more coarsely scaled than in other taxa; similarly, although it possesses a reduced number of plantar pads on the pes, the posterolateral pad is unusually small and not obviously an amalgam of two original pads as in the majority of other phalangeridans; and while *Tarsipes* is phalangeridan-like in having a squamosal epitympanic sinus leading off from the tympanic cavity, this is unusually small and lacks the usual internal complexity seen in the other taxa. Indeed, of its various "phalangeridan" features, only the morphology of the external ear actually compares closely to the condition seen in the "typical" families.

As with the acrobatids and macropodoids, *Tarsipes* also appears to be more plesiomorphic in some respects than the "typical" phalangeridans. In particular, *Tarsipes* lacks the characteristic postauricular hair reversal, it has simple, rounded pelage hairs, it lacks plasma membrane invaginations on the spermatozoan, it has a poorly developed squamosal epitympanic wing and it lacks bony contact in the middle-ear between the ectotympanic and the promontory of the petrosal.

Among the more striking autapomorphies of *Tarsipes*, special mention should be made of the extremely reduced dentition and the correspondingly ill-formed maxillae and dentaries, the numerous unique features of auditory and general basicranial anatomy, the unusual mode of construction of the medial orbital wall, the specialised nature of the tongue and palatal rugae, the complex nature of the stomach combined with a very simple gut which lacks a caecum, the extremely large size and peculiarly dasyurid-like morphology of the spermatozoa, the specialised nature of both male and female reproductive tracts and the highly derived 2N = 24 karyotype.

Comments on evolutionary history of Phalangerida: Compared with Vombatiformes, Phalangerida possesses a slightly higher overall familial diversity (11 families as against seven), but a far lower proportion of extinct to modern families (3:8 as against 5:2). While this difference might be due in part to differential ease of collection of the larger fossil taxa of vombatiformes, we consider this factor to be insufficient to account for such a striking difference in modern versus fossil representation. Similarly, the difference cannot be explained in terms of differential description of the larger taxa. At present we are aware of three additional, undescribed families of Vombatiformes compared with only one undescribed family of Phalangerida, despite the presence of a very large backlog of as yet unnamed taxa of early to mid-Miocene age.

In terms of ecological diversity, Phalangerida also differs from Vombatiformes inasmuch as it is comprised predominantly of very generalised animals, showing only minor variations in basic body plan, body size and lifestyle. By contrast, highly specialised taxa are scarce, with the only notable examples being *Tarsipes*, some macropodoids and the striped possums. Gigantic taxa are unknown within Phalangerida and the only moderately-large members of the group are restricted to one or two families (Macropodidae and Phalangeridae).

Overall, the contrasts between the two suborders are striking. As argued above, Neogene vombatiformes appear to be a relictual assemblage, representing the "tail-end" of an as yet largely unknown Palaeogene radiation. By contrast, phalangeridans give every impression of being in the process of progressive radiation at the present time. Unlike vombatiformes, most phalangeridan families are taxonomically diverse at generic and species level and most are comprised of ecologically, very generalised animals. Moreover, although the number of entirely extinct families now stands at 3, this could reasonably be interpreted as "normal" familial turnover and there is no real evidence of extensive Neogene extinctions, despite the clear indications that this has occurred within Vombatiformes.

To speculate that phalangeridans are currently in the process of a vigorous radiation is not of course to suggest that they are of recent origin. To the contrary, both phalangeridans and vombatiformes as currently recognised are too specialised to have given rise to the other; by implication, the phalangeridan lineage must be at least as ancient as the vombatiforme lineage. Based on arguments presented earlier, we would thus anticipate an earliest Tertiary origin for both groups.

Superfamilial groupings within Phalangerida: As reviewed in an earlier section, taxa presently included within Phalangerida have, in the past, generally been divided into a "possum" group and a "kangaroo" group, with many recent workers recognising *Tarsipes* as a third major division. Based on evidence reviewed herein, each of these latter groups may now be regarded as "probable" phalangeridans in the sense that they share some, though not all, of the characteristic features of the group. In the case of the macropodoids, the relationship to the remaining phalangeridans would appear to be a genuinely distant one, suggesting one possible means of taxonomically subdividing the group. On the other hand, each of the acrobatids and *Tarsipes* present highly contradictory indications of affinity and there can be no surety whatsoever about the nature of their relationship to the other families.

In view of these conclusions, we consider that any attempt to delimit major suprafamilial taxa

within Phalangerida would be premature and might well result in the recognition of decidedly unnatural taxa. On the other hand, to recognise no formal interfamilial groupings among the 11 families of phalangeridans would not, in our view, accurately reflect the current level of understanding regarding the phylogenetic history of the group. A midway position between these extremes is to recognise a larger number of less inclusive superfamilies, with each cluster representing a strictly monophyletic unit inasmuch as this can be determined from all available evidence. One advantage of this approach is that the infraordinal slot remains vacant, thereby leaving scope for future formalisation of more fundamental groupings within this highly diverse and taxonomically difficult group of diprotodontians.

Following this procedure we recognise a total of four primary clusters of families, each of which can be validly recognised at superfamilial level. Two taxa cannot be confidently allocated to any of these groups. Of these, the living family Burramyidae appears to be genuinely distant from, as well as taxonomically coordinate with, each of the other superfamilial clusters. Accordingly, we consider it to represent a fifth superfamilial unit. By contrast, the family Pilkipildridae appears to share features with members of two different superfamilies (Petauroidea and Phalangeroidea), yet cannot be confidently allocated to either group. It is classified herein as Phalangerida: incertae sedis.

Superfamily Macropodoidea (Gray, 1821): The monophyly of the rat-kangaroos with the true kangaroos is supported by a massive and rapidly-growing body of data from comparative morphology (Ride 1961; Case 1984 and Flannery 1987), comparative serology (Kirsch 1977a and Baverstock in prep.) and allozyme electrophoresis (Richardson and McDermid 1971). Although the essential unity of the group is thus universally accepted, the question of whether the macropodids and potoroids should be classified in separate families (e.g., Pearson 1950, informal suggestion only; Archer and Bartholomai 1978; Archer and Flannery 1985; Flannery 1987 and earlier papers cited therin), or should be placed as subfamilies within one family (Macropodidae; e.g., Ride 1964; Kirsch 1977a; Marshall 1981; Woodburne 1984a, b and Case 1984), remains undecided.

In a phylogenetic sense, the issue is a trivial one. However, in terms of the general aim of maintaining approximate levels of distinctiveness among taxa recognised at the same rank level, some comment on the matter is required. As indicated in Table 2, we have opted for a two-family arrangement. There are two primary reasons for this. First, although the degree of differentiation between a potoroid and a macropodid is probably less than that between some of the other phalangeridan taxa (e.g., Phalangeridae and Ektopodontidae within Phalangeroidea), we suspect that the basic macropodoid pattern may be fairly tightly constrained as a result of the unique masticatory and locomotary adaptations of the group. Looking to less obviously constrained systems such as reproductive anatomy, we note that consistent differences between macropodids and potoroids are present, as in the gross organisation of the female reproductive tract (Pearson 1950 and Case 1984) and in spermatozoan morphology (Harding 1987a). Secondly, palaeontological and biochemical evidence both suggest an early divergence of the modern families, perhaps sometime during the Oligocene.

The earliest and most plesiomorphic macropodoids occur in the ?middle Miocene Tarkarooloo Local Fauna (Flannery and Rich 1985). These taxa are all referrable to the modern groups Macropodidae, Potoroinae and Hypsiprymnodontinae. Isolated pedal elements of macropodoids from these sediments indicate that these dentally very primitive macropodoids were already terrestrially adapted by that time and moreover, might have been capable of saltatory progression.

As reviewed in general terms above, available evidence appears to favour placement of Macropodoidea as the immediate sister taxon to a clade containing all other known phalangeridans. On the other hand, conflicting evidence that would link the macropodoids to almost any other family of phalangeridans is not difficult to find. Hayman and Sharp (1982) and Renfree (1980), for example, suggested that macropodoids may be most closely related to Tarsipes, based on a combination of cytological and reproductive evidence. Likewise, Flannery (1984 and see Flannery and Rich 1986) has long pushed for a specific macropodoid-phalangerid relationship, based on the similarity of the molar teeth, although more recently he has backed away from this position (Flannery 1987).

Superfamily Phalangeroidea (Thomas, 1888): This superfamilial grouping unites the living family Phalangeridae with two, closely related fossil groups, namely Miralinidae and Ektopodontidae. Both of these groups are best known from deposits of Miocene age, although ektopodontids occur as well in a single early Pliocene locality.

Ektopodontids are remarkably autapomorphic in all known aspects of their dentition and were originally described as possible monotremes. However, subsequent discoveries of more complete materials and the recent description of the closely related but less specialised family Miralinidae, indicate the phalangeridan origins of this remarkable group of marsupials (Woodburne 1987). Although known primarily from isolated teeth and broken dentaries, one larger maxillary fragment of a Miocene ektopodontid shows that these taxa possessed a markedly foreshortened rostrum.

Miralinids are intermediate in dental complexity between the relatively simple-molared phalangerids and the extraordinarily complex-toothed ektopodontids (Woodburne *et al.* 1987). Because phalangerids and miralinids are autapomorphic in other aspects of their dentition, however, none of the three groups could be regarded as either structurally or actually annectant to any one of the others. Moreover, since each of the three phalangeroid taxa are fully distinct by middle Miocene times, any actual cladogenic events must have occurred well prior to this time.

Modern phalangerids are taxonomically diverse yet morphologically fairly homogeneous, with the Sulawesian *Ailurops ursinus* (distinguished at subfamilial level by Flannery *et al.* 1987) being perhaps the most distinctive and plesiomorphic of the living taxa. Among the diverse phalangerids of New Guinea and Australia, Flannery *et al.* (ibid) recognise two tribal-level taxa: the Phalangerini, including most of the New Guinean species and the Trichosurini, containing the Australian species of *Trichosurus* and *Wyulda* and a variety of New Guinean and Molluccan taxa (e.g., "*Strigocuscus*" *gymnotis* and "*S.*" *ornatus*). Apart from differing in various external and cranio-dental features, members of these groups also differ in chromosome number ($2N=20$ in *Trichosurus* spp. but $2N=14$ in *Phalanger* spp.). However, at least one member of Flannery *et al.*'s Trichosurini ("*S.*" *gymnotis*, S. Donnellan in prep.) possess a karyotype which is indistinguishable from that of Phalangerini. Interestingly, this species would probably also be placed within *Phalanger (sensu stricta)* on the basis of MC'F results currently in preparation by Baverstock.

The earliest known phalangerid is an undescribed species of early Miocene age from the Geilston Bay Local Fauna of Tasmania. Otherwise Tertiary phalangerids occur in the diverse Miocene age local faunas of the Riversleigh area of northwestern Queensland (Flannery and Archer 1987) and in the early Pliocene Hamilton Local Fauna from Victoria (Turnbull and Lundelius 1970).

What is known about the comparative anatomy and serology of living phalangerids (almost exclusively from *Trichosurus vulpecula*) appears to support their present separation from the other possums at superfamilial level. Anatomically, phalangerids are distinguishable from other possums in virtually all systems, including cranio-dental morphology (Flannery *et al.* 1987), external morphology (Pocock 1921 and Lyne 1959), visceral anatomy (MacKenzie 1918a, b and Perrot 1966), reproductive anatomy (Kean *et al.* 1964) and digestive anatomy (Lonnberg 1902 and MacKenzie 1918a). Serologically, Kirsch (1968b and 1977a) found the phalangerids to be among the more divergent of the living possum families. A similar pattern of relationships (or lack thereof) is also emerging from Baverstock's (1984 and Baverstock *et al.* 1987) MC'F studies of albumin.

Superfamily Petauroidea (Gill, 1872): This group contains the two living families Pseudocheiridae and Petauridae, the latter containing the well-marked subfamilies Petaurinae and Dactylopsilinae. Inclusion of these taxa within a single monophyletic taxon was first suggested by Kirsch (1968a and 1977a; Gill's Petaurinae included *Acrobates* as well), based chiefly on his serological results. Unlike in the present arrangement, however, Kirsch recognised a single, more inclusive Family Petauridae, in which he formalised three distinct subfamilies. As discussed by Archer (1984b), elevation of each of Kirsch's Pseudocheirinae and Petaurinae to full familial rank results in a more balanced classification of the possums as a whole and also allows for the formal expression of the close phylogenetic relationship between petaurines and dactylopsilines.

Modern pseudocheirids attain their greatest diversity in closed rainforests of tropical North Queensland and New Guinea, but they have also radiated in less spectacular fashion in the drier forests and woodlands of southern Australia. As might be expected given this modern distribution, pseudocheirids are well-represented in many Miocene to early Pliocene fossil localities (Turnbull and Lundelius 1970; Turnbull *et al.* 1987 and Woodburne *et al.* 1987), suggesting a decline in both the geographic range and the diversity of pseudocheirids towards the close of the Tertiary.

Along with the moister, rainforest habitats. Among the oldest pseudocheirines, the middle Miocene species *Paljara tirarensae* from the Kutjamarpu Local Fauna (Woodburne *et al.* 1987) is arguably also the most plesiomorphic member of the family. As discussed at length by Woodburne *et al.* (1987), there is little in the known dental morphology of either *Paldjara* spp. or any other Tertiary pseudocheirid to suggest an alternative familial relationship to that formalised here.

The two subfamilies of petaurids each contain a small number of ecologically highly specialised taxa. Modern petaurines are best known for their woodgouging activities and volplaning. As yet this group is poorly represented in the Tertiary fossil record, with the oldest reported material being that tentatively referred to modern species of *Petaurus* from the early Pliocene Hamilton Local Fauna (Turnbull and Lundelius 1970). However, recently collected material of middle Miocene age from the Riversleigh area of north-western Queensland appears to include definite petaurids marginally more plesiomorphic than the modern genus.

Dactylopsilines, with their elongate digits and powerful incisors, are often regarded as marsupial counterparts of the lemuriform primate *Daubentonia* (the Aye-aye). In most of their special craniodental features, however, they appear to represent a more extreme version of *Petaurus* spp. Fossil dactylopsilines are as yet unknown from any Australian Tertiary or even Pleistocene locality. However, given that they are currently found only in New Guinea and on Cape York, it seems likely that this group had its primary origin in New Guinea and perhaps entered northern Australia only recently.

The basis for formal recognition of a "petauroid" clade comes chiefly from serological data, although some morphological support is also available. Data presented by Kirsch (1968b, 1977a) indicate a very close serological relationship between species of *Petaurus* and *Pseudocheirus;* in Kirsch's view, the distance between these taxa was about that seen between typical marsupial subfamilies. Baverstock's (1984 and Baverstock *et al.* 1987) MC'F studies of albumin confirm this view inasmuch as petaurids and pseudocheirids form a well-supported (=long common branch length) cluster to the exclusion of most other phalangeridans (e.g. phalangerids, burramyids (present sense)) and macropodoids. On the other hand, Baverstock's results seemingly contradict those of Kirsch in that the MC'F results indicate that *Tarsipes* and the acrobatids also belong within this group.

In morphological terms, petaurids and pseudocheirids are at least superficially quite distinct (see Archer 1984b for illustrations covering external morphology, cranial shape, dental morphology). Without here wishing to go into extensive detail, we regard the majority of these differences to be autapomorphic specialisations related to the considerable ecological divergence between each of the families and between each of the subfamilies of petaurids. The striking selenodont dentition of the pseudocheirids, for example, although commonly now regarded to be a primitive feature in this group, might represent a secondary elaboration of an originally simpler pattern (Woodburne *et al.* 1987). In a similar fashion, many of the cranial and dental features that serve to distinguish members of the Petauridae appear to be related to specialisation of the anterior dentition for the purpose of woodgouging (e.g., Kay and Cartmill 1979); this is carried to an extreme degree in the dactylopsilines.

At a more detailed level of comparison, petaurids and pseudocheirids do appear to share numerous, more subtle features of cranial structure (e.g., orbital fossa, auditory region), general visceral anatomy, aspects of myology, detailed oral and lingual morphology etc. (mostly Aplin pers. observ.; but see Shrivastava 1962; Segall 1969b, 1970, 1971 and Haswell 1887). From the cladistic point of view, however, many of these features appear to be of little significance, either representing phalangeridan symplesiomorphies or being too poorly documented in other groups for adequate comparisons.

One feature which might constitute a valid synapomorphy of the Petauroidea is the presence of a single precava, as against the usual paired precavae of other marsupials.

The presence of a single precava in *Petaurus breviceps* was first noted by Forbes (1881), with subsequent confirmation by Sonntag (1921), Pearson (1940) and Amoroso *et al.* (1943). Although fairly widespread among placental mammals (see Amoroso *et al.* ibid. and Barnett *et al.* 1958), this represented the only documented occurrence of this undoubted specialised condition in a marsupial. During the course of wider anatomical investigations, Aplin (in prep.) has observed a single precava in one or more individuals of the following additional taxa: *Petaurus australis*, *Dactylopsila trivirgata*, *Pseudocheirus cupreus*, and *Petauroides volans*.

Among other marsupials examined to date, the more typical (and undoubtedly plesiomorphic) condition of paired precavae has been reported for a variety of macropodoids (including members of both families), *Trichosurus vulpecula*, the koala, a species of wombat, various bandicoots and dasyurids and several didelphids (see refs. cited above, McClure (1903) and Heighway (1939)) and has additionally been noted by Aplin in *Distoechurus pennatus*, several species each of *Cercartetus* and *Phalanger* and in one specimen of *Pseudocheirus forbesi*.

The presence of paired precavae in one species of *Pseudocheirus* does not in our opinion substantially detract from the potential cladistic significance of this feature. One observation relevant to this point is that the arrangement of the precavae is typically very stable within any given species; McClure (1903), for example, found only paired precavae in a survey of over 100 individual opossums. At least among placental mammals, this stability is carried over to the higher taxonomic levels; hence, most primates have a single precava, whereas most rodents have paired vessels. In all groups there are exceptions, however. Among the rodents, for example, there are several families in which all taxa possess a single precava, yet still other families in which this occurence is restricted to one or more genera (George 1981).

In the context of these wider observations, it thus seems reasonable to conclude that the presence of a single precava is synapomorphic for those phalangeridan taxa in which it occurs — the majority of Petauroidea — but with the qualification that the more plesiomorphic condition has presumably reappeared as an atavistic trait in at least one species of *Pseudocheirus*.

Baverstock's (1984 and Baverstock *et al.* 1987) suggestion that acrobatids and tarsipedids may be specially related to petaurids is discussed at length in the following section. Suffice it to say here that we feel that there are several reasons to doubt the correctness of this view and have thus chosen not to formalise such a concept.

As a final point of relevance to petauroid classification, we direct attention to the newly described fossil family Pilkipildridae (Archer *et al.* 1987). As discussed in detail in that paper, these poorly known Miocene possums are decidedly petaurid-like in some aspects of dental morphology, yet are similar in other features of their dentition to phalangerids and miralinids. In view of these contradictory indications of affinity, we herein include this group in Phalangerida: but list it as *incertae sedis* (see below).

Superfamily Tarsipedoidea (Gervais and Verreaux, 1842): The inclusion among the Phalangerida of *Tarsipes* and its phylogenetic association with the Feather-tailed Possums (herein described as a new family of extant marsupials) arguably represent the most radical and potentially controversial aspects of this new classification. Accordingly, in the the following pages we present an extended justification for the formal disbandment of the Family Burramyidae and for the establishment of a new familial group for the Feather-tailed Possums. We also present our reasons for including this new family within an expanded concept Tarsipedoidea.

The Feather-tailed possums (namely *Acrobates pygmaeus* of Eastern Australia and *Distoechurus pennatus* of New Guinea and nearby islands) are small to medium-sized diprotodontians which, prior to Kirsch's (1968a, 1968b) multi-familial classification of the late 1960's, were included by all workers in a broadly-conceived Phalangeridae. Compared with other members of this group, *Acrobates* and *Distoechurus* were generally regarded to be closely related to each other (e.g., Bensley 1903; Winge 1941; Ride 1959b) and to be somehow related to a larger grouping comprised of the other "bunodont" (i.e., possessing low-crowned molars) possums. One early dissenter was Ride (1959) who considered the Feather-tailed Possums to be more primitive than the other "phalangerids" on account of their triangular-shaped molars and unreduced, secodont premolars. Interestingly, most workers classified *Tarsipes* apart from the "bunodont" possums; Winge (1941), however, considered *Tarsipes* to be a highly specialised member of the bunodont group.

The first major challenge to this essentially very stable arrangement was issued by Gunson *et al.* (1968), following their karyotypic study of various small possums including the recently "resurrected" Mountain Pigmy Possum, *Burramys parvus*. They suggested that the 2N=14 karyotype found in each of *Burramys, Cercartetus* spp. and *Acrobates pygmaeus* was sufficiently distinct from the karyotypes of other possum genera to warrant the removal of these taxa to a family apart from the Phalangeridae.

Formal recognition of this new familial concept came later that year following presentation by Kirsch (1968b) of selected results of his extensive serological study of Australian marsupials. Because Kirsch was unable at that time to raise an antiserum against any of the smaller possums, his results do not include any "direct" demonstration of special affinity between *Acrobates* and either *Burramys* or *Cercartetus*. Nevertheless, he was able to show that each was individually distinct from the other major groups of possums and that they "form a group at least by exclusion from the other(s)" (1968b:44). As for *Distoechurus*, Kirsch noted its probable affinities with the "burramyids", but listed it conservatively as "Phalangeroidea incertae sedis". Subsequently, however, Kirsch and Calaby (1981) included *Distoechurus* within the Burramyidae based on its morphological resemblance to *Acrobates*.

Doubts about the monophyly of Burramyidae as then understood surfaced in 1982 during the course of two independent phylogenetic studies of Diprotodontia. Baverstock, working at the South Australian Muséum, had at that time begun to explore the molecular relationships of Australian marsupials using the technique of MC'F. One of the more surprising results of his early studies was the recognition that *Acrobates pygmaeus* fell unexpectedly far away from the other "burramyids". This suggested that Burramyidae *sensu* Kirsch was either far more diverse at the molecular level than any other family of marsupials, or that it was not a natural group. Subsequent results supported the latter interpretation, since *Acrobates* (by then, with *Distoechurus*) was found to group more closely with members of Petauridae and *Tarsipes* than with the remaining burramyids (Baverstock 1984 and Baverstock *et al.* 1987).

At the same time, Aplin and Archer in Sydney were investigating various aspects of cranial and dental morphology of diprotodontian marsupials, with particular emphasis upon the phylogenetically informative auditory region. During this study it was noted that *Acrobates* and *Distoechurus* shared a common basicranial and dental pattern that was strikingly divergent, not only from the specific morphology seen in the other "burramyids", but also from the more general pattern common to all of the "possum" families. Based on these observations, they were confident that the feather-tailed possums represented a distinct family-level grouping of diprotodonts, but were far less certain as to exactly where within Diprotodontia it might belong. Based on materials available at that time, there appeared to be no special resemblance either to *Tarsipes* or to petaurids.

In an attempt to shed further light upon the phylogenetic affinities of the feather-tailed possums, Aplin subsequently began a more wide-ranging comparative anatomical study of this and other potentially related groups. Several aspects of this work, undertaken in collaboration with relevant specialist workers, are reported in these volumes (Hughes *et al.* 1987 and Harding 1987a). Unfortunately, however, the larger part of this work (dealing with most aspects of the gross anatomy including external, cranio-dental, oro-lingual, visceral, cardiovascular, reproductive and muscular features) remains unfinished; it must now be published elsewhere. However, because the original suggestions of Aplin and Archer (1983) and Baverstock (1984) as to the probable familial distinction of the feather-tailed possums have proven acceptable to many workers, formal diagnosis of this group ahead of the larger comparative work is appropriate. This is undertaken below with an extended discussion of acrobatid and wider tarsipedoid affinities. This discussion should be regarded as the formal justification for the familial diagnosis.

FAMILY ACROBATIDAE APLIN, NEW

Type genus: Acrobates Desmarest, 1817.

Included genera: Distoechurus Peters, 1874.

Family diagnosis: Small to medium-sized phalangeridans differing from all other members of this group in possessing a distichous tail; a uniquely complex external ear with prominent anterior helical process, paired antitragal processes and a well-defined bursa; pelage hairs that lack noticeable distal "shields"; a unique hair follicle pattern (see Hardy 1947); "primitive", all caudal pattern of hair tracts; unusually numerous genal vibrissae; a separate post-marginal pad on the pes; a non-papillose circum-vallate region of the tongue and an enlarged posterior vallate papilla; reduced sublingual structures; a well-developed marginal "fringe" of conical papillae around the full perimeter of the papillated portion of the tongue; a distinct "lingual eminence" at the rear of the tongue; an elongate semi-tubular epiglottis; a long upper canine that projects well below the level of the upper incisors and premolars; a short cranio-incisive diastema; tall "secondont" upper premolars of which P^3 is either slightly or distinctly larger than P^2; only two non-vestigial lower premolars; loss of the posterior molar in both upper and lower dentition; upper molars with paracone taller than metacone and small metaconule; lower M_2 with tall anterobuccal "protostylid" and entoconid distinctly lower than metaconid; unusually elongate incisive foramina; greatly enlarged posterior palatal vacuities; extensive contact of the lacrimal and palatine elements in the medial orbital wall; very large inter-parietal element; reduced neurocranial lamina and zygomatic process of squamosal; unique foramen for temporo-masseteric ramus of mandibular nerve at anteromedial end of glenoid fossa; alisphenoid contribution to anteromedial margin of glenoid fossa; mandibular condyle lying close to level of cheek-tooth row; masseteric fossa of ascending ramus strongly developed but coronoid process reduced; auditory bulla lacking alisphenoid contribution and formed instead from petrosal, squamosal and ectotympanic elements; "bulla" underrun medially and posteriorly by "secondary" tympanic processes from basi- and exoccipital bones; primary tympanic cavity complexly compartmentalised by numerous septa; facial nerve canal complete through middle-ear; promontory of petrosal with strong transverse constriction at level of fenestra vestibuli; cochlear fossula deep and "winged" posteriorly; malleo-incudal joint fused in adult; anterior malleolar process reduced to splint; stapedial footplate very elongate; both ossicular muscles reduced to tendinous vestiges; ectotympanic tubular and giving rise to unique ectotympanic disc that narrowly occludes tympanic membrane; stomach with expanded fornical region; pancreas largely confined to lesser omentum; intestinal caecum moderately to very elongate; paired precavae; glans penis undivided at tip, blunt rather than tapering and lacking flagellum; urethral aperture shielded dorsally by a blunt process; levator penis muscle absent or very reduced; no dorsal suspensory ligament of penis; 1-2 pairs of bulbo-urethral glands and single(?) pair of paracloacal glands; tunica vaginalis of testis unpigmented; rete testis simple, duct-like; efferent duct single (?); spermatic cord containing separate bundles of arteries and veins; unique features of spermatozoal morphology (see Harding 1987); unique mode of placentation (see Hughes *et al.* 1987); females polyoestus, showing embryonic diapause.

Discussion of Acrobatidae:

Comparative remarks and an overview of Tarsipedoidea: A general comparison between Acrobatidae and the artificial grouping of "typical" phalangeridans was given in an earlier context from which it was concluded that acrobatids were phalangeridans, albeit somewhat "atypical" ones. Within Phalangerida, Acrobatidae need be compared specifically with members of three other families: Burramyidae *(sensu stricta)*, Petauridae and Tarsipedidae.

Comparison with burramyids: Burramyids differ from acrobatids in respect of virtually every feature cited in the familial diagnosis. With few exceptions, the more derived condition occurs in the acrobatids such that burramyids appear overwhelmingly plesiomorphic in the majority of features examined.

Three areas of somewhat more detailed resemblance relate to the mode of construction of the orbital mosaic, the "bunodont" molar morphology and loss of the posteriormost molar in each toothrow. In the orbital fossa, the feature of interest is the presence of extensive lacrimo-palatine contact which contrasts with the more usual involvement of the maxillary in this area in diprotodontians (Aplin in prep.). Other than in acrobatids and burramyids *(s.s.)*, a broad lacrimo-palatine contact in the orbit is found in several phalangerids and macropodoids (Flannery *et al.* 1987), in all petauroids and in most if not all Australian and American "polyprotodonts". On distributional grounds, the acrobatid-burramyid condition would thus appear to be symplesiomorphic.

The shared presence of "bunodont" molars is similarly uninformative since they are present also in petaurids and in primitive thylacoleonids (see earlier). Moreover, on detailed comparison, the bunodont molars of *Petaurus* spp. and *Cercartetus* spp. appear to resemble each other more closely than either resemble the molars of acrobatids.

Loss of the posteriormost molar, herein regarded as diagnostic of Acrobatidae, has occurred also in two of the four species of *Cercartetus*. Presumably, given the presence of the full molar row in two other *Cercartetus* spp. and in *Burramys parvus*, this similarity can be readily written off as convergent.

As reported by Gunson *et al.* (1968), acrobatids and burramyids *(s.s.)* are unique among the "possums" in having a 2N = 14 diploid comprised of a predominantly metacentric autosomal series and small sex chromosomes. To Kirsch (1968), this observation constituted critical support for Burramyidae *(sensu lato)*, which was otherwise only weakly supported by his serological data (see earlier comments).

Subsequently, Hayman and Martin (1974) suggested that the 2N = 14 "basic" karyotype, such as occurs in the small possums, was actually close to a hypothetical ancestral 2N = 14 shared with various other major groups of marsuials (e.g., vombatids, some peramelids, some didelphids). Recent work has confirmed this view, with the critical evidence coming from the G-banding studies of Rofe (1979, cited in McKay 1984) and Rofe and Hayman (1985). Interestingly, of the Australian marsupials successfully G-banded to date, it is a burramyid (*Cercartetus concinnus*) that appears to show least divergence from the reconstructed "ancestral" karyotype. Karyotypic similarities between acrobatids and burramyids may thus be understood as an example of symplesiomorphic retention and not of special affinity.

Comparison with petaurids: Petaurids differ from acrobatids in the majority of the features listed in the diagnosis. In general, however, they appear to be rather more similar to acrobatids than are the burramyids, possibly reflecting in part a similar dietary specialisation in these groups. Special features of the skull and teeth shared by petaurids and acrobatids include a similar orbital mosaic, a bunodont dentition, a similar modification of the posterior section of the dentary and a similar pattern of palatal rugae.

As noted above under burramyids, the shared features of bunodonty and the construction of the orbital fossa are probably of little or no phylogenetic significance. The close resemblance in the pattern of palatal rugae in these taxa is of uncertain status; work in progress by Aplin, however, suggests that the essential attributes of this resemblance, such as the number of ridges, general shape of palate, etc., are all diprotodontian symplesiomorphies.

One feature which is clearly apomorphic in these taxa is the unusual modification of the ascending ramus of the dentary. In both groups the articular condyle lies at about the level of the cheekteeth and the coronoid process is swept far back. In acrobatids, and to a lesser degree in petaurids, the coronoid process is weakly developed. Seemingly correlated with these features of the dentary, the facial skeleton in petaurids and acrobatids is deflected downwards relative to the basicranial axis. In both groups, these changes are probably related specifically to increased emphasis upon incisive function, perhaps as an adaptation to woodgouging and gumnivory.

Away from the skull, petaurids and acrobatids are similar in several features of visceral anatomy including the possession of a relatively long colon, an expanded fornical region of the stomach and a similarly elongate caecum. These resemblances do not hold if comparisons are extended to include Dactylopsilinae; in this group the whole gut is very simple and lacks all of the specialisations seen in the other petaurids and the acrobatids (Smith 1981). Again, dietary factors are strongly indicated.

As noted in the diagnosis, acrobatids display the plesiomorphic, paired precavae in contrast to the single precava of the petaurids and most other petauroids.

In terms of cytology, each of the petaurid genera is readily distinguishable from the acrobatids (McKay 1984), but this contributes nothing in the way of useful phylogenetic information.

Serological evidence presented by Kirsch (1968) is of considerable importance here inasmuch as his results appear to group petaurids and pseudocheirids to the exclusion of acrobatids. As noted by Baverstock (1984), some measure of special affinity between *Acrobates* and the petaurines might be inferred from the fact that preabsorption of an anti-*Petaurus* antiserum with *Burramys* removed all activity for *Cercartetus* but left some reaction for *Acrobates*. By itself, however, this result suggests only that *Acrobates* may fall closer to the combined petauroid clade than does *Cercartetus*.

As mentioned earlier, Baverstock's (1984 and Baverstock *et al.* 1987) results give a very different picture of affinities. In contrast to Kirsch's results, the MC'F data suggest that acrobatids are specially related to members of the general petauroid group and that, within this group, they might be closer to the petaurids than to the pseudocheirids.

Comparison with tarsipedids: The Honey Possum, *Tarsipes rostratus*, would appear on first glance to be least likely of these three groups to show any special morphological similarity to acrobatids; this was certainly our conclusion after initial comparisons of the basicranial region in the two groups (Aplin and Archer 1983). However, subsequent, more objective comparisons resulted in the recognition of a large number of apparently shared derived similarities between these groups, located not only in the basicranial region, but spread as well throughout many other parts of the body.

Among the more compelling of these apparent synapomorphies are the mode of enclosure of the posterior two-thirds of the tympanic cavity (e.g., presence of large, secondarily united caudal and rostral tympanic processes of petrosal), the detailed morphology of the promontory of the petrosal, the morphology of the auditory ossicles (e.g., fusion of malleo-incudal complex, reduction of ossicular muscles), the construction of the neurocranium and zygomatic arch (e.g., reduced squamosal contribution, greatly enlarged interparietal), various aspects of male reproductive anatomy (e.g., shape and detailed structure of glans penis, absence of levator penus muscle and of suspensory ligament, reduced number of bulbo-urethral glands) and of the spermatozoan (see Harding 1984, 1987), aspects of visceral anatomy (e.g., lobation of liver), several features of external anatomy (e.g., detailed rhinoglyphics, form of pinna), the morphology of the tongue (e.g., reduced sublingua, details of circumvallate region) and the shared practice of embryonic diapause of a similar kind (Renfree 1980 and Ward and Renfree 1986).

Cytological and serological evidence neither support nor contradict the concept of special affinity between *Tarsipes* and acrobatids. As reported by Hayman and Sharp (1981), the *Tarsipes* karyotype features a high number (2N=24) of graded telocentric autosomes and an enlarged X. In basic number, it is comparable to certain macropodids, but it differs from these in regard to autosomal morphology. Among the various other kinds of possums, the karyotype of *Tarsipes* is perhaps closest to that of *Gymnobelideus leadbeateri* (S. Donnellan unpubl. cited by McKay 1984) and *Trichosurus* spp., all have a lower basic number (2N=22 and 2N=20, respectively) but a similar, primarily telocentric autosomal morphology. Since no G-banding results are yet available for either *Tarsipes* or *Gymnobelideus*, however, the degree of karyotypic homology between these taxa remains unknown.

As discussed earlier, Kirsch's (1977) serological results for *Tarsipes* show only that it is a diprotodontian and a moderately distinct member of this group. However, since no direct tests were made between *Tarsipes* and *Acrobates*, or between *Tarsipes* and *Petaurus*, these data have little bearing on the present issue. On the other hand, the MC'F results provide a very clear picture of molecular affinities, grouping *Tarsipes*, the acrobatids and the petaurids in a relatively tight cluster; the molecular distance in each case is around 40 units (Baverstock *et al.* 1987).

In overview, immunological and morphological evidence are in good agreement on two points, but show a lack of congruency (or perhaps, of resolution) over a third. On the question of whether the feather-tailed possums are correctly placed within Burramyidae (*sensu* Kirsch 1968b), the answer must surely be that they are not. As it happens, of the three groups compared in the foregoing section, burramyids actually show the least special resemblance to the acrobatids. Other than for some retained phalangeridan symplesiomorphies and the fact that members of both families are small-bodied, acrobatids and burramyids are strikingly dissimilar in all aspects of their known morphology. Moreover, in the light of subsequent developments, the karyotypic and serological evidence cited by Kirsch (1968b) in support of Burramyidae,

can be rejected — the karyotypic similarity is almost certainly symplesiomorphic and the serological evidence of affinity was, as Kirsch emphasised at the time, very weak.

On the question of acrobatid affinities, the case for a special relationship to *Tarsipes rostratus* is undeniably strong; not only is the molecular evidence of affinity between these taxa relatively strong, but it is supported by a remarkably rich and varied body of morphological evidence. Importantly, this comes not from a single anatomical region or functional system but from a wide range of regions and systems including several functional components within the cranium, the reproductive system, the spermatozoan, the tongue and so on.

Considering the evidence of close affinity between the acrobatids and *Tarsipes,* one systematic option is to refer the feather-tailed possums to an expanded Tarsipedidae. An alternative and, in our opinion, more acceptable option is to recognise the extreme phenetic divergence between these taxa at familial rank, but to unite them within an expanded superfamilial taxon Tarsipedoidea.

In the event that the molecular picture of tarsipedoid affinities is supported (i.e., that tarsipedoids are shown to belong within Petauridae or Petauroidea), further taxonomic changes will clearly be in order. However, given that the molecular picture is currently unsupported by morphological evidence and appears to be actually contradicted by serological evidence, we regard the arrangement offered herein to be most in line with available evidence.

Superfamily Burramyoidea (Broom, 1898): Burramyidae as recognised herein is comprised of the species of only two genera, namely *Cercartetus* and *Burramys*. That these are indeed monophyletic to the exclusion of all other phalangeridan taxa is not certainly known, although such shared features as the characteristic pattern of premolar reduction (Archer 1984b) and the reduction in size of all digital claws (Anonymous 1966), suggest that they probably are. Molecular and serological data, though limited in scope, also appear to place *Burramys* with *Cercartetus* (Kirsch 1968; Baverstock 1984) although the MC'F results in particular suggest that the relationship may be a relatively distant one.

In terms of their general anatomical features, burramyids are perhaps the most plesiomorphic of known phalangeridans. As a consequence, morphological evidence of affinity to any other familial group is weak. Molecular data presented by Baverstock *et al.* (1987) are likewise consistent with a wide separation of burramyids from all other phalangeridan families.

Fossil evidence also indicates a relatively ancient phyletic separation of burramyids from other phalangeridans, inasmuch as there are now undoubted burramyids described from several local faunas of Miocene to Pliocene age (Pledge 1987c; Turnbull *et al.* 1987). In all cases known to us, these taxa appear to be referrable to the extant genus *Burramys.* Earlier references to *Cercartetus*-like forms in the Miocene of Australia (e.g., Tedford *et al.* 1977), actually refer to taxa now classified as pilkipildrids.

In our view an appropriate expression of the degree of morphological distinction of the burramyids and of their apparently wide phyletic separation from other possums is to place the revised, more restricted family in a separate superfamily Burramyoidea.

Phalangeridans of uncertain superfamilial position: As indicated in an earlier context, the Pilkipildridae, a newly described familial cluster of Miocene age taxa, are similar in aspects of their known morphology, both to the living petaurids and to members of the Miocene family Miralinidae (Archer *et al.* 1987). In the present classification, these latter mentioned families are placed in different superfamilies, Petauroidea and Phalangeroidea respectively. Since incorrect reference of Pilkipildridae to either of these taxa would have the effect of rendering one of them polphyletic and the other, presumably paraphyletic, we prefer instead to place them as Phalangerida: *incertae sedis.*

MARSUPIAL TAXA UNALLOCATED AT COHORT LEVELS

Because in the present classification we do not recognise any stem taxon for the Marsupialia (i.e., such as: Didelphidae of Kirsch 1977 and others; Pediomyidae of Szalay 1982; Microbiotherioidea of Reig *et al.* 1987 and Marshall 1987), we are left with a number of taxa which, for one reason or another, cannot be satisfactorily accommodated within either the Ameridelphia or the Australidelphia. These are listed in Table 2 as 'Cohort *incertae sedis*'. Here we provide brief comments about the three groups that we have placed in this category.

Peradectidae (Crochet, 1980)

This taxon was initially proposed by Crochet (as a Tribe within the Didelphidae) to encompass a variety of essentially plesiomorphic, late Cretaceous to Eocene 'didelphid' marsupials. Central among these are the species of *Alphadon,* known from the late Cretaceous of North America, and species of *Peradectes,* known from the Palaeogene of North America, Europe and possibly South America. The species of *Alphadon* in particular have commonly been taken to represent a basal grade of morphology from which all other known late Cretaceous and Tertiary marsupials could ultimately be derived (e.g., Clemens 1966, 1979; Archer 1976a; Crochet 1980). Crochet regarded the Peradectini to be the counterpart within the Didelphidae of the Didelphini which contained the more derived of the Cainozoic opossums.

Szalay (1982) evidently accepted the distinction between the peradectids and didelphids but in contrast to Crochet's approach, included the peradectids within his expanded concept of Pediomyidae. Among the various samples of 'pediomyid' tarsal elements examined by Szalay (1982), those from the Bitter Creek, Four Mile, Bridger, Green River and Teepee Trail faunal assemblages were likely to contain at least some elements referable to peradectids *(sens. strict.).*

Reig *et al.* (1987) confirmed the distinction of Crochet's Peradectini from their Didelphoidea (didelphids, sparassocynids and many of the Ibaborian genera) and took the additional step of boosting it up to family level. In regard to the phylogenetic significance of the peradectids, they suggested that it was "likely . . . to represent the ancestral stock of the Tertiary [didelphoid] radiation". This view is clearly expressed in their summary 'phylogram' in which Peradectidae is shown as a paraphyletic sister taxon (i.e., both a stem group and a persistent, collateral branch) to the Didelphidae.

A similar but perhaps more plausible scenario was developed for North American peradectids by Krishtalka and Stucky (1983). They suggested that Peradectini be restricted in composition to the species of *Peradectes, Nanodelphys, Mimoperadectes* and *Armintodelphys* and that together these represent the sister taxon of the Didelphini (i.e., our Didelphimorphia). They regard *Alphadon* (as currently conceived) to be a grade taxon, some members of which (e.g., *A. lulli*) are at least structurally annectant to a combined peradectine-didelphine clade. However, they also raise the possibility that specific *Alphadon* species may be specially related to peradectines (e.g., *A. russelli*) and others to didelphines (e.g., *A. marshi* and *A. rhaister*).

Several taxa currently included within Peradectidae clearly deserve further attention. One of these is *Alloedectes mcgrewi* (Russell 1984) from the lower Oligocene Cypress Hills Formation of Saskatchewan. This moderate-sized taxon (with molars about 5 mm long) was originally referred to the Peradectini on the basis of its subequal paracone and metacone and its poorly developed centrocrista. However, apart from these peradectid-like features, it is decidedly atypical in having an enlarged metastylar region, extreme reduction of stylar cusps C and D without concomitant reduction of the stylar shelf, a low stylar cusp B and reduced conules adjacent to the protocone. This combination of features sets this species well apart from other forms currently placed in the Peradectidae.

Another very unusual taxon currently placed in this family is *Peradectes marandati*, recently named by Crochet and Sige (1983) on the basis of several upper molars from the middle Palaeocene of Belgium. Interesting features of this species include the considerably larger size of the paracone over the metacone, the plesiomorphically straight centrocrista, the presence of well-developed stylar cusps B and C and a small stylar cusp D each of which are marginally positioned on the stylar shelf, the poorly developed metastylar corner of the tooth and its large parastylar corner. The last upper molar also displays a very early stage in reduction of the metacone and metaconule. While in these features it might well be annectant to later Tertiary European peradectids, its strikingly plesiomorphic morphology suggests that it was probably derived from an ancestor with a pre-*Alphadon* grade of dental morphology. Perhaps it is itself a form annectant between more typical late Cretaceous peradectids and the early Cretaceous *Holoclemensia*. One conclusion that might be drawn from the very plesiomorphic nature of this species is that mid- to late Cretaceous marsupials will eventually be found, perhaps in Europe itself or possibly in Africa. In this latter regard, we draw attention to the recently described possible marsupial *Garatherium* from the Eocene of Algeria (see below).

Overall, it is apparent that Peradectidae as recognised by Reig *et al.* (1987) probably requires significant revision. It seems likely that future phylogenetic studies will demonstrate that some of the forms currently placed in this family are indeed ameridelphians, while others probably belong outside of either of the major cohort-level taxa recognised here, or may even be basal or outside of the supercohort Marsupialia. Pending more detailed studies, we have maintained Peradectidae as conceived by Reig *et al.* (1987). However, we are convinced that it is at best a paraphyletic taxon of uncertain position within Marsupialia and at worst a polyphyletic grade.

Pediomyidae (Simpson, 1927)

Both the phyletic affinities and even the monophyly of this interesting group of late Cretaceous marsupials are currently under debate. There are at least 11 species currently divided between two genera, *Pediomys* and *Aquiladelphis*. The only certain records are from North America. Although Sige (1972) has tentatively suggested the presence of *Pediomys* from South America, it is possible that this material represents a caenolestoid or possibly even a microbiotherioid. None of the species are known to be represented by anything other then teeth and jaw fragments.

Traditionally, the Pediomyidae has been regarded to be a strictly monophyletic group,

with membership defined by a reduction of the stylar shelf, particularly the stylar cusps buccal to the paracone, and enlargement of the protoconal and talonid basins. These features have commonly been regarded to be specialised character states that were ultimately derived from a more *Alphadon*-like morphology (Clemens 1966 and Lillegraven 1969). *Aquiladelphis*, with its more complex stylar cusp B region, is at the same time more plesiomorphic and yet more derived than at least some species of *Pediomys*.

As argued by Fox (1987; this Volume), recent discoveries of older more plesiomorphic taxa suggest that reduction of the stylar shelf has occurred independently, and in slightly different ways, in several different groups of pediomyids. This has led him to challenge the idea that *Pediomys* is a monophyletic genus and to suggest that the family itself may be paraphyletic.

The traditional view of pediomyid evolution was that they became extinct at the close of the Cretaceous, leaving no Cainozoic descendants. Marshall (1978), however, suggested that they may have given rise to either borhyaenoids or thylacinids but this idea has not received subsequent support (Fox 1978 and Clemens 1979a). More recently, Reig *et al.* (1987) have suggested that pediomyids and microbiotheriids are closely related and formalised this suggestion by uniting the two taxa (along with glasbiines) in the single family Microbiotheriidae. This concept of Microbiotheriidae (but without glasbiines) was supported by Marshall (1987). This hypothesis serves simultaneously to identify ancestors for the microbiotheriines and descendants for the pediomyines. Reig *et al.* (1987) further suggest by way of their 'phylogram' that this group may also have had something to do with the ancestry of the Australian marsupials.

Although cheektooth morphology is undeniably similar between members of the Pediomyidae and the Microbiotheriidae, for several reasons, we hesitate to accept the conclusion that this is sufficient grounds for concluding that the two taxa are most closely related to each other. First, as detailed by Archer (1976a), the common dental morphotype of Australian dasyuroids and perameloids would have differed in only small details from an *Alphadon*-like pattern. Derivation of this pattern from a pediomyid/microbiotheriid dental pattern would necessitate the atavistic redevelopment of numerous plesiomorphic dental features including a broad stylar shelf, large stylar cusps B and D, prominent proto- and metaconlules, a smaller protocone and a more lingually directed cristid obliqua.

Secondly, it seems difficult to reconcile a special pediomyid/microbiotheriid relationship with the view proposed by Szalay (1982) and supported here that microbiotheriids are part of a monophyletic Australidelphia. Because pediomyids are represented in the earliest late Cretaceous faunas of North America, acceptance that pediomyids and microbiotheriids are each others' closest relatives implies that the divergence of the Australian marsupials from a common ancestor shared with the combined pediomyid/microbiotheriid taxon must have been a mid-Cretaceous event, something which although possible seems highly improbable.

Further, acceptance of a special pediomyid/microbiotheriid relationship means that the derived australidelphian tarsal features noted by Szalay (1982) to occur in the living microbiotheriid *Dromiciops australis* and all Australian marsupials must either have been independently acquired in microbiotheriids or have also been present in pediomyids. Unfortunately, it is not yet possible to determine with any certainty whether or not pediomyids had the australidelphian tarsal features. However, among the various samples of tarsal bones examined by Szalay (1982), that from the Lance Formation, in which "species of *Pediomys* form the most abundant group of marsupials" (Clemens 1966:34), might well have included pediomyid tarsal elements. If it did and they were among the elements examined by Szalay (1982), either the pediomyid/microbiotheriid dental similarity is illusory, presumably the result of convergence, or the microbiotheriid tarsus must be convergent on that of the other australidelphians. At present, we would favour the former conclusion because the concept of Australidelphia, as discussed above, is supported by another independent body of data (MC'F serology; Sarich, pers. comm. to Archer, 1986).

Stagodontidae Marsh, 1889

Stagodontids are a remarkably specialised group of marsupials which appear to be restricted to the late Cretaceous of North America. Nine species are known, representing three or four genera (Clemens 1979a) of which the recently described *Pariadens kirklandi* (Cifelli and Eaton 1987) is the oldest definite marsupial (as distinct from the oldest didelphian). All known stagodontids are relatively large by comparison with contemporary marsupials (*Alphadon* and *Pediomys* species) and all show characteristic dental specialisations involving enhanced trigonid shear with broad talonid-protocone contact. The premolars are also massively built and clearly suited for a crushing function, a structural adaptation that was carried to extreme in *Didelphodon vorax*.

Unlike all other Mesozoic marsupials, recognisable basicranial portions of stagodontids have been available for study (Clemens 1966). As with the dentition, the auditory region of *Didelphodon*

vorax is remarkably specialised. It exhibits substantial inflation of the primary tympanic cavity and a complex, derived pattern of middle ear enclosure involving an extensive contribution from the petrosal. As noted by Clemens (1966), similar auditory specialisations are seen among American marsupials only in microbiotheriids.

The relationships of stagodontids to other Mesozoic marsupials are uncertain although it has been usual to regard them as derived from an *Alphadon*-like form (Clemens 1966; Reig *et al.* 1987). This view was questioned by Fox (1971), but only insofar as this concept threatened the monophyly of *Alphadon* itself. Hoffstetter (1970; as cited in Fox 1971), however, suggested that stagodontids had been derived directly from an ancestral pappotheriid stock (presumably *sens.* Turnbull 1971). As Fox (1971) emphasised, stagodontids resemble later marsupials in all essential features of molar morphology (more open trigonid, broadened talonid with twinned entoconid and hypoconulid) whereas the similarities to pappotheriids all appear to be symplesiomorphies.

An alternative interpretation would be to group stagodontids with microbiotheriids on the basis of middle ear structure. Although dentally these groups differ substantially, with the two appearing to have specialised in different directions away from an *Alphadon*-like pattern, they nevertheless share derived features relating to protocone/talonid occlusion such as extreme buccal positioning of the cristid obliqua and an enlarged protocone with poorly developed conules. If these shared derived basicranial and dental features are genuinely synapomorphic rather than convergent, we would expect that when stagodontid tarsal bones are found, they will display the australidelphian tarsal pattern.

At present, we do not consider that available evidence enables confident placement of the Stagodontidae within either of the two cohorts recognised in our classification. Similar uncertainty may be the reasons why Reig *et al.* (1987) placed Stagodontidae within their paraphyletic stem taxon Microbiotherioidea. In the context of our present classification, this uncertainty induces us to classify Stagodontidae as 'Marsupialia: cohort *incertae sedis*'. As in the case of Pediomyidae, however, we consider this to be a temporary measure reflecting a combination of incomplete knowledge (e.g., of tarsal and basicranial morphology) and the very autapomorphic nature of the dentition in known members of the group. Further discoveries of older, less specialised taxa, together with increasing interest in the study of non-dental material, should lead to rapid resolution of the affinities of both of these groups.

TAXA THAT ARE ONLY QUESTIONABLY MARSUPIAL

While there may be among the more poorly known taxa listed above as marsupials a few that will prove with more knowledge not to belong within Marsupialia, there are only two genera known to us whose marsupial affinities are the subject of significant doubt: *Necrolestes* and *Garatherium*.

Necrolestidae Ameghino, 1894

This monotypic family is based on the single mole-like and zalamdodont taxon *Necrolestes patagonensis* from the early Miocene (Santacrucian) of Argentina. Although it is known from remarkably complete material, its taxonomic position remains uncertain. It was listed as a possible "Insectivore" by Simpson (1945). Patterson (1958) reviewed its status and concluded that it was a marsupial with possible special affinities to the borhyaenoids. Patterson's conclusions were followed by Kirsch (1977).

The common assumption that it was a marsupial was challenged by Archer (1984a). He suggested that its supposed marsupial features, such as its high molar number, were therian symplesiomorphies. Further, the lack of an alisphenoid tympanic process also suggested to Archer that *Necrolestes* was not a marsupial. However, some borhyaenoids also lack an alisphenoid tympanic process so that this absence in itself is not an adequate basis for arguing against marsupial affinities.

Before the phylogenetic affinities of this enigmatic animal can be satisfactorily determined, the material should be thoroughly reexamined. At present, we cannot decide one way or another about its possible status as a marsupial.

Garatherium mahboubii Crochet, 1984

This curious animal was described on the basis of a single upper molar from the early Eocene of Algeria. Crochet (1984) considered it to be a marsupial but with derivation direct from the mid-late Cretaceous — i.e., from a pre-*Alphadon* stock. He suggested that it was a late survivor of an endemic Palaeogene radiation of marsupials in Africa, one that had Gondwanan origins.

Based on the illustration, it is certainly marsupial-like in its moderately wide stylar shelf with distinct stylar cusps C and D, a metacone which although smaller than the paracone is nevertheless relatively large by comparison with that cusp in placentals. Compared with the species of *Alphadon,* it is unusual in being relatively short anteroposteriorly, in having a paracone larger than the metacone, in having a relatively larger protocone, a narrower stylar shelf, a V-shaped centrocrista and a connection between stylar cusp C and the buccal extremity of the centrocrista.

Clearly this African form is not readily derivable from a *Peratherium*-like didelphid of the kind that was relatively common at the same time in Europe, despite the presence of the V-shaped centrocrista. If it is a marsupial (and the size of the metacone leaves us in doubt), then Crochet (1984) is probably correct in suggesting derivation from a pre-*Alphadon* grade of animal. For the present, however, the Palaeogene record of Africa is simply too poorly known to rule out the possibility that it is not a peculiar placental or even a representative of an otherwise different group of eutherian mammals.

AFTERTHOUGHTS

We have none; this classification and its rationale are the best we can offer given present data and current phylogenetic interpretations. However, considering the high rate at which new primary data about marsupial systematics are being produced, we trust that the inevitable rotten spots in our summary of marsupial systematics will quickly be identified and carved out.

ACKNOWLEDGMENTS

We would like to thank many colleagues who have over the years shared their convictions and doubts with us about marsupial systematics and/ or systematic methodology. In particular, we have benefited from discussions with R. H. Tedford, L. G. Marshall, J. A. W. Kirsch, R. Pascual, R. Fox, M. McKenna, P. R. Baverstock, M. A. Adams, W. A. Clemens, R. A. Stirton, W. D. Turnbull, M. O. Woodburne, J. Case, G. G. Simpson, F. S. Szalay, A. Bartholomai, R. Molnar, J. Mahoney, S. J. Hand, T. F. Flannery, G. B. Sharman, G. M. McKay, D. Briscoe, B. Marlow, E. Lundelius, B. Richardson, J. Calaby, W. D. L. Ride, M. D. Plane, T. H. Rich, D. Merrilees, P. Murray, M. Springer, J. Haight, J. Nelson, B. Rauscher, R. H. Harding, L. Hughes, R. Wells, H. Cogger, A. Ritchie, N. S. Pledge, G. Sanson, N. Wakefield, L. Dawson, G. George, C. Groves, M. L. Augee, L. Hall, H. Godthelp, P. Temple-Smith, P. A. Woolley, R. Crozier, J. Humphrey-Smith, A. Baynes, J. Hope and D. Kitchener.

We would also like to thank Vincent Sarich for communicating the results of his MC'F analysis of *Dromiciops australis*.

REFERENCES

ABBIE, A. A., 1937. Some observations on the major subdivisions of the Marsupialia, with especial reference to the position of the Peramelidae and Caenolestidae. *Jour. Anat.* **71:** 424-36.

AMEGHINO, F., 1887. Enumeracion sistematica de las especies de mamiferos fosiles coleccionados por Carlos Ameghino en los terrenos eocenos de la Patagonia austral y depositados en el Museo La Plata. *Bol. Mus. de La Plata* **1:** 1-26.

AMEGHINO, F., 1889. Contribucion al conocimiento de los mamiferos fosiles de la Republica Argentina, obra escrita bajo los auspicios de la Academia Nacional de Ciencias de la Republica Argentina para presentarla a la Exposicion Universal de Paris de 1889. *Actas Acad. Cienc. Cordoba* **6**: 1-1027.

AMEGHINO, F., 1894. Enumeration synoptique des especies de mammiferes fossiles des formations eocenes de Patagonie. *Bol. Acad. Cienc. Cordoba, Buenos Aires* **13**: 259-452.

AMEGHINO, F., 1897. Mammiferes cretaces de l'argentine (Deuxieme contribution a la Connaissance de la faune mammalogique des couches a *Pyrotherium). Bol. Inst. Geog. Argent.* **18**: 406-521.

AMEGHINO, F., 1901. Notices prelininaires sur de ongules nouveaux de terrains cretaces de Patagonie. *Bol. Acad. Nac. Cien. Cordoba, Buenos Aires* **16**: 349-426.

AMEGHINO, F., 1904. Nuevas especies de mamiferos cretaceos y terciarios de la Republica Argentina. *An. Soc. Cien. Argent.* **56, 57, 58**: 1-142.

AMOROSO, E. C., BARCLAY, A. E., FRANKLIN, K. J. AND PRICHARD, M. M. L., 1943. Incidence of bilateral anterior venae cavae in a series of eutherian foetuses. *Proc. Zool. Soc. Lond.* **113**: 43-53.

ANDERSON, S. AND JONES, J. K. (eds), 1984. "Orders and families of recent mammals of the world". John Wiley & Sons: New York.

ANONYMOUS, 1966. A relict marsupial. *Nature (Lond.)* **212**: 225.

APLIN, K. P., 1987. Basicranial anatomy of the early Miocene diprotodontian *Wynyardia bassiana* (Marsupialia: Wynyardiidae) and its implications for wynyardiid phylogeny and classification. Pp. 369-91 *in* "Possums and opossums: studies in evolution" ed by M. Archer. Surrey Beatty & Sons and the Royal Zoological Society of New South Wales: Sydney.

ARCHER, M., 1976a. The dasyurid dentition and its relationships to that of didelphids, thylacinids, borhyaenids (marsupicarnivores) and peramelids (Peramelina, Marsupialia). *Aust. J. Zool. Suppl. Ser.* **39**.

ARCHER, M., 1976b. Phascolarctid origins and the potential of the selenodont molar in the evolution of diprotodont marsupials. *Mem. Qd Mus.* **17**: 367-71.

ARCHER, M., 1976c. The basicranial region of marsupicarnivores (Marsupialia), inter-relationships of carnivorous marsupials, and affinities of the insectivorous marsupial peramelids. *J. Linn. Soc. (Lond.)* **59**: 217-322.

ARCHER, M., 1976d. Miocene marsupicarnivores (Marsupialia) from central South Australia, *Ankotarinja tirarensis* gen. et sp. nov., *Keenuna woodburnei* gen. et sp. nov., and their significance in terms of early marsupial radiations. *Trans. Roy. Soc. S. Aust.* **100**: 53-73.

ARCHER, M., 1977. *Koobor notabilis* (De Vis), an ususual koala from the Pliocene Chinchilla Sand. *Mem. Qd Mus.* **18**: 31-35.

ARCHER, M., 1978a. The nature of the molar-premolar boundary in marsupials and a reinterpretation of the homology of marsupial cheekteeth. *Mem. Qd Mus.* **18**: 157-64.

ARCHER, M., 1978b. Koalas (phascolarctids) and their significance in marsupial evolution. Pp. 20-28 *in* "The Koala" ed by T. J. Bergin. Zoological Parks Board New South Wales: Sydney.

ARCHER, M., 1982a. Review of the dasyurid (Marsupialia) fossil record, integration of data bearing on phylogenetic interpretation, and suprageneric classification. Pp. 397-443 *in* "Carnivorous marsupials" ed by M. Archer. Royal Zoological Society of New South Wales: Sydney.

ARCHER, M., 1982b. A review of Miocene thylacinids (Thylacinidae, Marsupialia), the phylogenetic position of the Thylacinidae and the problem of apriorisms in character analysis. Pp. 445-76 *in* "Carnivorous marsupials" ed by M. Archer. Royal Zoological Society of New South Wales: Sydney.

ARCHER, M., 1984a. Origins and early radiations of marsupials. Pp. 585-631 *in* "Vertebrate zoogeography and evolution in Australasia" ed by M. Archer and G. Clayton. Hesperian Press: Perth.

ARCHER, M., 1984b. The Australian marsupial radiation. Pp. 633-808 *in* "Vertebrate zoogeography and evolution in Australasia" ed by M. Archer and G. Clayton. Hesperian Press: Perth.

ARCHER, M. AND BARTHOLOMAI, A., 1978. Tertiary mammals of Australia: a synoptic review. *Alcheringa* **2**: 1-19.

ARCHER, M., FLANNERY, T. F., RITCHIE, A. AND MOLNAR, R. E., 1985. First Mesozoic mammal from Australia — an early Cretaceous monotreme. *Nature* **318**: 363-66.

ARCHER, M. AND HAND, S. J., 1984. Background to the search for Australia's oldest mammals. Pp. 517-65 *in* "Vertebrate zoogeography and evolution in Australasia" ed by M. Archer and G. Clayton. Hesperian Press: Sydney.

ARCHER, M. AND KIRSCH, J. A. W., 1977. The case for the Thylacomyidae and Myrmecobiidae, Gill, 1872 or why are marsupial families so extended? *Proc. Linn. Soc. New South Wales* **102**: 18-25.

ARCHER, M., TEDFORD, R. H. AND RICH, T. H., 1987. The Pilkipildridae, a new family and four new species of ?petauroid possums (Marsupialia: Phalangerida) from the Australian Miocene. Pp. 607-27 *in* "Possums and opossums: studies in evolution" ed by M. Archer. Surrey Beatty & Sons and the Royal Zoological Society of New South Wales: Sydney.

ARCHER, M. AND WADE, M., 1976. Results of the Ray E. Lemley Expeditions, Part I. The Allingham Formation and a new Pliocene vertebrate fauna from northern Queensland. *Mem. Qd Mus.* **17**: 379-97.

BAGGETT, H. A., 1978. The tongue of *Tarsipes. Bull. Aust. Mammal Soc.* **5**: 44.

BARNETT, C. H., HARRISON, R. J. AND TOMLINSON, J. D. W., 1958. Variations in the venous system of mammals. *Biol. Rev.* **33**: 442-87.

BAVERSTOCK, P. R., 1984. The molecular relationships of Australian possums and gliders. Pp. 1-8 *in* "Possums and gliders" ed by A. Smith and I. Hume. Surrey Beatty & Sons and the Australian Mammal Society: Sydney.

BAVERSTOCK, P. R., ARCHER, M., ADAMS, M. AND RICHARDSON, B. J., 1982. Genetic relationships among 32 species of Australian dasyurid marsupials. Pp. 641-50 *in* "Carnivorous marsupials" ed by M. Archer. Royal Zoological Society of New South Wales: Sydney.

BAVERSTOCK, P. R., BIRRELL, J. AND KRIEG, M., 1987. Albumin immunological relationships among Australian possums: a progress report. Pp. 229-34 *in* "Possums and opossums: studies in evolution" ed by M. Archer. Surrey Beatty & Sons and the Royal Zoological Society of New South Wales: Sydney.

BENSLEY, B. A., 1903. On the evolution of the Australian Marsupialia: with remarks on the relationships of the marsupials in general. *Trans. Linn. Soc. Lond. (Zool.) Ser. 2* **9**: 83-217.

BIGGERS, J. D., CREED, R. F. S. AND DELAMATER, E. D., 1963. Conjugated spermatozoa in American marsupials. *J. Reprod. Fert.* **6**: 324.

BIGGERS, J. D. AND DELAMATER, E. D., 1965. Marsupial spermatozoa pairing in the epididymis of American forms. *Nature (Lond.)* **208**: 402-404.

BOARDMAN, W., 1943. The hair tracts in marsupials. Part I. Description of species. *Proc. Linn. Soc. NSW* **68:** 95-113.

BOARDMAN, W., 1946. The hair tracts in marsupials. Part II. Description of species, continued. *Proc. Linn. Soc. NSW* **70:** 179-202.

BOARDMAN, W., 1950a. The hair tracts in marsupials. Part IV. Evolution and genetics of tract pattern. *Proc. Linn. Soc. NSW* **75:** 267-72.

BOARDMAN, W., 1950b. The hair tracts in marsupials. Part VII. A system of nomenclature. *Proc. Linn. Soc. NSW* **75:** 273-78.

BOWN, T. M. AND ROSE, K. D., 1979. *Mimoperadectes,* a new marsupial, and *Worlandia,* a new dermopteran, from the lower part of the Willwood Formation (early Eocene), Bighorn Basin, Wyoming. *Contrib. Mus. Paleont. Univ. Michigan* **25:** 89-104.

BRADLEY, A. J., 1982. Steroid binding proteins in the plasma of dasyurid marsupials. Pp. 651-57 *in* "Carnivorous marsupials" ed by M. Archer. Royal Zoological Society of New South Wales: Sydney.

BROOM, R., 1896. On a small fossil marsupial with large grooved premolars. *Proc. Linn. Soc. NSW* **10:** 563-67.

BROOM, R., 1898a. On the affinities and habits of *Thylacoleo. Proc. Linn. Soc. NSW* **23:** 57-74.

BROOM, R., 1898b. On the affinities of *Caenolestes. Proc. Linn. Soc. NSW* **23:** 315-20.

BONAPARTE, C. L. J. L., 1838. Synopsis vertebratorum systematis. *Nuovi Ann. Sci. Nat., Bologna* **2:** 105-33.

BROTHERS, D. J., 1983. Nomenclature at the ordinal and higher levels. *Syst. Zool.* **32:** 34-42.

BRUNNER, H. AND COMAN, B. J., 1974. "The identification of mammalian hair". Inkata Press: Melbourne.

BURNETT, G. T., 1830. Illustrations of the Quadrupeda, or quadrupeds, being the arrangement of the true four-footed beasts indicated in outline. *Quart. J. Sci. Lit. Art.* **28:** 336-53.

BUTLER, P. M., 1977. Evolutionary radiation of the cheek teeth of Cretaceous placentals. *Acta Palaeontologica Polonica* **22:** 241-71.

BUTLER, P. M., 1978. A new interpretation of the mammalian teeth of tribosphenic pattern from the Albian of Texas. *Breviora* **446.**

CALABY, J. H., CORBETT, L. K., SHARMAN, G. B. AND JOHNSON, P. G., 1974. The chromosomes and systematic position of the marsupial mole, *Notoryctes typhlops. Aust. J. Biol. Sci.* **27:** 529-32.

CARLSSON, A., 1904. Zur Anatomie des *Notoryctes typhlops. Zool. Jb Abt. Anat.* **20:** 81-122.

CASE, J. A., 1984. A new genus of Potorinae (Marsupialia: Macropodidae) from the Miocene Ngapakaldi Local Fauna, South Australia, and a definition of the Potoroinae. *J. Paleont.* **58:** 1074-86.

CHOW, M. AND RICH, T. H. V., 1982. *Shuotherium dongi,* n. gen. and sp., a therian with pseudo-tribosphenic molars from the Jurassic of Sichuan, China. *Aust. Mammal.* **5:** 127-42.

CIFELLI, R. L. AND EATON, J. G., 1987. Marsupial from the earliest late Cretaceous of Western US. *Nature (Lond.)* **325:** 520-2.

CLEMENS, W. A., 1966. Fossil mammals of the type Lance Formation, Wyoming. Part 2. Marsupialia. *Univ. Calif. Publ. Geol. Sci.* **66:** 1-122.

CLEMENS, W. A., 1979a. Marsupialia. Pp. 192-220 *in* "Mesozoic mammals: the first two-thirds of mammalian history" ed by J. A. Lillegraven, Z. Kielan-Jaworowska and W. A. Clemens. Univ. Calif. Press: Berkeley.

CLEMENS, W. A., 1979b. Notes on the Monotremata. Pp. 309-11 *in* "Mesozoic mammals: the first two-thirds of mammalian history" ed by J. A. Lillegraven, Z. Kielan-Jaworowska and W. A. Clemens. Univ. Calif. Press: Berkeley.

CLEMENS, W. A., AND LILLEGRAVEN, A., 1986. New Late Cretaceous, North American advanced therian mammals that fit neither the marsupial nor eutherian moulds. *Contr. Geol. Univ. Wyom. Spec. Pap.* **3:** 55-85.

CLEMENS, W. A., AND MARSHALL, L. G., 1976. "Fossilium catalogus: American and European Marsupialia. Pars 123". Junk: The Hague.

CLEMENS, W. A., AND PLANE, M., 1974. Mid-Tertiary Thylacoleonidae (Marsupialia, Mammalia). *J. Paleont.* **48:** 661-69.

CROCHET, J.-Y., 1979. Diversite systematique des Didelphidae (Marsupialia) Europeens Tertiares. *Geobios.* **12:** 365-78.

CROCHET, J.-Y., 1980. "Les marsupiaux du Tertiaire d'Europe". Editions de la Fondation Singer-Polignac: Paris.

CROCHET, J.-Y., 1984. *Garatherium mahboubii* nov. gen., nov. sp., marsupial de L'Eocene Inferieur D'el Kohl (Sud-Oranais, Algerie). *Annales de Paleontologie* **70:** 275-94.

CROCHET, J.-Y. AND SIGE, B., 1983. Les mamiferes montiens de Hainin (Paleocene Moyan de Belgique) Part III: Marsupiaux. *Palaeovertebrata, Montpellier* **13:** 51-64.

CUVIER, G., 1817. "Le regne animal". Deterville: Paris.

DE BAVAY, J. M., 1951. Notes on the female urogenital system of *Tarsipes spenserae* (Marsupialia). *Pap. and Proc. Roy. Soc. Tasmania* **1950:** 143-48.

DE BLAINVILLE, H. M. S., 1816. Prodrome d'une nouvelle distribution systematique du regne animal. *Bull. Sci. Soc. Phlom. Parts Ser. 3* **3:** 105-24.

DE BLAINVILLE, H. M. S., 1834. Cours de la faculte des sciences, 1834. *Fide* Palmer 1904.

DEDERER, P. H., 1909. Comparison of *Caenolestes* with Polyprotodonta and Diprotodonta. *Amer. Nat.* **43:** 614-18.

DOLLO, L., 1899. Les ancestres des marsupiaux etaient-ils arboricoles? *Trav. Stat. Zool. Wimereus* **7:** 188-203.

DORAN, A. H. G., 1878. Morphology of the mammalian *ossicula auditus. Trans. Linn. Soc. (Lond.)* **1:** 271-498.

FINCH, E., 1982. The discovery and interpretation of *Thylacoleo carnifex* (Thylacoleonidae, Marsupialia). Pp. 537-51 *in* "Carnivorous marsupials" ed by M. Archer. Royal Zoological Society of New South Wales: Sydney.

FLANNERY, T. F., 1984. Kangaroos: 15 million years of Australian bounders. Pp. 817-35 *in* "Vertebrate zoogeography and evolution in Australasia" ed by M. Archer and G. Clayton. Hesperian Press: Perth.

FLANNERY, T. F., 1987. The relationships of the Macropodoids (Marsupialia) and the polarity of some morphological features within the Phalangeriformes. Pp. 741-47 *in* "Possums and opossums: studies in evolution" ed by M. Archer. Surrey Beatty & Sons and the Royal Zoological Society of New South Wales: Sydney.

FLANNERY, T. F. AND ARCHER, M., 1987. *Strigocuscus reidi* and *Trichosurus dicksoni,* two new fossil phalangerids (Marsupialia: Phalangeridae) from the Miocene of north-western Queensland. Pp. 527-36 *in* "Possums and opossums: studies in evolution" ed by M. Archer. Surrey Beatty & Sons and the Royal Zoological Society of New South Wales: Sydney.

FLANNERY, T. F. AND RICH, T. H. V., 1986. Macropodoids from the middle Miocene Namba Formation, South Australia, and the homology of some dental structures in kangaroos. *J. Paleont.* **60:** 418-47.

FORBES, W. A., 1881. On some points on the anatomy of the Koala (Phascolarctos cinereus). Proc. Zool. Soc. Lond. **1881:** 180-95.

FOX, R. C., 1971. Marsupial mammals from the early Campanian Milk River Formation, Alberta, Canada. Pp. 145-64 in "Early mammals" ed by D. M. Kermack and K. A. Kermack. Suppl. No. 1 to the Zool. J. Linn. Soc. Lond. **50.**

FOX, R. C., 1972. A primitive therian mammal from the Upper Cretaceous of Alberta. Can. J. Earth Sci. **9:** 1479-94.

FOX, R. C., 1979a. Mammals from the Upper Cretaceous Oldman Formation, Alberta. I. Alphadon Simpson (Marsupialia). Can. J. Earth Sci. **16:** 91-102.

FOX, R. C., 1979b. Mammals from the Upper Cretaceous Oldman Formation, Alberta. II. Pediomys Marsh (Marsupialia). Can J. Earth Sci. **16:** 103-13.

FOX, R. C., 1979c. Ancestry of the 'dog-like' marsupials. J. Paleont. **53:** 733-35.

FOX, R. C., 1980. Picopsis pattersoni, n. gen. and sp., an unusual therian from the Upper Cretaccous of Alberta, and the classification of primitive tribosphenic mammals. Can. J. Earth Sci. **17:** 1489-98.

FOX, R. C., 1981. Mammals from the Upper Cretaceous Oldman Formation, Alberta. V. Eodelphis Matthew, and the evolution of the Stagodontidae. Can. J. Earth Sci. **18:** 350-65.

FOX, R. C., 1983. Notes on the North American Tertiary marsupials Herpetotherium and Peradectes. Can. J. Earth Sci. **20:** 1565-78.

FRECHKOP, S., 1930. Notes sur les mammiferes II. Caracteres distinctifs et phylogenie du wombat (Phascolomys) et du koala (Phascolarctos). Bull. du Musee Royal d'Histoire naturelle de Belgique 6(12): 1-34.

GADOW, H., 1892. On the systematic position of Notoryctes typhlops. Proc. Zool. Soc. Lond. **1892:** 361-70.

GEORGE, W., 1981. Blood vascular patterns in rodents: contributions to an analysis of rodent family relationships. J. Linn. Soc. Lond. **73:** 287-306.

GERVAIS, P. AND VERREAUX, J., 1842. On a new genus of marsupial animals, Tarsipes rostratus. Proc. Zool. Soc. Lond. **1842:** 1-5.

GILL, T., 1872. Arrangement of the families of mammals with analytical tables. Smithsonian Misc. Coll. **11:** 1-98.

GRAHAM, S. F. AND RIDE, W. D. L., 1967. Infraclass Metatheria. Pp. 767-68 in "The fossil record: a symposium with documentation". Geological Society of London: London.

GRAY, J. E., 1821. On the natural arrangement of vertebrose animals. London Med. Redeposit. **15:** 296-310.

GRAY, J. E., 1825. Outline of an attempt at the disposition of the Mammalia into tribes and families with a list of the genera apparently appertaining to each tribe. Ann. Philos. n.s. **10:** 336-44.

GREGORY, W. K., 1910. The orders of mammals. Bull. Amer. Mus. Nat. Hist. **27.**

GREGORY, W. K., 1947. The monotremes and the Palimsest theory. Bull. Amer. Mus. Nat. Hist. **88:** 3-52.

GUNSON, M. M., SHARMAN, G. B. AND THOMSON, J. A., 1968. The affinities of Burramys (Marsupialia: Phalangeroidea) as revealed by a study of its chromosomes. Aust. J. Sci. **31:** 40-41.

HALL, L. S., 1987. Syndactyly in marsupials — problems and prophecies. Pp. 245-55 in "Possums and opossums: studies in evolution" ed by M. Archer. Surrey Beatty & Sons and the Royal Zoological Society of New South Wales: Sydney.

HAIGHT, J. R. AND MURRAY, P. F., 1981. The cranial endocast of the early Miocene marsupial, Wynyardia bassiana: an assessment of taxonomic relationships based upon comparisons with recent forms. Brain, Behav., Evol. **19:** 17-36.

HARDING, H. R., 1987. Interrelationships of the families of the Diprotodonta — A view based on spermatozoan ultrastructure. Pp. 195-216 in "Possums and opossums: studies in evolution" ed by M. Archer. Surrey Beatty & Sons and the Royal Zoological Society of New South Wales: Sydney.

HARDING, H. R., CARRICK, F. N. AND SHOREY, C. D., 1979. Special features of sperm structure and function in marsupials. Pp. 289-303 in "The spermatozoan" ed by D. W. Fawcett and J. M. Bedford. Urban and Schwarzenberg Inc.: Baltimore.

HARDING, H. R., CARRICK, F. N. AND SHOREY, C. D., 1981. Marsupial phylogeny: new indications from sperm ultrastructure and development in Tarsipes spenserae? Search **12:** 45-47.

HARDING, H. R., WOOLLEY, P. A., SHOREY, C. D. AND CARRICK, F. N., 1982. Sperm ultrastructure, spermiogenesis and epididymal sperm maturation in dasyurid marsupials: phylogenetic implications. Pp. 659-73 in "Carnivorous marsupials" ed by M. Archer. Royal Zoologial Society of New South Wales: Sydney.

HARDING, H. R., CARRICK, F. N. AND SHOREY, C. D., 1984. Sperm ultrastructure and development in the Honey Possum, Tarsipes rostratus. Pp. 451-61 in "Possums and gliders" ed by A. Smith and I. Hume. Surrey Beatty & Sons and the Australian Mammal Society: Sydney.

HARDING, H. R., CARRICK, F. N. AND SHOREY, C. D., 1987. The affinities of the Koala, Phascolarctos cinereus (Marsupialia: Phascolarctidae) on the basis of sperm ultrastructure and development. Pp. 353-64 in "Possums and opossums: studies in evolution" ed by M. Archer. Surrey Beatty & Sons and the Royal Zoological Society of New South Wales: Sydney.

HARDY, M. H., 1947. The group arrangement of hair follicles in the mammalian skin. Proc. Roy. Soc. Qd **58:** 125-48.

HASWELL, W. A., 1887. On the myology of the flying phalanger (Petaurista taguanoides). J. Linn. Soc. NSW Second Series **1:** 176-82.

HAYMAN, D. L., KIRSCH, J. A. W., MARTIN, P. G. AND WALLER, P. F., 1971. Chromosomal and serological studies of the Caenolestidae and their implications for marsupial evolution. Nature (Lond.) **231:** 194-95.

HAYMAN, D. L. AND MARTIN, P. G., 1974. Mammalia I: Monotremata and Marsupialia. In "Animal cytogenetics. Vol. 4: Chordata 4" ed by J. Bernard. Gebruder Borntraeger: Berlin-Stuttgart.

HAYMAN, D. L. AND SHARP, P. J., 1982. The chromosomes of Tarsipes spencerae Gray (Marsupialia). Aust. J. Zool. **30:** 799-803.

HEIGHWAY, F. R., 1939. "The anatomy of Hypsiprymnodon moschatus." Unpubl. M. D. Thesis, University of Sydney: Sydney.

HILL, J. P., 1900. Contribution to the morphology and development of the female urogenital organs in the Marsupialia. Proc. Linn. Soc. NSW. **25:** 519-32.

HILL, W. C. O., 1954. Observations on marsupials in the Royal Scottish Museum with special reference to the foetal material. Trans. Roy. Soc. Edinb. **62:** 145-69.

HILL, W. C. O. AND REWELL, R. E., 1954. The caecum of monotremes and marsupials. *Trans. Soc. Zool. Lond.* **28:** 185-242.

HOFER, H., 1952. Uber das gegenwartige Bild der Evolution der Beuteltiere. *Zool. Yahrb. Abt. Anat. Ontog.* **72:** 365-437.

HOPE, J. H. AND WILKINSON, H. E., 1982. *Warendja wakefieldi,* a new genus of wombat (Marsupialia, Vombatidae) from Pleistocene sediments in McEachern's Cave, western Victoria. *Mem. Nat. Mus. Vict.* **43:** 109-20.

HUGHES, R. L., 1965. Comparative morphology of spermatozoa from five marsupial families. *Aust. J. Zool.* **13:** 533-43.

HUGHES, R. L., 1974. Morphological studies on implantation in marsupials. *J. Reprod. Fert.* **39:** 173-86.

HUGHES, R. L. AND MCNALLY, J., 1968. Marsupial foetal membranes with particular reference to placentation. *J. Anat.* **103:** 211.

HUGHES, R. L., HALL, L. S., APLIN, K. P. AND ARCHER, M., 1987. Organogenesis and fetal membranes in the New Guinea Pen-tailed Possum *Distoechurus pennatus* (Acrobatidae: Marsupialia). Pp. 715-24 *in* "Possums and opossums: studies in evolution" ed by M. Archer. Surrey Beatty & Sons and the Royal Zoological Society of New South Wales: Sydney.

HUME, I. D., 1982. "Digestive physiology and nutrition of marsupials". Cambridge University Press: Cambridge.

HUXLEY, T. H., 1880. On the application of the laws of evolution to the arrangement of the Vertebrata and more particularly of the Mammalia. *Proc. Zool. Soc. Lond.* **1880:** 649-62.

ILLIGER, C., 1811. "Prodromus systematis mammalian et avium additus terminus zoographicis utriudque classis". C. Salfeld: Berlin.

JOHNSON, P. M. AND STRAHAN, R., 1982. A further description of the Musky Rat-kangaroo, *Hypsiprymnodon moschatus* Ramsay, 1876 (Marsupialia, Potoroidae), with notes on its biology. *Aust. Zool.* **21:** 27-46.

JONES, F. W., 1923-25. "The mammals of South Australia". Parts 1-3. Government Printer: Adelaide.

JONES, F. W., 1931. A re-examination of the skeletal characters of *Wynyardia bassiana,* an extinct Tasmanian marsupial. *Pap. Proc. Roy. Soc. Tas.* **1930:** 96-115.

KEAN, R. I., MARRYATT, R. G. AND CARROLL, A. L. K., 1964. The female urogenital system of *Trichosurus vulpecula* (Marsupialia). *Aust. J. Zool.* **12:** 18-41.

KEMP, T. S., 1983. The relationships of mammals. *Zool. J. Linn. Soc. (Lond.)* **77:** 353-84.

KERMACK, D. M., KERMACK, K. A. AND MUSSETT, F., 1968. The Welsh pantothere *Kuehneotherium praecursoris. J. Linn. Soc. (Zool.)* **47:** 407-23.

KERMACK, K. A., LEES, P. M. AND MUSSETT, F., 1965. *Aegialodon dawsoni,* a new trituberculosectorial tooth from the lower Wealden. *Proc. R. Soc. Lond., Ser. B* **162:** 535-54.

KIELAN-JAWOROWSKA, Z., 1982. Marsupial-placental dichotomy and paleogeography of Cretaceous Theria. Pp. 367-83 *in* "Palaeontology, essential of historical geology" ed by E. M. Galitelli. Istituto di Paleontologia, Universita di Modena: Modena.

KIELAN-JAWOROWSKA, Z., EATON, J. G. AND BOWN, T. M., 1979. Theria of metatherian-eutherian grade. Pp. 182-91 *in* "Mesozoic mammals, the first two-thirds of mammalian history" ed by J. A. Lillegraven, Z. Kielan-Jaworowska and W. A. Clemens. University of California Press: Berkeley.

KIRSCH, J. A. W., 1968a. Prodromus of the comparative serology of Marsupialia. *Nature (Lond.)* **217:** 418-20.

KIRSCH, J. A. W., 1968b. The serological affinities of *Burramys* and related possums (Marsupialia: Phalangeroidea). *Aust. J. Sci.* **31:** 43-45.

KIRSCH, J. A. W., 1977a. The comparative serology of Marsupialia, and a classification of marsupials. *Aust. J. Zool. Suppl. Ser.* **52.**

KIRSCH, J. A. W., 1977b. The classification of marsupials. Pp. 1-50 *in* "The biology of marsupials" ed by D. Hunsaker. Academic Press: New York.

KIRSCH, J. A. W. AND ARCHER, M., 1982. Polythetic cladistics, or, when parsimony's not enough: the relationships of carnivorous marsupials. Pp. 595-619 *in* "Carnivorous marsupials" ed by M. Archer. Royal Zoological Society of NSW: Sydney.

KIRSCH, J. A. W. AND CALABY, J. H., 1977. The species of living marsupials: an annotated list. *In* "The biology of marsupials" ed by B. Stonehouse and D. Gilmore. Macmillan Press Ltd: New York.

KRAUSE, W. J., 1972. The distribution of Brunner's Gland in 55 marsupial species native to the Australian region. *Acta anat.* **82:** 17-33.

KRAUSE, W. J. AND LEESON, C. R., 1973. The stomach gland patch of the Koala *(Phascolarctos cinereus). Anat. Rec.* **176:** 475-88.

KRISHTALKA, L. AND STUCKY, R. K., 1983a. Revision of the Wind River faunas, early Eocene of central Wyoming. Part 3. Marsupialia. *Ann. Carnegie Mus.* **52:** 205-27.

KRISHTALKA, L. AND STUCKY, R. K., 1983b. Paleocene and Eocene marsupials of North America. *Ann. Carnegie Mus.* **52:** 229-63.

KRISHTALKA, L. AND STUCKY, R. K., 1984. Middle Eocene marsupials (Mammalia) from northeastern Utah and the mammalian fauna from Powder Wash. *Ann. Carnegie Mus.* **53:** 31-45.

KUHNE, W. G., 1972. The systematic position of monotremes reconsidered (Mammalia). *Zeit. fur Morph. Tiere* **75:** 59-64.

KUHNE, W. G., 1977. On the Marsupionta, a reply to Dr Parrington. *Jour. Nat. Hist.* **11:** 225-28.

LECHE, W., 1891. Beitrage zur Anatomie des *Myrmecobius fasciatus. Biologiska Foreningens Forhandlingar* **3:** 136-54.

LEE, A. K. AND COCKBURN, A., 1985. "Evolutionary ecology of marsupials". Cambridge University Press: Sydney.

LEE, A. K., WOOLLEY, P. AND BRAITHWAITE, R. W., 1982. "Life history strategies of dasyurid marsupials. Pp. 1-11 *in* "Carnivorous marsupials" ed by M. Archer. Royal Zoological Society of New South Wales: Sydney.

LILLEGRAVEN, J. A., 1969. Latest Cretaceous mammals of upper part of Edmonton Formation of Alberta, Canada, and review of marsupial-placental dichotomy in mammalian evolution. *Univ. Kansas Paleont. Contribs* **50.**

LILLEGRAVEN, J. A., 1972. Ordinal and familial diversity of Cenozoic mammals. *Taxon* **21:** 261-74.

LONNBERG, E., 1902. On some remarkable digestive adaptations in diprotodont marsupials. *Proc. Zool. Soc. Lond.* **1902:** 12-31.

LYNE, A. G., 1959. The systematic and adaptive significance of vibrissae in marsupials. *Proc. Zool. Soc. Lond.* **133:** 79-133.

MACALISTER, A., 1870. On the myology of the Wombat *(Phascolomys wombata)* and the Tasmanian Devil *(Sarcophilus ursinus). Ann. Mag. Nat. Hist.* **27:** 153-73.

MACKENZIE, C., 1934. Comparative anatomy and the Koala. *Vict. Nat.* **51:** 58-61.

MACKENZIE, W. C., 1918a. 'The comparative anatomy of Australian marsupials. Part I. The gastro-intestinal tract in monotremes and marsupials'. Critchley Parker: Australia.

MACKENZIE, W. C., 1918b. 'The comparative anatomy of Australian marsupials. Part II. The liver, spleen, pancreas, peritoneal relations, oral biliary system in monotremes and marsupials'. Critchley Parker: Australia.

MCCLURE, C. F. W., 1903. A contribution to the anatomy and development of the venous system of *Didelphys marsupialis* (L.). — Part I, anatomy. *Amer. J. Anat.* **2:** 371-404.

McKAY, G. M., 1984. Cytogenetic relationships of possums and gliders. Pp. 9-16 in "Possums and gliders" ed by A. P. Smith and I. D. Hume. Surrey Beatty & Sons and the Australian Mammal Society: Sydney.

MARSH, O. C., 1889. Discovery of Cretaceous Mammalia, Part ii. Am. J. Sci., Ser. 3 38: 177-80.

MARSHALL, L. G., 1972. Evolution of the peramelid tarsus. Proc. R. Soc. Vic. 85: 51-60.

MARSHALL, L. G., 1976a. Evolution of the Thylacosmilidae, fossil marsupial "saber-tooths" of South America. Paleo. Bios. 23: 1-31.

MARSHALL, L. G., 1976b. New didelphine marsupials from the La Venta Fauna (Miocene) of Colombia, South America. J. Paleont. 50: 402-18.

MARSHALL, L. G., 1976. Revision of the South American fossil marsupial subfamily Abderitinae (Mammalia, Caenolestidae). Publ. del Museo Municipal de Ciencias Naturales de Mar del Plata Lorenzo Scaglia 2: 57-90.

MARSHALL, L. G., 1977a. Cladistic analysis of borhyaenoid, dasyuroid, and thylacinid (Marsupialia: Mammalia) affinity. Syst. Zool. 26: 410-25.

MARSHALL, L. G., 1977b. Evolution of the carnivorous adaptive zone in South America. Pp. 709-21 in "Major patterns in vertebrate evolution" ed by M. Hecht, P. C. Goody and B. M. Hecht. Plenum Press: New York.

MARSHALL, L. G., 1978. Evolution of the Borhyaenidae, extinct South American predaceous marsupials. Univ. Calif. Publ. Geol. Sci. 117: 1-89.

MARSHALL, L. G., 1979. Review of the Prothylacyninae, an extinct subfamily of South American "dog-like" marsupials. Fieldianna Geol. n.s. 3: 1-50.

MARSHALL, L. G., 1980. Systematics of the South American family Caenolestidae. Fieldiana Geol. n.s. 5: 1-145.

MARSHALL, L. G., 1981. The families and genera of Marsupialia. Fieldiana Geol. n.s. 8: i-vi and 1-65.

MARSHALL, L. G., 1982. Systematics of the South American marsupial family Microbiotheriidae. Fieldiana Geol. n.s. 10: 1-75.

MARSHALL, L. G., DE MUISON, C. AND SIGE, B., 1983. Late Cretaceous mammals (Marsupialia) from Bolivia. Geobios 16: 739-45.

MARSHALL, L. G., 1987. Systematics of Itaboraian (Middle Paleocene) Age "Opossum-like" marsupials from the limestone quarry at São José De Itaboraí, Brazil. Pp. 91-160 in "Possums and opossums: studies in evolution" ed by M. Archer. Surrey Beatty & Sons and the Royal Zoological Society of New South Wales: Sydney.

MAXON, L. R., SARICH, V. M. AND WILSON, A. C., 1975. Continental drift and the use of albumin as an evolutionary clock. Nature (Lond.) 225: 397-400.

McKAY, G. M., 1984. Cytogenetic relationships of possums and gliders. Pp. 9-16 in "Possums and gliders" ed by A. Smith and I. Hume. Surrey Beatty & Sons and the Australian Mammal Society: Sydney.

McKENNA, M. C., 1975. Toward a phylogenetic classification of the Mammalia. Pp. 21-46 in "Phylogeny of the Primates" ed by W. P. Luckett and F. S. Szalay. Plenum Press: New York.

VAN MOELLER, H., 1968. Zur frage der Parallelerscheinunge bei Metatheria und Eutheria. Z. fur Wiss. Zool. 177: 283-92.

MURRAY, P. D., WELLS, R. AND PLANE, M., 1987. The cranium of the Miocene thylacoleonid, Wakaleo vanderleuri: Click go the shears — A fresh bite at thylacoleonid systematics. Pp. 433-66 in "Possums and opossums: studies in evolution" ed by M. Archer. Surrey Beatty & Sons and the Royal Zoological Society of New South Wales: Sydney.

NOVACEK, M. J., 1986. The skull of lepticid insectivorans and the higher-level classification of eutherian mammals. Bull. Amer. Mus. Nat. Hist. 183: 1-112.

OGILBY, J. D., 1892. "Catalogue of Australian Mammalia, with introductory notes on general mammalogy". The Australian Museum: Sydney.

OSBORN, H. F., 1910. "The age of mammals in Europe, Asia and North America". Macmillan and Co.: New York.

OSGOOD, W. H., 1921. A monographic study of the American marsupial Caenolestes. Field Mus. Nat. Hist., Zool. Ser. 14: 1-162.

OWEN, R., 1839. Outlines of a classification of the Marsupialia. Proc. Zool. Soc. Lond. 7: 5-19.

OWEN, R., 1866. "On the anatomy of vertebrates". Vol. II. Longmans, Green: London.

PALMER, T. S., 1904. "Index generum mammalium: a list of the genera and families of mammals". US Dept. Agr., North American Fauna 23: 1-984.

PARKER, T. J. AND HASWELL, W. A., 1897. "A text-book of zoology". Macmillan and Co.: London.

PASCUAL, R., 1980. Nuevos y singulares tipos ecologicos de marsupiales extinguidos de America del Sur (Paleoceno tardio o Eoceno temprano) del Noreste Argentino. Actas II Congr. Arg. Paleont. y Biestr. y I Congr. Latinoamer. Paleont. 2: 151-73.

PASCUAL, R., 1981a. Prepidolopidae, nueva familia de Marsupialia Didelphoidea del Eoceno Sudamericano. Ameghiniana 17: 216-42.

PASCUAL, R., 1981b. Adiciones al conocimiento de Bonapartherium hinakusijum (Marsupialia, Bonapartheriidae) del Eoceno temprano del noroeste Argentino. Anais II Congresso Latin-Americano Paleontologia, Porto Alegre 1981: 507-20.

PATTERSON, B., 1952. Un nuevo y extraordinario marsupial Deseadiano. Rev. Mus. Munic. Cien. Nat. Trad. de Mar del Plata 1: 39-44.

PATTERSON, B., 1958. Affinities of the Patagonian fossil mammal Necrolestes. Breviora 94.

PATTERSON, B., 1965. The auditory region of the borhyaenid marsupial Cladosictis. Breviora 217.

PAULA COUTO, C. DE, 1952. Fossil mammals from the beginning of the Cenozoic in Brazil. Marsupialia: Polydolopidae and Borhyaenidae. Novitates 1559.

PEARSON, J., 1940. Notes on the blood system of the Marsupialia. Pap. & Proc. Roy. Soc. Tasm. 1939: 77-94.

PEARSON, J., 1949. Placentation of the Marsupialia. Proc. Linn. Soc. Lond. 161: 1-9.

PEARSON, J., 1950. The relationships of the Potoroidae to the Macropodidae (Marsupialia). Pap. & Proc. Roy. Soc. Tasm. 1949: 211-29.

PEARSON, J. AND DE BAVAY, J. M., 1959. The female urogenital system of Pseudocheirus peregrinus. Aust. J. Zool. 7: 13-21.

PERROT, J. W., 1966. The peritoneum of Trichosurus vulpecula. Anat. Rec. 154: 295-304.

PLEDGE, N. S., 1987a. *Muramura williamsi*, a new genus and species of ?wynyardiid (Marsupialia: Vombatoidea) from the Middle Miocene Etadunna Formation of South Australia. Pp. 393-400 *in* "Possums and opossums: studies in evolution" ed by M. Archer. Surrey Beatty & Sons and the Royal Zoological Society of New South Wales: Sydney.

PLEDGE, N. S., 1987b. *Kuterintja ngama*, a new genus and species of primitive vombatoid marsupial from the Medial Miocene Ngama Local Fauna of South Australia. Pp. 419-22 *in* "Possums and opossums: studies in evolution" ed by M. Archer. Surrey Beatty & Sons and the Royal Zoological Society of New South Wales: Sydney.

PLEDGE, N. S., 1987c. A new species of *Burramys* Broom (Marsupialia: Burramyidae) from the Middle Miocene of South Australia. Pp. 725-28 *in* "Possums and opossums: studies in evolution" ed by M. Archer. Surrey Beatty & Sons and the Royal Zoological Society of New South Wales: Sydney.

POCOCK, R. I., 1921. The external characters of the Koala (*Phascolarctos*) and some related marsupials. *Proc. Zool. Soc. Lond.* **1921:** 591-607.

PROTHERO, D. R., 1981. New Jurassic mammals from Como Bluff, Wyoming, and the interrelationships of the non-tribosphenic Theria. *Bull. Amer. Mus. Nat. Hist.* **167:** 277-362.

RAUSCHER, B., 1987. *Priscileo pitikantensis*, a new genus and species of the thylacoleonid marsupial (Marsupialia: Thylacoleonidae) from the Miocene Etadunna Formation, South Australia. Pp. 423-32 *in* "Possums and opossums: studies in evolution" ed by M. Archer. Surrey Beatty & Sons and the Royal Zoological Society of New South Wales: Sydney.

REIG, O. A., 1955. Noticia preliminar sobre la presencia de microbiotherinos vivientes en la fauna Sudamericana. *Invest. Zool. Chil.* **2:** 121-30.

REIG, O. A., 1958. Notas para una actualizacion del conocimiento de la fauna de la formacion Chapadmalal. I. Lista faunistica preliminar. *Acta geol. Lilloana* **2:** 241-53.

REIG, O. A., KIRSCH, J. A. W. AND MARSHALL, L. G., 1987. Systematic relationships of the living and neocenozoic American "Opossum-like" marsupials (Suborder Didelphimorphia), with comments on the classification of these and of the cretaceous and paleogene new world and European metatherians. Pp. 1-89 *in* "Possums and opossums: studies in evolution" ed by M. Archer. Surrey Beatty & Sons and the Royal Zoological Society of New South Wales: Sydney.

REIG, O. A. AND SIMPSON, G. G., 1972. *Sparassocynus* (Marsupialia, Didelphidae), a peculiar mammal from the late Cenozoic of Argentina. *J. Zool., Lond.* **167:** 511-39.

RENFREE, M. B., 1980. Embryonic diapause in the Honey Possum, *Tarsipes spencerae*. *Search* **11:** 81.

RENFREE, M. B., RUSSELL, E. M. AND WOOLLER, R. D., 1984. Reproduction and life history of the Honey Possum, *Tarsipes rostratus*. Pp. 427-37 *in* "Possums and gliders" ed by A. Smith and I. Hume. Surrey Beatty & Sons and the Australian Mammal Society: Sydney.

RICH, T. H. V. AND ARCHER, M., 1979. *Namilamadeta snideri*, a new diprotodontan (Marsupialia, Vombatoidea) from the medial Miocene of South Australia. *Alcheringa* **3:** 197-208.

RICH, T. H. V., ARCHER, M., PLANE, M., FLANNERY, T., PLEDGE, N. S., HAND, S. J. AND RICH, P. V., 1982. Australian Tertiary mammal localities. Pp. 525-72 *in* "The fossil vertebrate record of Australasia" ed by P. V. Rich and E. M. Thompson. Monash Offset Printing Unit: Clayton.

RICHARDSON, B. J. AND MCDERMID, E. M., 1978. A comparison of genetic relationships within the Macropodidae as determined from allozyme, cytological and immunological data. *Aust. Mammal.* **2:** 43-51.

RICHARDSON, K. C., WOOLLER, R. D. AND COLLINS, B. G., 1986. Adaptations to a diet of nectar and pollen in the marsupial *Tarsipes rostratus* (Marsupialia: Tarsipedidae). *J. Zool. Lond. (A)* **208:** 285-97.

RIDE, W. D. L., 1959. Mastication and taxonomy in the macropodine skull. Pp. 33-59 *in* "Function and taxonomic importance" ed by A. J. Cain. Systematics Association: London.

RIDE, W. D. L., 1961. The cheek-teeth of *Hypsiprymnodon muschatus* Ramsay 1876 (Macropodidae: Marsupialia). *J. Proc. R. Soc. West. Aust.* **44:** 53-60.

RIDE, W. D. L., 1962. On the evolution of Australian marsupials. Pp. 281-306 *in* "The evolution of living organisms" ed by G. W. Leeper. Melbourne University Press: Melbourne.

RIDE, W. D. L., 1964. A review of Australian fossil marsupials. *J. Proc. R. Soc. West. Aust.* **47:** 97-131.

RIGGS, E. S., 1933. Preliminary description of a new marsupial saber-tooth from the Pliocene of Argentina. *Field Mus. Nat. Hist., Geol. Ser.* **6:** 61-66.

ROFE, R. AND HAYMAN, D., 1985. G-banding evidence for a conserved compliment in the Marsupialia. *Cytogenet. Cell Genet.* **39:** 40-50.

RUSSELL, E. M., 1984. Social behaviour and social organisation of marsupials. *Mamm. Rev.*

RUSSELL, L. S., 1984. Tertiary mammals of Saskatchewan. Part VII: Oligocene marsupials. *Roy. Ont. Mus. Life Sci. Contr.* **139.**

SARICH, V., LOWENSTEIN, J. M. AND RICHARDSON, B. J., 1982. Phylogenetic relationships of the Thylacine (*Thylacinus cynocephalus*, Marsupialia) as reflected in comparative serology. Pp. 707-9 *in* "Carnivorous marsupials" ed by M. Archer. Roy. Zool. Soc. New South Wales: Sydney.

SEGALL, W., 1969a. The middle ear region of *Dromiciops*. *Acta Anat.* **72:** 489-501.

SEGALL, W., 1969b. The auditory ossicles (malleus, incus) and their relationships to the tympanic: in marsupials. *Acta Anat.* **73:** 176-91.

SEGALL, W., 1970. Morphological parallelisms of the bulla and auditory ossicles in some insectivores and marsupials. *Fieldianna Zool.* **51:** 169-205.

SERNIA, C., BRADLEY, A. J. AND MCDONALD, I. R., 1979. High affinity binding of adrenocorticol and gonadal steroids by plasma proteins of Australian marsupials. *Gen. Com. Endocr.* **38:** 496-503.

SHARMAN, G. B., 1961. The embryonic membranes and placentation in five genera of diprotodont marsupials. *Proc. Zool. Soc. Lond.* **137:** 197-220.

SHARMAN, G. B., 1982. Karyotypic similarities between *Dromiciops australis* (Microbiotheriidae, Marsupialia) and some Australian marsupials. Pp. 711-14 *in* "Carnivorous marsupials" ed by M. Archer. Roy. Zool. Soc. New South Wales: Sydney.

SHRIVASTAVA, R. K., 1962. The deltoid musculature of the marsupialia. *Amer. Midl. Nat.* **67:** 305-20.

SIGE, B., 1971. Les Didelphoidea de Laguna Umayo (Formation Vilquechico, Cretace Superieur, Perou), et le Peuplement marsupial d'Amerique du Sud. *C. R. Acad. Sc. Paris, t.* 273: 2479-481.

SIMONS, E. L. AND BOWN, T. M., 1984. A new species of *Peratherium* (Didelphidae; Polyprotodonta): the first African marsupial. *J. Mamm.* **65**: 539-48.

SIMPSON, G. G., 1927. Mammalian fauna of the Hell Creek Formation of Montana. *Novitates* **267**.

SIMPSON, G. G., 1928. Affinities of the Polydolopidae. *Novitates* **323**.

SIMPSON, G. G., 1941. The affinities of the Borhyaenidae. *Novitates* **1118**.

SIMPSON, G. G., 1945. The principles of classification and a classification of mammals. *Bull. Amer. Mus. Nat. Hist.* **85**.

SIMPSON, G. G., 1948. The beginning of the Age of Mammals in South America. Part 1. *Bull. Amer. Mus. Nat. Hist.* **91**: 1-232.

SIMPSON, G. G., 1970. The Argyrolagidae, extinct South American marsupials. *Bull. Mus. Comp. Zool.* **139**: 1-86.

SINCLAIR, W. J., 1906. Mammalia of the Santa Cruz Beds. *Rept. Princeton Univ. Exped. Patagonia* **4**: 333-460.

SLAUGHTER, R. H., 1968. Earliest known marsupials. *Science* **162**: 254-55.

SLAUGHTER, R. H., 1971. Mid-Cretaceous (Albian) therians of the Butler Farm Local Fauna, Texas. Pp. 131-43 *in* "Early mammals" ed by D. M. Kermack and K. A. Kermack. Academic Press: London.

SMITH, A. P., 1981. Is the Striped Possum (*Dactylopsila trivirgata;* Marsupialia, Petauridae) an arboreal anteater? *Aust. Mammal.* **5**: 229-34.

SMITH, A. P., 1985. Stomach contents of the Long-tailed Pygmy-possum (*Cercartetus caudatus*) (Marsupialia: Burramyidae). *Aust. Mammal.* **9**: 135-37.

SONNTAG, C. F., 1921a. Contributions to the visceral anatomy and myology of the Marsupialia. *Proc. Zool. Soc. Lond.* **1921 (B)**: 851-82.

SONNTAG, C. F., 1921b. The comparative anatomy of the Koala (*Phascolarctos cinereus*) and Vulpine Phalanger (*Trichosurus vulpecula*). *Proc. Zool. Soc. Lond.* **1921**: 547-77.

SONNTAG, C. F., 1922. On the myology and classification of the wombat, koala, and phalangers. *Proc. Zool. Soc. Lond.* **1922**: 863-96.

SPENCER, W. B., 1901. A description of *Wynyardia bassiana*, a fossil marsupial from the Tertiary beds of Table Cape, Tasmania. *Proc. Zool. Soc. Lond.* **1900**: 776-94.

STIRTON, R. A., 1967. The Diprotodontidae from the Ngapakaldi Fauna, South Australia. *Bull. Aust. Bur. Min. Res.* **85**: 1-44.

STIRTON, R. A., TEDFORD, R. H. AND WOODBURNE, M. O., 1967. A new Tertiary formation and fauna from the Tirari Desert, South Australia. *Rec. S. Aust. Mus.* **15**: 427-62.

STIRTON, R. A., WOODBURNE, M. O. AND PLANE, M. D., 1967. A phylogeny of the Tertiary Diprotodontidae and its significance in correlation. *Bull. Aust. Bur. Min. Res.* **85**: 149-60.

STRAHAN, R., 1978. What is a koala? Pp. 3-19 *in* "The Koala" ed by T. J. Bergin. Taronga Zoological Park Board, New South Wales: Sydney.

SZALAY, F., 1977. Phylogenetic relationships and a classification of the eutherian Mammalia. *In* "Major patterns in vertebrate evolution" ed by M. K. Hecht, P. C. Goody and B. M. Hecht. *Nato Advanced Study Institute Series A* **14**. Plenum Publ. Co.: New York.

SZALAY, F., 1982. A new appraisal of marsupial phylogeny and classification. Pp. 621-40 *in* "Carnivorous marsupials" ed by M. Archer. Roy. Zool. Soc. New South Wales: Sydney.

TATE, G. H. H., 1945. Results of the Archbold Expeditions. No. 55. Notes on the squirrel-like and mouse-like possums (Marsupialia). *Novitates* **1305**.

TATE, G. H. H., 1948. Results of the Archbold Expeditions. No. 56. On the anatomy and phylogeny of the Macropodidae (Marsupialia). *Bull. Amer. Mus. Nat. Hist.* **91**: 237-351.

TEDFORD, R. H., BANKS, M. R., KEMP, N. R., McDOUGALL, I. AND SUTHERLAND, F. L., 1975. Recognition of the oldest known fossil marsupials from Australia. *Nature (Lond.)* **255**: 141-42.

TEDFORD, R. H., ARCHER, M., BARTHOLOMAI, A., PLANE, M. D., PLEDGE, N. S., RICH, T. H. V., RICH, P. V. AND WELLS, R. T., 1977. The discovery of Miocene vertebrates, Lake Frome area, South Australia. *BMR J. Aust. Geo. Geophys.* **2**: 53-57.

TEDFORD, R. H. AND WOODBURNE, M. O., 1987. The Ilariidae, a new family of vombatiform marsupials from Miocene strata of South Australia and an evaluation of the homology of molar cusps in the Diprotodonta. Pp. 401-18 *in* "Possums and opossums: studies in evolution" ed by M. Archer. Surrey Beatty & Sons and the Royal Zoological Society of New South Wales: Sydney.

TEMPLE-SMITH, P., 1987. Sperm structure and marsupial phylogeny. Pp. 171-93 *in* "Possums and opossums: studies in evolution" ed by M. Archer. Surrey Beatty & Sons and the Royal Zoological Society of New South Wales: Sydney.

THOMAS, O., 1888. "Catalogue of the Marsupialia and Monotremata in the collection of the British Museum (Natural History)". British Museum (Natural History): London.

THOMAS, O., 1895. On *Caenolestes*, a still existing survivor of the *Epanorthidae* of Ameghino, and the representative of a new family of recent marsupials. *Proc. Zool. Soc. Lond.* **1895**: 870-78.

TROUSSART, E. L., 1898. "Catalogus Mammalium tam viventium quam fossilium". Part 5. New ed. Berlin.

TROUGHTON, E. LeG., 1923. The "Honey Mouse", *Tarsipes spenserae* Gray. *Aust Zool.* **3**: 148-56.

TURNBULL, W. D., 1971. The Trinity therians: their bearing on evolution in marsupials and other therians. Pp. 151-79 *in* "Dental morphology and evolution" ed by A. A. Dahlberg. The University of Chicago Press: Chicago.

TURNBULL, W. D. AND LUNDELIUS, E. L., JR, 1970. The Hamilton Fauna, a late Pliocene mammalian fauna from the Grange Burn, Victoria, Australia. *Fieldiana Geol.* **19**: 1-163.

TURNBULL, W. D., RICH, T. H. V. AND LUNDELIUS, E. L., JR, 1987a. The petaurids (Marsupialia: Petauridae) of the early Pliocene Hamilton Local Fauna, southwestern Victoria. Pp. 629-38 *in* "Possums and opossums: studies in evolution" ed by M. Archer. Surrey Beatty & Sons and the Royal Zoological Society of New South Wales: Sydney.

TURNBULL, W. D., RICH, T. H. V. AND LUNDELIUS, E. L., JR, 1987b. Burramyids (Marsupialia: Burramyidae) of the early Pliocene Hamilton Local Fauna, southwestern Victoria. Pp. 729-39 *in* "Possums and opossums: studies in evolution" ed by M. Archer. Surrey Beatty & Sons and the Royal Zoological Society of New South Wales: Sydney.

TURNER, V., 1982. Marsupials as pollinators in Australia. Pp. 55-66 *in* "Pollination and evolution" ed by J. A. Armstrong, J. M. Powell and A. J. Richards. Royal Botanical Gardens: Sydney.

VALENTE, A. AND WOOLLEY, P. A., 1982. Hair structure of some Western Australian mammals. *J. Proc. R. Soc. West. Aust.* **64**: 101-32.

VAN VALEN, L., 1963. The origin and status of the mammalian order Tillodontia. *J. Mamm.* **44**: 364-73.

VAN VALEN, L., 1971. Adaptive zones and the orders of mammals. *Evolution* **25**: 420-28.

WARD, S AND RENFREE, M. B., 1986. Some aspects of reproduction in *Acrobates pygmaeus*. Abst. 32nd Sci. Mtg Aust.
Mamm. Soc. Adelaide 1986.

WATERHOUSE, G. R., 1838. "Catalogue of the mammalia preserved in the museum of the Zoological Society". Second
ed. Richard and John E. Taylor: London.

WATROUS, L. E. AND WHEELER, Q. P., 1981. The out-group
comparison method of character analysis *Syst. Zool.* **30**:
1-11.

WEBER, M., 1928. "Die Saugetiere". Vol. 2. systematischer
Teil. Gustav Fischer: Jena.

WELLS, R. T. AND NICHOL, B., 1977. On the manus and pes of
Thylacoleo carnifex Owen (Marsupialia). *Trans. Roy. Soc. S.
Aust.* **101**: 139-46.

WINGE, H., 1893. Jordfunde og nulevende Pungdyr (Marsupialia) fra Lagoa Santa, Minas Geraes, Brasilien. *Med.
Udsigt over Pungdyrenes Slaegtskab. E. Mus. Lundii* **11**(2):
1-149.

WINGE, H., 1941. "The interrelationships of mammalian
genera. Vol. 1. Monotremata, Marsupialia, Insectivora,
Chiroptera, Edentata". (A translation by Deichmann and
Allen of "Pattedyr-Slaegter", 1923, an edited and anno-
tated collection of Winge's earlier works). C. A. Reitzels
Forlag: Copenhagen.

WOLFF, R. G., 1984. A new early Oligocene argyrolagid
(Mammalia: Marsupialia) from Salla, Bolivia. *J. Vert.
Paleo.* **4**: 108-13.

WOLLER, R. D., RUSSELL, E. M. AND RENFREE, M. B., 1984.
Honey Possums and their food plants. Pp. 439-43 *in*
"Possums and gliders" ed by A. Smith and I. Hume.
Surrey Beatty & Sons and the Australian Mammal
Society: Sydney.

WOODBURNE, M. O., 1967. The Alcoota Fauna, central Australia, an integrated palaeontological and geological
study. *Bull. Aust. Bur. Min. Res.* **87**: 1-187.

WOODBURNE, M. O., 1984a. Families of marsupials: relation-
ships, evolution and biogeography. Pp. 48-71 *in*
"Mammals: notes for a short course" ed by T. W.
Broadhead. *Univ. Tenn. Dept. Geol. Sci. Studies in Geol.* **8**.

WOODBURNE, M. O., 1984b. *Wakiewakie lawsoni*, a new genus
and species of Potoroinae (Marsupialia: Macropodidae)
of medial Miocene age, South Australia. *J. Paleont.* **58**:
1062-73.

WOODBURNE, M. O., 1987. The Ektopodontidae, an unusual
family of Neogene phalangeroid marsupials. Pp. 603-06
in "Possums and opossums: studies in evolution" ed by

M. Archer. Surrey Beatty & Sons and the Royal Zoological Society of New South Wales: Sydney.

WOODBURNE, M. O. AND CASE, J. A., 1984. Carnivorous marsupials: an essay review. *J. Vert. Paleont.* **4**: 155-63.

WOODBURNE, M. O., PLEDGE, N. S. AND ARCHER, M., 1987.
The Miralinidae, a new family and two new species of
phalangeroid marsupials from Miocene strata of South
Australia. Pp. 581-602 *in* "Possums and opossums:
studies in evolution" ed by M. Archer. Surrey Beatty &
Sons and the Royal Zoological Society of New South
Wales: Sydney.

WOODBURNE, M. O., TEDFORD, R. H., ARCHER, M. AND
PLEDGE, N. S., 1987. *Madakoala*, a new genus and two
species of Miocene Koalas (Marsupialia: Phascolarctidae)
from South Australia, and a new species of *Perikoala*. Pp.
293-317 *in* "Possums and opossums: studies in evolution" ed by M. Archer. Surrey Beatty & Sons and the
Royal Zoological Society of New South Wales: Sydney.

WOODBURNE, M. O., TEDFORD, R. H. AND ARCHER, M., 1987.
New Miocene ringtail possums (Marsupialia:
Pseudocheiridae) from South Australia. Pp. 639-79 *in*
"Possums and opossums: studies in evolution" ed by M.
Archer. Surrey Beatty & Sons and the Royal Zoological
Society of New South Wales: Sydney.

WOODS, H. E., 1924. The position of the "sparassodonts":
with notes on the relationships and history of the Marsupialia. *Bull. Amer. Mus. Nat. Hist.* **51**: 77-101.

WOODS, J. T., 1956. The skull of *Thylacoleo carnifex. Mem. Qd
Mus.* **13**: 125-40.

WOOLLEY, P. A., 1982. Phallic morphology of the Australian
species of *Antechinus* (Dasyuridae, Marsupialia): a new
taxonomic tool? Pp. 767-81 *in* "Carnivorous marsupials"
ed by M. Archer. Roy. Zool. Soc. New South Wales:
Sydney.

YADAV, M., 1973. The presence of the cervical and thoracic
thymus lobes in marsupials. *Aust. J. Zool.* **21**: 285-301.

YOUNG, A. H., 1981. Anatomy of the Koala (*Phascolarctos
cinereus*). *J. Anat. and Phys.* **1881**: 466-74.

YOUNG, G. J., GRAVES, J. A. M., BARBIERI, I., WOOLLEY, P. A.,
COOPER, D. W. AND WESTERMAN, M., 1982. The chromo-
somes of dasyurids (Marsupialia). Pp. 783-95 *in*
"Carnivorous marsupials" ed by M. Archer. Royal
Zoological Society of New South Wales: Sydney.

Section 2

INTERCONTINENTAL PHYLOGENETICS AND NON-AUSTRALIAN MARSUPIALS

PLATE 1: The living American didelphid *Didelphis virginiana*. The reconstructions are based on the drawings of the skeleton in W. K. Gregory's (1910) "The Orders of Mammals". (*Bull. Amer. Mus. Nat. Hist.* **27**). Drawings by Peter Murray.

SYSTEMATIC RELATIONSHIPS OF THE LIVING AND NEOCENOZOIC AMERICAN "OPOSSUM-LIKE" MARSUPIALS (SUBORDER DIDELPHIMORPHIA), WITH COMMENTS ON THE CLASSIFICATION OF THESE AND OF THE CRETACEOUS AND PALEOGENE NEW WORLD AND EUROPEAN METATHERIANS

OSVALDO A. REIG[1], JOHN A. W. KIRSCH[2] and LARRY G. MARSHALL[3]

Studies of aspects of crania, dentitions, reproductive tracts, and other anatomical features of extant and Neogene South American "opossum-like" marsupials are integrated with cytogenetic and macromolecular data on living forms in a phylogenetic study employing cladistic and Wagner-tree analyses. The phylogenetic affinities indicated by serology are similar to those derived from a cladistic interpretation of morphology and are consistent with karyotypes, permitting inferences to be drawn about the relationships of fossil genera based on the arrangement of extant ones. Wagner trees showed a high degree of homoplasy among the dental and osteological characters. The most informative cladogram in a mathematical sense was one requiring several modifications of the Wagner trees, including removal of the convergent *Peratherium* from the group of murine opossums (*Marmosa, Micoures, Thylamys, Lestodelphys, Monodelphis, Thylatheridium*) and of *Sparassocynus* from within the 2n=22 large-opossum lineage (*Philander, Didelphis, Chironectes, Lutreolina, Hyperdidelphis, Thylophorops, Hondadelphys*). The large and murine opossums are considered tribes (Didelphini and Marmosini, respectively) within the subfamily Didelphinae; *Metachirus* represents a third tribe, Metachirini. The didelphid subfamily Caluromyinae (*Caluromys, Caluromysiops, Glironia, Pachybiotherium*) is also recognized. The Paleocene-Middle Miocene subfamily Herpetotheriinae includes *Herpetotherium, Peratherium*, and *Amphiperatherium*. The Late Miocene-Early Pleistocene subfamily Sparassocyninae (*Sparassocynus* only) is raised to family rank. The Didelphidae and Sparassocynidae comprise a restricted superfamily Didelphoidea, conveniently characterized by the joint possession (except in *Honda delphys*) of the derived V-shaped centrocrista (the "didelphoid dilambdodonty" of Crochet). The remaining genus of extant opossums, *Dromiciops*, is placed along with *Microbiotherium* (subfamily Microbiotheriinae) in a second superfamily Micro-biotherioidea new, which represents a clade distinct since the Upper Cretaceous. The Microbiotheriinae, the North American *Glasbius*, the Eocene South American *Reigia*, and the probably allied *Protodidelphis* from the Paleocene of South America (subfamily Glasbiinae), and the Cretaceous New World *Pediomys* and allies (subfamily Pediomyinae) constitute one of four families (Microbiotheriidae) included in the Microbiotherioidea: two other families are the carnivorous Stagodontidae of the North American Late Cretaceous (*Didelphodon, Eodelphis*), and the Late Cretaceous-Early Eocene South American Caroloameghiniidae (*Roberthoffstetteria, Procaroloameghinia, Caroloameghinia*) which may have had a contemporaneous origin with Microbiotheriidae. The tribe Peradectini (comprised of the North American Late Cretaceous *Alphadon* and *Albertatherium; Mimoperadectes* and *Alloeodectes* from the North American Lower Eocene and Lower Oligocene, respectively; *Peradectes* from the Late Cretaceous of South America, Paleocene to Oligocene of North America, and Paleocene to Eocene of Europe, and probably *Ankotarinjia* from the Miocene of Australia, *Bobbschaefferia* from the Paleocene of South America and *Garatherium* from the Eocene of North Africa) is recognised as the fourth and the most primitive microbiotherioid family, Peradectidae. Peradectidae were probably ancestral to Didelphoidea (*s.s.*), Borhyaenoidea, and other microbiotherioid taxa; Australian marsupials and remaining members of the South American radiation (caenolestoids, polydolopoids) may have had successively earlier origins from a peradectid morphotype. The majority of fossil opossums of the Paleocene fauna at Itaborai, Brazil are considered *incertae sedis*, but we accept Crochet's conclusion that *Bobbschaefferia* is a peradectid and *Protodidelphis* a glasbiine.

Pages 1-89 *in* POSSUMS AND OPOSSUMS: STUDIES IN EVOLUTION ed by M. Archer. Surrey Beatty & Sons and the Royal Zoological Society of New South Wales: Sydney, 1987.

INTRODUCTION

THE OPOSSUM-LIKE American marsupials, superfamily Didelphoidea in current classifications, comprise a sizable array of living and fossil mammals. They are of particular interest to students of evolution because some of the living taxa are thought to approximate, in aspects of some features at least, early therians; and they include the ancestral stem of the Metatheria, representing the early stock from which the diversification of American, European, and Australasian marsupials took place (Tedford 1974).

Living and fossil opossums have long been grouped (e.g., Clemens and Marshall 1976) into three families — Didelphidae, Pediomyidae, and Stagodontidae; the latter two being known only as fossils (Marshall 1981b). A fourth currently recognized family, Microbiotheriidae (Kirsch 1977b; Marshall 1982a; Reig 1981) includes

[1]Departamento de Ciencias Biologicas, F.C.E. yN., Universidad de Buenos Aires, Ciudad Universitaria Nuxez 1428 Buenos Aires, Argentina.
[2]University of Wisconsin Zoological Museum Madison, Wisconsin 53706, U.S.A.
[3]Berkeley Geochronology Center at the Institute of Human Origins, 2453 Ridge Road, Berkeley, California, U.S.A. 94709.

extant (*Dromiciops*) and fossil (*Microbiotherium*) taxa; and a fifth, Caroloameghiniidae, includes only the Late Cretaceous, Paleocene, and early Eocene genera *Roberthoffstetteria, Procaroloameghinia,* and *Caroloameghinia,* respectively (Marshall 1982d; Marshall *et al.* 1983).

The families of Borhyaenoidea, Necrolestidae, Thylacinidae, and Prepidolopidae have been included in the Didelphoidea by some workers (e.g., Simpson 1970, 1971 for the first two taxa) although their phylogenetic relationships are a matter of controversy (Archer 1976a, 1976b, 1982; Kirsch 1977b; Marshall 1977d; Reig 1981). We believe, as suggested by McKenna (1975, 1981), that the Necrolestidae may not be marsupials as thought by Patterson (1958) and others, and that the Thylacinidae and Borhyaenoidea warrant placement elsewhere within the Marsupialia (see also Reig 1981:58).

For the purpose of this paper we include in the Didelphimorphia all families which have been currently classified as strictly didelphoids, and that, as will be shown as a result of the present analysis, are probably best arranged in two superfamilies, namely Didelphoidea and Microbiotherioidea. We thus avoid the divergence of opinion among the present authors as to the inclusion within the Didelphimorphia either of borhyaenoids (for instance, Kirsch 1977b) or Dasyuroidea (Reig 1981). As for the Prepidolopidae (Pascual 1980b), two of the present authors (OAR and JAWK) agree, and will indicate in the appropriate part of this paper, that these marsupials can hardly be considered didelphoids. As so restricted, Didelphimorphia appears to be a relatively tightly united array by reason of (a) closer serological affinity (Kirsch 1977b), (b) biogeographic relationship, and (c) so far as is known, pairing of the epididymal sperm. Thus, although the concept of Didelphimorphia adopted here will not satisfy a strict cladist in the sense that not all derivative taxa are necessarily included, at least our restricted definition allows us to proceed effectively to demonstrate the phylogenetic relationships of those groups which we have chosen to include. Opinions contrary to our concept of Didelphimorphia expressed by Szalay (1982a and b) and Sharman (1982) will be considered in the appropriate parts of this paper.

Didelphidae are by far the most diverse of all recognized families of Didelphimorphia. Included as didelphids in current classifications are the earliest undoubted marsupials from the Late Cretaceous of North America and their generalized living descendants, the New World opossums. As a result, didelphids have the longest biochron (spanning about 75 million years) of any known family of Mammalia. However, many taxa included in Didelphidae are conservative in structure and their association within the family is based largely on retention of primitive characters. As is demonstrated below, the extremely long temporal range of this group is a reflection of the coarser scale of earlier classifications.

The primary aim of this paper is to establish the phylogenetic relationships of extant genera of opossum-like marsupials and of their South American Neocenozoic fossil relatives. As a secondary consideration we compare these taxa with fossil opossums from the Late Cretaceous and Tertiary of North America, and from the Tertiary of Europe. The relationships of extant and fossil didelphoids are poorly known. One reason for this is that paleontologists are often unfamiliar with the structural diversity of extant taxa, while neontologists pay little attention to the fossils. There has also been limited interest in the suprageneric relationships of living and fossil didelphoids. A first attempt to unravel their relationships was undertaken by one of us 30 years ago (Reig 1955). A substantial and concordant amount of data accumulated since then now permits association of all extant and Neogene didelphoid genera into groups, each considered to represent a monophyletic unit. The newer information comes from increased knowledge of anatomy, primarily of skull and dental characters, and from cytogenetics and comparative serology of living taxa.

Each of us is actively involved in one or more of these lines of research. Kirsch is responsible for information on serological relationships; Kirsch and Reig have synthesized present knowledge of cytogenetics; and Marshall and Reig are engaged in the study and description of Neogene taxa, and in the study of dental and cranial anatomy of living and fossil forms. The wealth of available data in these areas permits a multidisciplinary approach to phylogenetic inference, providing an exceptional opportunity to evaluate the merits of different sources of systematic data and to explore the relationships between biochemical, cytological, and morphological evolution (e.g., Miyamoto 1981; Solbrig 1970). Cladistics is used herein more as a tool to infer systematic relationships than as a framework to construct classifications, and we incidentally demonstrate some of the theoretical and practical problems (and advantages) afforded by this method of phylogenetic analysis.

MATERIALS AND METHODS

Most information on skull and dental morphology comes from study of specimens in institutional collections. Catalogue numbers preceded by the following institutional abbreviations are given in several instances to substantiate the factual evidence: AMNH, American Museum of Natural History, New York; BMNH, British Museum (Natural History), London; DGM, Dirección de Minas e Geología, Brasil; EBRG,

Estación Biológica de Rancho Grande, Venezuela; FMNH, Field Museum of Natural History, Chicago; FSM, Florida State Museum, Gainesville, Florida; INDERENA, Instituto de los Recursos Naturales Renovables y del Ambiente, Bogotá, Colombia; MACN, Museo Argentino de Ciencias Naturales "Bernardino Rivadavia," Buenos Aires; MCZ, Museum of Comparative Zoology, Harvard University, Cambridge; MBUCV, Museo de Biología, Instituto de Zoología Tropical, Universidad Central de Venezuela, Caracas; MLP, Museo de La Plata, La Plata, Argentina; MMP, Museo Municipal de Ciencias Naturales "Lorenzo Scaglia", Mar del Plata, Argentina; MVZ, Museum of Vertebrate Zoology, University of California, Berkeley; PV, Princeton University; UCMP (or UCVP), Museum of Paleontology, University of California, Berkeley; UKMNH, University of Kansas, Museum of Natural History, Lawrence; UMVP, University of Michigan Museum of Paleontology, Ann Arbor; USB, Universidad Simón Bolívar, Caracas, Venezuela; USNM, National Museum of Natural History (formerly, the United States National Museum), Washington, D.C.

Anatomical observations were made on dry skeletons with the aid of Wild M-5 stereomicroscopes. Some drawings (labeled OAR) were made from sketches taken with the drawing tube of this microscope. Illustrations of other specimens were made directly by an artist from original materials or epoxy casts (those labeled EL are by Elizabeth Liebman; MHW are by Marlene Hill Werner; and ZD are by Zorica Dabich). References to the techniques employed in karyotypic studies are found in Reig *et al.* (1977), serological methods are described by Kirsch (1977b), and chronology and use of South American Land Mammal Ages follows Marshall (1982c).

For reasons of uniformity in this volume, the authors agreed to use the molar nomenclature proposed by Archer (1978). In this system, the M1 is the deciduous tooth usually called dP4. Thus, the adult molar formula is M2-5. The conventional adoption of this nomenclature in this paper does not imply that the authors agree with Archer's interpretation and they may use the classical one in forthcoming papers.

METHODOLOGICAL FOUNDATIONS

We employ cladistic methodology to infer phylogenetic relationships within a taxon for which various sorts of data are available, and use Wagner-tree analysis to quantify those relationships. A statement of the foundations of these tools follows; later we evaluate the usefulness of the different data sets (morphological, chromosomal, and serological).

Cladistic Analysis

Cladistic analysis is a procedure for inferring branching sequences (Ashlock 1974; Brundin 1966 and 1968; Hennig 1966; Kavanaugh 1972;

Wiley 1981). Its fundamental premise is that relatedness is demonstrated by shared derived character-states, not by shared primitive ones. Thus, branching sequences are documented by the successive changes in characters associated with those branchings. It is vital to the method that sound procedures for inferring the direction, or *polarity*, as well as the ordering of changes be developed and applied. The means used here for determining if a character-state is *plesiomorphic* (primitive, ancestral) or *apomorphic* (advanced, derived), and whether taxa are united by *symplesiomorphy* (possession in common of primitive states) or *synapomorphy* (possession in common of derived states) largely conform to those of Schaeffer *et al.* (1972) and Stevens (1980). Use of such terms as *sister-group, stem species, polarity, convergence,* and *homoplasy* follow the definitions employed by Hecht (1976), Kirsch (1977b), and Simpson (1961).

One criterion for determining whether a character-state is plesiomorphic or apomorphic for a particular taxon is by application of the *principle of commonality* (Crisci and Stuessy 1980; Ride 1962:302; Schaeffer *et al.* 1972). If a character-state is found in the majority or in all members of the group under consideration, that character-state was probably the condition present in their ultimate common ancestor, and can be considered primitive for that group; any variation from the inferred primitive state is then regarded as derived. Obviously, what is primitive for one group may be derived with respect to its ancestor, so the terms "primitive" and "derived" are always relative.

Doubts, sometimes well-founded, have been expressed about the validity of the "common is primitive" postulate (Estabrook 1977). Most cladists prefer to rely on the criterion of *out-group comparison* which involves consideration of the distribution pattern of character-states among taxa of different levels of cladistic relatedness (Watrous and Wheeler 1981). In applying this principle the assumption must be made that some taxon is cladistically removed from all those under consideration. If the character-state in the out-group is the same as that in some members of the considered group, then that state was probably the condition in the stem species which gave rise to both groups (i.e., it is the primitive state).

A limitation of the method arises when the nearest out-group of the group under consideration is so divergent that its characters have themselves undergone change from the plesiomorphous to the apomorphous states; such is often found to be the case in attempting to determine the ancestral condition in marsupials by using contemporary placentals or monotremes as the out-group.

Decisions about the relative plesiomorphy or apomorphy of character-states may also be tempered by knowledge of what sorts of changes are possible for a given character. For example, when inferring directions of changes in karyotypic characters, some authors allege that alterations in chromosome number due to fission are less likely than those due to fusion of chromosomes or their arms (a doubtful contention given the present state of knowledge). Or, some polarities, such as the resurrection of complex structures from lost or simplified ones, can be given low probability of occurrence (a restatement of Dollo's Law). The validity of these criteria depends on the strength of the theoretical arguments on which they are founded.

Finally, it is often argued that some sequences of change are intrinsically more likely than others, such as successive reduction in numbers of mammalian premolars or of the digits. Yet, those arguments are frequently based simply on prior knowledge of what *has* occurred, or can be shown to reduce to application of out-group or commonality criteria. We attempt to avoid such reasoning here. (For a review of the criteria for inferring polarity of character-states, see Stevens 1980.)

Once the plesiomorphic state of a character is established and the apomorphous states ordered in a sequence corresponding to the likely pathway of change, this series is used to infer the relationships of the taxa in which various derived states co-occur. A character-state sequence may be linear or branching, and both kinds are represented in our data-sets. Each character may serve as the basis for a cladogram relating the taxa, and concordance or discordance among characters will suggest the extent of convergent or parallel changes in those characters. Ordinarily one begins an analysis with three or four taxa thought to be closely related and then determines their cladistic relatedness by means of the best-established character-state sequence; that scheme is then tested against less-certainly analyzed characters and extended to more taxa. For more than a few characters the number of possible final, integrative cladograms can be large, particularly when some character-state sequences do not give concordant arrangements due to convergence or otherwise mistaken homologies. The greater the concordance between different characters, the more likely that the original hypothesis of relatedness is correct, if "parsimony" is any sort of guide. In the GENERAL DISCUSSION we show how some cladograms may be eliminated or modified when biochemical data are available, and we note some limitations of the popular concept of parsimony.

Wagner-tree Analysis

An alternative to the tedious analysis of characters by hand is to use a "numerical cladistic" method, which, through application of an heuristic algorithm, simultaneously analyzes many characters with the aim of reconstructing a tree representing the shortest evolutionary pathway. Three basic approaches are in use: 1) historically pre-eminent is that of Camin and Sokal (1965), which performs a sequential examination for concordance or discordance between character-state pathways and does not allow for reversals in the evolution of characters in the final trees; 2) popular among botanists is *clique analysis*, a related tool for partitioning a group into the largest subsets concordant in their character-states (e.g., Meacham 1980); and 3) the *Wagner-tree method* formalized from the principles of W. H. Wagner (e.g., 1969) by Kluge and Farris (1969; see also Farris 1970 and 1972; Wiley 1981) and which we adopt here. We give a general introduction to this method below and discuss some Wagner trees generated in and relevant to our study.

SURVEY OF STATUS OF SOME "KEY" GENERA

Early opossum-like marsupials, representing four distinct lineages (see below), are known in the Late Cretaceous of North and South America and in the Early Eocene of South America (Archibald 1982; Clemens 1979; Crochet 1980; Pascual 1983; Sigé 1972). *Alphadon, Peradectes, Pediomys, Glasbius, Reigia, Didelphodon*, and *Eodelphis* are well-known, while *Albertatherium, Aquilodelphis*, and *Boreodon* are more poorly known (Clemens 1979; Pascual 1983). Additional information on these taxa is provided later.

Roberthoffstetteria, Procaroloameghinia, and *Caroloameghinia*, with bunodont dental specializations, are known respectively from the Late Cretaceous, Paleocene, and Early Eocene of South America (Marshall 1982d; Marshall *et al.* 1983; Marshall *et al.* 1984); we will comment on these taxa in more detail below. Contemporaneous with *Procaroloameghinia* are diverse opossum taxa described from the Paleocene (Riochican Age) fissure-fills at Itaborai, Brazil (Paula Couto 1952, 1962, 1970). They belong to ten genera:* *Derorhynchus, Didelphopsis, Eobrasilia, Gaylordia* (including *Xenodelphis*), *Guggenheimia, Marmosopsis, Minusculodelphis, Mirandatherium, Monodelphopsis*, and *Sternbergia*. These genera are in dire need of systematic revision, and for this reason we do not include them in this study. An eleventh didelphoid genus from this fauna, *Itaboraidelphys*, was recently named by Marshall and de Muizon (1984).

Prepidolops, the single known genus of the family Prepidolopidae (Pascual 1980a and b) is also not included here because two of the present authors (OAR and JAWK) disagree with the third concerning its placement as a didelphoid (but see also Marshall 1982e). In addition, the

---

* *Eobrasilia* Simpson 1947, originally described as a didelphoid or a borhyaenoid is clearly a didelphoid (Marshall 1984), and *Ischyrodidelphis* Paula Couto 1952 represents the lower dentition of *Patene* and is thus a borhyaenoid (Marshall 1981a:6). Moreover, *Bobbschaefferia* and *Protodidelphis* are, following Crochet (1980), classified respectively as peradectid and glasbiine (see below).

Late Paleocene-Early Eocene bonapartheriids are not treated in this paper. Bonapartheriidae were, on the basis of their molar teeth, originally regarded as polydolopoids (Pascual 1980a). Later, the discovery by Galileo Scaglia of a nearly complete skull showing several didelphimorphian character-states led Pascual (1981) to assign them to the Didelphoidea. More detailed study is certainly necessary to establish the relationships of these peculiar marsupials. But if they are eventually demonstrated to be opossumlike, they should be placed in their own superfamily within Didelphimorphia.

Opossums are also known from the Early Paleocene to Middle Miocene of North America and Europe. *Alloeodectes*, *Mimoperadectes*, *Thylacodon* and *Herpetotherium* in North America, *Peratherium* and *Amphiperatherium* in Europe, and *Peradectes* in both continents are currently recognized (Crochet 1977a and b, 1980; Bown and Rose 1979; Russell 1984). Recently, Crochet (1984) reported the presence of *Garatherium* in the Eocene of North Africa. This genus was referred to the peratherines (see later). We will also claim below peratherine relationships for the Miocene Australian *Ankotarinja*.

Living opossums are a conspicuous element of the Recent Neotropical mammalian fauna. One monotypic genus, *Dromiciops*, long considered a didelphid, is now placed in the family Microbiotheriidae (Kirsch 1977b; Marshall 1982a; Reig 1981). This family also includes species of *Microbiotherium* from the Late Oligocene and Early Miocene of southern Argentina (Marshall 1982a).

The wealth of living Neotropical opossums belong to the family Didelphidae, comprising about 76 species which Kirsch, Pine, and Van Gelder (1982) group in 11 genera. We here recognize *Thylamys* and *Micoures* as distinct genera (considered as subgenera of *Marmosa* by some previous workers), raising the number to 13, grouped here into two subfamilies. The Didelphinae include *Marmosa*, *Micoures*, *Thylamys*, *Monodelphis*, *Lestodelphys*, *Metachirus*, *Philander*, *Didelphis*, *Lutreolina*, and *Chironectes*; the Caluromyinae include *Caluromys*, *Caluromysiops*, and *Glironia* (Kirsch 1977b; Reig 1981). Sixty-four species belonging to the 13 genera occur exclusively in South America; seven species also occur in South and North America, and four species of three genera are reported exclusively from Central America and Mexico. One species, *Didelphis virginiana*, has successfully invaded temperate North America (Gardner 1973; Reig *et al.* 1977:192).

Several fossil didelphids described from the Neogene of South America are closely related to living genera. We recognize *Hyperdidelphis* (=*Paradidelphis*), *Thylophorops*, *Thylatheridium*, and *Hondadelphys* as Didelphinae. No fossil caluromyines have been described from the Neogene, but the Late Oligocene *Pachybiotherium*

is considered the first known member of this subfamily (Reig 1981). *Sparassocynus*, long placed in a subfamily Sparassocyninae Reig 1958b, is here raised to family rank.

This study concentrates on the 19 recognized genera of living and Neogene South American opossums, which are characterized as follows:

Didelphis Linnaeus 1758

Common opossums, chuchas comunes, zarigueyas, rabipelados, or comadrejas, are the largest of living didelphines (325-500 mm in combined head and body length), and constitute one of the most widely distributed mammalian genera in the New World, ranging from southern Canada to central Argentina. Three living species are recognized: *D. virginiana* in Central and North America; *D. marsupialis* in lowlands of tropical and subtropical Mexico, Central America, and South America; and *D. albiventris* in temperate South America and highlands of tropical South America. *Didelphis marsupialis* and *D. albiventris* are more closely related to each other than either of them is to *D. virginiana* (Gardner 1973). The latter is closely comparable to the fossil genus *Hyperdidelphis* (OAR personal observation). Both temperate and tropical populations are referred to *D. albiventris* mainly because they have white ears. All species are terrestrial, good climbers, omnivorous and eurytopic, and have a well-developed marsupium or abdominal pouch. Fossil species of *Didelphis* are first known from the Middle Pleistocene (Ensenadan) of Argentina (*D. mutilata* Ameghino 1889), and fossil specimens of *D. virginiana* (FSM 11826, 11827) are first known from the Middle Pleistocene (Irvingtonian) of Florida in North America (Marshall and Reig in prep.).

Philander Brisson 1762

Grey or black four-eyed opossums, or cuicas, are slimmer and smaller (250-350 mm in combined head and body length) than species of *Didelphis*, and range from southern Mexico to northern Argentina. Two living species are recognized: *P. opossum* (range as for the genus), and *P. mcilhennyi* (Gardner and Patton 1972), restricted to Amazonian Peru. They are carnivorous and frugivorous, are mostly terrestrial but are also good climbers and swimmers (Crespo 1950:6), and inhabit tropical forests. A well-developed abdominal pouch is present. Simpson (1972) and Pine (1973) commented that there is little reason to consider *Philander* more than a subgenus of *Didelphis*. We disagree. *Philander* is distinctive in anatomy, ecology, and in other characters as demonstrated below. A fossil species is reported from the Pliocene (Montehermosan) of Argentina (*P. entrerrianus*; see Reig 1957b), and *P. opossum* is found in Late Pleistocene-Holocene cave deposits of Brazil (Winge 1893). As for the availability of this generic name for the grey and for the black four-eyed opossum, see Gardner (1981) and Hershkovitz (1977, 1981).

Metachirus Burmeister 1854

Metachirus nudicaudatus, the single species of brown four-eyed opossum or cuica, is similar in shape, size (about 265 mm in combined head and body length), skull, and dental morphology to species of *Philander*, but differs in having relatively longer hindlegs (Hildebrand 1961), a tendency toward an incipiently bipedal stance, and in chromosome number. It ranges from Nicaragua to northern Argentina and is mostly terrestrial, but is also a good swimmer. It inhabits tropical forests close to rivers and is omnivorous. Females are pouchless. Fossils are known from the Late Pleistocene-Holocene cave deposits at Lagoa Santa, Brazil (Winge 1893).

Lutreolina Thomas 1910

Lutreolina crassicaudata, the lutrine opossum or comadreja colorada, is a short-legged, weasel-like, reddish opossum, slightly larger (210-445 mm in combined head and body length) than species of *Philander*. It ranges from the southern pampas of Argentina to south of the Amazon Basin in Brazil and Bolivia and, discontinuously, in the Guianas and lowlands of southern Venezuela and Colombia. It is mostly terrestrial, but is also a good swimmer and occasionally climbs. Carnivorous, insectivorous, and at times herbivorous in diet, it inhabits pampas and savannas, usually near water, but is also found in gallery woodland. Despite reports to the contrary, females have a pouch. Fossils from the Late Miocene (Huayquerian) of Argentina have been referred by Simpson (1974) to *L.* cf. *crassicaudata*. The fossil species *L. tracheia* from the Pliocene (Montehermosan) of Argentina may prove to be synonymous with *L. crassicaudata* (Marshall 1977c; Reig 1958b); and *"Didelphis" lujanensis* from the Late Pleistocene (Lujanian) of Argentina is referrable to *L. crassicaudata* (Marshall and Reig in prep.), as are specimens reported from beds of Late Pleistocene-Holocene age in the cave deposits at Lagoa Santa, Brazil and from Tarija, Bolivia (Clemens and Marshall 1976:23).

Chironectes Illiger 1811

Chironectes minimus, the water opossum, perrito de agua, yapo, or yapock, is amphibious, robust, short-legged, and somewhat smaller (270-340 mm in combined head and body length) than living species of *Didelphis*. It ranges from Oaxaca, Mexico to northern Argentina. The aquatic adaptations include webbed hindfeet which serve as the primary means of propulsion in the water; a dense, water-repellent pelage; long, stout, supernumerary facial vibrissae, and sensitive forepaws with enlaged pisiform bones which act as tactile organs under water; a "water-proof" pouch in the females; and (*fide* Enders 1937) a scrotal pouch in the males (Kirsch 1977a; Marshall 1978c). These animals are excellent swimmers and divers (Stein 1981). They eat fish, crabs, insects, and frogs (Collins 1973). Fossils referrable to the living species are recorded from the Pliocene (Montehermosan) of Argentina (Marshall 1977a; Reig 1958b:262n) and from the Late Pleistocene-Holocene cave deposits of Lagoa Santa, Brazil (Winge 1893).

Hyperdidelphis (=*Paradidelphis*) Ameghino 1904

Hyperdidelphis is a fossil genus of large and robust, *Chironectes*- or *Didelphis*-like opossums. Four species are recognized: *H. pattersoni* from the Late Miocene (Huayquerian) and Pliocene (Montehermosan) of Catamarca Province, Argentina; *H. inexpectata* from the Pliocene of Buenos Aires Province, Argentina; *H. biforata* also from the Pliocene of Buenos Aires Province; and *H. brachyodonta* from the Late Pliocene (Chapadmalalan) of Buenos Aires Province. The use of the name *Hyperdidelphis* for what has usually been called *Paradidelphis* results from the fact that the recently found type specimen of *Hyperdidelphis acutidens* is inseparable in morphology from *Didelphis inexpectata* Ameghino, the type species of *Paradidelphis* (Marshall and Reig in prep.). The genotypic species is thus *H. inexpectata*. Simpson (1972, 1974) did not recognize *Hyperdidelphis* (*Paradidelphis* in his paper), but included its species in *Didelphis*. As is demonstrated below, *Hyperdidelphis* is a valid genus, albeit very similar to *D. virginiana*, and is distinctive in dental and cranial characters (see also Marshall and Reig in prep.; Reig 1952 and 1958b).

Thylophorops Reig 1952

Thylophorops, a *Lutreolina*-like fossil genus with carnivorous dental specializations, is the largest Late Tertiary didelphid. It is known from lower jaws and fragmentary skulls belonging to a single species, *Thylophorops chapalmalensis*, from the upper levels of the Late Pliocene Chapadmalal Formation (Chapadmalalan) and lower part of the Late Pliocene-Early Pleistocene (Uquian) Barranca de Los Lobos Formation of southeastern Buenos Aires Province, Argentina (Simpson 1972:17).

Marmosa Gray 1821

Murine opossums, or achocayas, constitute a polytypic genus ranging from tropical Mexico to southern Brazil and Bolivia. Species live in tropical and subtropical humid and dry forests, thorn scrubs and "matas" of the savanna. Many are arboreal and are insectivorous, but some eat fruits and small vertebrates. As generally recognized (e.g., Pine *in* Collins 1973), *Marmosa* is a composite genus of small, pouchless opossums, including forms which are as different from each other as are the species placed in such genera as *Philander*, *Didelphis*, and *Chironectes*. We here restrict the concept of *Marmosa* to species placed by Tate (1933) in his *murina* and *noctivaga* groups, adding the more recently named *Marmosa (Stegomarmosa) andersoni* (Pine 1972), and *Marmosa cracens* and *M. xerophila* (Handley and Gordon 1979).

Species of *Marmosa* are characterized by their small or intermediate size (120 to 160 mm in combined head and body length); medium-sized feet; lack of pectoral mammae but occasional presence of inguinal teats; lack of a pouch; moderately long tail; skull with bullar region of the "tripartite" type (see below), with a well-inflated alisphenoid process; upper molars with moderately reduced paracone, well-developed metacrista and metastylar spur; M^4 longer than M^3; and anterioposteriorly compressed M^4 with moderately reduced metacone.

Subgeneric distinctions may prove convenient, in which case *Marmosa s.s.* should be restricted to the *murina* group of Tate (*murina, rubra, tyleriana, robinsoni, xerophila, mexicana,* and *canescens*); *Marmosops*, erected by Matschie (1916) with *incana* as the type species, is the appropriate name for Tate's *noctivaga* group (*noctivaga, yungasensis, leucastra, ocellata, incana, scapulata, invicta, fuscata, cracens,* and *parvidens*); and *Stegomarmosa*, created by Pine (1972), would include only the species *andersoni*.

A fossil representative of *Marmosa*, probably of the subgenus *Marmosops*, was described from the Late Pliocene (Chapadmalalan) of Argentina by Reig (1958b). *Marmosa laventica* from the Middle Miocene (Friasian) of Colombia (Marshall 1976) seems also to belong here.

Micoures Lesson 1842

These large murine opossums are a polytypic genus of predominantly arboreal, medium-sized opossums, ranging from Belize (British Honduras) to northern Argentina. They are typically inhabitants of tropical rain and montane forests, and are insectivorous, frugivorous, and/or carnivorous in diet. As discussed below, there are cogent reasons for separating the large murine opossums of Tate's *cinerea* group into a genus of their own. The name *Micoures* became available when Thomas (1888) designated *cinerea* as the type species. We include in *Micoures* the species *cinereus* (including *alstoni* and *demararae*), *constantiae, dominus, mapirensis, rapposus, germanus, reginus,* and *phaeus*. These are distinguished from other murine opossums by their relatively large size (130 to 215 mm in combined head and body length); large and strong feet; lack of pectoral and inguinal mammae; long tail with a woolly basal portion; skull without a "tripartite" type of bullar region but with a moderately inflated

alisphenoid process; upper molars with moderately reduced paracone and moderately developed metacrista and metastylar spur; protocone not compressed anteroposteriorly; M^4 shorter than M^3; and M^5 moderately compressed anteroposteriorly and with a well-developed metacone. Fossil representatives of *Micoures* have been recorded only from Late Pleistocene-Holocene cave deposits at Lagoa Santa, Brazil (Winge 1893).

Thylamys Gray 1843

Small murine opossums are mostly terrestrial-scansorial, and some arboreal animals, ranging from northern Venezuela to the pampas of southern Argentina. They live in open grassland, rockeries, thorn-scrub bushes, semi-desert mountain areas, paramos, and adjacent tropical forests. Mostly insectivorous-carnivorous in diet, they are at times frugivorous.

Thylamys was long considered a subgenus of *Marmosa* (e.g., Cabrera 1919 and 1957; Gilmore 1941; Kirsch and Calaby 1977). Species are characterized by their small size (68 to 139 mm in combined head and body length), although larger species (e.g. *T. elegans*) may attain the size of the smallest species of *Marmosa*. Most typical of *Thylamys* are their very small feet, relatively short and frequently incrassated tails,* (e.g., *T. elegans, T. pusillus, T. dryas,* and *T. velutinus*), and presence of pectoral mammae. As in *Marmosa* and *Micoures*, a pouch is lacking. The skull normally lacks supraorbital crests, has an enlarged braincase, and a large "tripartite" type of bulla with a swollen tympanic process of the alisphenoid. The teeth are distinctive owing to the subequal P^2 and P^3; upper molars have reduced paracones, compressed protocones, well-developed metacristae and metastylar spurs; M^4 longer and wider than M^3; and strongly compressed M^5 with metacone either much reduced or absent.

We include in *Thylamys*, following Kirsch and Calaby (1977), 14 species (*aceramarcae, agilis, agricolai, dryas, elegans, formosus, griseus, karimii, maricus, microtarsus, pusillus, tatei, unduaviensis,* and *velutinus*); the type species is *T. elegans* (Cabrera 1919:39). Also probably included is *lepida*, which Tate (1933) considered a member of the *microtarsus* group, and which Kirsch and Calaby (1977:14) included in the subgenus *Marmosa*. The species *emiliae* may also be placed here, but Pine (*in* Kirsch and Calaby 1977:18) suggested that it could belong elsewhere. Pine (pers. comm.) believes that the *Marmosa elegans* group of Tate (i.e., those with incrassated tails — *elegans, formosus?, griseus, karimii, pusillus, tatei, velutinus*) may at the least deserve subgeneric recognition.

---

* Incrassated tails contain large fat deposits at the base which function in food storage (Morton 1980). It is known that incrassated tails developed independently in various lineages of marsupials, namely in species of *Thylamys*, in *Lestodelphys halli*, and in *Dromiciops australis* among didelphimorphians; in *Rhyncholestes raphanurus* among caenolestoids; and in species of *Sminthopsis* and *Cercartetus* among Australasian groups (Kirsch 1977a; Morton 1980).

Late Pleistocene-Holocene remains from caves in Lagoa Santa, Brazil (Winge 1893), and in Cordoba, Argentina (i.e., *"Didelphis" incerta* Ameghino — see Reig 1958b:276), are referrable to *Thylamys*. A lower jaw with M_4 and M_5 (MACN 9963) from Pliocene (Montehermosan) beds at Rio Quequen Salado, Buenos Aires Province, and *"Marmosa" contrerasi* from Pliocene beds at Monte Hermoso are apparently referrable to *Thylamys* (Mones 1980:163; Reig unpubl.).

Lestodelphys Tate 1934

The Patagonian opossum, *Lestodelphys halli*, is a small to intermediate (132-220 mm in combined head and body length), pouchless (based on examination of MACN 47.24 and MACN 47.25 by LGM), terrestrial opossum restricted to central Patagonia, Argentina (Marshall 1977b), where it inhabits semi-desert pampas and thorn-scrub bushes. It shows advanced carnassial adaptations in skull and dentition, and feeds on mice and small birds. *Lestodelphys* shares with some species of *Thylamys* an incrassated tail and an enormous "tripartite" type of bulla.

A specimen (MLP 67-VIII-11-3) found in the Late Pleistocene Junin Formation in northwest Buenos Aires Province demonstrates that the living species recently inhabited the Pampa Region (Odreman Rivas and Zetti 1969). In addition, an undescribed specimen (FMNH 1262) from the latest Pliocene-Early Pleistocene Vorohue Formation near Necochea, Buenos Aires Province is referrable to *Lestodelphys* (Reig unpubl.).

Monodelphis Burnett 1830

Short-tailed opossums are a polytypic genus of small (110-140 mm in combined head and body length), pouchless, terrestrial animals, ranging from Panama to the southern pampas of Argentina. They inhabit open grasslands, savanna, and dense forests; at least one species, *M. domestica*, frequents human habitations. They feed on small rodents, insects, fruits, seeds, and carrion (Hunsaker 1977). Following Kirsch and Calaby (1977) and Pine (1975, 1976, 1977), we tentatively recognize 17 living species (*adusta, americana, brevicaudata, dimidiata, domestica, henseli, iheringi, kunsi, maraxina, orinoci, osgoodi, scalops, sorex, theresa, touan, umbristriata,* and *unistriata*).

Monodelphis has been reported from beds of Late Miocene (Chasicoan) age of Buenos Aires Province (Reig 1957a), and *"Didelphis" juga* Ameghino, from the Late Pleistocene (Lujanian) of Córdoba, Argentina, probably should be included (Reig 195Bb). *Monodelphis* is also represented by an undescribed specimen (MACN 6408) from Middle Pleistocene (Ensenadan) beds at Miramar, Buenos Aires Province (Reig unpubl.)

Thylatheridium Reig 1952

Thylatheridium is a genus of medium-sized, *Monodelphis*-like opossums, found in beds of Late Miocene (Huayquerian) to Early Pleistocene (Uquian) age in Argentina. The three described species (*T. cristatum, T. pascuali,* and *T. dolgopoli*; Reig 1952, 1958a and b; Simpson 1972) are larger than species of *Monodelphis* and show advanced carnassialization of the dentition, distinctive skull features, and a moderately inflated auditory bulla.

Hondadelphys Marshall 1976

Hondadelphys is known from a single species, *H. fieldsi*, from the Middle Miocene (Friasian) La Venta fauna of Colombia (Marshall 1976). It is quite large, similar in size to and sharing dental characters with *Thylophorops chapalmalensis*. In *H. fieldsi* shear specializations have evolved by reduction of the metaconid and elongation of the metastylar ridge, while crushing specializations are evidenced by enlargement of the protocone and talonid basin. There is no evidence of an ossified auditory bulla, beyond the weakly developed tympanic process of the alisphenoid. *H. fieldsi* differs from *T. chapalmalensis* in relative size differences of the lower premolars, morphology of the talonid of M_5, weaker stylar cusps B and C, orientation of P^1, weakly developed anterior basal cingulum on the lower molars, and greater reduction of the metaconid. Several of the cranial and dental specializations in *H. fieldsi* are also shared with members of the Borhyaenoidea, including the structure of the basisphenoid and basioccipital, absence of a tympanic process on the petrosal, enlargement of posterior carotid foramina, emphasis on shear, and great reduction of the metaconid (Marshall 1976).

Caluromys Allen 1900

Woolly opossums or comadrejas include three species (*C. philander, C. lanatus,* and *C. derbianus*) of medium size (180-290 mm in combined head and body length), which are strictly arboreal and occur throughout tropical and subtropical America from southern Mexico to northern Argentina (Bucher and Hoffmann 1980). They inhabit rain- and deciduous forests, and are mostly frugivorous in diet but also feed on insects, seeds, and small vertebrates. Females of *C. lanatus* and *C. derbianus* have pouches, but females of *C. philander* are apparently pouchless (Pine 1973). Only *C. derbianus* (listed as *"D." laniger*) is reported as fossil, from the Late Pleistocene-Holocene cave deposits of Lagoa Santa, Brazil (Winge 1893).

Caluromysiops Sanborn 1951

Caluromysiops irrupta, the black-shouldered opossum, is arboreal, medium-sized (ca. 217 mm in combined head and body length), and restricted to wet, dense tropical forests of Peru (Sanborn 1951) and southern Colombia (Simonetta 1979:247). Females have a well-developed pouch (Pine pers. comm.).

Caluromysiops is reported to be similar to *Caluromys* in diet and habits, and some authors consider *irrupta* a species of *Caluromys*. Cabrera (1957:4), for example, suggested that *Caluromysiops* may prove to warrant recognition only as a subgenus of *Caluromys*, and Hershkovitz (pers. comm.) regards *irrupta* as a fourth species of *Caluromys*, doubting that even subgeneric recognition is deserved. As we demonstrate below, generic distinction seems well-warranted. No fossils have been reported.

Glironia Thomas 1912

The bushy-tailed opossum is a poorly known genus of presumably arboreal, medium-sized (160-205 mm in combined head and body length) animals which inhabit dense, humid forests. Two species have been described: *Glironia venusta*, from Ecuador and northern Bolivia, and *G. criniger* from Peru; the latter apparently represents a local race of the former. The diet is unknown, but from what can be inferred from the dentition it is probably omnivorous-insectivorous and frugivorous. Females (e.g., FMNH 41440) have four abdominal mammae and are pouchless. Fossils are unknown (Marshall 1978a).

Dromiciops Thomas 1894

Dromiciops australis, the monito del monte, is a small (85-130 mm in combined head and body length), arboreal opossum inhabiting the Valdivian temperate forests and Chiloe Island, southern Chile, and adjacent areas in Argentina. It lives in dense, humid forests on trees, in bamboo thickets, or in holes in the ground, and feeds mostly on insects and other invertebrates. Females have a small but distinct pouch with four mammae (Marshall 1978d). As demonstrated by Marshall (1982a), Reig (1955), and Segall (1969a), *Dromiciops australis* is very similar in dentition and skull morphology to species of the fossil genus *Microbiotherium* Ameghino 1887, from beds of Late Oligocene-Early Miocene age in Patagonia, southern Argentina.

Sparassocynus Mercerat 1898

Sparassocynus is a fossil genus of medium-sized opossums. Three species are recognized: *S. bahiai* and *S. derivatus* from the Pliocene and Early Pleistocene (Montehermosan, Chapadmalalan, and Uquian) of Argentina, and *S. heterotopicus* from the Late Miocene of Bolivia (Villarroel and Marshall 1983). Fossils from beds of Late Miocene (Huayquerian) age in Catamarca Province, Argentina (Simpson 1974) may represent a fourth species. *Sparassocynus* is unique among didelphoids in the possession of enormous alisphenoid bullae with epitympanic sinuses, a palate without vacuities, and advanced carnassialization of the molar teeth (Reig and Simpson 1972). It is currently placed in a subfamily of its own, Sparassocyninae Reig 1958b. As discussed below, we regard the peculiarities of *Sparassocynus* as warranting familial distinction for the genus.

CHARACTER-ANALYSIS

General Statement

Our character-analysis is not, obviously, thorough in the sense that we do not use all the available evidence on different organ-systems. Unfortunately, much of this evidence is fragmentary, and our aim was to compile data on all, or most, taxa in order to make comprehensive comparisons. At the anatomical level we have mostly used dental and cranial evidence which we have observed and analyzed ourselves. Additions from other character-sets have been selected from published summaries. The reader will probably be surprised that we have not used the tarsal characters which are the basis of a recent reassessment of marsupial relationships by Szalay (1982a and b). The reason is that we lacked access to relevant material and that we consider the evidence presented by Szalay to be in need of careful checking. However, we shall mention this source of data in our discussions throughout the text.

Size

Body size is traditionally considered a poor character for systematic studies. It has been informally recognized, however, that living opossums are either "large" or "small", though there is some overlap between these categories with a few species of intermediate size. We can conventionally define as "small", "medium", and "large" those groups ranging in body-size from 70-150, 150-300, and greater than 300 mm, respectively. These classes correspond to dental categories given by measurements of lower molar series of 4.5-8, 8-15, and greater than 15 mm in length. *Didelphis*, *Chironectes*, *Metachirus*, *Philander*, and *Lutreolina* among the living genera comprise the large-size group, measuring from about 300 to 500 mm in combined head and body length. The fossil genera *Didelphodon*, *Eodelphis*, *Thylophorops*, *Hyperdidelphis*, and *Hondadelphys* also belong here. The large species of *Alphadon*, *A. rhaister*, should also be included, but the type species of this genus, *A. marshi*, and other referred species such as *A. lulli*, *A. wilsoni*, and *A. presagious*, belong to the medium-sized group as judged on dentition. Also, some *Pediomys*, such as *P. florencae*, are large, whereas the type species,

P. elegans, and also *P. cooki* and *P. crece*, are really of small to medium size. *Thylamys*, *Marmosa*, *Monodelphis*, *Lestodelphys*, and *Dromiciops* represent the small opossums, ranging from 68 to 180 mm in combined head and body length. The fossil genera *Alloeodectes*, *Mimoperadectes*, *Peradectes*, *Glasbius*, *Herpetotherium*, *Microbiotherium*, and *Zygolestes* are also considered small. A group of medium-sized animals, represented by species of *Micoures*, *Caluromys*, *Caluromysiops*, and *Glironia* range from 130 to 310 mm in combined head and body length; the fossils *Peratherium*, *Amphiperatherium*, *Caroloameghinia*, *Procaroloameghinia*, *Roberthoffstetteria*, *Sparassocynus*, *Thylatheridium*, and *Pachybiotherium* are also medium-sized. The size data given above have been taken mainly from Archibald (1982), Clemens (1966), Crochet (1980), Tate (1933), Walker (1975), and our own measurements.

As we will demonstrate below, there is a general correlation between body size and other features, although *Metachirus nudicaudatus* and *Caluromys philander* represent striking exceptions. No apparent size-related (allometric) linkages exist among the characters studied. Cope's rule (see Stanley 1973) would favor a small-to-large direction of change, but this rule has many exceptions (Marshall and Corruccini 1978). That species of the structurally ancestral Late Cretaceous opossum *Alphadon* belong in the large-size group might, for example, be used as counterevidence against the application of Cope's rule to didelphimorphians. Moreover, out-group comparisons with dasyurids suggest that the character-state "medium-sized" has more chance of representing the plesiomorphic condition. This conclusion is substantiated by the fact that most representatives of the small- or large-size classes share significant derived character-states, and that most of the medium-sized genera are plesiomorphic for other character-states. For the purpose of coding size, we decided to classify variable and possibly composite genera, such as *Alphadon* and *Pediomys*, according to the sizes of their type species.

Tooth Morphology

Didelphoids are conservative in aspects of their dentition (i.e., all genera show the same dental formula and basic pattern in tooth morphology, and lack major adaptive departures for masticatory specializations). This fundamental uniformity is probably a reflection of a highly structured and coadapted genome, organized in fairly uniform karyotypes (Reig *et al.* 1977). However, the teeth of didelphoids are far from being identical, and a considerable amount of inter-taxon variation exists in many characters. These character-variations are the chief basis for study of fossil taxa. Unfortunately, many paleontologists are not fully acquainted with dental variation in extant taxa, and their comparisons have usually been limited to *Didelphis* on the implicit assumption that the remaining genera repeat the dental character-states found in species of that genus. That this assumption is unwarranted was clearly demonstrated when Reig (1955) showed that the living *Dromiciops* was more similar in dentition to the fossil *Microbiotherium* than to *Didelphis*.

Attempts to study comparatively the variation in dental characters of living didelphids have been made by Archer (1976a), Bensley (1903), Winge (1893), and Wood (1924). Unfortunately, these works focus on but a few characters, and include consideration of only a limited sample of living taxa. As regards the fossil taxa, great progress in understanding the molar-pattern variation in didelphimorphians has come from the work of Crochet (1980); but even more recently, Szalay (1982a) disparaged the value of molar characters in classifying didelphimorphs, claiming that these features do not provide reliable evidence for allocating fossil taxa within the group or to define major subtaxa. As demonstrated below, Szalay's contention can be shown to be incorrect, as molars *do* provide several and in some cases sufficient characters to establish cladistic groupings consistent with those made on the basis of other sets of features.

As implied and as is true for most mammalian groups, the most important source of information regarding the dentition comes from study of the gross morphology of the molar teeth, other evidence being afforded by premolars and incisors, canine variation not yet having been studied. There is no comparative information on the microstructure of the enamel or other components of the teeth, although the enamel structure of *Alphadon* and *Pediomys* is known (Clemens 1979; Moss 1969).

It is conventional to regard *Alphadon* as approximating the "proto-marsupial" in dental structure (Clemens 1966, 1968, 1979; Fox 1979a), and we mostly agree with this judgment. However, it is possible that *Alphadon* is advanced in some features, as will become clear below. Marsupials must have originated earlier than the Upper Cretaceous, so we consider also the states of characters of earlier mammals related to the therian lineage in establishing the directions of character-change. Whatever the exact relations of *Holoclemensia*, *Pappotherium*, deltatheriids, and other mammals of the "metatherian-eutherian grade" (Kielan-Jaworowska *et al.* 1979; Turnbull 1971), at least some are at, or near, the condition of primitive marsupials. We therefore consider Cretaceous marsupials and earlier therians in arriving at our assessments of the primitive state for each character. For the sake of uniformity, we adopt (Fig. 1) the terminology of molar tooth structure recently proposed by Crochet (1980).

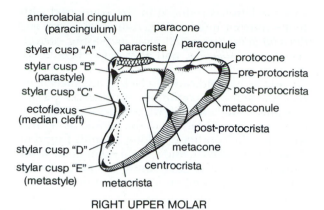

anterolabial cingulum (paracingulum)
paracone
paraconule
stylar cusp "A"
paracrista
protocone
stylar cusp "B" (parastyle)
pre-protocrista
stylar cusp "C"
post-protocrista
ectoflexus (median cleft)
metaconule
post-protocrista
stylar cusp "D"
metacone
stylar cusp "E" (metastyle)
centrocrista
metacrista

RIGHT UPPER MOLAR

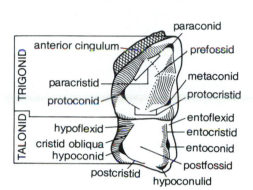

TRIGONID
anterior cingulum
paraconid
prefossid
paracristid
metaconid
protoconid
protocristid
entoflexid
TALONID
hypoflexid
entocristid
cristid obliqua
entoconid
hypoconid
postfossid
postcristid
hypoconulid

LEFT LOWER MOLAR

Fig. 1. Nomenclature for molar teeth used in this study (modified after Crochet 1980). (ZD).

An excellent starting point for evaluating the relative primitiveness of didelphoid molars is Crochet's distinction between *pre-dilambdodonty* and *didelphoid dilambdodonty*, which refers to the condition of the centrocrista connecting metacone and paracone. This crista is straight in *Peradectes* and *Alphadon* (Figs 2, 3), whereas it forms a V-shaped pattern with the vertex directed towards the stylar shelf in most others

(cf. Figs 2a and 2b). The primitiveness of the first condition is corroborated by the fact that Lower Cretaceous therians like *Pappotherium* and *Holoclemensia* display this state, as do all known Cretaceous eutherians (Kielan-Jaworowska *et al.* 1979). Other important features include the relative size of para- and metacones, and the presence of conules on the upper molars. Both *Alphadon* and *Peradectes* show subequal para- and metacones and possess conules, while most later didelphoids show a reduction of the paracone compared with the metacone and obsolescence of the conules. The presence of conules in the dasyurids is variable, as shown by Archer (1976a).

The anatomy of the molar teeth in Lower Cretaceous therians and in *Alphadon* and *Peradectes* indicates that a well-developed stylar shelf with a complete series of stylar cusps, but without the enlargement of the metastylar area, is the primitive condition. The shelf is retained in most later forms, which usually also show an enlarged metastylar area, but a few genera show a derived reduction of the shelf and stylar conules.

A well-developed paracrista united to the stylocone (stylar cusp B) is present in *Holoclemensia* and *Pappotherium*, a condition inherited from eupantotheres (Bown and Kraus 1979). This character-state is maintained in *Alphadon* and most later genera, although in some of them, and in *Albertatherium* (Fox 1971), and *Peradectes* (Crochet and Sigé 1983) the labial ending of the paracrista may lay a bit anterior to cusp B or between cusp A and B. Lack of this crest-connection is considered a derived state.

A large and broad protocone is found in *Alphadon* and *Peradectes*, whereas a less-developed protocone seems to be the common condition in *Pappotherium* and *Holoclemensia*. In this instance the inference of the primitive state is less certain: for Late Cenozoic didelphimorphs, with an anteriorly-posteriorly reduced protocone, it seems obvious that such a structure is associated

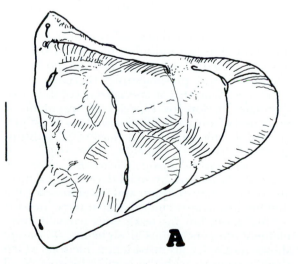

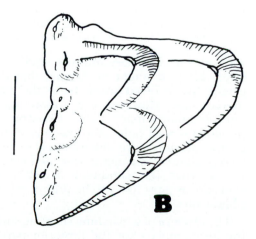

Fig. 2. Occlusal views of upper molars of A. *Peradectes* sp.; B., *Metachirus nudicaudatus.* Scale = 4 mm. (OAR).

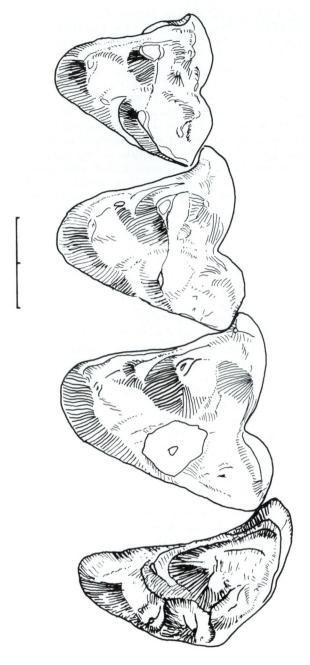

Fig. 3. Alphadon rhaister, UCMP 50292 (type), left M^{2-4};
AMNH 59513, a left M^5 (composed after Clemens
1966, by OAR). Both in occlusal view. Scale = 2 mm.

with advanced carnassialization, itself a derived
state. Thus, it might be argued that the out-
group has evolved away from the primitive
therian state. However, it is a more reasonable
interpretation that Late Cenozoic didelphimorphs with reduced protocones converged
on the primitive pre-didelphimorph condition,
as represented by *Pappotherium* and *Holoclemensia*, so that the condition in *Alphadon* and
Peradectes is most likely to be primitive for didelphimorphs.

The absence of a matching shearing surface on
the lower molars for the posterior part of the
didelphimorph last upper molar is the reason

why the latter is reduced in size compared with
more anterior molars. This reduction in size is a
derived state associated with increasing carnassialization. *Alphadon* and *Peradectes* (Figs 2, 3),
however, show a normal amount of reduction in
size of the M^5, indicating a lack of advanced carnassialization, a condition considered plesiomorphic. Advanced carnassialization is also
expressed by a reduction of the talonid of M_5 and
we judge the condition of long talonids found in
Alphadon, Peradectes, and several other genera not
showing carnassial adaptations as the primitive
state. This judgement is corroborated by the
occurrence of long talonids in Lower Cretaceous
therians. At the same time, a trigonid broader
than long, with subequal para- and metaconids —
a condition here considered primitive — is found
in *Pappotherium, Holoclemensia*, and in *Alphadon,
Peradectes*, their allies, and in a few later didelphimorphs. Reduction of metaconids, lengthening of the trigonid, and enlargement of the
paracristid connecting paraconid and protoconid
are also associated with carnassialization, and are
an indication of apomorphy.

These considerations provide a legitimate
point of departure for inferring polarity in
opossum molar teeth, and in making judgments
about symplesiomorphy or synapomorphy
among fossil and living genera.

Similarly, it seems likely that the primitive condition for upper premolars is of well-developed,
trenchant, narrow teeth increasing in size from
the first to the third. Tendencies toward reduction in the size of the first premolar and for
bulbousness in the third are here considered
derived. This is not necessarily in good keeping
with the presence of a rather bulbous P^3 in *Alphadon marshi* and *A. wilsoni* (cf. Lillegraven 1969:32,
Fig. 14 and p. 43, Fig. 20) and in other Upper
Cretaceous marsupials (e.g., *Eodelphis, Didelphodon*); but it seems to be a permissable inference on
the basis of sister-group comparisons (i.e., with
borhyaenoids and dasyuroids), and from the lack
of normal-sized P^3 in *Alphadon* (see Lillegraven
1969).

Less certain is the determination of polarity in
the states of upper and lower incisors, which are
missing from the known specimens of most early
marsupials and pre-marsupial therians. However, it is clear that a specialized condition of
spatulate incisors or especial enlargement or
reduction of any particular incisor may be judged
derived. In this context, Takahashi (1974) distinguished *Didelphis, Philander, Chironectes*, and
Lutreolina from *Caluromys, Monodelphis, Marmosa
(s.l.)*, and *Metachirus* by small details in the form
and relative sizes of upper and lower incisors. We
find his distinctions a bit subjective and in need of
confirmation, but his work does partially complement our conclusions given below. *Dromiciops*
shows mesio-distally broadened incisors which
sharply contrast with the peglike incisors of all
living didelphids. The former condition is most

Table 1. List and brief description of characters used to generate the data of Table 2. Character-pairs 6 and 7, 25 and 26, 28 and 29, and 34 and 35 represent branching features. See CHARACTER ANALYSIS for further discussion.

No. and name of character	Primitive (plesiomorphic) state	Derived (apomorphic) states(s)
1. Centrocrista shape	centrocrista rectilinear (predilamb-dodont) (0)	centrocrista V-shaped (didelphoid dilamb-dodont) (1)
2. Presence/absence, M^x para- and metaconules	M^x para- and/or metaconules present (0)	M^x para- and/or metaconules absent (1)
3. Relative sizes, M^3 and M^4	M^3 longer or subequal in length to M^4 (0)	M^3 shorter than M^4 (1)
4. Relative lengths, M^4 borders	outer border of M^4 subequal or longer than anterior border (0)	outer border of M^4 noticeably shorter than anterior border (1)
5. Relative sizes, M^x para- and metacones	M^x with subequal, separated para- and metacones (0)	M^x with separated paracone smaller (1), or with approximated paracone smaller (2) than metacone
6. Stylar shelf size and cusps I.	stylar shelf with usually five regular-sized stylar cusps (0)	stylar shelf narrow with reduced cusps (1), or with very much reduced cusps (2)
7. Stylar shelf size and cusps II.	stylar shelf with usually five regular-sized stylar cusps (0)	stylar shelf broadened, with enlarged B and C (or D) stylar cusps (1)
8. Union, M^x paracrista and either stylar cusps A or B	paracrista of M^x united to stylar cusps (0)	paracrista not united to stylar cusps (1)
9. Development, M^x metastylar spur and cristae	metastylar spur and metacrista of M^x moderately developed, metacrista not longer than postprotocrista (0)	spur and metacrista well-developed, metacrista longer than postprotocrista (1), or spur and metacrista highly developed, especially in M^4 (2)
10. Shape, M^x protocone	protocone of M^x bulbous, not compressed anteroposteriorly (0)	protocone moderately compressed (1), or protocone highly compressed (2)
11. Relative sizes, M^4 and M^5	M^5 reduced in length as regards M^4 but width of M^5 as large as or subequal to width of M^4 (0)	M^5 reduced in size and width as regards M^4 (1), or very reduced in size and width (2)
12. Shape and relative development of M^5 para- and metacone	M^5 subtriangular, its length more than half its width, with subequal para- and metacones (0)	M^5 moderately compressed, its length about half its width, with metacone moderately smaller than paracone (1), or M^5 compressed anteroposteriorly, its length less than half its width and metacone noticeably reduced (2), or M^5 much compressed, its length about 2/5 of its width, with metacone highly reduced or absent (3)
13. Shapes, M_3-M_5 trigonids	trigonids of M_3-M_5 wider than long (0)	trigonids of M_3-M_5 as wide as long (1), or longer than wide (2)
14. Relative sizes and shapes, M_4 and M_5	M_5 subequal in length or longer than M_4 (0)	M_5 narrower and moderately shorter than M_4 (1), or M_5 markedly shorter and narrower than M_4 (2)
15. Relative lengths, M_x talonid and trigonid	talonid of M_x longer than trigonid (0)	talonid of M_x shorter than trigonid (1)
16. Relative sizes and shapes, M_4 and M_5 talonids	talonid of M_5 as that of M_4 in size and shape (0)	talonid of M_5 reduced and narrower than that of M_4 (1)
17. Relative sizes, M_x para- and metaconids	metaconid of M_x subequal to or slightly reduced relative to paraconid (0)	metaconid of M_x noticeably reduced (1)
18. Development, M_x protoconid and paraconids	protoconid of M_x moderately larger and higher than paraconid (0)	protoconid much higher than paraconid (1), or both protoconid and paraconid high (2)
19. Development, M_x anterior cingulum	M_x with anterior cingulum reduced or moderately developed (0)	M_x with anterior cingulum well-developed (1), or highly developed (2)

Table 1. Continued.

No. and name of character	Primitive (plesiomorphic) state	Derived (apomorphic) state(s)
20. Shape, I^x	I^x normal (0)	I^x spatulate (1)
21. Size and shape, Px	Px normally sized, usually compressed anteroposteriorly (0)	Px enlarged and broadened (1)
22. Size, P^1	P^1 of normal size (0)	P^1 reduced in size (1)
23. Relative sizes, P2 and P3	P3 longer than P2 (0)	P3 shorter than P2 (1)
24. Shape, P^3	P^3 normal shape (0)	P^3 bulbous (1)
25. Body size I	body size medium (0)	body size large (1), or very large (2)
26. Body size II	body size medium (0)	body size small (1)
27. Presence/absence, palatal vacuities	palatal vacuities present (0)	palatal vacuities absent (1)
28. Otic region I	otic region with alisphenoid and perioticum not or little expanded (0)	"open tripartite bulla" with alisphenoid and perioticum moderately expanded and ectotympanic contributing the floor of tympanic cavity (1), or "bipartite bulla" with alisphenoid and perioticum closing the floor of tympanic cavity (2), or "closed tripartite bulla" with alisphenoid, pars petrosal, and pars mastoidea united to form a complete bulla (3), or "complete alisphenoid bulla" with alisphenoid enormously expanded and contributing alone to form complete bulla (4)
29. Otic region II	otic region as in state 1 for character 28 (0)	otic region with increasing inflation of alisphenoid, but still incomplete tripartite bulla (1)
30. Size, paraoccipital process	paroccipital process long (0)	paroccipital process reduced or absent (1)
31. Size, ectotympanic process	ectotympanic without process (0)	ectotympanic with process of increasing length (1, 2, 3)
32. Size, pars mastoidea	pars mastoidea normal (0)	pars mastoidea expanded (1)
33. Presence/absence, sinus epitympanicus	sinus epitympanicus absent (0)	sinus epitympanicus present (1), or sinus highly developed (2)
34. Serology I	antigens as shared by all didelphoids (0)	antigens as shared by all except *Ancestor* (1), by all didelphids (2), by all didelphids except Caluromyinae (3), by *Chironectes*, *Didelphis*, and *Philander* (4), or by *Didelphis* and *Philander* alone (5)
35. Serology II	antigens as in state 3 for character 34 (0)	antigens as specially shared by *Micoures*, *Marmosa*, *Thylamys*, and *Monodelphis* (1)
36. Diploid chromosome number	karyotype 2n=14 (0)	karyotype 2n=18 (1), or 2n=22 (2)
37. Lateral vaginal canals	short, simple loops (0)	large, U-shaped loops (1), or long, complex loops (2)
38. Median vaginal septum	complete (0)	incomplete (1), or absent (2)
39. Lining of median vagina	smooth (0)	folded (1)
40. Pseudo-vaginal canal	absent (0)	rudimentary (1), or present (2)
41. Anterior prolongation of median vagina	absent (0)	present (1)
42. Position of mammae	abdominal, inguinal, and pectoral (0)	abdominal and inguinal (1), or abdominal only (2)
43. Number of teats	greater than 9 (0)	less than 9 (1)
44. Presence/absence, pouch	pouch absent (0)	pouch present (1)
45. Sperm morphology	as for didelphids generally (0)	as for *Caluromys* (1)

Table 2. Data used in computing Wagner trees discussed herein; character-states are defined in Table 1. Underlined scores were inferred from incomplete or ambiguous material, and blanks indicate that scoring was not done on those taxa. All computations were performed on data read into the program in the order shown, by columns from left to right.

Characters	Thylatheridium	Monodelphis	Lestodelphys	Thylamys	Marmosa	Micoureus	Chironectes	Hondadelphys	Hyperdidelphis	Didelphis	Philander	Thylophorops	Lutreolina	Metachirus	Caluromys	Caluromysiops	Glironia	Borhyaena	Arminiheringia	Lycopsis	Sipalocyon	Patene	Sparassocynus	Dromiciops	Microbiotherium	Caroloameghinia	Glasbius	Pediomys	Peratherium	Didelphodon	Peradectes	Alphadon	Ancestor
1. Centrocrista shape	1	1	1	1	1	1	1	0	1	1	1	1	1	1	1	1	1	1	1	1	1	1	1	0	0	0	0	0	1	0	0	0	0
2. Presence/absence Mˣ para- and metaconules	1	1	1	1	0	0	0	0	1	1	1	0	1	1	1	1	0	1	1	0	1	0	1	0	0	0	0	0	0	0	0	0	0
3. Relative sizes, M³ and M⁴	1	1	0	1	1	1	1	1	1	1	1	0	1	0	0	0	0	0	0	0	0	0	0	0	0	0	0	1	0	1	1	1	0
4. Relative lengths, M⁴ borders	1	1	2	2	1	1	0	1	1	1	0	2	2	1	0	0	1	0	2	2	2	2	2	2	2	0	1	2	1	1	1	1	0
5. Relative sizes, Mˣ para- and metacones	2	2	2	2	1	1	1	1	1	1	1	2	2	0	0	0	0	2	2	2	2	2	2	2	2	0	0	2	0	0	0	0	0
6. Stylar shelf size and cusps I.	1	0	1	1	0	0	0	1	0	0	0	2	1	0	0	0	0	2	2	0	0	1	2	0	0	0	1	0	0	1	0	0	0
7. Stylar shelf size and cusps II.	0	0	0	0	1	1	0	1	0	0	0	0	0	0	0	0	0	2	2	0	0	0	0	0	2	0	0	2	0	0	0	0	0
8. Union, Mˣ paracrista and stylocone	0	0	0	0	0	0	0	0	0	0	0	0	0	0	0	0	0	1	1	0	0	0	0	0	0	0	0	0	0	0	0	0	0
9. Development, Mˣ metastylar spur and cristae	2	1	2	2	1	1	0	0	2	0	0	2	2	0	0	0	0	2	2	2	2	2	2	2	2	0	1	2	0	2	0	0	0
10. Shape, Mˣ protocone	2	1	2	2	1	1	0	2	1	0	0	1	1	0	0	0	0	2	2	1	1	1	0	2	0	0	0	0	0	0	0	0	0
11. Relative sizes, M⁴ and M⁵	0	0	2	2	2	2	1	0	3	0	0	3	3	1	0	0	0	2	2	2	2	2	3	0	1	0	2	2	1	0	0	0	0
12. Shape and relative development, M⁵ para- and metacone	3	3	3	2	2	2	0	2	2	2	2	3	2	0	0	0	0	3	3	3	3	2	2	2	0	0	0	0	0	0	0	0	0
13. Shapes, M₃-M₅ trigonids	0	0	0	2	1	1	0	1	0	0	0	0	1	0	0	0	0	2	2	2	2	2	0	2	2	0	0	0	0	0	0	0	0
14. Relative sizes and shapes, M₄ and M₅	1	1	1	1	1	1	1	0	1	0	0	0	0	0	0	0	0	1	1	1	0	1	0	0	1	0	1	1	1	0	0	1	0
15. Relative lengths, Mₓ talonid and trigonid	0	1	0	0	0	0	0	0	0	0	0	1	0	0	0	0	0	1	1	1	1	1	0	2	2	0	0	1	1	0	0	0	0
16. Relative sizes and shapes, M₄ and M₅ talonids	0	0	0	0	0	0	0	0	0	0	0	0	0	0	0	0	0	1	1	0	0	0	0	0	1	0	0	0	1	0	0	1	0
17. Relative sizes, Mₓ para- and metaconids	1	1	1	2	2	2	1	0	1	0	0	2	2	0	0	0	0	1	1	0	0	0	0	0	3	0	0	0	0	2	1	1	0
18. Development, Mₓ proto- and paraconids	0	2	0	0	0	0	0	0	0	0	0	0	0	0	2	2	0	1	1	0	0	0	0	1	3	0	0	0	1	0	0	0	0
19. Development, Mₓ anterior cingulum	0	1	0	0	0	0	0	0	0	0	0	0	0	0	0	0	0	0	0	0	0	0	0	0	0	0	0	0	0	1	0	0	0
20. Shape, Iˣ	0	0	0	0	0	0	0	0	0	0	0	0	0	0	0	0	0	0	0	0	0	0	0	1	0	0	0	0	0	1	0	0	0
21. Size and shape, Px	0	0	0	0	0	0	0	0	0	0	0	0	0	0	0	0	0	0	0	0	0	0	0	0	0	0	0	0	0	0	0	0	0
22. Size, P¹	1	1	1	1	1	1	1	1	1	1	1	1	1	1	0	0	0	1	1	1	1	1	1	0	0	0	0	0	0	0	0	1	0
23. Relative sizes, P2 and P3	0	0	0	0	0	0	0	0	0	0	0	0	0	0	1	1	1	0	0	0	0	0	0	1	0	0	0	0	0	0	0	0	0
24. Shape, P³	1	1	1	2	2	2	1	0	1	1	1	2	2	1	0	0	0	2	2	2	2	2	0	0	1	0	0	0	0	0	1	0	0
25. Body size I.	0	1	1	1	1	1	1	1	1	1	1	1	1	1	1	1	0	1	1	1	1	1	1	1	1	1	1	1	1	1	1	0	0
26. Body size II.	0	0	0	0	0	0	0	0	0	0	0	0	0	0	0	0	0	0	0	0	0	0	0	0	0	0	0	0	0	0	0	0	0
27. Presence/absence, palatal vacuities	1	1	1	1	1	1	1	0	1	1	1	1	1	1	1	1	1	0	0	0	0	0	0	1	1	0	0	1	1	1	1	1	0
28. Otic region I.	0	1	1	1	1	1	1	0	0	0	0	0	1	1	0	0	0	0	0	0	0	0	0	0	1	0	0	0	0	0	0	0	0
29. Otic region II.	1	2	1	2	2	2	0	1	0	0	0	1	0	0	2	2	0	0	0	0	0	0	0	3	3	0	0	0	0	0	0	0	0
30. Size, paroccipital process	1	0	0	0	0	0	0	0	0	0	0	0	0	0	0	0	0	0	0	0	0	0	0	1	0	0	0	0	0	1	0	0	0
31. Size, ectotympanic process	0	0	0	0	0	0	0	0	0	0	0	0	0	0	0	0	0	0	0	0	0	0	0	0	1	0	0	0	0	0	0	0	0
32. Size, pars mastoidea	0	0	0	0	0	0	0	0	0	0	0	0	0	0	0	0	0	0	0	0	0	0	0	1	0	0	0	0	0	0	0	0	0
33. Presence/absence, sinus epitympanicus	0	0	0	0	0	0	0	0	0	0	0	0	0	0	0	0	0	0	0	0	0	0	1	1	1	1	0	0	0	0	0	0	0
34. Serology I.	0	0	0	0	3	3	4	0	0	5	5	0	4	3	2	2	2	0	0	0	0	0	4	0	3	0	0	0	0	0	0	0	0
35. Serology II.	0	1	1	3	1	1	0	0	0	0	0	0	0	2	0	0	0	0	0	0	0	0	0	1	0	0	0	0	0	0	0	0	0
36. Diploid chromosome number	0	0	0	0	2	2	2	0	2	2	2	0	2	3	0	0	0	0	0	0	0	0	0	0	0	0	0	0	0	0	0	0	0
37. Lateral vaginal canals	0	1	0	1	1	1	1	0	0	0	0	0	0	1	1	1	0	0	0	0	0	0	1	2	2	0	0	0	0	0	0	0	0
38. Median vaginal septum	0	0	0	0	0	0	0	0	0	0	0	0	0	0	0	0	0	0	0	0	0	0	0	0	0	0	0	0	0	0	0	0	0
39. Lining of median vagina	0	0	0	0	0	0	0	0	0	0	0	0	0	0	0	0	0	0	0	0	0	0	0	1	0	0	0	0	0	0	0	0	0
40. Pseudo-vaginal canal	1	1	1	1	1	1	1	0	1	1	1	1	1	1	1	1	1	0	0	0	0	0	0	2	2	0	0	0	0	0	0	0	0
41. Anterior prolongation of median vagina	0	0	0	0	0	0	0	0	0	0	0	0	0	0	1	1	0	0	0	0	0	0	0	0	0	0	0	0	0	0	0	0	0
42. Position of mammae	0	0	0	0	2	2	2	1	2	2	2	0	0	2	2	2	2	0	0	0	0	0	0	2	0	0	0	0	0	2	0	0	0
43. Number of teats	0	0	0	0	0	0	1	0	1	1	1	1	1	1	1	1	1	0	0	0	0	0	0	1	1	0	0	0	0	0	0	0	0
44. Presence/absence, pouch	0	0	0	0	0	0	1	0	0	0	0	0	0	0	1	1	0	0	0	0	0	0	0	1	1	0	0	0	0	0	0	0	0
45. Sperm morphology	0	0	0	0	0	0	0	0	0	0	0	0	0	0	1	1	0	0	0	0	0	0	0	0	0	0	0	0	0	0	0	0	0

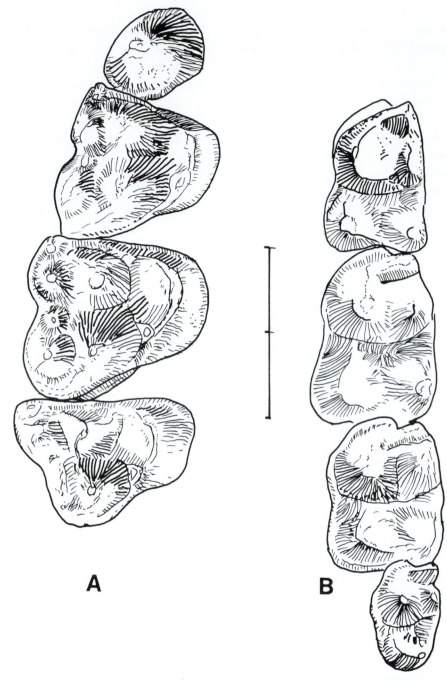

Fig. 4. Glasbius intricatus, A. Right P³-M⁴ (UMVP 1593). B. Left M₂₋₅ (AMNH 58736). Both in occlusal view (redrawn and composed after Clemens 1966, by OAR). Scale = 2 mm.

probably a derived character-state, which can be thought of as an adaptation towards functionally improved grasping in small insectivorous animals. Szalay (1982a) pointed out that dasyurids and peramelids share this condition with *Dromiciops*, and gave a particularly strong phylogenetic and taxonomic weight to the resemblance.

Table 1 includes descriptions of the 24 dental characters we studied, and their inferred primitive and derived states. Table 2 summarizes the scoring of dental and other character-states in the 27 opossum and five borhyaenoid genera

examined. A hypothetical totally primitive ancestral genus is included, from which various departures are shown in the different didelphimorphs. These departures we interpret as reflecting various feeding adaptations which evolved from an omnivorous ancestral condition toward more frugivorous, insectivorous, or carnivorous habits.

Microbiotheres, as represented by *Dromiciops* (Figs 5, 6) and *Microbiotherium* (Figs 7 and 8B), are the only taxa excepting *Hondadelphys* which agree with the Late Cretaceous genera *Alphadon*, *Pediomys* (Fig. 8A), *Glasbius*, *Didelphodon* (and

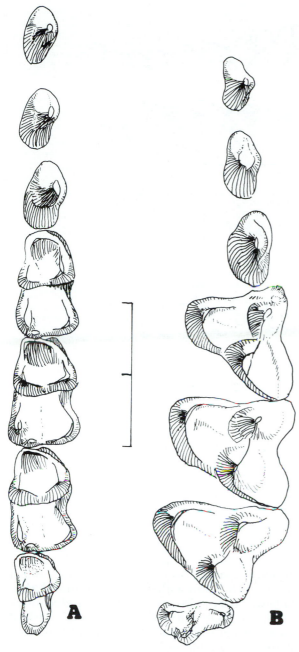

Fig. 5. Dromiciops australis, MMP. [1233] A. Right P_1-M_5. B. Left P^1-M^5. Both in occlusal view. Scale = 2 mm. (OAR).

Eodelphis), the Late Cretaceous-Eocene *Peradectes*, the Eocene *Mimoperadectes*, the Early Oligocene *Alloeodectes*, and the Late cretaceous-Early Eocene *Roberthoffstetteria*, *Procaroloameghinia*, and *Caroloameghinia* (Fig. 9) in retaining the primitive rectilinear centrocrista, this condition being obscured in the latter by advanced bunodonty; all other didelphimorph taxa, as well as all known dasyuroids (Archer 1976c) and borhyaenoids, have, to a greater or lesser degree, adopted the familiar V-shaped approximation of this crista to the stylar shelf (but the Australian Miocene "dasyurid" genus *Ankotarinja*, as judged from Archer's illustration, has a rectilinear centrocrista, this condition together with others being the basis of our belief that it is better

considered as a microbiotherioid, as explained below). This is true of *Glironia* (Figs 10A, 11A-C), *Caluromys* (Figs 10B, 12A-C), and *Caluromysiops*, which in other respects (discussed below) share with *Alphadon* and allies, and with *Microbiotherium* and *Dromiciops*, several plesiomorphic character-states, either as a group or in different subgroup combinations. *Dromiciops*, *Microbiotherium*, *Glironia*, and *Caluromys* are nonetheless synapomorphous in their common reduction of the stylar shelf and stylar cusps, though a group synapomorphous for paracrista not united to stylocone would join *Dromiciops*, *Microbiotherium*, *Caluromysiops*, and *Caluromys*, but would not include *Glironia*. *Caluromysiops* has relatively well-developed stylar cusps and a rather wide stylar shelf, although it has, as do the other genera, a reduced metacrista. *Dromiciops*, *Microbiotherium*, *Caluromysiops*, and *Caluromys* are also united in that the outer borders of their M^4's are shorter than the anterior borders, and in the M_5's being much smaller than the M_4's, two characters in which *Glironia* maintains the plesiomorphic states. The reduction of the M_5's in *Dromiciops* and *Microbiotherium* is more advanced than in *Caluromys* and *Caluromysiops*.

These comparisons suggest a closer relatedness between *Dromiciops*, *Microbiotherium*, *Caluromys*, and *Caluromysiops* than of any of these with *Glironia*. However, *Dromiciops* stands apart from the other three genera in the structure of its upper incisors (Fig. 6A-C), which are larger, spatulate, and form a continuous series. *Dromiciops* (Fig. 6A-C) and *Microbiotherium* (Fig. 7A-C) also have simple premolars, transversely compressed and continous in position. *Glironia* (Fig. 11A-C), *Caluromys* (Fig. 12A-C), and *Caluromysiops* have unmodified upper incisors, the first pair being separated from the succeeding ones by a short diastema. All three genera share the derived condition of a reduced P^1 (Fig. 10A-C; Fig. 12) separated from the P^2 by a diastema. *Glironia* has simple P^2's and P^3's, but those of *Caluromys* and *Caluromysiops* are bulbous (Fig. 10C).

Caluromys and *Caluromysiops* are also synapomorphously united by a lack of conules in the upper molars, but *Glironia* has small metaconules (Fig. 10A). Though *Dromiciops* also lacks conules, its nearest relative *Microbiotherium* shows distinctive conules (Figs 7A-C, 8B; Marshall 1982a), albeit ones reduced in size. In other respects *Dromiciops* is almost indistinguishable from *Microbiotherium* (Marshall 1982a). *Caluromys*, *Caluromysiops*, and *Glironia* have also been alleged to be microbiotheres on the basis of molar structure (Reig 1955), but this contention has been rebutted by Segall (1969a) on the basis of basicranial structure (see below). As noted above, many of the characters shared by *Dromiciops*, *Microbiotherium*, *Glironia*, *Caluromys*, and *Caluromysiops* are plesiomorphic and are thus not indicative of cladistic affinity. Additionally,

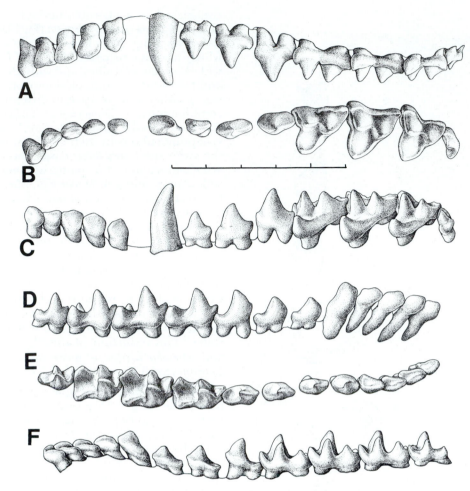

Fig. 6. Dromiciops australis, FMNH 22671, A-C, left I^1-M^5: A. Labial. B. Occlusal. C. Lingual
views. D-F. Right I_1-M_5: D. Labial. E. Occlusal. F. Lingual views. Scale = 5 mm. (EL).

the apomorphic character-states of *Dromiciops*
and *Caluromysiops* are not shared by the three
remaining genera as a whole, and the states of
their incisor and premolar anatomy are not
shared by any of the three. These data suggest
that a great deal of parallelism and mosaicism
occurred independently in the evolution of
Glironia, *Caluromys*, and *Caluromysiops* on the one
hand, and in that of *Dromiciops* and *Microbiotherium* on the other. At the same time,
the Late Oligocene (Colhuehuapian) genus
Pachybiotherium (Fig. 13) is structurally more
similar to *Glironia* and *Caluromys* than to *Microbiotherium* (Reig 1981).

Pediomys (Figs 8A, 14A), *Glasbius* (Fig 4), *Didelphodon* (and *Eodelphis*), and *Caroloameghinia* (Figs
9, 15) share with *Microbiotherium*, *Dromiciops*,
Alphadon, and *Peradectes*, *Mimoperadectes* and
Alloeodectes several primitive characters, including similar-sized para- and metacones (except in
Didelphodon), an undeveloped or little-developed
metastylar spur, presence of well-developed
conules (except in *Dromiciops*), unreduced protocones, and uncompressed trigonids with unreduced metaconids. These same conditions are
also present, surprisingly and probably significantly, in the Miocene "dasyurid" *Ankotarinja*

(Archer 1976c), reinforcing our belief that this
genus is not a dasyurid but a microbiotheriid (see
later). *Glasbius* is autapomorphous in its greatly
developed stylar cones and stylar shelf, but it
shares with microbiotheres much-reduced M^5's.
Pediomys shares with microbiotheres reduction of
the stylar shelf, reduced size of the M^4, lack of
union of paracone and stylocone, and size and
shape of trigonids and talonids. *Caroloameghinia*
approaches *Glasbius* in bunodont development of
molar cusps, but is more primitive in having a
normal stylar shelf and unreduced M_5's. *Didelphodon* (see Clemens 1979, Figs 11.5 and 11.6) and
Eodelphis are distinctive among this group in their
large size, crushing premolars, unreduced
stylocones not connected with the paracrista,
relatively small paracones, high shearing paracristid, and reduced metaconids.

The remaining genera are united by common
possession of various apomorphic character-
states (e.g., a full didelphoid dilambdodonty,
again excepting *Hondadelphys*; reduction of
paracone relative to metacone; enlargement of
metastylar spur and metacrista; non-bulbous,
more or less compressed protocones; absence of
paraconule and metaconule; talonids shorter
than trigonids). The more primitive situation

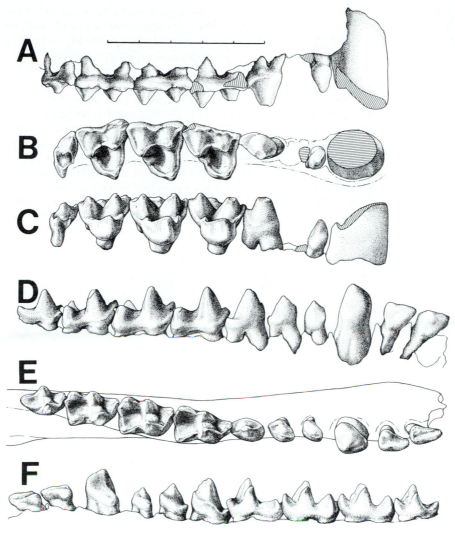

Fig. 7. A-C. *Microbiotherium tehuelcum*, MLP-58-IX-3-4, a right upper jaw with base of C, P¹ complete, roots of P², and with P³-M⁵ complete: A. Labial. B. Occlusal. C. Lingual views. D-F. *Microbiotherium praecursor*, MACN 10799 (type), a right mandibular ramus with alveoli of I₁₋₂ and I₃-M₅ complete: D. Labial. E. Occlusal. F. Lingual views. Scale = 5 mm. (EL).

within genera sharing these character-states seems to be represented by *Metachirus* (Figs 16C-D, 17), *Philander* (Figs 16A-B, 18), and *Micoures* (Fig. 19), which have uncompressed, subtriangular, and unreduced M⁵'s, unbroadened M⁴'s, and several other plesiomorphic character-states (see Table 2). Character-states of *Metachirus* and *Philander* on the one hand, and of *Micoures* on the other, may represent the starting-points of several primitive-to-derived sequences within this group of genera. One of the sequences is represented by character-states in *Micoures-Marmosa-Thylamys-Lestodelphys* (Figs 19, 20, 21, 22, 23C-D); another by character-states in *Micoures-Monodelphis-Thylatheridium* (Figs 19, 23A-B, 24, 25, 26). Both sequences, from *Marmosa* or *Monodelphis* onwards, show the apomorphic conditions of approximation of para- and metacones, well-developed metacrista, anteroposteriorly compressed M⁵'s, and

broadened M⁴'s. These character-states determine sharper and longer cutting edges and less-developed crushing surfaces, in correlation with carnivorous habits. They also occur in *Lutreolina* (Fig. 27), *Thylophorops* (Fig. 28), and most markedly in *Sparassocynus* (Fig. 29), which probably evolved them independently. *Thylamys* (Fig. 20C) and *Lestodelphys* (Fig. 23C-D) show the highly derived condition of upper molars markedly wider than long, resulting from the linguad expansion of the protocone and the more laterally projecting metacrista; and compressed M⁵'s with the metacone being highly reduced or absent. *Monodelphis* and *Thylatheridium*, although showing laterally expanded upper molars, are less advanced in this direction and have less-compressed M⁵'s, with reduced but distinct metacones. *Monodelphis* and *Thylatheridium* have a high paraconid, which contributes to a sharp and elevated paracristid, whereas *Marmosa*, *Thylamys*,

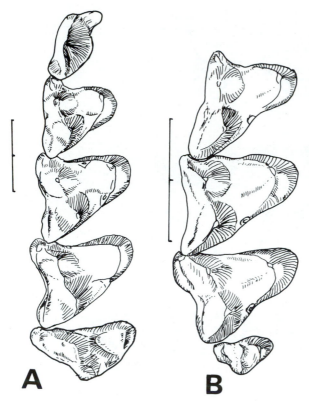

Fig. 8. Occlusal views of right upper cheek teeth of: A. *Pediomys cooki*, composed from UCMP 47738 (type), P³-M⁴, and UCMP 46911, M⁵ (redrawn after Clemens 1966). B. *Microbiotherium patagonicum*, MACN 8471, M²⁻⁵ (original). Scale = 2 mm. (OAR).

and *Lestodelphys* maintain a paraconid as low as the metaconid. *Lestodelphys*, *Monodelphis*, and *Thylatheridium* share the apomorphic condition of a reduced and narrow talonid on the M_5, which remains a plesiomorphic feature in *Micoures*, *Marmosa*, and *Thylamys*.

Polarity is more difficult to establish among the "large body-sized" group of opossums. *Metachirus* (Figs 16C-D, 17) and *Philander* (Figs 16A-B, 18) appear to be the most primitive, but differ from each other primarily in the lower molars. *Metachirus* uniquely among didelphines maintains the primitive condition of a trigonid wider than long, with a convex lateral surface, but shows a narrower M_5 which is moderately shorter than M_4. Relative to *Metachirus* and *Philander*, *Didelphis* (Figs 30, 31) and *Chironectes* (Figs 30, 31) have larger, more bulbous molars and moderately compressed M⁵'s with a reduced metacone. A peculiar feature of *Chironectes* is the lack of a connection of the paracrista with either stylar cusps A or B, and the marked development of stylar cusps B, C, and D, the first being even higher than the paracone. *Hyperdidelphis* (Fig. 30B) differs from *Chironectes* in these respects, though these genera are similar in the general structure of their molar teeth. *Hyperdidelphis* has, however, more advanced carnassial specializations (e.g., more-compressed protocones; better-

developed metacrista on M⁴; more-compressed M⁵; developed protoconid; and, in some species, a reduced talonid on M_5 — see Fig. 33B). These comparisons suggest a three-way evolution from a *Philander*-like ancestor to *Didelphis*, *Chironectes*, and *Hyperdidelphis*. Another sequence may be represented by *Metachirus-Philander* toward *Thylophorops* but bypassing *Lutreolina*. The latter two genera share several apomorphies and parallel some of the modifications found in *Lestodelphys* and *Thylatheridium* (e.g., paracone and metacone are approximated on M³ and M⁴; protocone is highly compressed on M³⁻⁵; M⁵ is wide and blade-like, with a very reduced metacone; protoconids are high; metaconids are reduced

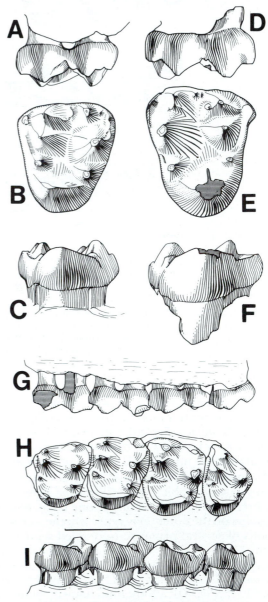

Fig. 9. A-F. *Caroloameghinia mater*, AMNH 28441a (A-C), and isolated left M³: A. Labial. B. Occlusal. C. Lingual views. AMNH 28441b (D-F), and isolated right M⁴: D. Labial. E. Occlusal. F. Lingual views. G-I. *Caroloameghinia tenuis*, MACN 10345 (type), a fragment of a left maxillary with M²⁻⁵ complete: G. Labial. H. Occlusal. I. Lingual views. Scale = 2.5 mm. (MHW).

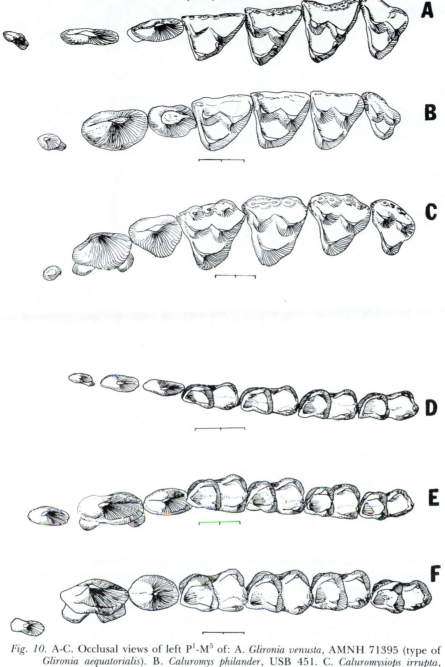

Fig. 10. A-C. Occlusal views of left P^1-M^5 of: A. *Glironia venusta*, AMNH 71395 (type of *Glironia aequatorialis*). B. *Caluromys philander*, USB 451. C. *Caluromysiops irrupta*, AMNH 208101. D-F. Right P_1-M_5 of: D. *Glironia venusta*, AMNH 71395 (type of *Glironia aequatorialis*). E. *Caluromys philander*, USB 451. F. *Caluromysiops irrupta*, AMNH 208101. (OAR).

relative to paraconids; and talonid of M_5 is narrower and shorter than that of M_4).

Hondadelphys (Fig. 34) belongs to the group of large-sized didelphines on the basis of most of its molar characters, with the exception of the straight centrocrista. *Hondadelphys* shares with *Lutreolina* and *Thylophorops* an elongated metastylar spur, but it has less-reduced protocones and more-reduced metaconids and stylar cusps.

The peculiar *Zygolestes* (Fig. 35) from the Pliocene of Argentina (see Reig 1957b) was not included in the above comparisons. Its phylogenetic relationships are obscure, and it is unique in the combination of reduced birooted P_3, buccally displaced mesial paraconid on M_2, and reduced M_5.

Sparassocynus (Fig. 29) has a highly modified dentition paralleling that of borhyaenoids and some carnivorous dasyuroids (Reig and Simpson 1972). These features place it quite apart from the Late Cenozoic and living didelphids considered here.

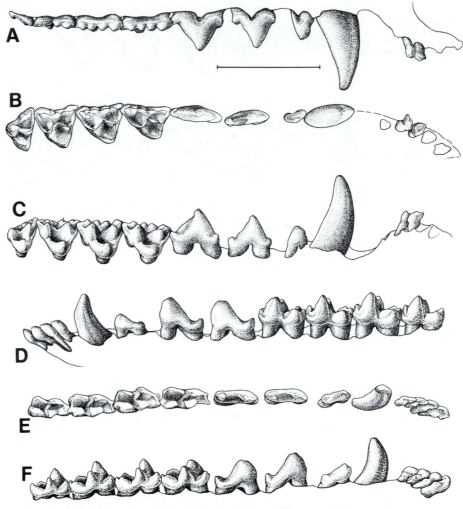

Fig. 11. Glironia venusta, FMNH 41440♀. A-C. Right I$^{2-3}$ and C-M^5: A. Labial. B. Occlusal. C. Lingual views. D-F. Left I$_1$-M$_5$: D. Labial. E. Occlusal. F. Lingual views. Scale = 5 mm. (EL).

The general picture of relationships that emerges from analysis of these dental character-states is presented in the Wagner trees of Figures 36 and 37. *Marmosa* and *Thylamys*, generally considered primitive didelphids (e.g., Bensley 1903; Gregory 1910:217), are relatively highly derived. This is evident after our placing the more primitive *cinerea* group in a distinct genus, *Micoures*, a conclusion corroborated by characters of the otic region of the skull discussed below. Also, *Metachirus* and *Philander*, which are so distinct cytologically (see under Karyotypes below), share many molar characters, a relatedness likewise indicated by details of the otic region.

MORPHOLOGY OF THE AUDITORY REGION

The auditory region of didelphids is discussed and comparatively described by Archer (1976b), van Kampen (1905), Reig and Simpson (1972), Segall (1969a and b), and Winge (1893, 1941). Archer gave a detailed analysis of species of *Didelphis, Marmosa, Monodelphis, Metachirus*, and

Caluromys, providing a basis for the following discussion, which is supplemented by unpublished notes of Reig covering species of genera not considered by Archer. Archer described the ear region of *Marmosa* as "typical" for didelphids, yet we conclude that *Marmosa* (along with *Thylamys* and *Lestodelphys*) has derived character-states in the otic region.

From out-group comparisons with thylacinids, borhyaenoids, and dasyurids some authors (Archer 1976b; van Kampen 1905) concluded that the anatomy of the ear region of didelphimorphs is primitive compared to these other marsupial groups. However, considerable variation exists within didelphimorphs and some lineages show modifications which developed independently in other marsupial taxa.

The primitive condition from which we infer that all other living and fossil opossums were derived is best approximated in species of *Metachirus* and *Philander*. In *Metachirus* (Fig. 38) the ossified bulla is small and incomplete, and much of the middle ear in the macerated skull is

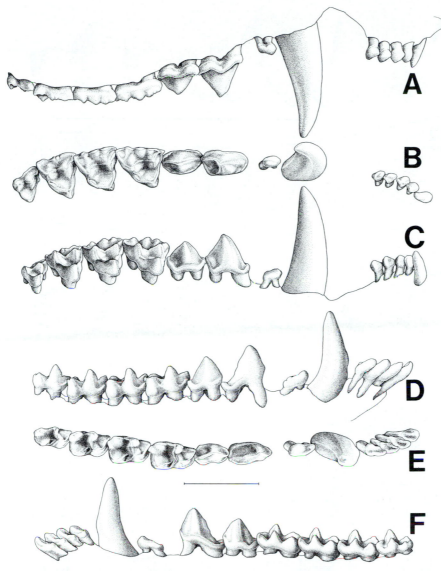

Fig. 12. Caluromys philander, FMNH 92030♀. A-C. Right I^1-M^5: A. Labial. B. Occlusal. C. Lingual views. D-F. Right I$_1$-M$_5$: D. Labial. E. Occlusal. F. Lingual views. Scale = 5 mm. (EL).

open ventrally. The tympanic wing of the alisphenoid contributes to formation of the bulla anteriorly, and posteriorly a tympanic process of the *pars petrosa* of the perioticum helps form the bullar wall. The tympanic wing of the alisphenoid is little inflated and merely represents a ledge of bone forming part of the anterior floor of the tympanic cavity. The periotic contribution to the wall is also very limited, and an open space extends between the two bones exposing the lower border of the ectotympanic. The perioticum is roughly triangular and is widely exposed ventrally, with the tip of the triangle reaching anteriorly to the basisphenoid. A relatively long paraoccipital process occurs behind the perioticum. The ectotympanic is narrow, horseshoe-shaped, lacks a tympanic process, and is visible from the lateral side along the whole extension of the thickened lateral rim. The relatively wide ectotympanic incisura opens

posterodorsally. There are no distinct epitympanic sinuses (van Kampen 1905; Reig and Simpson 1972:526).

From this condition minor modifications occur in *Didelphis, Hyperdidelphis, Lutreolina, Thylophorops,* and *Chironectes,* leading to a more complete ventral closure of the tympanic cavity and increasing inflation of the alisphenoid bulla. *Philander* (Fig. 39A-B) is essentially like *Metachirus* in this respect, but shows a less-inflated alisphenoid wing and a less well-developed tympanic process of the perioticum. In *Didelphis* (Fig. 39C-D) the condition is similar, but there is a larger ventral exposure of the mastoid part of the petrosum and the tympanic process of the alisphenoid is slightly larger. *Chironectes* (Fig. 40A-B) and *Hyperdidelphis* are similar but show an even better-developed tympanic wing of the alisphenoid which, even when

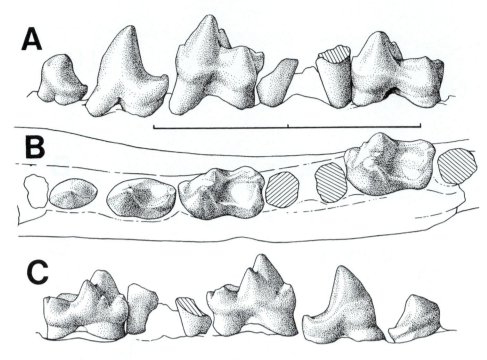

Fig. 13. *Pachybiotherium acclinum*, MACN 52-370a (type). A fragment of a left mandibular ramus with alveolus of C, roots of P_1 and M_3, anterior root and portion of posterior alveolus of M_5, and P_2-M_2 and M_4 complete: A. Labial. B. Occlusal. C. Lingual views. Scale = 10 mm. (EL).

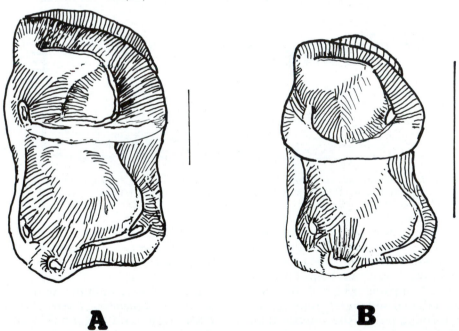

Fig. 14. A. *Pediomys hatcheri*, UCMP 3690, right M_4 (redrawn after Lillegraven 1969). B. *Dromiciops australis*, MMP 1233, right M_4. Both in occlusal view. Scales = 1 mm. (OAR).

not inflated, contributes to formation of the anterior half of the ventral floor of the tympanic cavity. In *Hyperdidelphis* the development of the perioticum is as in *Didelphis*, the tympanic process of the *pars petrosa* being rather wide and well-separated from the alisphenoid wing. In *Chironectes* this process is wider and projects anteriorly to a position very close to the tympanic wing of the alisphenoid, so that there is little open space in the floor of the tympanic cavity. In *Lutreolina* (Fig. 40C-D) and *Thylophorops* the periotica are very much alike, and are essentially as in *Didelphis;* but in them the tympanic wing of the alisphenoid is moderately inflated, forming a small swelling which enlarges the tympanic cavity. In *Hondadelphys* there is no evidence of an ossified auditory bulla, beyond the weakly developed tympanic process of the alisphenoid.

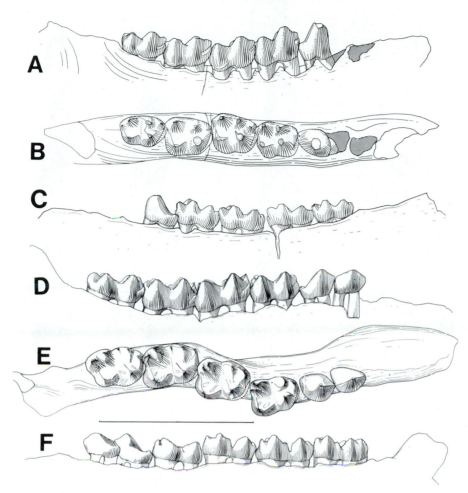

Fig. 15. A-C. *Procaroloameghinia pricei,* DGM 805-M (type), a fragment of a right mandibular ramus with alveoli of C-P$_2$ and with P$_3$-M$_3$ complete; and DGM 924-M, a fragment of a right mandibular ramus with M$_{4-5}$ complete (both specimens are of same individual): A. Labial. B. Occlusal. C. Lingual views. D-F. *Caroloameghinia mater,* MACN 10348, a nearly complete right mandibular ramus with P$_2$-M$_5$ complete: D. Labial. E. Occlusal. F. Lingual views (note, ramus is bent as result of preservation). Scale = 10 mm. (MHW).

In all the above genera the observed variations in the ear region are minor departures from the plesiomorphic condition typically represented in *Metachirus* and *Philander*. A major departure from the primitive state occurs in *Marmosa, Thylamys,* and *Lestodelphys.* In *Marmosa* (Fig. 41A-B) and *Thylamys* (Fig. 42A-B) the tympanic process of the alisphenoid is highly inflated and encloses the anterior half of the tympanic cavity, forming a hemispheric swelling which opens posteroexternally. The tympanic process of the *pars petrosa* is expanded and is well-separated from the posteroventral border of the alisphenoid bulla. The intermediate space is partially closed ventrally by a laterally expanded posterior end of the ectotympanic, which forms a tympanic process well-exposed on the ventral side of the ear region and contributes to the floor of the tympanic cavity. Therefore, there is a sort of "tripartite" bulla, more or less closed ventrally, with an anterior portion formed by the inflated alisphenoid wing, a medial portion formed by the tympanic process of the ectotympanic, and a posterior portion formed by the tympanic process of the *pars petrosa* of the perioticum. In *Marmosa* the alisphenoid is perforated by a *formen ovale* (through which the trigeminal or fifth cranial nerve exits the skull; see MacIntyre 1967:834) mesial to the anterior end of the tympanic wing; the lateral rim of the ectotympanic is thick and horseshoe-shaped, with the incisura in a posterodorsal position. In *Thylamys* the tympanic process of the ectotympanic is a little broader than in *Marmosa,* forming a larger enclosure of the tympanic cavity. *Lestodelphys* (Fig. 42C-D; Fig. 50) shows essentially the same character-states of the ear region as do *Marmosa* and *Thylamys.* However, in *Lestodelphys* the alisphenoid bulla is slightly more swollen and extends more posteriorly, so that the space between its posterior border and the anterior border of the tympanic process of the *pars petrosa* is shorter. Therefore, the tympanic process of the ectotympanic is less well-exposed ventrally. It is

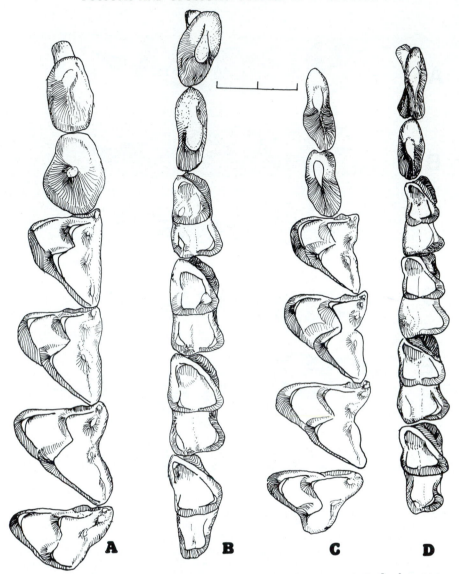

Fig. 16. Occlusal views of: A-B. *Philander opossum*, BMNH 3.3.4.51; A. Left P^2-M^5. B. Right
P$_2$-M$_5$. C-D. *Metachirus nudicaudatus*, MMP 20; C. Left P^2-M^5. D. Right P$_2$-M$_5$. Scale =
1 mm. (OAR).

also less developed, and does not reach the tympanic wing of the perioticum, leaving a small
open space in the ventral surface of the macerated skull. A *foramen ovale* is present in the same
position as in species of *Marmosa*.

Micoures (Fig. 43C-D), *Monodelphis* (Fig. 41C-
D), and *Thylatheridium* share characters of the ear
region, showing a combination of the features
found also in the *Metachirus*-like and the *Marmosa*-like groups of genera. The perioticum is
essentially as in *Metachirus*, with a poorly
developed and not laterally expanded tympanic
process of the *pars petrosa*. The perioticum is not
well-represented in the available specimens of
Thylatheridium, but the alisphenoid is similar to
that of *Monodelphis* and *Micoures*. In the latter two
genera, the tympanic wing of the alisphenoid is
swollen, forming a small hemisphere, but it is
much less developed than in *Marmosa, Thylamys,*
and *Lestodelphys*. According to Archer (1976b), a

foramen ovale is lacking, but a *foramen pseudovale*
occurs in its place. Due to the less-ventral extension of the tympanic wing of the alisphenoid and
the less-developed tympanic process of the *pars
petrosa*, a considerable space exists in-between
which is open mesially, but which is laterally
occupied by a well-exposed contribution of the
ectotympanic. This bone is similar in shape to the
ectotympanic of *Marmosa*, but it is narrower and
has a smaller tympanic process and an unthickened lateral rim. The stapes of *Monodelphis* are
imperforate (an autapomorphy), whereas in
Marmosa, in most other didelphids, and in most
marsupial groups, a stapedial foramen is present.

Micoures seems to represent a point of departure for two transformation series. One is conservative and leads to *Monodelphis* and
Thylatheridium. The other, by attainment of the
"tripartite" type of bulla and increased inflation
of the alisphenoid, is synapomorphous and

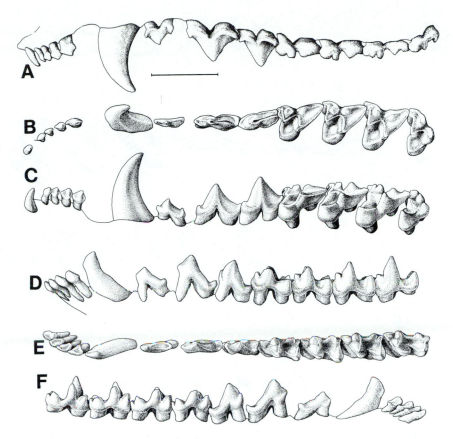

Fig. 17. Metachirus nudicaudatus, FMNH 94285♂. A-C. Left I^1-M^5: A. Labial B. Occlusal. C. Lingual views. D-F. Left I$_1$-M$_5$: D. Labial. E. Occlusal. F. Lingual views. Scale = 5 mm. (EL).

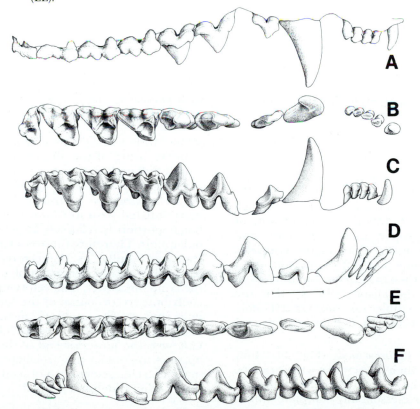

Fig. 18. Philander opossum, FMNH 114708♂. A-C, right I^1-M^5: A. Labial. B. Occlusal. C. Lingual views. D-F, right I$_1$-M$_5$: D. Labial. E. Occlusal. F. Lingual views. Scale = 5 mm. (EL).

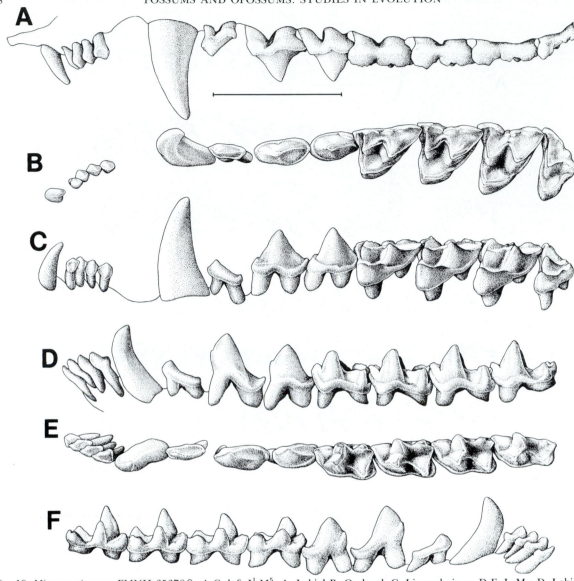

Fig. 19. Micoures cinereus, FMNH 65678♀. A-C, left I^1-M^5: A. Labial B. Occlusal. C. Lingual views. D-F, I$_1$-M$_5$: D. Labial. E. Occlusal. F. Lingual views. Scale = 5 mm. (EL).

follows a progression series from *Marmosa* to *Lestodelphys*, bypassing *Thylamys*.

A condition comparable to that of *Micoures* and *Monodelphis* occurs in *Glironia* (Fig. 43A-B). *Glironia* shows, however, a more advanced stage of bullar enclosure in that the tympanic wing of the alisphenoid projects more posteriorly, and contributes more to the floor of the tympanic cavity. The tympanic wing of the perioticum is better-developed than in *Micoures*, but is still well-separated from the alisphenoid, so that the ectotympanic is visible on the ventral side, though it does not show an expansion of its posterior end.

Caluromys and *Caluromysiops* (Fig. 44) show modifications in their basicrania which can be considered derived from the states found in *Glironia*. They have essentially complete bullae, with most of the ventral floor of the tympanic cavity covered by a tympanic wing of the alisphenoid and a tympanic process of the *pars*

petrosa, with a broad contact zone between. The tympanic wing of the alisphenoid is only moderately inflated; it forms a cone with the apex pointing ventrad and is somewhat swollen laterally. A *foramen ovale* is present. The perioticum is more inflated than in *Glironia*, and the periotic bullar portion is relatively larger than in other didelphids. Therefore there is a kind of complete "bipartite" bulla, as the ectotympanic is not exposed ventrally, being completely included in the tympanic cavity. The ectotympanic does not contribute to formation of the lower wall of the bulla nor to the auditory meatus (Reig and Simpson 1972:525). Archer (1976b) notes that in *Caluromys* the posterior end of the ectotympanic appears to contact the orbicular process of the malleus, a character-state not seen in *Glironia* nor in any other didelphid studied.

The observations of the bullar structure reported here agree with those of Segall (1969a and b), who reported that the auditory regions of

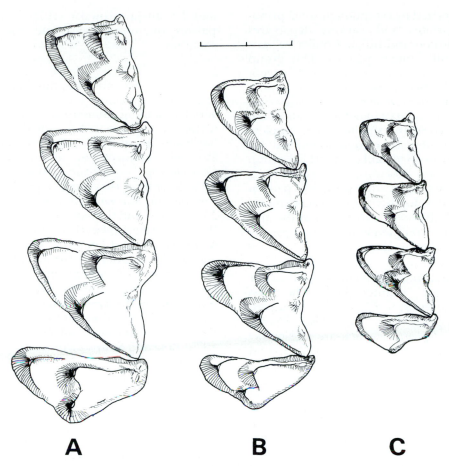

Fig. 20. Occlusal views of left M$^{2-5}$ of: A. *Micoures cinereus*, MACN 57761. B. *Marmosa fuscata*, MBUCV 1-1409. C. *Thylamys pusillus*, USB 447. Scale = 2 mm. (OAR).

Caluromys, *Caluromysiops*, and *Glironia* are more similar to each other than any of these is to *Didelphis* or other didelphines; but all of these genera are very distinct from *Dromiciops* and *Microbiotherium* (see below). "In fact, as far as the auditory region is concerned the similarities [of *Caluromys*, *Caluromysiops*, and *Glironia*] with *Didelphis* are so close that I cannot agree with Reig [1955] who considered them specialized Microbiotheriinae rather than Didelphinae" (Segall 1969a:497-498).

The most significant departures from the plesiomorphic ear region of *Metachirus* and *Philander* occur in *Dromiciops*, *Microbiotherium*, and *Sparassocynus*. *Dromiciops* (Fig. 45) and *Microbiotherium* (Fig. 46) have complete, well-ossified and greatly inflated bullae (Segall 1969a and b). The highly inflated tympanic process of the alisphenoid contributes about one-third of the bullar wall anteriorly. The posterior two-thirds are formed not only by an inflated wing of the *pars petrosa* of the perioticum, but also by a swollen wing of the *pars mastoidea* (Patterson 1965), which Segall (*opp. cit.*) considers a large inflated entotympanic. There is, then, a "tripartite" bulla, but distinct from the *Marmosa* type in that the ectotympanic does not form part of it. Rather, in *Dromiciops* the ectotympanic is completely enclosed within the bulla and is thus hidden from lateral and ventral view; there is no open space on the ventral side of the bulla as in most didelphines. The tympanic cavity is divided by two septa into anterior, medial, and posterior chambers (Fig. 46; Segall 1969a), and there is no paraoccipital process. *Dromiciops* is also unique among didelphimorphs in lacking a lamina between the follian process and the manubrium of the malleus (Segall 1969a).

Sparassocynus (Fig. 47; see Reig and Simpson 1972) has a well-developed, highly inflated, and complete ossified bulla, formed solely by the alisphenoid. This situation is not found in any other carnivorous marsupial, but does occur in some genera of phalangers and macropodids. *Sparassocynus* also has an enormous epitympanic sinus above the middle ear, a structure absent or at best incipiently developed in didelphids (Van Kampen 1905; however, see Clemens 1966 for its presence in *Didelphodon*), but present in some dasyurids (e.g., *Dasycercus* and *Dasyuroides*, according to Reig and Simpson 1972:527, although Archer 1976b compared it more closely to *Sarcophilus*).

Three main groups of opossum genera can be recognized from this analysis of the ear region.

One is characterized by retention of some primitive character-states and various departures toward complicated and inflated bullar regions within a basically similar pattern. This group comprises all the genera discussed except *Dromiciops*, *Microbiotherium*, and *Sparassocynus*. The second is the *Dromiciops-Microbiotherium* group, which represents a major alteration of the primitive pattern in the shaping of a complete "tripartite" alisphenoid-petrosal-mastoideal (or entotympanic?) ossified bulla. The third group comprises only *Sparassocynus*, and represents a radical departure from the primitive pattern through the development of an enormous, inflated bony bulla made exclusively of the alisphenoid and complemented by greatly developed epitympanic sinuses.

Within the first and most inclusive group, four subgroupings are distinguished. The first includes *Metachirus*, *Philander*, *Didelphis*, *Chironectes*, *Hyperdidelphis*, *Lutreolina*, *Hondadelphys*, and *Thylophorops*, which are plesiomorphous in most character-states. The second includes *Micoures*, *Monodelphis*, *Thylatheridium*, and *Glironia*, which show moderately inflated alisphenoid wings, but simple ectotympanics and only a moderately expanded tympanic process of the *pars petrosa*. The third includes *Marmosa*, *Thylamys*, and *Lestodelphys*, which possess a highly modified "tripartite" but incomplete bulla. The fourth is represented by *Caluromys* and *Caluromysiops*, which show an almost complete but not much inflated "bipartite", alisphenoid-periotic bulla. The first three subgroupings represent the combined variations of the alisphenoids, perioticum, and ectotympanic; while in *Caluromys* and *Caluromysiops* the ectotympanic is excluded from participation in enclosure of the tympanic cavity, a condition derivable from that found in *Glironia*.

Palatal Vacuities

In most marsupial groups (e.g., many didelphimorphs, dasyuroids, diprotodontans, perameloids, caenolestoids, and polydolopoids) the palate is perforated by large vacuities or fenestrae, often in two pairs. Because of their widespread occurrence, these vacuities are generally regarded (e.g., by Marshall 1977d and Tyndale-Biscoe 1973) as plesiomorphous for the Metatheria. Moreover, some of the genera lacking such openings are derived in other character-states. Palatal vacuities are also present in various placentals (e.g., rabbits, some rodents, macroscelidids, hedgehogs, and *Carpolestes*), but are absent in most. Some multituberculates have distinct palatal vacuities (e.g., *Sloanbaatar* and *Ptilodus*), but others do not (e.g., *Kamptobaatar* and *Kuehneodon*); see Clemens

and Kielan-Jaworowska (1979, Fig. 6-6). The palates of monotremes and therapsids lack vacuities.

The presence, absence, number, and/or relative sizes of palatal vacuities is in fact quite variable among marsupial groups and even within a species. No vacuities occur in borhyaenoids or in *Sparassocynus* (Fig. 47); they are very reduced or absent in *Caluromys*, *Caluromysiops*, *Glironia* (Fig. 48), and *Thylatheridium*; they are typically large but variable in *Marmosa*, *Micoures*, *Thylamys*, and *Monodelphis*; and they are very large in *Dromiciops* (Fig. 49), *Lestodelphys* (Fig. 50), *Metachirus*, *Lutreolina* (Fig. 51), *Chironectes* (Fig. 52), *Philander*, and *Didelphis*. Among Australasian taxa vacuities are vestigial or absent in some species of Dasyuridae (e.g., *Dasycerus*, *Dasyuroides*, and in some species-groups of *Antechinus* and *Sminthopsis*), in *Myrmecobius*, and in the phalangeroids *Petaurus*, *Dactylopsila*, and *Dactylonax* (Marshall 1979; Reig and Simpson 1972:523).

W. K. Parker (1886:270) found that in the developing skull of marsupials the palatal plates of the maxillary and palatine bones generally were at first not fenestrated, but became so by the gradual resorption of bone in certain areas. These observations suggest that the palatal vacuities in marsupials and placentals are independently derived. A solid, unfenestrated palate would thus be regarded as plesiomorphous for marsupials, placentals, and for their common therian ancestors; but the infrequent occurence of unfenestrated palates in some marsupials may represent truncation of the typical ontogenetic sequence rather than retention of the ancestral therian state. That truncation is descriptively a derived condition, and for this reason we consider fenestration as primitive for marsupials (Table 1) in those species which have it. In this as in other cases where something is known of ontogeny, a distinction must therefore be made between the character-state tree and the occurence of the various states in the phylogeny of the taxa.

Pouch

There are two related questions regarding the evolution of the female's pouch or marsupium: first, whether a pouch was present in the common ancestor of marsupials and placentals; and second, if it was not, then whether the pouch might nonetheless be a primitive feature of marsupials or, alternatively, have evolved independently in various marsupial lineages.

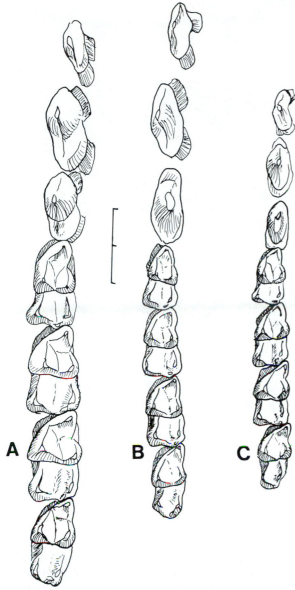

Fig. 21. Occlusal views of left P_1-M_5 of: A. *Micoures cinereus*, MACN 30-297. B. *Marmosa murina*, MACN 50-29. C. *Thylamys pallidior*, MACN 28-127. Scale = 2 mm. (OAR).

No one to our knowledge has suggested that any living eutherian mammal possesses a pouch or a rudiment thereof. Among the monotremes, which can be considered an out-group with respect to therian mammals, a pouch occurs seasonally in the echidnas (*Tachyglossus*, *Zaglossus*) but is absent in the nest-building platypus (*Ornithorhynchus*). The homology of the pouch in the echidnas and marsupials is debated; it has been declared both homologous (Long 1972:142) and non-homologous (Griffiths 1968:195) in the two groups. Whether the homology is true, or whether the pouch was independently derived in the two taxa, the abdominal area anterior

to the female genitalia in early mammals was apparently receptive to modifications involving the formation of a pouch-like structure for protection of the young (Hopson 1973:449; Lillegraven 1969:105). While these points hardly settle the question, it does seem unlikely to us that there was a pouch in the therian ancestor of marsupials and placentals, and studies of the taxonomic distribution and anatomy of the pouch in those marsupials possessing it (see below) suggest that marsupials were also primitively pouchless.

To begin with, not all female marsupials have pouches. Many small-bodied taxa such as some dasyurids (e.g., *Antechinus*) and the great majority of didelphids (e.g., species of *Glironia*, *Metachirus*, *Monodelphis*, *Thylamys*, probably *Micoures*, and *Marmosa*) are without pouches (Marshall 1977c, 1978a; Tate 1933). These absences are generally regarded (e.g., Lillegraven 1969:106; Reig *et al.* 1977:206; Tate 1933:14) as secondary losses from a pouched ancestor. If this were the case, it might be expected that the teats in these pouchless forms would be confined to the area occupied by the pouch in their predecessors. However, some species of *Thylamys* have up to 19 and some species of *Monodelphis* up to 27 teats distributed over the pectoral and inguinal as well as abdominal areas, as in placentals, suggesting that pouchlessness was the ancestral condition. Winge (1923:112-113) reviewed early literature on pouch structure and development and reached a similar conclusion.

Woolley (1974) documented a remarkable variety of pouch development in various members of Dasyuridae. She described four basic types, but noted that the typical conditions may not always be apparent. Some were evident only during the breeding season, or were absent in immature females. In addition, the appearance and size of the pouch is known to change during estrus, pregnancy, and lactation. No obvious correlation was observed between pouch type, habitat of the animal, adult body-size, number of young per litter, duration of period of continuous suckling, or total period of lactation (Woolley 1974:14).

On the hypothesis that pouchlessness is primitive, marsupials which possess pouches for only part of the year could represent an intermediate stage toward development of a true and permanent pouch (Kirsch 1977a), as might also be the case for those species which have a rudimentary pouch. *Caluromys philander*, for example, apparently has only two anteroposteriorly extended folds of skin, one on each side of the belly (Pine 1973).

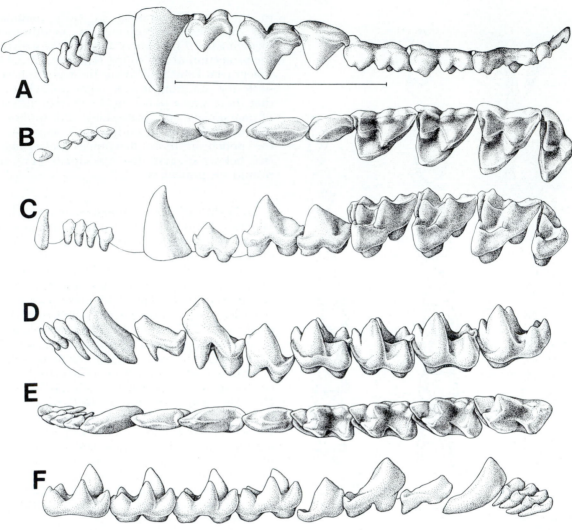

Fig. 22. Marmosa mexicana, FMNH 13806♂. A-C, left I^1-M^5: A. Labial. B. Occlusal. C. Lingual views. D-F, left I$_1$-M$_5$: D. Labial. E. Occlusal. F. Lingual views. Scale = 5 mm. (EL).

In addition, the gross structure of the pouch in permanently pouched forms is highly variable between species. In kangaroos it is a great pendulous bag which opens anteriorly and transversely, but pouches may open posteriorly as in fossorial bandicoots, or along a central slit as in burrowing wombats. The pouch is best-developed in forms which climb (phalangers and *Tarsipes*), hop (kangaroos), or dig (bandicoots, wombats).

In view of the variation remarked above it is probable that the earliest marsupials, as well as the immediate ancestors of both marsupials and placentals, did not have a pouch (Winge 1923:112; Carlsson 1903), but that a pouch evolved independently in various marsupial lineages as an adaptation for special mechanical and/or locomotory needs (Kirsch 1977a; Marshall 1979; Sharman 1976:63). The conclusion that pouches could have evolved independently is somewhat strengthened by the occurence of epipubic bones in several groups of primitive mammals and even (possibly) in *Oligokyphus*, a

tritylodont. These structures have been suggested as preadaptations for having a pouch. Thus, unrelated groups of mammals characterized by possession of epipubes could have separately evolved pouches, as, apparently, have some monotremes and various marsupials. However, arguments about the functional significance of marsupial bones (or their homology with other structures, such as the os penis; see Jellison 1945) have been so close to bald speculation that we are skeptical about their value in support of the hypothesis of multiple evolution of the pouch.

Among Didelphimorphia, a well-developed pouch is present in *Caluromysiops irrupta*, in two species of *Caluromys* (*C. derbianus* and *C. lanatus*), and in species of *Philander*, *Didelphis* (Carlsson 1903), and *Chironectes* (Enders 1937). *Lutreolina* is generally reported to have an undeveloped or absent pouch (Cabrera 1919; Marshall 1977c; and Tate 1933 — but the latter contradicts himself on this point); however, Krieg (1924) described and figured a pouch in *Lutreolina*, and

Fig. 23. A-B. *Monodelphis orinoci*, UKMNH 123941 ♀ (Kirsch field nr. 222): A. Left P^1-M^5; B. Right P_1-M_5. C-D. *Lestodelphys halli*, BMNH 21.6.7.19 (type): C. Left P^1-M^5. D. Right P_1-M_5. All in occlusal view. Scale = 2 mm. (OAR).

this is confirmed by specimen INDERENA No. 1322 (Pine pers. comm.). All other didelphimorphs lack pouches, although some authors refer to a pouch in at least some species of "*Marmosa*" as "rudimentary" or "almost nonexistent".

Thus the presence of a pouch is not always certain, but on the ground of commonality it seems unlikely that possession of a pouch is plesiomorphic for the taxon. Yet, a pouch is present in *Dromiciops* (see Mann 1955; Marshall 1978d; Osgood 1943:49), which, as the nearest extant relative of didelphids, represents a possible outgroup for determining the primitive state at least in Didelphidae.

There is some disagreement about the occurrence of pouches in Caenolestidae, another likely out-group. Tomes (1860) wrote that his specimen of "*Hyracodon*" (= *Caenolestes*) *fuliginosus* (as he later designated it) possessed "a small and rudimentary pouch," a feature which has not been remarked by any subsequent writers, even though lactating females have been examined

(Kirsch and Waller 1979; Osgood 1921). The measurements given in Tomes' (1863) description of the species indicate that the type specimen is a juvenile, suggesting to Osgood that the pouch may be a juvenile feature in caenolestids. That would seem adaptively pointless, but might represent recapitulation of a condition ancestral at least for Caenolestidae.

While it is superficially unparsimonious to conclude that the pouch evolved independently several times, the evidence seems strongly to favor this conclusion, especially in view of cases like *Caluromys* where some species have the pouch and others apparently do not. Moreover, the conclusion that presence of a pouch is a derived state is inescapable in terms of the cladogram indicated by the majority of other anatomical and biochemical features, with which a single evolution of the pouch is inconsistent. We thus agree with Winge (1923:112) that a pouch evolved repeatedly within Didelphoidea.

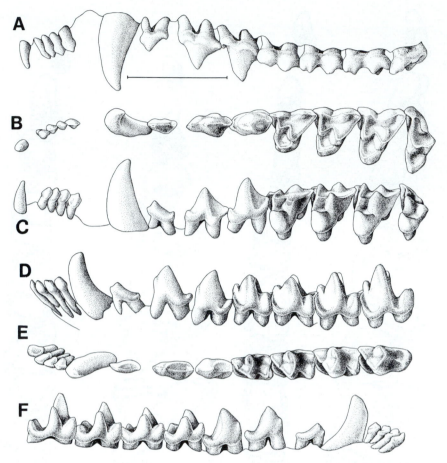

Fig. 24. *Monodelphis domesticus*, FMNH 19504♀. A-C, left I^1-M^5: A. Labial. B. Occlusal. C. Lingual views. D-F, left I$_1$-M$_5$: D. Labial. E. Occlusal. F. Lingual views. Scale = 5 mm. (EL).

Teat Number and Distribution

The number of teats varies greatly among didelphoid genera, and it may also vary within a genus and a species. Unfortunately published data do not exist for all the living taxa. The following comments are based mainly on Bresslau (1912, 1920), Tate (1933), and Walker (1975).

If we start from the assumption (see above) that pouchlessness is the plesiomorphic condition for opossums, then, as noted, the concentration of teats in the abdominal region must be considered a derived character-state, since one would otherwise expect the teats to show a distribution related to the location (or previous location) of the protective pouch (Kirsch 1977a; Tate 1933). Similarly, a low number of mammae may also be considered a derived feature, because it is associated with smaller litter-size, and linked with advanced maternal care.

Thylamys, Monodelphis, and probably *Lestodelphys* (Tate 1933) are the only didelphids that show pectoral mammae in addition to abdominal and, in at least some cases, inguinal ones. *Thylamys* has around 15 mammae, whereas *Monodelphis* has been reported to have from eight

to 27 (Hunsaker 1977; Walker 1975). All the remaining genera lack pectoral mammae and have a variable number of teats concentrated in the abdominal, and in a few instances also the inguinal, region. *Micoures* shows from nine to 15 abdominal and inguinal teats, while in *Marmosa* the number is from 11 to 18, all abdominal in position.

Didelphis has typically about 13 abdominal mammae, whereas both *Metachirus* and *Philander* have from five to nine abdominal ones. *Lutreolina* has been reported (Krieg 1924) as having nine abdominal mammae, and *Chironectes* (Mondolfi and Medina-Padilla 1958) as showing four or five abdominal teats. *Glironia* (Marshall 1978a) and *Caluromys* (FMNH 24142) possess only four abdominal mammae, a condition which suggests a low number for *Caluromysiops*, in which teat number has not yet been recorded. *Caluromys* was reported as having only three young in the pouch (Hall and Dalquest 1963) and FMNH 21725 (*C. philander*) was taken with four well-developed young. *Dromiciops* has only four abdominal mammae (Mann 1958).

In scoring these characters (Table 2) we consider fewer than nine teats to represent the

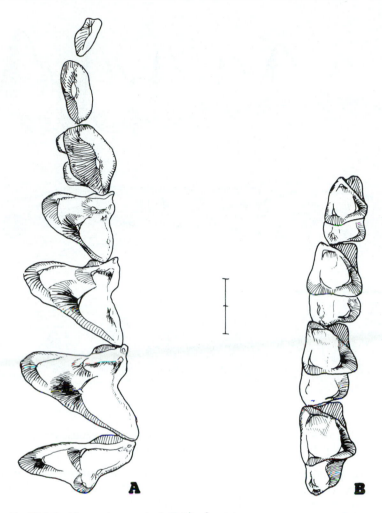

Fig. 25. Thylatheridium cristatum. A. Left P^1-M^5 MACN 6442 (type). B. Right M$_{2-5}$, MACN 6444. Both in occlusal view. Scale =2 mm. (OAR).

derived state, and their concentration in the abdominal region likewise derived (distribution over the pectoral, abdominal, and sometimes inguinal regions is regarded as primitive). The condition in *Micoures* (where mammae are inguinal and abdominal) is considered intermediate, though it may represent a divergent situation.

Female Reproductive Tract

Hill and Fraser (1925) examined the gross structure and serial sections of the female urogenital system of 11 species and nine genera of didelphids (employing the nomenclature adopted here, these included *Didelphis marsupialis, Philander opossum, Metachirus nudicaudatus, Caluromys philander, Micoures cinereus, Marmosa murina, Thylamys pusillus, Monodelphis americana, Monodelphis domestica, Monodelphis theresa,* and *Chironectes minimus*). Their paper provides the only available broad survey of the female reproductive tract in Didelphidae, demonstrating the female genital organs (specifically the lateral vaginal canals and their fusion into median vaginal cul-de-sacs) to be quite variable in structure. We found that the character-variation described by Hill and Fraser provided data capable of clustering the included genera, though those groupings are in partial contradiction to other information (see Fig. 53). Additional published descriptions exist for *Dromiciops australis* (Mann 1955, 1958) and *Caenolestes obscurus* (Osgood 1921), the latter providing a point of reference for determination of character-polarities along with the extensive information available on Australian marsupials.

The determination of advanced and primitive character-states is not as straightforward for reproductive organs as for other characters. However, it seems certain that a permanent pseudovaginal canal is apomorphic for any marsupial: in most marsupials the canal is formed under hormonal influences near the time of birth, and usually regresses afterwards; in only some marsupials (e.g., kangaroos, *Tarsipes,* and possibly some *Didelphis*; Barbour 1977), does the structure remain. Likewise, hypertrophy of any of the regular components of the female reproductive tract may be considered apomorphic. This may be the case for elaborate looping of the

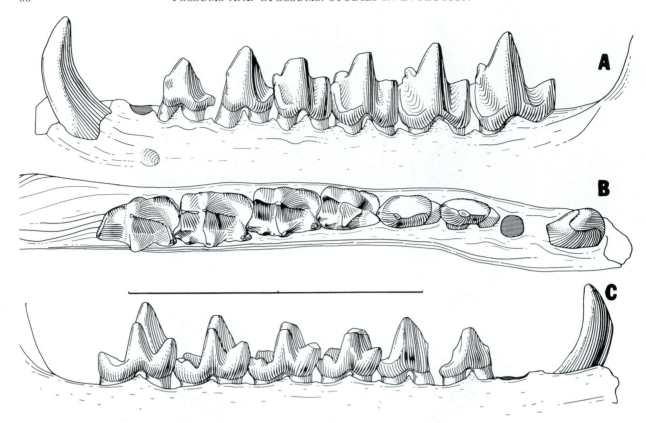

Fig. 26. A-C. *Thylatheridium pascuali*, MLP 46-V-13-52 (type), a nearly complete left mandibular ramus with C, roots of P$_1$, and P$_2$-M$_5$ present: A. Labial. B. Occlusal. C. Lingual views. Scale = 10 mm. (MHW).

vaginal canals or anterior prolongations of the median cul-de-sacs. However, it is less clear, but probable, that the folded lining of the median vaginae is an apomorphic state. Perforation or absence of a septum is surely an advanced character-state, but since the septum breaks at first parturition in those species that possess it, one needs complete life-history information in order to assess the state of this character.

Micoures cinereus appears to be primitive for all of these characters. It has no trace of a persistent pseudovaginal canal, and while only an anterior fragment of the median vaginal septum was present in specimens examined by Hill and Fraser, they regarded the septum as having originally been complete. The median vaginae are small, and their linings smooth; the lateral vaginal canals are short, and they have simple loops.

Marmosa murina also lacks a persistent pseudovaginal canal, and has two short median vaginae without smooth linings separated by a thick septum. The median vaginae have the form of simple cul-de-sacs extending posteriorly independently from each other. The lateral vaginal canals are narrow and elongate, forming two double or S-shaped loops.

The organs of *Thylamys pusillus* are similar to those of *Marmosa murina*: the median vaginae are in the form of minute cul-de-sacs, but the lateral canals form simple loops and are occluded

posteriorly for about two-fifths of their entire length; there is no trace of the pseudovaginal canal.

In species of *Monodelphis*, the median vaginae are wide transversely, are paired in front and separated by a thick septum, but are unpaired behind where they terminate in a short common canal. There is no evidence of a pseudovaginal canal, and the lateral vaginal canals are simplest in *Monodelphis americana*, forming simple loops projecting slightly outward. The loops are more prominent in *M. domestica* and *M. theresa*.

Metachirus nudicaudatus has no definite trace of a pseudovaginal canal. There are two median vaginae, a complete septum, and the lining is slightly folded, while the lateral vaginal canals form large and prominent U-shaped loops.

Hill and Fraser compared *Didelphis marsupialis* with *D. virginiana* and found them very similar. In *D. marsupialis* the median vaginal septum is complete, the median vaginae are of equal length, the lining is markedly folded, the pseudovaginal canal is present, and the lateral vaginal canals form prominent, freely projecting U-shaped loops.

Philander opossum is readily distinguishable from *Didelphis marsupialis* by characters of the vaginal loops, which are relatively much smaller and project freely forward and downwards in

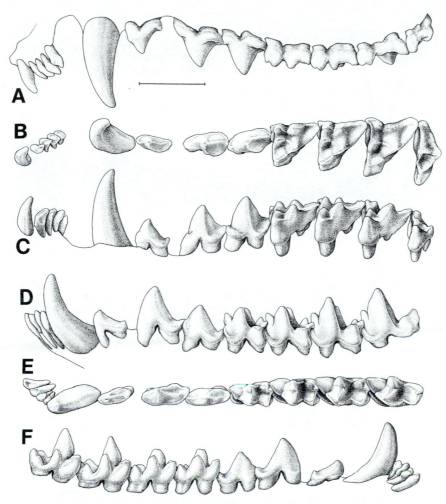

Fig. 27. Lutreolina crassicaudata, FMNH 53944♀. A-C, left I^1-M^5: A. Labial. B. Occlusal. C. Lingual views. D-F, left I$_1$-M$_5$: D. Labial. E. Occlusal. F. Lingual views. Scale = 5 mm. (EL).

front of the uterine necks, while their ascendant and descendant limbs are more equal in size. The vaginal septum is incomplete, and is perforated in front and behind, where there is thus a short common median vagina, the lining of which is folded. There is a trace of a pseudovaginal canal.

In *Chironectes minimus* the median vaginae are well-marked cul-de-sacs, dorsoventrally flattened, but not so wide transversely as in *Monodelphis*; they are separated by a thick septum and terminate some distance behind the openings of the ureter into the neck of the bladder. The lateral vaginal canals run outwards and forwards, and then bend posteriorly to form simple loops. Hill and Fraser's drawing (p. 207) shows no evidence of a pseudovaginal canal, and their discussion implies that this structure is absent.

Caluromys philander is distinguishable from all other opossums studied by the presence of a median cylindrical structure situated between the uterine cervices and the origins of the lateral vaginal canals. This structure represents the drawnout anterior portions of the median vaginae into which the uterine necks open. The

vaginal septum is incomplete, as it is thin and is perforated in two places, so that there is a short common vagina; the lining is markedly folded. There is only a trace of a pseudovaginal canal, and the lateral vaginal canals form well-marked, closed loops.

The characteristics of the female reproductive organs of *Dromiciops* are poorly known. The following is our interpretation of the rather vague information provided by Mann (1955, 1958), supplemented by unpublished observations of Hiriarte Cea (1975). The median vaginae are short (Hiriarte says they are also wide), and seemingly single; no trace of the septum is seen in the transverse section. The section illustrated by Mann is rather posterior in position, and it is not possible to assess the presence or absence of the septum in a more anterior position. However, Hiriarte confirms that a septum is absent, while cautioning that she was unsure whether her specimens were virginal. The lining is gently folded; the pseudovaginal canal seems to be present; and the lateral vaginal canals are short, with very simple U-shaped loops, although

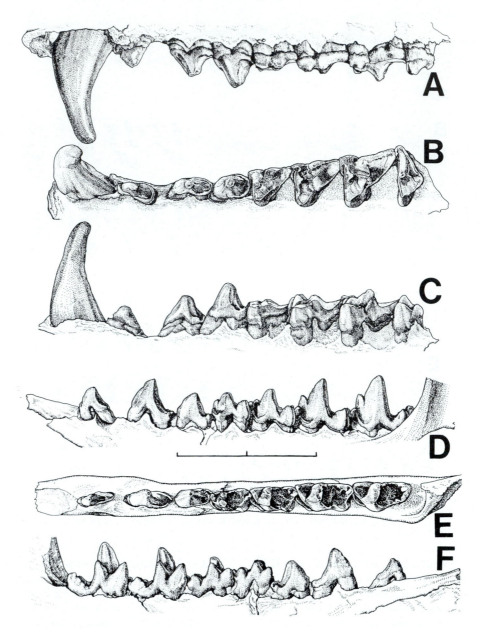

Fig. 28. Thylophorops chapalmalensis. A-C (MMP 1037M), a left maxillary with C-M⁵ complete: A. Labial. B. Occlusal. C. Lingual views. D-F (MMP 354S), mandibular ramus with alveolus of C, and P₁-M₅ complete: D. Labial. E. Occlusal. F. Lingual views. Scale = 20 mm. (ZD).

Hiriarte remarks that the left canal is more complex than the right. Anterior expansions of the median vaginae are clearly absent.

Male Reproductive Tract

Biggers (1966) studied the glans penis in four species of didelphids (*Didelphis marsupialis, Philander opossum, Marmosa mexicana,* and *Caluromys derbianus*). In these four, as in all other marsupials except the macropodids, *Tarsipes,* and *Notoryctes* (Barbour 1977:241), the glans penis is cleft, bifid, and divided; and the urethra is continued as a groove along the inner aspect of each half of the split glans. The glans penis is also bifid in most reptiles, and is grooved in the echidna

(*Tachyglossus*), giving the impression of being bifid in the latter (Griffiths 1968:131; but this may be autapomorphous for monotremes). In placentals, however, the glans is typically simple and non-bifid (Marshall 1979). The bifid glans is thus considered the plesiomorphic condition in didelphimorphs and in marsupials as a whole.

The glans penis is deeply cleft in *D. marsupialis, P. opossum,* and *M. mexicana*; it is more deeply divided in *Caluromys derbianus* and *Caluromysiops irrupta* (Pine pers. comm.) and in the caenolestid *Caenolestes* (Osgood 1921:64). In the former three species, the halves of the glans are tapered and pointed, which is apparently also true in *Dromiciops* (Hiriarte Cea 1975); while in *C.*

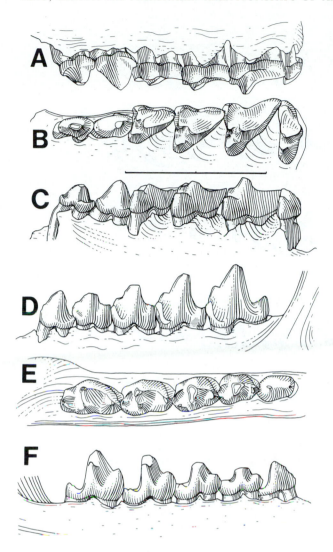

Fig. 29. A-C. *Sparassocynus* sp., MLP 11-92, a left maxillary with P²-M⁵: A. Labial. B. Occlusal. C. Lingual views. D-F. *Sparassocynus bahiai*, MACN 17909, a fragment of a left mandibular ramus with P₃-M₅ complete: D. Labial. E. Occlusal. F. Lingual views. Scale = 10 mm. (MHW).

derbianus and *C. irrupta* the halves are long and slender, and the distal ends markedly rounded. In addition, the urethral grooves in *D. marsupialis* and *P. opossum* terminate some distance from the tips of the glans, while in *M. mexicana*, *C. derbianus*, and *Dromiciops* the grooves continue to the ends.

These incomplete data on didelphoid penis morphology clearly indicate the separation of *Caluromys* and *Caluromysiops* from the remaining didelphid genera, and, within the latter, they show that *Didelphis* and *Philander* are closer to each other than to *Marmosa*, a pattern of relationships which also emerges from analysis of other characters. *Dromiciops* shows features of both major subgroupings.

Sperm Morphology

The morphology of spermatozoa from eight species of didelphids *Didelphis marsupialis*, *Philander opossum*, *Chironectes minimus*, *Monodelphis brevicaudata*, *Metachirus nudicaudatus*, *Marmosa mexicana*, *Micoures cinereus* (= *M. demararae*), and *Caluromys derbianus* was studied by Biggers (1966) and by Biggers and Delamater (1965). Two distinct types were observed. The first, found in all except *C. derbianus*, " . . . is characterized by a flattened, hook-shaped head with one thickened limb and one longer and tapered. The acrosome lies on the anterior part of the thicker limb. The mid-piece is attached by a very fine filament to the base of the cleft separating the two limbs of the head. A complex system of granules is present near the point of attachment and is presumably part of the system of centrioles" (Biggers 1966:269). In *Caluromys* the spermatozoan " . . . is characterized by a saucer-shaped head. The acrosome lies in the centre of the concave side of the head, and the mid-piece is inserted into the centre of the convex side" (Biggers 1966:270). Once again, the distinctness of *Caluromys* is demonstrated. The broad

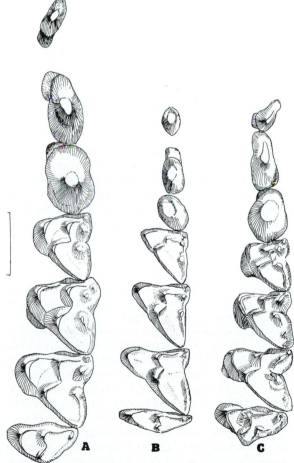

Fig. 30. Occlusal views of left P¹-M⁵ of: A. *Didelphis marsupialis*, USB 456. B. *Hyperdidelphys biforata*, MACN 7920. C. *Chironectes minimus*, ERBG 2371. Scale = 5 mm. (OAR).

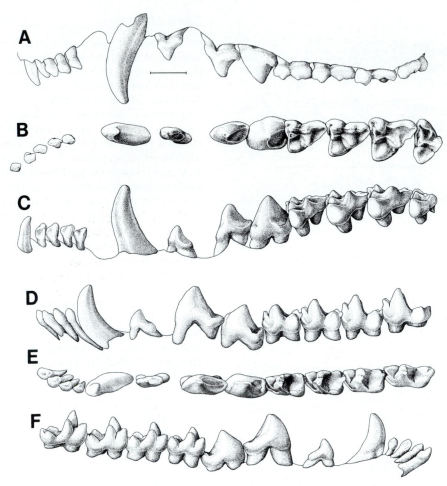

Fig. 31. *Didelphis marsupialis*, FMNH 89217♀. A-C, left I^1-M^5: A. Labial. B. Occlusal. C. Lingual views. D-F, left I$_1$-M$_5$: D. Labial. E. Occlusal. F. Lingual views. Scale = 5 mm. (EL).

distribution of the didelphine type of sperm is suggestive that it is the plesiomorphic condition, though this must be a tentative conclusion until more careful comparisons with Australasian marsupials are made. *Caenolestes* has, according to Biggers and Delamater, yet a third type of sperm.

Karyotype

A considerable amount of information is available for number, gross structure, and variation in the chromosomes of 24 species representing 11 genera of opossums (Hayman and Martin 1974; Kirsch 1977b; Schneider 1977; Sharman 1973a and b; Reig *et al.* 1977; Wainberg *et al.* 1979). For only a few of these species do we know chromosome banding patterns (Yonenaga-Yassuda *et al.* 1982; Seluja *et al.* 1984 and citations therein). Opossums are conservative in chromosomal complements, affording an exceptional opportunity to study inter-taxon relationships based on relatively small numbers of chromosome types. All of the studied species can be sorted into one of three patterns: 2n=14 (four species of *Marmosa*, two species of *Micoures*, two species of *Thylamys*, three species of *Caluromys*, and the single species

of *Metachirus* and of *Dromiciops*), 2n=18 (three species of *Monodelphis*), and 2n=22 (three species of *Didelphis*, two species of *Philander*, and the single species of *Lutreolina* and of *Chironectes*).

Marsupial chromosome numbers in general have a bimodal distribution around 2n=14 and 2n=22, with a range from 2n=10 to 2n=32 (Hayman 1977:28). It is assumed, as for anatomical characters, that some species are conservative in chromosome evolution and retain the chromosome number and morphology of the stem form from which the extant Australasian and South American radiations arose, or a karyotype much like it. Thus, following the principle of commonality, either 2n=14 or 22 could represent the "basic marsupial karyotype" (Sharman 1973b:508-509). Inasmuch as both modal numbers occur within the Didelphimorphia, identification of the stem karyotype is important for determining the primitive pattern and in assessing polarity in didelphimorph chromosome evolution.

Several authors discuss the question of the original marsupial karyotype, and it is now generally agreed that a diploid number of 14 is basic

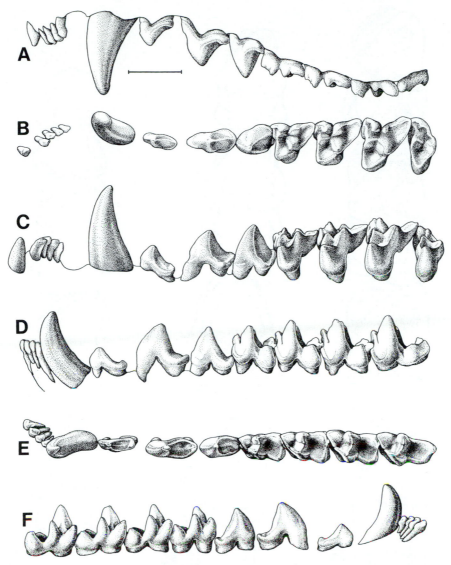

Fig. 32. *Chironectes minimus*, FMNH 69328♂. A-C, left I^1-M^5: A. Labial B. Occlusal. C. Lingual views. D-F, left I_1-M_5: D. Labial. E. Occlusal. F. Lingual views. Scale = 5 mm. (EL).

for metatherians (e.g., Hayman and Martin 1974; Reig and Bianchi 1969; Reig *et al.* 1972; Reig *et al.* 1977). However, Sharman (1973a and b) and Matthey (1973) express dissent and argue for some higher number. The chief reasons for their disagreement are the occurrence of higher basic numbers in placentals (wherein the mean is around 2n=48, with a range from 2n=6 male or 2n=7 female in *Muntiacus munjac* to 2n=96 in *Anotomys*), monotremes (2n=54 in *Ornithorhynchus* and 2n=64 in *Tachyglossus*), and some reptiles; and a reluctance to admit the general occurrence of chromosome increase through fission, since it is alleged that there is no known mechanism for this.

The relevance of the first argument is doubtful if one admits that eutherians and metatherians diverged from a common ancestor from 80 to 100 mybp (Lillegraven 1969; and chapters 2, 9, 10, and 11 in Lillegraven *et al.* 1979), and

followed two different patterns of chromosome evolution — one (in metatherians) of retention of uniformity in karyotypes, probably related to selection for evolutionary stability of their regulatory gene-action mechanisms (Reig *et al.* 1977:208-210); and the other (in eutherians) of a high rate of chromosomal evolution, probably linked to selection for new regulatory mechanisms (Wilson *et al.* 1974a and b).

As for the second argument, it depends on an outdated, classical view of chromosomal evolution. Current evidence supports the view that chromosome fissioning at the level of the centromere is cytologically possible and perhaps even a common occurrence, and that fissioning played a significant role in karyotypic evolution in various groups of organisms (John and Hewit 1968; John and Lewis 1968; White 1978). Even if 2n=22 be admitted as the primitive number, at least some occurrences of fission must have taken place

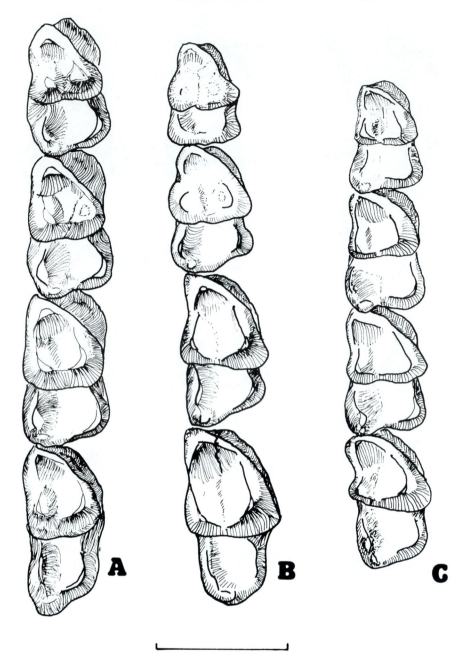

Fig. 33. Right M_{2-5} of: A. *Didelphis marsupialis*, USB 456. B. *Hyperdidelphys inexpectata*, MACN 1615. C. *Chironectes minimus*, EBRG 2371. Scale = 5 mm. (OAR).

(Hayman 1977). In any case, the higher number in marsupials is far removed from the modal values of any of the possible out-groups.

A further point against the 2n=22 hypothesis is that the morphologies of chromosomes comprising that number are frequently different in their various occurrences (e.g., in Didelphinae compared with the various phalangers and kangaroos which have 2n=22). In contrast, there is a striking uniformity in the appearance of the 2n=14 karyotypes possessed by many didelphimorphs, all dasyurids, caenolestids, burramyids, vombatids, most peramelids, some phalangerids (e.g, *Phalanger*), and macropodids (e.g., some species of *Dendrolagus*). Only among the last three is there reasonable doubt that their

karyotypes are "the same", although Sharman (1973b:146) did suggest that the burramyid constitution might also have been separately derived.

The overall uniformity of the 2n=14 pattern from group to group thus favors its choice as the primitive one; and in at least the genera of dasyurids (Young *et al.* 1982) and didelphids (Seluja *et al.* 1984; Yonenaga-Yassuda *et al.* 1982) known in this respect, this uniformity involves not only gross morphology but also G-banding patterns, although minor differences appear in C-banding. That pattern (Fig. 54) consists of three pairs of large metacentric or submetacentric autosomes; one pair of medium-sized metacentric autosomes; two pairs of small autosomes variously described as acrocentric, submetacentric, or

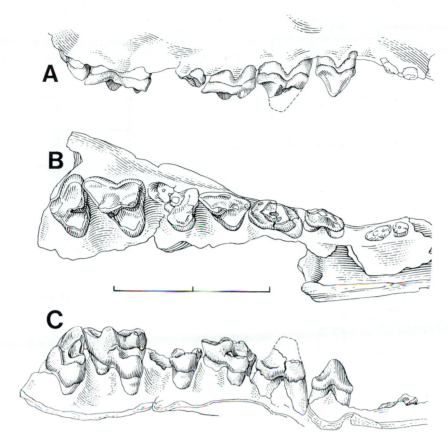

Fig. 34. *Hondadelphys fieldsi*, UCMP 37960 (type). A fragment of a right maxilla with alveolus of C roots P^1, P^2 complete, P^3 missing tip of crown (restored from left P^3 of type), M^2 complete, M^3 broken, M$^{4-5}$ complete: A. Labial. B. Occlusal. C. Lingual views (after Marshall 1976). Scale = 20 mm.

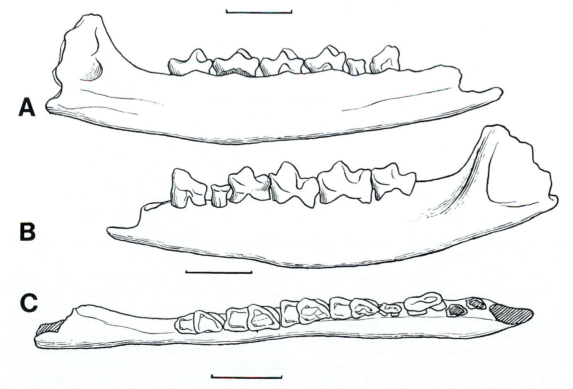

Fig. 35. *Zygolestes paranensis*, MACN 8889 (type). Left mandibular ramus with alveoli of C, roots of P$_1$, P$_2$-M$_5$, complete: A. Lingual. B. Labial. C. Occlusal views (X5). D. Occlusal view of M$_{2-5}$ (X20). (OAR).

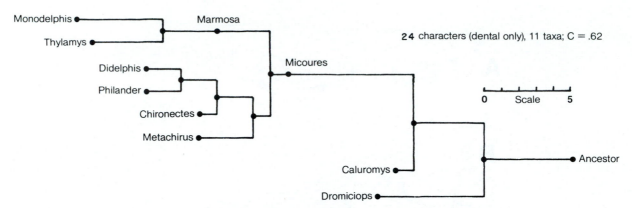

Fig. 36. Wagner tree of 11 taxa computed on the basis of the 24 dental characters listed in Table 2, with overall consistency of .62. Horizontal lines between linkage levels are drawn in proportion to evolutionary distances, in increments of one unit; the scale may be used to determine the exact number of steps separating nodes. As the all-primitive *Ancestor* is to the right, taxa to the left are more derived.

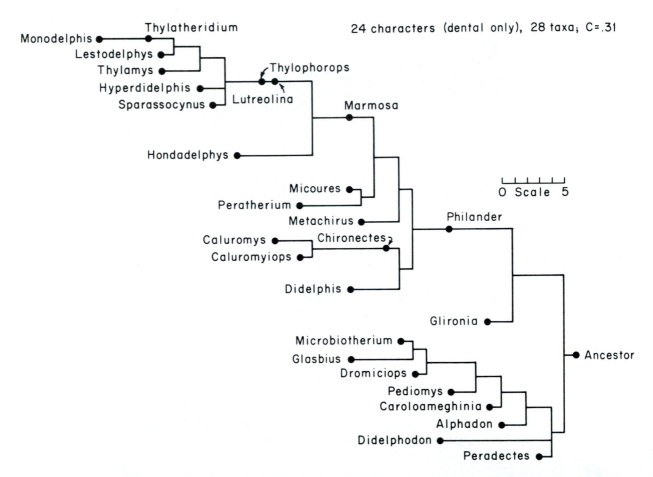

Fig. 37. Wagner tree of 28 taxa computed on the basis of the 24 dental characters listed in Table 2, with overall consistency of .31. Conventions as for Figure 36.

subtelocentric; and a pair of variable sex-chromosomes (Hayman and Martin 1974:41). As the pattern is shared by distantly related taxa (Fig. 55), joint possession of this karyotype implies not close affinity but symplesiomorphy (Reig *et al.* 1977:207). On the other hand, 2n=22 may well be the basic number for Phalangeridae and Macropodidae, since it is only in these taxa that 2n=14 ever takes on an appearance markedly different from that in other families, and because most of the

phalangerids and macropodids have higher numbers. That is not the case for didelphoids, since all studied species can be sorted into one of three groups *vis-à-vis* karyotypes: 2n=14, 2n=18, and 2n=22. Moreover, each of these karyotypes appears to be stable in chromosomal structure, showing only minor variations within each basic pattern; and, in the known cases, a quite similar G-banding pattern (Yonenaga-Yassuda *et al.* 1982; Seluja *et al.* 1984). This stability suggests that the

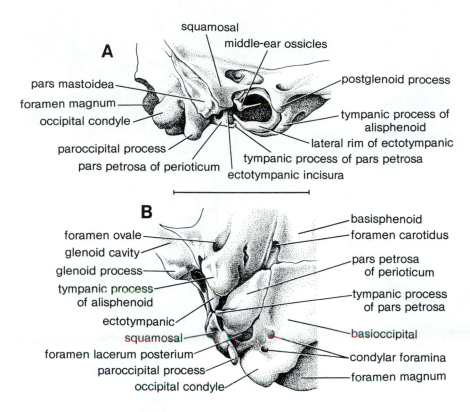

squamosal
middle-ear ossicles
A
pars mastoidea
foramen magnum
occipital condyle
postglenoid process
tympanic process of alisphenoid
lateral rim of ectotympanic
paroccipital process
tympanic process of pars petrosa
pars petrosa of perioticum
ectotympanic incisura

B
foramen ovale
glenoid cavity
glenoid process
tympanic process of alisphenoid
ectotympanic
squamosal
foramen lacerum posterium
paroccipital process
occipital condyle
basisphenoid
foramen carotidus
pars petrosa of perioticum
tympanic process of pars petrosa
basioccipital
condylar foramina
foramen magnum

Fig. 38. A., lateral, and B., ventral view of otic region of *Metachirus nudicaudatus* (composite of FMNH 20800♀, and FMNH 24787♀). Scale = 10 mm. (EL).

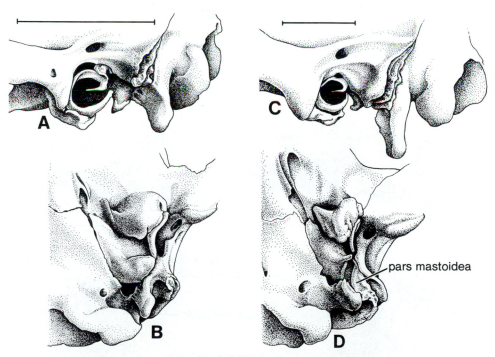

A

C

B

D

pars mastoidea

Fig. 39. A., lateral, and B., ventral view of left otic region of *Philander opossum* (composite of FMNH 114710♀, and FMNH 24790♀). C., lateral, and D., ventral view of left otic region of *Didelphis marsupialis* (FMNH 89217♀). Scale = 10 mm. (EL).

three are the main, if not only, patterns established during the evolution of the superfamily (Reig *et al.* 1977:205). Yet the small variations may be important for understanding transitions between each number.

Secondary but significant variation exists mainly in the 2n=14 karyotype. *Marmosa fuscata* (Fig. 55E) is the most divergent, with chromosome pairs C-1 and C-2 (Reig and Sonnenschein 1970; Reig *et al.* 1977), corresponding to pairs

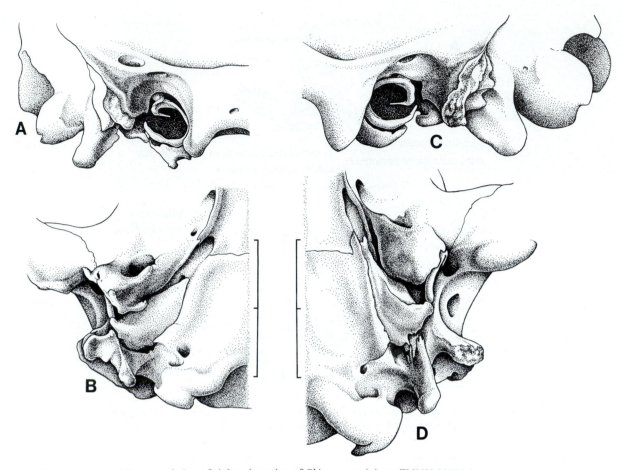

Fig. 40. A., lateral, and B., ventral view of right otic region of *Chironectes minimus*, FMNH 69328♂. C., lateral and D., ventral view of left otic region of *Lutreolina crassicaudata*, FMNH 24375♂. Scale = 10 mm. (EL).

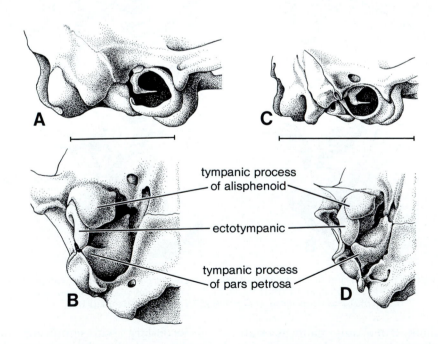

Fig. 41. A., lateral, and B., ventral view of right otic region of *Marmosa mexicana*, FMNH 13806; scale = 5 mm. C., lateral, and D., ventral view of right otic region of *Monodelphis domesticus* (composite of FMNH 20257♂, and FMNH 20255♀); scale = 10 mm. (EL).

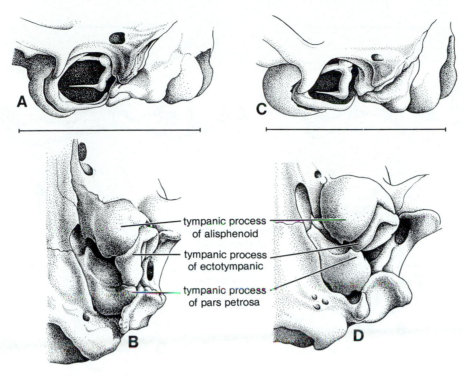

tympanic process
of alisphenoid

tympanic process
of ectotympanic

tympanic process
of pars petrosa

Fig. 42. A., lateral, and B., ventral view of left otic region of *Thylamys elegans*, FMNH 22667♂.
C., lateral, and D., ventral view of left otic region of *Lestodelphys halli*, BMNH
23.12.11.206♀. Scales = 10 mm. (EL).

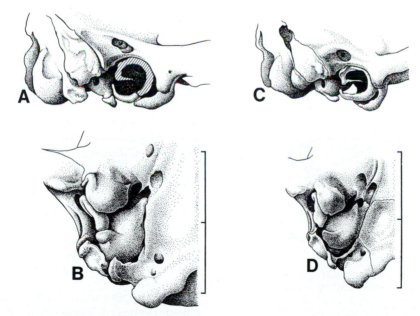

Fig. 43. A., lateral, and B., ventral view of right otic region of *Glironia venusta*, FMNH
41440♀. C., lateral, and D., ventral view of right otic region of *Micoures cinereus*,
FMNH 75102♂. Scale =10 mm. (EL).

five and six of Hayman and Martin (1974), almost perfectly metacentric. *Marmosa robinsoni* (Fig. 55D) and M. *murina* have subtelocentric C-1 and C-2 autosomes, whereas pair C-1 is acrocentric (with centromeric index greater than seven) in *Thylamys pusillus* (Fig. 55A). *Thylamys elegans* (Fig. 55B), *T. aguilis*, and *T. microtarsus* also show more-terminal centromeres in pairs C-1 and C-2,

whereas *Micoures cinereus* (Fig. 55C) has minute short arms which are difficult to measure in those pairs. This is a feature that *Micoures cinereus* shares with *Metachirus nudicaudatus; Micoures cinereus* also differs strikingly from *Marmosa murina* in nucleolar organizers, supporting the generic distinction advocated here (Yonenaga-Yassuda *et al.* 1982). Absence or extreme minuteness of

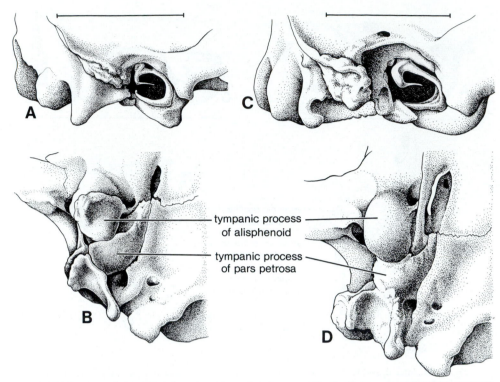

Fig. 44. A., lateral, and B., ventral view of right otic region of *Caluromys philander* (composite of FMNH 92030♀, and FMNH 61850♂). C., lateral, and D., ventral view of right otic region of *Caluromysiops irrupta* (composite of USNM 396160♂, and FMNH 84426♀). Scale = 10 mm. (EL).

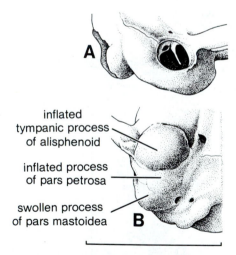

Fig. 45. A., lateral, and B., ventral view of right otic region of *Dromiciops australis*, FMNH 22672♀). Scale = 10 mm. (EL).

short arms in the two pairs of small-sized autosomes is a character-state expected in the particular 2n=14 karyotype which gave rise by autosomal fissioning to the higher numbers — otherwise the resulting number could be 2n=26, which does not occur among marsupials (Hayman and Martin 1974:11).

Thus, starting from a form like either *Micoures cinereus* or *Metachirus nudicaudatus*, the major course of chromosomal evolution within didelphoids appears to have been by ascending

meroaneuploidy (*sensu* Jackson 1971) through the mechanism of Robertsonian increase (i.e., centric fissions or dissociations), producing uniarmed chromosomes from the four pairs of biarmed autosomes, and an increase in the diploid number to 22. Superimposed upon this basic mechanism for karyotypic repatterning has been the incorporation of pericentric inversions and/or translocations in some lineages (e.g., *Marmosa, Thylamys, Caluromys, Dromiciops,* and *Didelphis*; Reig *et al.* 1977:207). However, the primitive karyotype for didelphimorphs as a whole is more likely to have been one like that found in *Caluromys philander* or in species of *Thylamys*, where pairs C-1 and C-2 both have measurable short arms, since these features are also found in the microbiothere *Dromiciops*, in caenolestids, and even in some Australasian groups.

Recently Sharman (1982) argued that the chromosomes of *Dromiciops* were more like those of the diprotodontans *Cercartetus* and *Vombatus* and the peramelid *Isoodon* than those of *Marmosa, Caenolestes,* and dasyurids in arm-ratios and positions of possible nucleolar organizers, thus suggesting that these patterns of similarity support Szalay's Australidelphian hypothesis (Szalay 1982a and b). However, it is clear to us that the similarities remarked by Sharman are based on probably misleading conclusions from incomplete data. There is neither sufficient evidence from banding patterns nor from the extent and position of nucleolar organizers in

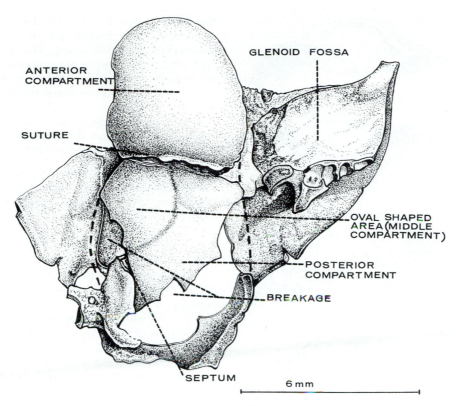

Fig. 46. Microbiotherium tehuelchum, PU 15038. Ventral view of left auditory region (after Segall 1969a, Fig. 5). Scale = 6 mm.

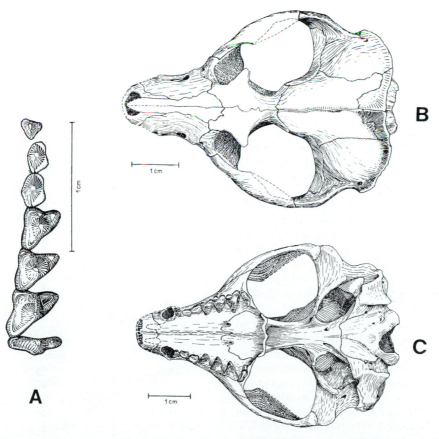

Fig. 47. Sparassocynus derivatus Reig and Simpson, MMP 1725 (type). A. Occlusal view of right P¹-M⁵. B. Skull in dorsal view. C. Skull in ventral view. Figure from Reig and Simpson (1972). Scale = 10 mm.

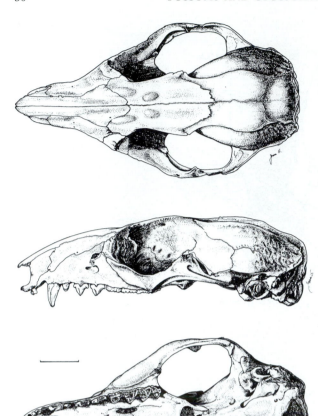

Fig. 48. Skull of *Glironia venusta*, FMNH 41440, shown from top to bottom in dorsal, lateral and ventral views (after Marshall 1978a). Scale = 5 mm.

most of the 2n=14 didelphoids to support his contentions. Moreover, in giving significance to differences in arm-ratios between *Dromiciops* and *Thylamys elegans* and the greater similarity between the former and some Australian marsupials, Sharman ignores arm-ratio differences found *among* marmosines and other didelphoids. In addition, even if the similarities noted by Sharman in support of the Australidelphian concept were true, they could also be regarded as symplesiomorphies and hence of no phylogenetic significance: that is, they would be *consistent* with Szalay's hypothesis but not *supportive* of it.

The 2n=22 karyotype is less variable than the basic 2n=14 pattern (Fig. 56). Species of *Didelphis*, *Philander*, and *Chironectes* share an identical karyotype of 10 pairs of uniarmed autosomes with apparently terminal centromeres, except for some of those of the exclusively North American *D. virginiana* which have recognizable short arms (Gardner 1973; Shaver 1962), plus an acrocentric X, and a minute Y chromosome (Reig *et al.* 1977:196). In *Lutreolina crassicaudata* the autosomal set is the same, but the X differs from

the X of all other 2n=22 didelphids in being fully metacentric and longer than the smallest autosomes (Reig *et al.* 1977:196). The X of the karyotype of *L. crassicaudata* was considered to be an isochromosome derived from the standard acrocentric X through centromeric misdivision by Reig *et al.* (1977:199), but this hypothesis has been refuted by Yonenaga-Yassuda *et al.* (1982) and by Seluja *et al.* (1984) on the basis of the differences found in G- and C-banding in the two arms.

The 2n=18 pattern occurs only in species of *Monodelphis*, where it is almost identical in the three known species. This pattern is conceivably derived either from 2n=14 by fissions or from 2n=22 by fusions. Sharman (1973a:145) indicated the mechanism and chromosome pairs which could be involved in each case, and stressed the difficulty of choosing between these alternatives due to uncertainty as to whether any marsupial chromosomes are truly telocentric. Because all autosomes in the 2n=18 karyotypes have short arms, Hayman and Martin (1974) favored derivation from 2n=22. Two fusions and pericentric inversions in the remaining acrocentric chromosomes would be sufficient to

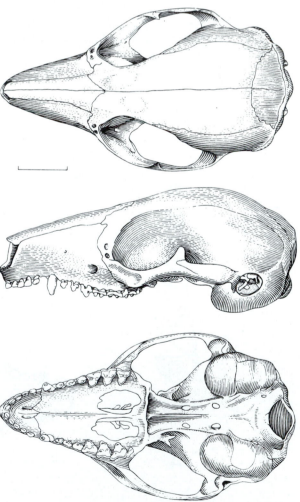

Fig. 49. Skull of *Dromiciops australis*, AMNH 92147, shown from top to bottom in dorsal, lateral, and ventral views (after Marshall 1978d). Scale = 5 mm.

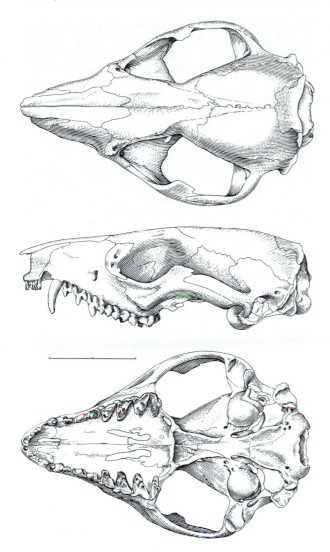

Fig. 50. Skull of *Lestodelphys halli*, BMNH 28.12.11.206, shown from top to bottom in dorsal, lateral and ventral views (after Marshall 1977b). Scale = 10 mm.

into the 2n= 18 karyotype of *Monodelphis*. Given the invariable intergeneric pattern of the 2n=22 karyotype in the large opossums, the six inversions in *D. virginiana* are indeed impressive, and suggest that it will be of interest to explore in more depth the generic relationships of that species.

In summary, a chromosome complement of 2n=14 is regarded as primitive for Didelphidae. An increase to 2n=18 occurred in one lineage, and a probably independent increase to 2n=22 occurred in another. Therefore, species of *Didelphis, Philander, Chironectes*, and *Lutreolina* share the derived character-state of 2n=22, and are considered cladistically related by synapomorphy in this character. On the other hand, species of various genera sharing the primitive 2n=14 chromosome pattern are symplesiomorphous, so that karyotypes give no evidence of relatedness among them. It is especially noteworthy that *Metachirus* possesses the primitive karyotype and is distinct from *Philander* in this respect.

Serology

The data of serology are usually given as single-number similarity or distance measurements between pairs of taxa, derived through any

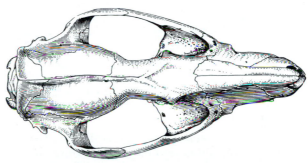

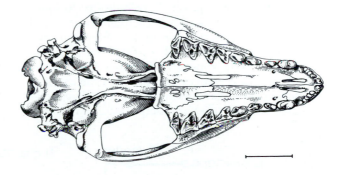

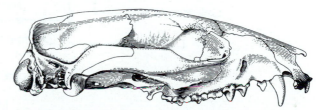

Fig. 51. Skull of *Lutreolina crassicaudata*, MVZ 134223, shown from top to bottom in dorsal, ventral and lateral views (after Marshall 1977c). Scale = 10 mm.

account for the changes. But there seems to be no good reason why inversions producing short arms could not occur in some elements of the 2n=14 pattern after fissioning of two pairs of biarmed elements. Reig and Bianchi (1969) and Reig *et al.* (1977:204-205) favor this means of derivation of 2n=18 from 2n=14.

There remains the question of whether 2n=18 might represent an actually intermediate stage between 2n=14 and 2n=22. Serological results (see below) indicate a special relation among (at least) *Marmosa robinsoni, Micoures cinereus*, and *Monodelphis orinoci*, a relation also substantiated by skull and dental characters and suggesting that the derivation of 2n=18 was instead a separate event (Kirsch 1977b). In this respect it is of interest to note that *Didelphis virginiana* differs in karyotype from the "standard" 2n=22 forms by 6 pericentric inversions, which involve a larger number of chromosomal mutations than those required to transform a "standard" 2n=14 form

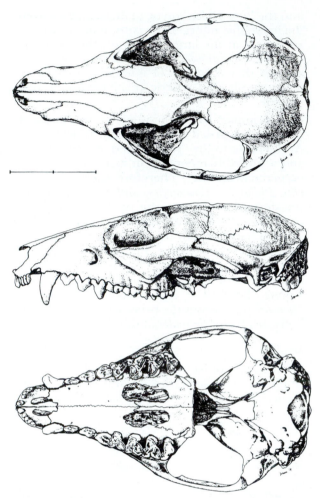

Fig. 52. Skull of *Chironectes minimus*, FMNH 75092, shown from top to bottom in dorsal, lateral and ventral views (after Marshall 1978c). Scale = 20 mm.

of a number of techniques. Classically these were estimates of antibody-antigen reactions in a liquid medium, sometimes measured nephelo-metrically; contemporary serological techniques of choice involve micro-complement fixation and immune-precipitation in gels. Whatever the methods, the results are difficult to interpret according to the canons of cladistics, because the similarity (or distance) indices are recalcitrant to ordering and polarization. In fact, serological distances, which are phenetic statements, are

usually interpreted as being directly proportional to cladistic separation. Such interpretation assumes that the characters measured have evolved in a constant and divergent manner, and a large literature deals with the evidence for (and theoretical likelihood of) "molecular clocks" (e.g., Ayala 1977; Fitch 1976; Wilson *et al.* 1977).

If it is indeed true that structural genes and their products evolve time-dependently, a powerful tool is available for establishing cladistic relationships. In the GENERAL DISCUSSION we show that serological distances approximate cladistic affinities among opossums and provide the basis for choosing among other hypotheses of relationship. Here we note only that the assumption of rate-constancy underlies most considerations of opossum biochemical systematics.

Some marsupials, but no American species, were included in the first comprehensive application of serology to phylogeny, that by Nuttall (1904). Weymss (1953), using a more sophisticated nephelometric technique, compared *Didelphis virginiana* serum with sera from five Australian marsupials and concluded that it was less similar to those species than they were to each other, a finding that has been confirmed by other workers using comparable and also quite distinct methods. Kirsch (1968, 1977b), however, obtained stronger nephelometric reactions between *Didelphis* and Australasian marsupials, suggesting caution in using different and evidently variable-strength antisera to make precise statements about categorical level. Using more qualitative gel-diffusion methods (double-diffusion, immunoelectrophoresis) and an haemagglutination technique, Kirsch (1977b) demonstrated that the distances *among* didelphids appear greater than are usual within Australasian marsupial families; this conclusion was also reached by Maxson *et al.* (1975) through study of marsupial albumins employing the currently favored micro-complement-fixation method. Their results do not show clear groupings within Didelphidae (*Didelphis, Metachirus, Caluromys,* and *Marmosa* spp. were studied) beyond the fact that *Caluromys* and *Marmosa*

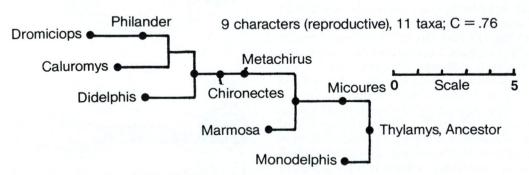

Fig. 53. Wagner tree of 11 taxa computed on the basis of the nine reproductive features listed in Table 2, with overall consistency of .76. Conventions as for Figure 36.

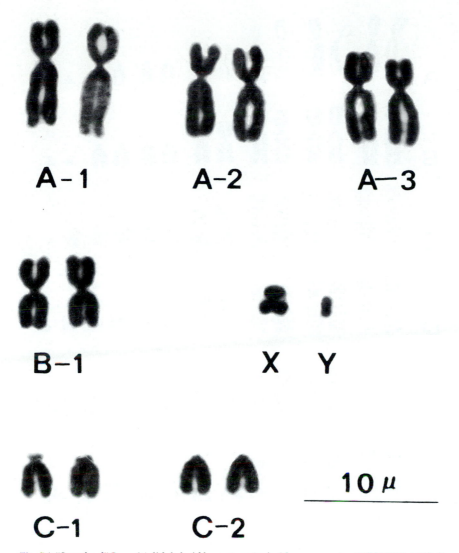

Fig. 54. "Standard" 2n = 14 didelphoid karyotype (as in *Marmosa murina*, MBUCV 1-1657♂).
Scale = 10 mm.

apparently are closer to each other than to the other two genera.

More discrimination of generic groupings was derived from Kirsch's data, by inspection and numerical analysis (clustering and principle-factor analysis) of immunoelectrophoretic tests of whole sera reacted with absorbed and unabsorbed antisera. Figure 57 is a phenogram of serological resemblances indicating that *Didelphis*, *Philander*, and *Chironectes* form a unit with the first two genera somewhat closer to each other than to the latter; *Marmosa robinsoni* and *Micoures cinereus* stand fairly close to *Monodelphis* spp., while both triplets of genera and *Metachirus*, representing a third taxon, are about equally separated from each other. Limited experiments with *Thylamys elegans* indicated that it reacted much as did *Marmosa*.

Kirsch also maintained that all of the above genera are more similar to each other than any is to *Caluromys*, a conclusion in superficial opposition to that of Maxson *et al.* (1975). However, the

latter workers examined just albumin, which may have an individual rate of change different from that of serum proteins as an averaged group. Kirsch also noted (1977b:91) special resemblances between *Caluromys* and *Marmosa*, but concluded that the similarities were not greater than those between *Marmosa* and the other genera. The implication is that *Marmosa* and *Caluromys* have experienced a relatively slower rate of change in at least some serum proteins than have other didelphids. Since the constant-rate model is a stochastic one, this kind of result is not unexpected.

Finally, *Dromiciops* is less like all the genera mentioned above than those genera are like each other, and caenolestids and Australian marsupials are still more dissimilar. Thus the serological results suggest a relatively early divergence of *Dromiciops* from other opossums, and a later branching between lineages represented respectively by *Caluromys* and the didelphines.

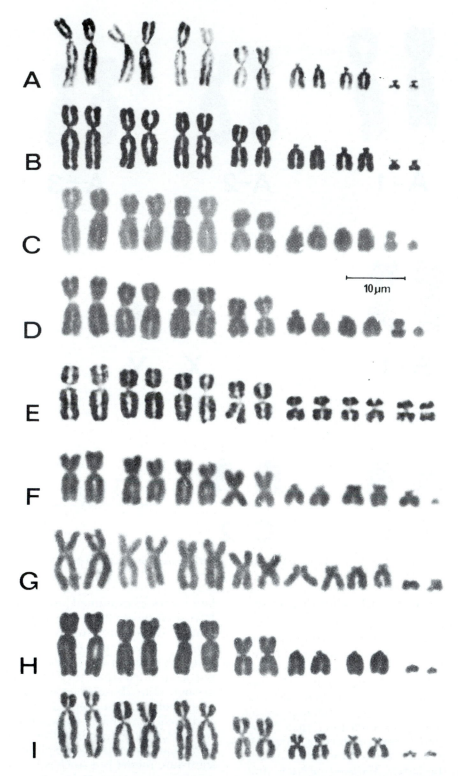

Fig. 55. Representative 2n = 14 didelphoid karyotypes: A. *Thylamys pusillus*. B. *Thylamys elegans*. C. *Micoures cinereus*. D. *Marmosa robinsoni*. E. *Marmosa fuscata*. F. *Caluromys philander*. G. *Caluromys lanatus*. H. *Metachirus nudicaudatus*. I. *Dromiciops australis* (after Reig *et al.* 1977).

In utilizing the serological results in our cladistic analysis we have provisionally accepted the dogma of rate-constancy, taking the serological phenogram as an approximation to a cladogram which was then compared to cladograms generated by anatomical data. Serology thus provides a kind of "initial cladistic hypothesis" which can be directly tested by morphological and other characters. Our attitude has been somewhat similar for the data of karyology, and this being so, we have not relied greatly on either serology or karyotypes in drawing conclusions from the

Fig. 56. Representative 2n = 22 didelphoid karyotypes: A. *Didelphis albiventris* (Perú). B. *Didelphis albiventris* (Argentina). C. *Didelphis marsupialis*. D. *Lutreolina crassicaudata*. E. *Philander mcilhennyi* (after Reig *et al.* 1977).

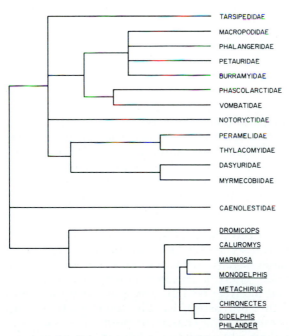

TARSIPEDIDAE
MACROPODIDAE
PHALANGERIDAE
PETAURIDAE
BURRAMYIDAE
PHASCOLARCTIDAE
VOMBATIDAE
NOTORYCTIDAE
PERAMELIDAE
THYLACOMYIDAE
DASYURIDAE
MYRMECOBIIDAE

CAENOLESTIDAE

DROMICIOPS
CALUROMYS
MARMOSA
MONODELPHIS
METACHIRUS
CHIRONECTES
DIDELPHIS
PHILANDER

Fig. 57. Phenogram showing the serological resemblances among most of the extant marsupial families, with additional detail for the genera of didelphimorphs, collated from several sets of experiments (see Kirsch 1977b). *Marmosa* should here be understood to include *Micoures* and *Thylamys*, the three genera and *Monodelphis* being about equally close. Only the horizontal axis of this figure has dimension, with linkages of groups being successively less close from right to left (from Kirsch 1977c).

Wagner analysis; the necessarily restricted coverage also militated against inclusion of biochemical and chromosomal data in the largest (28- or 33-taxon) trees. However, for a limited range of experiments on 9, 11, or 12 taxa, codings for both karyotypes and serology were established and added to or subtracted from the data-set to determine their effects on the dendrograms.

Biochemical features were coded as two characters, one relating all genera in a linear series and another representing a separate branch grouping *Marmosa*, *Thylamys*, *Micoures*, and *Monodelphis*. While no direct comparisons of *Thylamys* to other murine opossums were made, it is a reasonable prediction that they will group with the latter; similarly, the judgment that *Lutreolina* will prove to be close to the other 2n=22 didelphines was made so that it also could be included in the calculations. Admittedly this two-character coding does not indicate all branchings and has a distinct polarity with *Dromiciops* near the base (thus formally contradicting the rate-constancy hypothesis); but a coding fully representing all branchings would have involved at least as many characters as taxa, quite overbalancing other features in representing distances on the trees. The purpose of including serology was just to see if the *groupings* indicated by chemistry would significantly alter trees based on anatomy.

WAGNERIAN AND INFORMATIONAL ANALYSIS

Introduction

The Wagner method is named after the botanist W. H. Wagner, who showed (e.g., 1969) that the character-states of a group of taxa could be used as the basis for a diverging network relating more primitive taxa to ones more derived. Comparison of the states of taxa often shows, however, that any three of them can be joined through a real or hypothetical ancestor that may differ in some ways from any of them. Thus, a Wagner diagram connecting a group of taxa will often interpose several phylogenetically necessary taxa. A Wagner tree differs in this respect from a simple nearest-neighbour network, and resembles a cladogram except that structural ancestors sometimes exist among the studied taxa and may be placed at the appropriate notes.

The Wagner algorithm developed by Farris (Farris 1970, 1972; Kluge and Farris 1969) simply formalizes the logic of such analysis. As the method allows for reversals in state, parallelisms, and convergences (collectively, *homoplasy*), it reflects well the likely complexity of evolutionary change.

The algorithm aims for, but does not always achieve, the most parsimonious arrangement. Since there is usually significant homoplasy in any real data-set, it is helpful to have a record of where changes take place in the tree; the programme documents these changes. Additionally, a measure of the evolutionary lability of each character can be calculated. A "perfect" character ought to change only so many times as is necessary to account for its observed states; the *range* of a character is thus one less than the number of its character-states. If, however, a character changes more frequently than this minimum, its *length* on the network (the sum of all its changes) will exceed the range. The *consistency* of a character is then the ratio of range to length. For example, a two-state character that changes twice, instead of only once, has a consistency of ½ or .5. Since the length of the entire tree is just the sum of all calculated character-state changes, the consistency of the *tree* is similarly the ratio of the sum of the ranges of all characters to the tree's length.

Once a tree has been constructed, and consistencies have been computed for it and the characters, an objective decision can be made about the cladistic usefulness of the particular characters. Those with poor consistencies might be eliminated from the data-set, or more reliable ones might be weighted for subsequent re-analysis. The Wagner method is thus a useful tool for examining the structure of data-sets.

Of course, a Wagner network is not truly phylogenetic unless some orientation or general direction of evolution is assumed. The most obvious way to *root* a Wagner tree is on the most plesiomorphous taxon; if this is not surely known, use of an out-group or rooting between the most divergent taxa (tantamount to assuming roughly equal rates of change in the group) are alternatives.

As already noted, the Wagner algorithm is but one heuristic approach designed to construct a "parsimonious" or minimum-pathlength dendrogram from many characters. Like all such except linear programming (Fitch and Smith 1982), it is not guaranteed to find *the* most parsimonious tree but only an approximation to it. It achieves this aim by several shortcuts, one effect of those being that the order in which the data are entered may affect the shape of the final network. Knowing this, we make no special claims of parsimony for the following trees. They are valid only for the sequence of taxa shown in Table 2, although some limited experiments with altering the ordering of input did produce similar trees. The results of these computations should be considered as a *guide* to relationships among the taxa (and characters), and not taken too literally.

Another point about these trees is that we largely avoided encoding autapomorphies because they provide no information about branching sequences; by definition only shared, derived character-states can do that. However, neither did we make a fetish about eliminating autapomorphies, as they do add to stem-lengths and therefore give some estimate of the phenetic divergence of taxa possessing them (see also Archer 1982). Yet it is useful to note which those characters are; and of course, they will be different for differing subsets of taxa, since in smaller sets of taxa only one may possess a state which two or more share in a more extensive sampling. Similarly, some characters are invariant and thus non-informative on small subsets of taxa. For convenience in comparing trees, however, they are referred to as based on (for example) "33" or "24" characters whether or not some of those features are invariant or autapomorphous (Figs 36, 37, 53, 58-67; Tables 3-5).

A third point is that each taxon must join the tree at some cycle of clustering however few the states it shares with other taxa represented at the node where it attaches; joining may occur late and in effect be a grouping by exclusion, hence rather untrustworthy. This interpretation is supported by the change in position of such late-joiners when more or fewer taxa are analyzed in different trees. Accordingly, we view late-joining taxa as being of uncertain relationship.

Table 3 lists some of the trees computed. The largest set for which we had information on all 45 characters, including reproduction, karyotypes, and serology, was 11. The initial section of description concerns trees based on the full

Table 3. List of Wagner trees computed from the data of Table 2. Columns indicate the taxa included, and the rows give the characters used. Consistency ratios (range/length; see WAGNERIAN AND INFORMATIONAL ANALYSIS for explanation) are given for each tree, as are the numbers of non-informative (or invariant; "nc"), autapomorphous ("aut."), and perfect (or non-homoplastic; "pc") characters in each case. All trees were rooted on an hypothetical *Ancestor* primitive in every character, included in each set of taxa. The symbol "Ø" means that computation could not be performed, usually because of incomplete scoring for those taxa; * means that *Lutreolina* was assumed "4" in character 33. All trees based on fewer than 45 characters omitted reproductive features, except for trees based on just those nine characters.

Chs. \ Taxa	9	11	12	28	28 OPT	33
73: all opossum inc. repro., and dasyuroid	83/134 = .62 15 nc 10 aut. 27 pc	Ø	Ø	Ø	Ø	Ø
45: all opossum inc. repro.	62/ 97 = .64 4 nc 7 aut. 19 pc	65/107 = .61 4 nc 6 aut. 15 pc	Ø	Ø	Ø	Ø
36: all opossum less repro.	49/ 71 = .69 4 nc 6 aut. 18 pc	52/ 79 = .66 4 nc 5 aut. 15 pc	53/88 = .60* 3 nc 6 aut. 12 pc	Ø	Ø	Ø
35: all opossum less repro. and karyotypes	47/ 68 = .69 4 nc 6 aut. 18 pc	50/ 76 = .66 4 nc 5 aut. 16 pc	51/85 = .60* 3 nc 6 aut. 14 pc	Ø	Ø	Ø
34: all opossum less repro. and serology	43/ 63 = .68 4 nc 6 aut. 17 pc	46/ 73 = .63 4 nc 5 aut. 13 pc	47/82 = .57 3 nc 6 aut. 10 pc	Ø	Ø	Ø
34: all opossum less repro., 21, and 23	48/ 68 = .71 3 nc 6 aut. 17 pc	51/ 75 = .68 3 nc 5 aut. 15 pc	52/84 = .62* 2 nc 6 aut. 13 pc	Ø	Ø	Ø
33: all opossum less serology and karyotypes	41/ 61 = .67 4 nc 6 aut. 16 pc	44/ 71 = .62 4 nc 5 aut. 12 pc	45/79 = .57 3 nc 6 aut. 11 pc	50/153 = .33 0 nc 1 aut. 4 pc	50/175 = .29 0 nc 1 aut. 4 pc	51/170 = .30 0 nc 1 aut. 3 pc
31: all opossum less serology kary., 21, and 23	40/ 57 = .70 3 nc 6 aut. 15 pc	43/ 64 = .67 3 nc 5 aut. 14 pc	44/72 = .61 2 nc 6 aut. 12 pc	48/143 = .34 0 nc 0 aut. 3 pc	48/166 = .29 0 nc 0 aut. 3 pc	49/163 = .30 0 nc 0 aut. 2 pc
24: opossum dental only	30/ 48 = .63 3 nc 2 aut. 10 pc	32/ 52 = .62 3 nc 2 aut. 9 pc	33/57 = .58 2 nc 3 aut. 9 pc	35/113 = .31 0 nc 1 aut. 2 pc	35/142 = .25 0 nc 1 aut. 3 pc	35/128 = .27 0 nc 1 aut. 2 pc
22: opossum dental less 21 and 23 only	29/ 43 = .67 2 nc 2 aut. 11 pc	31/ 49 = .63 2 nc 2 aut. 9 pc	32/53 = .60 1 nc 3 aut. 9 pc	33/102 = .32 0 nc 0 aut. 2 pc	33/133 = .25 0 nc 0 aut. 2 pc	33/121 = .27 0 nc 0 aut. 1 pc
9: reproduction only	13/ 19 = .68 0 nc 1 aut. 4 pc	13/ 17 = .76 0 nc 1 aut. 5 pc	Ø	Ø	Ø	Ø

data-set. Scores for 28 additional characters on nine of these 11 taxa were available from a previous study (Kirsch and Archer 1982), so trees of nine taxa are considered along with those of 11. In addition, *Lutreolina* was scored on 33 of the 45 characters, so that 12-genus trees including *Lutreolina* are also discussed in this section. In order to calculate trees from as many characters as possible, serological affinities were assumed in three instances.

The states of 33 characters were also known or inferred for an additional 16 didelphimorph genera and five borhyaenoids, providing a basis for calculating trees of 28 and 33 taxa, respectively. Also, the 33 characters or subsets of them

were fitted to a predetermined tree (approximating our final estimate of relationships) using that option of the Wagner 78 programme package in order to determine the length of the "preferred" cladogram.

Finally, some of the 11- and 28-genus trees were assessed for their information contents using various subsets of the 45 characters. This procedure (Kirsch 1983; Smith and Kirsch in prep.) gives some indication of how well differing dendritic arrangements distribute the potential information in variable characters.

Descriptions of Trees

The conventions of the drawings (Figs 36, 37, 53, 58-68) are that the most plesiomorphous taxa are to the right and/or bottom. Distances from the nodes in all figures except Figures 58 and 68 are proportional to the number of "evolutionary steps", and it is helpful to remember that branches may be rotated around the node from which they emanate.

Trees of 9, 11, and 12 Genera: Figure 58 shows, in schematic form, 29 trees calculated on various subsets of characters and taxa; Figures 36, 53, and 59-63 depict five of the 11-taxon trees and one each of nine and 12 genera from Figure 58, drawn with distances proportional to evolutionary steps. Figure 59 represents the 11-taxon tree computed on all 45 characters. The arrangement is typical and expected from considerations of certain individual characters. *Didelphis* and *Philander* are paired, with *Chironectes* and *Metachirus* successively less close; the four are the sister-group of the four "small" didelphines, of which *Monodelphis* and *Thylamys* are closest, with *Marmosa* and then *Micoures* less so. All eight together are the sister-group of *Caluromys* with *Dromiciops*. These relationships are stable throughout considerable expansion and contractions of the data-set, except that *Philander* and *Chironectes* are sometimes interchanged, as are *Monodelphis* and *Marmosa*, and *Metachirus* sometimes joins the group of small opossums rather than the large (Fig. 58L-U). In some instances *Caluromys* and *Dromiciops* are not linked and then the latter is the more distant from didelphines. Allowing for the absence of two of the small opossums, the tree of nine taxa based on this character-set augmented by 28 dasyurid features (Kirsch and Archer 1982, Fig. 60) is identical cladistically to that of 11 genera; the calculated distances are, of course, much greater.

Omission of the nine reproductive characters yields the tree of Figure 61, where *Caluromys* and *Dromiciops* have separate origins. The nine reproductive features alone give a poorly differentiated tree (Fig. 53), with almost exactly the opposite sequence of derivation. (Except that *Chironectes* is closer to *Caluromys*, the tree of nine genera based on reproductive features is similar; cf. Fig. 58K.) Note, however, that *Dromiciops* and *Caluromys* are among the most derived of the eleven. Thus, reproductive characters are probably one reason for their association in the 45- or 73-character trees (Figs 59, 60). At the same time, other characters must also bring about their association, since *Dromiciops* and *Caluromys* also remain together in the 36-character tree of nine taxa (Fig. 58C). However, a 36-character tree including *Lutreolina*, where serum of that genus is presumed to give serological reactions like those of *Chironectes*, separates *Caluromys* and *Dromiciops* (Fig. 58V). In this and other features that tree is like the one of 11 genera; *Lutreolina*, *Chironectes*, and the *Didelphis-Philander* pair form an unresolved trichotomy.

When karyotypes are eliminated from the scoring there is only one change in both the 11- and 12-genus trees (Fig. 58N, W): *Marmosa*, rather than *Monodelphis*, is the nearest relative of *Thylamys*. (*Thylamys* and *Micoures* are absent from the nine-genus set, so that cladistic relationships on that tree — Fig. 58D — are unaltered from ones calculated on more characters.) Once again, serological affinities of *Lutreolina* have been assumed for this calculation.

When serology is eliminated from the set of characters, *Monodelphis* is again the sister-group of *Thylamys* (in 11- and 12-genus trees; Figs 62, 63). *Chironectes* rather than *Philander* is now the closest relative of *Didelphis*; the latter relationship holds for all three taxonomic sets (cf. Fig. 58E). *Caluromys* and *Dromiciops* are united in a single lineage, suggesting that serology is critical in maintaining their independent cladistic origins. Note that this is the largest set of characters for which a tree including *Lutreolina* could be calculated without inferring any data.

Another set of trees was calculated from 34 characters, but in this instance eliminating character 21 (size and shape of premolars, which is autapomorphous for *Didelphodon*, and therefore invariant in this set of taxa) and the highly inconsistent character 23 (relative sizes of P2's and P3's — chosen for elimination retrospectively, after it was found to be the "worst" character on 28- and 33-taxon trees, changing eight times). Trees of nine, 11, or 12 genera were changed from previous ones calculated on 34 only in that *Marmosa* was again the closest relative of *Thylamys* (Fig. 58F, P, Y).

Elimination of *both* karyotypes and serology, however, caused major changes (Fig. 58G, Q, Z). In all three cases *Metachirus* became the sister-group of the small didelphines rather than of the large, and *Didelphis* and *Philander* were again each other's nearest relatives. In addition, *Chironectes* became the out-group to all other didelphines in the 11- and 12- but not 9-taxon trees. *Lutreolina* was paired with *Monodelphis* in the 12-taxon tree, a feature of all other 12-taxon cases discussed below as well.

When serology, karyotypes, invariant character 21, and inconsistent character 23 were all eliminated, leaving just 31 characters (Fig. 58H, R, AA), *Chironectes* again became the sister-group of *Didelphis*, and *Metachirus* became the sister-group of all other large opossums in 11- and 12-taxon trees; *Lutreolina* remained with *Monodelphis*. The 9-taxon tree was considerably more altered — *Philander* became the sister-group of the pair of pairs *Didelphis-Chironectes* and *Monodelphis-Marmosa*, and *Metachirus* was the sister-taxon of all five.

One example of the next set of trees, based just on the 24 dental characters and drawn with

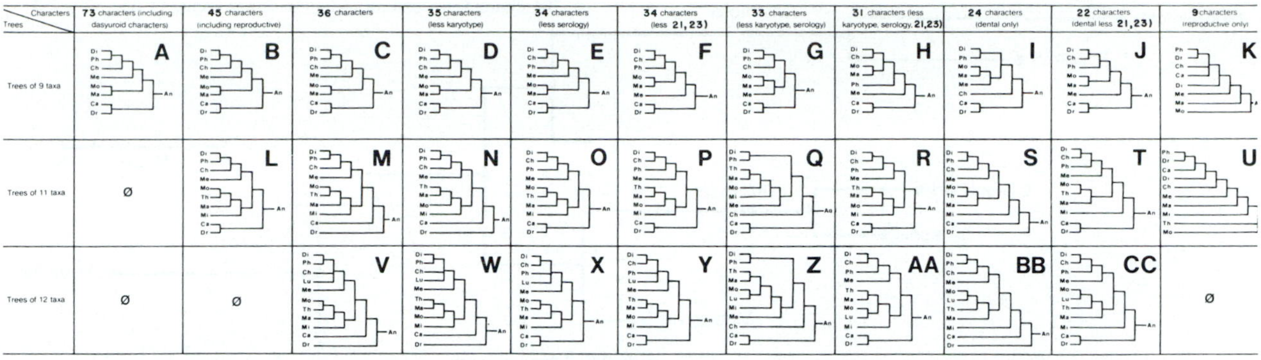

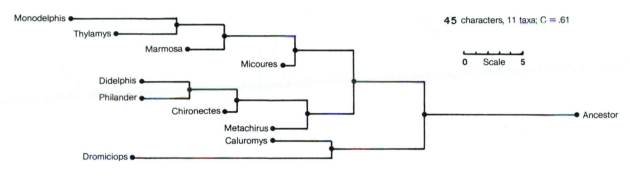

Fig. 58. Cladograms of 9, 11, and 12 taxa under the various regimes of characters indicated by the column-headings, and presented as phenograms for clarity. Consistencies and other details are given in Table 3. Figures 58 L, A, M, U, O, X, and S are shown as Wagner trees in Figures 59-61, 53, 62, 63 and 36, respectively. Abbreviations: Di =*Didelphis*, Ph = *Philander*, Ch = *Chironectes,* Me = *Metachirus*, Lu =*Lutreolina*, Ma = *Marmosa*, Th =*Thylamys*, Mi = *Micoures*, Mo =*Monodelphis*, Ca = *Caluromys*, Dr =*Dromiciops*, An = *Ancestor*.

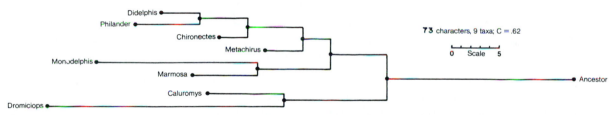

Fig. 59. Wagner tree of 11 taxa computed on the basis of all 45 characters listed in Table 2, with overall consistency of .61. Conventions as for Figure 36.

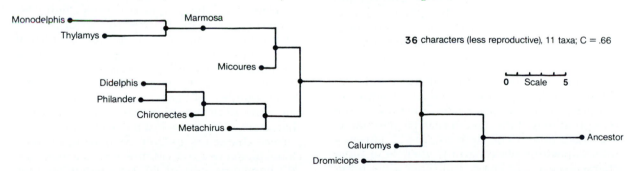

Fig. 60. Wagner tree of nine taxa computed on the basis of all 45 characters listed in Table 2 and the 28 "dasyuroid codings" of Kirsch and Archer (1982), with overall consistency of .62. Conventions as for Figure 36.

Fig. 61. Wagner tree of 11 taxa computed on the basis of the first 36 characters listed in Table 2, omitting nine reproductive features, with overall consistency of .66. Conventions as for Figure 36.

proportional distances, is shown in Figure 36 (for 11 genera). Here, *Didelphis* and *Philander* are paired, with *Chironectes* and *Metachirus* successively less close. *Monodelphis* and *Thylamys* are also paired, but *Micoures* is the sister-group not just of these, but of all other didelphines. *Caluromys* and *Dromiciops* have separate origins. All these points hold for the 12-genus tree (Fig. 58BB), and

Lutreolina remains the closest associate of *Monodelphis*. In the 9-taxon tree, however, *Metachirus* is the sister-group of *Marmosa* plus *Monodelphis*, and *Chironectes* is that of all other didelphines (Fig. 58I). *Caluromys* and *Dromiciops* are united, so that the bearings of just dental characters on their relationships are unclear.

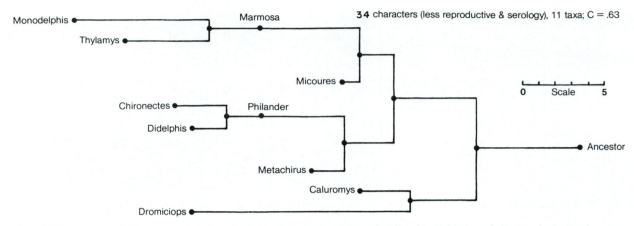

Fig. 62. Wagner tree of 11 taxa computed on the basis of characters 1-33 and 36 listed in Table 2, omitting serological and reproductive features, with overall consistency of .63. Conventions as for Figure 36.

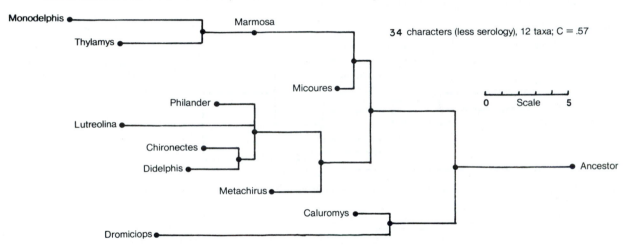

Fig. 63. Wagner tree of 12 taxa computed on the basis of characters 1-33 and 36 listed in Table 2, omitting serological and reproductive features, with overall consistency of .57. Conventions as for Figure 36.

That ambiguity is emphasized when characters 21 and 23 are eliminated, since one result is that *Caluromys* and *Dromiciops* are united in all three trees (Fig. 58J, T, CC). The only other change in the 11- and 12-taxon trees is that *Didelphis* and *Chironectes* are again paired. Because that is true in the 9-genus tree as well, *Chironectes* is no longer the sister-group of the remaining didelphines; instead, *Metachirus* occupies that position.

In sum, experiments with a limited number of extant opossum genera showed a fair degree of uniformity in the placement of these, even though subsets of the data were in partial conflict with each other. The derived, 2n=22 chromosomes opossums always group when 34 or more characters are employed, and often when fewer are used. Similarly, the small opossums *Monodelphis, Marmosa, Micoures,* and *Thylamys* form a group that is stable on as few as 31 characters. *Metachirus* is generally associated with other large opossums, but occasionally with the small; in either event, it is relatively primitive with respect to its sister-group. It is either the case that *Caluromys* and *Dromiciops* are associated apart from didelphines or represent two lineages increasingly removed from true didelphines.

Dental characters by themselves preserve most features of the tree, except that *Chironectes* or *Micoures* is then the most distinct of the didelphines.

Technically, these trees had quite high consistencies, averaging .64 over the set of 29 considered here. While there is no clear pattern of change in consistencies among the elements of the columns of Table 3, a trend is clear among the cells of each row. Consistencies always decrease as the number of taxa increases, except in the single case of the two trees based on reproductive characters. Only in two instances is the number of perfect characters greater than half the total character-set (for trees of nine taxa calculated on 22 characters, and of 11 taxa based on nine reproductive features); but again the proportion decreases with an increase in the number of taxa.

The relationships depicted were in close agreement with inferences from serology and, so far as the limited variation permits, from chromosome numbers. The repeated placement of *Metachirus* near the base of the dichotomy within didelphines is consistent with the trifurcation indicated by serological comparisons. Where the serological phylogeny and Wagner trees of anatomy differed most strikingly was in the (sometime) association of *Caluromys* and *Dromiciops*. The reasons that these genera separated or associated are not obvious, but must include critical features of the dentition. Evidently also the presence of additional genera of small opossums is important, because *Caluromys* and *Dromiciops* were never separate in 9-genus trees (Figs 58A-K, 60), and the integrity of both small and large opossums is threatened in such trees based on fewer than 33 characters. This fact indicates that the number, and identity, of the taxa under consideration may be as important as the characters themselves, a point made even more clearly by trees relating many more genera.

Trees of 28 or 33 Genera: Figures 37 and 64-67 show trees calculated for 28 or 33 genera. The additional taxa include the two other caluromyines, *Lestodelphys*, and representatives of all major taxa of fossil opossums plus (in the case of 33-genus trees) five borhyaenoids. Since no data on softpart anatomy were available for the fossils, and only in fragmentary form for the three additional living genera, the maximum number of characters that could be examined was 33. It will be recalled that it was with restriction of the data-set to this number that groupings began to change in the smaller trees.

The first of the large trees to be described is that of 28 taxa calculated on 33 features (Fig. 64). In many respects it is similar to smaller trees based on a like number of characters. The small opossums, now including *Lestodelphys* and *Thylatheridium*, form a discrete group, as do of the large opossums augmented by various fossil taxa. *Metachirus* is the sister-group of the large opossums, and *Chironectes* is between it and the remainder.

Unlike in the smaller 33-character trees, the caluromyines and *Dromiciops* have distinct origins. In fact, *Dromiciops* is but one member of a branch representing the most basic dichotomy within didelphoids, and is (with *Microbiotherium*) the most derived member of a sequence running from *Alphadon* and *Didelphodon* through *Peradectes, Caroloameghinia, Pediomys,* and *Glasbius*.

There are thus a few unexpected relationships: *Peradectes* is not the sister-taxon to *Alphadon*. *Peratherium* might also be expected to lie near the base of the tree, but is instead the sister-group of didelphines including *Sparassocynus*, which is part of the 2n=22 chromosomes large opossum group.

Neither of those last two relationships are materially altered when character 21 and character 23 are eliminated from the data-set (Fig. 65).

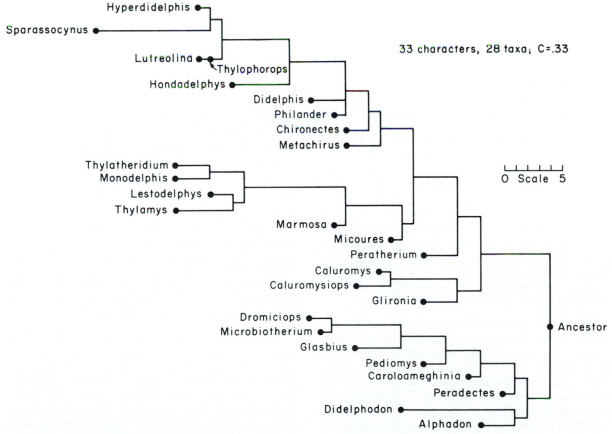

Fig. 64. Wagner tree of 28 taxa computed on the basis of characters 1-33 listed in Table 2, with overall consistency of .33. Conventions as for Figure 36.

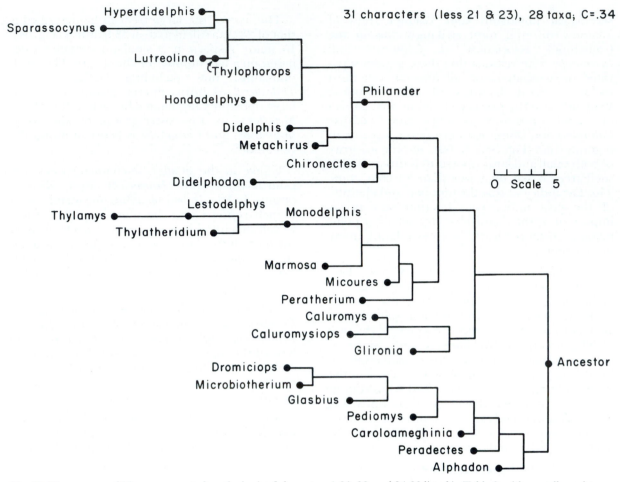

Fig. 65. Wagner tree of 28 taxa computed on the basis of characters 1-20, 22, and 24-33 listed in Table 2, with overall consistency of .34. Conventions as for Figure 36.

However, *Didelphodon* is then paired with *Chironectes*, and *Peratherium* is more closely associated with the small living opossums; also, *Metachirus* is the closest taxon to *Didelphis*. Other relationships remain much as in Figure 64.

As was true for smaller trees when only dental characters were used (Figs 36; 58I, S, BB), the integrity of most didelphid groups breaks down in the larger tree (Fig. 37) based on dentition. Nevertheless, the lineage including *Dromiciops* remains unified, as one would hope since most of the non-dental characters of included taxa are primitive or were assumed so if not known; in particular, *Didelphodon* is part of that group. Moreover, the didelphoid branch of the tree at least clearly reflects the increasing carnivory typical of opossum evolution.

The theme of ever-increasing carnivory is also borne out in the largest trees, of 33 taxa, which include five borhyaenoids (Figs 66, 67). As might be expected, *Sparassocynus* is closely associated with these five; less expected is that *Lutreolina* and other large opossums with which *Sparassocynus* was associated in Figures 37, 64, and 65 move to a derived position among the more carnassialized of the small opossums (*Lestodelphys*, *Thylatheridium*, and *Monodelphis*) in the tree based

on 33 characters (Fig. 66). While the unity of the large didelphines is thus compromised, those remaining apart from the small opossums are quite closely related, and include both *Chironectes* and *Metachirus*, but also *Didelphodon* (which is, not unexpectedly in view of their shared pre-dilambdodonty, paired with *Hondadelphys*). *Peratherium* remains with the small opossums.

Elimination of characters 21 and 23 from the set produces the tree of Figure 67, the form of which is close to that of Figure 65. That is to say, both the large and small opossum groups retain their integrity (the borhyaenoids and *Sparassocynus* derive from the former, and *Chironectes* (along with *Didelphodon*) is the most primitive member of it, a sister-group to all the rest). The structure of the lineage culminating in *Dromiciops* is unchanged from that in Figure 66.

The arrangement of taxa in the tree of 33 genera based only on dental characters (not shown) is similar to that in the corresponding 28-taxon tree (in that the large and small opossums are intermingled), except that the five borhyaenoids are added to the *Sparassocynus* lineage. Neither of the 28- and 33-genus trees based on dental characters without 21 and 23 is depicted here. The basic dichotomy of previously

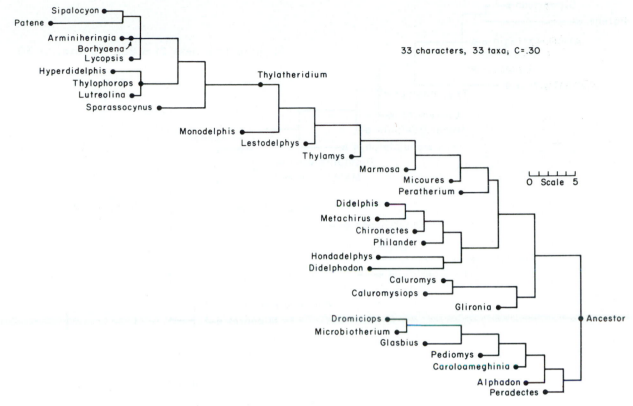

33 characters, 33 taxa; C=.30

Fig. 66. Wagner tree of 33 taxa computed on the basis of characters 1-33 listed in Table 2, with overall consistency of .30. Conventions as for Figure 36.

described trees is preserved, with the same internal structure to the lineage including *Dromiciops* (*Didelphodon* is part of that branch); didelphoids again show only increasing carnivory and not the anticipated subgroupings.

In summary, the larger trees extend most of the conclusions drawn from trees of nine, 11, or 12 taxa. Most significantly, they confirm the separation of *Dromiciops* and demonstrate a coherence of several fossil taxa in its clade. The most difficult results to accept are the placement of *Peratherium* with the small opossums and of *Sparassocynus* (with the borhyaenoids) generally close to the large ones, together with the tendency of *Didelphodon* to alter its allegiance. Once again the consistencies of these trees (averaging about .30) drop as more taxa are added; so too do the numbers of perfect characters, which are never more than four in 28- or 33-taxon trees.

Tree-fitting: The tree shown in Figure 68, which, as we shall indicate in the GENERAL DISCUSSION is our preferred overall cladogram of didelphoid relationships, was fitted with 31, 33, 24, and 22 characters. Since the tree-fitting option of Wagner 78 will only accept strictly bifurcating trees, it was necessary to resolve the polychotomies of Figure 68 somewhat arbitrarily. *Didelphodon* was placed closer to *Pediomys* than was *Caroloameghinia*. *Didelphis* became the nearest relative of *Hyperdidelphis*, with *Chironectes*, *Philander*, and *Hondadelphys*

successively less close and all forming the sister-group of *Lutreolina* plus *Thylophorops*; *Metachirus* was the sister-group to all of these. An all-plesiomorphic Ancestor was added as the outgroup.

The programme found no characters to distinguish levels of relatedness among *Peratherium*, *Sparassocynus*, and the didelphids, considering these three lineages members of an unresolved trichotomy (see also the section on *Information-assessment*). In all cases the preferred tree proved somewhat longer than the comparable tree calculated from those characters by the Wagner 78 programme itself, as indicated by lower overall consistencies (see Table 3). However, character 23 is not the most inconsistent character on these trees; character 16 is.

Information-assessment: While it is true that the preferred cladogram is longer in each case tested than the tree calculated from that data-subset by the Wagner programme, it is also true that in all cases the preferred tree was more informative by a wide margin (whether it was strictly bifurcating or with multifurcations as shown in Figure 68); this fact is indicated in Table 4, where the information contents of 28-taxon trees are given for various subsets of characters. It is interesting to note that trees based on a greater number of characters are better at recovering dental information than the dental tree itself, paralleling results described in Kirsch (1983) and Smith and Kirsch (in prep.) for other data-sets.

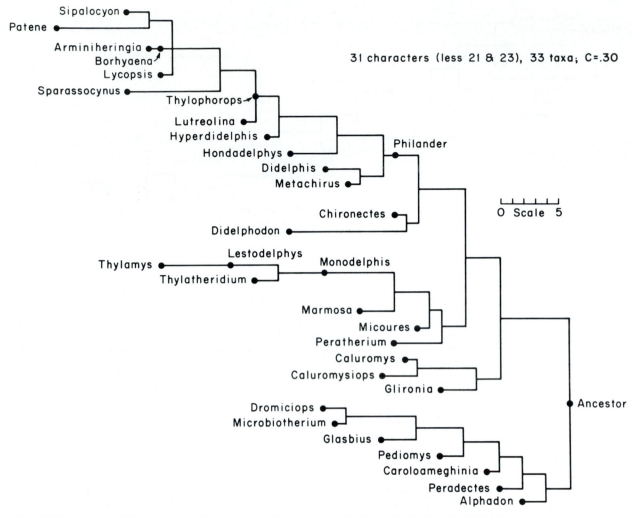

Fig. 67. Wagner tree of 33 taxa computed on the basis of characters 1-20, 22, and 24-33 listed in Table 2, with overall consistency of .30. Conventions as for Figure 36.

A similar point is made in Table 5, which records the informativeness of some 11-taxon trees, including that based just on reproduction. The last is extremely uninformative even about reproductive features, while the best trees are those which unite *Caluromys* and *Dromiciops*. These are ones based on 45 or 33 characters, identical cladistically except in the pairing of *Didelphis* with *Philander* in the first but with *Chironectes* in the second.

Discussion

The results described above are encouraging because some relationships seem real and stable throughout all or most regimes of coding and taxa, but they are also discouraging because even small alterations in scoring or the elimination of characters can sometimes re-place taxa. The combination of robustness and sensitivity considerably undermines confidence in abandoning judgment to the computer. This being the case, we follow the approach of Carleton (1980), who chose to look for commonalities among his trees but not to follow slavishly the indications of any one of them.

In fact, some reasons that are intrinsic to the algorithm, as well as some that are biological, can be suggested to account for our difficulties; in the end the assessment of those reasons is as interesting as any taxonomic discoveries that have been made employing the Wagner programme.

To begin with, there are the above-mentioned problems of constructing a parsimonious tree. Given any moderate degree of homoplasy, or a large number of taxa, the computational shortcuts will have different effects depending on the sequence in which taxa are encountered in the iterative procedure. This is why there is an ordering effect, which we have minimized by always submitting the data in the same way. A second point is that, since the programme operates by sequential addition of taxa, the last taxa to join are more likely to be mis-placed because a greater number of compromises will normally have accumulated. As Fitch and Smith (1982) have shown, a truly minimum-pathlength tree *must* have zero homoplasy between terminal sister-taxa; the proportion of homoplasy, and hence the uncertainty of the branching sequence, thus

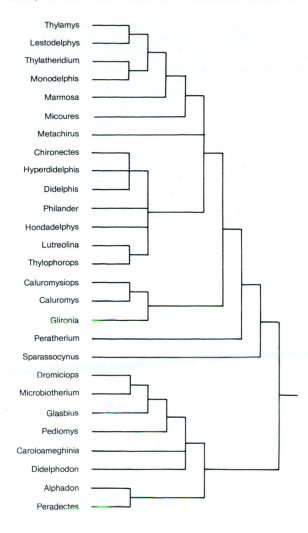

Thylamys
Lestodelphys
Thylatheridium
Monodelphis
Marmosa
Micoures
Metachirus
Chironectes
Hyperdidelphis
Didelphis
Philander
Hondadelphys
Lutreolina
Thylophorops
Caluromysiops
Caluromys
Glironia
Peratherium
Sparassocynus
Dromiciops
Microbiotherium
Glasbius
Pediomys
Caroloameghinia
Didelphodon
Alphadon
Peradectes

Fig. 68. Final cladogram of 27 non-borhyaenoid didelphimorph genera, based in part on Wagnerian analysis of the characters listed in Table 2 but taking into account informal consideration of other features, presented as a phenogram for clarity. Although the line-segments do not have dimension, more derived taxa are in general toward the top of the figure.

increases as more nodes are interposed, and the last taxa to join are accordingly of most uncertain position. Since these are normally also the most distinct (such as *Dromiciops* or *Sparassocynus*), *any* affinities they show to other taxa on the tree are bound to be based on proportionally fewer synapomorphies. A related point is that the most primitive taxa (here, the putative ancestor and ones like *Alphadon*) will be mostly symplesiomorphous, and the branching sequence in their part of the tree (involving the early, and often most interesting, furcations) will be based on relatively few differences. The same problem can afflict derived taxa as well, which may be separated from each other by as few differences as are more plesiomorphic relatives; this is why certain close pairs of genera often changed position with minor alterations in the number of characters used.

These, and other, technical problems with parsimony methods remove any reason for feeling embarrassed about rejecting *specific* results of an analysis, at least if there are good biological considerations mandating acceptance of an alternative relationship.

Actually, the notion of a shortest pathlength (whether achievable or not) is a strange one to equate with Ockham's philosophical concept of parsimony. As further discussed below in the GENERAL DISCUSSION, there is a "higher-level" parsimony which depends more on the integration of character-sets which one has good reason to believe are under different selective regimes, or to be subject to characteristically different modes of evolution. Thus, congruence among characters is less likely to be significant when the characters are all of the same kind (or perhaps are correlated, as are the shapes of occluding upper and lower molar teeth) than when there are grounds for believing them to be independent in their evolution.

In fact, one of the points of these experiments was to test for congruence between the serological and anatomical phylogenies. As will be noted below, that congruence is close, yet the assumption underlying the use of biochemistry in phylogeny-reconstruction is that the characters evolve time-dependently and divergently, while anatomical ones are manifestly irregular in their rates of change and subject to much homoplasy.

Yet, even among the latter characters there is some evidence for congruence, in that similar associations of, for example, the small opossums are indicated by dental and reproductive characters. At the same time, of course, the relative *position* of these didelphids is different when based on one set or the other. In dentition the murine group is among the most derived; their reproductive anatomy, however, is primitive. There is really no incompatibility here — the fundamental heterobathmy (mosaic evolution) of the opossums involves just this sort of disjunction between soft and hard parts — but there is more than a hint that trees based on all of the characters may represent a compromise among different groupings.

Thus, in spite of the fact that few, if any, of the anatomical characters are reliable in the usual sense (most of them are mathematically inconsistent in the larger trees), nearly all seem necessary to give a tree that accords with a "Gestalt" appreciation of the animals and the conclusions suggested by certain sorts of data (e.g., serology and karyotypes). Nevertheless, often the closest approximations to expectation are achieved when characters 21 and 23 are eliminated (Figs 65, 67). No special justification can be given for disregarding them: character 21 is autapomorphous and irrelevant in any case, and while 23 is highly inconsistent, other characters

Table 4. Informativeness of selected Wagner trees of 28 taxa and two versions of the "final cladogram" (rows) on various subsets of the characters listed in Table 2 (columns). Consistencies of trees are given (first column), as are the number of internal nodes ("M") in each tree. All figures are in bits; numbers in parentheses are the average information per character per internal cluster.

Tree	C	M	all 33 chs.	31 chs.	Subset 24 chs.	22 chs.	2 chs.
On all 33 chs.	50/153 = .33	24	331.58 (.419)	325.04 (.437)	308.26 (.535)	301.73 (.571)	6.54 (.136)
On 31 chs. (-21, 23)	48/143 = .34	25	348.29 (.422)	347.46 (.448)	324.90 (.542)	324.07 (.589)	0.83 (.017)
On 24 chs. (dental)	35/113 = .31	24	115.33 (.146)	111.93 (.150)	204.13 (.354)	200.74 (.380)	3.39 (.071)
On 22 chs. (dental -21, 23)	33/102 = .32	23	213.06 (.281)	211.85 (.297)	266.83 (.483)	265.63 (.525)	1.20 (.026)
Bifurcating Final Cladogram (fitted)	33: 50/175 = .29 31: 48/166 = .29 24: 35/142 = .25 22: 33/133 = .25	26	286.37 (.334)	283.39 (.352)	250.60 (.402)	247.62 (.433)	2.97 (.057)
Final Cladogram with Multifurcations	–	20	369.47 (.560)	362.44 (.585)	308.35 (.642)	301.33 (.685)	7.02 (.176)
Maxima	.34	26	876.71	843.39	682.95	649.63	33.32

Table 5. Informativeness of selected Wagner trees of 11 taxa (rows), on various subsets of the characters listed in Table 2 (columns). Consistencies of trees are given (first column), as are the number of internal nodes ("M") in each tree. All figures are in bits; numbers in parentheses are the average information per character per internal cluster.

Tree	C	M	all 45	36 chs.	Subset 34 chs.	24 chs.	21 and 23	9 chs.
Tree 1 (all 45)	65/107 = .61	9	228.36 (.604)	175.77 (.610)	149.80 (.555)	107.76 (.570)	3.96 (.440)	44.19 (.488)
Tree 2 (36 chs.)	52/79 = .66	9	185.49 (.491)	138.91 (.482)	116.86 (.433)	88.32 (.467)	3.93 (.437)	33.96 (.377)
Tree 3 (34 chs. −serology)	46/73 = .63	9	228.36 (.604)	171.77 (.596)	147.80 (.547)	105.76 (.560)	1.96 (.218)	48.19 (.532)
Tree 4 (9 chs., repro.)	13/17 = .76	7	-35.49 (-.122)	-9.93 (-.044)	-19.09 (-.091)	-32.65 (-.222)	1.62 (.231)	4.91 (.058)
Maxima	.76	9	455.89	357.66	322.68	228.69	10.34	99.81

are nearly as labile on the smaller trees. More importantly, and as Kirsch and Archer (1982) observed, it appears that the very structures of the trees are dependent on homoplasy; rejection of a character for inconsistency alone is thus not logical. Close examination of the "synapomorphies" of most taxa reveals that those states are only rarely unique. Rather, monophyletic groups are characterized by *combinations* of states which have each evolved independently in other portions of the tree. These combinations are the *true* synapomorphies, and even so groups are often of a polythetic nature.

One might then validly question why the groupings are as intuitively pleasing as they are, given that virtually the entire data-set is homoplastic. The answer most likely involves the fundamental taxonomic problem of definition and choice of characters: we have simply dissected the animals, and particularly their dentitions, too finely. An analogy from biochemical systematics may clarify that point. Little is learned about the relationships of living organisms from compiling

percentages of individual amino acids or nucleotides in their proteins or DNA, but a great deal is gained from considering *sequences* of such acids or bases. Similarly, the presence or size of an individual tooth-cusp rarely means much by itself, while a combination of such details making up a functioning tooth or dentition provides the true (and, we suspect, less-often-convergent) character. It is no accident that our rather simple codings of the complex otic and karyotypic features appear much more consistent: they denote *real* characters.

Nonetheless, it is not easy to recognize the appropriate level of character-definition *a priori*. In any case, our entire set of 45 characters appears to provide a delicate balance of opposing multiply convergent features, which only gives stable structures if fewer than a dozen are eliminated (most of those being reproductive). It is easy to understand that a few associations at the termini (of *Didelphis* with *Philander* or *Chironectes*, and of *Monodelphis* or *Marmosa* with *Thylamys*) are labile, since these genera are closely related in any case. But the association or lack of it between *Dromiciops* and *Caluromys* dramatizes how a high-level relationship can also be uncertain.

There are several reasons why *Dromiciops* and *Caluromys* associate: the reproductive characters provide one; others include a few superficial resemblances in the molar teeth; and finally, the ear regions have been naively coded as part of a single sequence, though the conditions in *Dromiciops* and *Caluromys* are most probably of independent origin. On the other hand, serology and character 23 provide sufficient separation to produce independent origins for *Dromiciops* and *Caluromys* in trees of fewer taxa. Moreover, *Dromiciops* is, on the small trees at least, the last taxon to join the network. That would seem to indicate that when it is being grouped with *Caluromys* the association is by exclusion — resembling nothing much at all, it nonetheless resembles *Caluromys* the most.

In marked contrast to its problematic placement on smaller trees, *Dromiciops* is always distinct from caluromyines in larger trees. The reason is that independent sequences of taxa leading to similar derived conditions in *Dromiciops* and *Caluromys* are documented by the included taxa. Thus, the tree-building algorithm is able to demonstrate that the similarities of *Dromiciops* and *Caluromys* (here at the level of combinations, not individual characters) have been arrived at by distinct if somewhat parallel routes. This is exactly like the situation for the remarkably alike *Thylacinus* and *Borhyaena* (Kirsch and Archer 1982): when only the convergently similar endpoints of lineages are present the Wagner program has no choice but to link them. Avoiding that result is a strong argument for including as many fossil taxa in an analysis as possible. Of course, this argument presupposes that the taxa of Late Cretaceous opossums and *Caroloameghinia* really are related to *Dromiciops* and *Microbiotherium*. There seems, in terms of the dental evidence at least, to be little doubt that they are. (However, the labile positioning of the later-joining *Didelphodon* may reflect instead that taxa near the base of the tree are almost entirely plesiomorphic, as a change in the state of a single character can alter allegiance from one branch of the dichotomy to another.)

But even when the entire panoply of taxa is included, some relationships remain problematic. Descriptively, it is certainly true that the borhyaenoids and *Sparassocynus* are highly carnivorous marsupials that can be derived from certain didelphines. Something similar may be said of *Peratherium*, always positioned close to the small opossums. But neither relationship is likely to be strictly true. Why is it that in some instances the programme is able to sort out parallelism but not for these taxa?

Part of the explanation lies in the recognition that increasing carnivory is a theme among didelphoids. As documented by the fossil record and analyzed by Marshall (1978b), there is a "relay" of various carnivorous marsupials through time in South America. While the borhyaenoids were apparently replaced first by phorusrhacoid birds and later by placental carnivores, at least some of their niches were filled by *Sparassocynus* and carnivorous didelphines. Any tree which attempts to relate a sampling of these several rather ecologically similar taxa drawn from a long chronological sequence (as do our larger trees) is almost bound to conflate the clades, and certainly *Sparassocynus* is unlikely to be *both* a derivative of some living opossum as well as the progenitor of borhyaenoids!

That confusion of lineages is exactly what we have demonstrated in the set of trees described above. Dendrograms including the fossil taxa have lower consistencies and depict doubtful relationships, while trees of the living taxa are much more consistent and probable. The difference is that the living opossums represent a single time-horizon, and probably a distinctly parcelled-out environment with little niche overlap; the fossil taxa represent sequential refilling of the same niches over a long time-period. Moreover, some of the taxa included in the larger trees are geographic as well as ecological or temporal vicars, providing a further source of confusion and convergence.

This explanation provides some rationale for prefering certain groupings over others. Given their compatibility with serology and other independent sources of data, as well as their chronological and geographic unity, the basic groupings of living didelphines indicated in the 11-genus (and most other) trees should inspire some confidence. On the other hand, we are not sure of the association of *Peratherium* with small opossums, since *Peratherium* and allies are known only from outside South America and from much earlier. Similarly, *Sparassocynus* and the borhyaenoids represent successively more remote vicars in time of the *Lutreolina*-like South American opossums with which they usually appear to be linked.

In short, we have considerable faith in the results of the Wagner analyses involving a chronologically uniform sampling (significantly, it is the sole persistent microbiothere, *Dromiciops*, whose relationship is most ambiguous here), which are confirmed by a truly parsimonious comparison of anatomical, biochemical, karyological, and other independent sources of data. The relationships depicted on the trees become increasingly less certain as we include taxa representing a greater temporal and/or geographical span, or fossils providing only fragmentary data. Technical constraints of the computer programme cast additional doubt on the sequence of branchings near the base of the tree or at late cycles of clustering. Nevertheless, it is hard to deny the possibility that some of the more surprising of these associations may be real.

It remains to comment on the information-assessment of these trees. The most striking results of those calculations is that informativeness (the ability to recover the information implicit in the data-set or subsets) does not track with consistency, for the shorter trees are not the most informative ones. That seems usually to be the case when data-sets have a large amount of homoplasy (Smith and Kirsch in prep.). The reason is that the "same" characters can be used more than once to provide discrimination, and hence information, in more than one portion of the tree. The Wagner algorithm, however, creates a tree that reduces, if it does not necessarily minimize, homoplasy, often subverting the discriminatory potential of homoplastic features. Departures from minimum-pathlength, such as our "preferred" cladogram (Fig. 68), may thus be more information-rich.

But the *basis* of our preferred arrangement, where both *Sparassocynus* and *Peratherium* are removed from the didelphids *s.s.*, is in fact additional data — data on anatomical characters not included in the codings as well as on chronological or geographical distributions. Though it may be something of a sophistry to argue so, there is a sense in which it is appropriate that longer trees should have a higher recovery of the information in a data-set, because additional information has been used to bring about the departure from "parsimony" which such trees represent.

GENERAL DISCUSSION

Introduction

The primary aim of this paper is to analyze and evaluate different sorts of data in order to gain a better understanding of the phylogenetic and taxonomic relationships of living and Neocenozoic didelphoid marsupials. Meeting this objective obliged us to review also the relationships of Cretaceous North American and Tertiary European and American representatives of the didelphimorphs, and furthermore resulted in a need to reassess not only the data but some theoretical and practical problems of phylogenetic inference.

The phylogenetic enterprise aims to discover and reconstruct the pathways of evolutionary change which have resulted in the diversity of living and extinct organisms. Cladistic analysis is one method of phylogenetic inference, which emphasizes the sequences of alteration in characters and the differential importance of derived over primitive states. Cladistics has contributed much to the formalization and sharpening of

what once was (and often still is) a sometimes imprecise endeavor. The distinction between primitive and advanced is not new [see for example the distinction between "habitus and heritage" of Gregory (1910), or between "paleotelic and cenotelic" of Camp (1923)]; what is new is the consistent application of such a distinction, and the insistence on avoiding paraphyletic or plesiomorphous taxa.

In principle one should need to discover only one reliable sequence of character-change in order to delineate a cladistic group and to document its evolution. In that respect the cladistic method is reminiscent of the nineteenth-century search for the one, true, and universally useful character, a search which was not notably successful. The reason that it was not, and the Achilles heel of cladistic (or any other phylogenetic) analysis, is the extreme vulnerability of single-character analysis to homoplasy. In fact, most conventional taxonomic characters are subject to repetitive evolution of similar character-states.

Indeed it could scarcely be otherwise, given the paradigm of natural selection, than that distinct lineages may undergo similar selective pressures, resulting in the appearance of apparently identical states in distinct lines that began from the same or even different initial conditions.

The pervasiveness of homoplasy suggests two alternative solutions to the problem it creates. The first is the adoption, implicitly or explicitly, of the epistemological principle of parsimony — the view that one should avoid the multiplication of explanations for observations beyond the minimum needed. In systematics, that usually means not inferring more independent appearances of the same state than necessary, and translates into the perhaps pious hope that a consistent pattern of relationships will emerge from the characters; when this is true not only for several characters of the same sort but also for highly independent *kinds* of characters (see below), the solution through parsimony can be effective indeed. Unfortunately, adoption of this methodological principle too often leads to a rigid insistence that the simplest hypothesis must accurately describe nature. The commonness of the difficulty from which parsimony represents one escape — homoplastic resemblance — should be adequate caution that nature is, far from being parsimonious, often quite profligate (see also Crisci 1982 and Kirsch 1983).

The other means of dealing with homoplasy is through reliance on one sort of data thought to be especially useful for indicating true cladistic relationships. This alternative amounts to another hopeful return to the nineteenth-century programme, and in the twentieth has taken the particular form of dependence on biochemical or karyotypic data. In the case of biochemistry, equation of resemblance with relationship is done with the faith that such characters evolve divergently (so that chemical similarities are not the result of convergence or parallelism) and time-dependently (so that degree of relationship is proportional to similarity). The evidence for both presumptions is fairly good (Wilson *et al.* 1977), and is supported by the growing awareness of the importance of regulatory changes to anatomy (e.g., Britten and Davidson 1969; Wilson *et al.* 1974a and b), which has freed us to regard changes in structural genes and their products as a minor element in adaptive evolution. The theoretical reasons that biochemical characters might not often converge follow mainly from the neutralist argument that most observed (fixed) changes in the genetic coding for proteins (and hence their primary sequences and immunological properties) are equally acceptable from the point of view of selection; or, alternatively, from the argument that the selective pressures on proteins are themselves so numerous and randomly distributed through time that the biochemical evolution of organisms mimics a stochastic, continually divergent, and homoplasy-free process.

But even the most enthusiastic supporters of the molecular clock hypothesis admit that it is a probabilistic model (e.g., Kimura and Ohta 1971). Some deviation from constancy should be expected, particularly at taxonomic levels representing genera within the same family, and Wilson *et al.* (1977) conclude that the variance is about twice that expected on the basis of a simple Poisson model. Moreover, at high levels of similarity the immunological data in particular have a large standard error (Nei 1977).

The objective of cladistic analysis is less to establish the absolute times of branchings than their sequences, and short-term variations in rates should not greatly obstruct inferences of cladistic relatedness providing that such variation does not obscure the correct *order* of branchings. Unfortunately, it may not always be true that biochemical data indicate the true branching sequence — in maximum parsimony analyses of amino acid sequences, for example, there is always at least one unlikely branching (e.g., Fitch 1976; Goodman 1976). Some of these discrepancies from "known" phylogeny may be due to homoplasy, but a simple model of stochastic change *excluding* homoplasy also gives results not exactly congruent with known branching sequences (Kirsch 1969). In addition, Farris (1972, 1981) indicates that the dendritic structure is sensitive to assumptions about the geometry of taxonomic space adopted in the interpretation of biochemical distance data.

While biochemical comparisons may not provide an unequivocal indication of branching sequences in all cases, chemical evidence ought nonetheless to be particularly useful to the cladistic enterprise. Yet the phylogenies suggested by such comparisons have not always been accepted, or even acknowledged, by taxonomists. Many cladists, especially, seem uncomfortable with the inability to order the characters of immunology or primary sequences such that some living taxa are recognizably more primitive than others, as is usually the result of a cladistic analysis of anatomical data. Indeed, if the assumption of time-dependent change is correct, there should be no such ordering. More important than this methodological consideration may be the obvious limitation of biochemical comparisons to living taxa; only a few data have yet accrued from study of fossils. Nevertheless, a biochemically based phylogeny certainly provides a framework in which fossils may be considered, and which, if correct, makes some demands for consistency among and limitations on speculations provoked by the anatomical data which are usually the sole source of information on extinct taxa.

This last point suggests what may be the chief theoretical importance of biochemical characters — that they provide an initial phyletic hypothesis against which the indications of other, non-biochemical, characters may be tested. An appropriate and efficient way of utilizing biochemical similarities for cladistic purposes would be to treat the chemically based phenogram as a cladogram which is then tested with other kinds of characters interpreted by orthodox cladistic analysis. To the extent that biochemical information gives correct branchings, the cladistic analysis ought to confirm those branchings. We emphasize that the significance of this procedures goes beyond the usual one of simply combining characters in one minimum-length tree, since the biochemical phylogeny is constructed on the basis of a different assumption about character-evolution than that used for interpreting anatomical features (i.e., regular as opposed to irregular rates of change). Moreover, by making the validation of biochemical phylogenies dependent on an explicitly cladistic interpretation of anatomical characters, one also avoids the confusion of earlier efforts to confirm biochemical phylogenies which involved comparison with dendrograms sometimes based on *phenetic* analyses of anatomical data.

We did in fact adopt this biochemistry-testing approach in the initial stages of our analysis of didelphimorphs. It quickly became evident that assumptions about the directions of changes in certain anatomical features might need to be altered if the biochemical data were given supremacy. Yet it also became clear that beginning with any cladogram and judging other characters against it is potentially or actually circular at best, and at worst is prejudiced in favour of the data which gave the initial tree. Those pitfalls may not be completely avoidable (for example, the determination of any character-state morphocline depends on an assumption of out-group relationship, itself an initial cladistic hypothesis), but the dangers can be minimized by independent evaluation of distinct character-sets and their later comparison.

We have largely avoided circularity through (usually) analyzing the morphological data by themselves, using the Wagner-tree method for dealing with a large number of data simultaneously, and so escaping the additional but related difficulty of cladistic analysis that the first character studied often unduly influences the interpretation of others. It should be noted that assessments of Wagner trees and of the usefulness (i.e., freedom from homoplasy) of constituent or contributing characters are themselves judgments made on the basis of parsimony. However, it remains a separate question how well trees generated by anatomical data match the ones suggested by serology. The hope and expectation is that they *will* agree, and if this is the case, that agreement constitutes a "higher-level" parsimony judgment, since the data-sets of anatomy and biochemistry are surely to a greater extent independent of each other than are partitions of either one. We might, for example, expect congruence among several dental characters, since they all form part of the same functional complex; but there is no obvious reason why these should give the same phyletic indications as does serology. This higher-level comparison more adequately meets Ruse's (1979) rather stringent interpretation of Whewell's principle of the consilience of inductions than does the usual taxonomic procedure of playing off one anatomical feature against another.

We have, in fact, considered three quite distinct data-sets in this analysis, namely macromolecular, anatomical, and chromosomal. Though some attempts have been made to interpret the interrelationships among biochemical, chromosomal, and morphological evolution (e.g., Wilson *et al.* 1974a and b), we are still mostly ignorant of the mutual dependence of these factors — if any — in evolution; and therefore a sound theoretical framework for comparatively evaluating these quite different sorts of data is almost nonexistent. Any congruence among the data may thus be more surprising than its lack would be; at the very least, it is essential to reason from an understanding of the very different modes of change which may apply to each kind of character. This being so, we have had to begin with a lengthy statement on the relative value of the sets of data as they provide independent evidence for phylogenetic and systematic inference.

The foregoing remarks may be sufficient as regards the independence of morphological and

biochemical evolution; what remains unclear, or perhaps even unlikely, is the basis for any correlation between chromosomal and anatomical evolution. As pertains to the didelphimorphs, the extent of that interdependence will emerge from the ensuing discussion, and from reference to the treatments of characters and presentation of Wagner trees in the preceding sections.

Our procedure in the following discussion will make it necessary to keep always in mind the indications of serology and the general outlines of the quantitative trees. Figure 57 shows the serological phenogram. To the extent that the branching sequences of the Wagner trees match the linkages of that phenogram, the serological affinities may be said to be confirmed; discrepancies may be due to homoplasy in either case, and evaluation of those discrepancies carries the burden of much of the next several sections of this GENERAL DISCUSSION.

Inferences on Systematic Relationships among Extant and Fossil Opossums

Relationships among Extant and Neocenozoic American Opossums: Inasmuch as a greater number of character-sets apply to the living opossums than to the fossils, we begin our discussion with an assessment of the pattern of relationships among extant didelphoids and related Neocenozoic forms. We later extend the discussion to more distantly related fossil taxa, where the implications of serology and/or karyotypes can only be drawn by extension from the living forms.

1. The Position of *Dromiciops*

The primary question regarding the phylogenetic relationships of living opossums concerns *Dromiciops*. On the basis of comparisons with the original description and illustrations of *Microbiotherium tehuelchum* by Sinclair (1906:410, Pl. 62, Fig. 7), Reig (1955) placed *Dromiciops* in the Microbiotheriinae, a group until then recognized only from fossil forms. Reig pointed out strong resemblances in the molar teeth and the similarity in the swelling of the periotic (entotympanic of Segall 1969b) and its participation in the formation of the ossified bulla posteriorly in both *Dromiciops* and *Microbiotherium*. Reig also noted similarities in the development of elongated premaxillae and in the form of the incisors. He added that *Caluromys*, *Caluromysiops*, and *Glironia* agree more closely in dental structure and in some particularities of the skull with *Dromiciops* and *Microbiotherium* than with "typical" didelphines. These statements were supported by earlier observations of Sinclair (1906:412), who noted that "Among living forms . . . [*Microbiotherium*] approaches most closely some of the subspecies of *Caluromys laniger*." Accordingly, and based largely on dental similarities, Reig tentatively included these genera among the microbiotheres.

This last conclusion is not supported by our present analysis. Though *Dromiciops* differs from *Microbiotherium* in having lost the para- and metaconules, the two genera are easily distinguished from most didelphids in their molar teeth. The M^5's in microbiotheres are strongly reduced in size relative to M^2-M^4 (Figs 5-8); the trigonid is wider than long, the trigonid cusps are typically low, and the talonids are longer than the trigonids and include a well-developed basin for occlusion with an enlarged protocone; the protocone is large and broad and the para- and metacones are low, rounded, and of subequal height; the centrocrista uniting these two cusps is straight and low, and does not form a V-shaped pattern; the stylar shelf and cusps and the metastylar spur are greatly reduced or absent. *Caluromys*, *Caluromysiops*, and *Glironia* (Figs 10-12) share with the microbiotheres a few primitive character-states, such as the relative development of the main molar cusps, reduced metastylar area, and enlarged talonids. Additionally, they share a portion of the derived character-states of the microbiotheres in pairwise combinations; but each pair of genera does not agree upon the same set of characters. Thus, *Caluromys* and *Caluromysiops* agree with the microbiotheres in having reduced M^5's, and *Caluromys* and *Glironia* in reduction of the stylar shelf. However, *Glironia* has fully developed and large M^5's, and *Caluromysiops* shows a normally developed stylar shelf. Moreover, the three caluromyine genera agree with each other in having trigonids longer than wide and a V-shaped centrocrista. Thus, and only on the grounds of molar character-states, there are good reasons to assume that the resemblances in molar pattern between microbiotheres and caluromyines are a consequence of convergence combined with a few cases of symplesiomorphy. Indeed, the latter condition should be invoked to explain the sharing by *Dromiciops* and *Caluromys* of the 2n=14 chromosomal set.

These conclusions are fully corroborated when we look at other characters, both anatomical and biochemical. The latter clearly indicate that there is no special affinity between *Caluromys* and *Dromiciops*, and that the former is more closely related to the didelphines (Kirsch 1977b). Moreover, caluromyines and microbiotheres are quite distinct in the morphology of the ear region, as Segall (1969a) pointed out. *Dromiciops* and *Caluromys* also differ in the absence in the former of an anterior extension of the median vaginae (among other, and perhaps more important, features of the reproductive systems). And although a pouch is present in *Dromiciops*, in two species of *Caluromys*, and in *Caluromysiops irrupta*, it is quite reasonable to conclude that a pouch

evolved independently in the two lineages. This is especially likely since *Glironia* is pouchless, a condition that can be regarded as primitive for the caluromyines, as for many marsupial groups.

Thus the major and most obvious discrepancy between the biochemical findings and some of the Wagner trees — in the relationship or lack of it between *Caluromys* and *Dromiciops* — can be considered due to several convergent anatomical features. When several fossil taxa were included in the trees, caluromyines and *Dromiciops* never clustered together (Figs 64-67). Recent evidence presented by Szalay (1982a and b) on the tarsal anatomy of *Glironia* as compared with *Dromiciops* strongly supports the same conclusions, although we do not necessarily agree with his contention that the didelphid tarsus — as represented by *Glironia* — is not a primitive condition.

We also claim as untenable Szalay's extreme position that *Dromiciops* stands quite apart from the American marsupials (his cohort Ameridelphia) and should be included in an alternative major division (cohort Australidelphia) together with all Australasian marsupials. This position is allegedly supported by the fact that *Dromiciops* shares with Australian marsupials strong, and phyletically important, similarities in tarsal morphology which he considers as uniquely derived. This complex consists primarily in a continuous lower ankle joint, typified by the continuous state of astragalar and calcaneal articular facets. The association is also supported (Szalay claims) by the distally broadened incisors in *Dromiciops*, dasyurids, and peramelids.

We can only say here that this conclusion is not supported by the whole array of characters studied in the present paper and discussed below. *Dromiciops* is clearly a didelphimorph, although belonging to a taxon within Didelphimorphia well-differentiated on the basis of several dental, cranial, softpart anatomical, and biochemical features; it shows no special similarities to Australian marsupials and indeed represents the most derived taxon in a lineage that is continuously documented from the very earliest didelphimorphs onward. Szalay's argument is considered in more detail in the section "Assessment of the Clade of Early Opossums," below.

2. Relationships of *Caluromys* and Allied Genera

In the previous section we showed that *Caluromys*, *Caluromysiops*, and *Glironia* are not microbiotheres, but are more closely related to living didelphines *s.s.* At the same time, these three genera show some differences in molar structure and in other features. We must, then, determine whether they are closely related to each other and, if so, what their relationships to the remaining didelphoid genera are.

Unfortunately, serological, karyological, and several anatomical data are lacking for *Glironia* and *Caluromysiops*. However, study of skull morphology (Reig pers. data) clearly indicates that these two genera are more closely related to each other and to *Caluromys* than any of them is to the remaining didelphids. Apart from characters already analyzed, significant synapomorphies linking the three genera are the broad supra-occipital process, which forms winglike ledges over the orbits; the sharpened, low, and narrow rostrum; the weakly developed and upright canines; the broad and rather inflated braincase; and the enlarged orbits (Fig. 48). We have also seen that these genera are united by having big, broad palates with reduced or absent palatal vacuities, a reduced P^1, dilambdodonty, a change in the proportion of the paracrista which is not united to the parastyle, trigonids longer than wide, and several other pairwise resemblances in molar teeth and the ear region. The discrepancies of *Glironia* in showing a primitive ear region and unreduced M4's, and of *Caluromysiops* in having a well-developed stylar shelf and a complete set of stylar cusps, as regard each other and one or the other of these two with respect to *Caluromys*, are quite swamped by the many other synapomorphies. These discrepancies may be considered good examples of heterobathmy in the evolution of molar character-states, and of the polythetic nature of the concept of the taxon which unites the three genera. It is obvious that a common ancestor of all of them should show the primitive condition of the stylar shelf displayed by *Caluromysiops* and the unreduced M4's and unmodified ear region possessed by *Glironia*. The three genera are also united in showing strictly arboreal, nocturnal, and lemur-like habits.

The conclusion that *Caluromys*, *Caluromysiops*, and *Glironia* are closely related thus seems well-warranted. Moreover, in the majority of character-states uniting them, they are equally separable from the remaining didelphids. Association of *Caluromys* with the large, pouched, 2n=22 chromosomes group of opossums is contradicted by its plesiomorphic karyotype, not to mention the lack in *Caluromys* and allies of the several synapomorphies which unite *Didelphis*, *Philander*, *Chironectes*, and *Lutreolina*. The same argument about chromosomes applies to a comparison of *Caluromys* with *Monodelphis*. Additionally, in her study of didelphoid osteology, Works (1950) concluded that *Caluromys* is widely separated from *Didelphis*, *Philander*, and *Chironectes*.

A relationship with *Metachirus* is not contradicted by the chromosomal data, but it is evident that this genus does not share any important set of derived character-states with *Caluromys* and its allies, and it is also quite distinct from them serologically.

Regarding the remaining, small-sized living didelphids, it is important that in the structure of the male and female reproductive tract and morphology of the spermatozoa, *Caluromys* is clearly separate from those forms studied. Also, in locomotor behavior *Caluromys* has a different gait from that of species of *Marmosa* (Bucher and Fritz in prep.); and, the unique features of the bipartite bullae and the lack of several synapomorphies of dentition and the ear-region shared by *Marmosa*, *Thylamys*, *Micoures*, *Monodelphis*, and *Lestodelphys* show quite clearly that the caluromyines cannot be closely related to the small-sized living didelphids.

Lastly, it appears that *Caluromys* and *Marmosa* have experienced a slower rate of change in some serological characters than have other opossums (Kirsch 1977b:91; Maxson *et al.* 1975). In spite of the coincidental slowdown in these genera, the majority of the serological tests show *Caluromys* to be distinct from other didelphids, in clear confirmation of the many morphological characters considered here. All the evidence therefore indicates that *Caluromys*, *Caluromysiops*, and *Glironia* are united and distinct from the remaining didelphids. We take here the position, advanced by one of us (Kirsch 1977b; see also Reig 1981), of grouping these genera in a subfamily Caluromyinae as opposed to the true Didelphinae (see below).

3. Relationships Within Didelphinae

As for the remaining genera of living didelphids, the serological tests do not clearly indicate any hierarchy among the three groups, but instead show an unresolved trichotomy. One major branch comprises only *Metachirus*; another links *Marmosa*, *Micoures*, *Thylamys*, and *Monodelphis*; the third associates *Philander*, *Didelphis*, and *Chironectes*, the last standing apart from the other two. Unfortunately we lack much serological information on *Thylamys* and *Lutreolina*, and have none on *Lestodelphys*. We now examine how well the pattern of resemblances among didelphines which *are* known biochemically matches the evidence afforded by chromosomal and anatomical data.

Metachirus shows the plesiomorphic 2n=14 karyotype (Fig. 55H), as do many of the small genera, and is traditionally associated with *Philander*, as suggested by its common English name — the brown four-eyed opossum. However, in spite of the similar markings and size, the two genera are otherwise rather distinct. *Philander* has a reduced number of teats and possesses a pouch, while *Metachirus* lacks a pouch; in the former the fur is soft and the pelage is practically all underfur, while the latter has a rough-textured coat of guard hairs (Kirsch 1977b). *Metachirus* retains a primitive trigonid wider than long (Fig. 17), whereas the trigonid of *Philander* (Fig. 18) shows the derived longer-than-wide condition; the two genera also display differences in the female reproductive tract. Most of the similarities in the skull region shown by the two genera are actually primitive character-states unreliable for inferring relationships. *Philander* is apomorphous in showing the derived 2n=22 karyotype (Fig. 56E), which distinctly relates it to *Didelphis, Chironectes* and *Lutreolina*. On the other hand, *Metachirus* does not share with the other 2n=14 chromosomes didelphines any of the several cranial and dental synapomorphies which characterize them. Thus, no inconsistency exists among the three kinds of data as regards the separate position of *Metachirus* within the didelphines, and Wagner-analysis strongly suggests a primitive position of this genus *vis-à-vis* either of the two groups of the remaining genera.

The relatedness of *Philander, Didelphis*, and *Chironectes* is strongly corroborated by their sharing, together with *Lutreolina*, the derived 2n=22 karyotype (Fig. 56). In morphology they are also closely connected by having a well-developed pouch (or one somewhat weakly developed in *Lutreolina*), large body-size, and some features of the female reproductive system. Their grouping is also indicated by the Wagner trees, which describe a clearcut assemblage on several kinds of morphological character-states, none of which is unique to the group or possessed by all its members, but which in their association support the conclusion that we are dealing with a closely interrelated group of genera for which suprageneric distinction is desirable.

The Wagner-tree character-analysis suggests that *Philander* is the more primitive living genus within this group, although *Chironectes* sometimes occupies that position. Serologically, *Philander* is more similar to *Didelphis* than either of them is to *Chironectes*. However, the anatomical analysis suggests that *Didelphis* and *Chironectes* are cladistically more closely related and that *Lutreolina* differentiated from them extensively. *Lutreolina* is the only large didelphine showing the derived, highly metacentric X-chromosome and a weasel-like body form, two autapomorphous details not included in the coding. To this group clearly belong the fossil genera *Hyperdidelphis* (=*Paradidelphis*) and *Thylophorops*. *Hyperdidelphis* is distinct from *Didelphis marsupialis* and *D. albiventris*, showing several autapomorphies which indicate it to be more advanced than either. The relationships of *Hyperdidelphis* and *D. virginiana*, which shows also a distinct karyotype, are open to further investigation. *Thylophorops* is closely related to *Lutreolina*, from which it can be distinguished by its greater size and more advanced carnassialization of the molar teeth (Marshall and Reig in prep.). *Hondadelphys* also seems to belong to this group, although Crochet (1980:96) remarked that it hardly seems to belong to the Didelphinae, mainly on account of its straight centrocrista and reduction of the

anterior portion of the stylar shelf on M². Yet, in most other characters analyzed by us, *Hondadelphys* approaches the group of large-bodied, 2n=22 chromosomes opossums, and in spite of the peculiar and quite unexpected feature of seemingly pre-dilambdodont molar structure, it is clearly part of the large-bodied group in Wagner trees (Figs 37, 64-67).

The grouping of *Micoures, Marmosa, Monodelphis*, and probably *Thylamys* obtained by serological analysis is not fully confirmed by the chromosomal evidence, as *Monodelphis* possesses a derived 2n=18 karyotype, whereas *Micoures* and *Marmosa* share with *Thylamys* the plesiomorphic 2n=14 (Fig. 55). A case can be made that the karyotype of *Monodelphis* is most parsimoniously considered representative of an intermediate stage in the derivation of 2n=22 from 2n=14. However, the serological findings suggest a separate derivation from 2n=14 (were 2n=18 an intermediate condition or else derived from 2n=22, *Monodelphis* ought to show greater serological resemblances with some 2n=22 opossums). We feel that the serological similarities among *Marmosa, Micoures*, and *Monodelphis* are more likely to be real than not, despite the separate derivation of 2n=18 required.

This conclusion is also consistent with several autapomorphies shown by *Monodelphis* when compared to the remaining small didelphines. *Monodelphis* is derived in its terrestrial habits as opposed to the arboreal or semi-arboreal original adaptation of most didelphines, and consequently in its shrew-like habitus and shorter tail, characters not included in the scoring for Wagner analysis. *Monodelphis* also shows some specializations in dental and skull characters.

Lestodelphys has been considered a close relative of *Monodelphis* (Simpson 1972). However, our character-analysis usually shows that it is closer to *Thylamys*, with which it also shares an incrassated tail. This affinity was noted by Tate (1933), who said that *Notodelphys* (=*Lestodelphys*) may be derived from his *elegans* group of *Marmosa* (=*Thylamys*). The specialized bullar region of *Lestodelphys* is comparable to that of species of *Thylamys*, and clearly separates the former from *Monodelphis*. As Reig (1958a and b) pointed out, a close ally of *Monodelphis* is the fossil *Thylatheridium*, which shares similar specializations of the palate and molar teeth, yet deserves generic distinction because of its larger size and more advanced carnassialization.

One of the new proposals in this paper is the separation of *Micoures, Marmosa*, and *Thylamys* as three genera, which are regarded as a single genus, *Marmosa*, in most current classifications. We believe that this separation is necessary to achieve a more reasonable balance in the classification of didelphines. The three taxa differ from each other at least as much as do *Philander* and *Didelphis*, or, perhaps more pertinently, as does *Monodelphis* from any of the three genera once included in *Marmosa*. Reasons for this separation are provided by our character-analyses, which show a distinctive distribution of character-states in each of the three taxa. A common feature of the Wagner trees is that *Marmosa* and *Thylamys* are closer to each other than either is to *Micoures*, which frequently takes a primitive position. This is underscored by the fact that *Monodelphis* sometimes occupies a position in the trees which separates *Micoures* from *Thylamys* and *Marmosa*. The distinctness of *Thylamys* is further confirmed by its peculiar mixture of primitive and advanced character-states: it is most primitive among didelphines in possessing pectoral mammae, and in all aspects of the female reproductive tract, whereas it is advanced in molar and bullar morphology. Comparative anatomy suggests that *Micoures* may be the ancestral genus for the murine group, but the derivations of *Monodelphis*, *Thylatheridium*, *Marmosa*, *Thylamys*, and *Lestodelphys* may have occurred through different pathways.

Thus, the small-opossum group is unified by serological resemblances (so far as they are known) and anatomical data, and the chromosomal derivation of *Monodelphis* is explained as one autapomorphy among many shown by this genus. As indicated by the Wagner trees and supported by serology, the group deserves the same level of distinction within the didelphines as do *Metachirus* on the one hand and *Didelphis*, *Lutreolina*, *Chironectes*, and *Philander* on the other (see Table 6).

Relationships of Early Fossil Opossums: An astonishing result of our Wagner-tree analysis is the clustering of early fossil opossum groups with living and fossil microbiotheres as a sister-group to the clade of living and Neocenozoic didelphoids (Figs 64-67). This result entails a radical reassessment of the classification of the major opossum groups, and renders necessary a more detailed discussion of the relationships of the included taxa.

1. Relationships of *Pediomys*

The microbiotheres stand apart from the remaining living didelphimorphs, as indicated by serological distance and many autapomorphies in skull and dental characters. An unexpected conclusion of our study is that in all derived molar characters, microbiotheres are closely comparable to the Upper Cretaceous Pediomyidae, a resemblance not commented upon

by authors who have studied pediomyids (Clemens 1966 and 1968; Fox 1979b; Lillegraven 1969; Shani 1972).

All known species of *Pediomys* share upper molars with enlarged protocones; subequal para- and metacones united by a straight centrocrista; and a reduced stylar shelf always lacking or showing a merely vestigial parastyle and stylar cusp C, and displaying a variable development of stylar cusps A and D, which are at most only moderately developed. In all small species of *Pediomys*, such as *elegans*, *krejcii*, *cooki*, *prokrejcii*, and *clemensi* (see measurements in Clemens 1966; Fox 1979b; Lillegraven 1969), the M^3 is smaller than or subequal to the M^4 in anteroposterior length. However, M^4 is definitely slightly larger than M^3 in the large species *florencae* and *hatcheri*, a striking difference from the former that is related to their noticeably larger size, and which is probably an indication of advanced carnassialization. M^4 is never noticeably larger than M^3 but is usually subequal or smaller, as in microbiotheres;* and the M^5 is reduced, although this condition is less advanced than in *Dromiciops* or *Microbiotherium* (Fig. 8). Pediomyids also show an M_5 smaller than the M_4, an anteroposteriorly compressed trigonid, enlarged and basined talonid with crista obliqua oriented anteriorly and opposed to the base of the protoconid, and hypoconulid twinned with entoconid (Fig. 14). Thus, *in combination* the matching of molar character-states among pediomyids and microbiotheres is almost complete, the only significant differences being the still-further reduced stylar cusps and M5 in the latter, characteristics not unexpected if microbiotheres are later representatives of the same stock. Pediomyids also agree with *Caluromys* and its allies (Fig. 10) in the same character-states as do the microbiotheres, although these partial resemblances are justifiably attributed to homoplasy and symplesiomorphy. But the extent and detailed nature of the similarities between pediomyids and microbiotheres require too much of a departure from the principle of parsimony to be judged as convergences, and there is no reason not to consider their shared apomorphies as indications of patristic affinity. The Wagner-tree analyses agree with this assignment (Figs 64-67), so far as the available evidence permits. This conclusion would be stronger if corroborated by other kinds of anatomical evidence, especially of the ear region, which is so markedly derived in *Dromiciops* and *Microbiotherium*. Unfortunately

---

*This difference in size and structure among species suggests that *Pediomys* may be a composite genus, a conclusion which Fox (this volume) has independently reached. As the type of *Pediomys* is the small species *elegans*, we restrict here our concept of this genus to the species that have an M^3 of less than 2.3 mm in length. "*Pediomys*" *hatcheri* and *florencae* have a mean M^3 length of 3.34 and 4.04 or 4.1 mm (depending on the sample), respectively. No known living genus of didelphid shows such an extreme in absolute size and structure as do these two groups of *Pediomys*.

this information is not yet available for pediomyids.*

Thus, microbiotheres and pediomyids represent a single taxon, with pediomyids occupying a relatively ancestral position with respect to microbiotheres. This conclusion may appear contrary to biogeographical expectations, despite its attractiveness in specifying the phyletic origins of microbiotheres. However, Sigé (1972) has, on the basis of fragmentary evidence, noted the apparent presence of pediomyids in the Late Cretaceous of South America. Given the suggested occurrence of *Peradectes* (= *Alphadon*) *austrinum* in the same deposits (Crochet 1980; Sigé 1972), and Glasbiinae in the Paleocene (Crochet 1980) and in the Eocene (Pascual 1983) of South America, there are several cases of the southern occurence of didelphimorph groups previously thought to occur only in North America.

2. Relationships of *Glasbius* to the Microbiotheres

Glasbius is a Late Cretaceous genus known from upper and lower molar teeth (Archibald 1982; Clemens 1966). The distinctness of these molars from those of the remaining opossums was the basis for the genus' separation in a subfamily of its own, Glasbiinae (Clemens 1966). Clemens dismissed similarities between *Glasbius* and the fossil South American microbiotheres, but Reig and Bianchi (1969) found that it was so similar to microbiotheres as to suggest that it should be considered closely related to the latter. *Glasbius* (Fig. 4) is distinguished by having a very reduced M_5 and bunodont molars, with subequal para- and metacones, enlarged protocones, well-developed para- and metaconules, and enlarged stylar cusps. The bunodonty does not mask many similarities shared with the teeth of microbiotheres other than reduction of the M_5's, such as the primitively straight centrocrista, M^4 smaller than M^3 and with metacrista shorter than postprotocrista, reduced metastylar area, an anteroposteriorly compressed trigonid, enlarged and basined talonid (which, however, is shorter in length than the trigonid), lower anterior cingulum tending to have a transverse position, and anteriorly oriented cristid obliqua. *Glasbius'* enlarged stylar shelf with bunodont stylar cusps is interpreted as an adaptation to an omnivorous diet, evolving from a common ancestor with microbiotheres not showing a strong reduction of the shelf. The striking resemblance in other molar features between *Glasbius* and microbiotheres may be an indication that they shared such a common ancestor more recently than did microbiotheres and pediomyids. If the three taxa are placed in the same family, *Glasbius* ought, on strict cladistic principles, to be included in the same subfamily with microbiotheres, with only tribal distinction between them; however, we prefer to retain subfamilial distinction for glasbiines. These conclusions are further supported by the recent finding of an obviously *Glasbius*-like genus in the Eocene of Argentina, namely *Reigia* (Pascual 1983). This discovery makes less suprising the possible glasbiine relationship of *Protodidelphis* (Crochet 1980).

3. Relationships of *Alphadon, Peradectes,* and Allies

Alphadon has classically been considered "the" ancestral didelphid, from which extant opossums and the various Cretaceous marsupial lineages originated (Clemens 1966, 1979). The several species of *Alphadon* are Late Cretaceous in age and North American in distribution. The great size differences between the type species (*A. marshi*) and other relatively small species compared with the much larger *A. rhaister*, suggests that, as in the case of *Pediomys*, the genus *Alphadon* may be composite. We here restrict our discussion to the typical medium-sized *Alphadon*.

Peradectes is a genus of fossil opossums which occurs in the Paleocene and Eocene of North America and Europe, and in the Late Cretaceous of South America (Crochet 1980). Provoked by Crochet's conclusion that *Alphadon* and *Peradectes* were closely related and constitute a separate group relative to other living and fossil didelphids (Crochet 1980), we reviewed the published information on those genera and their allies *Nanodelphys, Mimoperadectes, Alloeodectes* and *Albertatherium* and we included *Alphadon* and *Peradectes'* character-states in our Wagner-tree analysis. The results of this analysis show that *Alphadon* and *Peradectes* closely approximate a hypothetical didelphimorph ancestor, having small differences which do not preclude their grouping, as proposed by Crochet (1980). We give full family status to this group, Peradectidae. Crochet placed *Nanodelphys* McGrew 1937 as a subgenus of *Peradectes*, a conclusion with which we tentatively agree (but see Russell 1984). Along with Fox (1979a), we consider *Albertatherium* to be allied with *Alphadon; Alloeodectes* and *Mimoperadectes* are two genera recently added to the same group (see Russell 1984 and Bown and Rose 1979 respectively). As suggested in the previous paragraph, it is possible also that *A. rhaister* should be placed in a genus of its own.

---

*However, Szalay (1982 a and b) recently claimed that pediomyids strikingly differed from the living microbiothere *Dromiciops* in tarsal structure. His conclusion is based on the identification of pediomyid proximal tarsal bones among material from the Lance and Bitter Creek Formations. Actually, Szalay does not provide any supporting evidence for this taxonomic allocation, and indeed, as a student of pediomyids recently stated, "dentitions and jaws are the only fossils of *Pediomys* known" (Fox 1979b:103; see also Clemens 1979:207, from which it is clear that as far as associated dental and postcranial remains are concerned, this assertion extends to all known Late Cretaceous didelphimorphs). In fact, Szalay illustrates as pediomyid an astragalus and a calcaneus from the Bitter Creek Formation, but this Paleogene deposit is not one of the mammal-bearing beds where pediomyids have been found (according to the summary by Clemens 1979:174)! Szalay probably assigned isolated tarsal elements to Pediomyidae found in deposits which contained pediomyid and other didelphoid dental remains, in view of similarities of those bones with corresponding parts of living didelphids, and then inferred that similar bones in deposits *lacking* pediomyids also belonged to that family. Nothing he writes, however, explains why he chose pediomyids and not *Alphadon*, for instance, as the primitive stock. While Szalay claims to have a broader concept of pediomyids (1982a:625), which might ameliorate these criticisms, one can only wonder what the factual basis of such a definition might be.

Our analysis indicates that *Alphadon* and *Peradectes* are cladistically closer to microbiotheres than to Caluromyinae or Didelphinae. *Alphadon* (Figs 2, 3) and *Peradectes* and their allies *Alloeodectes*, *Mimoperadectes* and *Albertatherium* share many primitive character-states with microbiotheres, pediomyids, and *Glasbius*, such as the straight centrocrista, presence of conules, absence of a well-developed metastylar region, relatively broad protocones, and lack of reduction of the paracone. At the same time, they do not show most of the derived characters of those taxa, as they have unreduced M_5's, a normal stylar shelf with fairly well-developed but not enlarged styles, and a stylocone united to paracone. It can be argued that their association with the microbiotheres is spurious, as it is mainly based on symplesiomorphy and not the synapomorphies which distinguish the latter as a cohesive clade. We are ready to accept the objection, but are also willing to depart from an orthodox cladistic arrangement, accepting the possibly paraphyletic grouping of *Alphadon* and *Peradectes* and allies as a family within a major taxon including microbiotheres and relatives. Peradectidae is considered a lineage within an early radiation which included phylads more specialized in food habits; microbiotheres are the only surviving representatives of this early radiation. That clade probably also included specialized carnivorous taxa such as Stagodontidae and omnivorous forms like Caroloameghiniidae. The plesiomorphic conditions of most if not all of their dental character-states also make peradectids ideal ancestors for both didelphines and caluromyines, and for such specialized Cenozoic groups as sparassocynids and borhyaenoids. This hypothesis of ancestry is in good keeping with the finding of *Peradectes* in the Late Cretaceous of Bolivia and of *Bobbschaefferia* in the Middle Paleocene of Brazil (Crochet 1980).

The Peradectidae are also likely to be proposed as the direct ancestors of the Australian marsupials. This idea is strongly suggested by the discovery of *Ankotarinja* in the Miocene of central South Australia, a genus which was described by Archer (1976c) as a dasyurid. In fact, and as indicated above, *Ankotarinja* does not show any of the derived dasyurid molar characters connected with carnassialisation, and in all details of molar teeth it fully agrees with the character states of peradectids and is so placed in our formal classification.

With representatives in the Late Cretaceous of North and South America, the Paleocene of Europe, and the Miocene of Australia, the peradectids show a world-wide distribution pattern. This pattern is reinforced by the discovery of an upper molar of a supposed peradectid in the Eocene of El Kohol (Alger), announced first by Mahboubi *et al.* (1983), and later described by Crochet (1984) as *Garatherium mahboubii*. The assignment of this fossil to the peradectids, and even to metatherians is, however, not quite convincing, as recently argued by Simons and Bown (1984) who claimed that *Garatherium* may be a minodectid that also resembles some microchiropterans.

4. Relationships of Stagodontids

As typically represented by *Eodelphis* and *Didelphodon* of the Late Cretaceous of North America (Clemens 1979), stagodontids are a specialized group of opossums which include the largest known Cretaceous marsupials. They show enlarged and crushing premolars, reduced metaconids, high and enlarged paraconids, a well-developed and medially notched stylar shelf with an enlarged stylocone, and the derived condition of large epitympanic sinuses in the squamosal. Most of these characters are autapomorphies and separate the stagodontids as a distinct family, but do not help to clarify their relationships.

We included in our Wagner-tree analysis characters of *Didelphodon* taken from Clemens (1966, 1968, 1979) and also considered descriptions of *Eodelphis*, mostly from Fox (1981). *Didelphodon* usually clustered with *Alphadon*, *Peradectes*, microbiotheres, and allied forms (Figs 64-67), but sometimes was associated with the didelphid branch. This first placement results mostly from the predilambdodont condition, well-developed conules, relative length of the M^4 borders, broadened stylar cusps, and disunited paracones and stylocone — a list including both primitive and derived character-states. Stagodontids are distinct from the more evolved Cenozoic and modern didelphoids in the possession of numerous primitive characters, though showing some convergent derived characters (e.g., the high paracristid and enlarged paraconid) in which they resemble some modern genera — hence the sometimes anomalous placement of *Didelphodon* with modern large opossums. Stagodontids are thus considered direct derivatives of the peradectids, as proposed by Clemens (1966) and supported by Fox (1981).

5. Relationships of Caroloameghiniids

Procaroloameghinia and *Caroloameghinia* (Figs 9, 15) are Early Cenozoic South American opossums, known from upper and lower molar teeth and mandibles (Marshall 1982d; Simpson 1948). Ameghino placed *Caroloameghinia* in a family of its own, Caroloameghiniidae, based on its broadly cuspidated and crushing molars, interpreted as an adaptation to an omnivorous diet. Wagner analysis of the dental character-states of *Caroloameghinia* clustered this Early Eocene genus as a sister-group of pediomyines, glasbiines, *Microbiotherium*, and *Dromiciops* together, and in-between that group of taxa and *Alphadon, Peradectes,* and (sometimes) *Didelphodon* (e.g., Fig. 64). Four synapomorphies support this placement: two (relative length of the M^4 border and disunion of paracone and stylocone) are shared by all six genera; one (size and shape of talonids) is shared by all except *Didelphodon* and is also present in *Alphadon*; the fourth (large body-size) unites only *Caroloameghinia, Didelphodon,* and some *Alphadon* within this clade. In all other respects *Caroloameghinia* is inferred to be primitive. It is of interest that *Caroloameghinia* shares with *Glasbius* an advanced trend towards omnivory in molar structure (Figs 4, 9, 15), as reflected in large and low bunodont cusps, well-developed conules, broad protocone, large basined talonids, and well-developed stylar cusps. *Glasbius* differs, however, in the more enlarged stylar shelf and reduction of M4's. *Roberthoffstetteria* has been recently added to the group of caroloameghiniids (Marshall *et al.* 1983; Muizon *et al.* 1984). It is based on good material from the Late Cretaceous of Bolivia, and it reinforces our previous conclusions.

6. Assessment of the Clade of Early Opossums

The clustering of microbiotheres, pediomyines, glasbiines, stagodontids, caroloameghiniids, and peradectids is a reasonable conclusion from our analysis, providing we depart from strict cladistic philosophy and recognize the utility of paraphyletic groupings in classification. Synapomorphies supporting this entire grouping are few, a point illustrated by the changing position of *Didelphodon* in the various trees (cf. Figs 37, 64-67), but we are impressed with the *sequence* of derivation that leads from the relatively plesiomorphous peradectids to the highly derived microbiotheriids. These several groups constitute half of a major dichotomy in most of our trees, the other being the remaining Cenozoic and extant didelphoids.

This is perhaps the place to comment on the interpretation of Pascual (1980b) regarding a taxon purportedly intermediate between didelphoid and polydolopoid marsupials. *Prepidolops* is a genus with two species from the Early and probably Middle Eocene of Salta Province, Argentina known from several lower and upper jaws, including most of the dentition, found in the Lumbrera Formation. *Prepidolops* is indeed a distinct genus, showing autapomorphies which justify placing it in a separate family, Prepidolopidae, as proposed by Pascual. What two of the present authors (OAR and JAWK) do not understand is his further proposal that prepidolopids belong in the Didelphoidea, since *Prepidolops* is in all respects a typical, though primitive, polydolopoid. The characters allegedly intermediate between those of didelphoids and polydolopoids are just primitive polydolopoid features not in contradiction with

current definitions of that taxon (Marshall 1982b). Moreover, to place the prepidolopids in the Didelphoidea would necessitate a thorough modification of our concept of that taxon and be quite at odds with the current logic of classification.

It is similarly necessary to remark here more fully on the recent papers of Szalay (1982a and b), in which he makes the radical suggestion that *Dromiciops* be considered phylogenetically close to the Australian marsupials. Szalay's main conclusions (Szalay 1982a) are:

a) Pediomyids are the most primitive marsupials, placed in a new order Didelphida and a new suborder Didelphiformes. Szalay supports an expanded concept of this family, implying the inclusion therein of stagodontids, glasbiines, and peradectines. The family distinctions of each on dental characters are, he says, not warranted. Moreover, he accepts that the grouping may be paraphyletic.

b) Szalay includes under Borhyaeniformes (within the Didelphida) such diverse groups as borhyaenoids, caenolestoids, polydolopoids, groeberioids, and argyrolagoids. These taxa are alleged to be united by tarsal morphology and to be direct derivatives of the pediomyids.

c) The Didelphidae are considered to have a separate derivation from pediomyids.

d) Didelphida constitutes the basis of the monotypic cohort Ameridelphia, mostly characterized by the retention of an allegedly primitive primitive proximal tarsal morphotype — specifically, separate lower-ankle, transversely oriented astragalar-navicular, and ball-and-socket calcaneal-cuboid joints.

e) Szalay opposes to the Ameridelphia the cohort Australidelphia, which includes the orders Dromiciopsia (monotypic for the microbiotheriids), Dasyurida (dasyuroids), and Syndactyla (all other Australasian marsupials, including *Notoryctes*). The fundamental similarities joining these three orders are the fusion of the facets of the lower ankle joint, the absence of the proximal sustenacular facet between the astragalus and calcaneus, and a modified calcaneal-cuboid joint, as well as other tarsal features. Szalay relates these putatively derived tarsal modifications to a greater degree of arboreality in *Dromiciops* than has supposedly been recognized heretofore, implying an arboreal ancestry for Australian marsupials. He claims also the presence of broadened incisors in microbiotheres, dasyurids, and peramelids as an additional synapomorphy bolstering the conclusion of relatedness from tarsal anatomy. Thus the microbiotheres would be completely separated from the remaining American marsupials and represent the common ancestor of the Australasian radiation.

f) The Syndactyla includes an hypothetical ancestral taxon as well as the Notoryctoidea, Perameliformes, and all Australian diprotodont forms. Evidence that the marsupial mole is close to bandicoots and diprotodontans, as previously maintained by Bensley (1903), is a useful contribution; but no new or cogent evidence for the single origin of syndactyly is presented to justify the association of syndactyl marsupials.

Most of these points are at odds with the results presented in this paper, as well as with previous phylogenetic arrangements of marsupials. We believe that Szalay's conclusions come from a selective interpretation of limited evidence and neglect of other well-founded sources of information. He claims at the beginning of his paper that he evaluated cranial and dental, as well as pedal (particularly tarsal) character-clines; yet, he only *secondarily* uses cranial or dental characters and plainly disregards previous contributions from these sources (e.g., Archer 1976a and b), so that his conclusions are fundamentally based on an intepretation of tarsal morphology. Our present analysis, as well, shows that dental characters *are* able to provide sound "morphoclines" of primitive and derived states when used to build Wagner trees along with other kinds of evidence. Szalay thus completely ignores the conflicting results which do not support his views at all.*

Moreover, and as stated previously, Szalay's characterization of primitive tarsal structure is tied to his unclear concept of Pediomyidae, and there are, to say the least, serious doubts that his "pediomyid" material has been correctly classified.

At the same time, the ear region of *Dromiciops*, even though peculiar for didelphimorphs, is derivable from the tripartite kind of bulla in some didelphids, but is not comparable with the complete bulla of dasyurids and other Australian marsupials; we thus regard it as unlikely to be ancestral to or even derivable from the primitive pattern in some dasyurids, as would be necessary to corroborate Australidelphian affinities. Additionally, and as noted in the section above on chromosomes, the supposed agreement of karyotypic evidence with Szalay's placement of *Dromiciops* (Sharman 1982) is a matter of weak intepretation.

Szalay claims that his weighting of tarsal anatomy is sound because it is based on functional considerations not evident in the cladistic use of other (e.g., dental) character-sets. However, he overlooks the point that the molar structure of didelphimorphs *in particular* has been the subject of thorough functional analyses at levels of sophistication far beyond his own pedal studies (for instance, see Crompton and Hiiemae 1969,

1970). In our opinion, Szalay's "functional morphology" is stained by typological thinking and a simplistic characterization of what it means to be "arboreal" or "terrestrial": for small animals in particular, all rough surfaces are equivalent (Jenkins and Parrington 1976), and the issue is really one of freedom of movement in the tarsus, not degree of incline of the substrate. We cannot see, moreover, why Szalay claims that the complex of bones of the proximal tarsus are an aspect of the ancestral morphological matrix more constrained by heritage, and therefore less subject to convergence, than are other character-sets.

Szalay's union of *Dromiciops* with the Australasian marsupials is not only at odds with analysis of more extensive anatomical data, as stated, but also seems contradicted by his own evidence from the fusion of the facets of the lower ankle joint itself. That feature shows intermediacy, if not actual separation or repartition of the facets, in several of Szalay's own diagrams illustrating the conditions in various Australidelphians (i.e., *Perameles, Distoechurus, Hypsiprymnodon*, and *Myrmecobius* in his Fig. 7, p. 632 of Szalay 1982a), strongly suggesting that this condition could have been acquired separately in different lineages. Some of these intermediate taxa, such as the bandicoots, are placed at the *end* of the sequence by Szalay because of presumed derivedness in other characters, in disregard of the principle of mosaic evolution. Perhaps Szalay believes that the apparent intemediacy represents a secondary approximation to the Ameridelphian condition, but this unstated belief is weakened by the actual separation which seems to be the true situation in some syndactylans.

An additional controversial conclusion of Szalay's is his concept of the Borhyaeniformes. We really believe that putting together such different marsupials as caenolestoids, borhyaenoids, polydolopoids, and such peculiar (and not certainly marsupial!) taxa as groeberioids and argyrolagoids (McKenna 1981; Reig 1981) is going too far beyond what is permissable in scientific inference on the basis of a few supposedly shared characters.

Thus, the whole scheme presented by Szalay seems quite questionable, and the sometimes dubious similarities on which he bases his conclusions are better interpreted as convergent adaptive responses. This conclusion is particularly strengthened by the more comprehensive character-analysis presented in this paper. Contrary to what Szalay maintains about teeth, his discovery of a difference in pedal features of *Dromiciops* as compared with the remaining

---

*All this aside, it might still be asked if the dichotomous arrangement of didelpimorphs suggested by the present analysis does not allow Australian marsupials to be derived from something like a microbiothere. The answer is that it does not: an earlier study utilizing many of the characters analysed here (Kirsch and Archer 1982), but including Australian taxa, showed these to form a cluster derived from within a paraphyletic didelphoid assemblage — but via *Monodelphis*-like forms, not microbiotheres.

didelphimorphs may be considered supportive of the separation of microbiotheres from the remaining living didelphids concluded herein, whatever the implications of pedal characters for the origin and relationships of Australian marsupials.

The overall phylogeny of opossum-like marsupials we propose (Fig. 69) has some implications for the broader relationships and biogeography of marsupials as well, points on which the present authors do not necessarily agree and which we have decided not to treat extensively in this paper (see instead Kirsch 1984); in fact, it may be that supposed geographic constraints will be cited as the strongest criticism of our scheme.

One implication of our work is that the divergence between the microbiothere and didelphoid groups occurred very early, since the oldest undoubted marsupials are allied with the former assemblage. Serologically, Australasian marsupials are more distinct from didelphoids than is *Dromiciops*, and the caenolestoids are even more so. These facts suggest divergences from the ancestral form possibly predating the separation of the southern continents which Maxson *et al.* (1975) correlated with albumin differences between didelphoids and Australasian marsupials.

Yet, the similarities of *Dromiciops* to didelphoids are only slightly greater than those between didelphoids and any other marsupial taxon (Kirsch 1977b:89). Such earlier divergences as the serological hierarchy necessitates cannot be much separated in time. Moreover, statistical variation in rates of protein evolution could affect the details of the branching pattern among taxa so distinct, and a more conservative intepretation of the serological phenogram (Fig. 57) would be as an unresolved four-way split amongst Australasian marsupials, didelphids, caenolestids, and microbiotheres.

But taking the biochemical results at face value, they suggest that caenolestids and polydolopoids, Australasian marsupials, microbiotheres, and didelphoids (with borhyaenoids) may represent successive derivations from a moderately persistent peradectid-like ancestral taxon. Such a series of cladogenic events are more easily envisaged as occurring in South America, with dispersals from there of the Australasian marsupials and some members of the didelphoid and microbiotherioid lineages. An equally valid alternative is origin of metatherians in Australia and emigration of the didelphimorph stock to South America, with further differentiation there and eventual emigration of the various microbiotherioids and didelphoids to the Northern continents, not disregarding the possibility of still-further invasions of Africa from Europe or directly from South America (for instance, by the peradectids alleged to have lived in Africa (see Mahboubi *et al.* 1983; Crochet 1984).

Relationships of Cenozoic Fossil Opossum Groups:

1. Relationships of *Peratherium* and Allies

Peratherium, Amphiperatherium, and *Herpetotherium* are closely related Early Tertiary didelphimorphs of Holarctic distribution. As redefined by Crochet (1977a and b, 1980), the first two are found only in Early Eocene to Middle Miocene deposits of Europe, and *Peratherium* has been recently (Simons and Bown 1984) reported from the Oligocene of Egypt, whereas *Herpetotherium* occurs only in the Early Eocene to Early Miocene of the United States and Canada, references to *Peratherium* in Nearctic deposits actually being to this genus (Crochet 1977a, 1977b; but see Simons and Bown 1984).

We scored *Peratherium* as a representative genus of this group for our Wagner analysis, using as a basis Crochet's (1980) descriptions and illustrations. In most trees *Peratherium* linked closely with didelphines, more particularly with the murine opossums. It shares with didelphines several derived character-states (lack of conules, didelphoid dilambdodonty, reduction in size of the paracones, well-developed metastylar spur, trigonid shape, and open tripartite bulla). *Peratherium* is linked more closely with the murine opossums by its small body-size, presence of a well-developed paracrista, shortened outer border of M^4, and elevated protoconid.

If the shared, derived states between *Peratherium* and didelphines, and the Marmosini in particular, are due to patristic affinity, then *Peratherium* and allies evolved from a South American didelphine after the divergence of the murine opossums, arriving in North America by rafting in the Early Eocene. An alternative recently proposed by Simons and Bown (1984), is that African *Peratherium* arrived from South America via a South Atlantic sweepstake route. It is at least conceivable that *Peratherium* and allies spread from Africa to Europe, and from Europe to North America. The probable time of murine opossum cladogenesis, most likely one of the last events in an evolutionary sequence which occurred during the mid-Tertiary, is however, at odds with this interpretation.

It is more reasonable to ascribe to homoplasy the resemblances between *Peratherium* and allies and didelphines, and to imagine that *Peratherium* and allies originated from a peradectid North American ancestor, and that they evolved advanced dental characters in parallel with or convergent upon the Marmosini in adaptation towards insectivory and small body-size. This being the more probable explanation of the results of our Wagner analysis, it would also be convenient to separate *Peratherium, Herpetotherium,* and *Amphiperatherium* in a taxon of their own parallel to the South American Didelphinae, i.e., the Herpetotheriinae.

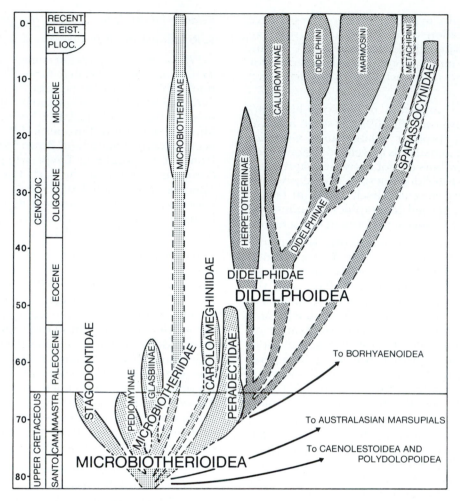

Fig. 69. Phylogram of the non-borhyaenoid Didelphimorphia. See GENERAL DIS-
CUSSION for reasoning behind the depicted associations and order of branchings.
Solid line segments indicate periods for which fossils are known, while broken lines
are conjectural. Widths of lineages represent our subjective judgments of relative
importance (diversity) of any lineage at the time shown. Extension of Glasbiinae into
the Paleocene is based on Crochet's (1980) suggestion that *Protodidelphis* is a glasbiine,
and of peradectids into the Eocene similarly on his belief that *Bobbschaefferia* belongs
to the Peradectidae.

2. Relationships of *Sparassocynus*

Sparassocynus (Fig. 29), from the Late Miocene-
Early Pleistocene, has variously been considered
of borhyaenoid, dasyuroid, or didelphid relationship. Revisionary studies by Reig (1958b) and
Reig and Simpson (1972) provided stronger
arguments in favor of didelphid affinities. This
genus is outstanding in its possession of large
epitympanic sinuses, a peculiar type of closed,
inflated, fully alisphenoid bulla, advanced
carnassialization of upper and lower molars
(Figs 29. 47), broadened palate without palatal
vacuities, wide occiput, shortened rostrum, and
laterally expanded braincase. *Sparassocynus* separates from borhyaenoids because of its five upper
incisors and broad frontal-maxillary contact, two
character-states consistent with didelphid
relationship. On the other hand, carnassialization is highly advanced by didelphid standards,
even as compared with *Lutreolina* and
Thylophorops, and together with the closed palate

is suggestive of borhyaenoid affinities. However,
Sparassocynus is primitive for borhyaenoids in
carnassialization, somewhat inconsistently with
its late position in the fossil record. *Sparassocynus*
also lacks conules, present in some of the early
borhyaenoids. It is probable, therefore, that
Sparassocynus represents a lineage of Cenozoic
didelphoids which converged on borhyaenoids in
some characters, while retaining the emblems of
a didelphoid relationship demonstrated by the
frontal-maxillary suture and dental formula. In
any case, its carnassial molars and peculiar
autapomorphies in the ear region indicate that it
belongs to a separate phylad within the didelphoids. The type of bulla is suggestive of a
separation in Cenozoic didelphoid cladogenesis
prior to the split of caluromyines and didelphines, which evolved different transformations
of the typical open tripartite bullae. It thus seems
reasonable to give full familial status to *Sparassocynus*, raising the rank of the Sparassocyninae
as proposed by Reig (1981).

Table 6. A classification of the non-borhyaenoid didelphimorph marsupials, to genus. The abbreviations used for geologic occurrence include: L., lower; M., middle; U., upper; Cret., Cretaceous; Paleoc., Paleocene; Eoc., Eocene; Olig., Oligocene; Mioc., Miocene; Plioc., Pliocene; Pleist., Pleistocene; R., Recent. The abbreviations used for geographic distributions include: Cent. Amer., Central America (México through Panamá); N. Amer., North America; S. Amer., South America. Only major synonymies are quoted. For a detailed synonomy at the generic level, see Marshall 1981b.

Order POLYPROTODONTA Owen 1866.

Suborder DIDELPHIMORPHIA Gill 1872.

Superfamily Microbiotherioidea (Ameghino 1887), new rank.

†Family Peradectidae (Crochet 1979), new rank.
  †*Albertatherium* Fox 1972. U. Cret., N. Amer.
  †*Alloeodectes* Russell 1984. L. Oligoc., N. Amer.
  †*Alphadon* Simpson 1927. U. Cret., N. Amer.
  †*Ankotarinja* (?) Archer 1976. M. Mioc., Australia
  †*Bobbschaefferia* (?) Paula Couto 1970. M. Paleoc., S. Amer.
  †*Garatherium* (?) Crochet 1984. L. Eoc., Africa
  †*Mimoperadectes* Bown and Rose 1979. L. Eoc., N. Amer.
  †*Peradectes* Matthew and Granger 1921 (including *Didelphidectes* Hough 1961; *Nanodelphys* McGrew 1937, and *Thylacodon* Matthew and Granger 1921, as junior synonyms. *Nanodelphys* may be considered as a valid subgenus.). U. Cret., S. Amer.; L. Paleoc. -M. Oligoc., N. Amer.; M. Paleoc. -L. Eoc., Europe.

Family Microbiotheriidae Ameghino 1887 (including Clenialitidae Amegh. 1909).
  Subfamily Microbiotheriinae Simpson 1929.
    Dromiciops Thomas 1894. Rec., S. Amer.
    †*Microbiotherium* Ameghino 1887 (including *Oligobiotherium* Amegh. 1902; *Clenia* Amegh. 1904; *Clenialites* Amegh. 1906; *Microbiotheridion* Ringuelet 1953; *Hadrorhynchus* Amegh. 1891; *Eodidelphys* Amegh. 1891; *Prodidelphus* Amegh. 1891; *Proteodidelphys* Amegh. 1898; *Stylognathus* Amegh. 1891; *Phonocdromus* Amegh. 1894 (*partim*). L. Olig. -E. Mioc., S. Amer.

  †Subfamily Glasbiinae Clemens 1966.
    †*Glasbius* Clemens 1966. U. Cret., N. Amer.
    †*Protodidelphis* (?) Paula Couto 1952. M. Paleoc., S. Amer.
    †*Reigia* Pascual 1983. M. (?) Eoc., S. Amer.

  †Subfamily Pediomyinae Simpson 1927.
    †*Aquiladelphis* Fox 1971. U. Cret., N. Amer.
    †*Pediomys* Marsh 1899 (incl. *Protolambda* Osborn 1898 and *Synconodon* Osborn 1898, as junior synonyms.). U. Cret., N. Amer.
    †*Gen. indet.* U. Cret., S. Amer.

†Family Stagodontidae Marsh 1889 (= Thlaeodontidae Cope 1892; Didelphodontinae Simpson 1927; Thlaeodontinae Hay 1930).
    †*Boreodon* Lambe 1902. U. Cret., N. Amer.
    †*Delphodon* Simpson 1927. U. Cret., N. Amer.
    †*Didelphodon* Marsh 1889 (including *Didelphops* Marsh 1899; *Stagodon* Marsh 1889; *Thlaeodon* Cope 1892; *Ectocondon* Osborn 1898; and *Diaphorodon* Simpson 1927, as junior synonyms). U. Cret., N. Amer.
    †*Eodelphis* Matthew 1916. U. Cret., N. Amer.

†Family Caroloameghiniidae Ameghino 1901.
    †*Caroloameghinia* Ameghino 1901. L. Eoc., S. Amer.
    †*Procaroloameghinia* Marshall 1982d. M. Paleoc., S. Amer.
    †*Roberthoffstetteria* Marshall, deMuizon and Sigé 1983. U. Cret., S. Amer.

Superfamily Didelphoidea (Gray 1821) Osborn 1910.

Family Didelphidae Gray 1821.

Subfamily Didelphinae (Gray 1821), Simpson 1927.

Tribe Didelphini (Gray 1821), Crochet 1979.
    Chironectes Illiger 1811 (including *Memina* Fisher 1813; *Cheironectes* Gray 1821; and *Gamba* Liais, as junior synonyms). L. Plioc. -Rec., S. Amer.; Rec., M. Amer.
    Didelphis Linnaeus 1758 (including *Didelphys* Schreber 1778; *Philander* Tiedemann 1808 *partim; Opossum* Schmid 1818; *Sarigua* Muirhead 1819; *Dasyurotherium* Liais 1872; *Gambatherium* Liais 1872; *Thylacotherium* Lund 1839; *Micoures* Lesson 1842 *partim; Leucodelphis* Ihering 1914; *Leucodidelphys* Krumbiegel 1941; †*Dimerodon* Amegh. 1889, as junior synonyms). M. Pleist. -Rec., S. Amer.; L. Pleist. -Rec., N. Amer.; Rec., M. Amer.
    †*Hondadelphys* Marshall 1976. M. Mioc., S. Amer.
    †*Hyperdidelphys* Ameghino 1904 (including *Paradidelphys* Amegh. 1904, and *Cladodidelphys* Amegh. 1904, as junior synonyms). U. Mioc. -U. Plioc., S. Amer.
    Lutreolina Thomas 1910 (including *Didelphis* Desmarest 1804, *partim; Peramys* Matschie 1916; *Metachirus* Hensel 1872, *partim,* as junior synonyms). L. Mioc. -Rec., S. Amer.
    Philander Tiedemann 1808 (including *Metachirops* Matschie 1916; *Holothylax* Cabrera 1919; *Metacheirus* Sanderson 1949, as junior synonyms). L. Plioc. -Rec., S. Amer.; Rec., M. Amer.
    †*Thylophorops* Reig 1952. U. Plioc. -L. Pleist. S. Amer.

Tribe Marmosini, new tribe.
    Lestodelphys Tate 1934 (including *Notodelphys* Thomas 1921; *nec* Allman 1847; *nec* Lichtenstein and Weynland 1854, as a synonym). L. Pleist. -Rec., S. Amer.
    Marmosa Gray 1821 (including *Asagis* Gloger 1842; *Notogogus* Gloger 1842; *Grymaeomys* Burmeister 1854; *Cuica* Liais 1872; and *Quica* Cabrera 1957, as synonyms; *Marmosops* Matschie 1916, and *Stegomarmosa* Pine 1972, stand as valid subgenera). M. Mioc. -Rec., S. Amer.; Pleist. -Rec., M. Amer.
    Micoures Lesson 1842 (including *Caluromys* Matschie 1916). U. Pleist. -Rec., S. Amer.; Rec., M. Amer.

Table 6. Continued.

Tribe Marmosini, new tribe — continued.

Monodelphis Burnett 1830 (including *Peramys* Lesson 1842; *Hemiurus* Gervais 1855; *Microdelphys* Burmeister 1856; and *Monodelphiops* Matschie 1916, as synonyms. *Minuania* Cabrera 1916, may be recognized as a subgenus). U. Mioc. -Rec., S. Amer.; Rec., M. Amer.

Thylamys Gray 1843. L. Plioc. -Rec., S. Amer.

†*Thylatheridium* Reig 1952. U. Mioc. -L. Pleist., S. Amer.

Tribe Metachirini, new tribe.

Metachirus (Burmeister 1854) Burmeister 1856 (including *Philander* Gray 1843, *partim,* as a synonym). Rec., S. Amer.; Rec., M. Amer.

Subfamily Caluromyinae Kirsch and Reig 1977 (in Kirsch 1977).

Caluromys Allen 1900 (including *Philander* Burmeister 1856, *nec* Tiedemann 1808; *Micoureus* Matschie 1916, *nec* Lesson 1842, as synonyms. *Mallodelphys* Thomas 1920, may be considered as a valid subgenus). U. Pleist. -Rec., S. Amer.; Rec., M. Amer.

Caluromysiops Sanborn 1951. Rec., S. Amer.

Glironia Thomas 1912. Rec., S. Amer.

†*Pachybiotherium* Ameghino 1902. U. Oligoc., S. Amer.

Subfamily Herpetotheriinae Trouessart 1879.

†*Amphiperatherium* Filhol 1879 (including *Oxygomphius* Meyer 1846, *nomen oblitum,* and probably *Ceciliolemur* Weigelt 1933; and *Microtarsoides* Weigelt 1933, as synonyms). L. Mioc. -M. Mioc., Europe.

†*Herpetotherium* Cope 1837 (including *Miothen* Cope 1893; *Spalacodon* Charlesworth 1844, as synonyms). L. Eoc.-L. Mioc., N. Amer.

†*Peratherium* Aymard 1850 (including *Alacodon* Quinet 1964, as a synonym). L. Eoc. -U. Oligoc., Europe; Oligoc., Africa.

Family Sparassocynidae (Reig 1958), new rank.

†*Gen. innom.* U. Mioc., S. Amer.

†*Sparassocynus* Mercerat 1898 (including *Perazoyphium* Cabrera 1928, and the misprint *Gerazoyphus* L. Kraglievich 1934, as synonyms). L. Plioc. -L. Pleistoc., S. Amer.

Superfamily Didelphoidea *incertae sedis.*

†*Coona* Simpson 1938. L. Eoc., S. Amer.

†*Derorhynchus* Paula Couto 1952. M. Paleoc., S. Amer.

†*Didelphopsis* Paula Couto 1952. M. Paleoc., S. Amer.

†*Eobrasilia* Simpson 1947. M. Paleoc., S. Amer.

†*Eomicrobiotherium* Marshall 1982a. L. Eoc., S. Amer.

†*Gaylordia* Paula Couto 1952 (including *Xenodelphis* Paula Couto 1962). M. Paleoc., S. Amer.

†*Guggenheimia* Paula Couto 1952. M. Paleoc., S. Amer.

†*Itaboraidelphys* Marshall and de Muizon 1984. M. Paleoc., S. Amer.

†*Marmosopsis* Paula Couto 1962. M. Paleoc., S. Amer.

†*Minusculodelphis* Paula Couto 1962. M. Paleoc., S. Amer.

†*Mirandatherium* Paula Couto 1952 (= *Mirandaia* Paula Couto 1952). M. Paleoc., S. Amer.

†*Monodelphopsis* Paula Couto 1952. M. Paleoc., S. Amer.

†*Sternbergia* Paula Couto 1970. M. Paleoc., S. Amer.

†*Zygolestes* Ameghino 1898. L. Plioc., S. Amer.

Overall Phylogenetic Relationships

The phylogram in Fig. 69 represents our best estimate of the phylogenetic relationships of the living and extinct major groups of opossum-like marsupials. That figure corresponds in most details to the Wagner-tree results, with the important exceptions of the placements of *Sparassocynus* and the *Peratherium*-like group of genera. Note also that the Peradectidae are regarded as the probable ancestors for the didelphoid group, and were most likely also the ancestors of Borhyaenoidea, Australasian marsupials, and American diprotodonts.

CLASSIFICATION

Most of the systematic and phylogenetic inferences in the previous section have been based on an unorthodox utilization of cladistic tools and concepts, which were demonstrated to be of great value in adumbrating relationships among the studied taxa. Needless to say, the pattern of relationships inferred by those means is much more reliable in the case of the taxa with living representatives, because of both the scope and

meaning of the characters used. The Wagner-tree method helped to evaluate relationships of taxa known exclusively on the basis of a more limited set of characters, and therefore on a less reliable foundation.

But even though we profited from cladistic analysis, we did not build a purely cladistic classification from our systematic conclusions. We were not tied to the mandatory correspondence between branching sequences and taxonomic grouping or of branching levels and categorical rank (e.g., McKenna 1975); nor did we feel compelled to exclude paraphyletic groupings or to disregard degrees of divergence. The main reason for departing from cladistic rules is that otherwise we could not benefit from the important information provided by fossil taxa, which made decisions on probable relatedness and phylogeny more direct. Thus, our classification is strongly cladistic when dealing with the living taxa, but departs from the tenets of cladistics in establishing ancestor-descendant relations among the fossil and living groups, and in the recognition of degrees of morphological differentiation. We could have taken the course, nevertheless, of grouping some of the fossil taxa in *plesions* (see, for instance, Wiley 1981). However, we avoided doing so because we strongly believe that placing fossil taxa in such groups is a subtle way of avoiding the effort required to integrate them with living taxa in a cogent and useful overall scheme.

The most important departure in our classification from previous ones, including our own (Kirsch 1977b; Reig 1981) is the recognition that living and fossil opossum taxa are grouped in two separate superfamilies, Microbiotherioidea and Didelphoidea (Table 6). In this formalization we attempt to be consistent with our systematic conclusions and the repeated results of Wagnerian analysis, and with current practice in marsupial systematics, which tends to recognize several restrictive superfamilial groupings (e.g., Kirsch 1968 and 1977b; Marshall 1981b; Marshall and Clemens 1976). Needless to say, the Didelphoidea are more clearly defined by synapomorphies than are the Microbiotherioidea, and it may be objected that the latter is mainly based on shared primitive states and therefore can be considered paraphyletic. Our answer will be "so what?" It is clear that this grouping, even if eventually demonstrated to be artificial by new evidence, has already great heuristic value; and the separation of the mainly Late Cretaceous clade of Microbiotherioidea from the Cenozoic Didelphoidea has the additional advantage of providing an ancestral group for the remaining polyprotodont superfamilies, the Didelphoidea, Borhyaenoidea, Perameloidea, and Dasyuroidea being thus rooted to a common generalized ancestral group. Moreover, although no obvious unique shared, derived features may unite, for example, *Alphadon* and *Dromiciops*, there is a clearly demonstrated *sequence* of character-changes which tie them together through a series of roughly intermediate taxa — a sequence not duplicated elsewhere within Didelphimorphia and thus constituting a kind of "super"-synapomorphy. This demonstration of a series of linking characters is, to us, as persuasive a reason for establishing formal taxonomic status for this group as would be the sharing of a larger set of derived features.

For this early opossum clade we chose the name Microbiotherioidea to emphasize the meaningful relationships of stagodontids, pediomyines, glasbiines, caroloameghiniids, and peradectids to the living *Dromiciops* and the closely allied fossil genus *Microbiotherium*, and in due respect for the priority of the family name Microbiotheriidae Ameghino 1887 over other available nomina.

Within the Microbiotherioidea, we recognize full family status for Peradectidae, Stagodontidae, Microbiotheriidae, and Caroloameghiniidae. The former name was proposed as a

tribal one, Peradectini, by Crochet (1979, 1980), but the group deserves full family rank within the framework of relationships ensuing from our analysis. Stagodontids are usually given family rank, which is here also confirmed by their separation from other related microbiotherioid taxa. The same reasoning supports familial categorical rank for the Caroloameghiniidae. As for the Microbiotheriidae, we include as subfamilies the South American Microbiotheriinae and the North and South American Glasbiinae and Pediomyinae. To reflect more strictly the cladistic pattern of relationships emerging from the Wagner-tree analysis, we could recognize two subfamilies, Pediomyinae and Microbiotheriinae, and split the latter into two tribes. However, with due allowance for the fact that we cannot be completely confident of the degrees of affinity indicated by a restricted set of characters, we prefer to maintain the tradition of recognizing Glasbiinae as a separate subfamily.

The remaining opossums are all grouped under a restricted concept of the superfamily Didelphoidea, the extension of which covers just the families Sparassocynidae and Didelphidae. The latter is divided into only three subfamilies — Caluromyinae, Didelphinae, and Herpetotheriinae. As discussed earlier in this paper, the recognition — or rather revival — of the Herpetotheriinae Trouessart 1879 as a separate taxon is based more on phylogenetic and biogeographic considerations than on the analytical results. In stringent adherence to phylogenetics, herpetotheres should be considered a family parallel to Didelphidae. However, as they are closely similar anatomically to the true opossums, we place them in the same family. The tribal arrangement within the Didelphinae, which has been advanced in a recent paper (Reig 1981), is based on the combined results of our chromosomal, serological, and morphological analyses. We recommend that three tribes be recognized within Didelphinae — the Marmosini, Metachirini, and Didelphini. Subtribes could possibly be recognized within these, with their included taxa obvious from the cladograms and accompanying discussions, but we do not formally propose subtribes at this time.

Certain fossil taxa currently affiliated with Didelphinae may warrant recognition at higher levels. The diversity of the 13 described genera of didelphoids from fissure fillings of Paleocene (Riochican) age at São José de Itaborai, Brazil, surpasses that of living didelphines and documents a well-advanced adaptive radiation for the group. In fact, those Riochican forms are so diverse that subfamilial (or familial?) division appears to be needed, but has not yet been attempted (Simpson 1971:112). These taxa have not been adequately compared with either living or other fossil forms, and the phylogenetic relationships of the Riochican marsupials among themselves are not clear. This is the reason why we classify most of them as Didelphoidea *incertae sedis*. It is our hope that the classification presented in this paper may help clear the way toward fuller understanding of the Riochican as well as other opossum-like marsupials.

ACKNOWLEDGMENTS

We thank P. Bottjer, P. Hershkovitz, R. H. Pine, K. Padian, R. Pascual, J. E. Rodman, and G. G. Simpson for helpful comments on early drafts of this paper; and two anonymous reviewers of the penultimate version. M. Renfree provided helpful criticisms of the sections on reproductive anatomy; J. D. Archibald and W. A. Clemens aided with the coding of *Alphadon* characters; J. S. Farris made available the Wagner 78 program; and the information-assessment program was written by T. F. Smith and W. Fitch. O. A. R. thanks the Guggenheim Foundation for supporting his studies at the British Museum (Natural History), London (1972-73), and the curators at that museum — G. Corbet, J. Bishop, and J. Ingles — for providing facilities and many appreciated courtesies. J. A. W. K.'s contribution benefitted from a grant from the Milton Fund, Harvard University and earlier grants from the National Science Foundation and other sources. Study of museum specimens by L. G. M. in Argentina was made possible by research grants 1329 and 1698 from the National Geographic Society, Washington, D.C., and completion of the work was facilitated by National Science Foundation grant DEB-7901976. For providing access to study of specimens in Argentine museums we all wish to acknowledge the assistance of the late Guillermo del Corro (MACN), Rosendo Pascual (MLP), and Galileo J. Scaglia (MMP). Figures 34, 46, and 47 are reprinted by permission of the *Journal of Paleontology, Acta Anatomica*, and the *Journal of Zoology*, respectively; Figures 48-52 are reproduced with the permission of the American Society of Mammalogists; Figure 55 is presented through the courtesy of the *Journal of the Linnean Society*; and Figure 57 is reprinted with the permission of Academic Press.

REFERENCES

ARCHER, M., 1976a. The dasyurid dentition and its relationship to that of didelphids, thylacinids, borhyaenids (Marsupicarnivora) and peramelids (Peramelina, Marsupialia). *Aust. Jour. Zool. Suppl. Ser. No. 39.*

————., 1976b. The basicranial region of marsupicarnivores (Marsupialia), relationships of carnivorous marsupials, and affinities of the insectivorous marsupial peramelids. *Zool. Jour. Linn. Soc. Lond.* 59: 217-322.

————., 1976c. Miocene marsupicarnivores (Marsupialia) from Central South Australia, *Ankotarinja tirarensis* gen. et sp. nov., *Keeuna woodburnei* gen. et sp. nov., and their significance in terms of marsupial radiations. *Trans. Roy. Soc. South Australia* 100: 53-73.

————., 1978. The nature of the molar-premolar boundary in marsupials and reinterpretation of the homology of marsupial cheekteeth. *Mem. Qd Mus.* 18: 157-64.

————., 1982. A review of Miocene thylacinids (Thylacinidae, Marsupialia), the phylogenetic position of the Thylacinidae and the problem of apriorisms in character analysis. Pp. 445-76 *in* "Carnivorous marsupials" ed by M. Archer. Royal Zool. Soc. New South Wales: Sydney.

ARCHIBALD, J. D., 1982. A study of Mammalia and geology across the Cretaceous-Tertiary boundary in Garfield County, Montana. *Univ. Calif. publ. Geol. Sci.* 122: v-xvi, 1-286.

ASHLOCK, P. D., 1974. The uses of cladistics. *Ann. Rev. Ecol. Syst.* 5: 81-99.

AYALA, F., 1977. Phylogenetics and Macromolecules. Pp. 262-73 *in* "Evolution" ed by T. Dobzhansky, F. Ayala, G. L. Stebbins and J. A. Valentine. W. H. Freeman and Co.: San Francisco.

BARBOUR, R. A., 1977. Anatomy of marsupials. Pp. 237-72 *in* "The Biology of Marsupials" ed by B. Stonehouse and D. Gilmore. University Park Press: Baltimore.

BENSLEY, B. A., 1903. On the evolution of the Australian Marsupialia: with remarks on the relationships of the marsupials in general. *Trans. Linn. Soc. London, Zool.* (2)9: 83-217.

BIGGERS, J. D., 1966. Reproduction in male marsupials. Pp. 251-80 *in* "Comparative biology of reproduction in mammals" ed by I. W. Rowlands. *Symp. Zool. Soc. Lond.* 15: 1-559.

———— AND DELAMATER, E. D., 1965. Marsupial spermatozoa pairing in the epididymis of American forms. *Nature* 208: 402-04.

BOWN, T. M. AND KRAUS, M. J., 1979. Origin of the tribosphenic molar and metatherian and eutherian dental formulae. Pp. 172-81 *in* "Mesozoic mammals" ed by J. A. Lillegraven, Z. Kielan-Jaworowska and W. A. Clemens. University of California Press: Berkeley.

BOWN, T. M. AND ROSE, K. D., 1979. *Mimoperadectes*, a new marsupial, and *Worlandia*, a new dermopteran, from the lower part of the Willwood Formation (Early Eocene), Bighorn Basin, Wyoming. *Univ. Michigan Contrib. Mus. Paleont.* 25: 89-104.

BRESSLAU, E. L., 1912. Die Entwicklung des Mammarapparates der Monotremen, Marsupialier u. einiger Placentalier. III. Semon, Forschungsreisen, Bd. 4.

————., 1920. "The mammary apparatus of the Mammalia (in the light of ontogenesis and phylogenesis)". Methuen and Co.: London.

BRITTEN, R. J. AND DAVIDSON, E. H., 1969. Gene regulation for higher cells: a theory. *Science* 165: 349-57.

BRUNDIN, L., 1966. Transantarctic relationships and their significance, as evidenced by chironomid midges with a monograph of the subfamilies Pondonominae and Aphroteniinae and the austral Heptagyinae. *K. Svenska Vetenskapsakad. Handl., ser. 4* 11: 1-472.

————., 1968. Application of phylogenetic principles in systematics and evolutionary theory. Pp. 473-85 *in* "Current problems of lower vertebrate phylogeny" ed by T. Orvig. Nobel Symposium 4. John Wiley and Sons: New York.

BUCHER, J. E. AND HOFFMANN, R. S., 1980. *Caluromys derbianus*. Mammalian Species. *The Amer. Soc. Mammal.* 140: 1-4.

CABRERA, A., 1919. "Genera Mammalium. Monotremata, Marsupialia". Madrid.

CABRERA, A., 1957. Catalogo de los mamíferos de America del Sur, I. (Metatheria-Unguiculata-Carnivora). *Rev. Mus. Argent. Cienc. Nat. "Bernardino Rivadavia", Cienc. Zool.* **4:** 1-307.

CAMIN, J. H. AND SOKAL, R. R., 1965. A method for deducing branching sequences in phylogeny. *Evolution* **19:** 311-26.

CAMP, C. L., 1923. Classification of the lizards. *Bull. Amer. Mus. Nat. Hist.* **48:** 289-481.

CARLETON, M. D., 1980. Phylogenetic relationships in neotomine-peromyscine rodents (Muroidea) and a reappraisal of the dichotomy within New World Cricetinae. *Misc. Publ. Mus. Zool., Univ., Mich.* No. 157.

CARLSSON, A., 1903. Beiträge zur Anatomie der Marsupialregion bei den Beutelthieren. *Zool. Jahrb., Abth. für Anat.* **18**(4): 489-506.

CLEMENS, W. A., 1966. Fossil mammals of the Type Lance Formation, Wyoming. Part II. Marsupialia. *Univ. Calif. Publ. Geol. Sci.* **62:** 1-122.

————., 1968. Origin and early evolution of marsupials. *Evolution* **22:** 1-18.

————., 1979. Marsupialia. Pp. 192-220 *in* "Mesozoic mammals" ed by J. A. Lillegraven, Z. Kielan-Jaworowska and W. A. Clemens. University of California Press: Berkeley.

———— AND KIELAN-JAWOROWSKA, Z., 1979. Multituberculata. Pp. 99-149 *in* "Mesozoic mammals" ed by J. A. Lillegraven, Z. Kielan-Jaworowska and W. A. Clemens. University of California Press: Berkeley.

———— AND MARSHALL, L. G., 1976. Fossilium Catalogus: American and European Marsupialia. Pars **123:** 1-114. W. Junk: The Hague.

COLLINS, L. R., 1973. "Monotremes and marsupials: a reference for zoological institutions". Smithsonian Ins. Publ. **4888:** 1-323.

CRESPO, J. A., 1950. Nota sobre mamíferos de Misiones nuevns para Argentina. *Com Mus. Argentina Cien. Nat. "Bernardino Rivadavia", Cien. Zool.* **1**(14): 1-14.

CRISCI, J. V., 1982. Parsimony in evolutionary theory: law or methodological prescription? *Jour. Theor. Biol.* **97:** 35-41.

CRISCI, J. V. AND STUESSY, T. F., 1980. Determining primitive character states for phylogenetic reconstruction. *Syst. Bot.* **5:** 112-35.

CROCHET, J. Y., 1977a. Les Didelphidae (Marsupicarnivora, Marsupialia) holarctiques tertiaires. *C.R. Acad. Sci. Paris*, ser. D. **284:** 357-60.

————., 1977b. Les Didelphidés Palèogénes Holarctiques: historique et tendances évolutives. *Géobios. Mem. special* **1:** 127-34.

————., 1979. Diversité systématique des Didelphidae (Marsupialia) Européens Tertiares. *Géobios* **12:** 365-78.

————., 1980. "Les Marsupiaux du Tertiare D'Europe". Éditions de la Foundation Singer-Polignac: Paris.

————., 1984. *Garatherium mahboubii* nov. gen., n. sp., marsupial de l'Eocène inférieur, d'El Kohol (Sud-Oranais, Algerie. *Ann. Paleont. (Vért.-Invért.)* **70:** 275-94.

CROMPTON, A. W. AND HIIEMAE, K. M., 1969. Functional occlusion in tribosphenic molars. *Nature* **222:** 678-79.

————., 1970. Molar occlusion and mandibular movements in the American opossum, *Didelphis marsupialis, L. Linn. Soc. Zool. Jour.* **49:** 21-47.

ENDERS, R. K., 1937. Panniculus carnosus and formation of the pouch in didelphids. *Jour. Morph.* **61:** 1-26.

ESTABROOK, G. F., 1977. Does common equal primitive? *Syst. Bot.* **2:** 36-42.

FARRIS, J. S., 1970. Methods for computing Wagner trees. *Syst. Zool.* **19:** 83-92.

————., 1972. Estimating phylogenetic trees from distance matrices. *Amer. Nat.* **106:** 645-68.

————., 1981. Distance data in phylogenetic analysis. Pp. 3-23 *in* "Advances in cladistics: proceedings of the first meeting of the Willi Hennig Society" ed by V. A. Funk and D. R. Brooks. The New York Botanical Garden: Bronx.

FITCH, W. M., 1976. Molecular evolutionary clocks. Pp. 160-78 *in* "Molecular evolution" ed by F. J. Ayala. Sinauer Associates: Sunderland.

———— AND SMITH, T. F., 1982. Implications of minimal length trees *Syst. Zool.* **31:** 68-75.

FOX, R. C., 1971. Marsupial mammals from the Early Campanian Milk River Formation, Alberta, Canada. Pp. 145-64 *in* "Early mammals" ed by D. M. and K. A. Kermack. *Jour. Linnean Soc. (Zool.) Suppl.* **50.**

FOX, R. P., 1979a. Mammals from the Upper Cretaceous Oldman Formation, Alberta. I. *Alphadon* Simpson (Marsupialia). *Can. J. Earth Sci.* **16:** 91-102.

————., 1979b. Mammals from the Upper Cretaceous Oldman Formation, Alberta. II. *Pediomys* Marsh (Marsupialia). *Can. J. Earth Sci.* **16:** 103-13.

————., 1981. Mammals from the Upper Cretaceous Oldman Formation, Alberta. V. *Eodelphis* Matthew, and the evolution of the Stagodontidae (Marsupialia). *Can. J. Earth Sci.* **18:** 350-65.

GARDNER, A. L., 1973. The systematics of the genus *Didelphis* (Marsupialia: Didelphidae) in North and Middle America. *Spec. Publ. Mus. Texas Tech. Univ.* **4:** 1-81.

————., 1981. (review of) The mammals of Suriname, by A. M. Husson. *J. Mamm.* **62:** 445-48.

———— AND PATTON, J. L., 1972. New species of *Philander* (Marsupialia: Didelphidae) and *Mimon* (Chiroptera: Phyllostomidae) from Peru. *Occ. Pap. Mus. Zool. Louisiana State Univ.* **43:** 1-12.

GILMORE, R. M., 1941. Zoology. Pp. 314-19 *in* "The susceptibility to yellow fever of the vertebrates of eastern Colombia" ed by J. C. Bugher *et al. Amer. J. Trop. Med.* **21**(2): 314-19.

GOODMAN, M., 1976. Protein sequences in phylogeny. Pp. 141-59 *in* "Molecular evolution" ed by F. J. Ayala. Sinauer Associates: Sunderland.

GREGORY, W. K., 1910. The orders of mammals. *Bull. Amer. Mus. Nat. Hist.* **27:** 1-524.

GRIFFITHS, M., 1968. "Echidnas". Pergamon Press: Oxford.

HALL, E. R. AND DALQUEST, W. W., 1963. The mammals of Veracruz. *Univ. Kansas Publ. Mus. Nat. Hist.* **14:** 165-362.

HANDLEY, C. A. AND GORDON, L. K., 1979. New species of mammals from Northern South America: mouse possums, genus *Marmosa* Gray. Pp. 65-72 *in* "Vertebrate ecology in the northern neotropics" ed by J. F. Eisenberg. Smithsonian Inst. Press: Washington.

HAYMAN, D. L., 1977. Chromosome number — constancy and variation. Pp. 27-48 *in* "The biology of marsupials" ed by B. Stonehouse and D. Gilmore. Macmillan Press: London.

HAYMAN, D. L. AND MARTIN, P. G., 1974. "Animal Cytogenetics, 4, Chordata. Mammalia I: Monotremata and Marsupialia". Gebruder Borntraeger: Berlin — Stuttgart.

HECHT, M. K., 1976. Phylogenetic inference and methodology as applied to the vertebrate record. *Evolutionary Biology* **9:** 335-63.

HENNIG, W., 1966. "Phylogenetic Systematics". Univ. Illinois Press: Urbana. [1968. Elementos de una sistematica filogenetica. Editorial Universitaria de Buenos Aires (Eudeba), 353 pp.].

HERSHKOVITZ, P., 1977. Comments on generic names of four-eyed opossums (Family Didelphidae). *Proc. Biol. Soc. Wash.* **89:** 295-304.

————., 1981. *Philander* and four-eyed opossums again. *Proc. Biol. Soc. Wash.* **93:** 943-46.

HILDEBRAND, M., 1961. Body proportions of didelphid (and some other) marsupials, with emphasis on variability. *Amer. Jour. Anat.* **109:** 239-49.

HILL, J. P. AND FRASER, E. A., 1925. Some observations on the female urogenital organs of the Didelphidae. *Proc. Zool. Soc. Lond. for* **1925:** 189-219.

HIRIARTE CEA, C. L., 1975. "La morfologia del aparato genital de *Dromiciops australis* (Marsupialia) y sus consecuencias sistematicas". Unpublished thesis, Universidad Austral de Chile – Facultad de Letras y Educación, Escuela del Biologia y Quimica.

HOPSON, J. A., 1973. Endothermy, small size and the origin of mammalian reproduction. *Amer. Natur.* **107:** 446-52.

HUNSAKER, D., 1977. Ecology of New World marsupials. Pp. 95-156 *in* "The biology of marsupials" ed by D. Hunsaker. Academic Press: New York.

JACKSON, R. D., 1971. The karyotype in systematics. *Ann. Rev. Ecol. Syst.* **2:** 327-68.

JELLISON, W. L., 1945. A suggested homolog of the os penis or baculum of mammals. *Jour. Mammal.* **26:** 146-47.

JENKINS, F. A. AND PARRINGTON, F. R., 1976. The postcranial skeletons of the Triassic mammals *Eozostrodon, Megazostrodon* and *Erythrotherium. Roy. Soc. London Philos. Trans., B (Biol. Sci.)* **273:** 387-431.

JOHN, B. AND HEWITT, G. M., 1968. Patterns and pathways of chromosome evolution within the Orthoptera. *Chromosoma* **25:** 40-74.

———— AND LEWIS, K. R., 1968. The chromosome complement. Protoplasmatologia 4(A): 1-206. Springer Verlag: Wien-N.Y.

KAMPEN, P. N. VAN., 1905. Die Tympanalgegend des Säugetierschähels. *Morph. Jahrb.* **34:** 321-722.

KAVANAUGH, D. H., 1972. Hennig's principles and methods of phylogenetic systematics. *Biologist* 54: 115-27.

KIELAN-JAWOROWSKA, Z., BOWN, T. M. AND LILLEGRAVEN, J. A., 1979. Eutheria. Pp. 221-58 *in* "Mesozoic mammals" ed by J. A. Lillegraven, Z. Kielan-Jaworowska and W. A. Clemens. University of California Press: Berkeley.

KIMURA, M. AND OHTA, T., 1971. Protein polymorphism as a phase of molecular evolution. *Nature* **229:** 467-69.

KIRSCH, J. A. W., 1968. Prodromus of the comparative serology of Marsupialia. *Nature* **217:** 418-20.

————., 1969. Serological data and phylogenetic inference: the problem of rates of change. *Syst. Zool.* **18:** 296-311.

————., 1977a. The six-percent solution: second thoughts on the adaptedness of marsupials. *Amer. Sci.* **65:** 276-88.

————., 1977b. The comparative serology of Marsupialia and a classification of marsupials. *Aust. Jour. Zool., Suppl. Ser.* **52:** 1-152.

————., 1977c. The classification of marsupials, with special reference to karyotypes and serum proteins. Pp. 1-50 *in* "The biology of marsupials" ed by D. Hunsaker. Academic Press: New York.

————., 1983. Informativeness and the problem of parsimony. *The Biologist* **65:** 65-77.

————., 1984. Marsupial origins: taxonomic and biological considerations. Pp. 627-31 *in* "Vertebrate zoogeography and evolution in Australasia" ed by M. Archer and G. Clayton. Perth Hesperian Press: Perth.

———— AND ARCHER, M., 1982. Polythetic cladistics or, when parsimony's not enough: the relationships of carnivorous marsupials. Pp. 595-619 *in* "Carnivorous marsupials" ed by M. Archer. Royal Zool. Soc. New South Wales: Sydney.

———— AND CALABY, J. H., 1977. The species of living marsupials: an annotated list. Pp. 9-26 *in* "The biology of marsupials" ed by B. Stonehouse and D. Gilmore. Macmillan: London.

————., PINE, R. H. AND VAN GELDER, R. G., 1982. Family Didelphidae. Pp. 18-76 *in* "Mammal species of the world: a taxonomic and geographic reference" ed by J. H. Honacki, K. E. Kinman, and J. W. Koeppl. Allen Press Inc.: Lawrence.

———— AND WALLER, P. F., 1979. Notes on the trapping and behavior of the Caenolestidae (Marsupialia). *Jour. Mammal.* **60:** 390-95.

KLUGE, A. G. AND FARRIS, J. S., 1969. Quantitative phyletics and the evolution of anurans. *Syst. Zool.* **18:** 1-32.

KRIEG, H., 1924. Biologische Reisestudien in Südamerika. III. Chilenische Beutelratten. *Zeitsch. Morphol. Ökol. der Tiere, Berlin* **3:** 169-76.

LILLEGRAVEN, J. A., 1969. Latest Cretaceous mammals of upper part of Edmonton Formation of Alberta, Canada and review of marsupial-placental dichotomy of mammalian evolution. *Univ. of Kansas, Paleont. Contrib., Art.* **50**(Vert. 12): 1-22.

————., KIELAN-JAWOROWSKA, Z. AND CLEMENS, W. A. (eds.), 1979. "Mesozoic Mammals: the first two-thirds of mammalian history". University of California Press: Berkeley.

LONG, D. A., 1972. Two hypotheses on the origin of lactation. *Amer. Nat.* **106:** 141-44.

McKENNA, M. C., 1975. Toward a phylogenetic classification of the Mammalia. Pp. 21-46 *in* "Phylogeny of the primates" ed by W. P. Luckett and F. S. Szalay. Plenum: New York.

McKENNA, M. C., 1981. Early history and biogeography of South America's extinct land mammals. Pp. 43-77 *in* "Evolutionary biology of the New World monkeys and continental drift" ed by H. L. Ciochon and A. H. Chiarelli. Plenum Publishing Corp.: New York.

MACINTYRE, G. T., 1967. Foramen pseudovale and quasi-mammals. *Evolution* **21:** 834-41.

MAHBOUBI, M., AMEUR, R., CROCHET, J.-Y. AND JAEGEH, J.-J., 1983. Première dé-couvente d'un marsupial en Afrique. *C. R. Acad. Sci., Paris,* **297**(7 Novembre 1983), ser. II: 691-99.

MANN, G. F., 1955. Monito del monte *Dromiciops australis* Philippi. *Invest. Zool. Chil.* **2:** 159-66.

————., 1958. Reproduccion de *Dromiciops australis* (Marsupialia, Didelphydae). *Invest. Zool. Chil.* **4:** 209-13.

MARSHALL, L. G., 1976. New didelphine marsupials from the La Venta Fauna (Miocene) of Colombia, South America. *Jour. Paleont.* **50**(3): 402-18.

MARSHALL, L. G., 1977a. First Pliocene record of the water opossum, *Chironectes minimus* (Didelphidae, Marsupialia). *J. Mammal.* **58**(3): 434-36.

————., 1977b. *Lestodelphys halli*. Mammalian Species. *The Amer. Soc. Mammal.* **81**: 1-3.

————., 1977c. *Lutreolina crassicaudata*. Mammalian Species. *Amer. Soc. Mammal.* **91**: 1-4.

————., 1977d. Cladistic analysis of borhyaenoid, dasyuroid, didelphoid, and thylacinid (Marsupialia: Mammalia) affinity. *Syst. Zool.* **26**: 410-25.

————., 1978a. *Glironia venusta*. Mammalian Species. *The Amer. Soc. Mammal.* **107**: 1-3.

————., 1978b. Evolution of the Borhyaenidae, extinct South American predaceous marsupials. *Univ. Calif. Publ. Geol. Sci.* **117**: 1-89.

————., 1978c. *Chironectes minimus*. Mammalian Species. *The Amer. Soc. Mammal.* **109**: 1-4.

————., 1978d. *Dromiciops australis*. Mammalian Species. *The Amer. Soc. Mammal.* **99**: 1-5.

————., 1979. The evolution of metatherian and eutherian (mammalian) characters: a review based on cladistic methology. *Zool. Jour. Linn. Soc. Lond.* **66**: 369-410.

————., 1981a. Review of the Hathlyacininae, an extinct subfamily of South American "dog-like" marsupials. *Fieldiana: Geology, n. s.*, no. **7**: 1-120.

————., 1981b. The families and genera of Marsupialia. *Fieldiana: Geology, n. s.*, no. **8**: 1-65.

————., 1982a. Systematics of the South American family Microbiotheriidae. *Fieldiana: Geology, n. s.*, no. **10**: 1-75.

————., 1982b. Systematics of the extinct South American marsupial family Polydolopidae. *Fieldiana: Geology, n. s.*, no. **12**: 1-109.

————., 1982c. Calibration of the Age of Mammals in South America. *Géobios, mém. spec.* **6**: 427-37.

————., 1982d. A new genus of Caroloameghiniinae (Marsupialia: Didelphoidea: Didelphidae) from the Paleocene of Brazil. *Jour. Mammal.* **63**: 709-16.

————., 1982e. Evolution of South American Marsupialia. Pp. 251-72 *in* "Mammalian biology in South America" ed by M. M. Mares and H. H. Genoways. Pittsburgh: Spec. Publ. Ser. Pymatuning Laboratory of Ecology, Univ. Pitts., 6.

————., 1984. The lower jaw of *Eobrasilia coutoi* Simpson, 1947, a unique didelphoid (not borhyaenoid) marsupial from the Paleocene of Brazil. *Jour. Paleont.* **58**: 173-77.

———— AND CORRUCCINI, R. S., 1978. Variability, evolutionary rates and allometry in dwarfing lineages. *Paleobiology* **4**(2): 101-19.

————., de MUIZON, Ch. AND SIGE, B., 1983. Late Cretaceous mammals (Marsupialia) from Bolivia. *Géobios* **16**: 739-45.

———— AND de MUIZON, Ch. 1984. A new didelphid marsupial (*Itaboraidelphys camposi* gen. et sp. nov.) from middle Paleocene (Itaboraian) age fissure fillings at the limestone quarry of Sao Jose de Itaborai, Brazil. *C. R. Acad. Sci., Paris.*

————., de MUIZON, Ch. AND HOFFSTETTER, R. L., 1984. Fossil Marsupialia in the Muséum National d'Histoire naturelle collected by André Tournouër from Patagonia, southern Argentina. *Bull. Mus. natn. Hist. nat., Paris, 4e sér., sec. C*, **6**: 33-58.

MATSCHIE, P., 1916. Bemerkungen über die Gattung *Didelphis* L. *Sitzber. Gesellsch. Naturf. Freunde, Berlin* **1916**(1): 259-72.

MATTHEY, R., 1973. The chromosome formulae of eutherian mammals. Pp. 531-616 *in* "Cytotaxonomy and vertebrate evolution" ed by A. B. Chiarelli and E. Capanna. Academic Press: London.

MAXSON, L. R., SARICH, V. M. AND WILSON, A. C., 1975. Continental drift and the use of albumin as an evolutionary clock. *Nature* **225**: 397-400.

MEACHAM, C. A., 1980. Phylogeny of the Berberidaceae with an evaluation of classifications. *Syst. Bot.* **5**: 149-72.

MIYAMOTO, M. M., 1981. Congruence among character sets in phylogenetic studies of the frog genus *Leptodactylus*. *Syst. Zool.* **30**: 281-90.

MONDOLFI, E. AND MEDINA-PADILLA, C. M., 1958. Contribución al conocimiento del "Perrito de Agua" (*Chironectes minimus* Zimmermann). *Mem. Soc. Cienc. Nat. La Salle* **17**: 141-49.

MONES, A., 1980. Sobre una colección de vertebrados fosiles de Monte Hermosa (Plioceno superior) Argentina, con la descripción de una neueva especie de *Marmosa* (Marsupialia: Didelphidae). *Com. Pal. Mus. de Hist. Nat. de Montevideo* **1**(8): 159-69.

MORTON, S. R., 1980. Ecological correlates of caudal fat storage in small mammals. *Aust. Mammal.* **3**: 81-86.

MOSS, M. L., 1969. Evolution of mammalian dental enamel. *Amer. Mus. Novit.* **2360**: 1-39.

MUIZON, C. de, MARSHALL, L. G. AND SIGÉ, B., 1984. The mammal fauna from El Molino Formation (Late Cretaceous, Maestrichtian) at Tiupampa, Southcentral Bolivia. *Bull. Mus. natn. Hist. nat. Paris (4e série).* **6, sect. C**: 327-51.

NEI, M., 1977. Standard error of immunological dating of evolutionary time. *Jour. Molecular Evolution* **9**: 203-11.

NUTTALL, G. H. F., 1904. "Blood immunity and blood relationship". Cambridge University Press: Cambridge.

ODREMAN RIVAS, O. AND ZETTI, J., 1969. Addenda paleontológica. Pp. 291-92 *in* "Caracteres geológicos de los depósitos eólicus del Pleistoceno superior de Juni (Provin de Buenos Aires)" ed by O. E. de Salvn, J. H. Ceci and A. Dillon. Actas IV Jornadas Geol. Argent., Mendoza, April 6-16. Vol. **1**: 269-92. *Assoc. Geol. Argent. Buenos Aires.*

OSGOOD, W. H., 1921. A monographic study of the American marsupial, *Caenolestes. Publ. Field Mus. Nat. Hist., Zool. Ser.* **14**(1): 1-156.

————., 1943. The mammals of Chile. *Field Mus. Nat. Hist., Zool. Ser.* **30**: 1-268.

PARKER, W. K., 1886. On the structure and development of the skull in the Mammalia. *Phil. Trans. Roy. Soc. Lond. for* **1885**: 1-275.

PASCUAL, R., 1980a. Nuevos y singulares tipos ecológicos de marsupiales extinguidos de América del Sur (Paleoceno tardió u Eoceno Temprano), del Noroeste Argentino: Act. II Cong. Arg. Paleont. y Bioestr., y I Congr. Latinoam. Paleont. Buenos Aires, 1978 **2**: 151-73.

————., 1980b. Prepidolopidae, nueva familia de Marsupialia Didelphoidea del Eoceno sudamericano. *Ameghiniana* **17**(3): 216-42.

————., 1981. Adiciones al conocimiento de *Bonapartherium hinakusijum* (Marsupialia, Bonapartheriidae) del Enceno temprano del Noroeste argentino. Anals II Congresso Latino-Americano Paleontológia, Porto-Alegre, Abril 1981 **2**: 507-20.

————., 1983. Novedosos marsupiales paleogenos de la Formación Pozuelos (Grupo Pastos Grandes) de la Puna, Salta, Argentina. *Ameghiniana* **20**: 265-80.

PATTERSON, B., 1958. Affinities of the Patagonian fossil mammal, *Necrolestes. Breviora* **94**: 1-14.

————., 1965. The auditory region of the borhyaenid marsupial *Cladosictis. Breviora* **217**: 1-9.

PAULA COUTO, C. de., 1952. Fossil mammals from the beginning of the Cenozoic in Brazil. Marsupialia: Didelphidae. *Amer. Mus. Novit.* **1567:** 126.

————., 1962. Didelfideos fosiles del Paleoceno de Brasil. Rev. Mus. Argent. Cien. Nat. "Bernardino Rivadavia", *Cien. Zool.* 8(112): 135-66.

————., 1970. News on the fossil marsupials from the Riochican of Brazil. *An. Acad. Bras. Ciênc.* **42**(1): 19-34.

PINE, R. H., 1972. A new subgenus and species of murine opossum (genus *Marmosa*) from Peru. *J. Mamm.* **53:** 279-82.

————., 1973. Anatomical and nomenclatural notes on opossums. *Proc. Biol. Soc. Wash.* **86:** 391-402.

————., 1975. A new species of *Monodelphis* (Mammalia: Marsupialia: Didelphidae) from Bolivia. *Mammalia* **39**(2): 321-22.

————., 1976. *Monodelphis umbristriata* (A. De Miranda-Ribeiro) is a distinct species of opossum. *J. Mamm.* **57**(4): 785-87.

————., 1977. *Monodelphis iheringi* (Thomas) is a recognizable species of Brazilian opossum (Mammalia: Marsupialia: Didelphidae). *Mammalia* **41**(2): 235-37.

REIG, O. A., 1952. Descripción previa de nuevos ungulados y marsupiales fósiles del Plioceno y del Eocuartario argentinos. *Rev. Mus. Munic. Cienc. Nat. Mar del Plata* **1:** 119-29.

————., 1955. Noticia preliminar sobre la presencia de microbiotherinos vivientes en la fauna Sudamericana. *Invest. Zool. Chil.* **2:** 121-30.

————., 1957a. Nota previa sobre los marsupiales de la formación Chasicó. *Ameghiniana* 1(3): 27-31.

————., 1957b. Sobre la posición sistematica de *Zygolestes paranensis* Amegh. y de *Zygolestes entrerrianus* Amegh., con una reconsideración sobre la edad y la correlación del mesopotamiense. *Holmbergia* **5:** 209-26.

————., 1958a. Comunicacion preliminar sobre nuevas especies del genero *Thylatheridium* Reig (Mammalia, Didelphidae). *Neotropica* **4**(15): 89-95.

————., 1958b. Notas para una actualización del conocimiento de la fauna de la formación Chapadmala. II. Amphibia, Reptilia, Aves, Mammalia (Marsupialia: Didelphidae, Borhyaenidae). *Acta geol. Lilloana* **2:** 255-83.

————., 1981. Teoria del origen y desarrollo de la fauna de mamíferos de América del Sur. Monographiae Naturae. *Mus. Munic. Cien. Natur.* "*Lorenzo Scaglia*" **1:** 1-162.

———— AND BIANCHI, N. O., 1969. The occurrence of an intermediate didelphoid karyotype in the short-tailed opossum (genus *Monodelphis*). *Experientia* **25:** 1210-11.

———— AND SONNESCHEIN, C., 1970. The chromosomes of *Marmosa fuscata* Thomas from Northern Venezuela (Marsupialia, Didelphidae). *Experientia* **26:** 199-200.

———— AND SIMPSON, G. G., 1972. *Sparassocynus* (Marsupialia, Didelphidae), a peculiar mammal from the Late Cenozoic of Argentina. *Jour. Zool., London* **167:** 511-39.

————., FERNANDEZ, D. R. AND SPOTORNO, O. A., 1972. Further occurrence of a karyotype of 2n=14 in two species of Chilean marsupials. *Z. Saugetierkd.* **37:** 37-42.

————., GARDNER, A. L., BIANCHI, N. O. AND PATTON, J. L., 1977. The chromosomes of the Didelphidae (Marsupialia) and their evolutionary significance. *Biol. Jour. Linn. Soc. London* **9:** 191-216.

RIDE, W. D. L., 1962. On the evolution of Australian marsupials. Pp. 281-306 *in* "The evolution of living organisms" ed by G. W. Leeper. University Press: Melbourne.

RUSE, M., 1979. Falsifiability, consilience and systematics. *Syst. Zool.* **28:** 530-36.

RUSSELL, L. S., 1984. Tertiary mammals of Saskathewan. Part VII: Oligocene marsupials. *Life Sci. Contrib. Royal Ontario Mus.* **139:** 4-13.

SANBORN, C. C., 1951. Two new mammals from southern Peru. *Fieldiana: Zoology* **31:** 473-77.

SCHAEFFER, B., HECHT, M. K. AND ELDREDGE, N., 1972. Phylogeny and paleontology. *Evolutionary Biology* **6:** 31-46.

SCHNEIDER, L. K., 1977. Marsupial chromosomes, cell cycles and cytogenetics. Pp. 51-93 *in* "The biology of marsupials" ed by D. Hunsaker. Academic Press: New York.

SEGALL, W., 1969a. The middle ear region of *Dromiciops. Acta Anat.* **72:** 489-501.

————., 1969b. The auditory ossicles (malleus, incus) and their relationships to the tympanic in marsupials. *Acta Anat.* **73:** 176-91.

SELUJA, G. A., DI TOMASO, M. V., BRUM-ZORRILLA, N. AND CARDOSO, H., 1984. Low karyotypic variation in two didelphids (Marsupialia): karyogram and chromosome banding analysis. *Jour. Mammal.* **65:** 702-7.

SHANI, A., 1972. The vertebrate fauna of the Judith River Formation, Montana. *Bull. Amer. Mus. Nat. Hist.* **147:** 321-412.

SHARMAN, G. B., 1973a. Marsupial taxonomy and phylogeny. *Aust. Mammal.* **1:** 137-54.

————., 1973b. The chromosomes of non-Eutherian mammals. Pp. 485-530 *in* "Cytotaxonomy and vertebrate evolution" ed by A. B. Chiarelli and E. Capanna Academic Press: London.

————., 1976. Evolution of viviparity in mammals. Pp. 32-70 *in* "Reproduction in mammals. 6. The Evolution of Reproduction" ed by C. R. Austin and R. V. Short.

————., 1982. Karyotypic similarities between *Dromiciops australis* (Microbiotheriidae, Marsupialia) and some Australian marsupials. Pp. 711-14 *in* "Carnivorous marsupials" ed by M. Archer. Royal Zoological Society of New South Wales: Sydney.

SHAVER, E. L., 1962. The chromosomes of the opossum, *Didelphis virginiana. Canadian Jour. Genet. and Cytol.* **4:** 62-68.

SIGÉ, B., 1972. La faunule de mamifères du Crétacé supérieur de Laguna Umayo (Andes péruviennes). *Bull. Mus. Natl. Hist. (3) No. 99 Sci. Terr.* **19:** 375-409.

SIMONETTA, A. M., 1979. First record of *Caluromysiops* from Colombia. *Mammalia* **43**(2): 247-78.

SIMONS, E. L. AND BOWN, T. M., 1984. A new species of *Peratherium* (Didelphidae, Polyprotodonta): the first African marsupial. *J. Mammal.* **65:** 539-48.

SIMPSON, G. G., 1948. The beginning of the Age of Mammals in South America. *Bull. Amer. Mus. Nat. Hist.* **91:** 1-232.

————., 1961. "Principles of animal taxonomy". Columbia Univ. Press: New York.

————., 1970. The Argyrolagidae, extinct South American marsupials. *Bull. Mus. Comp. Zool.* **139:** 1-86.

————., 1971. The evolution of marsupials in South America. *Ann. Aca. Ciênc.* **43:** 103-18.

SIMPSON, G. L., 1972. Didelphidae from the Chapadmalal Formation in the Museo Municipal de Ciencias Naturales of Mar del Plata. *Mar del Plata Mus. Cienc. Natur., Publ.* **2:** 1-40.

—————., 1974. Notes on Didelphidae (Mammalia, Marsupialia) from the Huayquerian (Pliocene) of Argentina. *Amer. Mus. Novit.* **2259:** 1-15.

SINCLAIR, W. J., 1906. Marsupials of the Santa Cruz beds. *Repts Princeton Univ. Expeds Patagonia* **4:** 330-460.

SOLBRIG, O., 1970. The phylogeny of Gutierrezia: an eclectic approach. *Brittonia* **22:** 217-29.

STANLEY, S. M., 1973. An explanation for Cope's Rule. *Evolution* **27:** 1-26.

STEIN, B. R., 1981. Comparative limb myology of two opossums, *Didelphis* and *Chironectes*. *J. Morph.* **169:** 113-40.

STEVENS, P. F., 1980. Evolutionary polarity of character states. *Ann. Rev. Ecol. Syst.* **11:** 333-58.

SZALAY, F. 1982a. A new appraisal of marsupial phylogeny and classification. Pp. 621-40 *in* "Carnivorous marsupials" ed by M. Archer. Royal Zool. Soc. New South Wales: Sydney.

—————., 1982b. Phylogenetic relationships of the marsupials. Pp. 177-90 *in* "Philogenie et palengeographie: livre jubilaire en l'honeur de Robert Hoffstetter" ed by E. Buffetant, P. Janvier, J. C. Rage and P. Tassy. Lyons: Géobios Memoire Spécial 6.

TAKAHASHI, F., 1974. Vara≷ao morfologica de incisives em Didelfideos (Marsupialia, Didelphinae). *An. Acad. Bras. Ciên.* **46**(3-4): 413-16.

TATE,, G. H. H., 1933. A systematic revision of the marsupial genus *Marmosa*. *Bull. Amer. Mus. Nat. Hist.* **66:** 1-250.

TEDFORD, R. H., 1974. Marsupials and the new paleogeography. Pp. 109-26 *in* "Paleogeographic provinces and provinciality" ed by C. A. Ross. *Soc. Econ. Paleont. Miner., Spec. Publ. 21.*

THOMAS, O., 1888. "Catalogue of the Marsupialia and Monotremata in the collection of the British Museum (Natural History)". British Museum (Nat. Hist.): London.

TOMES, R. F., 1860. Notes on a second collection of Mammalia made by Mr Fraser in the Republic of Ecuador. *Proc. Zool. Soc. London for* **1860:** 211-21.

—————., 1863. Notice of a new American form of marsupial. *Proc. Zool. Soc. London for* **1863:** 50-1.

TURNBULL, W. D., 1971. The Trinity therians: their bearing on evolution in marsupials and other therians. Pp. 151-79 *in* "Dental morphology and evolution" ed by A. A. Dahlberg. Chicago University Press: Chicago.

TYNDALE-BISCOE, H., 1973. "Life of Marsupials". Edward Arnold Ltd.: London.

VILLARROEL, C. AND MARSHALL, L. G., 1983. Two new late Tertiary marsupials (Hathlyacyninae and Sparassocyninae) from the Bolivian Altiplano. *Jour. Paleont.*

WAINBERG, R. L., GENTILE DE FRONZA, T. AND GARCIA, J. G., 1979. Cromosomas marsupiales del generos *Marmosa: M. pusilla bruchi, M. agilis chacoensis* y *M. microtarsus* (Marsupialia: Didelphidae). *Physis* **38:** 33-38.

WAGNER, W. H., 1969. The construction of a classification. Pp. 67-90 *in* "Systematic biology" ed by C. G. Sibley. National Academy of Sciences U.S.A.: Washington.

WALKER, E. P., 1975. "Mammals of the World, Vol. I.", 3rd Ed. The John Hopkins Press: Baltimore.

WATROUS, L. E. AND WHEELER, Q. D., 1981. The out-group comparison method of character analysis. *Syst. Zool.* **30:** 1-11.

WEYMSS, C. T., 1953. A preliminary study of marsupial relationships as indicated by the precipitin test. *Zoologica (New York)* **38:** 173-81.

WHITE, M. J. D., 1978. "Modes of speciation". W. H. Freeman and Co.: San Francisco.

WILEY, E. O., 1981. "Phylogenetics the theory and practice of phylogenetic systematics". John Wiley and Sons: New York.

WILSON, A. C., MAXSON, L. R. AND SARICH, V. M., 1974a. Two types of molecular evolution. Evidence from studies of interspecific hybridization. *Proc. Nat. Acad. Sci.* **71:** 2843-3847.

—————., SARICH, V. M. AND MAXSON, L. R., 1974b. The importance of gene rearrangements in evolution: evidence from studies on rates of chromosomal, protein and anatomical evolution. *Proc. Nat. Acad. Sci.* **71:** 3028-30.

—————., CARLSON, S. S. AND WHITE, T. J., 1977. Biochemical evolution. *Ann. Rev. Biochem.* **46:** 573-639.

WINGE, H., 1893. Jordfundne og nulevende Pungdyr (Marsupialia) fra Lagoa Santa, Minas Gerais, Brasilien. Med. Udsigt over Pungdyrenes Slaegtskab. *E. Mus. Lund.* **11**(2): 1-149.

—————., 1923. "Pattedyr-Slaegeter. I. Monotremata, Marsupialia, Insectivora, Chiroptera, Edentata". Copenhagen.

—————., 1941. "The interrelationships of the mammalian genera. Vol. 1. Monotremata, Marsupialia, Insectivora, Chiroptera, Edentata". C.A. Reitzels Forlag: Copenhagen.

WOOD, H. E., 1924. The position of the "sparassodonts": with notes on the relationships and history of the Marsupialia. *Bull. Amer. Mus. Nat. Hist.* **51:** 77-101.

WOOLLEY, P., 1974. The pouch of *Planigale subtilissima* and other dasyurid marsupials. *Jour. Roy. Soc. West. Aust.* **57**(1): 11-15.

WORKS, M. E., 1950. "A morphological comparison of the central American genera of Didelphidae (opossums)". Unpubl. M.A. Thesis, Univ. Kansas.

YONENAGA-YASSUDA, Y., KASAHARA, S., SOUZA, M. J. AND L'ABBATE, M. L., 1982. Constitutive heterochromatin, G-bands and nucleolus-organizer regions in four species of *Didelphidae* (Marsupialia). *Genetica* **58:** 71-7.

YOUNG, G. J., MARSHALL-GRAVES, J. A., BARBIERI, I., WOOLLEY, P. A., COOPER, D. W. AND WESTERMAN, M., 1982. The chromosomes of dasyurids (Marsupialia) Pp. 783-95 *in* Carnivorous marsupials" ed by M. Archer. Royal Zoological Society of New South Wales: Sydney.

SYSTEMATICS OF ITABORAIAN (MIDDLE PALEOCENE) AGE "OPOSSUM-LIKE" MARSUPIALS FROM THE LIMESTONE QUARRY AT SÃO JOSÉ DE ITABORAÍ, BRAZIL

LARRY G. MARSHALL[1]

A systematic revision of the "opossum-like" marsupials from the Itaboraian (Middle Paleocene) age fissure fillings in the Itaborai Formation at the limestone quarry of São José de Itaboraí, Brazil permits recognition of 16 genera and 17 species: *Bobbschaefferia fluminensis, Derorhynchus singularis, Didelphopsis cabrerai, Eobrasilia coutoi, Gaylordia doelloi, G. macrocynodonta, Guggenheimia brasiliensis, Itaboraidelphys camposi, Marmosopsis juradoi, Minusculodelphis minimus, Mirandatherium alipioi, Monodelphopsis travassosi, Procaroloameghinia pricei, Protodidelphis vanzolinii, Robertbutleria mastodontoidea* gen. et sp. nov., *Sternbergia itaboraiensis* and *Zeusdelphys complicatus* gen. et sp. nov.

The systematics of all non-Australian Marsupialia is reconsidered in order to clarify the taxonomic affinities of these taxa. This overview results in recognition of two non-Australian suborders, the members of which are as follows (only Itaborai "opossum-like" taxa are listed): suborder Didelphimorphia, superfamilies Microbiotherioidea [families Peradectidae, Stagodontidae, Microbiotheriidae (subfamilies Pediomyinae, *Monodelphopsis*; and Microbiotheriinae, *Mirandatherium*], Borhyaenoidea [families Borhyaenidae, Thylacosmilidae], and Didelphoidea [families Didelphidae (subfamilies Didelphinae, *Itaboraidelphys, Marmosopsis, Sternbergia*; Caluromyinae; Herpetotheriinae; Derorhynchinae nov., *Derorhynchus, Minusculodelphis*; Eobrasiliinae nov., *Eobrasilia, Didelphis, Gaylordia*) and Sparassocynidae]; and suborder Polydolopimorphia nov., superfamilies Caroloameghinoidea nov. [family Caroloameghiniidae (subfamilies Caroloameghiniinae, *Procaroloameghinia, Robertbutleria*; and Protodidelphinae nov., *Bobbschaefferia, Guggenheimia, Protodidelphis, Zeusdelphys*)], Caenolestoidea [family Caenolestidae], Argyrolagoidea [families Gashterniidae nov., Groeberiidae, Argyrolagidae], and Polydolopoidea [families Prepidolopidae, Bonapartheriidae, Polydolopidae (subfamilies Epidolopinae, Polydolopinae)].

It is demonstrated that: 1) Borhyaenoidea and Didelphoidea share no special affinity (i.e. they are not sister-groups as long believed) and evolved independently in South America from Late Cretaceous peradectid-like ancestors. 2) The Early Tertiary families Prepidolopidae and Bonapartheriidae are sister-groups, and together represent the sister-group of Polydolopidae. They are thus placed with Polydolopidae in the superfamily Polydolopoidea, and not in the Didelphoidea as recently done by Pascual (1980b, 1981). 3) The Late Paleocene (Riochican) age *Gashternia* is placed in a new family Gashterniidae, regarded as structurally ancestral to *Groeberia* (Groeberiidae), which in turn is regarded as a sister-taxon of Argyrolagidae. These three families represent a monophyletic unit and are placed in the superfamily Argyrolagoidea. 4) A new superfamily Caroloameghinoidea includes the family Caroloameghiniidae, and the subfamilies Caroloameghiniinae (including Glasbiinae Clemens 1966) and Protodidelphinae nov. The Caroloameghinoidea shared a common ancestor with Microbiotheriidae in South America, and one lineage (*Glasbius*) dispersed to North America in latest Campanian or earliest Maestrichtian time. The Caroloameghinoidea are the plesiomorphic sister-group of Caenolestoidea, Argyrolagoidea and Polydolopoidea, and the four superfamilies are placed in the new suborder Polydolopimorphia. The Polydolopimorphia are characterized by a brachydont to bunodont molar structure, large and subequal sized stylar cusps B and D, and numerous other similarities in lower molar structure. 5) The family Microbiotheriidae includes the subfamilies Microbiotheriinae and Pediomyinae. The initial dichotomy of these subfamilies is possibly attributed to a vicariant or dispersal event of a common microbiotheriid ancestor that resulted in isolation of one stock (Pediomyinae) in North America and another (Microbiotheriinae) in South America. The presence of a pediomyine in the Middle Paleocene of South America is attributed to dispersal of a member of this lineage from North to South America in latest Cretaceous time. 6) Of the living groups of South American marsupials, Microbiotheriinae (*Dromiciops*) and Caenolestinae (*Caenolestes, Lestoros, Rhyncholestes*) are more closely related than they are with Didelphidae. 7) Microbiotheriinae have been a distinct lineage since at least Early Campanian time. It is therefore conceivable that a member of this subfamily may represent the ancestral stock for at least some Australian marsupials (Szalay 1982a, b). This affinity is supported by the close similarities in upper molar structure of the Middle Miocene age *Ankotarinja* of Australia and the Middle Paleocene age *Mirandatherium* of South America. 8) The initial known distributions of Peradectidae and Microbiotheriidae may be explained by vicariance or dispersal, while the known distributions of all other non-Australian marsupial groups resulted from dispersal. 9) It is concluded that South America was the *primary* area for early cladogenesis of Marsupialia, and that all basic aspects of marsupial evolution and distribution can be explained by envisioning initial dispersal of stocks from, and not to, that continent.

Pages 91-160 *in* POSSUMS AND OPOSSUMS: STUDIES IN EVOLUTION ed by M. Archer. Surrey Beatty & Sons and the Royal Zoological Society of New South Wales: Sydney, 1987.

INTRODUCTION

THE most taxonomically diverse marsupial fauna of Early Tertiary age known in South America is from the limestone quarry of São José de Itaboraí, located about 21 km east of Niterói, capital of the state of Rio de Janeiro, Brazil. The fossils were collected from marls filling channels and caves in the Early Paleocene age Itaborai Formation (Rodrigues Francisco and Souza Cunha 1978). These fissure fillings accumulated during Middle Paleocene time and the mammals in them represent the type fauna of the Itaboraian Land Mammal Age (*sensu* Marshall

[1]Berkeley Geochronology Center at The Institute of Human Origins, 2453 Ridge Road, Berkeley, California, U.S.A. 94709.

1985) (Fig. 1). This fauna provides most of our present knowledge of aspects of marsupial evolution in South America at the beginning of the Age of Mammals.

Among the known marsupials are the earliest undoubted representatives of the families Polydolopidae (*Epidolops ameghinoi* Paula Couto 1952a, includes *E. gracilis* Paula Couto 1952a, subfamily Epidolopinae, see Marshall 1982d) and Borhyaenidae (cf. *Nemolestes* sp., subfamily Borhyaeninae, see Marshall 1978; and *Patene simpsoni* Paula Couto 1952a, subfamily Hathliacyninae, see Marshall 1981a).

Sixteen other genera were referred initially and most tentatively to the family Didelphidae: *Bobbschaefferia fluminensis* (Paula Couto 1952b), *Derorhynchus singularis* Paula Couto 1952b, *Didelphopsis cabrerai* Paula Couto 1952b, *Eobrasilia coutoi* Simpson 1947, *Gaylordia macrocynodonta* Paula Couto 1952b, *G. mendesi* Paula Couto 1970, *Guggenheimia brasiliensis* Paula Couto 1952b, *Ischyrodidelphis castellanosi* Paula Couto 1952b, *Itaboraidelphys camposi* Marshall and de Muizon 1984, *Marmosopsis juradoi* Paula Couto 1962, *Minusculodelphis minimus* Paula Couto 1962, *Mirandatherium alipioi* (Paula Couto 1952c), *Monodelphopsis travassosi* Paula Couto 1952b, *Procaroloameghinia pricei* Marshall 1982c, *Protodidelphis vanzolinii* Paula Couto 1952b, *Sternbergia itaboraiensis* Paula Couto 1970, and *Xenodelphis doelloi* Paula Couto 1962.

In the following discussion I refer collectively to these 16 genera as Itaborai or "opossum-like" taxa. Their systematic affinities are clarified in parts 5 and 6 of the PHYLOGENETICS section (see below). Taxonomic ranks (i.e. subfamilies, families) used by previous workers which are shown to be either invalid or greatly modified in concept as a result of this study are placed in quotation marks.

The phylogenetic relationships of the majority of these 16 genera among themselves and with other marsupial groups are either totally unknown or only tentatively established. Paula Couto suggested possible ancestral-descendant relationships for two of these taxa with living Didelphidae [i.e. *Marmosopsis* with *Marmosa* (Paula Couto 1962:157) and *Protodidelphis* with *Didelphis* (Paula Couto 1952b:6)] and noted close "resemblances" of three others [i.e. *Bobbschaefferia* and *Guggenheimia* with *Philander* (Paula Couto 1952b:6), and *Monodelphopsis* with *Monodelphis* (Paula Couto 1952b:24)]. He also noted close structural similarity between *Mirandatherium* and the Casamayoran (Early Eocene, Fig. 1) genus *Coona* Simpson 1938, and suggested that they may prove synonymous. Tedford (1974:119) proposed that *Bobbschaefferia* (as *Schaefferia*, holotype only) may possibly be a microbiothere, while Marshall (1982c:715)

regarded *Procaroloameghinia* as the probable ancestor of the Casamayoran age genus *Caroloameghinia* Ameghino 1901 and placed that taxon in the subfamily Caroloameghiniinae. Based on a previously undescribed specimen of *Eobrasilia*, Marshall (1984) demonstrated that this taxon is a good "didelphid" and not a possible "didelphid-borhyaenid" transition group as suggested by Simpson (1947:7). It is also now known that *Ischyrodidelphis* was erected on the lower dentition of the borhyaenid *Patene simpsoni* (see Marshall 1981a:7), and that *Xenodelphis* is a junior synonym of *Gaylordia* (Marshall 1981b:20, see below).

Paula Couto (1952b:3) commented that P3 is the largest premolar in Itaborai taxa and in Late Cretaceous "Pediomyidae" of North America, in contrast to living Didelphinae in which P2 is the largest premolar (this generalization is erroneous, see below, Reig *et al.* 1985, 1987). Clemens (1966:21) observed many similarities in molar structure of *Didelphopsis*, *Marmosopsis* and *Minusculodelphis* which are also found in the Late Cretaceous "didelphid" *Alphadon marshi* of North America.

Based on a detailed comparison of the Itaborai taxa with Late Cretaceous and/or Early Tertiary "didelphids" from Europe and North America, Crochet (1979a, b, 1980) tentatively included *Bobbschaefferia* in the "didelphine" tribe "Peradectini" with *Alphadon*, *Albertatherium*, *Nanodelphis* and *Peradectes*. He also (1980:221) championed Sigé's (1972:397) observation regarding the similarities between *Protodidelphis* and *Glasbius* ("Glasbiinae") from the Late Cretaceous of North America and tentatively included *Protodidelphis* in that subfamily. Crochet included the remaining Itaborai taxa in the tribe "Didelphini" along with living and other fossil "didelphids".

Most recently, Reig *et al.* (1985, 1987) reanalyzed the systematics of most "opossum-like" marsupials. They tentatively assigned *Bobbschaefferia* to the family Peradectidae and *Protodidelphis* to the family Microbiotheriidae; while *Procaroloameghinia* they securely assigned to the family Caroloameghiniidae. The remaining Itaborai taxa, along with the Casamayoran age *Coona* and *Eomicrobiotherium* Marshall 1982a and the Montehermosan (Late Miocene to Early and Middle Pliocene, Fig. 1) age *Zygolestes* Ameghino 1898, they listed as Didelphimorphia *incertae sedis*.

This overview clearly demonstrates the need for a synthetic study of these Itaborai taxa. They have never been comprehensively compared with living or other fossil taxa, and their interrelationships are either obscure or unknown. Consequently, their suprageneric affinities have yet to be realized. It has long been

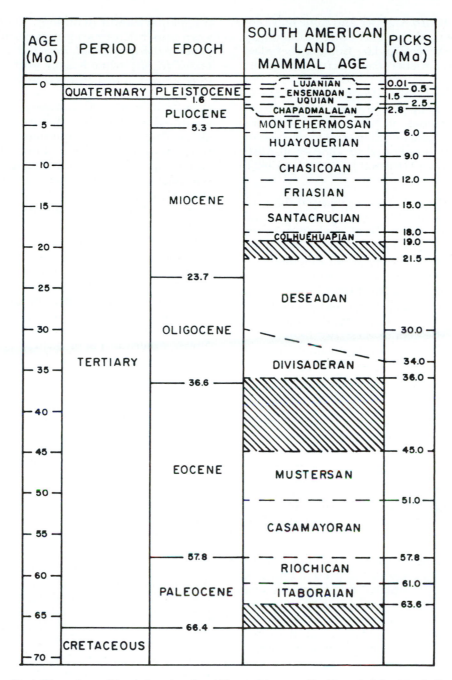

Fig. 1. Chronology of South American Land Mammal Ages used in this study (after Marshall 1985, Fig. 1). Hatching denotes hiatuses in knowledge of land mammal faunas.

recognized (e.g. Simpson 1971:112) that these taxa are so diverse that subfamilial, or higher, division appears warranted, ". . . but that has not yet been achieved." These taxa seem to be more varied than the "didelphoids" from the Late Cretaceous of North America which Clemens (1979:193; largely following Simpson 1935b and Clemens 1966) arranges into three families, one with two subfamilies. In terms of number of genera, the Itaborai taxa are more diverse (16 genera, see below) than either those from the Late Cretaceous of North America (9 genera, see Clemens 1979:193) or those living today in South America (14 genera, see Reig *et al.* 1985, 1987).

Furthermore, they exceed both of those groups in absolute size and morphological variation (see below). They thus document a diverse and well advanced radiation of "opossum-like" forms by Middle Paleocene time in South America.

It is universally agreed that the Cenozoic radiation of marsupials in South America stemmed from "opossum-like" or peradectid-like ancestors (e.g. Simpson 1970a:58; Reig *et al.* 1985, 1987). The Itaborai taxa are thus of potential importance in clarifying the phylogenetic relationships of other South American marsupial groups. They are the only "opossum-like" taxa of undoubted Paleocene age known in South America and some may represent members, direct ancestors, or ancestral-like forms of Argyrolagidae, Bonapartheriidae, Caenolestidae, Groeberiidae, Microbiotheriinae, Prepidolopidae, Sparassocynidae and/or later Didelphidae. Some may also prove referrable to groups presently or primarily known only from the Late Cretaceous of North America ("Glasbiinae", Pediomyinae, Stagodontidae) and/or prove structurally ancestral to some or all Australian groups as well (Russell 1976, Szalay 1982a, b).

Tedford (1974:119) has emphasized the existence of a remarkable overlap in structural "types" between the Itaborai taxa and North American Late Cretaceous "didelphids". Among the Itaborai taxa are generalized "didelphid" or *Alphadon*-like forms (e.g. *Didelphopsis, Marmosopsis*), bunodont *Glasbius*-like forms (e.g. *Guggenheimia, Mirandatherium*), and stagodontid-like forms with large crushing P3's (e.g. *Gaylordia*). Thus, the taxa in these two faunas ". . . show a similar range of dental morphologies and presumed adaptations. The extent to which this represents parallel or direct phyletic relationship is not presently known, but close comparison suggests that many, if not all, of these resemblances could have arisen as parallel trends of adaptation on both continents".

SCOPE AND ORGANIZATION OF STUDY

This study is organized into three principal sections: MARSUPIAL DISTRIBUTION and PALEOBIOGEOGRAPHY, SYSTEMATICS, and PHYLOGENETICS.

The first section is provided to familiarize the reader with knowledge of the present and past distribution of marsupials, and of some diverse views that exist which attempt to explain this distribution. This section focuses on the specific importance of the Itaborai fauna and how it has been interpreted to explain aspects of marsupial paleobiogeography.

The SYSTEMATICS section was initiated in 1979 when I was given permission by Dr. Diogenes Campos to describe the large collection of fossil marsupials made in recent years by personnel of the Divisão de Geologia e Mineralogia do Departamento Nacional da Produção Mineral (DGM) in Rio de Janeiro. During the course of this study I examined original and/or epoxy casts of all pertinent materials, including type and most referred specimens. This part of the study includes the description and naming of two new genera and species, and the description of many new specimens of taxa previously known only from partial lower dentitions. Prior to this study most "opossum-like" taxa from Itaborai were known only or primarily from partial, often times

fragmentary lower dentitions. Little to medium worn upper dentitions were known in only four genera (*Didelphopsis, Marmosopsis, Itaboraidelphys, Protodidelphis*), while heavily worn upper molars were known in two others (*Eobrasilia, Gaylordia*). Partial upper dentitions of two additional genera (*Mirandatherium, Monodelphopsis*) are here identified and described. A reappraisal is made of all previously known specimens and some of those assigned by Paula Couto (opt. cit.) to various species are here reassigned to others. I was unable to relocate some of the specimens studied by Paula Couto and for these I designate them "*fide* Paula Couto" and assign them to the same taxa as did he.

With one exception (i.e. the type of *Gaylordia macrocynodonta*), none of the upper and lower dentitions were surely found in direct association. I therefore follow Clemens (1966:83) and the "four lines of evidence" he used to associate "unassociated" upper and lower molars of "opossum-like" taxa from the Late Cretaceous of North America. These include relative size of upper and lower molars, height and width of dentary, relative abundance of species, and knowledge of premolar structure. In all cases I am conservative in identifying associations and have made them only when I felt secure that no other alternatives were possible. The primary associations are based only on multi-toothed jaws. In some rare cases I identify isolated teeth, but only where these elements are already represented among more complete specimens (i.e. in multi-toothed jaws). Thus, I make no attempt to assign all of the several hundred isolated teeth in the DGM collection to named taxa.

In the SYSTEMATICS section I begin with an introduction of abbreviations and techniques used. I then discuss, in alphabetical order, the genera and species. All of the specimens are from the Middle Paleocene (Itaboraian) age fissure fillings in the Itaborai Formation at the limestone quarry of São José de Itaboraí. For each taxon I provide information on authorship, previous taxonomic usage, type, hypodigm, diagnosis-description, and include figures of all type and some referred specimens. As some of the taxa were never adequately diagnosed, I include characters in the *diagnosis-description* whose utility have yet to be fully tested. This approach is the preferred alternative to omitting characters which are potentially of diagnostic significance. In the PHYLOGENETICS section I identify some characters for each taxon which are of particular value in phylogenetic inference. In the *comments* part I include information relating to the phylogenetic relationships and/or "habitus" of a species as reported by earlier workers. A detailed analysis of the interrelationships of these taxa, and of their affinities with other fossil and living marsupials is presented in the

PHYLOGENETICS section (see below). Thus, the information in the SYSTEMATICS section is primarily and intentionally of descriptive, diagnostic and/or historical nature, and serves as the data base for the PHYLOGENETICS section of this study.

The PHYLOGENETICS section is divided into six parts: 1) interrelationships of the Itaborai taxa; 2) their affinities with other South American marsupials; 3) their affinities with North American, European, and African marsupials; 4) their affinities with Australian marsupials; 5) a proposed classification and phylogeny of non-Australian taxa based on knowledge of data presented in parts 1-4; and 6) a reconsideration of the paleobiogeographic history of marsupials based on the phylogenetic conclusions presented in part 5.

MARSUPIAL DISTRIBUTION AND PALEOBIOGEOGRAPHY

Native marsupials occur today in Australia (about 58 genera), South America (17 genera), Central America (8 genera), and North America (one genus) (Marshall 1981b). Fossil marsupials are known from Africa (two genera, Late Eocene of Algeria, Mahboubi *et al.* 1983; Oligocene of Egypt, Bown and Simons 1984), Antarctica (one genus, Late Eocene, Seymour Island, Woodburne and Zinsmeister 1984), Australia (about 110 living and fossil genera, Late Oligocene to Recent, Archer 1981), Europe (three genera, Middle Paleocene through Middle Miocene, Crochet 1980; Crochet and Sigé 1983), North America (14 genera, Late Cretaceous through Middle Miocene, Fox 1979a, b, c, 1981; Slaughter 1978; Middle Pleistocene through Recent, Gardner 1973; Martin 1974), and South America (about 100 living and fossil genera, Late Cretaceous through Recent, Marshall *et al.* 1983; de Muizon *et al.* 1984; Sigé 1971, 1972; Marshall 1982b; Reig *et al.* 1985, 1987).

The paleobiogeographic history of marsupials, and particularly the role played in that history by the present continent of South America, has long been debated (Hoffstetter 1975). Many models have been proposed which purport to explain the distribution of marsupials based both on a stable continental configuration and on a mobile configuration that incorporates plate tectonic theory (for a review of this literature see Marshall 1980a).

For South America the assorted models fall into three categories: 1) some or all marsupial groups evolved *in situ* in South America from Late Cretaceous ancestors; 2) some or all groups came from stocks that dispersed to South America from elsewhere (i.e. North America, Africa, or Australia *via* Antarctica); or 3) the distribution of some groups is best attributed to vicariance and fragmentation of a broader

distribution resulting primarily from continental separations due to plate movements. Thus, these categories do not necessarily dictate an either/or choice for all marsupials, but entertain the possibility that each group (i.e. genus, subfamily, family etc.) may have had a paleobiogeographic history that is independent of and distinct from that of any other.

Some workers believe that North America was the area of origin of marsupials because the oldest known taxa occur there. Other workers believe that at least some marsupials evolved in South America or Australia, while others include Antarctica and/or Africa in their models. The debate continues, yet in reality ". . . we simply do not know where the marsupials evolved" (Simpson 1978:323). However, there is now concensus that marsupial distribution has been influenced by plate movements, and most recent workers agree that their early biogeographic history involved only what are now the continents of North America, South America, Antarctica and Australia. Unfortunately, the available evidence does not permit favoring any one or any combination of these continents as an area of origin over that of any other(s) (Simpson 1978:323).

It is also now evident that any dispersal(s) in what ever direction(s) which may have occurred between Australia and South America was *via* Antarctica, but not necessarily by a continuous land connection (Woodburne and Zinsmeister 1984). The same is apparently true for inferred Late Cretaceous and/or Early Tertiary dispersals between North and South America. "There is no evidence for any complete land connection between North and South America via the Greater and/or Lesser Antilles throughout later Mesozoic or Tertiary time. Nor is there any evidence for complete land connection via Central America and the Isthmus of Panama before Neogene time" (Coney 1982:432). Nevertheless, there are numerous groups of vertebrates found in Late Cretaceous and/or Early Tertiary faunas in North and South America which demonstrate close biotic affinity, and suggest that dispersals north and/or south occurred at this time (Rage 1978, 1981).

Prior to the discovery of the Itaborai fauna, no Early Tertiary didelphids were known in South America. The only surely identifiable didelphid known prior to Colhuehuapian (Early Miocene, Fig. 1) time was the Casamayoran (Early Eocene) age *Coona* Simpson 1938 from southern Argentina, once thought to be an ancestral microbiothere or "pediomyid" (Simpson 1948:36). The apparent absence of pre-Pliocene Didelphinae in South America led Patterson (1937), McGrew (1937:451), and Riggs and Patterson (1939:148) to the erroneous conclusion that Late Cenozoic didelphines on that continent were descendant from North American late Tertiary waif immigrants. ". . . the *Didelphinae* were of Holarctic if not North American origin; they went through a large part of their radiation in the north; and they reached South America probably in the late Miocene, there displacing the *Microbiotheriinae*. The absence of didelphines in the late Tertiary plains deposits of North America cannot be regarded as evidence of their absence from the entire continent. They were almost certainly abundant and diversified in the forested areas" (Riggs and Patterson 1939:148-49). It is now known that the Late Cenozoic and Recent South American Didelphinae are descendant from autochthonous Cretaceous and/or Paleogene ancestors (Reig *et al.* 1985, 1987).

Subsequent to the discovery of the Itaboraí fauna, the biogeographic history of South American marsupials was seriously reconsidered. The principal model that received most support in the mid-1950's through the mid-1970's was that all South American marsupials were descendant from a North American Late Cretaceous immigrant stock represented by *Alphadon* or an *Alphadon*-like animal. This model was largely developed and championed by Clemens (e.g. 1966: 21-22):

"A review of the published descriptions indicates the didelphines from São José de Itboraí could all be descendants of a Late Cretaceous didelphine immigrant from North America. Where comparisons of upper molars of these marsupials and the species of *Alphadon* could be made, distinct similarities in coronal outline and morphology of stylar cusps were found. The larger size of P3's in relation to P2's of many North American Mesozoic marsupials and didelphines from São José de Itaboraí is another point of resemblance. The only consistent differences appear to be the small size of the paracone, relative to the metacone, and the reduction or loss of conules on the molars of the South American species. These would be in line with the suggestion that South American didelphines were descendants of North American didelphines such as *Alphadon*, and that their upper dentitions had been modified along patterns similar to those found in the *Alphadon-Peradectes-Peratherium* lineage."

The discovery of fragmentary but undoubted Late Cretaceous age marsupials at Laguna Umayo in Peru (Sigé 1971, 1972) prompted some workers to once again reconsider the biogeographic history of South American marsupials. These new fossils and the occurrence of diverse "opossum-like" taxa, borhyaenids, and polydolopids at São José de Itaboraí was interpreted to indicate that marsupials were a very ancient group in South America and may indeed have arisen there.

"The scope of the South American marsupial radiation, as far as evidenced by the Late Cretaceous and early Tertiary record, suggests that marsupials may have been on that continent for a long time, possibly from early in the Cretaceous. In contrast the unified structure of the North American Late Cretaceous marsupicarnivores indicates a radiation of limited scope and more recent origin . . . evolving in isolation, possibly from a southern didelphine migrant close to, or congeneric with, *Alphadon*" (Tedford 1974:120).

The recent discovery of a diverse marsupial (and placental) fauna in the Late Cretaceous (Maestrichtian) age El Molino Formation of Bolivia has rekindled interest in South America's contribution to marsupial evolution and paleobiogeography (Marshall *et al.* 1983; de Muizon *et al.* 1984). At least 12 "opossum-like" taxa are known in this fauna (Marshall and de Muizon in preparation), although only one has thus far been named (*Roberthoffstetteria nationalgeographica* Marshall *et al.* 1983). Two others show special affinity to taxa from Itaboraí and are described by de Muizon *et al.* (1984) as "*Gaylordia*-like" and "*Steinbergia*-like" forms. A preliminary study of this new late Cretaceous marsupial fauna has demonstrated that before the taxa can be meaningfully described and named, the systematics of the "opossum-like" taxa from Itaboraí must first be clarified. This paper is thus a necessary prerequisite to the study of that important fauna. It is also an extension of the study of Reig *et al.* (1985, 1987) and attempts to identify the phylogenetic affinities of the 14 genera which those workers list as Didelphimorphia *incertae sedis*.

SYSTEMATICS

Abbreviations used in text, figure captions, and tables of measurements are: C, canine; ~, approximate measurement; I, incisor; L, length; M, molar; P, premolar; and W, width.

The following abbreviations are used for specimens from institutional collections: AMNH, American Museum of Natural History, New York; DGM, Divisão de Geologia e Mineralogia do Departmento Nacional da Produção Mineral, Rio de Janeiro, Brazil; MACN, Museo Argentino de Ciencias Naturales "Bernardino Rivadavia", Buenos Aires, Argentina; MLP, Museo de La Plata, La Plata, Argentina; MNHN, Muséum national d'Histoire naturelle, Paris, France; MNRJ, Museu Nacional e Universidade Federal do Rio de Janeiro, Brazil.

The nomenclature for molar tooth structure (Fig. 2) follows that employed by Reig *et al.* (1987, Fig. 1). The classic serial designation for cheek tooth number is based on the dental formula $P_{1-2-3}^{1-2-3} M_{1-2-3-4}^{1-2-3-4}$. The predecessor of the P_3^3 (i.e. the milk tooth) is designated dP_3^3. It must be noted that a different nomenclature for the dental series is used in this volume and is attaining wide acceptance. This alternative dental formula recognizes the permanent molar number as $M_{2-3-4-5}^{2-3-4-5}$ and is based on the ontogenetically demonstrated fact that the "dP_3^3" is homologous with the true M_1^1. This new serial designation for cheek tooth number is followed here.

Specimens were measured to the nearest 0.1 mm when possible using dial calipers. All measurements are in millimetres (mm).

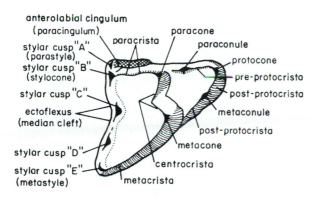

RIGHT UPPER MOLAR

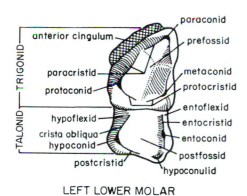

LEFT LOWER MOLAR

Fig. 2. Occlusal views of an anterior upper and lower molar of a generalized marsupial showing nomenclature of tooth structure used in this study (after Reig *et al.* 1985b, Fig. 1).

Bobbschaefferia (Paula Couto 1952b)

Schaefferia Paula Couto 1952b:12, *nec* Absolon 1900 (Collemb.), *nec* Houlbert 1918 (Lepidoptera).

Bobbschaefferia Paula Couto 1970: 20.

Type: Bobbschaefferia fluminensis (Paula Couto 1952b:13).

Diagnosis-Description: As for type and only known species.

Bobbschaefferia fluminensis (Paula Couto 1952b).

(Figs 3, 4; Table 1)

Schaefferia fluminensis Paula Couto 1952b:13, Fig. 4; 1962:150, Figs 7, 8.

Bobbschaefferia fluminensis Paula Couto 1970:20.

Type: MNRJ 1350-V, a fragment of a left mandibular ramus with alveoli of I_{1-4} and C, roots of P_1, alveoli of M_2, roots of M_3, and with P_{2-3} and M_{4-5} complete (figured by Paula Couto 1952b, Fig. 4B; 1962, Fig. 7A-C.

Hypodigm: Type and DGM 314-M (paratype), a relatively complete left mandibular ramus with alveoli of I_1-M_2, and M_{3-5} complete (figured by Paula Couto 1952b, Fig 4A; 1962, Fig. 8A, B, and part of C); DGM 315-M (paratype), anterior part of a left mandibular ramus with alveoli of C-M_3, M_4 complete, and anterior alveolus of M_5 (figured by Paula Couto 1962, Fig. 8D, E, part of C); MNRJ 2899-V, two right lower molars (*fide* Paula Couto 1970:21).

Diagnosis-Description: Medium size; dental formula ?/4 ?/1 ?/3 ?/5; mandibular ramus robust, relatively deep; symphysis extends posteriorly to point below posterior root of P_2; well developed mental foramen below anterior root of P_2 and smaller one below posterior root of P_3; as seen in MNRJ 1350-V, lingual-most incisor (I_1) very large with second tiny incisor alveolus situated ventrolabial to it, and equal sized alveoli of two additional incisors situated one above the other posterolabial to I_1; C large, well developed; in length $P_1 < P_2 \approx P_3$, in width $P_1 < P_2 < P_3$; P_1 set at slight oblique angle relative to rest of tooth row; small diastema separates P_1 and P_2; crowns of P_2 and P_3 very high (P_3 higher than P_2 and both much higher than molars) with sharp, pointed primary cusp, and small but distinct posterobasal heel; molars brachydont, small relative to premolars and ramus; in length $M_3 \approx M_4 \approx M_5$, in width $M_3 \approx M_4 > M_5$; trigonids and talonids subequal in width; trigonids higher, but not markedly so, than talonids; talonid basins filled largely by inflated bases of hypo- and entoconids; anterobasal cingulum distinct but weakly developed on M_{3-5}; in size protoconid > metaconid > paraconid; hypoconid and entoconid large, subequal in size; hypoconulid small but distinct on M_{3-4}, large on M_5.

Comments: Paula Couto (1952b:12) compared *Bobbschaefferia* (as *Schaefferia*) only with *Guggenheimia*, noting that among the known taxa from Itaborai they are the most similar. He further noted that among living genera, *Bobbschaefferia* and *Guggenheimia* most closely resemble *Philander* in aspects of their dental and mandibular morphology. Tedford (1974:120) suggested that *Bobbschaefferia* (as *Schaefferia*, holotype only) may possibly be a microbiothere. Crochet (1980:222) tentatively assigned

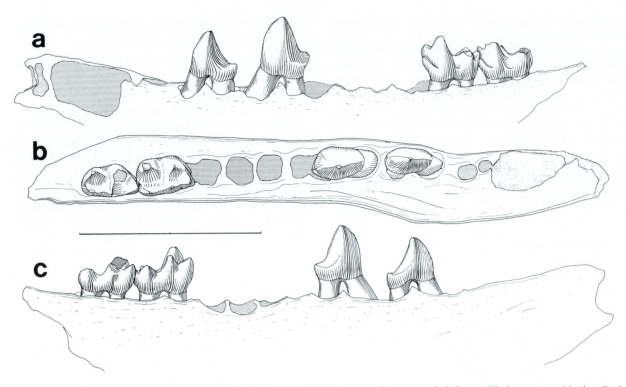

Fig. 3. Bobbschaefferia fluminensis (Paula Couto 1952b). MNRJ 1350-V (type), a fragment of a left mandibular ramus with alveoli of I$_{1-4}$ and C, roots of P$_1$, alveoli of M$_2$, roots of M$_3$, and with P$_{2-3}$ and M$_{4-5}$ complete: a, labial; b, occlusal; c, lingual views. Scale = 10 mm.

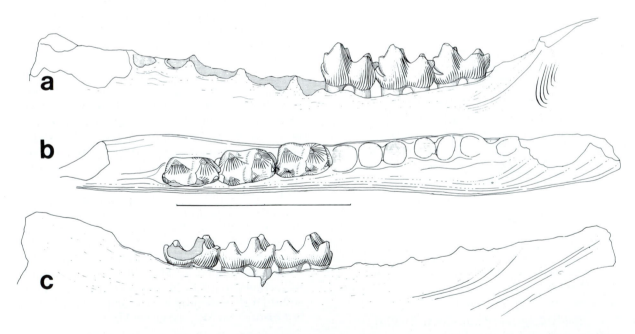

Fig. 4. Bobbschaefferia fluminensis (Paula Couto 1952b). DGM 314-M, a relatively complete left mandibular ramus with alveoli of I$_1$-M$_2$, and M$_{3-5}$ complete: a, labial; b, occlusal; c, lingual views. Scale = 10 mm.

Table 1. Measurements of lower cheek teeth of *Bobbschaefferia fluminensis*.

Specimen	P$_2$ L	P$_2$ W	P$_3$ L	P$_3$ W	M$_3$ L	M$_3$ W	M$_4$ L	M$_4$ W	M$_5$ L	M$_5$ W	M$_{2-5}$ L	P$_3$-M$_5$ L
DGM 314-M	–	–	–	–	3.0	2.0	3.0	2.0	3.0	1.7	–	–
DGM 315-M	–	–	–	–	–	–	3.2	2.1	–	–	–	–
MNRJ 1350-V	2.6	1.5	2.6	1.7	–	–	3.0	2.0	3.1	1.7	~12.5	15.2

Bobbschaefferia to the "didelphine" tribe "Peradectini". Most recently, Reig *et al.* (1985, 1987) raised "Peradectini" to the rank of family, Peradectidae, and tentatively included *Bobbschaefferia* as a member.

Derorhynchus Paula Couto 1952b

Derorhynchus Paula Couto 1952b:13.

Type: Derorhynchus singularis Paula Couto 1952b:15.

Diagnosis-Description: As for type and only known species.

Derorhynchus singularis Paula Couto 1952b

(Fig. 5; Table 2)

Derorhynchus singularis Paula Couto 1952b:15, Figure 5; 1962:147, Figure 5.

Type: MNRJ 1348-V, a relatively complete right mandibular ramus with roots of I_{1-2} and P_1-M_2, alveoli of I_3 and C, and M_{3-5} complete (figured by Paula Couto 1952b, Fig. 5; 1962, Fig. 5A-C).

Hypodigm: Type and AMNH 49826 (= MNRJ 1349-V, paratype), a fragment of a right mandibular ramus with roots or alveoli of P_1-M_2, and M_{3-5}, complete; DGM 810-M, a fragment of a right mandibular ramus with M_{3-5} complete.

Diagnosis-Description: Small size; dental formula ?/3 ?/1 ?/3 ?/5; mandibular ramus extremely elongate, especially anteriorly, shallow; symphysis extends posteriorly to point below P_3; a mental foramen occurs below anterior edge of P_1, another below posterior edge of P_3 or M_2; I_{1-2} (judging from alveoli in MNRJ 1348-V) procumbent, well developed, apparently elongate; I_3 very reduced and situated lateral to I_{1-2}; C large, well developed, inclined anteriorly (procumbent), apparently overhangs incisors to some degree; premolars increase sharply but evenly in size from P_1 to P_3, aligned along anteroposterior axis of jaw; P_1 separated from C and P_2 by distinct diastems; P_2 separated slightly from P_3 by small diastem; trigonids of M_{3-5} well elevated above talonids; talonid wider than trigonid on M_3, subequal in width to talonid on M_4, notably narrower than trigonid on M_5 anterobasal cingulum well developed on M_{3-5}; metaconid intermediate in size between higher protoconid and lower paraconid; entoconid slightly higher than hypoconid; hypoconulid very small but distinct; talonid with shallow basin.

Comments: Paula Couto (1952b:14) noted that in structure of the lower molars *Derorhynchus* differs little from some small living Didelphinae (e.g. *Marmosa*), while in structure of the elongate mandibular ramus and procumbent nature of the incisors and canine it resembles Caenolestidae. He concluded (ibid.) that *Derorhynchus*

". . . was almost certainly an insectivorous didelphid . . . [and] . . . is convergent with caenolestids. . ."

Table 2. Measurements of lower cheek teeth of *Derorhynchus singularis.*

Specimen	M3		M4		M5	
	L	W	L	W	L	W
AMNH 49826	1.7	1.2	1.7	1.1	1.7	1.0
DGM 810-M	1.9	1.4	1.9	1.4	1.9	1.3
MNRJ 1348-V	1.7	1.2	1.7	1.1	1.7	1.0

Didelphopsis Paula Couto 1952b

Didelphopsis Paula Couto 1952b:7.

Type: Didelphopsis cabrerai Paula Couto 1952b:8.

Diagnosis-Description: As for type and only known species.

Didelphopsis cabrerai Paula Couto 1952b

(Figs 6-9; Table 3)

Didelphopsis cabrerai Paula Couto 1952b:8; 1962:144, Figure 3; 1970:26.

Type: MNRJ 1429-V, a nearly complete left maxilla with partial alveolus of C, roots of P^{1-2}, alveoli of P^3, and M^{2-5} present but very worn (figured by Paula Couto 1962, Fig. 3C).

Hypodigm: Type and DGM 243-M (paratype), a nearly complete right mandibular ramus with roots of I_4? and C, base of P_1, and P_2-M_5, complete but very worn (figured by Paula Couto 1962, Fig. 3A, B); DGM 642-M, an isolated left M^4; MNRJ 1345-V, an isolated right M^5 (*fide* Paula Couto); MNRJ 1347-V, an isolated left M_5 (*fide* Paula Couto); MNRJ 1426-V, an isolated left M^3; MNRJ 2341-V, an isolated left M_5 (*fide* Paula Couto); MNRJ 2883-V, two right lower molars (*fide* Paula Couto); MNRJ 2884-V, an incomplete right maxilla with P^{1-2} and M^5, and alveoli of C and P^3-M^4; MNRJ 2885-V, posterior part of a right mandibular ramus with alveoli of M_5, and posterior root of M_4, (*fide* Paula Couto).

Diagnosis-Description: Large size; dental formula ?/4 1/1 3/3 5/5; mandibular ramus well developed; large mental foramen below posterior root of P_1 and smaller one below middle of M_2; symphysis extends posteriorly to below P_2; C's large and well developed; premolars aligned straight in jaw, increase sharply in size (length and width) from P^1_1 to P^3_3; P_1 separated from P_2 by small diastema; P_2 and P_3 with distinct posterobasal heel; P^3_3 enormous and bulbous; in length $M_2 < M_3 \approx M_4 > M_5$, in width $M_2 < M_3 < M_4 > M_5$; lower molars with well developed anterobasal cingulum; trigonid well elevated

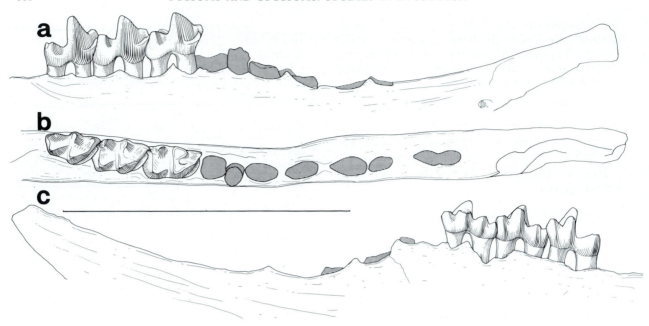

Fig. 5. Derorhynchus singularis Paula Couto 1952b. MNRJ 1348-V (type), a relatively complete right mandibular ramus with roots of I_{1-2}, alveoli of I_3 and C, roots of P_1-M_2, and with M_{3-5} complete: a, labial; b, occlusal; c, lingual views. Scale = 10 mm.

above talonid; metaconid intermediate in size between larger protoconid and smaller paraconid; talonid broader than trigonid on M_2, subequal in width on M_{3-4}, narrower on M_5; talonid with shallow basin; hypoconid slightly larger than entoconid; hypoconulid very small but distinct; preorbital canal located above middle of P^3; paroccipital process well developed, conical, located on same sagittal plane as tips of occipital condyles well behind precondyloid foramen; in length $M^2<M^3\approx M^4>M^5$, in width $M^2<M^3<M^4>M^5$; protocone large, talon deeply basined, paracone and metacone large, paracone smaller than metacone on M^{2-4} but larger on M^5; metacrista and paracrista sharp, blade-like, moderately to well developed; paracingulum small but distinct; paraconule tiny; metaconule usually distinct, often notably cuspate (giving appearance of small hypocone); stylar shelf broad, forming deep trough between para- and metacones and stylar cusps; ridge formed by stylar cusps sharp, serrated with supernumary styles on M^{2-4}; stylar cusp A small, B and D larger, C distinct but smaller than B and D, E very small; in size on M^3 B > D > C ≈ A > E; M^5 with prominent stylar cusp B only, stylar spur inflected anteriorly; ectoflex shallow or absent on M^{2-3}, shallow but present on M^4.

Comments: Paula Couto (1952b:8) noted that *Didelphopsis* is ". . . very similar to *Protodidelphis* and consequently to *Didelphis*". He also emphasized the presence in *Didelphopsis* of a large bulbous P3, a characteristic shared with members of the family Stagodontidae in the Late Cretaceous of North America. In 1970 (p. 25) Paula Couto gave a lengthy comparison of *Didelphopsis*

with *Didelphis*, and concluded that "*Didelphopsis* was relatively more robust than the modern *Didelphis* . . . [and that] . . . it was perhaps more of a carrion eater . . .". Clemens (1966:20-21) pointed out that ". . . in occlusal outline and morphology of stylar cusps, including the presence of a distinct stylar cusp C on M^4, upper molars of *Didelphopsis* resemble those of *Alphadon marshi*" from the Late Cretaceous of North America.

Eobrasilia Simpson 1947

Eobrasilia Simpson 1947:2.

Type: Eobrasilia coutoi Simpson 1947:2.

Diagnosis-Description: As for type and only known species.

Eobrasilia coutoi Simpson 1947
(Figs 10, 11; Table 4)

Eobrasilia coutoi Simpson 1947:2, Figures 1, 2; Paula Couto 1962:138; Marshall 1978:70; 1984:174, Figures 1-3.

Type: AMNH 39424, incomplete facial part of skull with right P^{1-3}, left P^3 and M^4 (latter crushed and fragmented); all teeth, except P^1, have advanced stages of occlusal wear (figured by Simpson 1947, Figs 1, 2; Marshall 1984, Fig. 1).

Hypodigm: Type and DGM 919-M, anterior half of a right mandibular ramus with alveoli of I_{1-2}, and C-P_1, P_{2-3} complete and heavily worn; anterior edge of M_2 present, alveoli of rest of M_2 and of M_3, on lingual side only (figured by Marshall 1984, Fig. 2).

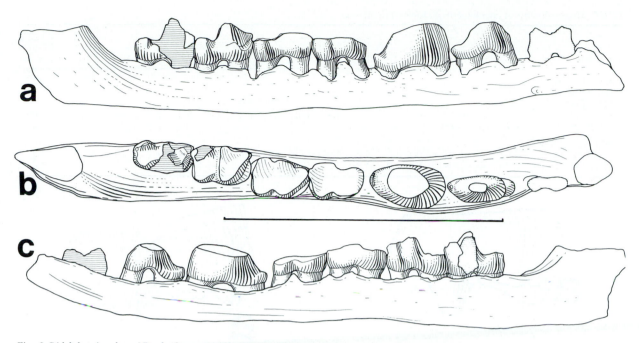

Fig. 6. *Didelphopsis cabrerai* Paula Couto 1952b. DGM 243-M, a nearly complete right mandibular ramus with roots of last incisor and C, base of P_1, and with P_2-M_5 present but worn (M_5 broken): a, labial; b, occlusal; c, lingual views. Scale = 20 mm.

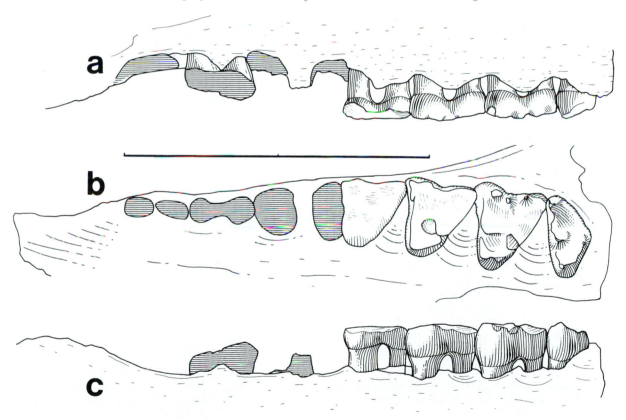

Fig. 7. *Didelphopsis cabrerai* Paula Couto 1952b. MNRJ 1429-V (type), a nearly complete left maxilla with partial alveolus of C, roots of P^{1-2}, alveoli of P^3, and with M^{2-5} present but very worn: a, labial; b, occlusal; c, lingual views. Scale = 20 mm.

Diagnosis-Description: Very large size; dental formula ?/2? 1/1 3/3 5/5; facial part of skull described and figured by Simpson (1947); short, blunt postorbital process, palate with one small pair of vacuities, small frontomaxilla contact; upper C large; premolars increase sharply in size from P^1 to P^3; P^1 small, single rooted, crown set obliquely in jaw; crown of P^1 consists of primary central cusp with small anterocuspule and slightly larger posterobasal heel, all three

Table 3. Measurements of cheek teeth of *Didelphopsis cabrerai.*

Specimen	P1 L	P1 W	P2 L	P2 W	P3 L	P3 W	M2 L	M2 W	M3 L	M3 W	M4 L	M4 W	M5 L	M5 W	M2-4 / M2-5 L	P3-M4 / P3-M5 L
Upper cheek teeth																
DGM 642-M	–	–	–	–	–	–	–	–	–	–	5.2	6.1	–	–	–	–
MNRJ 1426-V	–	–	–	–	–	–	–	–	4.9	5.4	–	–	–	–	–	–
MNRJ 1429-V	–	–	~4.4	–	–	–	4.4	4.4	4.7	5.1	4.8	5.6	2.7	5.4	13.7	~20.0
MNRJ 2884-V	1.7	1.0	4.0	–	–	–	–	–	–	–	–	–	2.8	4.8	–	–
Lower cheek teeth																
DGM 243-M	3.0	–	4.5	1.9	5.5	3.2	4.0	2.7	4.4	2.8	4.3	2.9	4.1	~2.4	17.0	22.7
MNRJ 2878-V(c)	–	–	–	–	5.0	3.5	–	–	–	–	–	–	–	–	–	–

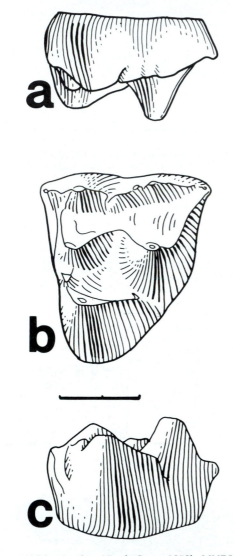

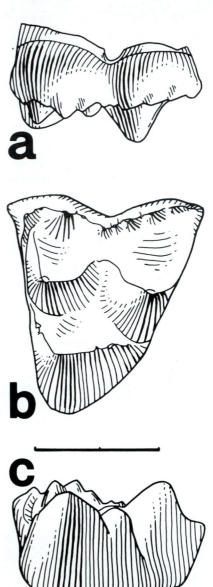

Fig. 8. *Didelphopsis cabrerai* Paula Couto 1952b. MNRJ 1426-V, an isolated left M³: a, labial; b, occlusal; c, lingual views. Scale = 2.5 mm.

connected by continuous ridge; P² (and P³) two rooted, ovoid in shape in occlusal view, small anterocuspule, broad posterior heel, crown broadest over posterior root; P³ larger than P², shape ovoid in occlusal view, widest part of crown at anterior end of posterior root, small tubercle on ventrolabial surface (seen in left P² only); in size M²<M³|M⁴>M⁵; M⁴ larger than P³;

Fig. 9. *Didelphopsis cabrerai* Paula Couto 1952b. DGM 642-M, an isolated left M⁴: a, labial; b, occlusal; c, lingual views. Scale = 2.5 mm.

mandibular ramus narrow, pointed anteriorly, extremely bulbous below P₃; well developed mental foramen below anterior root of P₁; symphysis deep, ligamentous, extending posteriorly to point below anterior root of P₃;

lower incisors and C semiprocumbent; lower C large, narrow transversely, deep dorsoventrally; P_1 two rooted, set obliquely in jaw, separated from C and P_2 by small diastema; P_2, set tightly against anterior edge of P_3; P_3 enormous, incipiently tri-rooted, shape ovoid to rectangular in occlusal view, middle part of crown slightly wider than either end (Marshall 1984:175).

Comments: In the original description of *Eobrasilia coutoi*, Simpson (1947) made most comparisons of skull and tooth structure with specimens of *Didelphis*. He also made comparisons with some Borhyaenidae (*Arminiheringia*, *Borhyaena*), Dasyuridae (*Dasyurus*) and "Didelphidae" (*Caroloameghinia*). Simpson noted that in structure of the large bulbous P_3 and rounded upper molars, a "habitus resemblance" existed between *Eobrasilia* and members of the family Stagodontidae (=Thlaeodontinae) from the Late Cretaceous of North America. He emphasized that the lack of a nasal-lacrimal contact in *Eobrasilia* was a distinct didelphoid character, as opposed to a broad nasal-lacrimal contact in borhyaenoids (Marshall 1978:70). Furthermore, the presence of a single rooted P^1 in *Eobrasilia* indicated that it was too specialized to be ancestral to any known later didelphoid or borhyaenoid. Simpson concluded his study by formally, but tentatively placing *Eobrasilia* in the family Didelphidae. "I suspect, nevertheless, that it represents a line derived from the earliest borhyaenids or from a didelphid-borhyaenid transition group" (Simpson 1947:7). Subsequent workers (e.g. Ride 1962:290, Fig. 5) followed this view, and *Eobrasilia* was long regarded as a potential "missing link" between borhyaenoids and didelphoids (Marshall 1984:173). In fact, Archer (1976:24) discussed *Eobrasilia* in his section on Borhyaenidae and not Didelphidae.

The controversial superfamily position of *Eobrasilia* was resolved by the recent discovery of a lower dentition (DGM 919-M; Marshall 1984). In fact, based on the comparable similarities of occlusal wear in the two known specimens, they apparently represent the upper (AMNH 39424-type) and lower (DGM 919-M) dentitions of the same individual (Marshall 1984, Fig. 3). The lower dentition has an enormous incipiently tri-rooted P_3 and is clearly didelphoid in aspect. The structure and relative size of P_{1-3} are similar to the living didelphid *Didelphis* "*paraguayensis*" (Clemens 1966, Fig. 73), members of the family Stagodontidae (Clemens 1966, 1979), and to other didelphoids from Itaborai (i.e. *Didelphopsis*, *Gaylordia*). Thus, *Eobrasilia* is a large unique didelphoid (a conclusion also reached by Paula Couto 1962:139) and "... bears no relation to the ancestry of that group or to any of its members" (Marshall 1984:176).

Paula Couto (1962:163-164) described a specimen (MNRJ 2505-V) as a right P_3 which based on its size (L = 5.3 mm, W = 3.7 mm *fide* Paula Couto 1962:164) he believed may be referrable to *Eobrasilia coutoi*. However, given its size and structure, as described by Paula Couto, this specimen is almost certainly referrable to *Didelphopsis*. Marshall (1984:176, Fig. 4) also described and figured an isolated right M^4 or, less likely, an M^3 (DGM 896a-M) which he tentatively referred to *Eobrasilia*. Below, this tooth is made the type of a new genus and species.

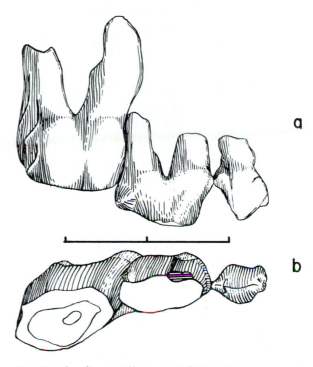

Fig. 10. Eobrasilia coutoi Simpson 1947. AMNH 39424 (type), right P^{1-3}: a, labial; b, occlusal views. Scale = 10 mm (adopted from Marshall 1984, Fig. 1).

Table 4. Measurements of cheek teeth of *Eobrasilia coutoi*.

Specimen	P1 L	P1 W	P2 L	P2 W	P3 L	P3 W
Upper cheek teeth						
AMNH 39424 (left)	–	–	–	–	7.2	4.7
AMNH 39424 (right)	3.1	2.2	5.3	3.6	7.3	4.7
Lower cheek teeth						
DGM 919-M	–	–	4.5	3.4	7.4	4.5

Gaylordia Paula Couto 1952b

Gaylordia Paula Couto 1952b:16; 1962:151.

Xenodelphis Paula Couto 1962:160.

Type of Gaylordia: G. macrocynodonta Paula Couto 1952b:17.

Type of Xenodelphis: X. doelloi Paula Couto 1962:160.

Diagnosis-Description: Small size; dental formula ?/? 1/1 3/3 5/5; rostral region of skull

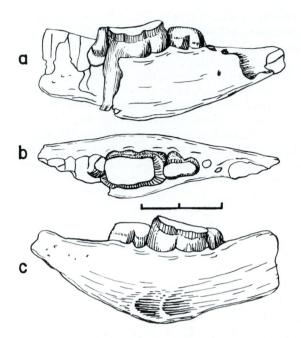

Fig. 11. Eobrasilia coutoi Simpson 1947. DGM 919-M, a fragment of a right mandibular ramus with alveoli of two incisors, C and P_1, P_{2-3} complete, anterior part of M_2 present: a, labial; b, occlusal; c, lingual views. Scale = 10 mm (adopted from Marshall 1984, Fig. 2).

described by Paula Couto (1952b:17-20); mandibular ramus short, low, robust; small mental foramen below anterior end of P_2, smaller one below M_2; symphysis extends posteriorly to point below anterior edge of P_3; lower C, judging from alveoli, well developed; upper C enormous, long, gently curved, laterally compressed, and "saber-like"; premolars two-rooted, increasing sharply in size from P_1^1 to P_3^3; P_1^1 very small, set at slight oblique angle in jaw, separated from P_2^2 by small diastema; P^2 ovate in occlusal view with bulbous primary cusp and low broad posterobasal heel; P^3 enormous, higher than M^2 or P^2, ovate in occlusal view with bulbous primary cusp and low broad posterobasal heel, crown inclined toward M^2; in length $M^2 \approx M^3 \approx M^4$, in width $M^2 < M^3$; protocone on M^{2-3} very narrow anteroposteriorly, crowns much wider than long; P_{2-3} ovate in occlusal view with broad posterobasal heel; P_3 enormous, much higher than P_2 or M_2, inclined slightly toward M_2; in length $M_2 < M_3 \approx M_4 \approx M_5$, in width $M_2 \approx M_3 \approx M_4 > M_5$; trigonids much larger, wider and elevated above very reduced talonids; protoconid large, paraconid slightly smaller than metaconid; talonids very narrow (compressed) anteroposteriorly, distinctly basined; hypoconid, entoconid, and hypoconulid subequal in size on M_{2-4}, hypoconulid larger than other two cusps on M_5 and located medially on posterior edge of talonid; talonid of M_5 more reduced, but still basined, than on M_{2-4}; oblique anterobasal cingulum well developed on M_{2-5}.

Comments: Paula Couto (1952b:17) suggested that based on tooth structure, *Gaylordia* should be regarded as insectivorous rather than omnivorous in diet. He later (1962:160) erected the genus *Xenodelphis* and made cursory comparisons only with *Marmosa*. Still later (1970:22) he noted that in structure of the large bulbous P3, *Gaylordia*, and *Didelphopsis*, resembled members of the family Stagodontidae (= Didelphodontinae) from the Late Cretaceous of North America.

The synonymy of *Gaylordia* and *Xenodelphis* was first formally recognized by Marshall (1981b:20). A "*Gaylordia*-like" taxon is reported from the Late Cretaceous El Molino Formation of Bolivia (de Muizon *et al.* 1984).

Gaylordia doelloi (Paula Couto 1962)
(Figs 12, 13; Table 5)

Xenodelphis doelloi Paula Couto 1962: 160; 1970: 22, Figure 1.

Gaylordia mendesi Paula Couto 1970:23; Figure 3.

Type of Xenodelphis doelloi MNRJ 2504-V, a fragment of a left mandibular ramus of a juvenile with alveoli of M_2 and M_{4-5}, and a complete M_3 (figured by Paula Couto 1970, Fig. 1).

Type of Gaylordia mendesi: MNRJ 2882-V, a relatively complete right mandibular ramus with P_2-M_5 complete (figured by Paula Couto 1970, Fig. 3).

Hypodigm: Two types and MNRJ 2487-V, a partial left mandibular ramus with alveoli of P_1-M_2 and M_5, and complete M_{3-4}.

Diagnosis-Description: Smallest known species of *Gaylordia*; linear tooth dimensions average 30% smaller than in *G. macrocynodonta*; P_3 typically smaller relative to M_2 than in *G. macrocynodonta*.

Comments: Paula Couto (1962:161) named *Xenodelphis doelloi* and in the discussion of the genus made cursory comparison only with *Marmosa*. Later (1970:23) he named *Gaylordia mendesi*, and in the diagnosis and discussion of that species compared it only with *Gaylordia macrocynodonta*. Thus, there is no evidence in his published works that he ever compared *X. doelloi* with *G. mendesi*. The two species are clearly synonyms and are inseparable in size and structure. Based on date of publication, the name for this taxon is *Gaylordia doelloi*.

Gaylordia macrocynodonta Paula Couto 1952b

(Figs 14-18; Table 5)

Gaylordia macrocynodonta Paula Couto 1952b:17, Figure 6; 1962:153, Figure 9; 1970:22, Figure 2.

Table 5. Measurements of cheek teeth of species of *Gaylordia*.

Specimen	P1 L	P1 W	P2 L	P2 W	P3 L	P3 W	M2 L	M2 W	M3 L	M3 W	M4 L	M4 W	M5 L	M5 W	M2-4/M2-5 L	P3-M4/P3-M5 L
Upper cheek teeth																
G. macrocynodonta																
DGM 329-M	–	–	1.2	0.8	2.0	1.6	~1.8	1.9	~1.8	2.2	~1.8	–	–	–	5.4	7.3
Lower cheek teeth																
G. doelloi																
MNRJ 2487-V	–	–	–	–	–	–	–	–	1.4	1.0	1.4	1.0	–	–	~5.5	~7.1
MNRJ 2504-V	–	–	–	–	–	–	–	–	1.3	1.0	–	–	–	–	–	–
MNRJ 2882-V	–	–	1.2	0.7	1.5	1.0	1.2	0.9	1.3	1.0	1.4	1.0	1.4	0.9	5.6	7.0
G. macrocynodonta																
AMNH 49801	–	–	2.0	1.1	3.0	1.8	1.7	1.5	1.8	1.5	–	–	–	–	–	–
DGM 186-M	–	–	–	–	3.5	2.0	1.9	1.6	–	–	–	–	–	–	~7.6	~11.0
DGM 329-M	–	–	–	–	2.3	1.3	–	–	–	–	–	–	–	–	–	–
DGM 644-M	–	–	–	–	–	–	1.8*	1.2*	2.0*	1.4*	~1.9*	–	2.0*	1.2*	–	–
DGM 645-M	–	–	–	–	1.8*	1.0*	1.8*	1.2*	2.0*	1.4*	~1.9*	–	2.0*	1.2*	–	9.6*
DGM 646-M	–	–	–	–	–	–	2.0*	1.5*	2.0*	1.6*	–	–	–	–	–	–
DGM 925-M	0.9	0.5	1.6	0.8	2.0	1.1	1.6	1.2	–	–	–	–	–	–	–	–
MNRJ 1355-V	–	–	1.6	0.9	2.8	1.6	1.7	1.4	1.8	1.4	1.9	1.4	1.9	1.2	7.0	9.8
MNRJ 1357-V	–	–	–	–	3.1	1.7	–	–	–	–	2.1	1.6	–	–	~8.0	~11.0
MNRJ 1358-V	–	–	–	–	3.0	1.7	–	–	–	–	–	–	–	–	~7.8	~10.6
MNRJ 1366-V	–	–	–	–	–	–	1.8	1.4	–	–	–	–	–	–	~8.0	~10.0

*After Paula Couto (1970: 24, Table 1).

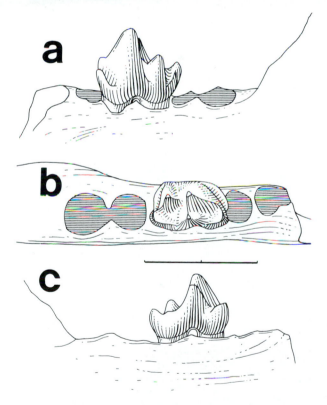

Fig. 12. Gaylordia doelloi (Paula Couto 1962). MNRJ 2504-V (type), a fragment of a left mandibular ramus with alveoli of M_2 and M_{4-5}, and with M_3 present: a, labial; b, occlusal; c, lingual views. Scale = 2.5 mm.

Type: DGM 329-M, rostrum of skull with right $C-M^4$ and left M^{2-3} present, roots of right P^1, anteroexternal root of right M^5, roots of left $C-P^3$ and M^4; medial part of left mandibular ramus with P_3 and M_2, roots of P_{1-2}, and crowns of M_{3-4} (figured by Paula Couto 1952b, Fig. 6; 1962, Fig. 9A-D).

Hypodigm: Type and AMNH 49801 (=MNRJ 1356-V), a fragment of a right mandibular ramus with P_2-M_3 complete and alveoli of M_4; MNRJ 1355-V, a nearly complete right mandibular ramus with alveoli of P_1 and complete P_2-M_5; MNRJ 1357-V, a left mandibular ramus with P_3 and M_4 complete, and alveoli of P_{1-2}, M_{2-3} and M_5; MNRJ 1358-V, a right mandibular ramus with P_3 present, alveoli of P_{1-2} and M_{2-5}; MNRJ 1366-V, a fragment of a left mandibular ramus with M_2 complete, alveoli of P_{2-3} and M_{3-5} (figured by Paula Couto 1962, Fig. 9E-G); DGM 186-M, a left mandibular ramus with P_3-M_2 complete and alveoli of rest of teeth (figured by Paula Couto 1962, Fig. 9H; 1970, Fig. 2); DGM 644-M, a fragment of a mandibular ramus with very worn M_{2-3} and M_5 complete, alveoli of M_4; DGM 645-M, a fragment of a mandibular ramus with very worn P_3-M_3 and M_5 complete, roots and fragments of crown of M_4; DGM 646-M, a fragment of a mandibular ramus with very worn M_{2-3} complete and roots of M_{4-5}; DGM 647-M, a partial right mandibular ramus with very worn M_4, alveoli of M_{2-3} and M_5; DGM 648-M, a fragment of a right mandibular ramus with very worn P_3 and M_2; DGM 925-M, a fragment of a left mandibular ramus with P_1-M_2 complete.

Diagnosis-Description: Largest known species of *Gaylordia*; averages about 30% larger in linear tooth dimensions than *G. doelloi*; P_3 typically larger relative to M_2 than in *G. doelloi*.

Comments: Paula Couto (1952b:17) compared *Gaylordia macrocynodonta* with specimens of the living didelphid *Monodelphis brevicaudata*. Clemens (1966:21) noted that the upper molars

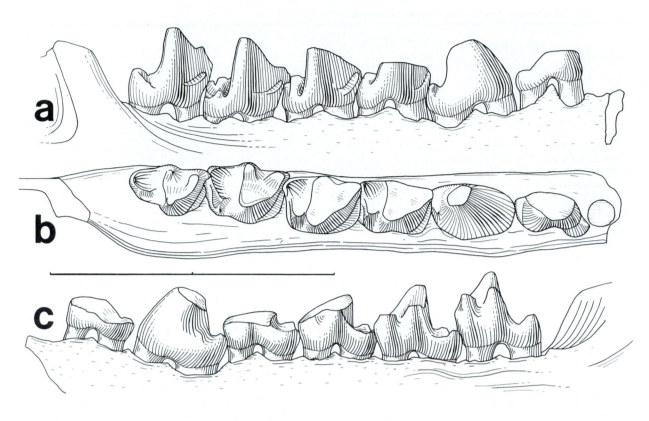

Fig. 13. Gaylordia doelloi (Paula Couto 1962). MNRJ 2882-V (type of *Gaylordia mendesi* Paula Couto 1970), a relatively complete right mandibular ramus with P$_2$-M$_5$ complete: a, labial; b, occlusal; c, lingual views. Scale = 5 mm.

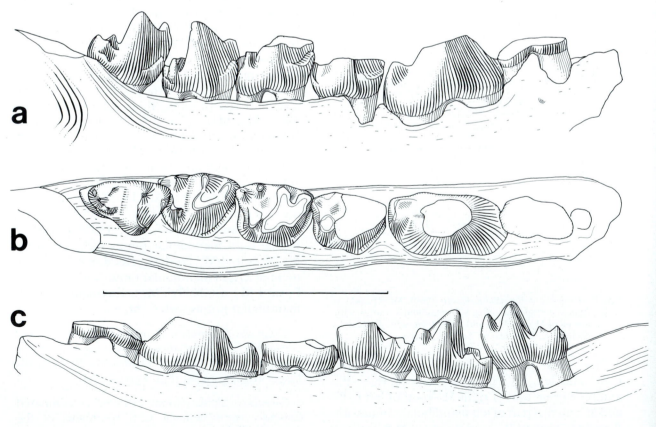

Fig. 14. Gaylordia macrocynodonta Paula Couto 1952b. MNRJ 1355-V, a nearly complete right mandibular ramus with alveoli of P$_1$ and with P$_2$-M$_5$ complete: a, labial; b, occlusal; c, lingual views. Scale = 5 mm.

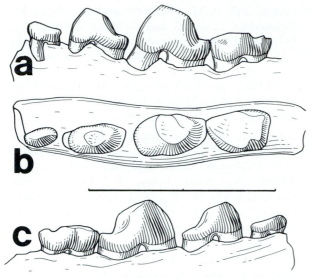

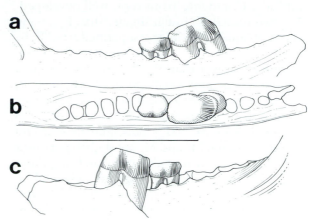

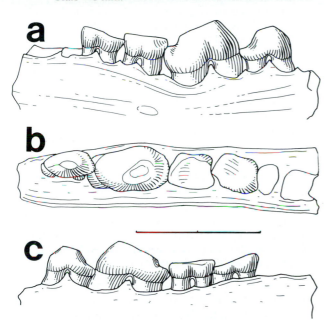

Fig. 15. *Gaylordia macrocynodonta* Paula Couto 1952b. DGM 925-M, a fragment of a left mandibular ramus with P$_1$-M$_2$ complete: a, labial; b, occlusal; c, lingual views. Scale = 5 mm.

Fig. 17. *Gaylordia macrocynodonta* Paula Couto 1952b. DGM 186-M, a left mandibular ramus with P$_3$-M$_2$ complete and alveoli of rest of teeth: a, labial; b, occlusal; c, lingual views. Scale = 10 mm.

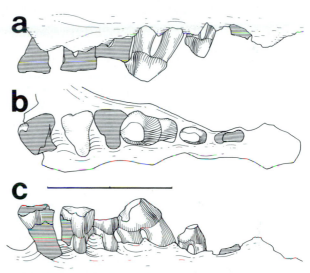

Fig. 16. *Gaylordia macrocynodonta* Paula Couto 1952b. AMNH 49801, a fragment of a right mandibular ramus with P$_2$-M$_3$ complete and alveoli of M$_4$: a, labial; b, occlusal; c, lingual views. Scale = 5 mm.

Fig. 18. *Gaylordia macrocynodonta* Paula Couto 1952b. DGM 329-M (type), palatal region of right side of skull with roots of P^1, P^2-M^4 present but worn and partially broken: a, labial; b, occlusal; c, lingual views. Scale = 5 mm.

in the type are heavily worn and ". . . little can be determined of their morphology. Their coronal outlines are similar to those of *Alphadon marshi* and *A. lulli*, suggesting presence of large stylar cusps".

Guggenheimia Paula Couto 1952b

Guggenheimia Paula Couto 1952b:11.

Type: Guggenheimia brasiliensis Paula Couto 1952b:12.

Diagnosis-Description: As for type and only known species.

Guggenheimia brasiliensis Paula Couto 1952b
(Fig. 19; Table 6)

Guggenheimia brasiliensis Paula Couto 1952b:12; 1962:149, Figure 6.

Type: DGM 297-M, a relatively complete left mandibular ramus with root of C, and complete P$_1$-M$_5$ (figured by Paula Couto 1962, Fig. 6).

Hypodigm: Type only.

Diagnosis-Description: Medium size; dental formula ?/? ?/1 ?/3 ?/5; mandibular ramus notably deep, robust; symphysis short, extending posteriorly to point below anterior edge of P$_3$; well developed mental foramen below anterior root of P$_2$, smaller one below anterior root of M$_2$

and M_3; C, judging from root, well developed; P_{1-3} two rooted; P_1 much smaller than P_{2-3}, set at slight oblique angle in jaw, separated from P_2 by small diastema; P_2 about same length as but slightly narrower than P_3; P_2 with large pointed primary cusp and small posterobasal heel; P_3 similar in structure to P_2 but with relatively larger posterobasal heel; tips of P_{2-3} as high as protoconid of M_2; M_{2-5} subequal in size and shape, although M_2 and M_5 are slightly narrower than M_{3-4}; M_{2-5} with weak anterobasal cingulum; trigonid only slightly higher or subequal to talonid in lateral view, but of similar size and width in occlusal view; protoconid and metaconid subequal in size in unworn teeth, paraconid considerably smaller but still well developed; hypoconid and entoconid very large, subequal in size, of about same size as protoconid and metaconid; hypoconulid very weakly developed, even on M_5.

Comments: Paula Couto (1952b:12) regarded *Guggenheimia* as "... of rather more primitive..." aspect than living Didelphinae, yet, suggested that in dental structure it shows closer affinity to omnivorous (i.e. derived) members of Didelphinae. He also noted that among the living genera of Didelphinae, *Guggenheimia* most closely resembles *Philander* in aspects of dental and mandibular morphology.

Itaboraidelphys Marshall and de Muizon 1984

Itaboraidelphys Marshall and de Muizon 1984:1297.

Type: Itaboraidelphys camposi Marshall and de Muizon 1984:1297.

Diagnosis-Description: As for type and only known species.

Itaboraidelphys camposi Marshall and de Muizon 1984

(Figs 20-23; Table 7)

Itaboraidelphys camposi Marshall and de Muizon 1984:1297 Figure 1.

Type: DGM 817-M, part of a right mandibular ramus with P_3-M_5 complete (figured by Marshall and de Muizon 1984, Fig. 1g-i).

Hypodigm: Type and DGM 804-M, a fragment of a right mandibular ramus with M_{3-4} complete, alveoli of M_2 and M_5; DGM 814-M, a fragment of a right mandibular ramus with M_{2-5} complete; DGM 923-M, a fragment of a right mandibular

ramus with P_2-M_2 complete and alveoli of M_3 (figured by Marshall and de Muizon 1984, Fig. 1j-l); DGM 926-M, a fragment of a left maxillary with M^{3-4} complete, alveoli of M^5 (figured by Marshall and de Muizon 1984, Fig. 1a-c); MNRJ 2878-V(a), an isolated right M^3 (figured by Marshall and de Muizon 1984, Fig. 1d-f); MNRJ 2878-V(b), an isolated right M^4.

Diagnosis-Description: Medium size; dental formula ?/? ?/? ?/3 5/5; mandibular ramus moderately well developed, deep; symphysis extends posteriorly to point below P_2; large mental foramen below M_2; P_{2-3} large, almost subequal in length and width although P_2 slightly smaller with trenchant primary cusp higher than protoconid of M_2; P_{2-3} with tiny cuspule on anteromedial edge, and large posterobasal heel (heel of P_3 relatively broader, more robust than that of P_2); in length $M_2 \lesssim M_3 \lesssim M_4 \approx M_5$, in width $M_2 < M_3 \approx M_4 > M_5$; anterobasal cingulum large; weak cingulum on hypoflexid and on posterior surface of hypoconid; trigonid high, well elevated above talonid in lateral view; trigonid and talonid subequal in size in occlusal view, trigonid narrower than talonid on M_{2-3}, subequal in width on M_4, wider on M_5; protoconid large, metaconid smaller, paraconid smallest; talonid basined; entoconid and hypoconid large, conical, subequal in size; hypoconulid large on M_{2-5}, subequal in size on all molars; M^4 subequal in length but wider than M^3; protocone large, talon basined; metacone about twice as large as paracone; para- and metaconules well developed; paracingulum well developed; paracrista well developed, blade-like, connects with stylar cusp A; metacrista moderately developed, somewhat blade-like; stylar cusps B and D well developed, D slightly smaller than B, C smaller than B and D, A and E very small; in size B>D>C>A>E.

Comments: Marshall and de Muizon (1984) regarded *Itaboraidelphys* as one of the most generalized "didelphids" in the Itaborai fauna. This conclusion is confirmed by the present study (see PHYLOGENETICS section below).

Marmosopsis Paula Couto 1962

Marmosopsis Paula Couto 1962:157.

Type: Marmosopsis juradoi Paula Couto 1962:158.

Diagnosis-Description: As for type and only known species.

Marmosopsis juradoi Paula Couto 1962

(Figs 24, 25; Table 8)

Table 6. Measurements of lower cheek teeth of *Guggenheimia brasiliensis.*

Specimen	P_1 L	W	P_2 L	W	P_3 L	W	M_2 L	W	M_3 L	W	M_4 L	W	M_5 L	W	M_{2-5} L	P_3-M_5 L
DGM 297-M	1.3	0.9	1.8	1.2	1.9	1.4	2.4	1.9	2.6	2.0	2.6	2.0	2.6	1.8	10.7	12.6

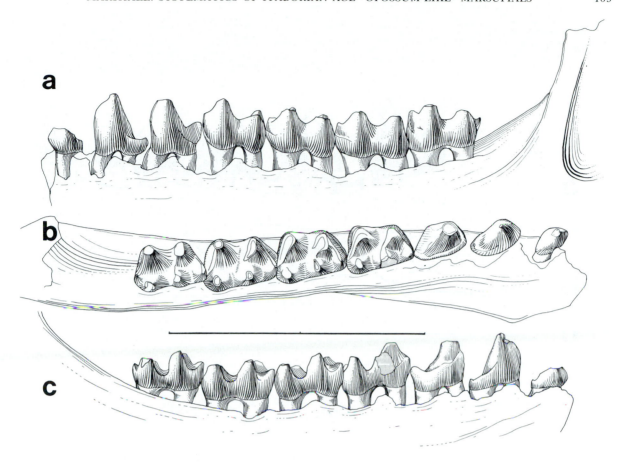

Fig. 19. Guggenheimia brasiliensis Paula Couto 1952b. DGM 297-M (type), a relatively complete left mandibular ramus with roots of C, and with P_1-M_5 complete: a, labial; b, occlusal; c, lingual views. Scale = 10 mm.

Table 7. Measurements of cheek teeth of *Itaboraidelphys camposi.*

Specimen	P^2 L	W	P^3 L	W	M^2 L	W	M^3 L	W	M^4 L	W	M^5 L	W	M^{2-5} L	P^3-M^5 L
Upper cheek teeth														
DGM 926-M	–	–	–	–	–	–	4.0	4.4	4.0	4.6	–	–	–	–
MNRJ 2878-V(a)	–	–	–	–	–	–	3.9	4.3	–	–	–	–	–	–
MNRJ 2878-V(b)	–	–	–	–	–	–	–	–	4.0	4.6	–	–	–	–
Lower cheek teeth														
DGM 804-M	–	–	–	–	–	–	3.7	2.3	3.8	2.4	–	–	–	–
DGM 814-M	–	–	–	–	3.3	1.9	3.4	2.0	3.5	2.0	3.5	1.9	14.0	–
DGM 817-M	–	–	4.0	1.8	3.6	2.2	3.6	2.4	3.6	2.4	3.6	2.1	14.7	19.3
DGM 923-M	3.6	1.6	3.7	1.7	3.6	2.3	–	–	–	–	–	–	–	–

Marmosopsis juradoi Paula Couto 1962:158, Figures 11, 12; 1970:27, Figures 4, 5.

Type: MNRJ 2343-V, a right mandibular ramus with alveoli of P_1 and complete P_2-M_5 (figured by Paula Couto 1962, Fig. 11A-C).

Hypodigm: Type and MNRJ 2344-V, a fragment of a left mandibular ramus with P_2-M_4 complete and talonid of M_5 (figured by Paula Couto 1970, Figs 4A, B, 5A); MNRJ 2345-V, a fragment of a right mandibular ramus with M_{2-5} (figured by Paula Couto 1970, Fig. 5B); MNRJ 2346-V, a fragment of a right mandibular ramus with M_{3-4} present, and alveoli of C-M_2 and M_5 (figured by Paula Couto 1970, Fig. 4C, D); MNRJ 2348-V, a fragment of a left mandibular ramus with M_{4-5} (figured by Paula Couto 1970, Fig. 5C); MNRJ 2349-V, a fragment of a right mandibular ramus with C-P_3 present, and roots of M_2 (figured by Paula Couto 1970, Fig. 4E); MNRJ 2350-V, a right mandibular ramus with P_3-M_5 complete and alveoli of C-P_2 (figured by Paula Couto 1962, Fig. 12A-C); MNRJ 2356-V, a fragment of a right mandibular ramus with M_{2-5}; MNRJ 2357-V, a fragment of a right mandibular ramus with P_{2-3} and M_5; MNRJ 2359-V, a fragment of a right mandibular ramus with P_3-M_4; MNRJ 2360-V, a fragment of a right mandibular ramus with trigonid of M_3 and complete M_4; MNRJ 2470-V, a fragment of a right mandibular ramus with P_2-M_4;

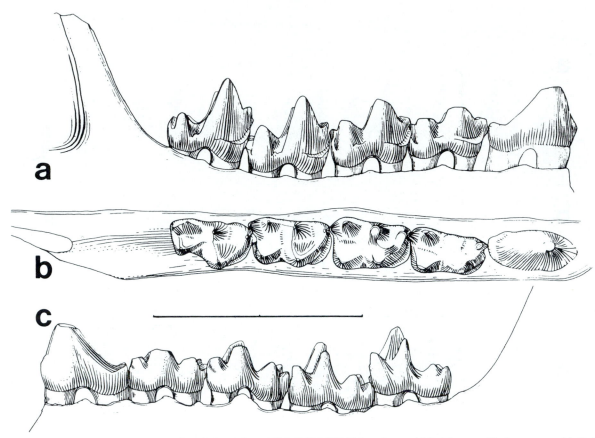

Fig. 20. Itaboraidelphys camposi Marshall and de Muizon 1984. DGM 817-M (type), a partial right mandibular ramus with P_3-M_5 complete: a, labial; b, occlusal; c, lingual views. Scale = 10 mm.

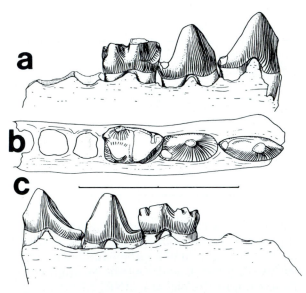

Fig. 21. Itaboraidelphys camposi Marshall and de Muizon 1984. DGM 923-M, a fragment of a right mandibular ramus with P_2-M_2 complete and alveoli of M_3: a, labial; b, occlusal; c, lingual views. Scale = 10 mm.

MNRJ 2471-V, a fragment of a right mandibular ramus with P_{1-2} and M_2; MNRJ 2472-V, a fragment of a right mandibular ramus with P_2 and M_2; MNRJ 2473-V, a fragment of a right mandibular ramus with M_3; MNRJ 2474-V, a fragment of a right mandibular ramus with M_2; MNRJ 2475-V, a fragment of a right mandibular ramus with roots of M_4 and complete M_5; MNRJ 2476-V, a fragment of a right mandibular ramus with M_5; MNRJ 2477-V, a fragment of a right mandibular ramus with M_2; MNRJ 2478-V, a fragment of a right maxillary with M^{2-3}; MNRJ 2479-V, a fragment of a right maxillary with M^{3-4}; MNRJ 2481-V, a fragment of a left maxillary with M^{2-3}; MNRJ 2482-V, a fragment of a left maxillary with M^{2-4}; MNRJ 2483-V, a fragment of a left maxillary with M^{3-4}; MNRJ 2484-V, a partial left mandibular ramus with M_{2-4}; MNRJ 2485-V, a partial left mandibular ramus with P_2-M_4; MNRJ 2486-V, a partial left mandibular ramus with P_2 and M_3; MNRJ 2495-V, a partial left mandibular ramus with M_{3-4}; MNRJ 2496-V, a partial left mandibular ramus with M_{2-3} and trigonid of M_4; MNRJ 2497-V, a partial left mandibular ramus with M_{3-5}; MNRJ 2498-V, a partial left mandibular ramus with P_2-M_3; MNRJ 2499-V, a partial left mandibular ramus with P_3 and M_4, and alveoli of rest of teeth; MNRJ 2500-V, a partial left mandibular ramus with talonid of M_3 and M_{4-5} present; MNRJ 2501-V, a partial left mandibular ramus with M_4 present and alveoli of M_5; MNRJ 2502-V, a partial left mandibular ramus with posterior half of P_1 and P_{2-3} present; MNRJ 2890-V, a fragment of a left mandibular ramus with alveoli of incisors, C-P_1 and P_3-M_2,

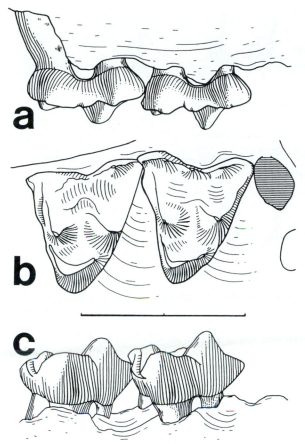

Fig. 22. *Itaboraidelphys camposi* Marshall and de Muizon 1984. DGM 926-M, a left maxilla with M^{3-4} complete, and alveoli of M^5: a, labial; b, occlusal; c, lingual views. Scale = 5 mm.

and with P_2 and M_{3-5} complete; DGM 649-M, a partial left mandibular ramus with M_{2-5}; DGM 650-M, a partial left mandibular ramus with alveoli of P_{1-2}, roots of P_3-M_2, and with M_{3-5} present; DGM 811-M, a fragment of a right mandibular ramus with M_{4-5}; DGM 931-M, a right mandibular ramus with alveoli and/or roots of P_{1-2} and M_{3-5}, and with P_3-M_2 present.

Diagnosis-Description: Small size; dental formula ?/4 ?/1 ?/3 5/5; mandibular ramus long, gracile; symphysis extends posteriorly to point below middle of P_2; mental foramen located below middle of P_2, another near point of M_{2-3} contact; lower C moderately to weakly developed; P_{1-3} two-rooted, all have large pointed primary cusp and a well developed posterobasal heel; P_1 smaller than P_{2-3}, separated from C and P_2 by small diastema, and with roots aligned in same anteroposterior axis as rest of cheek teeth; P_{2-3} subequal in length and width, sometimes separated by a small diastema; P_3 differs from P_2 in being set higher in jaw and inclined toward M_2 (tip of P_3 as high as protoconid of M_2, crown structure and size of P_{2-3} virtually identical; in length $M_2 < M_3 \approx M_4 \geqslant M_5$, in width $M_2 \approx M_3 \approx$

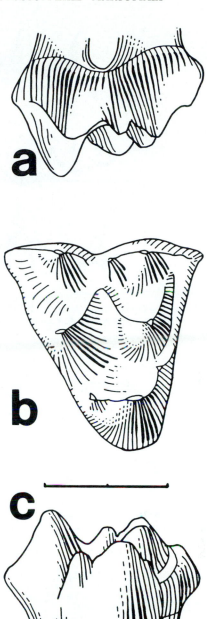

Fig. 23. *Itaboraidelphys camposi* Marshall and de Muizon 1984. MNRJ 2878(a), an isolated right M^3: a, labial; b, occlusal; c, lingual views. Scale = 2.5 mm.

$M_4 \approx M_5$; anterobasal cingulum moderately developed; trigonids large, well elevated above talonids; in size protoconid>metaconid>paraconid; paraconid small on M_2 increasing in size relative to metaconid from M_2 to M_5; trigonid and talonid of equal width on M_2, talonid narrower than trigonid on M_{3-5}; talonid decreases in size from M_3 to M_5; talonid short anteroposteriorly with shallow basin on M_{2-4}; hypo- and entoconid subequal in size on M_{2-4}; hypoconulid very reduced, indistinct, virtually fused to posterior end of entoconid; talonid of M_5

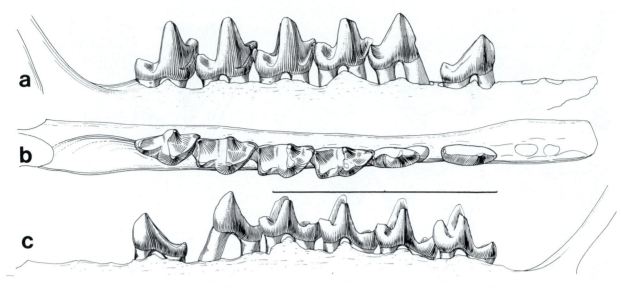

Fig. 24. Marmosopsis juradoi Paula Couto 1962. MNRJ 2343-V (type), a right mandibular ramus with alveoli of P₁, and P₂-M₅ complete: a, labial; b, occlusal; c, lingual views. Scale = 5 mm.

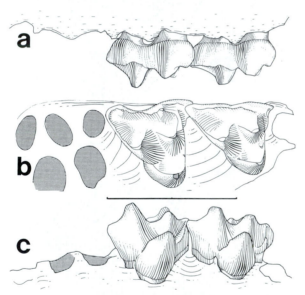

Fig. 25. Marmosopsis juradoi Paula Couto 1962. MNRJ 2478-V, a right maxilla with M²⁻³ complete, and with alveoli of M⁴⁻⁵: a, labial; b, occlusal; c, lingual views. Scale = 2.5 mm.

very reduced and narrow, slightly basined, cusps indistinguishable; in length $M^2 \approx M^3$, in width $M^2 < M^3$; judging from alveoli, M^4 similar in size to M^3; M^5 well developed, as wide as M^4, three-rooted; protocone large, well developed, talon basined; hint of paraconule and metaconule on M^{2-3}; metacone about twice as large as and higher than paracone; paracingulum short, broad; stylar shelf broad; stylar cusps A, C, D and E small but distinct; B largest of stylar cusps; in size B>D>A>C>E; paracrista moderately developed, blade-like, connects with anterior edge of stylar cusp B; metacrista moderately long, blade-like.

Comments: Paula Couto (1962:157) regarded *Marmosopsis* as the probable Paleocene ancestor

of the living *Marmosa*. He believed these taxa to be almost indistinguishable, and gave the fossil a different generic name simply because he found it inconceivable that a mammal genus could survive so long without change. He also noted that the reduction and structure of the M_5 talonid in some individuals of *Marmosopsis* resembled specimens of *Microbiotherium* from the middle Tertiary of Argentina.

Clemens (1966:21) observed that M^{2-4} of *Marmosopsis* have stylar cusps A through E, and according to Paula Couto they differ from molars of *Marmosa* only in the shorter and narrower protocones. "Thus upper molars of *Marmosopsis*, like . . . *Didelphopsis*, resemble those of *Alphadon marshi* in outline and stylar cusp pattern" (*ibid.*).

Minusculodelphis Paula Couto 1962

Minusculodelphis Paula Couto 1962:161.

Type: Minusculodelphis minimus Paula Couto 1962:162.

Diagnosis-Description: As for type and only known species.

Minusculodelphis minimus Paula Couto 1962

(Fig. 26; Table 9)

Minusculodelphis minimus Paula Couto 1962:162, Figure 13; 1970:21.

Type: MNRJ 2342-V, a left mandibular ramus with P_{1-2} and M_{2-5} complete, and roots of P_3 (figured by Paula Couto 1962, Fig. 13A-C). The M_2 of the type was lost subsequent to Paulo Couto's study of this specimen. This tooth is illustrated in his Figure 13A-C (1962), but was not

Table 8. Measurements of cheek teeth of *Marmosopsis juradoi.*

Specimen	P_2 L	W	P_3 L	W	M_2 L	W	M_3 L	W	M_4 L	W	M_5 L	W	M2-4 / M2-5 L	P3-M4 / P3-M5 L
Upper cheek teeth														
MNRJ 2478-V	–	–	–	–	1.5	1.4	1.5	1.6	–	–	–	–	–	–
MNRJ 2481-V	–	–	–	–	1.5	1.3	1.5	1.6	–	–	–	–	–	–
Lower cheek teeth														
DGM 649-M	–	–	–	–	1.2	0.8	1.3	0.9	1.3	0.9	1.2	0.8	5.2	–
DGM 650-M	–	–	–	–	–	–	1.3	1.0	1.3	1.0	1.2	0.8	–	–
DGM 811-M	–	–	–	–	–	–	–	–	1.4	1.0	1.3	0.9	–	–
DGM 931-M	–	–	1.1	0.5	1.2	0.7	–	–	–	–	–	–	–	–
MNRJ 2343-V	1.1	0.4	1.1	0.5	1.1	0.6	1.2	0.8	1.2	0.8	1.2	0.8	5.1	6.2
MNRJ 2344-V	1.1	0.5	1.1	0.5	1.3	0.8	1.4	0.9	1.4	0.8	–	–	5.5	6.5
MNRJ 2346-V	–	–	–	–	–	–	1.3	0.8	1.3	0.8	–	–	–	–
MNRJ 2348-V	–	–	–	–	–	–	–	–	1.3	0.9	1.3	0.8	–	–
MNRJ 2349-V*	1.1	0.5	1.2	0.5	–	–	–	–	–	–	–	–	–	–
MNRJ 2350-V	–	–	1.1	0.5	1.2	0.8	1.4	0.9	1.4	0.9	1.3	0.9	5.5	6.6
MNRJ 2357-V	1.2	0.5	1.1	0.5	–	–	–	–	–	–	1.4	0.8	~5.5	6.7
MNRJ 2359-V	–	–	1.1	0.5	1.3	0.8	1.4	0.8	1.4	0.8	–	–	–	–
MNRJ 2470-V	1.2	0.5	1.2	0.6	1.4	0.8	1.4	0.8	1.4	0.8	–	–	~5.2	~6.2
MNRJ 2471-V**	1.1	0.5	–	–	1.3	0.8	–	–	–	–	–	–	–	–
MNRJ 2472-V	1.2	0.5	–	–	1.4	0.8	–	–	–	–	–	–	–	–
MNRJ 2474-V	–	–	–	–	1.3	0.8	–	–	–	–	–	–	–	–
MNRJ 2475-V	–	–	–	–	–	–	–	–	–	–	1.2	0.8	–	–
MNRJ 2476-V	–	–	–	–	–	–	–	–	–	–	1.3	0.8	–	–
MNRJ 2477-V	–	–	–	–	1.4	0.9	–	–	–	–	–	–	–	–
MNRJ 2484-V	–	–	–	–	1.3	0.7	1.4	0.9	1.4	0.9	–	–	–	–
MNRJ 2485-V	1.2	0.5	1.2	0.5	1.2	0.8	1.3	0.9	1.3	0.9	–	–	~5.0	~6.2
MNRJ 2486-V	1.1	0.5	–	–	–	–	1.3	0.8	–	–	–	–	–	–
MNRJ 2495-V	–	–	–	–	–	–	1.3	0.8	1.3	0.8	–	–	–	–
MNRJ 2496-V	–	–	–	–	1.3	0.8	1.4	0.8	–	–	–	–	–	–
MNRJ 2497-V	–	–	–	–	–	–	1.4	0.8	1.3	–	1.3	0.8	–	–
MNRJ 2499-V	–	–	1.1	0.5	–	–	–	–	1.3	0.8	–	–	–	–
MNRJ 2500-V	–	–	–	–	–	–	–	–	1.4	0.9	1.3	–	–	–
MNRJ 2501-V	–	–	–	–	–	–	–	–	1.3	0.9	–	–	–	–
MNRJ 2502-V	1.1	0.4	1.0	0.5	–	–	–	–	–	–	–	–	–	–
MNRJ 2890-V	–	–	–	–	–	–	1.4	1.0	1.4	1.0	1.3	0.9	–	–

*P_1 L=1.0 mm, W=0.4 mm. **P_1 L=1.0 mm, W=0.4 mm.

present when I made my study and is not shown in my Figure 27.

Hypodigm: Type and MNRJ 2503-V, a nearly complete right mandibular ramus with alveoli of I_1-C and M_2, and with complete P_{1-3} and M_{3-5}; MNRJ 2901-V, an incomplete left mandibular ramus with M_{3-5} very broken.

Diagnosis-Description: Very small size, smallest species of didelphoid known; dental formula ?/4 ?/1 ?/3 ?/5; mandibular ramus extremely gracile, shallow, elongate; mental foramina occur below P_1 and anterior root of M_2; I_1-C, judging from alveoli and root, well developed, markedly procumbent; P_{1-3} structurally similar, subequal in length and width, very reduced in size, all two-rooted with roots aligned in same antero-posterior axis as molar series; P_2 separated from P_1 and P_3 by small diastema; crowns of P_{1-3} low, greatly elongate anteroposteriorly with very small cuspule anteriorly which becomes larger from P_1 to P_3, higher primary cusp medially and distinct but small posterobasal cuspule; M_{3-4} subequal in length, M_4 slightly wider than M_3; M_2 judging from Paula Couto's figure and the roots,

and M_5 slightly shorter and narrower than M_{3-4}; thus in length $M_2<M_3\approx M_4>M_5$, in width $M_2<M_3<M_4>M_5$; trigonid and talonid subequal in width on M_3 talonid slightly narrower than trigonid on M_4 and considerably narrower on M_5; weak anterobasal cingulum present; trigonid larger and greatly elevated above talonid; protoconid largest of trigonid cusps, metaconid somewhat smaller and situated directly linguad of protoconid, both connected by a very high blade-like protocristid; paraconid reduced relative to other two trigonid cusps and inclined anterolingually; talonid very low, distinctly basined even on M_5; hypoconid very large on M_{3-4}, prodont; entoconid indistinct and fused into exceedingly large blade-like structure that is continuous with metaconid; hypoconulid small but distinct; talonid of M_5 reduced; hypoconulid well developed; hypoconid and entoconid indistinct.

Comments: In the original description of *Minusculodelphis*, Paula Couto (1962:162) made no direct comparison with any other marsupial, apart from noting that it was much smaller than any known living species of *Marmosa*.

Table 9. Measurements of cheek teeth of *Minusculodelphis minimus*.

Specimen	P_1 L	W	P_2 L	W	P_3 L	W	M_2 L	W	M_3 L	W	M_4 L	W	M_5 L	W	M_{2-5} L
MNRJ 2342-V	0.5	0.2	0.6	0.2	–	–	0.7*	0.4*	0.8	0.5	0.8	0.6	0.7	0.5	~3.0
MNRJ 2503-V	0.5	0.1	0.5	0.2	0.5	0.2	–	–	0.8	0.5	0.8	0.6	0.7	0.5	~3.0

*Taken from Figure 13 of Paula Couto (1962).

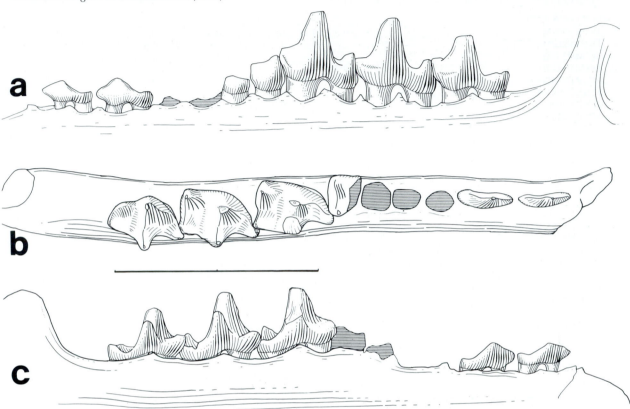

Fig. 26. *Minusculodelphis minimus* Paula Couto 1962. MNRJ 2342-V (type), a left mandibular ramus with roots of P_3 and anterior root of M_2, and with talonid of M_2, and P_{1-2} and M_{3-5} complete: a, labial; b, occlusal; c, lingual views. Scale = 2.5 mm.

Mirandatherium Paula Couto 1952c

Mirandaia Paula Couto 1952b:22, *nec* Travassos 1937:360 (Nematodia).

Mirandatherium Paula Couto 1952c:503.

Type: Mirandatherium alipioi (Paula Couto 1952b:23).

Diagnosis-Description: As for type and only known species.

Mirandatherium alipioi (Paula Couto 1952c)

(Figs 27-29; Table 10)

Mirandaia ribeiroi Paula Couto 1952b:23, *nec Mirandaia ribeiroi* (Nematodia) Travassos 1937:360.

Mirandatherium alipioi Paula Couto 1952c:503; 1962:145, Figure 4; 1970:28, Figure 6.

Type: MNRJ 1359-V, a fragment of a right mandibular ramus with P_2-M_3 complete, and alveoli of P_1 and M_4 (figured by Paula Couto 1962, Fig. 4).

Hypodigm: Type and AMNH 49802 (= MNRJ 1361-V), a fragment of a right mandibular ramus with M_3 present, anterior root and talonid of M_2, and roots of M_4; AMNH 49859 (= DGM 360-M, paratype), a partial right mandibular ramus with alveoli of M_2 and complete M_{3-5}; DGM 361-M (paratype), a fragment of a right mandibular ramus with alveoli of M_2, and complete M_{3-5} (figured by Paula Couto 1970, Fig. 6A-C); DGM 362-M, posterior fragment of a right mandibular ramus with M_{3-4}, posterior edge of P_3 and roots of M_2 and M_5; DGM 922-M, a fragment of a right mandibular ramus with M_{2-5}; DGM 929-M, a fragment of a left mandibular ramus with broken C, complete P_{2-3}, and alveoli of P_1 and M_{2-3}; MNRJ 1362-V, a fragment of a right mandibular ramus with M_{3-4} complete, alveoli of M_2, and anterior root of M_5; MNRJ 1363-V, a fragment of a right mandibular ramus with M_3, and talonid and anterior root of M_2; MNRJ 1364-V, an isolated right M_4 (*fide* Paula Couto); MNRJ 2340-V, a fragment of a right mandibular ramus with alveoli of I_1-C, and P_1-M_3 present; MNRJ 2506-V, left maxilla with M^{2-4} complete, and alveoli of P^3 and M^5; MNRJ 2900-V, an isolated right M_3 (*fide* Paula Couto).

Diagnosis-Description: Small size; dental formula ?/4 ?/1 ?/3 5/5; mandibular ramus deep, short; large mental foramen below M_2 or posterior edge of P_3; symphysis extends posteriorly to point below anterior edge of P_2; C very well developed; premolars increase in size from P_1 to P_3; P_1 markedly smaller than P_{2-3}, set obliquely in jaw; P_{2-3} with sharp, high, blade-like primary cusp, distinct posterobasal heel (larger on P_3); P_2 only slightly smaller than P_3, crown less elevated; P_3 crown same height as M_2 protoconid; molars brachydont; in length and width $M_2 < M_3$ $\approx M_4 > M_5$; talonid broader than trigonid on M_{2-4}, narrower on M_5; trigonid low and only moderately elevated above talonid; trigonid slightly smaller than talonid on M_{2-4} in occlusal view; well developed anterobasal cingulum; trigonid cusps well developed, slightly inflated; in size protoconid>metaconid>paraconid; protoconid has broad vertical convex crest on lingual surface extending from tip to between para- and metaconid in prefossid; talonid large, distinctly basined; hypoconid and entoconid large, subequal in size; hypoconulid also well developed, situated labiad of entoconid; small shelf situated below hypoflexid and posterobasal cingular shelf on posterior surface of hypoconid on M_{2-4}; in length $M^2 > M^3 > M^4$, in width $M^2 < M^3 \approx M^4$; protocone large, talon basined; metacone about twice as large and higher than paracone; para- and metaconules well developed on M^{2-4}; paracingulum well developed; centrocrista weak, rectilinear; metacrista moderately developed; paracrista weak, connects with anterior edge of stylar cusp B; stylar shelf of moderate width; stylar cusp B largest, D smaller, A and C smaller still and subequal in size, E virtually absent; in size B>D>C $\approx$A>E on M^{2-4}; ectoflexus absent on M^2, shallow on M^3, moderately developed on M^4.

Comments: In the original description of *Mirandatherium alipioi* (as *Mirandaia ribeiroi*) Paula Couto (1952b:23) did not compare it with any other taxon. However, he later (1970:28) compared it with the Casamayoran genus *Coona* Simpson 1938, concluding that "It is possible that the differences, . . ., between *Mirandatherium* and *Coona*, are only of specific significance, and that *Mirandatherium* is a synonym of *Coona*". As I will demonstrate in the PHYLOGENETICS section below, this animal represents the earliest known member of the subfamily Microbiotheriinae.

Monodelphopsis Paula Couto 1952b

Monodelphopsis Paula Couto 1952b:24.

Type: Monodelphopsis travassosi Paula Couto 1952b:25.

Diagnosis-Description: As for type and only known species.

Monodelphopsis travassosi Paula Couto 1952b

(Figs 30-32; Table 11)

Monodelphopsis travassosi Paula Couto 1952b:25, Figure 7; 1962:155, Figure 10; 1970:29, Figure 7.

Type: MNRJ 1365-V, a fragment of a left mandibular ramus with P_3-M_2 complete (figured by Paula Couto 1952b, Fig. 7; 1962, Fig. 10A-C).

Hypodigm: Type and MNRJ 1368-V, a fragment of a right mandibular ramus with alveoli of M_4 and complete M_5 (figured by Paula Couto 1970, Fig. 7); MNRJ 2347-V, a fragment of a left mandibular ramus with M_5 (figured by Paula Couto 1962, Fig. 10D-F); MNRJ 2354-V, a partial left mandibular ramus with base of P_3 and complete M_{2-5}; MNRJ 2355-V, a partial left mandibular ramus with unerupted P_3 and M_2, alveoli of I_{1-4}, C-P_2, M_1 and M_3; MNRJ 2886-V, a partial right mandibular ramus with P_2 and M_{3-4} complete, anterior root of M_2, alveoli of P_1, P_3-M_2 and M_5; MNRJ 2887-V, a partial right mandibular ramus with P_3 and M_3 complete, anterior root and talonid of M_2, anterior root of M_4, alveoli of P_{1-2}, and anterior part of posterior alveolus of M_4; MNRJ 2888-V, an isolated left M_4 (*fide* Paula Couto); MNRJ 2889-V, a fragment of a left mandibular ramus with M_{4-5}; MNRJ 2891-V, a lower premolar, and a right and two left molars (*fide* Paula Couto); DGM 808-M, a fragment of a left maxilla with M^{3-5}; DGM 812-M, a left mandibular ramus with P_3 and M_{4-5} complete, and alveoli of rest of dentition; DGM 651-M, a partial left mandibular ramus with M_{2-5}; DGM 652-M, a partial left mandibular ramus with talonid of M_4 and complete M_5; DGM 928-M, a right mandibular ramus with M_{4-5} complete and alveoli of rest of dentition; DGM 932-M, a fragment of a right mandibular ramus with M_{2-4} complete and alveoli of M_5.

Diagnosis-Description: Small size; dental formula ?/4 ?/1 ?/3 5/5; mandibular ramus of moderate robustness and depth; symphysis extends posteriorly to point below P_2; mental foramina below P_2 and M_2; I_{1-4}, judging from alveoli, subequal in size and moderately developed; C, judging from alveolus, moderately developed; P_{1-3} two-rooted and in contact, not

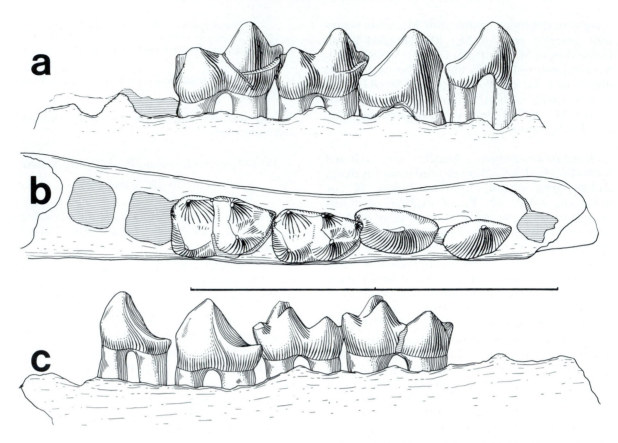

Fig. 27. Mirandatherium alipioi Paula Couto 1952c. MNRJ 1359-V (type), a fragment of a right mandibular ramus with P_2-M_3 complete and alveoli of P_1 and M_4: a, labial; b, occlusal; c, lingual views. Scale = 10 mm.

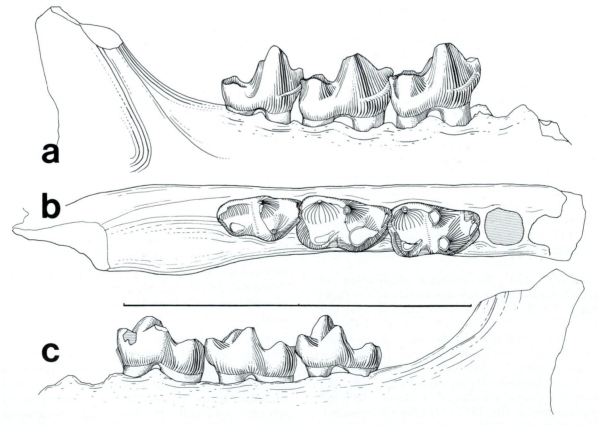

Fig. 28. Mirandatherium alipioi Paula Couto 1952c. DGM 361-M, a fragment of a right mandibular ramus with alveoli of M_2, and with M_{3-5} complete: a, labial; b, occlusal; c, lingual views. Scale = 10 mm.

Table 10. Measurements of cheek teeth of *Mirandatherium alipioi.*

Specimen	P2 L	P2 W	P3 L	P3 W	M2 L	M2 W	M3 L	M3 W	M4 L	M4 W	M5 L	M5 W	$\frac{M^{2-4}}{M_{2-5}}$ L
Upper cheek teeth													
MNRJ 2506-V	–	–	–	–	2.1	1.9	1.9	2.4	1.7	2.4	–	–	6.2
Lower cheek teeth													
AMNH 49859	–	–	–	–	–	–	2.2	1.5	2.4	1.6	2.3	1.3	–
DGM 361-M	–	–	–	–	–	–	2.4	1.7	2.5	1.6	2.3	1.4	–
DGM 362-M	–	–	–	–	–	–	2.2	1.4	2.2	1.4	–	–	–
DGM 922-M	–	–	–	–	2.2	1.3	2.3	1.5	2.3	1.5	2.2	1.3	9.1
DGM 929-M	2.2	1.0	2.4	1.2	–	–	–	–	–	–	–	–	–
MNRJ 1359-V	2.0	1.0	2.2	1.2	2.2	1.4	2.4	1.7	–	–	–	–	–
MNRJ 1362-V	–	–	–	–	–	–	2.3	1.6	2.3	1.5	–	–	–
MNRJ 1363-V	–	–	–	–	–	–	2.3	1.5	–	–	–	–	–
MNRJ 2340-V*	2.0	1.0	2.1	1.1	2.1	1.4	2.2	1.5	–	–	–	–	–

*P_1 L=1.4 mm, W=0.8 mm.

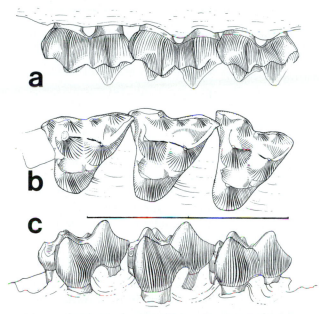

Fig. 29. Mirandatherium alipioi (Paula Couto 1952c). MNRJ 2506-V, a left maxilla with M^{2-4} complete: a, labial; b, occlusal; c, lingual views. Scale = 5 mm.

separated by diastema; P_1, judging from alveoli, well developed, not much smaller than P_2 and with roots aligned in same anteroposterior axis as rest of cheek teeth; P_2 either subequal in size to, or slightly larger than, P_3; P_2 and P_3 with very high, trenchant primary cusp and well developed posterobasal heel; P_2 differs from P_3 in being of greater height, tip of crown higher than protoconid of M_2 and with proportionately narrower posterobasal heel; P_2 about same height as M_2 protoconid; molars brachydont; in length $M_2 \approx M_3 \approx M_4 > M_5$, in width $M_2 \approx M_3 \approx M_4 > M_5$; M_{2-5} with weakly developed anterobasal cingulum; M_{2-4} with very weakly developed cingulum on posterior surface of hypoconid; trigonid low, only moderately elevated above talonid; trigonid narrower than talonid on M_{2-3}, trigonid and talonid subequal in width on M_{4-5}; all trigonid cusps well developed; in size protoconid>metaconid>paraconid; talonid large,

basined; entoconid considerably higher and larger than hypoconid; hypoconulid well developed on M_{2-4}, subequal in size to entoconid on M_5; M_5 talonid large, with well developed and distinct hypoconid, hypoconulid, and entoconid; in length $M^3 \approx M^4 > M^5$, in width $M^3 < M^4 \approx M^5$; M^{3-5} exceptionally narrow anteroposteriorly; protocone enormous, talon basined; para- and metaconules distinct but weakly developed; metacone about twice as large and higher than paracone on M^{3-4}, paracone about twice as large as metacone on M^5; paracingulum small, does not reach lingually beyond point below paracone; centrocrista rectilinear, weakly developed, sharp; paracrista short, moderately developed, blade-like,connects with stylar cusp A; metacrista very short, blade-like; stylar shelf narrow, stylar cusps small or absent; stylar cusp B largest of stylar cusps but small and only slightly larger than A on M^{3-4}, vestigial on M^5, A small but distinct on M^{3-5}, C distinct on M^4 only, D and E visible on M^{3-4} only; on M^3 B>A>D>E; ectoflexus deep on M^{3-5}.

Comments: Paula Couto (1952b:24) noted that of all the "Didelphids" he described in his 1952b paper, *Monodelphopsis* was most similar in size and dental structure to some living genera, especially *Monodelphis*. This similarity was particularly evident in the relative size of P_{2-3} in which P_2 was larger than P_3, a character he observed as shared with living didelphids as opposed to the other Itaborai genera in which P_3 was as large as or larger than P_2 (but see below).

DGM 808-M, a fragment of a left maxilla with M^{3-5} (Fig. 32) is the only specimen known of the upper dentition of this species. It has multiple similarities shared with North American Late Cretaceous pediomyines (see PHYLO-GENETICS section), and in size it falls within the range of measurements of the population of *Pediomys cooki* described by Clemens (1966:42, Table 9) from the Maestrichtian age Lance Formation. The secure referral of this specimen

Table 11. Measurements of cheek teeth of *Monodelphopsis travassosi.*

Specimen	P2 L	P2 W	P3 L	P3 W	M2 L	M2 W	M3 L	M3 W	M4 L	M4 W	M5 L	M5 W	M2-4 / M2-5 L	P3-M5 L
Upper cheek teeth														
DGM 808-M	–	–	–	–	–	–	1.9	2.4	1.9	2.7	1.0	2.7	–	–
Lower cheek teeth														
DGM 651-M	–	–	–	–	1.7	1.0	1.7	1.1	1.7	1.2	1.6	1.0	6.7	–
DGM 652-M	–	–	–	–	–	–	–	–	–	–	1.8	1.0	–	–
DGM 812-M	–	–	1.4	0.7	–	–	–	–	1.8	1.1	1.6	1.0	~7.0	8.0
DGM 928-M	–	–	–	–	–	–	–	–	1.8	1.2	1.6	1.0	~7.0	~8.2
DGM 932-M	–	–	–	–	1.8	1.0	1.9	1.1	1.9	1.1	–	–	~7.0	–
MNRJ 1365-V	1.6	1.0	1.5	0.9	1.9	1.2	–	–	–	–	–	–	–	–
MNRJ 1368-V	–	–	–	–	–	–	–	–	–	–	1.9	1.1	–	–
MNRJ 2347-V	–	–	–	–	–	–	–	–	–	–	1.9	1.1	–	–
MNRJ 2354-V	–	–	–	–	1.8	1.2	1.8	1.2	1.8	1.2	1.7	1.0	7.1	–
MNRJ 2355-V	–	–	–	–	2.2	1.3	–	–	–	–	–	–	–	–
MNRJ 2886-V	1.6	0.7	–	–	1.8	1.2	1.8	1.2	–	–	–	–	–	–
MNRJ 2887-V	–	–	1.6	0.8	–	–	2.0	1.3	–	–	–	–	–	
MNRJ 2889-V	–	–	–	–	–	–	–	–	1.8	1.2	1.7	1.1	–	

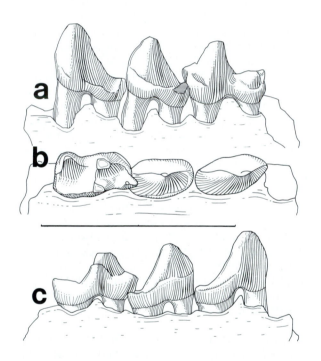

Fig. 30. Monodelphopsis travassosi Paula Couto 1952b. MNRJ 1365-V (type), a fragment of a left mandibular ramus with P$_2$-M$_2$ complete: a, labial; b, occlusal; c, lingual views. Scale = 5 mm.

to *Monodelphopsis travassosi* led me to reconsider aspects of structure of the lower dentition and hence of its possible affinities to pediomyines. As expected, there are indeed numerous pediomyine characters present in the lower dentition of this species, and these are identified and discussed in the PHYLOGENETICS section, below.

Procaroloameghinia Marshall 1982c

Procaroloameghinia Marshall 1982c:711.

Type: Procaroloameghinia pricei Marshall 1982c:711.

Diagnosis-Description: As for type and only known species.

Procaroloameghinia pricei Marshall 1982c

(Fig. 33; Table 12)

Procaroloameghinia pricei Marshall 1982c:711, Figure 1A-C.

Type: DGM 805-M, a fragment of a right mandibular ramus with alveoli of C-P$_2$ and complete P$_3$-M$_3$; and DGM 924-M, a fragment of a right mandibular ramus with M$_{4-5}$ (both specimens are of the same individual; figured by Marshall 1982c, Fig. 1A-C).

Hypodigm: Type only.

Diagnosis-Description: Medium size; dental formula ?/? ?/? ?/3 ?/5; mandibular ramus robust; symphysis extends posteriorly to point below anterior edge of P$_2$; C, judging from alveoli, weakly developed; premolars increase slightly in size from P$_1$ to P$_3$, with no diastems separating them from each other or from C or M$_2$; P$_3$ somewhat bulbous, of the same height as M$_2$ trigonid, and with low broad posterobasal heel; M$_{2-5}$ bunodont; in length M$_2$<M$_3$≈M$_4$>M$_5$, in width M$_2$<M$_3$≈M$_4$>M$_5$; anterobasal cingulum weakly developed, increasing slightly in relative size from M$_2$ to M$_5$; M$_5$ reduced in size relative to M$_{2-4}$; trigonids slightly narrower than talonids on M$_{2-4}$, subequal in width on M$_5$; trigonids only slightly higher than talonids; talonids distinctly basined, hypoconulid large and distinct on M$_{2-5}$; overall size intermediate between larger *Caroloameghinia mater* and smaller *C. tenuis*; compared to *C. mater*, trigonid cusps more distinct, trigonids absolutely higher in lateral view and larger in occlusal view relative to talonids, lower molars lack wrinkling of enamel especially in talonid basin, lack metastylid, and hypoconulid (especially on M$_5$) much smaller relative to rest of crowns (after Marshall 1982c:711, 713).

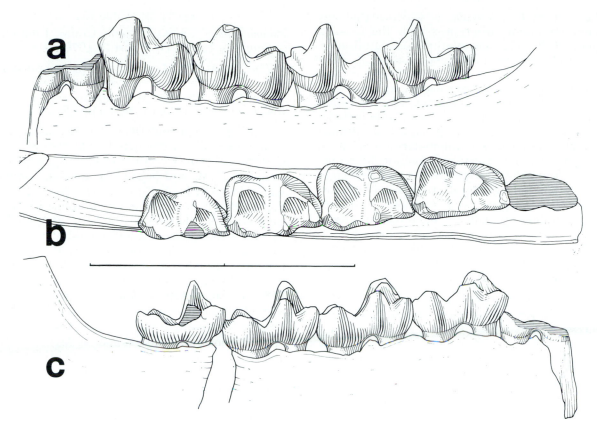

Fig. 31. *Monodelphopsis travassosi* Paula Couto 1952b. MNRJ 2354-V, a partial left mandibular ramus with base of P$_3$, and M$_{2-5}$ complete: a, labial; b, occlusal; c, lingual views. Scale = 5 mm.

Comments: Procaroloameghinia is an ideal structural ancestor for the Casamayoran genus *Caroloameghinia* Ameghino 1901 from southern Argentina. In the original description (Marshall 1982c) it was accordingly placed with *Caroloameghinia* in the subfamily Caroloameghiniinae.

Three species of *Caroloameghinia* are recognized: *C. mater* Ameghino 1901 (genotype); *C. intermedia* Marshall, de Muizon and Hoffstetter 1984; and *C. tenuis* Ameghino 1901. *Caroloameghinia mater* averages 20% larger in linear tooth dimensions than *C. intermedia; C. intermedia* is about the same size, or slightly larger in some dimensions, as *Procaroloameghinia pricei*; while *C. tenuis* averages 20% smaller than *C. intermedia* (for list of tooth dimensions see Marshall 1982c, Table 1; Marshall, de Muizon and Hoffstetter 1984:38). Unfortunately, direct comparison of many important aspects of dental morphology among these species is not yet possible. *C. intermedia* is known only from a partial right mandible with deeply worn P$_3$-M$_3$ (type, MNHN CAS 27); *C. tenuis* is best known by its type (MACN 10345), a left maxilla with M$^{2-5}$; *C. mater* is known from nearly complete lower dentitions and some isolated upper molars; while *Procaroloameghinia pricei* is known only from a partial right mandible with little worn P$_3$-M$_5$ (Fig. 33).

In overall size, *Procaroloameghinia pricei* compares best with *Caroloameghinia intermedia*. Yet,

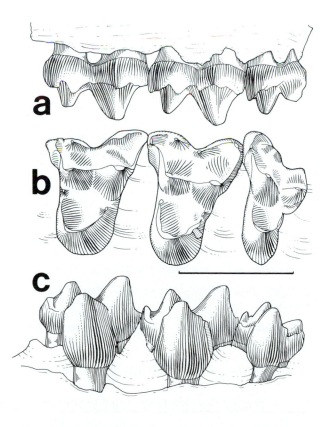

Fig. 32. *Monodelphopsis travassosi* Paula Couto 1952b. DGM 808-M, a left maxilla with M$^{3-5}$: a, labial; b, occlusal; c, lingual views. Scale = 2.5 mm.

the type of *C. intermedia* is so worn that any observed minor differences with the unworn teeth of *P. pricei* are of dubious significance. Relatively unworn lower dentitions of *C. mater* are available (see Marshall 1982c:710, Fig. 1D-I) and these permit meaningful comparison with *P. pricei*. "The primary changes required in a lineage leading from *P. pricei* to *C. mater* include: increase in size . . ., acquisition of distinct wrinkling of enamel on the molars (especially in talon and talonid basins) and a metastylid; reduction in distinction of trigonid cusps, and in height (in lateral view) and overall area (in occlusal view) relative to talonids; and increase in size of hypoconulid (especially on M_5) relative to rest of crown surface. Most importantly, there are no features in *P. pricei* which would negate such a proposed ancestral-descendant relationship" (Marshall 1982c:715).

Thus, the available evidence suggests that *Procaroloameghinia* is the Middle Paleocene ancestor of the Early Eocene genus *Caroloameghinia*. The exact phylogenetic relationships of the three species of *Caroloameghinia* are, however, unclear.

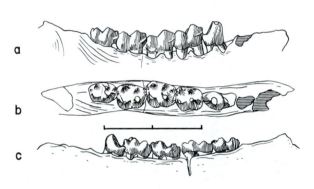

Fig. 33. *Procaroloameghinia pricei* Marshall 1982c. DGM 805-M (type), a fragment of a right mandibular ramus with alveoli of C-P_2, and with P_3-M_3 complete; and DGM 924-M, a fragment of a right mandibular ramus with M_{4-5} complete (both specimens are of same individual): a, labial; b, occlusal; c, lingual views. Scale = 10 mm (after Marshall 1982c, Fig. 1A-C).

Protodidelphis Paula Couto 1952b

Protodidelphis Paula Couto 1952b:5.

Type: *Protodidelphis vanzolinii* Paula Couto 1952b:6.

Diagnosis-Description: As for type and only known species.

Protodidelphis vanzolinii Paula Couto 1952b

(Figs 34-38; Table 13)

Protodidelphis vanzolinii Paula Couto 1952b:6, Figure 3; 1962:141, Figure 2; 1970:20.

Type: DGM 271-M, rostral and palatal portion of left side of skull with alveoli of P^1, roots of P^2, complete P^3-M^2 and M^5, and alveoli of M^{3-4} (figured by Paula Couto 1962, Fig. 2E).

Hypodigm: Type and DGM 250-M (paratype), greater part of a right mandibular ramus with posterior edge of C alveolus, roots of P_{1-2} and M_3, complete P_3-M_2 and M_4 and M_5 missing lingual side (figured by Paula Couto 1952b, Fig. 3; 1962, Fig. 2A-B); DGM 303-M, a right maxilla with M^{2-5} (figured by Paula Couto 1962, Fig. 2D); DGM 319-M, greater part of a left mandibular ramus with alveoli of last incisor and C, P_1 and P_3 present, and alveoli or roots of P_2 and M_{2-5} (figured by Paula Couto 1962, Fig. 2C); AMNH 49857 (= DGM 320-M), a nearly complete right mandibular ramus with alveolus of C, P_1-M_2 present (P_3 missing posterior edge), roots of M_{3-4} and anterior root and talonid of M_5; AMNH 49803 (= MNRJ 1370-V), an isolated right M_2; AMNH 49858, a fragment of a left mandibular ramus with P_2-M_3 complete; MNRJ 1371-V, an isolated left M_2; MNRJ 1427-V, an isolated right M^5; MNRJ 2897-V, a left upper molar and three right M^4's (*fide* Paula Couto); MNRJ 2898-V, an isolated left M_4?.

Diagnosis-Description: Large size; dental formula ?/? 1/1 3/3 5/5; mandibular ramus elongate, deep, massive; symphysis extends posteriorly to point below anterior edge of P_2; large mental foramina below posterior edge of P_1 and anterior end of M_2; upper and lower C, judging from alveoli, exceptionally large; lower premolars increase sharply in size from P_1 to P_3; P_1 tiny, two-rooted with roots aligned in same anteroposterior axis as rest of cheek teeth, separated from P_2 by large diastema, crown with primary cusp only; crowns of P_{2-3} with large trenchant, pointed primary cusp (which is considerably higher than protoconid of M_2) with tiny cuspule on antero-medial edge and well developed posterobasal heel, notably larger on P_3 than on P_2; in length $M_2<M_3<M_4>M_5$; in width $M_2<M_3<M_4 \approx M_5$; molars brachydont; anterobasal cingulum well developed; talonids broader than trigonids on all molars, more so on M_2, becoming less so to M_5; trigonids very low, only slightly higher than talonids; trigonids narrow anteroposteriorly, in occlusal view trigonid slightly (M_2) to markedly (M_5) smaller than talonid; trigonid cusps nearly equal in size and

Table 12. Measurements of lower cheek teeth of *Procaroloameghinia pricei*.

Specimen	P_3 L	W	M_2 L	W	M_3 L	W	M_4 L	W	M_5 L	W	M_{2-5} L	P_3-M_5 L
DGM 805-M+924-M	2.0	1.6	2.5	2.2	2.7	2.4	2.7	2.4	2.6	1.9	11.0	13.0

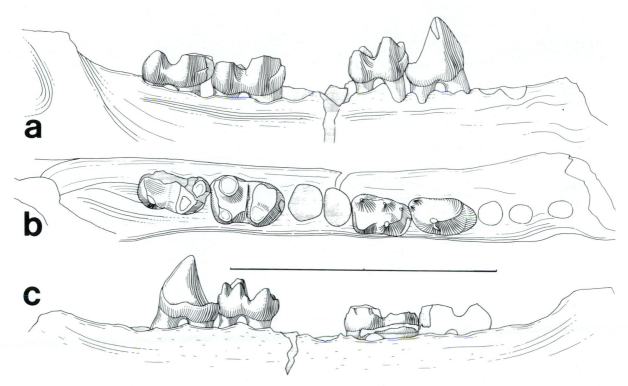

Fig. 34. Protodidelphis vanzolinii Paula Couto 1952b. DGM 250-M, greater part of a right mandibular ramus with posterior edge of C alveolus, roots of P_{1-2} and M_3, P_3-M_2 and M_4 complete, M_5 missing lingual side: a, labial; b, occlusal; c, lingual views. Scale = 20 mm.

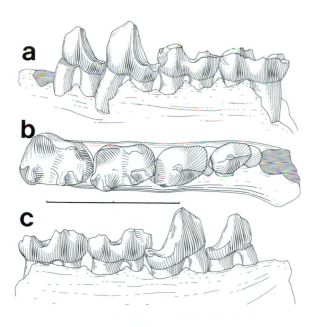

Fig. 35. Protodidelphis vanzolinii Paula Couto 1952b. AMNH 49858, a fragment of a left mandibular ramus with P_2-M_3 complete: a, labial; b, occlusal; c, lingual views. Scale = 10 mm.

height, although protoconid (is slightly) >metaconid>paraconid; talonid large, basined; hypoconid and entoconid very large, only slightly smaller than trigonid cusps, and subequal in size; hypoconulid very reduced, represented only by tiny cuspule immediately posterior to entoconid; upper premolars increase sharply in size from

P^1 to P^3; P^1 tiny, separated from P^2 by large diastema; P^3 large, proodont, ovoid in occlusal view, primary cusp extends considerably below occlusal surface of molars, with only hint of posterobasal cingulum; in length $M^2<M^3<M^4>M^5$, in width $M^2<M^3<M^4\approx M^5$; molars rounded, cusps bulbous and all about subequal in height, although paracone is slightly lower; protocone large; only trace of para- and metaconules; metacone about twice as large and only slightly higher than paracone; paracrista very reduced, extending labially to between stylar cusps A and B; paracingulum well developed; stylar shelf broad; stylar cusps B and D very large (B slightly larger than D), subequal in size to paracone; no trace of stylar cusp C; E weakly developed or absent; in size B>D>A>E; no trace of ectoflexus.

Comments: Paula Couto (1952b:6) noted that of the living didelphids, *Protodidelphis* showed closest affinity in size and tooth structure to *Didelphis*, and that the two ". . . may be placed in the same phyletic line." He thus regarded *Protodidelphis* as the Paleocene ancestor of *Didelphis*. Later, Paula Couto (1962:141) stated that *Protodidelphis* was more omnivorous than *Didelphis*.

Clemens (1966:20) commented that comparison of the molars of *Protodidelphis* and *Alphadon rhaister* from the Late Cretaceous of North America ". . . reveals gross similarities in both shape and morphology of stylar cusps".

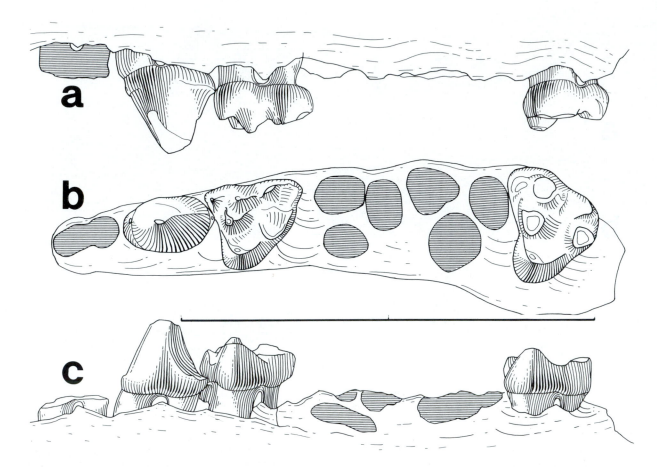

Fig. 36. Protodidelphis vanzolinii Paula Couto 1952b. DGM 271-M (type), rostral and palatal portion of left side of skull with alveoli of P^1, roots of P^2, P^3-M^2 and M^5 complete, and alveoli of M$^{3-4}$: a, labial; b, occlusal; c, lingual views. Scale = 20 mm.

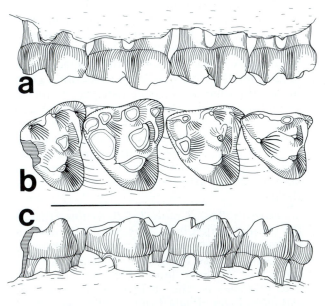

Fig. 37. Protodidelphis vanzolinii Paula Couto 1952b. DGM 303-M, a right maxilla with M$^{2-5}$: a, labial; b, occlusal; c, lingual views. Scale = 10 mm.

Sigé (1972:397) noted marked similarity in the dentitions of *Protodidelphis* and *Glasbius* from the Late Cretaceous of North America. Crochet (1980:222) was also impressed by these similarities

and tentatively placed *Protodidelphis* in the "didelphid" subfamily "Glasbiinae". This arrangement was followed by Reig *et al.* (1985, 1987).

Robertbutleria gen. nov.

Etymology: Named in honour of Robert F. Butler in recognition of his numerous contributions to the knowledge of the magnetostatigraphy and geochronology of the mammal-bearing Cenozoic of South America (e.g. Marshall, Butler, Drake and Curtis 1981).

Type: Robertbutleria mastodontoidea sp. nov.

Diagnosis-Description: As for type and only known species.

Robertbutleria mastodontoidea sp. nov.
(Fig. 39)

Etymology: mastodontoidea is given in reference to the mammary-like appearance of stylar cusps B and D in particular, and to those of the trigon cusps in general.

Type: DGM 896a-M, an isolated left M^2.

Hypodigm: Type and DGM 896b-M, an isolated left M^3.

Measurements: M^2 L=4.4 mm, W=4.6 mm; M^3 L=5.0 mm, W=5.1 mm.

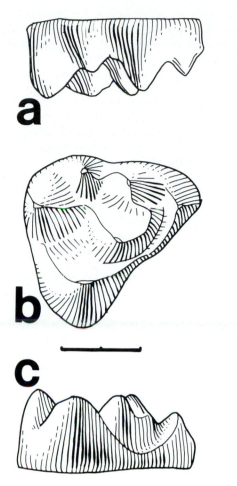

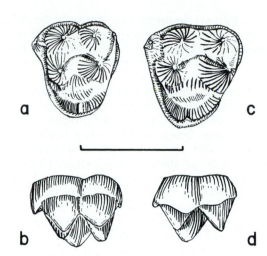

Fig. 39. *Robertbutleria mastodontoidea* n. gen., n. sp. a-b, DGM 896a-M (type), an isolated left M²; and c-d, DGM 896b-M, an isolated left M³: a + c, occlusal; b + d, labial views. Scale = 5 mm.

Fig. 38. *Protodidelphis vanzolinii* Paula Couto 1952b. MNRJ 1427-V, an isolated right M⁵: a, labial; b, occlusal; c, lingual views. Scale = 2.5 mm.

Diagnosis-Description: Large size; dental formula unknown; molars distinctly ovoid in occlusal view, slightly wider than long, cusps large, rounded, bulbous; dentition markedly bunodont; enamel wrinkled, especially on M^3; M^{2-3} in occlusal view with protocone about twice as large as metacone, metacone about twice as large as paracone; in lateral view, metacone higher than paracone, protocone and metacone about same height; hint of para- and metaconule on M^{2-3}; no trace of centrocrista; paracrista short, rounded, joins with anterior edge of stylar cusp B; talon basin virtually absent, position filled by swollen labial portion of protocone; distinct basal cingulum extends almost continuously from stylar cusp A along anterior surface of tooth, around protocone to point below posterior edge of metacone; cingulum stronger on M^3, and on M^{2-3} it is absent between posterior edge of protocone and medial point of metacone; stylar shelf broad, stylar cusps B and D very large, about size of paracone, with B slightly larger in occlusal view and higher in lateral view than D; A small; E very small and connected to metacone by short metacrista; no trace of cusp C; in size B>D>A>E; a weak shelf extends labial to bases

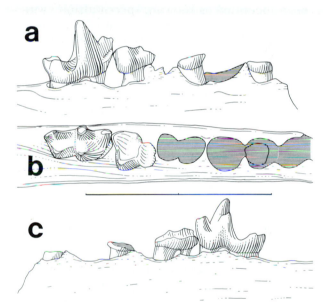

Fig. 40. *Sternbergia itaboraiensis* Paula Couto 1970. MNRJ 2892-V (type), a fragment of a right mandibular ramus with roots of M_3 and parts of crowns of M_{4-5}: a, labial; b, occlusal; c, lingual views. Scale = 5 mm.

of stylar cusps B and D, and connects stylar cusps A and E; slight hint of ectoflexus at point between stylar cusps B and D on M^2, no trace of ectoflexus on M^3; M^2 slightly smaller than M^3; similar in size and general structure to *Protodidelphis vanzolinii*, but differs from that species in having upper molars more ovoid in occlusal view, protocone set more mediolingual on tooth and not inflected anteriorly, cusps more bulbous, bunodont, presence of labial and lingual cingula, and a less well developed metacrista.

Comments: Robertbutleria mastodontoidea shares many similarities with *Glasbius intricatus* Clemens 1966 from the Late Cretaceous of North America (see Clemens 1966, Figs 14, 16). In fact, the similarities are so numerous that only the differences warrant comment. These taxa differ in *Robertbutleria* being about 60% larger in linear tooth dimensions, having smaller para- and metaconules, lacking a stylar cusp C (although one is present in *G. intricatus*, see Clemens 1966), having slightly stronger and more nearly continuous cingulum around lingual side of crown, cusps being slightly more bulbous and rounded, and in centrocrista, metacrista and paracrista being absent or less well developed. In most part these differences represent degree of development or relative expression of features found in both taxa; while only the absence of a centrocrista in *Robertbutleria* appears to represent a significant difference between these two taxa. Thus, in the vast majority of comparable features these taxa are virtually indistinguishable, and *Glasbius* represents an ideal structural ancestor for *Robertbutleria*.

Two other fossil taxa from South America have been identified as showing special affinity with *Glasbius*. *Protodidelphis* Paula Couto 1952b from Itaborai was tentatively referred by Crochet (e.g. 1980) to the "Glasbiinae", and *Reigia* Pascual 1983 of Middle or Late Eocene age was assigned securely by Pascual (1983) to this subfamily. These taxa are discussed, compared, and their familial and subfamilial affinity reassessed in the PHYLOGENETICS section, below.

Sternbergia Paula Couto 1970

Sternbergia Paula Couto 1970:30.

Type: Sternbergia itaboraiensis Paula Couto 1970:31.

Diagnosis-Description: As for type and only known species.

Sternbergia itaboraiensis Paula Couto 1970

(Figs 40-44; Table 14)

Sternbergia itaboraiensis Paula Couto 1970:31, Figure 8.

Type: MNRJ 2892-V, a fragment of a right mandibular ramus with roots of M_3 and parts of crowns of M_{4-5} (figured by Paula Couto 1970, Fig. 8).

Hypodigm: Type and DGM 803-M, a left maxilla with roots of P^3 and M^4, and complete M^{2-3} and M^5; DGM 807-M, a left maxilla with roots of P^2-M^2, and complete M^{3-5}; DGM 812-M, a left mandibular ramus with complete P_3 and M_{4-5}, and alveoli of rest of dentition; DGM 920-M, a left mandibular ramus with roots of P_{1-2} and M_2, and complete P_3 and M_{3-5}; DGM 927-M, a fragment of a right mandibular ramus with P_2-M_4 complete, P_3 erupting.

Diagnosis-Description: Small size; dental formula ?/4 ?/1 ?/3 5/5; mandibular ramus moderately developed; symphysis extends posteriorly to point below P_2; mental foramina below P_1 and M_2; C, judging from alveolus, moderately developed; P_1, judging from roots, considerably smaller than P_{2-3}, set at slight oblique angle in jaw; P_{2-3} subequal in size and structure, with prominent central cusp and small cuspate posterobasal heel; in length $M_2 < M_3 < M_4 > M_5$, in width $M_2 < M_3 \approx M_4 > M_5$; trigonids well elevated above talonids, and subequal in size in occlusal view; anterobasal cingulum moderately developed, and on M_{2-4} it continues posteriorly along labial side of tooth and connects with moderately developed cingulum on posterior surface of hypoconid; in size protoconid > metaconid > paraconid; talonid well developed and basined; entoconid larger and higher than hypoconid; hypoconulid small on M_{2-4}; on M_5 hypoconulid, entoconid and hypoconid subequal in size, first two considerably higher than latter; in length $M^2 \approx M^3 > M^4 > M^5$, in width $M^2 < M^3 \approx M^4 > M^5$; protocone large, talon basined; metacone about twice as large and higher than paracone; hint of paraconule on M^2 only, metaconule distinct on M^{2-4}; paracingulum well developed; centrocrista weak, apex pointing labially; paracrista weak, connects with anterior edge of stylar cusp B; metacrista moderately developed; stylar shelf broad; stylar cusp B largest, A and C smaller, D next smallest, E virtually absent; in size B > C > A > D > E; ectoflexus absent on M^2, shallow on M^{3-4}; talon of M^5 not basined.

Comments: Paula Couto (1970:31) regarded *Sternbergia* as "closer" to *Monodelphopsis* ". . . than to anyone of the other Itaboraian genera of didelphids." A "*Sternbergia*-like" taxon is reported from the Late Cretaceous (Maestrichtian) age El Molino Formation of Bolivia (de Muizon *et al.* 1984).

Zeusdelphys gen. nov.

Etymology: Zeus-, Greek god of gods, in this context to mean the largest and most important of; -delphys, in reference to the relationship of this taxon with "opossum-like" marsupials.

Type: Zeusdelphys complicatus sp. nov.

Diagnosis-Description: Same as type and only known species.

Zeusdelphys complicatus sp. nov.

(Fig. 45)

Etymology: complicatus, name given in reference to the complicated, wrinkled appearance of the enamel.

Type: DGM 896a-M, an isolated right M^4.

Hypodigm: Type only.

Measurements: L = 8.3 mm, W = 7.6 mm.

Diagnosis-Description: Very large size; dental formula unknown; M^4 L > W, brachydont,

Table 13. Measurements of cheek teeth of *Protodidelphis vanzolinii*.

Specimen	P2 L	P2 W	P3 L	P3 W	M2 L	M2 W	M3 L	M3 W	M4 L	M4 W	M5 L	M5 W	M2-4 / M2-5 L	P3-M4 / P3-M5 L
Upper cheek teeth														
DGM 271-M	~3.8	–	4.2	2.8	4.6	4.2	–	–	–	–	4.4	5.8	~14.8	~19.0
DGM 303-M	–	–	–	–	4.5	4.3	5.0	5.2	5.2	5.8	4.4	5.9	15.0	–
MNRJ 1427-V	–	–	–	–	–	–	–	–	–	–	4.2	5.5	–	–
Lower cheek teeth														
AMNH 49803	–	–	–	–	4.8	3.3	–	–	–	–	–	–	–	–
AMNH 49857*	3.5	2.0	–	–	4.8	3.0	–	–	–	–	–	3.5	19.6	~24.5
AMNH 49858	3.5	2.4	4.4	3.0	4.8	3.3	5.3	4.0	–	–	–	–	–	–
DGM 250-M	–	–	4.2	2.8	4.6	2.9	–	–	5.0	3.7	5.3	–	20.3	24.7
DGM 319-M	–	–	4.2	2.6	–	–	–	–	–	–	–	–	–	–
MNRJ 1371-V	–	–	–	–	4.8	3.1	–	–	–	–	–	–	–	–
MNRJ 2898-V	–	–	–	–	–	–	–	–	4.6	3.4	–	–	–	–

*P_1 L = 2.1 mm, W = 1.3 mm.

Table 14. Measurements of cheek teeth of *Sternbergia itaboraiensis*.

Specimen	P2 L	P2 W	P3 L	P3 W	M2 L	M2 W	M3 L	M3 W	M4 L	M4 W	M5 L	M5 W	M2-4 / M2-5 L	P3-M5 L
Upper cheek teeth														
DGM 803-M	–	–	–	–	1.8	1.7	1.8	2.1	–	–	1.0	2.2	~5.0	–
DGM 807-M	–	–	–	–	–	–	1.7	1.9	1.5	1.9	0.8	1.7	–	–
Lower cheek teeth														
DGM 812-M	–	–	1.5	0.6	–	–	–	–	1.9	1.2	1.8	1.0	–	8.1
DGM 920-M	1.6	–	1.7	0.6	–	–	2.0	1.1	2.1	1.2	1.9	1.0	~8.9	9.5
DGM 927-M	1.7	0.7	–	–	1.7	1.0	1.8	1.2	1.9	1.2	–	–	–	–
MNRJ 2892-V	–	–	–	–	–	–	–	–	–	–	1.7	0.9	–	–

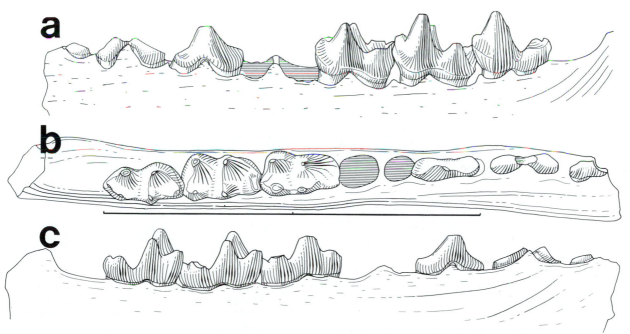

Fig. 41. Sternbergia itaboraiensis Paula Couto 1970. DGM 920-M, a left mandibular ramus with roots of P_{1-2} and M_2, and with P_3 and M_{3-5} complete: a, labial; b, occlusal; c, lingual views. Scale = 10 mm.

enamel heavily wrinkled; protocone very large, talon somewhat basined but position filled by swollen labial side of protocone; metacone about three times as large and higher than paracone; no trace of para- or metaconule; paracingulum small but distinct; centrocrista very weak, apex pointing labially; paracrista short, rounded, connects with stylar cusp A; metacrista short, rounded, virtually absent; stylar shelf broad; stylar cusps B and D enormous, subequal in size; A and E smaller than B and D, no trace of C; in size B≈D>A≈E; ectoflexus very shallow; small shelf-like tubercle on posterolabial face of metacone.

Comments: Earlier I (Marshall 1984:175, Fig. 4) tentatively referred this tooth to *Eobrasilia coutoi*,

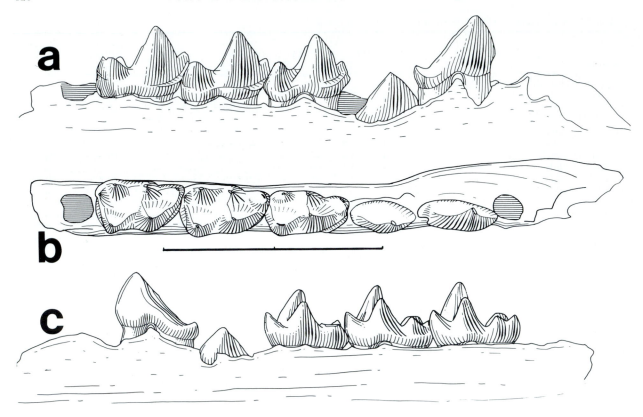

Fig. 42. Sternbergia itaboraiensis Paula Couto 1970. DGM 927-M, a fragment of a right mandibular ramus with P_2-M_4 complete (P_3 erupting): a, labial; b, occlusal; c, lingual views. Scale = 5 mm.

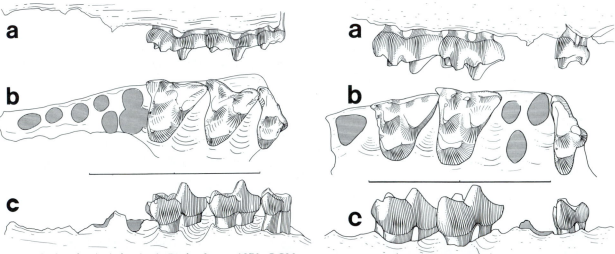

Fig. 43. Sternbergia itaboraiensis Paula Couto 1970. DGM 807-M, a left maxilla with roots of P^2-M^2, and M^{3-5} complete: a, labial; b, occlusal; c, lingual views. Scale = 5 mm.

Fig. 44. Sternbergia itaboraiensis Paula Couto 1970. DGM 803-M, a left maxilla with roots of P^3 and M^4, and with M^{2-3} and M^5 complete: a, labial; b, occlusal; c, lingual views. Scale = 5 mm.

although noted that if it ". . . is not referrable to *Eobrasilia* then it clearly represents a new and distinct taxon, for based solely on its size it cannot possibly be referred to any other known fossil or living didelphoid" (*ibid.*:176). During the course of this study and based on additional knowledge of the complementary structure of upper and lower dentitions of Itaborai taxa, it became apparent that this tooth and the lower dentition of *Eobrasilia* are not structurally compatible. For example, *Gaylordia* and *Didelphopsis* have large

bulbous P2's and P3's as in *Eobrasilia*, but the upper molars either definitely (*Didelphopsis*) or apparently (*Gaylordia*) have reduced stylar cusps. On the other hand, taxa which definitely have large stylar cusps B and D (i.e. *Protodidelphis*) do not have large crushing P2's or P3's in their surely referred lower dentitions. Thus, DGM 896a-M represents a taxon other than *Eobrasilia*, and for this reason I have made it the type of a new genus and species. As will be demonstrated in the

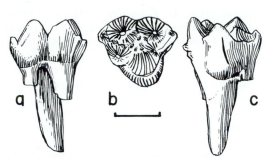

Fig. 45. *Zeusdelphys complicatus* n. gen., n. sp. DGM 896a-M (type), an isolated right M^4: a, labial; b, occlusal; c, lingual views. Scale = 5 mm (after Marshall 1984, Fig. 4).

PHYLOGENETICS section below, the features represented in *Eobrasilia* and *Zeusdelphys* indicate that they are members of different suborders. I have also reconsidered the identity of this tooth and now believe that it almost certainly represents an M^4.

Marsupialia *incertae sedis*

The specimens listed and/or described above are among the most complete and most diagnostic of the material yet known from Itaborai. In fact, all specimens represented by multi-toothed jaws, as well as some isolated teeth and one-toothed jaws, are referred to one of the 17 named species. There are, however, several hundred isolated teeth and one-toothed jaws in the DGM collection that remain to be identified, and most of these are as yet uncatalogued. The identity of all these specimens was not, however, an objective of this study. It is very probable that new taxa are represented among these specimens. This probability is demonstrated by the fact that one named taxon (*Zeusdelphys*) is known from only one upper molar, while another (*Robertbutleria*) is known by only two isolated upper molars. Two other taxa (*Guggenheimia, Procaroloameghinia*) are known only by their types, partial multi-toothed lower jaws. At the very least, some of these unidentified specimens will surely be shown to represent elements presently unknown in named taxa.

Paula Couto (opt. cit.) described numerous fragmentary specimens as Didelphidae indeterminate or Didelphinae *incertae sedis*. In his 1952b paper (p. 25) he compared two specimens (MNRJ 1367-V, MNRJ 1368-V; both fragments of right mandibular rami with M_5), apparently of the same unnamed taxon, with living species of *Marmosa*. MNRJ 1367-V he noted as showing special affinity with *Marmosa elegans*. MNRJ 1368-V I refer above to *Monodelphopsis travassosi*.

In 1962, Paula Couto described five additional specimens and figured two. MNRJ 2352-V, a fragment of a left maxilla with M^2, he figured (1962, Fig. 14E) and suggested that it may belong to a taxon already named on the basis of lower

dentitions only. MNRJ 2505-V, described as an enormous right P_3, he compared primarily with *Protodidelphis (sic,* he almost certainly meant *Didelphopsis*), but suggested possible affinity with *Eobrasilia*. Given the measurements of this specimen and its description as provided by Paula Couto, it is almost certainly referrable to *Didelphopsis cabrerai* (see *Comments* under *Eobrasilia coutoi*). A third specimen (MNRJ 2506-V, a left maxilla with M^{2-4}) he suggested may represent the upper dentition of *Mirandatherium* or *Monodelphopsis*. Above, I refer this specimen to *Mirandatherium alipioi*. The fourth and fifth specimens apparently represented the same species and included MNRJ 2351-V, a fragment of a right mandibular ramus with M_5 (see Paula Couto 1962, Fig. 14A-D), and MNRJ 2353-V, a fragment of a right mandibular ramus with M_4 present and alveoli of M_3 and M_5. Both he compared with living species of *Monodelphis*.

In 1970, Paula Couto described and measured additional specimens and noted the existence of many others. However, he did not compare any of these with named taxa. These include MNRJ 2893-V, a crown of a right lower molar; MNRJ 2894a-V, two right upper molars, apparently M^{2-3}; MNRJ 2894b-V, a left M^5; MNRJ 2894c-V, a small right lower molar; MNRJ 2895-V, 94 fragmentary lower jaws of small size without teeth; MNRJ 2896-V, three fragmentary lower jaws with one tooth each; DGM 643-M, a fragment of a left maxilla with M^3 and alveoli of M^2 and M^{4-5}; and DGM 653-M, a fragment of a left mandibular ramus with a premolar (P_3?) and perhaps a deciduous tooth.

PHYLOGENETICS

1) INTERRELATIONSHIPS OF ITABORAI TAXA

Sixteen genera and 17 species of Itaborai "opossum-like" taxa are recognized: *Bobbschaefferia fluminensis, Derorhynchus singularis, Didelphopsis cabrerai, Eobrasilia coutoi, Gaylordia doelloi, G. macrocynodonta, Guggenheimia brasiliensis, Itaboraidelphys camposi, Marmosopsis juradoi, Minusculodelphis minimus, Mirandatherium alipioi, Monodelphopsis travassosi, Procaroloameghinia pricei, Protodidelphis vanzolinii, Robertbutleria mastodontoidea, Sternbergia itaboraiensis,* and *Zeusdelphys complicatus.*

These taxa vary greatly in absolute size and in aspects of dental morphology (Figs 46, 47). The absolute size distribution of various species are indicated by comparison of length versus width of P_3 (Fig. 48), length versus width of M_3 (Fig. 49), length of P_{1-3} (Fig. 50), and length of M_{2-5} (Fig. 51). These plots demonstrate the existence of a nearly continuous size gradient between the smallest (*Minusculodelphis minimus*) and the largest (*Eobrasilia coutoi*) species. These plots also show considerable overlap among small to medium size taxa, a feature indicating that size alone is insufficient to separate and distinguish some species.

As a starting point for comparing, contrasting, and ultimately inferring the affinities of these taxa, I provide a character analysis of some dental differences and similarities. I attempt to identify the primitive and derived states for each character, and to this end I follow with only minor degression the polarities recognized by Reig *et al.* (1987). This approach negates unnecessary repetition of aspects of that study and permits integration of the results generated in this study into that taxonomically more comprehensive work. I list 17 characters (not all and some in addition to those used by Reig *et al.* 1987), and for each list its probable primitive state then identify the distribution of character states among the Itaborai taxa. In a few cases I discuss alternative polarities for characters based on work done by colleagues during or subsequent to completion of the Reig *et al.* study. The abbreviation *ps* is used to denote the probable primitive state. Table 15 summarizes the conclusions of the following character analysis. Characters 7, 16 and 17 represent branching features, while the others are regarded as linear (i.e. are ordered series representing trends).

1). *Centrocrista* (*ps* rectilinear or predilamb-dodont, crista straight): The only two genera which retain the primitive state are *Miranda-therium* and especially *Monodelphopsis*. All of the other taxa known for this feature are derived (1, Table 15) in having a V-shaped centrocrista (didelphoid dilambdodont state) with the vertex directed labially toward the stylar shelf. As will become clear below, the derived state is not uniquely derived among those taxa possessing this state, but apparently was attained at least two times among South American groups.

2). *Para- and metaconules* (*ps*, both kinds of conules present): These conules are present, although in some cases are only weakly developed, in all taxa except *Sternbergia* in which paraconules are absent on M^{3-4} (1, Table 15) and *Zeusdelphys* which lacks both kinds of conules (2, Table 15). A consideration of other characters below demonstrates that this is apparently not a simple linear series, but that reduction of the conules occurred independently in these two taxa.

3). *Relative size of M^{3-4}* (*ps*, M^3 longer or subequal in length to M^4): Only *Protodidelphis*, of the taxa known for this feature, is derived in having M^3 shorter than M^4 (1, Table 15).

4). *Relative size of para- and metacones on M^{2-4}* (*ps*, subequal in size): All of the taxa are derived (1, Table 15) in having the metacone about twice as large as the paracone in occlusal view and higher

in lateral view. In *Zeusdelphys* the metacone is about three times as large as the paracone. Fox (1984) recently proposed that ancestral marsupials had the metacone larger and higher than the paracone. If so, then all the Itaborai taxa are primitive and not derived in this character. I follow Reig *et al.* (1987) for consistency and because the polarity recognized by those workers is also the one followed by Clemens (1966).

5). *Gross molar structure* (*ps*, molars tribosphenic, trigonid well elevated above talonid): Six taxa are derived (1, Table 15) in being brachydont (*Bobbschaefferia, Guggenheimia, Mirandatherium, Monodelphopsis, Protodidelphis, Zeusdelphys*), and two are further derived (2, Table 15) in being distinctly bunodont (*Procaroloameghinia, Robertbutleria*). This distinction is somewhat arbitrary and *Protodidelphis* can just as well be regarded as bunodont, but less so than the other two taxa.

6). *Stylar shelf (ps,* well developed, broad): All of the taxa are primitive in this character except *Mirandatherium* (1, Table 15) in which the shelf is only moderately developed, and *Monodelphopsis* (2, Table 15) in which it is very narrow.

7). *Stylar cusps* (*ps*, probably only four stylar cusps — A, B, D, E — with B largest, D next largest, A and E smaller, C absent): Clemens (1979) and Fox (1984) believe that stylar cusp C was not present in the ancestral marsupial, and this view is followed here. However, some earlier studies (e.g. Clemens 1966, Marshall 1979) were made with the belief that stylar cusp C was primitively present in the ancestral marsupial. In all the Itaborai taxa, stylar cusp B is large and is either the largest stylar cusp or is subequal in size to D. The other stylar cusps are all smaller than B, and C is present in some taxa but absent in others. All of the taxa show some departure from the inferred primitive state, and these can be grouped into a tripartite branching sequence (7a, 7b, 7c, Table 15).

Character 7a derived state 1 includes those taxa (*Didelphopsis, Itaboraidelphys, Marmosopsis, Mirandatherium, Sternbergia*) which most closely approximate the primitive state but have a small stylar cusp C. They all have B as the largest cusp, D is large but smaller than B, while A, C and E are smaller than D. Derived state 2 (*Monodelphopsis*) has all the stylar cusps reduced, although B is still largest followed by A, then D, then E on M^3; C is absent on M^3, present on M^4.

Character 7b derived state 1 (*Protodidelphis, Robertbutleria, Zeusdelphys*) has large to enormous stylar cusps B and D (B is either slightly larger than or subequal in size to D), while A and E are much smaller and subequal in size; there is no trace of C.

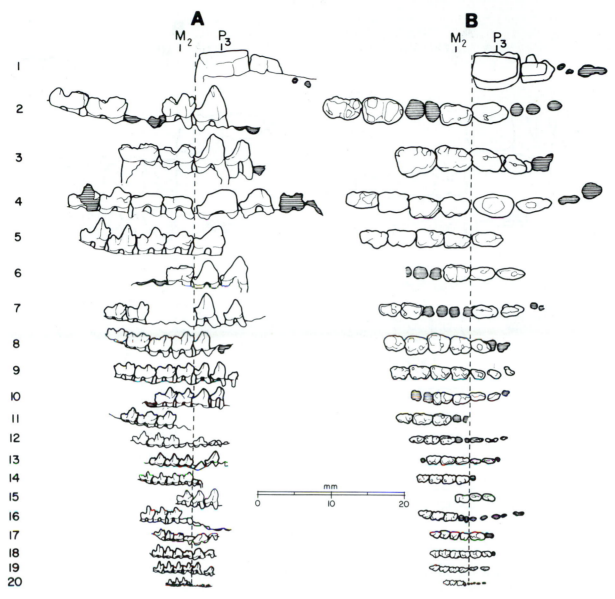

Fig. 46. Comparison of lower dentitions of various Itaborai taxa showing relative size and proportions of teeth. All illustrations are drawn to same scale, with largest size taxa at top and smallest at bottom. A, labial and B, occlusal views. 1, *Eobrasilia coutoi;* 2-3, *Protodidelphis vanzolinii;* 4, *Didelphopsis cabrerai;* 5-6, *Itaboraidelphys camposi;* 7, *Bobbschaefferia fluminensis;* 8, *Procaroloameghinia pricei;* 9, *Guggenheimia brasiliensis;* 10-11, *Mirandatherium alipioi;* 12-13, *Sternbergia itaboraiensis;* 14-15, *Monodelphopsis travassosi;* 16, *Bobbschaefferia fluminensis;* 17, *Gaylordia macrocynodonta;* 18, *Gaylordia doelloi;* 19, *Marmosopsis juradoi;* 20, *Minusculodelphis minimus.*

Character 7c derived state 1 includes only *Sternbergia* (also a character 7a derived state 1 taxon) which is uniquely derived in having stylar cusp B largest, C next largest, followed by A, D and E.

A schematic representation of the character 7 branching sequence is as follows:

character 7a ⊂②④⊃ character 7c
⊂①③⊃ character 7b

8). *Paracrista (ps,* moderately to well developed, united to stylar cusp A or B): The primitive state occurs in *Didelphopsis, Itaboraidelphys, Marmosopsis*

and *Monodelphopsis* (i.e. paracrista well developed, blade-like). In derived state 1 (*Mirandatherium, Sternbergia*) the paracrista is weakly developed but blade-like; in derived state 2 (*Protodidelphis, Robertbutleria, Zeusdelphys*) the paracrista is short, rounded and virtually absent (Table 15). This character may represent a branching and not linear sequence.

9). *Metacrista (ps,* moderately developed, blade-like): In derived state 1 (*Monodelphopsis*) the metacrista is short and blade-like; in derived state 2 (*Protodidelphis, Robertbutleria, Zeusdelphys*) the metacrista is very short, rounded and definitely not blade-like (Table 15). This character may also represent a branching and not linear sequence.

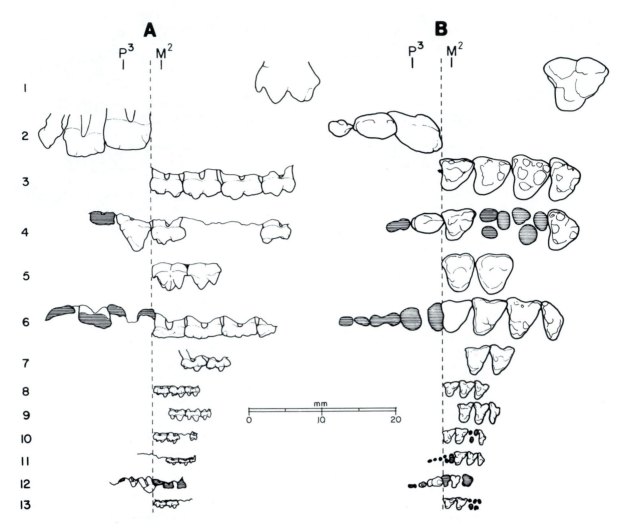

Fig. 47. Comparison of upper dentitions of various Itaborai taxa showing relative size and proportions of teeth. All illustrations are drawn to same scale, with largest size taxa at top and smallest at bottom. A, labial and B, occlusal views. 1, *Zeusdelphys complicatus;* 2, *Eobrasilia coutoi;* 3-4, *Protodidelphis vanzolinii;* 5, *Robertbutleria mastodontoidea;* 6, *Didelphopsis cabrerai;* 7, *Itaboraidelphys camposi;* 8, *Mirandatherium alipioi;* 9, *Monodelphopsis travassosi;* 10-11, *Sternbergia itaboraiensis;* 12, *Gaylordia macrocynodonta;* 13, *Marmosopsis juradoi.*

10). *Relative length of* M_4 *and* M_5 *(ps,* M_5 *subequal in length or longer than* M_4*):* The following taxa are derived in consistently having M_4 longer than M_5 — *Didelphopsis, Minusculodelphis, Mirandatherium, Monodelphopsis, Procaroloameghinia, Protodidelphis* and *Sternbergia.*

11). *Relative width of* M_4 *and* M_5 *(ps,* M_4 *and* M_5 subequal in width):* The following taxa are derived (1, Table 15) in having M_5 narrower than M_4 — *Bobbschaefferia, Derorhynchus, Didelphopsis, Gaylordia, Guggenheimia, Itaboraidelphys, Minusculodelphis, Mirandatherium, Monodelphopsis, Procaroloameghinia,* and *Sternbergia.*

12). *Relative size and development of hypo- and entoconids on* M_{2-4} *(ps, both distinctly developed, but not large and low relative to trigonid cusps):* Three taxa (*Mirandatherium, Monodelphopsis, Procaroloameghinia*) have *large* subequal sized hypo- and entoconids (derived state 1, Table 15). Three other taxa (*Bobbschaefferia, Guggenheimia, Protodidelphis*) have these conids *very large,*

subequal in size, bulbous, and only slightly lower than trigonid cusps (especially latter two genera; derived state 2, Table 15). The internal swelling of these conids in the latter three taxa results in loss of definition and decrease in depth of the talonid basin.

13). *Shape of* P_3 *(ps, well developed, trenchant, narrow transversely):* In derived state 1 (*Mirandatherium, Monodelphopsis, Procaroloameghinia*) P_3 is somewhat inflated and ovoid in occlusal view. In derived state 2 (*Guggenheimia, Protodidelphis*) P_3 is inflated, rounded in occlusal view with a broad posterobasal heel and crowns with anterior end directed labially. In derived state 3 (*Didelphopsis, Eobrasilia, Gaylordia*) P_3 is enormous, bulbous, and has a distinct crushing function (Table 15). As will be evident below, derived states 1 and 2 represent a linear sequence, while derived state 3 evolved directly and independently from a primitive state condition. This reality is apparent only when other characters are taken into consideration.

Table 15. Distribution of characters and states among Itaborai taxa. Character numbers (1-17) follow those listed in text. Missing data are indicated by blanks, primitive state by 0, derived states by numbers (1, 2, 3). Characters 7, 16 and 17 represent branching features; the other characters are here regarded as linear. The taxa are recorded by "groups" as recognized in text.

Character	Ancestor	Robertbutleria	Zeusdelphys	Procaroloameghinia	Bobbschaefferia	Guggenheimia	Protodidelphis	Mirandatherium	Monodelphopsis	Itaboraidelphys	Marmosopsis	Sternbergia	Didelphopsis	Eobrasilia	Gaylordia	Minusculodelphis	Derorhynchus
1. Centrocrista	0	1	1				1	0	0	1	1	1	1				
2. Para- and metaconules	0	0	2				0	0	0	0	0	1	0				
3. Size M^{3-4}	0	0					1	0	0	0	0	0	0				
4. Para- and metacones	0	1	1				1	1	1	1	1	1	1				
5. Molar structure	0	2	1	2	1	1	1	1	1	0	0	0	0	0	0	0	0
6. Stylar shelf	0	0	0				0	1	2	0	0	0	0				
7a. Stylar cusps	0	0	0				0	1	2	1	1	1	1				
7b. Stylar cusps	0	1	1				1	0	0	0	0	0	0				
7c. Stylar cusps	0	0	0				0	0	0	0	0	1	0				
8. Paracrista	0	2	2				2	1	0	0	0	1	0				
9. Metacrista	0	2	2				2	0	1	0	0	0	0				
10. Length M_4 and M_5	0			1	0	0	1	1	1	0	1	1	1		0	1	0
11. Width M_4 and M_5	0			1	1	1	0	1	1	1	0	1	1		1	1	1
12. Hypo- and entoconids	0			1	2	2	2	1	1	0	0	0	0	0	0	0	0
13. Shape P_3	0			1	0	2	2	1	1	0	0	0	3	3	3	0	0
14. Length P_2 and P_3	0			0	1	1	0	0	2	0	1	1	0	0	0	0	0
15. Width P_2 and P_3	0			0	0	0	0	0	2	0	1	1	0	0	0	0	0
16a. Structure I_1-C	0			0	0	0	0	0	0	0	0	0	0	0	0	1	1
16b. Structure I_1-C	0			0	0	0	0	0	0	0	0	0	0	1	0	0	0
17a. Height P_3 and M_2	0			0	1	0	1	0	0	1	0	0	1	1	1	0	0
17b. Height P_3 and M_2	0			0	0	0	0	0	0	0	0	0	0	0	0	1	1

14). *Length of P_2 and P_3 (ps, P_2 shorter than P_3):* In derived state 1 (*Bobbschaefferia, Guggenheimia, Marmosopsis, Sternbergia*) P_2 and P_3 are subequal in length. In derived state 2 (*Monodelphopsis*) P_2 is slightly longer than P_3.

15). *Width of P_2 and P_3 (ps, P_2 narrower than P_3):* In derived state 1 (*Marmosopsis, Sternbergia*) P_2 and P_3 are subequal in width. In derived state 2 (*Monodelphopsis*) P_2 is slightly wider than P_3.

16). *Orientation of I_1-C in jaw and structure of ramus anteriorly (ps, I_1-C situated more or less vertically or anterovertically, ramus decreases gradually in depth anteriorly):* In character 16a derived state 1 (*Derorhynchus, Minusculodelphis*) the ramus is elongate anteriorly, P_{1-3} are elongate and reduced, and I_1-C are procumbent. In character 16b derived state 1 (*Eobrasilia*) the ramus is narrow anteriorly and foreshortened, P_{1-3} are inflated, and I_1-C are semiprocumbent (Table 15).

17). *Relative height of P_3 and M_2 protoconid (ps, apparently subequal):* In character 17a derived state 1 (*Bobbschaefferia, Didelphopsis, Eobrasilia, Gaylordia, Itaboraidelphys, Protodidelphis*) P_3 is demonstrably or apparently higher. In character 17b derived state 1 (*Derorhynchus, Minusculodelphis*) P_3 is, based on knowledge of roots and/or alveoli, apparently lower.

Character analysis: In considering the distribution of the 17 characters and their states I will, where appropriate, note additional similarities and/or differences among the taxa. I will also indicate which characters and states I deem to be particularly useful in grouping these taxa, and thus introduce a subjective weighting of their probable importance. Since some of the taxa are represented by only a few teeth, it is not possible to compare all of them directly. Thus, taxon A may prove comparable with B and C on the basis of upper dentitions, while C may prove comparable with D and E on the basis of lower dentitions. It is then possible to indirectly compare A and B with D and E, using C as the "linking" intermediary. This is not an ideal situation, but is unavoidable given the lack of alternative options. I will begin with the most obvious and securely established groupings and proceed to the least obvious. As is demonstrated below, all of the taxa except *Bobbschaefferia, Procaroloameghinia* and *Protodidelphis* can readily be organized into one of six groups. In addition, *Zeusdelphys* shows initial affinity to one group, but is ultimately shown to be referrable to another.

Group 1: Protodidelphis, Robertbutleria and *Zeusdelphys* share three derived character states — stylar cusps B and D large to enormous and subequal in size with no trace of cusp C (character 7); paracrista short, rounded, and virtually absent (character 8); and metacrista very short and rounded (character 9). These taxa also have a V-shaped centrocrista (character 1); metacone about twice as large (or larger) than paracone (character 4); and are either brachydont or bunodont (character 5). Within this group I

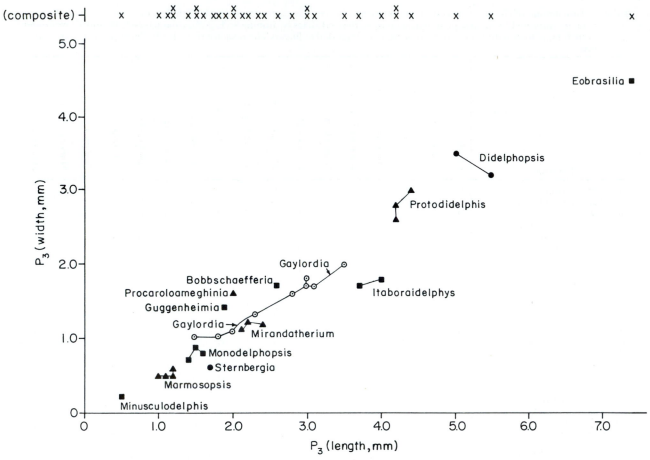

Fig. 48. Size distribution of Itaborai "opossum-like" taxa as indicated by P₃ length vs. width. Composite at top (x's) represents total number of specimens upon which plot is based.

tentatively include *Procaroloameghinia* because it is markedly bunodont (as is *Robertbutleria*), and because it shows special affinity with *Caroloameghinia* (see Marshall 1982c) and *Roberthoffstetteria* (see Marshall *et al.* 1983), both of which are known by upper molars that have large stylar cusps B and D. This primarily upper molar group is compared with taxa known only by lower dentitions using *Protodidelphis*, which is known by both upper and lower dentitions, as the "linking" or "bridging" intermediary.

Group 2: Protodidelphis compares uniquely with *Guggenheimia* in structure of the P₃ (character 13) in which, that tooth, and P₂, are somewhat rounded in occlusal view, have broad posterobasal heel, and the crown (especially of P₂) is oriented at a slight oblique angle in the jaw. In addition, both taxa have a very large M₅ talonid; rectangular shape of molars in occlusal view; trigonids and talonids subequal in height; hypo- and entoconids are large, subequal in size and inflated internally (character 12); lack of a distinct crista obliqua, and trigonid and talonid separated by broad transverse valley; proto- and metaconids subequal in size and height; and trigonid constricted somewhat anteroposteriorly. These taxa also compare favorably with *Bobbschaefferia* in the large size of the hypo- and

entoconids (character 12); rectangular shape of molars in occlusal view; and large size of M₅ talonid. The molars in all three taxa are also brachydont, although *Bobbschaefferia* has the trigonid slightly higher than the talonid and the crista obliqua is weakly developed (i.e. the distinction between trigonid and talonid is not as evident). The premolars of *Bobbschaefferia* differ from the other two taxa in being more trenchant and elongate. Overall, however, the molars of *Bobbschaefferia* display features that can be considered structurally antecedent to those found in *Protodidelphis* and *Guggenheimia*. Thus, this group securely includes *Guggenheimia* and *Protodidelphis*, and tentatively includes *Bobbschaefferia*.

Group 3: Didelphopsis, Eobrasilia and *Gaylordia* uniquely share an enormous, bulbous P₃ that is clearly modified for crushing (character 13). The premolars decrease sharply in size from P₃ to P₁, and P₂ is also bulbous and modified for crushing. In *Didelphopsis* and *Gaylordia* (and by inference in *Eobrasilia*) the protoconid is notably higher than the para- and metaconid, and the trigonids are foreshortened anteroposteriorly. Worn teeth demonstrate that trigonids share in the crushing function since they too wear to flat surfaces. The mandibular rami in *Didelphopsis* and *Gaylordia* are

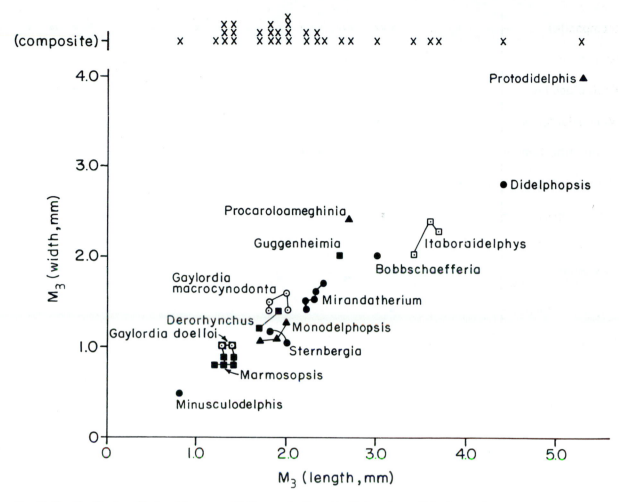

Fig. 49. Size distribution of Itaborai "opossum-like" taxa as indicated by M₃ length vs. width. Composite at top (x's) represents total number of specimens upon which plot is based.

robust anteriorly, while in *Eobrasilia* the anterior part is narrow, and the incisors and canine are semiprocumbent (character 16). Thus, *Eobrasilia* is derived in this regard relative to *Didelphopsis* and *Gaylordia*.

The upper molars of *Didelphopsis* are similar in structure to that expected in the ancestral marsupial, except that they have a small stylar cusp C (character 7). The worn upper teeth of *Gaylordia* suggest that the stylar area was similar to *Didelphopsis* and hence approaches the primitive condition. A similar conclusion was reached by Clemens (1966:21) who remarked that the coronal structure of *Gaylordia* suggests similarity to both *Alphadon marshi* and *A. lulli*. The important point is that the upper molars of *Didelphopsis, Gaylordia* and by inference *Eobrasilia* have a structure that appnoaches or approximates that expected in the ancestral marsupial. These three taxa are thus quite distinct from those in group 1, and by inference group 2, which have enormous stylar cusps B and D (see above).

Group 4: Derorhynchus and *Minusculodelphis* have the anterior part of the mandibular ramus

elongate; P$_{1-3}$ elongate and reduced; incisors and canine procumbent (character 16); P$_3$ apparently lower than M$_2$ (character 17); and M$_5$ shorter and narrower than M$_4$ (characters 10, 11). These taxa also have the posterior edge of the M$_{2-4}$ straight, a feature resulting from virtual loss of hypoconulid; talonid somewhat foreshortened anteroposteriorly; paraconid inclined somewhat anteriorly (more so in *Minusculodelphis*); and cristid connecting entoconid with metaconid well developed and somewhat blade-like (again, more so in *Minusculodelphis*).

Group 5: Mirandatherium and *Monodelphopsis* have trigonid cusps somewhat approximated; talonid distinctly basined and notably wider than trigonid; M$_5$ notably shorter and narrower than M$_4$ (characters 10, 11); molars brachydont (character 5); stylar shelf reduced relative to primitive state (more so in *Monodelphopsis*, character 6); hypo- and entoconids well developed and subequal in size (character 12); and P$_3$ somewhat inflated and ovoid in occlusal view (character 13). These taxa also resemble *Procaroloameghinia* in the approximation of the trigonid cusps; talonid distinctly basined and

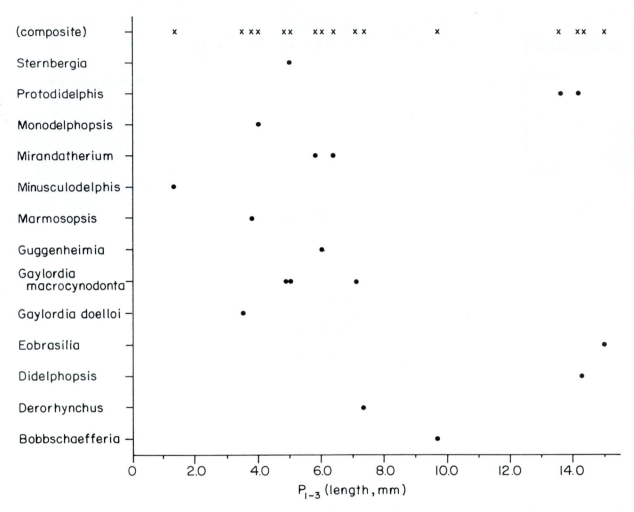

Fig. 50. Size distribution of Itaborai "opossum-like" taxa as indicated by length of P_{1-3}. Composite at top (x's) represents total number of specimens upon which plot is based.

broader than trigonid; well developed, subequal sized hypo- and entoconids; M_5 shorter and narrower than M_4; P_3 ovoid in shape; and molars brachydont or bunodont. Thus, the lower dentition of *Procaroloameghinia* suggests special affinity with group 5, while inference about its upper dentition (i.e. it probably has large stylar cusps B and D, see above) suggests affinity with groups 1 and 2. I will return to this apparent conflict below.

Group 6: Marmosopsis, Itaboraidelphys and *Sternbergia* represent generalised taxa that either departed little, if at all, from the inferred ancestral marsupial, or in some or all the same ways as did representatives of groups 1-5 (i.e. in characters 1, 4, 7, 10, 11, 14, 15, 17). The only derived feature not found in a contemporaneous taxon is the presence of a labial cingulum on M_{2-4} in *Sternbergia*; although this feature occurs in some species of *Alphadon* (see Clemens 1966, Figs 58, 62, 64) and in both species of *Glasbius* (see Clemens 1966, Figs 17-19; Archibald 1982, Figs 46-48) from the Late Cretaceous of North America. *Bobbschaefferia* can be referred to this group based on the generalized structure of its premolars, but possession of brachydonty (character 5) and large size of hypo- and entoconid (character 12) suggest affinity with group 2. I will return to the problem of *Bobbschaefferia* below.

With the exceptions of *Bobbschaefferia, Procaroloameghinia* and *Protodidelphis* the taxa in each of the above groups appear to represent discrete and cohesive taxonomic units. Furthermore, groups 3, 4 and 6 include taxa with tribosphenic molars, while groups 1, 2 and 5 include taxa with brachydont or bunodont molar specializations. On the other hand, groups 1 and 2 include taxa which have, or are inferred to have, enlarged stylar cusps B and D; while groups 3, 4, 5 and 6 have unspecialized or reduced stylar cusps (i.e. stylar cusp D is not large). Thus, group 5 is problematical because it includes taxa of "intermediate aspect" that have brachydont molars and reduced stylar cusps. Nevertheless, groups 1 and 2 appear to form an unambiguous and intergrading unit, while groups 3, 4 and 6 are all discrete. The interrelationships of these groups and the affinities of

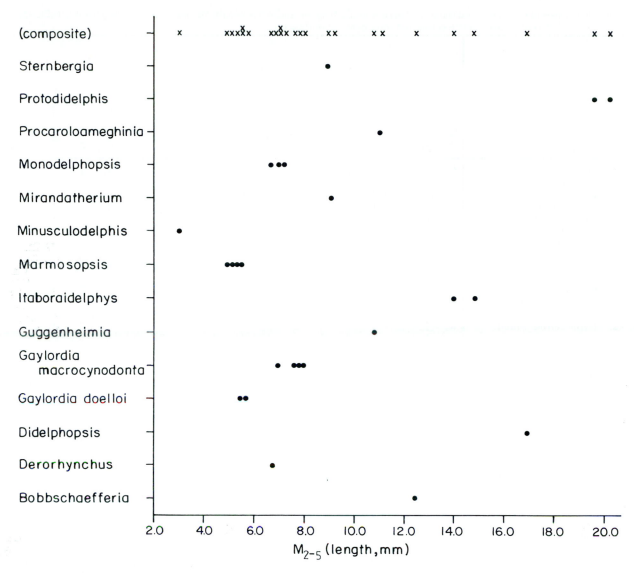

Fig. 51. Size distribution of Itaborai "opossum-like" taxa as indicated by length of M_{2-5}. Composite at top (x's) represents total number of specimens upon which plot is based.

Bobbschaefferia, Procaroloameghinia and *Protodidelphis* to these groups, will be clarified in part 5 of the PHYLOGENETICS section, below.

J. A. W. Kirsch, while reviewing this paper, tested the unity of the above six groups by computing Wagner trees (Fig. 52) using the coded data sets in Tables 15 and 16. A discussion of the methodology used in the preparation, computation, and interpretation of Wagner trees is provided by Reig *et al.* (1987). With his kind permission I include these trees here, along with some of his observations which make up the bulk of the following four paragraphs.

As noted by Kirsch, the Wagner program requires a complete data set "to do its thing", and a decision had to be made about the states of missing characters in Table 15. Two internally consistent but quite different options were possible. The first, and most conservative, was to consider all missing character-states as primitive; the

second, and more realistic, was to code missing states as the same as that of the most primitive member of the group to which an incomplete taxon belongs (*sensu* my groups and using my method of linking intermediates). "The second option smacks of circular reasoning, and still requires some bold inferences in particular cases".

Considering the first option (Wagner tree, Fig. 52A), Kirsch noted the surprise at how closely the Wagner programmes' matched mine. Groups 3, 4 and 5 are definite, and group 6 is nearly so. *Itaboraidelphys* is primitive with respect to group 6 plus a composite of groups 1 and 2. However groups 1 and 2 are neither together on the tree nor are they of the composition which I infer. The reason for this is that *Protodidelphis* possesses several states lacking in other group 2 members, while *Procaroloameghinia* represents the complementary situation (see Table 15). Kirsch noted that the lack of association of groups 1, 2 and 5 into what he calls supergroup A probably results

Table 16. Data set of characters and states used in computing Wagner tree in figure 52b. Missing characters (see blanks, Table 15) are coded as for nearest relative. Inferred codings are underlined.

Character	Ancestor	Robertbulleria	Zeusdelphys	Procaroloameghinia	Bobbschaefferia	Guggenheimia	Protodidelphis	Mirandatherium	Monodelphopsis	Itaboraidelphys	Marmosopsis	Sternbergia	Didelphopsis	Eobrasilia	Gaylordia	Minusculodelphis	Derorhynchus
1. Centrocrista	0	1	1	0	1	1	1	0	0	1	1	1	1	1	1	1	1
2. Para- and metaconules	0	0	2	0	0	0	0	0	0	0	0	1	0	0	0	0	0
3. Size M$^{3-4}$	0	0	0	0	1	1	1	0	0	0	0	0	0	0	0	0	0
4. Para- and metacones	0	1	1	1	1	1	1	1	1	1	1	1	1	1	1	0	0
5. Molar structure	0	2	1	2	1	1	1	1	1	0	0	0	0	0	0	0	0
6. Stylar shelf	0	0	0	0	0	0	0	1	2	0	0	0	0	0	0	0	0
7a. Stylar cusps	0	0	0	0	0	0	0	1	2	1	1	1	1	0	0	0	0
7b. Stylar cusps	0	1	1	1	1	1	1	0	0	0	0	0	0	1	1	0	0
7c. Stylar cusps	0	0	0	0	0	0	0	0	0	0	0	1	0	0	0	0	0
8. Paracrista	0	2	2	2	2	2	2	1	0	0	1	0	0	0	0	0	0
9. Metacrista	0	2	2	2	2	2	2	0	1	0	0	0	0	0	0	0	0
10. Length M$_4$ and M$_5$	0	1	1	1	0	0	1	1	1	0	0	1	1	0	0	1	0
11. Width M$_4$ and M$_5$	0	1	1	1	1	1	0	1	1	1	0	1	1	1	1	1	1
12. Hypo- and entoconids	0	1	1	1	2	2	2	1	1	0	0	0	0	0	0	0	0
13. Shape P$_3$	0	1	1	1	0	2	2	1	1	0	0	0	3	3	3	0	0
14. Length P$_2$ and P$_3$	0	0	0	0	1	1	0	0	2	0	1	1	0	0	0	0	0
15. Width P$_2$ and P$_3$	0	0	0	0	0	0	0	0	2	0	1	1	0	0	0	0	0
16a. Structure I$_1$-C	0	0	0	0	0	0	0	0	0	0	0	0	0	0	0	1	1
16b. Structure I$_1$-C	0	0	0	0	0	0	0	0	0	0	0	0	0	1	0	0	0
17a. Height P$_3$ and M$_2$	0	0	0	0	1	0	1	0	0	1	0	0	1	1	1	0	0
17b. Height P$_3$ and M$_2$	0	0	0	0	0	0	0	0	0	0	0	0	0	0	0	1	1

from the inconsistent distribution of the V-shaped centrocrista. A realization of Kirsch's observations is that a V-shaped centrocrista evolved independently at least two times among the Itaborai taxa, and the centrocrista character is therefore liable to convergent evolution of the derived state. In retrospect, I did recognize this situation in constructing my groups and in the phylogeny and classification which I present below, although I failed to recognize this situation above when coding this character. The convergent nature of this character is thus obvious when other characters are brought into consideration, and a single origin for a V-shaped centrocrista is simply inconsistent with the other characters considered here.

Considering the second option (Wagner tree, Fig. 52B), educated guesses were made about missing states based on knowledge of these states in nearest relatives (Table 16), and character 1 was eliminated from the computation of this tree for reasons noted above. As seen in this tree, the taxa are nicely clustered in my putative groups (except *Itaboraidelphys*), although the groups are rather strung out in a series rather than being lumped into two supergroups. There is also a problem with *Zeusdelphys* , which is incomplete in many characters critical to its secure association with either group 1 or 2. If characters 3, 10, 11, 12 and 13 for this taxon are recoded in Table 16 to match the most primitive codings of group members (i.e. codings changed respectively to 1, 0, 0, 2, 0,), then *Zeusdelphys* clusters with group 2

members. Thus, Kirsch's tests indicate that *Zeusdelphys* is probably a member of group 2 and not group 1. As demonstrated below, Kirsch is correct on this point, and in the classification proposed below, *Zeusdelphys* is securely associated with group 2 taxa.

Kirsch concluded his comments on these tests by noting that they appear to be "useful in showing (1) how securely based most of your primary groups are, (2) the intergrading nature of some of these, (3) the manner in which your method of linking intermediates works, and (4) the less-secure basis of the super groupings — *at least as indicated by this limited data-set*".

In parts 2-4 which follow, I compare the Itaborai taxa with other marsupials in South America, North America, Europe, Africa,and Australia. In all cases these comparisions are made simply to identify potential relationships. Except for a few obvious exceptions, I reserve judgement on the systematic implications of the noted similarities and differences for part 5 of the PHYLOGENETICS section.

2) AFFINITIES OF ITABORAI TAXA WITH OTHER SOUTH AMERICAN MARSUPIALIA

In this part I compare the groups of Itaborai taxa discussed in part 1 above with the following South American marsupials: *Coona, Eomicrobiotherium, Pachybiotherium,* Microbiotheriinae, *Gashternia, Zygolestes, Reigia, Caroloameghinia, Roberthoffstetteria, Peradectes, Prepidolops, Bonapartherium,* Caenolestidae, Argyrolagidae,

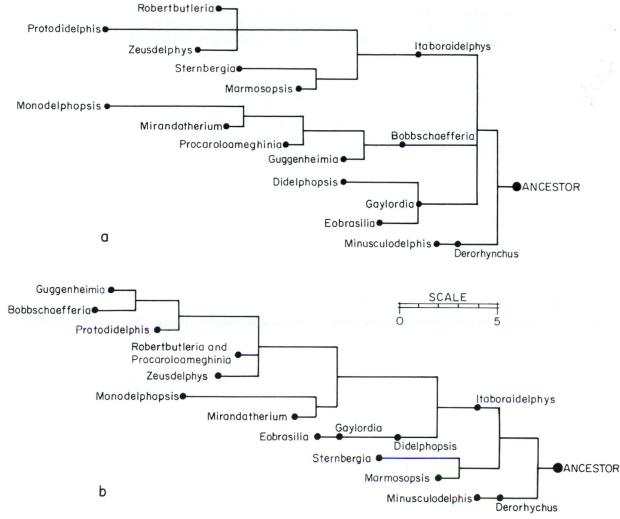

Fig. 52. A, Wagner tree of 17 taxa computed on the basis of the 21 characters listed in Table 15, with overall consistency of .49; all missing characters are assumed primitive. B, Wagner tree of 17 taxa computed on 20 characters (less character 1) listed in Table 16, with overall consistency of .58; all missing characters are coded as for nearest relative. Horizontal lines between linkage levels are drawn in proportion to evolutionary distances, in increments of one unit; the scale may be used to determine the exact number of steps separating nodes. As the all-primitive *Ancestor* is to the right, taxa to the left are more derived. Wagner trees provided courtesy of J. A. W. Kirsch.

Groeberiidae, Borhyaenoidea, and Didelphoidea.

Coona pattersoni Simpson 1938 of Casamayoran (Early Eocene) age is known only by its type, a left mandibular ramus with alveoli of P_3-M_3, M_4 complete, and talonid of M_5 (see Marshall 1982a, Fig. 24). Paula Couto (1970:28) noted similarities between *Coona* and *Mirandatherium*, and suggested that they may prove synonymous. These taxa are similar in size, but they are definitely distinct at the generic level. *Coona* differs from *Mirandatherium* in having the trigonid much higher than talonid (it is tribosphenic, not brachydont), a very reduced hypoconulid, and a much larger M_5 talonid. In all these features *Coona* compares only with members of Itaborai group 6 (i.e. it is generalized and retains states expected in or approximated by the ancestral marsupial). It is derived in having M_5 shorter and narrower than M_4, a state shared with *Sternbergia* among group 6 (see Fig. 52).

Eomicrobiotherium Marshall 1982a:57 is known from two species of Casamayoran age in southern Argentina; *E. gaudryi* Simpson 1964 and *E. gutierrezi* del Corro 1977 (see Marshall 1982a, Figs 25-27). These taxa are characterized by their small size, ovoid P_3, brachydonty, trigonid moderately elevated above talonid, paraconid reduced, talonid distinctly basined and broader than trigonid, and lack of prominent anterobasal cingulum. *Eomicrobiotherium* compares best with *Mirandatherium* in structure and relative size of trigonid and talonid, brachydonty, ovoid shape of P_3, and lack of prominent anterobasal cingulum. *Eomicrobiotherium* also compares favorably with *Pachybiotherium* Ameghino 1902 from the Colhuehuapian (Early Miocene) of Argentina (for a listing of these similarities see Marshall 1982a). Earlier (1982a) I argued that *Eomicrobiotherium* and *Pachybiotherium* may represent an ancestral-descendant lineage which paralleled microbiotheres in aspects of their dental structure and opted not to include these taxa in that group. Reig *et al.* (1985, 1987) regarded *Pachybiotherium* as a caluromyine didelphid, although I now feel confident that it, *Eomicrobiotherium*, and *Mirandatherium* (see below) are good microbiotheres. This change of position stems primarily from a slight broadening of my concept of the structural breadth of that group, and acknowledges that some early taxa need not be structurally ancestral to all later taxa. I thus formally recognize these three taxa as members of the Microbiotheriinae (*sensu* Reig *et al.* 1985, 1987). Del Corro's (1977) original referral of *gutierrezi* to *Microbiotherium* is appreciated in view of the present taxonomic positon of this species. Also, it is possible that when Paula Couto (1970:28) suggested affinity of *Mirandatherium* and *Coona*, he made his comparisons not with the type species of *Coona*, *C. pattersoni*, but with ?*Coona gaudryi* which I later (1982a) made the genotype of *Eomicrobiotherium*.

As noted above, *Mirandatherium* compares favorably with microbiotheres. This group is monographed by Marshall (1982a) who recognizes two genera: *Microbiotherium* Ameghino 1887 from the Colhuehuapian (Early Miocene) and Santacrucian (Early-Middle Miocene) of southern Argentina, and the living *Dromiciops* of Chile and western Argentina. The similarities are numerous and include ovoid structure of P_3, presence of small posterobasal heel on P_3, P_3 longer and wider than P_2, molars brachydont, trigonid only moderately elevated above talonid, talonid distinctly basined and wider than trigonid, M_5 shorter and narrower than M_4, absence of prominent anterobasal cingulum, stylar shelf only moderately developed, stylar cusps greatly reduced, centrocrista rectilinear, and para- and metaconules distinct in fossil taxa, but absent in *Dromiciops*. *Mirandatherium* is thus recognized as the earliest member of this subfamily.

Gashternia ctalehor Simpson 1935a is from the Riochican (Late Paleocene) of Argentina (see Simpson 1935a, Fig. 4; 1948, Fig. 9). The type and only known specimen is a partial lower jaw containing two molars or molariform teeth, which if a marsupial apparently represent M_{2-3} (Simpson 1935a, 1948). Simpson cautioned that *Gashternia* may not be a marsupial, while Clemens (1966:33) and Archer (1976:23) believed it was and suggested didelphoid affinity. The two teeth in *Gashternia* are bunodont, trigonid only slightly higher than talonid, presence of well developed anterobasal cingulum, paraconid very reduced and not joined to protoconid by paracrista, proto- and metaconid subequal in size, hypo- and entoconid subequal in size, talonid separated from trigonid by horizontal valley (i.e. crista obliqua weak and lacks distinct contact with protoconid), and hypoconulid apparently absent. I am confident that *Gashternia* is a

marsupial and it compares favorably in aspects of its dental structure with Itaborai group 2 taxa (i.e. in relative height of trigonid and talonid, large size of hypo- and entoconids, separation of talonid from trigonid, brachydonty to bunodonty, etc.). Absence of a hypoconulid does not warrant exclusion as a marsupial, since hypoconulids are virtually or totally absent in such taxa as *Minusculodelphis*, *Coona*, *Groeberia*, and *Bonapartherium*. The presence of a metastylid in *Gashternia* is a feature shared with *Caroloameghinia*, a taxon that is also related to Itaborai group 2 (see below).

Zygolestes paranensis Ameghino 1898 is from the Montehermosan (Pliocene) of Argentina. The type and only known specimen is a left mandibular ramus with P_2-M_5 (see Reig *et al.* 1987, Fig. 35). The affinities of this animal were most recently considered by Reig *et al.* (1987) who concluded by listing it as Didelphimorphia *incertae sedis*. Among didelphoids it is unique in having a small bi-rooted P_2, paraconid on M_2 displaced medially, and M_5 shorter and narrower than M_4. *Zygolestes* shows no special affinity to any of the Itaborai groups, except that it has a generalized molar structure as in group 6 taxa. *Sternbergia* has M_5 shorter and narrower than M_4, but so does *Didelphopsis*, *Minusculodelphis*, *Mirandatherium*, *Monodelphopsis* and *Procaroloameghinia* (i.e. one member in each of the Itaborai groups). The affinities of *Zygolestes* clearly lie with members of the subfamily Didelphinae (*sensu* Reig *et al.* 1987). In fact, it shows special affinity with members of the tribe Marmosini (i.e. size, most members have M_5 shorter and narrower than M_4, and most have P_3 smaller than P_2). The remaining features of *Zygolestes*, save for the displacement of the M_2 paraconid, occur widely among Didelphinae. I therefore recommend that *Zygolestes* be placed in a monotypic tribe, Zygolestini nov., in the Didelphinae and that its closest affinities lie with members of the Marmosini (*sensu* Reig *et al.* 1987).

Reigia punae Pascual 1983 is from the Middle or Late Eocene age Pozuelos Formation of Salta Province, northwest Argentina. The type and only known specimen is a fragment of a right maxilla with M^{4-5} (Pascual 1983, Fig. 1, Pl. 1). The M^4 is brachydont and has inflated cusps, small cingulum labial to stylar shelf, stylar cusps B and D enormous, C large but smaller than B and D, A and E small, metacone about twice as large and higher than paracone, centrocrista V-shaped, metaconule displaced posterolingually, and no trace of paraconule or paracrista. Pascual (1983) assigned *Reigia* to the subfamily Glasbiinae, and regarded it structurally more primitive than *Glasbius*. This assignment was championed by Reig *et al.* (1987). *Reigia* compares well with Itaborai group 1, and by inference group 2, taxa in having brachydont to bunodont molars, virtual lack of paracrista, short metacrista, and large size of stylar cusps B and D. *Reigia* also resembles *Robertbutleria* and *Glasbius* in having a cingulum labial to the stylar shelf.

Caroloameghinia Ameghino 1901 is known from three species (*C. mater, C. tenuis, C. intermedius*) from rocks of Casamayoran (Early Eocene) age of southern Argentina (see Marshall 1982c; Marshall, de Muizon and Hoffstetter 1984). The genus was placed by Ameghino (1901) in the monotypic family Caroloameghiniidae, which Clemens (1966) later recognized as a subfamily (Caroloameghiniinae) of the Didelphidae. *Caroloameghinia* was recently rediagnosed by Marshall (1982c) and the types and many referred specimens were figured. As seen by those figures, *Caroloameghinia* compares favorably with Itaborai groups 1 and 2, and especially with *Procaroloameghinia* which Marshall (1982c) regards as its probable direct ancestor. *Caroloameghinia* shares the following features with groups 1 and 2, or with some of those taxa: marked bunodonty (as in *Procaroloameghinia, Robertbutleria*), large stylar cusps B and D, C absent, P_3 ovoid to rounded, trigonid foreshortened anteroposteriorly, trigonid narrower and only slightly higher than talonid on M_{2-4}, proto- and metaconid subequal in size, hypo- and entoconid subequal in size and very large, trigonid separated from talonid by transverse valley, para- and metaconules distinct (absent in *Zeusdelphys*), paracrista small to absent, metacrista short and rounded. *Caroloameghinia* differs from the taxa in those groups in having the centrocrista rectilinear and the para- and metacones subequal in size (or paracone only slightly smaller).

I agree with Pascual (1983) that the tooth (AMNH 28928, an isolated right M^4) referred tentatively by Simpson (1948:38) to *Caroloameghinia tenuis* represents a new taxon. In fact, in aspects of its structure (i.e. brachydonty, absence of paraconule, metaconule displaced posterolingually, metacone about twice as large and higher than paracone) it agrees most favorably with *Reigia punae*. Two options appear likely, either make this specimen a new species of *Reigia*, or make it type of a new genus and species.

Roberthoffstetteria nationalgeographica Marshall, de Muizon and Sigé 1983 from the Late Cretaceous (Maestrichtian) of Bolivia also compares very favorably with *Caroloameghinia* and Itaborai groups 1 and 2. In fact, Marshall *et al.* (1983) placed this species in the Caroloameghiniinae along with *Caroloameghinia* and *Procaroloameghinia*. The principal features shared by *Roberthoffstetteria* and group 1 and 2 taxa include: bunodonty (shared with *Procaroloameghinia, Robertbutleria*), large size of stylar cusps B and D (especially on M^{2-3}), paracrista reduced or absent, metacrista short, metacone about twice as large and higher than paracone, trigonid cusps

foreshortened anteroposteriorly, trigonid narrower and only slightly higher than talonid, and hypo- and entoconids large and subequal in size. *Roberthoffstetteria* differs from those taxa in having a distinct stylar cusp C (a feature shared with *Glasbius intricatus*), having a metacingulum (found also in *Glasbius*), centrocrista rectilinear on M^2 (indistinct on M^{3-4}), and para- and metaconules displaced lingually.

Alphadon austrinum Sigé 1971 is based on an isolated right M^4, missing protocone, from the Late Cretaceous (Maestrichtian) of Peru. Sigé (1972) later referred additional fragmentary teeth to this species and figured most of them. Crochet (e.g. 1980) recently reassigned the species *austrinum* to *Peradectes*. The type agrees best with Itaborai group 6 taxa, except that the centrocrista is more rectilinear and the paracone is only slightly smaller than the metacone. Thus, this species retains features expected in the ancestral marsupial and overall displays more primitive states than do any of the Itaborai group 6 taxa.

Prepidolops Pascual 1980a is known by three species: *P. didelphoides* Pascual 1980a and *P. molinai* Pascual 1980b from the Casamayoran age Lumbrera Formation of northwest Argentina, and *P. alonsoi* Pascual 1983 from the Middle or Late Eocene age Pozuelos Formation of northwest Argentina. *Prepidolops* was originally placed by Pascual (1980a) in the superfamily Polydolopoidea, family Polydolopidae, and later (Pascual 1980b, 1983) in its own family Prepidolopidae in the superfamily Didelphoidea. The teeth of the various species of *Prepidolops* are generally brachydont, with a V-shaped centrocrista, large and subequal stylar cusps B and D united basally with para- and metacone, stylar cusp C absent, paracrista virtually absent, and metacrista very short. In the majority of these features *Prepidolops* resembles Itaborai groups 1 and 2. It agrees with Itaborai group 5 and *Procaroloameghinia* in having trigonid foreshortened anteroposteriorly, talonid basined and wider than trigonid, trigonid only slightly higher than talonid, and M_5 shorter and narrower than M_4. It also compares with group 3 taxa in having an enormous P_3, although in *Prepidoelops* the P_3 is modified for cutting while in group 3 taxa it is modified for crushing. As is demonstrated below, the conflicting array of features displayed by *Prepidolops* relative to the Itaborai taxa are easily explained by the realization that *Prepidolops* is a generalized polydolopoid and not a didelphoid.

Bonapartherium hinakusijum Pascual 1980a is from the Casamayoran age Lumbrera Formation of Argentina. Pascual (1980a) originally placed *Bonapartherium* in the monotypic family Bonapartheriidae and tentatively referred it to the superfamily Polydolopoidea. He later (1981) placed the Bonapartheriidae in the superfamily

Didelphoidea. The teeth of *Bonapartherium* are bunodont, M_{3-5} trigonids slightly narrower than talonids, trigonid only slightly higher than talonid, paraconid very reduced, protoconid and metaconid subequal in size, hypo- and entoconid very large and subequal in size, hypoconulid apparently absent, stylar cusps B and D very large and united basally to para- and metacone giving labial half of molars a selenodont appearance, hypocone present, P3 enormous, P^2 and P^3 with large buccal shelf, trenchant P_3 and small anvil-shaped P_2, and P_{1-2} very reduced relative to P_3. As with *Prepidolops*, this animal shows affinity with Itaborai groups 1 and 2 in the large size of stylar cusps B and D, absence of stylar cusp C, bunodonty, large and subequal size of hypo- and entoconids, and in trigonid only slightly higher than talonid on M_{3-5}. It also has a large P3 as in Itaborai group 6, although differs from that group in having a broad buccal shelf and being modified for a combined shearing-crushing function. I demonstrate below that *Bonapartherium* is a member of the Polydolopoidea and not a derived didelphoid.

The Caenolestidae are monographed by Marshall (1980b). A broken tooth from the Casamayoran apparently represents a member of this family (see Marshall 1980b, Fig. 3), although surely identifiable taxa are first known from the Deseadan (see Fig. 1). Three subfamilies (Caenolestinae, Palaeothentinae, Abderitinae) are recognized. Because the Caenolestinae represent or approximate the ancestral caenolestid stock, I restrict my comparisons to members of that subfamily. The upper molars of Caenolestinae are highly derived and the homology of the cusps is not readily apparent. In my 1980b monograph, I identified the cusps in the quadritubercular upper molars of the Santacrucian age *Stilotherium* (see Marshall 1980b, Figs 4, 5) as a protocone and hypocone lingually, and a paracone and metacone labially. I made this interpretation on the assumption that the stylar shelf and stylar cusps were lost in caenolestids. However, I now believe that the two large labial cusps represent stylar cusps B and D, and that the posterior lingual cusp is an enlarged metaconule. A cusp situated lingual to the posterior most labial cusp I designated an "intermediate conule" in my 1980b paper, although this is almost certainly homologous with the metacone. Thus, the paracone is absent in caenolestids. Based on this new interpretation of upper molar cusp homology, caenolestids compare favorably with Itaborai groups 1 and 2 in having large stylar cusps B and D, brachydonty, large and subequal sized protoconid and metaconid, hypo- and entoconids large and subequal in size, and trigonid only slightly higher than talonid. The lower molars of Caenolestinae also agree with Itaborai group 5 taxa in being brachydont,

trigonid cusps approximated and only slightly higher than talonid cusps, talonid distinctly basined and wider than trigonid, and M_5 shorter and narrower than M_4. Features shared with Itaborai group 4 include reduction of premolars (especially P_{1-2}) and incisors, and canine procumbent. Early Caenolestinae also have a weak labial cingular shelf on M_{2-4}, a feature shared with *Sternbergia* (group 6) and North American Late Cretaceous *Glasbius* and some *Alphadon* (see Clemens 1966). The upper molars of some members of each of the three subfamilies of Caenolestidae also have a weak cingulum labial to stylar cusps B and D, a feature shared with *Robertbutleria*, *Reigia*, and *Glasbius*.

Argyrolagidae range from Deseadan (Middle Oligocene-Early Miocene) to Uquian (Late Pliocene-Early Pleistocene) (Simpson 1970a; Wolff 1984). The dentition of argyrolagids is highly modified, with hypsodont cheek teeth and procumbent incisors. Simpson (1970a:15-16) believes that the two gliriform lower teeth are I_{1-2}, although below I suggest that they may be the canine and P_1. Argyrolagids show no special affinity with any Itaborai groups, although they do share a few similarities with group 4 taxa, especially *Minusculodelphis*: incisors and canine procumbent, trigonid cusps inclined forward and paraconid inflected labially and reduced, talonid foreshortened anteroposteriorly, hypoconulid very reduced or lost, M_5 shorter and narrower than M_4, sharp blade-like crest connects entoconid and metaconid, and blade-like protocrista connects protoconid and metaconid. I will demonstrate below that argyrolagids show special affinity with groeberiids, and the resemblances shared with Itaborai group 4 taxa are apparently attributable to convergence or parallism.

The family Groeberiidae is represented only by *Groeberia* Patterson 1952 from the Early Oligocene (Divisaderan) age Divisadero Largo Formation of Argentina. Two specimens are known, a partial lower dentition (Patterson 1952) and a partial skull and upper dentition (Simpson 1970b). *Groeberia* has a large gliriform tooth in the lower jaw separated by a large diastema from the cheek teeth which presumably represent P_3-M_5. The molars are bunodont, trigonid and talonids are subequal in height and width, and P_3 is rounded and subequal in height to M_2. The structure of the molars suggest affinities with *Gashternia*, *Bonapartherium*, some Caenolestidae (Abderitinae, Palaeothentinae), and Itaborai groups 1 and 2.

The Borhyaenoidea are the "dog-like" and "saber-tooth" marsupials of South America. Two families, Borhyaenidae and Thylacosmilidae, are recognized (Marshall 1978). Within the "dog-like" Borhyaenidae are four subfamilies (Hathliacyninae, Borhyaeninae,

Prothylacyninae, Proborhyaeninae), of which the Hathliacyninae are the most generalized and early members approximate the ancestral borhyaenid (Marshall 1981a). The earliest known hathliacynine, *Patene simpsoni* from Itaborai, is known from relatively complete upper and lower dentitions (see Marshall 1981a, Figs 3-8). *Patene* compares only with Itaborai group 6 taxa, and especially with *Itaboraidelphys* in the following features: para- and metaconules distinct on M^{2-4}, molars tribosphenic, stylar shelf well developed, stylar cusp B largest of stylar cusps but very small, paracrista well developed and blade-like, P_3 trenchant and narrow transversely, and metacone larger and higher than paracone. *Patene* is more primitive than group 6 taxa in having a rectilinear centrocrista, and with M_5 longer and wider than M_4 (only *Marmosopsis* is regarded primitive in this character). *Patene* is more derived than group 6 taxa in having a more reduced protocone, metacrista notably longer, stylar cusps very reduced (A, B and sometimes C incipiently developed, D and E absent), P_3 higher than M_2 (as in *Bobbschaefferia* and *Itaboraidelphys*), and metaconid smaller than paraconid. Thus, *Patene* shares multiple derived states with group 6 taxa, has additional states regarded more derived than those taxa, and retains two states regarded as more primitive than those taxa. These features jointly suggest that borhyaenoids evolved from a more generalized "opossum-like" stock than that represented by the taxa in Itaborai group 6.

South American Didelphoidea are represented by the families Didelphidae and Sparassocynidae (Reig *et al.* 1987). The latter is regarded as a derived sister-taxon of the former, and because of this I will restrict my comments to the Didelphidae, and more specifically to the structurally conservative subfamily Didelphinae. South American didelphoids are known from the Friasian (Middle Miocene — I here place the Colhuehuapian genus *Pachybiotherium* in the Microbiotheriidae, see below) and are common in later fossil and recent faunas on that continent (Reig *et al.* 1987). The Didelpinae compare best with Itaborai group 6 taxa, and there is virtually complete overlap in the distribution of primitive and derived states among both groups (see Reig *et al.* 1987, and above).

The question that warrants special consideration is — Can some or all group 6 taxa be regarded as ancestral to some or all Didelphidae? Collectively *Marmosopsis*, *Itaboraidelphys*, and *Sternbergia* potentially qualify as ancestral didelphoids, and they share with Didelphidae a V-shaped centrocrista, metacone larger and higher than paracone, unmodified protocone, tribosphenic dentition, and talonids shorter than trigonids. In all other features (i.e. relative size of P_{2-3}, relative size of M_{4-5}, structure of para- and

metacrista, structure of stylar shelf, and size of stylar cusps) these taxa can be regarded as ancestral to some or all later Didelphoidea. *Marmosopsis* and *Itaboraidelphys* have distinct para- and metaconules on M^{2-4}, and *Sternbergia* lacks paraconules on M^{3-4}, while the majority of Didelphidae (except *Glironia* and *Hondadelphys*) lack these conules. The distribution of these conules among these taxa suggest that they were present in the ancestral didelphoid. *Sternbergia* has a weak labial cingulum on M_{2-4}, and at least a hint hint of this structure occurs in many later Didelphidae (see numerous figures in Reig *et al.* 1987). Thus, this feature does not eliminate *Sternbergia* as a potential didelphid or as a didelphid ancestor. In conclusion, *Marmosopsis*, *Itaboraidelphys*, and *Sternbergia* represent ideal structural ancestors for some or all later didelphoids. Below I address the obvious question — Should some or all these group 6 taxa be classified as Didelphidae (i.e. generalized and/or ancestral didelphoids)?

3) Affinities Of Itaborai Taxa With North American, European, And African Marsupialia

The North American and European Marsupialia are described, illustrated, and/or reviewed by Clemens (1966, 1979), Lillegraven (1969), Archibald (1982), Fox (1979a, b, c, 1981), Crochet (1980), and references therein. I discuss the Late Cretaceous North American taxa first (Stagodontidae, Glasbiinae, Pediomyinae, Peradectidae), then the Cenozoic age North American, European, and African taxa (Peradectidae, Didelphidae).

Stagodontidae are known from the Late Cretaceous (Campanian-Maestrichtian) of North America (Clemens 1966, 1979, Fox 1981). They resemble Itaborai group 3 taxa in having large crushing P_{2-3}, a well developed stylar shelf, and a large stylar cusp B. The latter two features are regarded primitive for marsupials, and the only shared derived state is the size and crushing specialization of the premolars. These groups differ markedly in many features, including stagodontids having a rectilinear centrocrista, reduced metaconid, enlarged paraconid, sharp decrease in size from M_5 to M_2 and emphasis on postvallum-prevalid shear. Fox (1981) demonstrated that early stagodontids (i.e. *Eodelphis)* lack a large stylar cusp D, while one is present in later forms (i.e. *Didelphodon*). In fact, *Didelphodon* resembles Itaborai groups 1 and 2 in having large stylar cusps B and D although this occurrence clearly represents a state derived independently in that lineage. In summary, stagodontids show no special affinity to any Itaborai taxa.

The subfamily Glasbiinae Clemens 1966 is based on *Glasbius* Clemens 1966 from the Late Cretaceous (Maestrichtian) of North America.

Two species, *G. intricatus* Clemens 1966 and *G. twitchelli* Archibald 1982 are recognized. *Glasbius* shares numerous derived states with Itaborai groups 1 and 2: i.e. marked bunodonty (with *Procaroloameghinia*, *Robertbutleria*), stylar cusps B and D enormous, stylar cusp C absent in *G. twitchelli* but present in *G. intricatus*, metacone larger and higher than paracone, distinct para- and metaconules (absent in *Zeusdelphys*), paracrista reduced or absent, metacrista short, and P$_3$ ovoid to rounded. In all aspects, *Glasbius* compares exceptionally well with *Robertbutleria* and both even have a cingulum bordering most or all M$^{2-3}$, while a cingulum labial to the stylar cusps also occurs in *Reigia* and cingula antero- and posterior to the protocone are known in some specimens of *Pediomys cooki* (see Clemens 1966:40). *Glasbius* has a centrocrista that is rectilinear, while in *Robertbutleria* it is V-shaped. *Glasbius* compares favorably with Itaborai group 5 taxa (and *Procaroloameghinia*)in trigonid cusps approximated and only slightly higher than talonid cusps, talonid distinctly basined and wider than trigonid on M$_{2-4}$, hypo- and entoconids large and subequal in size, centrocrista rectilinear, M$_5$ shorter and narrower than M$_4$, and P$_3$ ovoid. *Glasbius* shares with *Sternbergia*, a group 6 taxon, a labial cingulum on M$_{2-4}$.

Pediomyinae (*sensu* Reig *et al.* 1987) are known primarily from the Late Cretaceous (Campanian-Maestrichtian) of North America (Clemens 1966, 1979, Fox 1979b, c, 1984), although Slaughter (1978) reports a possible member from the Late Eocene of Texas, and Sigé (1972) tentatively records another from the Late Cretaceous (Maestrichtian) of Peru. The systematics of this group, and even its unity, are uncertain. Fox (1984) considers *Pediomys* ". . . a form genus for evolutionary separate lineages that have suppressed the anterior part of the stylar shelf and the stylocone independently", and that the family comprises "a polyphyletic taxon". In view of this situation I make reference in the following discussion to particular species of *Pediomys*. In overall structure, *Pediomys* compares favorably with *Monodelphopsis*, a group 5 taxon. In size, *Monodelphopsis travassosi* compares with *Pediomys cooki* (see Clemens 1966, Table 9). *Monodelphopsis* shares with *Pediomys* ovoid shaped P$_{2-3}$ with distinct posterobasal heels and labial (compare *P. ?hatcheri*, Fig. 70 in Clemens 1966 with my Fig. 30) and lingual (compare *P. ?krejcii*, Fig. 65 in Clemens 1966 with my Fig. 30) cingula, M$_5$ slightly smaller than M$_4$, talonid distinctly basined with crista obliqua shifted labially and contacting back of trigonid at base of protoconid, enormous basined protocone, distinct para- and metaconules, centrocrista rectilinear, stylar shelf and cusps greatly reduced (although anterior part of stylar shelf less reduced and stylar cusp B somewhat larger in *Monodelphopsis* than in specimens figured by Clemens 1966), and stylar

cusp D present as in most species of *Pediomys*. These taxa differ in the paracone and metacone being subequal in size in *Pediomys*, while in *Monodelphopsis* the metacone is much larger and higher than the paracone.

Representatives of the family Peradectidae (*sensu* Reig *et al.* 1987; Peradectini of Crochet e.g. 1980) include *Alphadon*, *Albertatherium*, *Peradectes*, and *Mimoperadectes* which collectively range from Late Cretaceous (Campanian) to Early Eocene in North America, Europe, and Africa. *Alphadon* has long been regarded as representing or approximating the ancestral marsupial in aspects of its dental structure (see Clemens 1966, 1979; Fox 1979a). Peradectids compare best with Itaborai group 6 taxa in retaining many character states regarded primitive for marsupials. *Sternbergia* and *Alphadon ?rhaister* (see Figs 62 and 64 in Clemens 1966) have a labial cingulum on M$_{2-4}$. Peradectids differ from Itaborai group 6 taxa in the latter having a V-shaped centrocrista, metacone about twice as large and higher than paracone, various differences in size of stylar cusps, M$_5$ in *Sternbergia* shorter and narrower than M$_4$, P$_3$ in *Marmosopsis* and *Sternbergia* subequal in size, and talonids of M$_{2-4}$ shorter than trigonids. *Marmosopsis*, *Itaboraidelphys*, and *Sternbergia* differ from peradectids in structure of the centrocrista and relative size of para- and metacones. Thus, these group 6 taxa are derived in ways that suggest special affinity with Didelphidae (*sensu* Reig *et al.* 1987) rather than Peradectidae.

The didelphid subfamily Herpetotheriinae (*sensu* Reig *et al.* 1987) includes *Amphiperatherium*, *Herpetotherium*, and *Peratherium* which are known respectively from the Early Eocene-Late Miocene of Europe, Early Eocene-Early Miocene of North America, Early Eocene-Late Oligocene of Europe, and with a member in the Oligocene of Africa. These taxa compare favourably with Itaborai group 6 and differ from peradectids in the following features: lack para- and metaconules, V-shaped centrocrista, metacone about twice as large and higher than paracone, metacrista more elongate, talonid shorter than trigonid, and M$_5$ narrower than M$_4$. These features thus demonstrate a didelphid affinity for the Herpetotheriinae as already shown by Reig *et al.* (1987).

4) AFFINITIES OF ITABORAI TAXA WITH MARSUPIALIA IN AUSTRALIA

Archer (1976, 1982), and Kirsch and Archer (1982) provide summaries of dental characters in Australian carnivorous marsupials. As dasyurids are believed to approximate or represent the basal stock for the Australian radiation, I make my comparisons only with members of that group.

One comparison warrants special mention. The dasyurid *Ankotarinja* (see Archer 1982) from the Middle Miocene shows special affinity with Itaborai group 5 taxa, and especially with *Mirandatherium*. The lower molars of these taxa have trigonids moderately elevated above talonids, M_{2-4} talonids distinctly basined and wider than trigonids, and M_5 reduced relative to M_4. The upper molar of *Ankotarinja* described by Archer (1982, Fig. 1C) appears to be a left M^4? and it compares with *Mirandatherium* in having stylar shelf moderately developed, centrocrista nearly rectilinear, stylar cusps B and D moderately developed and subequal in size, cusp C present but smaller than stylar cusps B and D, ectoflexus on M^4 moderately deep, and metacrista (metastylar area) only moderately developed. These taxa differ in *Ankotarinja* being more primitive in having a larger paracrista, and in para- and metacones more subequal in size.

In more general terms, dasyurids as a whole compare best with Itaborai group 6 taxa in retaining many features regarded plesiomorphic for marsupials. Yet, both groups have a distinct V-shaped centrocrista (except *Ankotarinja*) and a metacone notably larger than paracone. *Sternbergia* and some dasyurids (e.g. *Dasyuroides burnei*, see Archer 1976, Fig. 7H) have a labial cingulum on M_{2-4}. Some dasyurids (see Archer 1976, Fig. 7) have stylar cusp D larger than B, while the reverse is true in Itaborai group 6 taxa. The marked size reduction of P_3 in some dasyurids occurs only in the didelphid *Zygolestes* (see above).

Peramelids compare with Itaborai group 4 taxa in reduction of anterior dentition, elongation of ramus anteriorly, and procumbent structure of incisors and canine. The upper molars of peramelids have large stylar cusps B and D, and short metacristae, features shared with Itaborai groups 1 and 2.

5) Phylogenetics Of Itaborai "Opossum-Like" Taxa

During the course of this study it became evident that the families Bonapartheriidae and Prepidolopidae are not didelphoids as emphatically concluded by Pascual (1980b, 1981), but are polydolopoids. In this part of the PHYLOGENETICS section I will first clarify the systematic position of these families within the Polydolopoidea. Second, I will identify the phylogenetic affinities of the Itaborai "opossum-like" taxa, building largely upon the work of Reig *et al.* (1985, 1987). Third, I will consider the phylogenetic relationships of these taxa with other marsupial groups. A classification for these and other non-Australian Marsupialia based on the following analysis is presented in Table 17, and a phylogeny is proposed in Figure 55.

The Polydolopoidea are monographed by Marshall (1982d); one family (Polydolopidae) and two subfamilies (Epidolopinae, Polydolopinae) are recognized. The Epidolopinae, represented by *Epidolops*, is regarded as the plesiomorphic subfamily, while the Polydolopinae (*Polydolops, Pseudolops, Amphidolops, Eudolops*) are collectively regarded as more derived. A fifth polydolopine genus, *Antarctodolops*, was recently described from the Late Eocene of Seymour Island, Antarctica (Woodburne and Zinsmeister 1984).

The primitive number of incisors in the proto-polydolopoid is not certain. Pascual (1981) infers (p. 509) that the number in *Bonapartherium* is "aparentemente 5/4", while on p. 511 he records that in the only specimen retaining evidence of upper incisors there are "fragmentos de ambos incisivos distales", meaning there is evidence for only two that he records. For the lower incisors he observes (p. 513) that in one specimen "puede observarse, aunque dificultosamente, las raíces de 4 incisivos." Thus, there appears to be evidence of only two upper and questionably four lower incisors in *Bonapartherium*. In *Prepidolops* the number of upper incisors is unknown, while there is evidence (Pascual 1980b:223–224) of "al menos tres incisivos" in the lower jaw. *Epidolops* (Epidolopinae) apparently has 3/2 incisors, while *Eudolops* (Polydolopinae) has clear evidence of only two lower incisors (Marshall 1982d). Although inconclusive, the proto-polydolopoid incisor number may thus have been as high as 5/4. What is certain, however, is that some reduction in this number occurred within several polydolopoid lineages.

Knowledge of size and structure of the canines is based on the following. In *Prepidolops* upper canines are unknown, while the lowers are procumbent and reduced relative to the molars in adults (Pascual 1980b:224). In *Bonapartherium* the lowers, based on knowledge of their roots, are large and "con la forma e implantación de los más comunes Didelphidae" (Pascual 1981:513); while the upper canine (see Pascual 1981, Pl. I) is not unlike that expected in *Epidolops* as demonstrated by alveoli preserved in a relatively complete skull (see Marshall 1982d, Fig. 62). In Polydolopidae the lower canine is large and semiprocumbent (Marshall 1982d). Marshall (1982d) demonstrates that the large semiprocumbent tooth in Polydolopidae is the canine, and not an incisor as believed by earlier workers. This conclusion is reinforced by knowledge of the lower anterior dentition of *Bonapartherium* and *Prepidolops* in which the incisors are apparently or actually reduced, and the canine well developed. Thus, the size and structure of the canine in *Bonapartherium* (and possibly the incisors as well) most closely approximate that expected in the proto-

Table 17. Classification of non-Australian Marsupialia proposed in this study. A detailed account of authorship, synonymies, and references can be found in Marshall (1981b) and Reig *et al.* (1987). Because of space limitations I do not here include, although I do recognize, the tribal ranks for Didelphinae (see Reig *et al.* 1985b), and Caenolestinae and Abderitinae (see Marshall 1980b). The stratigraphic ranges of South American genera are based on the time scale in Figure 1. Composite stratigraphic ranges of families and subfamilies with indication of their apparent affinities are shown in Figure 55. The genera are listed in alphabetical order.

Suborder Didelphimorphia Gill 1872

I follow the basic identity of this suborder as employed by Reig *et al.* (1985, 1987) and list genera only in those families and subfamilies that require emendation to that work based on data presented in this study.

Superfamily Microbiotherioidea (Ameghino 1887) Reig, Kirsch and Marshall 1987

Family Peradectidae (Crochet 1979) Reig, Kirsch and Marshall 1987
Albertatherium Fox 1972 (Campanian)
Alphadon Simpson 1927 (Campanian-Maestrichtian)
Mimoperadectes Bown and Rose 1979 (Clarkforkian-Wasatchian)
Peradectes Matthew and Granger 1921 (Maestrichtian-"middle Oligocene")

Family Stagodontidae Marsh 1889

Family Microbiotheriidae Ameghino 1887

Subfamily Microbiotheriinae (Ameghino 1887) Simpson 1929
Dromiciops Thomas 1894 (Recent)
Eomicrobiotherium Marshall 1982 (Casamayoran)
Microbiotherium Ameghino 1887 (Colhuehuapian-Santacrucian)
Mirandatherium Paula Couto 1952 (Itaboraian)
Pachybiotherium Ameghino 1902 (Colhuehuapian)

Subfamily Pediomyinae Simpson 1927
Aquiladelphis Fox 1971 (Campanian)
Monodelphopsis Paula Couto 1952 (Itaboraian)
Pediomys Marsh 1899 (Campanian-Maestrichtian)

Superfamily Borhyaenoidea (Ameghino 1894) Simpson 1930

Family Borhyaenidae Ameghino 1894

Subfamily Borhyaeninae (Ameghino 1894) Cabrera 1927

Subfamily Hathliacyninae (Ameghino 1894) Kirsch 1977
(The spelling of this family-group name follows that used by Petter and Hoffstetter 1983 : 207.)

Subfamily Proborhyaeninae (Ameghino 1897) Trouessart 1898

Subfamily Prothylacyninae (Ameghino 1894) Trouessart 1898

Family Thylacosmilidae (Riggs 1933) Marshall 1976

Superfamily Didelphoidea (Gray 1821) Osborn 1910

Family Didelphidae Gray 1821

Subfamily Didelphinae (Gray 1821) Simpson 1927
Includes the genera listed by Reig *et al.* (1987) plus the following:
Coona Simpson 1938 (Casamayoran)
Itaboraidelphys Marshall and de Muizon 1984 (Itaboraian)
Marmosopsis Paula Couto 1962 (Itaboraian)
Sternbergia Paula Couto 1970 (Itaboraian)

Tribe Zygolestini nov.
Zygolestes Ameghino 1898 (Montehermosan)

Subfamily Caluromyinae Kirsch and Reig 1977 in Kirsch 1977
Includes only the following genera:
Caluromys Allen 1900 (Recent)
Caluromysiops Sanborn 1951 (Recent)
Glironia Thomas 1912 (Recent)

Subfamily Herpetotheriinae Trouessart 1879

Subfamily Derorhynchinae nov.
Derorhynchus Paula Couto 1952 (Itaboraian)
Minusculodelphis Paula Couto 1962 (Itaboraian)

Subfamily Eobrasiliinae nov.
Didelphopsis Paula Couto 1952 (Itaboraian)
Eobrasilia Simpson 1947 (Itaboraian)
Gaylordia Paula Couto 1952 (Itaboraian)

Family Sparassocynidae (Reig 1958) Reig, Kirsch and Marshall 1987

Suborder Polydolopimorphia (Ameghino 1897) nov.

Superfamily Caroloameghinoidea (Ameghino 1901) nov.

Family Caroloameghiniidae Ameghino 1901

Subfamily Caroloameghiniinae (Ameghino 1901) Clemens 1966 (includes Glasbiinae Clemens 1966)
Caroloameghinia Ameghino 1901 (Casamayoran)
Glasbius Clemens 1966 (Maestrichtian)
Procaroloameghinia Marshall 1982 (Itaboraian)
Robertbutleria nov. (Itaboraian)
Roberthoffstetteria Marshall, de Muizon and Sigé 1983 (Maestrichtian)

Table 17. Continued.

Subfamily Protodidelphinae nov.
    Bobbschaefferia (Paula Couto 1952)(Itaboraian)
    Guggenheimia Paula Couto 1952 (Itaboraian)
    Protodidelphis Paula Couto (Itaboraian)
    Reigia Pascual 1983 (Mustersan? or Divisaderan?)
    Zeusdelphys nov. (Itaboraian)
Superfamily Caenolestoidea (Trouessart 1898) Osborn 1910
  Family Caenolestidae Trouessart 1898
    Subfamily Caenolestinae (Trouessart 1898) Sinclair 1906
    Subfamily Palaeothentinae Sinclair 1906
    Subfamily Abderitinae (Ameghino 1889) Sinclair 1906
Superfamily Argyrolagoidea (Ameghino 1904) Simpson 1970
  Family Gashterniidae nov.
    Gashternia Simpson 1935 (Riochican)
  Family Groeberiidae Patterson 1952
    Groeberia Patterson 1952 (Divisaderan)
  Family innominate Pascual and Carlini 1984 (Colhuehuapian)
  Family Argyrolagidae Ameghino 1904
    Argyrolagus Ameghino 1904 (Montehermosan-Chapadmalalan)
    Microtragulus Ameghino 1904 (Huayquerian-Uquian)
    Proargyrolagus Wolff 1984 (Deseadan)
Superfamily Polydolopoidea (Ameghino 1897) Clemens and Marshall 1976
  Family Prepidolopidae Pascual 1980
    Prepidolops Pascual 1980 (Casamayoran-Mustersan? or Divisaderan?)
  Family Bonapartheriidae Pascual 1980
    Bonapartherium Pascual 1980 (Casamayoran)
  Family Polydolopidae Ameghino 1897
    Subfamily Epidolopinae Pascual and Bond 1981
      Epidolops Paula Couto 1952 (Itaboraian)
    Subfamily Polydolopinae (Ameghino 1897) Pascual and Bond 1981
      Amphidolops Ameghino 1902 (Riochican-Casamayoran)
      Antarctodolops Woodburne and Zinsmeister 1984 ("Late Eocene")
      Eudolops Ameghino 1897 (Casamayoran)
      Polydolops Ameghino 1897 (Riochican-Mustersan)
      Pseudolops Ameghino 1902 (Casamayoran)

polydolopoid. The Prepidolopidae and Polydolopidae are thus apparently derived in aspects of incisor number and/or canine size and structure.

In *Bonapartherium* and *Prepidolops* the P_1 and P_2 are reduced to low anvil-shaped structures. In *Epidolops* the structure of P_{1-2} are similar but are more reduced in size; P_1 is sometimes lost and P_2 is single or double rooted. In Polydolopinae the P_1 is apparently lost, and P_2 is lost in some specimens of *Polydolops mayoi* (see Marshall 1982d:44).

The P_3 is large and trenchant in all polydolopoids. In Prepidolopidae, Bonapartheriidae, and Polydolopinae it is aligned in the same anteroposterior axis as the molars, while in Epidolopinae it is set at about a 30° angle relative to the axis of the molar series (a derived condition).

The P^{1-2} are unknown in *Prepidolops*, although given the known structure of P_{1-2} it is probable that these teeth were present and weakly to moderately developed. In *Bonapartherium* P^1 is lost, while P^2 and P^3 are large (the latter is enormous) and both have large buccal shelves. *Epidolops* has P^{1-2} very reduced and a very large sectorial P^3 that is set obliquely in the jaw as is P_3. P^1 is unknown in Polydolopinae, but P^{2-3} are large

sectorial blades. All polydolopoids are derived in having a tri-rooted P^3 (two labially, one posterolingually). The basic size and structure of P^{2-3} in *Bonapartherium* is, except for the buccal shelves, similar to that in Polydolopinae (see figures of *Polydolops* and *Eudolops* in Marshall 1982d). Thus, the basic size and structure of P^{2-3} in Polydolopinae is regarded as approximating that expected in the ancestral polydolopoid.

The M_2 trigonid is slightly elevated above the talonid in all polydolopoids and participates to some degree in the continued sectorial function of P_3. In *Prepidolops* the trigonid cusps are distinct and least specialized, representing the primitive condition for polydolopoids (although the paraconid is somewhat reduced). In *Epidolops* the trigonid cusps are slightly higher and the paraconid is virtually absent. In *Bonapartherium* the trigonid cusps are more elevated and fused into a blunt anteroposterior blade that is continuous with the posterior edge of P_3. In Polydolopinae the anterior most edge of the M_2 trigonid is very high and distinctly blade-like, and clearly shares in the sectorial function of P_3. The trigonid of *Bonapartherium* and Polydolopinae are thus modified in slightly different ways, although both modifications can easily be derived from a state represented by *Epidolops*, which in turn can be derived from a state represented by *Prepidolops*. In all polydolopoids the M_2 talonid is broad and distinctly basined.

In all polydolopoids M_{3-4} are brachydont or bunodont with trigonids and talonids subequal in height. *Prepidolops* is brachydont, trigonid cusps are distinct although the paraconid is reduced relative to larger proto- and metaconid, and the hypo- and entoconids are large and subequal in size. In the other taxa these molars are bunodont, the paraconid is very reduced or virtually lost, the teeth are rectangular in shape in occlusal view with a bilophodont appearance resulting from a transverse crest connecting proto- and metaconid and another connecting hypo- and entoconid. These features are best seen in *Bonapartherium* and *Epidolops*, while in Polydolopinae they are less distinct due to shallowing of the molar basins and addition of supernumary cusps.

In *Bonapartherium* the M_5 is slightly shorter and narrower than the M_4 but retains the general structure of M_{3-4}. In *Prepidolops* and *Epidolops* the M_5 is very reduced, ovoid, and lacks details of molar structure. In Polydolopinae the M_5 is lost.

In *Prepidolops* M^{2-4} are distinctly tritubercular. In the other taxa a hypocone is added and these teeth are quadritubercular. In *Prepidolops*, *Bonapartherium*, and *Epidolops* stylar cusps B and D are large and subequal in size and fused basally by crest-like connections with the para- and metacone respectively, giving this part of the molars a selenodont appearance. This arrangement is not obvious in Polydolopinae where addition of supernumary cusps and general structural rearrangement of the basic quadritubercular pattern obscure the ancestral selenodont condition. M^5 is reduced relative to M^4 in *Bonapartherium*, is very reduced in *Prepidolops* and *Epidolops*, and is lost in Polydolopinae.

Thus, polydolopoids share an array of specializations either not found jointly or at all in any other group of South American Marsupialia: i.e. P_{1-2} very reduced, anvil-shaped; P^3 tri-rooted and in part or in total blade-like; P_3 large, trenchant; molars brachydont to bunodont; trigonid of M_2 elevated above talonid and participates to some degree in sectorial function of P_3; M_{3-4} with paraconid reduced, trigonids and talonids subequal in height, hypo- and entoconid large and subequal in size; stylar cusps B and D large and basally fused by crest-like structures to para- and metacone respectively, giving labial side of tooth a selenodont appearance; and M5 reduced relative to M4. Variations from this proto-polydolopoid pattern occur in all taxa, although the inferred joint occurrence of these features primitively within this group permits recognition of its monophyly. My interpretation of the phylogenetic relationships of the families

and subfamilies of polydolopoids and of their relative branching pattern is shown in the cladogram in Figure 53.

The Riochican age *Gashternia ctalehor* Simpson 1935a compares favorably with some polydolopoids. As discussed above, the M_{2-3} in *Gashternia* are distinctly bunodont, trigonid and talonid subequal in height, paraconid very reduced, molars rectangular in occlusal view with bilophodont structure resulting from a transverse crest connecting proto- with metaconid and another crest connecting large and subequal sized hypo- with entoconid. "Alveoli indicate that there is a large semiprocumbent canine, followed by a very small, one-rooted, crowded P_1, a larger two-rooted P_2, closely crowded and planted obliquely so that its anterior root is posteroexternal to that of P_1, and a still larger two-rooted P_3." (Simpson 1948:69). Aspects of structure of the lower molars suggest a stage of polydolopoid evolution that approximates point 3 to 5 in the cladogram in Figure 53. The trigonid of M_2 is not elevated above the talonid, although this could be explained by a slight secondary reduction in its height, which as evidenced by *Prepidolops* was not great in early polydolopoids. The P_3, based on knowledge of alveoli, is smaller than that expected in polydolopoids at this point in their evolution, but this too could represent a simple secondary reduction in its size. In fact, reduction in height of the M_2 trigonid would predictably accompany size reduction of P_3. The semiprocumbent canine and reduced size of the P_{1-2} further suggest a stage of evolution that approximates point 5 in the Figure 53 cladogram. Below I present alternative interpretations of the possible affinities of *Gashternia* with early Tertiary age marsupial groups.

In terms of Middle Tertiary age taxa, *Gashternia* agrees most favorably with the Divisaderan (Early Oligocene) age *Groeberia*: i.e. molars bunodont, rectangular shaped in occlusal view, bilophodont, trigonid and talonid of M_{2-3} subequal in height, paraconid very reduced or absent, P_3 smaller than M_2, and subequal in height, canine semiprocumbent (possibly representing the gliriform tooth in *Groeberia*, see below), and P_{1-2} reduced in *Gashternia* and lost in *Groeberia*. In all aspects of its structure, *Gashternia* represents an ideal ancestor for *Groeberia*. I therefore propose that *Gashternia* be grouped with *Groeberia*, although not in the same family. The presence of a less specialized canine (i.e. the gliriform tooth in *Groeberia*) and retention of P_{1-2} in *Gashternia* represent notable differences from *Groeberia*, and these I interpret as warranting family distinction. I thus place *Gashternia* in its own family, Gashterniidae nov., in the same superfamily with the Groeberiidae. The numerous similarities shared by *Gashternia* and

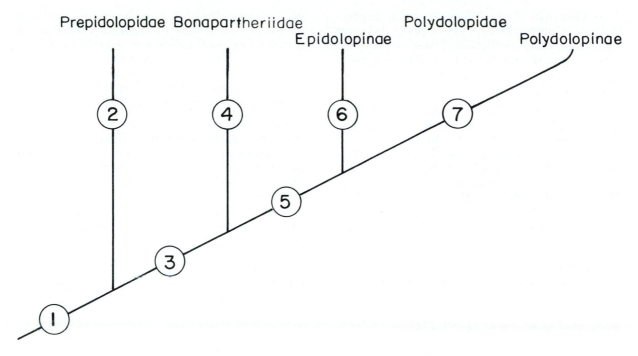

Fig. 53. Cladogram showing probable phylogenetic relationships of Polydolopoidea. Diagram depicts only relative position of common ancestor. Numbers refer to distribution and relative times of acquisition of derived character states (expanded and modified after Figure 74 in Marshall 1982d).

1. Dental formula possibly 5/4 1/1 3/3 5/5; C large; P^1 very reduced; P^2 well developed, two-rooted blade-like; P^3 three-rooted (two labial, one posterolingual); P_{1-2} very reduced, anvil-shaped; P3 trenchant, large and aligned in same anteroposterior axis as molar series; molars tritubercular, brachydont; trigonid of M_1 slightly elevated above talonid and participates slightly in shearing function of P_3; M_{3-4} trigonids and talonids subequal in height, paraconid reduced, talonid basined and broader than trigonid; stylar cusps B and D large, united basally to paracone and metacone by crest-like structures, giving labial side of M^{2-4} selenodont appearance; M5 reduced relative to M4.

2. Incisor number possibly reduced to ?/3; C reduced in size relative to molars; inferred possible reduction of P^2; size increase of P3; M5 greatly reduced relative to M4; lower incisors and C semiprocumbent.

3. M^{2-5} quadritubercular (addition of hypocone); molars bunodont; M_{3-4} rectangular in shape in occlusal view, paraconid very reduced (virtually lost), attain bilophodont appearance by transverse crest connecting subequal sized protoconid and metaconid, and another connecting subequal sized hypoconid and entoconid.

4. P^1 lost; P^{2-3} modified into partial crushing function by addition of large buccal shelf; trigonid of M_2 notably elevated above talonid, and trigonid cusps fused into blunt anteroposterior blade; P^{2-3}/P_3 and M_2 trigonid thus modified into shearing-crushing combination; M_{2-3} develop slight tendency toward hypsodonty, but not M_{4-5}.

5. Incisors reduced to 3/2; P_{1-2} greatly reduced in size; lower canine procumbent; molar basins shallow with very lightly wrinkled enamel; M5 greatly reduced in size relative to M4.

6. P_1 sometimes lost; P^2 very small, single or double rooted, crown without distinct cusps, flattened, irregularly oval in outline, loss of sectorial function; axis of P3 set at about 30° angle relative to axis of molar series; sectorial blades P^3/P_3 only.

7. Complete loss of P1 and M5; anterior edge of M_2 trigonid distinctly elevated above rest of molar series and modified into shear-blade; molars multituberculate; supernumary cusps develop along labial and lingual sides of M_2 talonid basin, M_{3-4} talonid basin and on labial sides of M^{2-3}.

some polydolopoids (see above) suggest affinity between the two, and I will return to their probable relationship below.

During the course of this study it has become evident that *Groeberia* also shares special affinity with members of the family Argyrolagidae. Argyrolagids are known from Deseadan (Wolff 1984) and Later Cenozoic (Simpson 1970a) faunas of South America. *Groeberia* shares several unique specializations with argyrolagids which demonstrate their affinity: i.e. two upper gliriform incisors separated from cheek teeth by long diastema; large lower gliriform tooth (the canine) separated from cheek teeth by long diastema ". . . with limited enamel and with intraalveolar part parallel with median line of the symphysis; subequal molars of similar size;

strong projecting coronal process; . . . absence of masseteric crest" (Simpson 1970a:42); reduction or loss of P^{1-2}; and similar relative size and structure of P_3 and M_2. I also believe that the lower gliriform tooth in argyrolagids is the canine and not an incisor as believed by Patterson (1952) and Simpson (1970a). This conclusion is based on the same reasoning I used to identify the large semiprocumbent tooth in polydolopoids as the canine (Marshall 1982d), and is supported by the identity of this tooth in *Gashternia* and hence in *Groeberia* (see above). Thus, the single rooted incisiform tooth posterior to the gliriform tooth in argyrolagids is probably the P_1, which is also small and single rooted in *Gashternia*. Pascual and Carlini (1984:14) report the occurrence of a new unnamed family of argyrolagoids from the

Colhuehuapian (Early Miocene) which displays features intermediate in aspect between argyrolagids and groeberiids. Those workers also remarked on the special affinity of these groups and championed their placement in the same superfamily.

For the above reasons I opt to place the Argyrolagidae (*sensu* Simpson 1970a) in the same superfamily as the Gashterniidae and Groeberiidae. The superfamily name available is Argyrolagoidea Simpson 1970a:3 which has priority of date of publication over the super-family Groeberioidea Clemens and Marshall 1976:10 (Table 17). The affinities of Argyrolagoidea with Polydolopoidea are discussed below.

Returning to the Itaboraian "opossum-like" taxa, four of them (*Mirandatherium, Monodelphopsis, Procaroloameghinia, Robertbutleria*) show unambiguous affinity with suprageneric groups recognized by Reig *et al.* (1985, 1987).

Mirandatherium is clearly a member of the sub-family Microbiotheriinae (*sensu* Reig *et al.* 1985, 1987) which includes the Casamayoran age *Eomicrobiotherium*, the Colhuehuapian age *Pachybiotherium*, the Colhuehuapian and Santa-crucian age *Microbiotherium* (with five species), and the living *Dromiciops australis* (see above, and Marshall 1982a). The Middle Miocene age *Ankotarinja* from Australia also shows marked affinity with *Mirandatherium*, the biogeographic implications of which are discussed below.

Monodelphopsis is a member of the subfamily Pediomyinae (*sensu* Reig *et al.* 1985, 1987), and shows striking similarity to species of *Pediomys* from the Late Cretaceous of North America (see above). This is the first secure record of a pediomyine in South America and the youngest known representative of this group anywhere (save for the questionable report of a "pediomyid" in the Miocene of Texas, Slaughter 1978). Sigé (1972) identified a possible "pediomyid" in the Late Cretaceous Laguna Umayo local fauna of Peru. This determination is based largely on the expansion of the talonid on a lower molar fragment, a character also shared with Microbiotheriinae. The identification of a "pediomyid" from Laguna Umayo seems reason-able given the presence of a member in the Middle Paleocene Itaborai fauna of South America. However, reference of this specimen to either the Microbiotheriinae or Pediomyinae is not possible to establish based on the fragmen-tary nature of the material, and this identification is best regarded as "microbiotheriid"?. The marked similarity in dental structure of *Miranda-therium* and *Monodelphopsis* further substantiates the referral of Pediomyinae and Microbiotheriinae to the family Microbiotheriidae (Reig *et al.* 1985, 1987) and explains their initial allocation to

group 5 (see above). It must be emphasized that the distinction of these two taxa and their referral to different subfamilies was greatly facilitated and largely based on knowledge of their upper dentitions. A revised and expanded classification of the Microbiotheriinae and Pediomyinae based on this study is presented in Table 17.

Procaroloameghinia represents an ideal structural ancestor for the Casamayoran age *Caroloameghinia*, and the two were placed together in the subfamily Caroloameghiniinae by Marshall (1982c) and later in the family Caroloa-meghiniidae by Reig *et al.* (1985, 1987). *Roberthoffstetteria* from the Late Cretaceous of Bolivia was tentatively assigned by Marshall *et al.* (1983) to the Caroloameghiniinae and later placed unquestionably in the Caroloameghini-idae by Reig *et al.* (1985, 1987). The Caroloameghiniidae were thus recognized only from the Late Cretaceous to Early Eocene of South America.

A structurally similar group, the "Glasbiinae", was placed by Reig *et al.* (1985, 1987) in the family Microbiotheriidae. This subfamily was long known only by the North American Late Cretaceous genus *Glasbius*, and was long recognized as a monotypic group (Clemens 1966, 1979). Crochet (1980) tentatively assigned *Proto-didelphis* to this subfamily, while Pascual (1983) regarded the South American Middle or Late Eocene genus *Reigia* as an unambiguous member. As noted above, *Robertbutleria* shows striking similarity to *Glasbius* and may be regarded as a structural descendant of that or of a closely related genus. Thus, *Robertbutleria* is unquestionably referrable to the "Glasbiinae", while at the same time it shares many features with group 1 and 2 taxa, some of which are unquestionably referrable to the subfamily Caroloameghiniinae.

Many workers (e.g. Clemens 1966; Marshall 1982c; Reig *et al.* 1987) have commented on the notable similarity in molar structure of Caroloa-meghiniinae and "Glasbiinae", although all have concluded that these bunodont groups represent classic examples of convergence or parallelism in evolution. These conclusions were based primarily on comparison of *Caroloameghinia* and *Glasbius*. As is demonstrated below, *Glasbius* is generalized in most aspects of its molar structure while *Caroloameghinia*, a closely related form, is simply more derived along the same structural lines as *Glasbius*. Thus, these taxa represent structural extremes within a monophyletic group, and primarily for this reason their true affinity has long been obscured. The discovery and/or recognition of additional Caroloa-meghiniinae and "Glasbiinae" provides for a broader interpretation of the structural limits of each group, such that their distinction is no longer justified. These additional taxa display an array of features that are intermediate between *Glasbius* and *Caroloameghinia*.

Clemens (1966:24) diagnosed the "Glasbiinae" as "stylar shelf broad with bulbous cusps; principal cusps of upper molars relatively low; prominent labial basal cingulum on lower molars; trigonid short, anteroposteriorly; last molar smaller than other molars". Of these "diagnostic" characters, two occur in members of unrelated groups. A labial cingulum on lower molars occurs in some species of *Alphadon* (see Clemens 1966, Fig. 62), in some Caenolestidae (see Marshall 1980b, Figs 6,7), and in *Sternbergia* (see Figs 41, 42 above). A reduced M_5, relative to M_4 occurs in at least one member of each of the six groups from Itaborai (Table 15), in microbiotheres, living Caluromyinae, caenolestids, polydolopoids, and numerous Australian lineages. These two characters are thus diagnostic of the genus *Glasbius*, while the other three "diagnostic" features occur also in Caroloameghiinae (see diagnosis in Marshall 1982c). Furthermore, all of the taxa assigned to both groups have large inflated stylar cusps B and D, a feature unknown in other "opossum-like" taxa (save Stagodontidae). Clemens (1966) emphasized the importance of these large stylar cusps as diagnostic features of *Glasbius*, although in retrospect and with increased knowledge of other taxa, the size and structure of these stylar cusps are now recognized as representing a diagnostic feature of suprageneric importance (see below).

Reig *et al.* (1987) recently assigned the "Glasbiinae" to the family Microbiotheriidae and regarded the Caroloameghiniidae as a sister-group. This distinction was based solely on a character state analysis of *Glasbius* and *Caroloameghinia* and on recognition of a larger stylar shelf and reduced M5's in *Glasbius*, characters which aligned that genus with *Pediomys*, *Microbiotherium*, and their allies. I now believe that we placed too much emphasis on these two characters and did not appreciate the significance, or lack thereof, of a reduced stylar shelf in *Caroloameghinia*. The stylar shelf in *Caroloameghinia* is indeed reduced relative to that of *Glasbius*, but based on knowledge of additional taxa it now appears that this reduction is simply a derived condition within that one taxon, while stylar cusps B and D are still large relative to the para- and metacone. The reduced size of the M_5 should have minimal weighting in phylogenetic inference (especially when it is only one of two primary characters used) since reduction of this tooth has occurred independently in many unrelated groups (see above).

For these reasons I offer the following emendations to the classification of Microbiotheriidae and Caroloameghiniidae proposed by Reig *et al.* (1987). I concur that caroloameghiniids are a sister-taxon of microbiotheriids, but only if the

Microbiotheriidae is redefined to include only the subfamilies Microbiotheriinae and Pediomyinae, and if the Glasbiinae are united within the Caroloameghiniidae The problematical position of *Procaroloameghinia* within Itaborai groups 1 and 5 is a clear reflection of the close structural similarity of Microbiotheriidae (Microbiotheriinae and Pediomyinae) and Caroloameghiniidae (see below). Thus, *Procaroloameghinia* approximates or retains many structural features expected in the ancestral caroloameghiniid.

Caroloameghiniidae are diagnosed as: molars brachydont to bunodont, cusps bulbous, trigonids subequal in height or only slightly higher than talonids, trigonid foreshortened anteroposteriorly, talonid broad with large hypo- and entoconid, protocone large, paracrista very reduced or absent (i.c. paracone not united with stylar cusp B), metacrista (metastylar area) short, stylar shelf with large subequal sized cusps B and D. Two subfamilies are recognized (Table 17). The Caroloameghiniinae (including Glasbiinae Clemens 1966) includes *Caroloameghinia*, *Glasbius*, *Procaroloameghinia*, *Roberthoffstetteria*, and *Robertbutleria*: distinctly bunodont, centrocrista rectilinear or nearly so (except in *Robertbutleria*), P_{2-3} ovoid and somewhat bulbous, and M_{2-4} talonid broader than trigonid. To emphasize the integrading nature of numerous derived character states among these taxa I note the following: large metacingulum developed along posterior surface of metacrista (*Glasbius*, *Roberthoffstetteria*), weak labial and lingual cingula on upper molars (*Glasbius*, *Robertbutleria*), stylar cusp C absent (*Caroloameghinia*, *Robertbutleria*, *Glasbius twitchelli*), stylar cusp C present (*Roberthoffstetteria*, *Glasbius intricatus*), M_5 longer and wider than M_4 (*Roberthoffstetteria*), M_5 narrower but subequal in length to M_4 (*Caroloameghinia*, *Procaroloameghinia*), and M_5 distinctly shorter and narrower than M_4 (*Glasbius*). The Protodidelphinae nov. include *Bobbschaefferia*, *Protodidelphis*, *Guggenheimia*, *Zeusdelphys*, and *Reigia*: brachydont, V-shaped centrocrista, and talonids of M_{2-4} subequal in width to trigonids.

The Caroloameghiniinae are derived relative to the Protodidelphinae in possessing marked bunodonty, having surely passed through a brachydont state best approximated by Microbiotheriinae. The Protodidelphinae are primitive relative to Caroloameghiniinae in their possession of brachydonty, but derived in having a V-shaped centrocrista and notably larger hypo- and entoconids; *Bobbschaefferia* represents the structurally most generalized taxon. The inferred polarities suggest that the M_{2-4} talonids in Protodidelphinae have undergone a slight secondary reduction in width relative to the trigonids.

Three of the Itaborai group 6 taxa (*Marmosopsis, Itaboraidelphys, Sternbergia*) have a V-shaped centrocrista and the metacone is about twice as large and higher than the paracone. In these features they represent generalized Didelphidae *sensu* Reig *et al.* (1987). *Marmosopsis* and *Sternbergia* are also derived relative to Peradectidae (*sensu* Reig *et al.* 1987) in having P_{2-3} subequal in length and width. These three taxa resemble Peradectidae in retaining distinct para- and metaconules, relative size of M^{3-4}, and in the moderately developed metacrista (i.e. metastylar region is unspecialized). Thus, these taxa collectively and individually display a spectrum of character states which suggest affinity with both Peradectidae and Didelphidae (*sensu* Reig *et al.* 1987). This situation is exactly what I would expect to find in generalized "opossum-like" taxa of Middle Paleocene age in South America, a time which apparently approximates or represents the final transitional stage in the evolution of didelphids from peradectid ancestors. This process apparently involved stochastic rates of change in multiple characters and character complexes. The mosaic distribution of typical peradectid and typical didelphid character states in these taxa simply documents the transitional nature of this process. Given this reality, the assignment of these taxa to Peradectidae or Didelphidae is made not by the last appearance of peradectid (primitive) character states but by the first appearance of didelphid (derived) character states. This situation is analogous to the biostratigraphic principal whereby first and not last appearances of taxa are of primary importance in defining time boundaries. Using this criteria these Itaborai taxa must be classified as generalized Didelphidae (i.e Didelphinae) because they have, among other features, a V-shaped centrocrista and a paracone reduced relative to the metacone. These taxa thus display two derived states that occur primitively in didelphoids, states regarded as structurally ancestral for didelphoids (e.g. relative size of P_{2-3}), and states that are lost in some or all later didelphoids (e.g. loss of para- and metaconules) (see Reig *et al.* 1987). These group 6 taxa I thus classify as Didelphinae, because they do not display features which permit their collective or individual separation from other members of that subfamily, and because they possess a suite of features that permit them to be regarded as ancestral to some or all later didelphoids. In the Didelphinae I also include the Casamayoran age *Coona* which (see above) compares favorably with these three group 6 taxa in being a structurally generalized didelphoid (Table 17). However, I make no formal attempt to group these four taxa into tribal ranks as was done for their living and Neocenozoic relatives by Reig *et al.* (1987). Nevertheless, the data do conclusively indicate that *Marmosopsis* and *Sternbergia* are more closely related to each other than they are to *Itaboraidelphys*, and these differences may prove to be of tribal significance. As a general observation, *Marmosopsis* and *Sternbergia* appear to show closest affinity with Marmosini, while *Itaboraidelphys* compares best with Didelphini (*sensu* Reig *et al.* 1987).

The group 3 taxa are also referrable to the Didelphoidea because of their possession, as seen in *Didelphopsis*, of a V-shaped centrocrista and reduced size of paracone relative to metacone. These taxa depart from the structure expected in generalized Didelphidae in having a large bulbous P3, in slight variations in molar structure of *Gaylordia* and *Didelphopsis*, and in the semi-procumbent structure of the lower anterior dentition in *Eobrasilia*. These taxa document a considerable radiation within this group and show variations in size and structure that center around a large bulbous P3. This latter feature I interpret as representing the key derived character state which establishes these taxa as a monophyletic unit. In recognition of this key specialization and of the diversity within the group, I opt to place *Didelphopsis, Gaylordia*, and *Eobrasilia* in a distinct didelphid subfamily, the Eobrasiliinae nov. (Table 17). *Didelphopsis* appears to represent the most generalized member of this subfamily, while *Eobrasilia* is the most derived (i.e. it differs from the other two taxa in having a semiprocumbent lower anterior dentition).

Secure placement of Itaborai group 4 (*Derorhynchus, Minusculodelphis*) within the existing systematic framework is somewhat tenuous. Neither of these taxa are known by upper molars, and these dental elements have provided many of the key features that permitted clarification of the affinities of other groups and taxa. Both Itaborai group 4 taxa are uniquely specialized in reduction in size of the premolars, elongation of the lower jaw anteriorly, and procumbent nature of the canine and incisors. The molars display common features that indicate deviation from the basic didelphine plan, although these features do not suggest special affinity with any of the other groups showing procumbency or semiprocumbency of the lower dentition (i.e. Caenolestoidea, Argyrolagoidea, Polydolopoidea). For this reason it is not possible to classify *Derorhynchus* and/or *Minusculodelphis* in an ancestral position within any of those groups. I thus opt to place these taxa in a separate didelphid subfamily, the Derorhynchinae nov. (Table 17). This arrangement best expresses their apparent didelphid affinity as represented by known lower molar structure, and their deviation from the basic didelphid plan as shown by structure of the premolars, canine, and lower incisors.

The Herpetotheriinae (*sensu* Reig *et al.* 1987; including *Amphiperatherium, Herpetotherium, Peratherium*) collectively range from the Early Eocene to the Middle Miocene of North America, Europe, and/or Africa. Based on a character analysis of *Peratherium* as a representative of this group, Reig *et al.* (1987) demonstrated that this subfamily linked closely with South American Didelphinae, and showed particular affinity with members of the tribe Marmosini. Those workers opted to regard the Herpetotheriinae as evolving in North America from a peradectid ancestor and paralleling or converging with South American didelphids which evolved independently on that continent from a different peradectid stock. I now believe that this interpretation is at odds with the character analysis, it dictates acceptance of the Didelphoidea as a paraphyletic taxon, and it is unnecessary given present knowledge of the fossil record of early Didelphinae as documented in the present study. There are several taxa in the Middle Paleocene of South America (*Marmosopsis, Itaboraidelphys, Sternbergia*) which represent ideal structural ancestors for all later Didelphoidea, including Herpetotheriinae. I therefore deem it more parsimonious to envision the transition from Peradectidae to Didelphidae as occurring in South America, from where a didelphine stock dispersed to North America and then to Europe either in or shortly before Early Eocene time. This interpretation is in strict agreement with the character analysis provided by Reig *et al.* (1987) which indicates affinity of Herpetotheriinae with South American Didelphinae, it is supported by knowledge of potential Middle Paleocene ancestors for later Didelphinae (and Herpetotheriinae), and it permits recognition of the Didelphoidea as a monophyletic unit. I must stress that I do not formally regard the similarities shared by Herpetotheriinae and Marmosini as real synapomorphies and hence as demonstrating any special affinity for these groups. The similarities shared by these groups are probably attributable in part to convergence and in part to them both including taxa that are small and generalized relative to other didelphid groups. In other words, these two groups include rather generalized didelphids, the other didelphid groups being overall more derived.

The earliest record of a potential sparassocynid ancestor is provided by two isolated teeth from rocks of Casamayoran age in Argentina. AMNH 28411 is an isolated right M_4? (see Simpson 1948:49, Fig. 6A) showing a moderately elevated trigonid, reduced metaconid, and well developed talonid with distinct hypoconid, entoconid, and hypoconulid. AMNH 28415 is an isolated left M^2? (see Simpson 1948:49, Fig. 6B) showing a reduced stylar shelf labial to the paracone, paracone reduced relative to metacone, metacrista (metastylar area) elongate, and protocone reduced. These specimens are good didelphoids and not borhyaenoids, and represent potential ancestors for later Sparassocynidae. If these specimens do indeed represent the earliest sparassocynids, then the cladogenetic event between Didelphidae and Sparassocynidae occurred before Early Eocene time, possibly in the Middle to Late Paleocene (Fig. 55).

Borhyaenoids (see Marshall 1978, 1981a, b, 1982b for a review of pertinent literature) were long believed to represent the derived sister-group of didelphoids (e.g. Marshall 1978). Features of the dentitions of early generalized borhyaenoids and knowledge of the dental structure of early didelphoids as presented in this study, now indicate that this interpretation is erroneous. As seen in *Patene simpsoni* from Itaborai (see Paula Couto 1952a, 1961) and in other early Hathliacyninae (see Marshall 1981a) the stylar shelf is broad, stylar cusp B is prominent but not connected to paracone by a paracrista, centrocrista is rectilinear, protocone is moderately developed, para- and metaconules are present, metacrista elongate, and a sharp size decrease occurs from M_5 to M_2 from M^4 to M^2, and from P3 to P1. The presence of a rectilinear centrocrista and of a large P3 relative to P2 are good peradectid characters, and not didelphid characters in which the centrocrista is V-shaped and the size difference between P3 and P2 is less distinct. These features demonstrate that borhyaenoids evolved directly from a peradectid ancestor, and that this event occurred independent of the peradectid-didelphid dichotomy. Thus, borhyaenoids share no special affinity with didelphoids, other than both having an apparent peradectid ancestor.

The affinities of Caenolestoidea are clarified by aspects of the present study. Generalized caenolestids of the subfamily Caenolestinae (see Marshall 1980b) have a lower molar structure that agrees well with Microbiotheriidae in having M_{2-4} brachydont, trigonid only moderately elevated above talonid, trigonid cusps somewhat approximated, and talonid basined and wider than trigonid. Caenolestines also agree with caroloameghiniids in having enlarged subequal sized hypo- and entoconids, and in the large size of the metaconid relative to the protoconid. In Caenolestidae the M_5 is greatly reduced relative to the M_4, a feature occurring in Microbiotheriinae and in some Caroloameghiniinae. The distribution of the size of M_5 relative to M_4 within and between these groups suggests that a potential common ancestor had an M_5 that was subequal in length to but narrower than M_4. Most importantly, I am certain that the two large labial cusps in caenolestids are homologous to the enlarged stylar cusps B and D in Caroloameghiniidae (see above). I also believe that the quadritubercular structure of the caenolestid upper molars is the result of enlargement and posterointernal expansion of a metaconule, and not by the addition of a hypocone as apparently occurred in polydolopoids. It is thus interesting to note the occurrence of a slightly enlarged and posterolingually expanded metaconule in *Reigia*, a condition which represents a structurally ancestral arrangement expected for caenolestids. Also, the similarities noted above in the lower dentitions of Caenolestidae and Microbiotheriidae represent or approximate the states of these characters as they apparently existed in early caroloameghiniids. These data thus demonstrate special affinity of Caenolestidae and Caroloameghiniidae.

The problem now arises of identifying the affinities of the large stylar cusp B and D groups (Caroloameghiniidae, Caenolestoidea, Polydolopoidea, and by inference of lower dental structure, the Argyrolagoidea). I believe that the derived character state of enlarged stylar cusps B and D evolved once in the early evolution of South American marsupials. [The presence of large stylar cusps B and D in the stagodontid *Didelphodon* clearly evolved independently in that lineage. As demonstrated by Fox (1981) *Didelphodon* evolved from an *Eodelphis*-like ancestor which lacked a large stylar cusp D.] I therefore recognize these groups as representing a monophyletic unit, a conclusion supported by the occurrence of multiple shared similarities in the lower molars of generalized members of each group (see above). For this group I propose a new suborder, Polydolopimorphia (Ameghino 1897) nov. I also propose the superfamily rank Caroloameghinoidea nov. for the family Caroloameghiniidae, and recognize it as representing a unit of evolution and diversification that is equal to that of the Caenolestoidea, Polydolopoidea, and Argyrolagoidea. I propose that the ancestry shared by Microbiotheriidae and Polydolopimorphia is best expressed by visualizing a dichotomy very early in the evolution of the Microbiotheriidae that predated the Microbiotheriinae- Pediomyinae dichotomy. This initial dichotomy occurred at a time when the stylar shelf was still well developed and the stylar cusps relatively unreduced, character states that closely approximated their peradectid or peradectid-like ancestors.

When I attempted to construct a cladogram for the branching sequence of polydolopimorph superfamilies, it became evident that viable alternative interpretations for the time of appearance and initial distribution of several key character states existed (Fig. 54). The only aspect of the branching sequence that is not contested is that the ancestral group within the Polydolopimorphia is the Caroloameghinoidea. The caenolestoid-polydolopoid dichotomy probably occurred early in the evolution of the Polydolopimorphia. These superfamilies may have shared a common ancestor that had tritubercular upper molars, an unspecialized P3, and developed a tendency toward procumbency. The stock leading to caenolestoids developed quadritubercular upper molars by enlargement and posterolingual expansion of the metaconule, experienced further reduction of P1-2, and developed marked procumbency of I_1 or I_2. The stock leading to polydolopoids attained a quadritubercular upper molar by addition of a hypocone, developed an enlarged P3, and attained semiprocumbency of the lower canine (Fig. 54a). Alternatively, both groups may have evolved independently from a caroloameghiniid ancestor, and all aspects of procumbency evolved separately in each lineage (Fig. 54b). The affinity of argyrolagoids is even less clear. If it is assumed that their lower gliriform tooth is a canine, then close affinity with polydolopoids is suggested (Fig. 54c). Such an affinity dictates secondary size reduction of P3 and height of M_2 trigonid. Alternatively, if the lower gliriform tooth is an incisor, then closer affinity with caenolestoids is possible (Fig. 54d). This affinity requires no secondary reduction of P3 or height of M_2 trigonid, and the structure of these teeth can easily be derived from a generalized caenolestoid. A third possibility is that all aspects of procumbency of the anterior lower dentition evolved independently in caenolestoids, polydolopoids, and argyrolagoids (Fig. 54d). If so, then the gliriform tooth in argyrolagoids could be the canine as in polydolopoids, the I_1 or I_2 as in caenolestoids, or even the I_3 or I_4.

The classification proposed here for non-Australian marsupials is summarized in Table 17. A phylogram of the families and subfamilies is shown in Figure 55, and my interpretation of the paleobiogeographic history (i.e. times and directions of dispersals) of marsupials is shown in Figure 56.

6) MARSUPIAL PHYLOGENY AND PALEOBIOGEOGRAPHY

The phylogeny of non-Australian Marsupialia proposed in this study (Fig. 55) is based solely on consideration of tooth characters. It is therefore of particular interest that many basic aspects of this phylogeny are corroborated by other data or character complexes.

First, the Borhyaenoidea and Didelphoidea show no special affinity, other than having evolved from a peradectid-like ancestor. This conclusion is supported by Szalay's (1982a, b) study of tarsal bone structure.

Second, living Microbiotheriidae (*Dromiciops*) and Caenolestidae (*Caenolestes, Lestoros, Rhyncholestes*) are more closely related to each other than either is with any living Didelphidae.

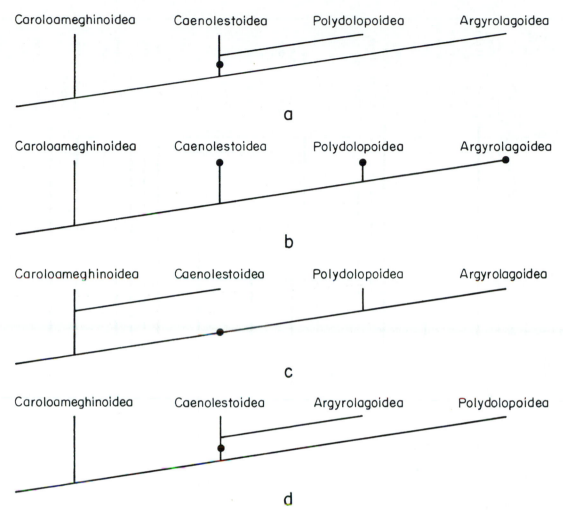

Fig. 54. Cladograms showing alternative phylogenetic relationships of polydolopimorph superfamilies. Solid circles mark relative times of acquisition of derived characters (if any) which would indicate unity of the various superfamilies (a, c, d), or the independent acquisition of derived characters by each superfamily (b). For discussion of the various characters, see text.

This affinity is supported by the serological studies of Kirsch (1977, Fig. 21).

Third, the similarities of tarsal structure noted by Szalay (1982a, b) between borhyaenoids, caenolestoids, and his "Pediomyidae" (i.e. Peradectidae, Stagodontidae, Caroloameghiniinae, and Pediomyinae of this study) is explained by the probable retention of primitive character states in all these groups. Szalay's Borhyaeniformes (i.e. his Borhyaenoidea, Caenolestoidea, Polydolopoidea, Argyrolagoidea) is thus a paraphyletic taxon based on plesiomorphic character states. Nevertheless, the monophyly of the Caenolestoidea, Polydolopoidea, and Argyrolagoidea as proposed by Szalay is supported by the present study.

Fourth, aspects of the present study are consistent with Szalay's (1982a, b) observations regarding special affinity of *Dromiciops* (Microbiotheriinae) and at least some Australian Marsupialia. The lineage leading to *Dromiciops* (i.e. the Microbiotheriinae) has been a distinct monophyletic unit since at least Early Campanian

time (Fig. 55). It is thus possible to envision an early member of this lineage as representing the basal stock for part or all of the Australian radiation (Fig. 56). This possibility is supported by the existence of the Middle Miocene genus *Ankotarinja* which I regard (see above) as displaying "Microbiotheriinae-like" tooth characters.

Fifth, further work is needed to clarity if any special affinity exists between Australian perameloids and South American Derorhynchinae (because of procumbent specializations of the anterior lower dentition) and Polydolopimorphia (because of the occurrence of large stylar cusps B and D, and procumbency of the lower anterior dentition, especially in caenolestoids).

The proposed phylogeny (Fig. 55) also dictates reconsideration of direction and timing of inferred dispersal events of various taxa (Fig. 56). The occurrence of Pediomyinae in the Late Cretaceous of North America and the Middle Paleocene (*Monodelphopsis*) of South America

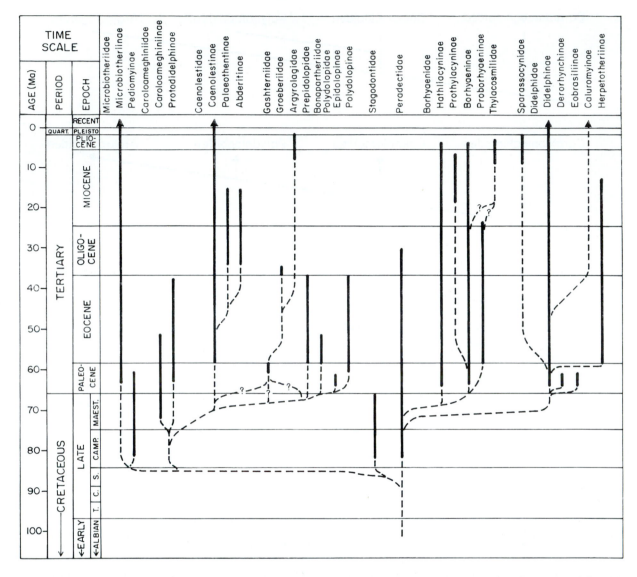

Fig. 55. Phylogram showing known stratigraphic ranges of the families and subfamilies of non-Australian Marsupialia with
      indication of their probable affinity.

suggests their initial development and radiation
in North America and subsequent dispersal of a
stock, possibly in the Latest Cretaceous, to South
America (Fig. 56). I also propose that the initial
Pediomyinae-Microbiotheriinae dichotomy is
best explained by envisioning either a vicariant
separation or a dispersal from South to North
America of a common ancestral stock of Microbiotheriidae. Thus, the stock in South America
gave rise to the Microbiotheriinae and that in
North America evolved into the Pediomyinae.

The notable diversity of Caroloameghiniidae
in the Middle Paleocene of South America and
their ancestral position relative to South
American Caenolestoidea, Argyrolagoidea, and
Polydolopoidea suggests that the Polydolopimorphia originated on that continent. The
occurrence of *Glasbius* in the Late Cretaceous
(Maestrichtian) of North America is most
parsimoniously explained by dispersal of a

caroloameghiine stock from South to North
America in Latest Campanian or Earliest
Maestrichtian time (Fig. 56). The occurrence of
Antarctodolops in the Late Eocene of Seymour
Island, Antarctica is attributed to probable
dispersal of a polydolopine stock from South
America to Antarctica during Early or Middle
Eocene time (Fig. 56).

The paleobiogeographic histories of Pera-
dectidae and Didelphidae are discussed by
Crochet (1980), and Crochet and Sigé (1983).
Initially, Peradectidae had a Late Cretaceous
(Earliest Campanian) distribution that included
at least North and South America, and possibly
Antarctica and Australia as well (Fig. 56). This
distribution is explained either by the occurrence
of a common ancestral stock on both continents,
or by dispersal of a stock from South to North
America or visa versa in the early part of the Late
Cretaceous. The distribution of *Peradectes* is best

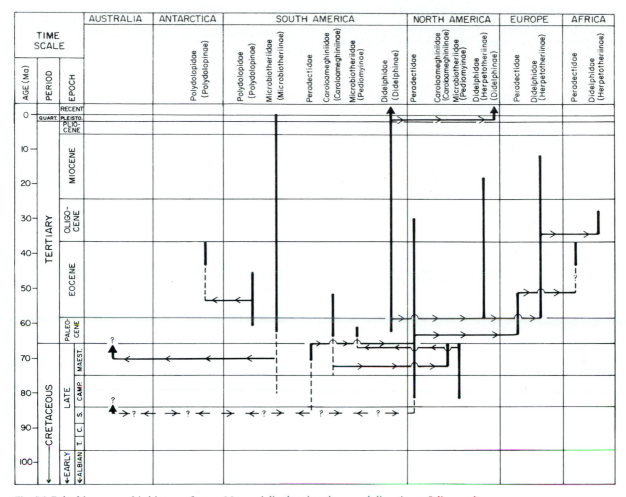

Fig. 56. Paleobiogeographic history of some Marsupialia showing times and directions of dispersal.

explained by envisioning dispersal of a member of this genus from South to North America around the Cretaceous-Paleocene boundary, and with subsequent dispersal from North America to Europe in the Early Paleocene and from there to Africa in the Early Eocene (Fig. 56). The distribution of Didelphidae apparently represents dispersal of a stock from South to North America and from there to Europe in Latest Paleocene time, and from Europe to Africa in Early Oligocene time (Fig. 56). However, it is also possible, although I think less probable, that the initial didelphid dispersal event was from South America to Africa, and from there to Europe and North America (see Crochet and Sigé 1983, Fig. 5). The presence of *Didelphis* (Didelphinae) in the Late Pleistocene to Recent of North America clearly represents dispersal of a member of this genus from South to North America after emergence of the Panamanian Land Bridge about 2.8 Ma (Marshall 1985) (Fig. 56). Thus, the only two marsupial groups for which their initial distribution is probably explained by vicariance are the Peradectidae and Microbiotheriidae. The distributions of all other groups and taxa are attributed solely to dispersal.

The phylogenetic (Fig. 55) and paleobiogeographic (Fig. 56) conclusions of this study either indicate or imply that South America was the primary area for early cladogenesis and diversification of Marsupialia. Furthermore, all basic aspects of marsupial evolution and distribution can be explained by envisioning at least initial dispersal of stocks from, and not to, this continent.

ACKNOWLEDGEMENTS

I thank the following individuals for permission to study specimens in their institutional collections: M. C. McKenna and R. H. Tedford,The American Museum of Natural History, New York; R. Pascual, Museo de La Plata, La Plata, Argentina; J. Bonaparte, Museo Argentino de Ciencias Naturales "Bernardino Rivadavia", Buenos Aires, Argentina; R. Hoffstetter, Muséum national d'Histoire naturelle, Paris, France; F. L. de Souza Cunha, Museu Nacional e Universidade Federal do Rio de Janeiro, Brazil; and D. Campos, Divisão de Geologia e Mineralogia do Departamento Nacional da Produção Mineral, Rio de Janeiro, Brazil. Initial stages of this study were facilitated by grants 1329, 1698 and 1943 from the National Geographic Society, Washington, D. C. and completion was made possible by National Science Foundation grant DEB-7901976. Special thanks to J. A. W. Kirsch, R. H. Tedford, and M. O. Woodburne for their constructive reviews of an early draft of this paper. I am particularly indebted to J. A. W. Kirsch for his assistance with coding characters and for providing me with the Wagner trees computed from the coded data set.

REFERENCES

ABSOLON, J., 1900. Vorläufige Mittheilung über einige neue *Collembolen* aus den Höhlen des märischen Karstes. *Zool. Anz.* **23**: 265-69.

AMEGHINO, F., 1887. Enumeración sistemática de los especies de mamíferos fósiles coleccionados por Carlos Ameghino en los terrenos eocenos de la Patagonia austral y depositados en el Museo de La Plata. *Bol. Mus. de La Plata* **1**: 1-26.

AMEGHINO, F., 1898. Sinopsis geológico-paleontológica. Segundo censo de la República Argentina. *Fol., Buenos Aires* **1**: 112-255.

AMEGHINO, F., 1901. Notices préliminaires sur des ongulés nouveaux des terrains crétacés de Patagonie. *Bol. Acad. Nac. Cien. Córdoba* **16**: 349-426.

AMEGHINO, F., 1902. Première contribution à la connaissance de la fauna mammalogique des couches à *Colpodon*. *Bol. Acad. Cienc. Córdoba* **17**: 71-141 (1902-04).

ARCHER, M., 1976. The dasyurid dentition and its relationship to that of didelphids, thylacinids, borhyaenids (Marsupicarnivora) and peramelids (Peramelina: Marsupialia). *Aust. J. Zool. Suppl. Ser.* No. **39**: 34p.

ARCHER, M., 1981. A review of the origins and radiations of Australian mammals. Pp. 1437-88 *in* "Ecological biogeography of Australia" ed by A. Keast. Dr. W. Junk Publ.: The Hague, London.

ARCHER, M., 1982. Review of the dasyurid (Marsupialia) fossil record, integration of data bearing on phylogenetic interpretation, and suprageneric classification. Pp. 397-443 *in* "Carnivorous marsupials" ed by M. Archer. Roy. Zool. Soc. New South Wales: Sydney.

ARCHIBALD, J. D., 1982. A study of Mammalia and geology across the Cretaceous-Tertiary boundary in Garfield County, Montana. *Univ. Calif. Publ. Geol. Sci.* **122**: 1-286.

BOWN, T. M. AND SIMONS, E. L., 1984. First record of marsupials (Metatheria: Polyprotodonta) from the Oligocene in Africa. *Nature* **308**: 447-49.

CLEMENS, W. A., 1966. Fossil mammals of the type Lance Formation, Wyoming. Part II. Marsupialia. *Univ. Calif. Publ. Geol. Sci.* **62**: 1-122.

CLEMENS, W. A., 1979. Marsupialia. Pp. 192-220 *in* "Mesozoic mammals, the first two-thirds of mammalian history" ed by J. A. Lillegraven, Z. Kielan-Jaworowska and W. A. Clemens. Univ. Calif. Press: Berkeley.

CLEMENS, W. A. AND MARSHALL, L. G., 1976. Fossilium Catalogus: American and European Marsupialia. W. Junk (The Hague), Pars **23**: 114p.

CONEY, P. J., 1982. Plate tectonic constraints on biogeographic connections between North and South America. *Ann. Missouri Bot. Garden* **69**: 432-43.

CORRO, G. DEL, 1977. Un nuevo microbiotherio (Marsupialia) del Eoceno del Patagonia. *Rev. Mus. Argent. Cien. Nat. "Bernardino Rivadavia", Paleont.* **2**(2): 31-33.

CROCHET, J.-Y., 1979a. Diversité systématique des Didelphidae (Marsupialia) Européens Tertiaires. *Geobios* **12**(3): 365-78.

CROCHET, J.-Y., 1979b. Donnees nouvelles sur l'histoire paléogéographique des Didelphidae (Marsupialia). *C. R. Acad. Sc. Paris* **288** (sér. D): 1457-60.

CROCHET, J.-Y., 1980. Les marsupiaux du Tertiaire D'Europe. Éditions de la fondation Singer-Polignac, Paris, 279 p.

CROCHET, J.-Y. AND SIGÉ, B., 1983. Les mammifères montinens de Hainin (Paléocène moyen de Belgique). Part III. Marsupiaux. *Palaeovertebrata* **13**(3): 51-64.

FOX, R. C., 1979a. Mammals from the Upper Cretaceous Oldman Formation, Alberta. I. *Alphadon* Simpson (Marsupialia). *Can. J. Earth Sci.* **16**: 91-102.

FOX, R. C., 1979b. Mammals from the Upper Cretaceous Oldman Formation, Alberta. II. *Pediomys* Marsh (Marsupialia). *Can. J. Earth Sci.* **16**: 103-13.

FOX, R. C., 1979c. Ancestry of the "dog-like" marsupials. *J. Paleont.* **53**: 733-35.

FOX, R. C., 1981. Mammals from the Upper Cretaceous Oldman Formation, Alberta. V. *Eodelphis* Matthew, and the evolution of the Stagodontidae (Marsupialia). *Can. J. Earth Sci.* **18**: 350-65.

FOX, R. C., 1984. Paleontology and the origin of marsupials. Abstract, *in* "Possums and opossums: studies in evolution", 1984 Symposium. Roy. Zool. Soc. NSW; Univ. NSW, 1984.

GARDNER, A. L., 1973. The systematics of the genus *Didelphis* (Marsupialia: Didelphidae) in North and Middle America. *Spec. Publ. Mus., Texas Tech Univ.* **4**: 1-81.

HOFFSTETTER, R., 1975. Les marsupiaux et l'histoire des mamifères : aspects phylogéniques et chorologiques. *Colloque International C. N. R. S.* **218**: 591-610.

HOULBERT, C., 1918. Revision monographique des Castniinae. in Oberthur Et. Lép. comp. 15. Rennes **1918**: 5-720.

KIRSCH, J. A. W., 1977. The comparative serology of Marsupialia, and a classification of Marsupialia. *Aust. J. Zool. Suppl. Ser.* **52**: 152p.

KIRSCH, J. A. W. AND ARCHER, M., 1982. Polythetic cladistics, or, when parsimony's not enough: the relationships of carnivorous marsupials. Pp. 595-619 *in* "Carnivorous marsupials" ed by M. Archer. Roy. Zool. Soc. New South Wales: Sydney.

LILLEGRAVEN, J. A., 1969. Latest Cretaceous mammals of upper part of Edmonton Formation of Alberta, Canada, and review of marsupial-placental dichotomy of mammalian evolution. *Univ. Kansas Paleont. Contrib.* **50** (vert. 12): 1-122.

MAHBOUBI, M., AMEUR, R., CROCHET, J.-Y. AND JAEGER, J.-J., 1983. Première découverte d'un marsupial en Afrique. *C. R. Acad. Sc. Paris*, sér. II, **297**: 691-94.

MARSHALL, L. G., 1978. Evolution of the Borhyaenidae, extinct South American predaceous marsupials. *Univ. Calif. Publ. Geol. Sci.* **117**: 1-89.

MARSHALL, L. G., 1979. Evolution of metatherian and eutherian (mammalian) characters : a review based on cladistic methodology. *Zool. J. Linn. Soc. Lond.* **66**: 369-410.

MARSHALL, L. G., 1980a. Marsupial paleobiogeography. Pp. 345-86 *in* "Aspects of vertebrate history" ed by L. L. Jacobs. Museum of Northern Arizona Press: Flagstaff.

MARSHALL, L. G., 1980b. Systematics of the South American marsupial family Caenolestidae. *Fieldiana, Geology* (n. s.) No. **5**: 1-145.

MARSHALL, L. G., 1981a. Review of the Hathlyacyninae, an extinct subfamily of South American "dog-like" marsupials. *Fieldiana, Geology* (n. s.) No. **7**: 1-120.

MARSHALL, L. G., 1981b. The families and genera of Marsupialia. *Fieldiana, Geology* (n. s.) No. **8**: 1-65.

MARSHALL, L. G., 1982a. Systematics of the South American marsupial family Microbiotheriidae. *Fieldiana, Geology* (n. s.) No. **10**: 1-75.

MARSHALL, L. G., 1982b. Evolution of South American Marsupialia. Pp. 251-72 *in* "Mammalian biology in South America" ed by M. Mares and H. Genoways. Special Publ. Ser. Pymatuning Lab. of Ecol., Univ. Pittsburgh Press, Vol. **6**: 539p.

MARSHALL, L. G., 1982c. A new genus of Caroloameghiniinae (Marsupialia: Didelphoidea: Didelphidae) from the Paleocene of Brazil. *J. Mammal.* **63**: 709-16.

MARSHALL, L. G., 1982d. Systematics of the extinct South American marsupial family Polydolopidae. *Fieldiana, Geology* (n. s.) No. **12**: 1-109.

MARSHALL, L. G., 1984. The lower jaw of *Eobrasilia coutoi* Simpson, 1947, a unique didelphoid (not borhyaenoid) marsupial from the Paleocene of Brazil. *J. Paleont.* **58**: 173-77.

MARSHALL, L. G., 1985. Geochronology and land-mammal biochronology of the transamerican faunal interchange. Pp. 49-85 *in* "The great American Biotic interchange" ed by F. G. Stehli and S. D. Webb. Plenum Press: New York.

MARSHALL, L. G., BUTLER, R. F., DRAKE, R. E. AND CURTIS, G. H., 1981. Calibration of the beginning of the Age of Mammals in Patagonia. *Science* **212**: 43-45.

MARSHALL, L. G. AND MUIZON, C. DE, 1984. A new didelphid marsupial (*Itaboraidelphys camposi* nov. gen., nov. sp.) from the Middle Paleocene (Itaboraian) of São José de Itaboraí (Brazil). *C. R. Acad. Sc. Paris*, sér. 2, no. 18, **299**: 1297-300.

MARSHALL, L. G., MUIZON, C. DE, AND HOFFSTETTER, R., 1984. Fossil Marsupialia in the Muséum national d'Histoire naturelle collected by André Tournouër from Patagonia, southern Argentina. *Bull. Mus. natn. Hist. nat. Paris* 4ᵉ ser., 6, sec. C, **1**: 33-58.

MARSHALL, L. G., MUIZON, C. DE, AND SIGÉ, 1983. Late Cretaceous mammals (Marsupialia) from Bolivia. *Geobios* **16**: 754-61.

MARTIN, R. A., 1974. Fossil mammals from the Coleman IIA fauna, Sumter County. Pp. 35-99 *in* "Pleistocene mammals of Florida" ed by S. D. Webb. Univ. Presses of Florida: Gainesville.

McGREW, P. O., 1937. New marsupials from the Tertiary of Nebraska. *Jour. Geol.* **45**: 448-55.

MUIZON, C. DE, MARSHALL, L. G. AND SIGÉ, B., 1984. The mammal fauna from the El Molino Formation (Late Cretaceous, Maestrichtian) at Tiupampa, southcentral Bolivia. *Bull. Mus. natn. Hist. nat. Paris* 4ᵉ sér., **6**, sec. C, no. **4**: 327-51.

PASCUAL, R., 1980a. Nuevos y singulares tipos ecológicos de marsupiales extinguidos de América del Sur (Paleoceno tardío o Eoceno temprano) del Noroeste Argentino. *Act. II. Congr. Argent. Paleont. y Bioestr. y I Congr. Latinoamer. Paleont. Buenos Aires* **2**: 151-73.

PASCUAL, R., 1980b. Prepidolopidae, nueva familia de Marsupialia Didelphoidea del Eoceno Sudamericano. *Ameghiniana* **17**: 216-42.

PASCUAL, R., 1981. Adiciones al conocimiento de *Bonapartherium hinakusijum* (Marsupialia, Bonapartheriidae) del Eoceno temprono del Noroesta Argentino. *II Congr. Latino-Amer. Paleont. Porto Alegre* **2**: 507-20.

PASCUAL, R., 1983. Novedosos marsupiales paleogenos de la Formación Pozuelos (Grupo Pastos Grandes) de la Puna, Salta, Argentina. *Ameghiniana* **20**: 265-80.

PASCUAL, R. AND CARLINI, A. A., 1984. Un nuevo Argyrolagoidea (Marsupialia) del Oligoceno tardío (Edad Colhuehuapense) de Patagonia (Gaiman, Chubut, Argentina). *I Jour. Arg. Paleont. Vert. res., La Plata*, p.14.

PATTERSON, B., 1937. Didelphines from the Pliocene of Argentina. *Proc. Geol. Soc. Amer.* **1936**: 379.

PATTERSON, B., 1952. Un nuevo y extraordinario marsupial Deseadiano. *Rev. Mus. Munic. Cienc. Nat. Trad. de Mar del Plata* **1**: 39-44.

PAULA COUTO, C. DE, 1952a. Fossil mammals from the beginning of the Cenozoic in Brazil. Marsupialia: Polydolopidae and Borhyaenidae. *Amer. Mus. Novit.* **1559**: 1-27.

PAULA COUTO, C. DE, 1952b. Fossil mammals from the beginning of the Cenozoic of Brazil. Marsupialia: Didelphidae. *Amer. Mus. Novit.* **1567**: 1-26.

PAULA COUTO, C. DE, 1952c. A new name for *Mirandaia ribeiroi* Paula Couto, 1952. *J. Mammal.* **33**: 503.

PAULA COUTO, C. DE, 1961. Marsupialis fósseis do Paleoceno do Brazil. *An. Acad. brasil Ciênc.* **33**: 312-33.

PAULA COUTO, C. DE, 1962. Didelfídeos fósiles del Paleoceno de Brazil. *Rev. Mus. Argent. Cien. Nat. "Bernardino Rivadavia", Cien. Zool.* **112**: 135-66.

PAULA COUTO, C. DE, 1970. News on the fossil marsupials from the Riochican of Brazil. *An. Acad. brasil Ciênc.* **42**: 19-34.

PETTER, G. AND HOFFSTETTER, R., 1983. Les marsupiaux du Déséadien (Oligocène Inférieur) de Salla (Bolivie). *Ann. Paleont.* **69**: 175-234.

RAGE, J. C., 1978. Une connexion ≷ontinentale entre Amérique du Nord et Amérique du Sud au Crétacé supérieur? l'example des vertébrés continentaux. *C. R. Somm. Soc. geol. Fr.* **6**: 281-85.

RAGE, J. C., 1981. Les continents Péri-atlantiques au Crétacé Supérieur : migrations des Faunes continentales et problèmes paléogéographiques. *Cret. Res.* **2**: 65-84.

REIG, O. A., KIRSCH, J. A. W. AND MARSHALL, L. G., 1985. New conclusions on the relationships of the opossum-like marsupials, with an annotated classification of the Didelphimorphia. *Ameghiniana*.

REIG, O. A., KIRSCH, J. A. W. AND MARSHALL, L. G., 1987. Systematic relationships of the living and Neocenozoic American "opossum-like" marsupials (suborder Didelphimorphia), with comments on the classification of these and of the Cretaceous and Paleogene New World and European Metatherians. Pp. 1-89 *in* "Possums and opossums: studies in evolution" ed by M. Archer. Surrey Beatty & Sons and the Royal Zoological Society of New South Wales: Sydney.

RIDE, W. D. L., 1962. On the evolution of Australian marsupials. Pp.281-306 *in* "The evolution of living organisms" ed by G. W. Leeper. Melbourne University Press: Melbourne.

RIGGS, E. S. AND PATTERSON, B., 1939. Stratigraphy of Late Miocene and Pliocene deposits of the province of Catamarca (Argentina) with notes on the faunae. *Physis* **14**: 143-62.

RODRIGUES FRANCISCO, B. H. AND SOUZA CUNHA, F. L. DE, 1978. Geologia e estratigrafia da Bacia de São José, Município de Itboraí, R. J. *An. Acad. brasil Ciênc.* **50:** 381-416.

RUSSELL, L. S., 1976. The palaeogeographic significance of the polyprotodont marsupials. *25th Inter. Geol. Congr. Sydney Australia, Abst.* **1** *(sec. 7c):* 333-34.

SIGÉ, B., 1971. Les Didelphoidea de Laguna Umayo (formation Vilquechico, Crétacé supérieur, Pérou), et le peuplement marsupial d'Amérique du Sud. *C. R. Acad. Sc. Paris* **273:** 2479-481.

SIGÉ, B., 1972. La faunule de mammifères du Crétacé supérieur de Laguna Umayo (Andes péruviennes). *Bull. Mus. nat. d'Hist. natur.*, 3ᵉ se., Sci. de la Terre **19:** 375-405.

SIMPSON, G. G., 1935a. Descriptions of the oldest known South American mammals from the Rio Chico Formation. *Amer. Mus. Novit.* **793:** 1-25.

SIMPSON, G. G., 1935b. Note on the classification of Recent and fossil opossums. *J. Mammal.* **16:** 134-37.

SIMPSON, G. G., 1938. A new marsupial from the Eocene of Patagonia. *Amer. Mus. Novit.* **989:** 1-5.

SIMPSON, G. G., 1947. A new Eocene marsupial from Brazil. *Amer. Mus. Novit.* **1357:** 1-7.

SIMPSON, G. G., 1948. The beginning of the Age of Mammals in South America. Part 1. *Bull. Amer. Mus. Nat. Hist.* **91:** 1-232.

SIMPSON, G. G., 1964. Los mamíferos casamayorenses de la Coleccion Tournouër. *Rev. Mus. Argent. Cien. Nat. Paleont.* **1:** 1-21.

SIMPSON, G. G., 1967. The beginning of the Age of Mammals in South America. Part 2. *Bull. Amer. Mus. Nat. Hist.* **137:** 1-259.

SIMPSON, G. G., 1970a. The Argyrolagidae, extinct South American marsupials. *Bull. Mus. Comp. Zool. Harvard* **139:** 1-86.

SIMPSON, G. G., 1970b. Addition to knowledge of *Groeberia* (Mammalia, Marsupialia) from the Mid-Cenozoic of Argentina. *Breviora* **362:** 1-17.

SIMPSON, G. G., 1971. The evolution of marsupials in South America. *An. Acad. brasil Ciênc.* **43:** 103-18.

SIMPSON, G. G., 1978. Early mammals in South America: fact, controversy, and mystery. *Proc. Amer. Phil. Soc.* **122:** 318-28.

SLAUGHTER, B. H., 1978. Occurrence of didelphine marsupials from the Eocene and Miocene of Texas Gulf coastal plain. *Jour. Paleont.* **52:** 744-46.

SZALAY, F. S., 1982a. A new appraisal of marsupial phylogeny and classification. Pp.621-640 *in* "Carnivorous marsupials" ed by M. Archer. Roy. Zool. Soc. New South Wales: Sydney.

SZALAY, F. S., 1982b. Phylogenetic relationships of the marsupials. *Geobios, mém. spéc.* **6:** 177-90.

TEDFORD, R. H., 1974. Marsupials and the new paleo-geography. Pp.109-126 *in* "Paleogeographic provinces and provinciality" ed by C. A. Ross. *Soc. Econ. Paleont. Miner., Spec. Publ.* No. 21.

TRAVASSOS, L., 1937. Revisão da familia Trichostrongylidae Leiper, 1921. *Monogr. Inst. Oswaldo Cruz, Rio de Janeiro* **1:** 1-512.

WOLFF, R. G., 1984. New early Oligocene Argyrolagidae (Mammalia, Marsupialia) from Salla, Bolivia. *Jour. Vert. Paleont.* **1:** 108-13.

WOODBURNE, M. O. AND ZINSMEISTER, W. J., 1984. The first land mammal from Antarctica and its biogeographic implications. *J. Paleont.* **58:** 913-48.

APPENDIX

Subsequent to completion of this paper a colleague suggested to me that it would be useful to include a summary table of the stratigraphic ranges of all families and genera of South American Marsupialia. It was further suggested that "spindle diagrams" depicting the generic diversity of each family (as summarized in that table) would also permit readers to quickly visualize present knowledge of the known record of marsupials on that continent. Following these suggestions, I include both the table and figure as an appendix to my paper on Itaboraian age marsupials.

Table 1. Stratigraphic ranges of all South American Marsupialia. The families and genera are listed in alphabetical order. A chronology of the South American Land Mammal Ages listed at the top of the table is shown in Fig. 1 of the main text. See opposite page for Table 1.

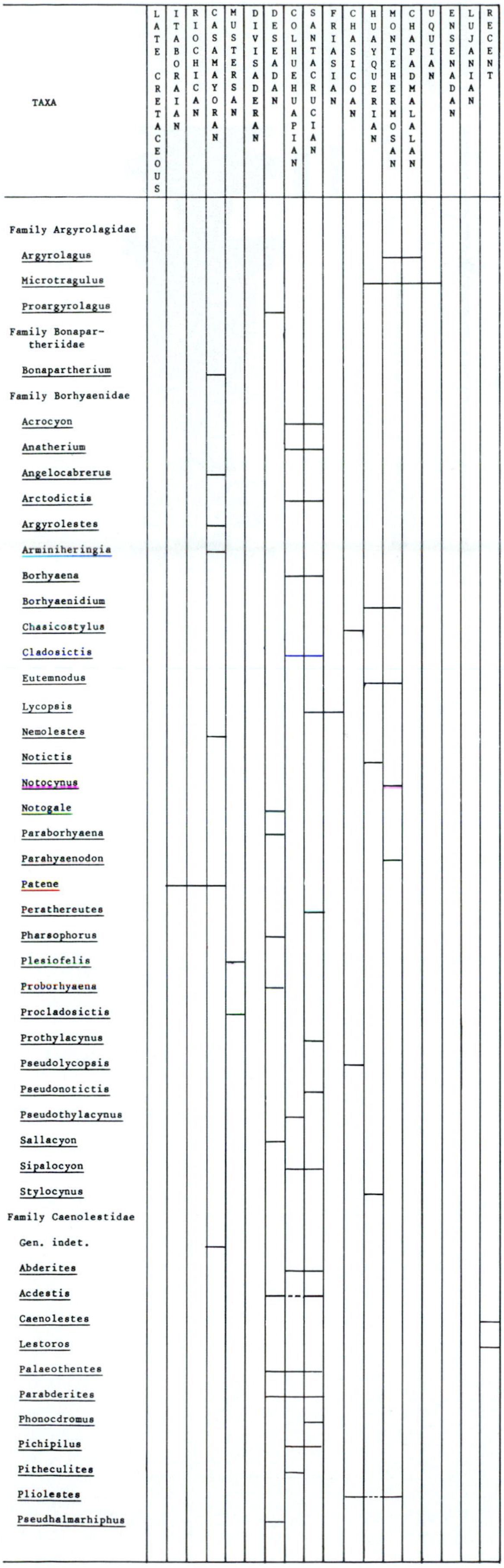
TAXA
LATE CRETACEOUS
ITABORAIAN
RIOCHICAN
CASAMAYORAN
MUSTERSAN
DIVISADERAN
DESEADAN
COLHUEHUAPIAN
SANTACRUCIAN
FRIASIAN
CHASICOAN
HUAYQUERIAN
MONTEHERMOSAN
CHAPADMALALAN
UQUIAN
ENSENADAN
LUJANIAN
RECENT
Family Argyrolagidae
Argyrolagus
Microtragulus
Proargyrolagus
Family Bonapartheriidae
Bonapartherium
Family Borhyaenidae
Acrocyon
Anatherium
Angelocabrerus
Arctodictis
Argyrolestes
Arminiheringia
Borhyaena
Borhyaenidium
Chasicostylus
Cladosictis
Eutemnodus
Lycopsis
Nemolestes
Notictis
Notocynus
Notogale
Paraborhyaena
Parahyaenodon
Patene
Perathereutes
Pharsophorus
Plesiofelis
Proborhyaena
Procladosictis
Prothylacynus
Pseudolycopsis
Pseudonotictis
Pseudothylacynus
Sallacyon
Sipalocyon
Stylocynus
Family Caenolestidae
Gen. indet.
Abderites
Acdestis
Caenolestes
Lestoros
Palaeothentes
Parabderites
Phonocdromus
Pichipilus
Pitheculites
Pliolestes
Pseudhalmarhiphus

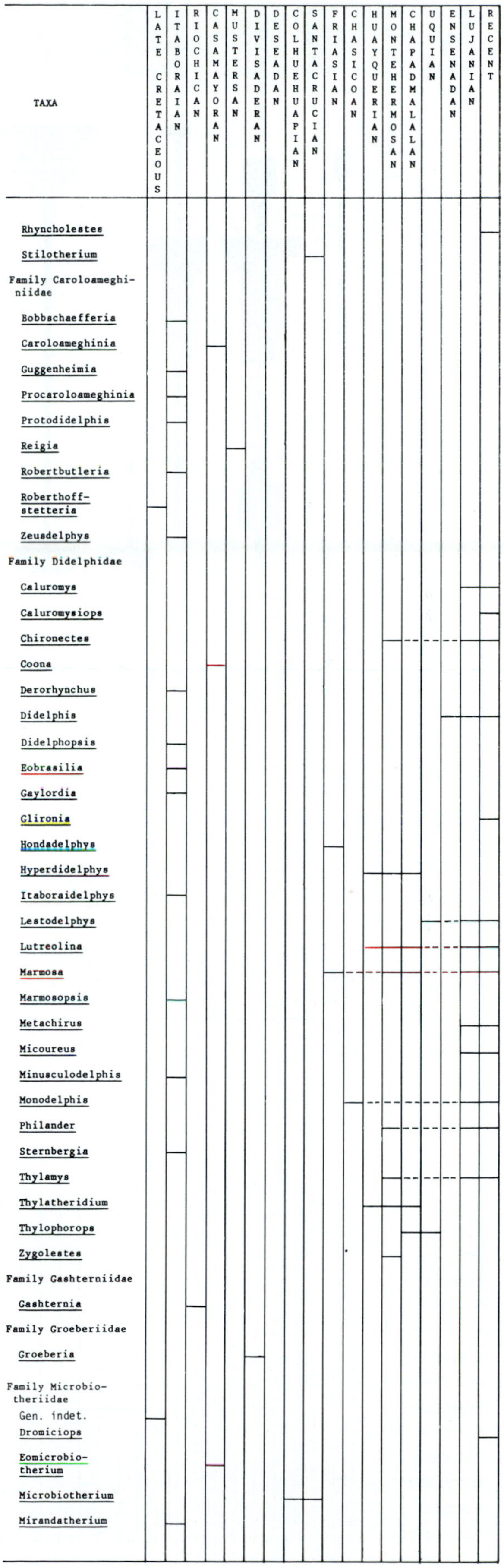
TAXA
LATE CRETACEOUS
ITABORAIAN
RIOCHICAN
CASAMAYORAN
MUSTERSAN
DIVISADERAN
DESEADAN
COLHUEHUAPIAN
SANTACRUCIAN
FRIASIAN
CHASICOAN
HUAYQUERIAN
MONTEHERMOSAN
CHAPADMALALAN
UQUIAN
ENSENADAN
LUJANIAN
RECENT
Rhyncholestes
Stilotherium
Family Caroloameghiniidae
Bobbschaefferia
Caroloameghinia
Guggenheimia
Procaroloameghinia
Protodidelphis
Reigia
Robertbutleria
Roberthoffstetteria
Zeusdelphys
Family Didelphidae
Caluromys
Caluromysiops
Chironectes
Coona
Derorhynchus
Didelphis
Didelphopsis
Eobrasilia
Gaylordia
Glironia
Hondadelphys
Hyperdidelphys
Itaboraidelphys
Lestodelphys
Lutreolina
Marmosa
Marmosopsis
Metachirus
Micoureus
Minusculodelphis
Monodelphis
Philander
Sternbergia
Thylamys
Thylatheridium
Thylophorops
Zygolestes
Family Gashterniidae
Gashternia
Family Groeberiidae
Groeberia
Family Microbiotheriidae
Gen. indet.
Dromiciops
Eomicrobiotherium
Microbiotherium
Mirandatherium

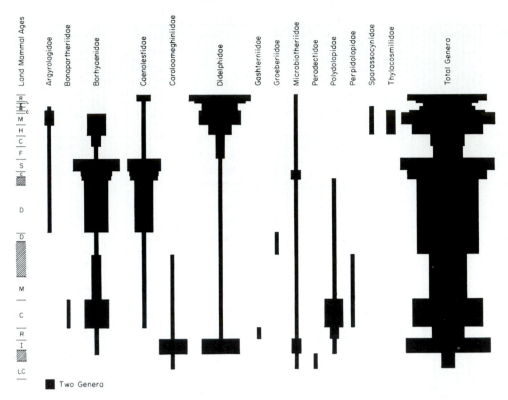

■ Two Genera

Figure 1. "Spindle diagrams" showing generic diversity of the known South American families of Marsupialia based on data presented in Appendix Table 1. The widths of "spindles" depict total (cumulative) generic diversity in each age. In cases where a "hiatus" exists between two ages (i.e. between Mustersan and Divisaderan), the diversity of the ages on both sides of this hiatus are extended to the middle of that hiatus to permit continuity of the "spindles". If a genus or family is recorded in sub- and superjacent ages, but is not recorded in an intermediate age(s), it is recognized as present in that intermediate age(s).

Stratigraphic ranges of all South American Marsupialia. The families and genera are listed in alphabetical order. A chronology of the South American Land Mammal Ages listed at the top of the table is shown in Fig. 1 of the main text.

PALAEONTOLOGY AND THE EARLY EVOLUTION OF MARSUPIALS

RICHARD C. FOX[1]

The geologically oldest undoubted marsupials so far described come from the middle Upper Cretaceous of Alberta, Canada, from rocks approximately 83 my old, and Aquilan (early Campanian) in age. These fossils, which occur in the upper member of the Milk River Formation, include species dentally resembling species of *Pediomys* in younger Cretaceous rocks, but which have a stylocone and lack stylar cusp C on the upper molars. The configuration of the stylar area in these marsupials is comparable to that in early eutherians, suggesting that the ancestral marsupial upper molar may have had enlarged stylar cusps at only the B and D positions, and not at the B, C, and D positions, as had previously been supposed. The Milk River marsupials also imply that *Pediomys* is polyphyletic, and includes several lineages of opossum-like marsupials that independently suppressed the anterior parts of the stylar shelf and anterior stylar cusps during the Late Cretaceous. Some potential implications of this new interpretation of early marsupial evolution are discussed.

Key Words: Pediomyidae; Didelphidae; Marsupialia; Cretaceous; Milk River Formation; Evolution.

Pages 161-69 *in* POSSUMS AND OPOSSUMS: STUDIES IN EVOLUTION ed by M. Archer. Surrey Beatty & Sons and the Royal Zoological Society of New South Wales: Sydney, 1987.

INTRODUCTION

THE geologically oldest undoubted marsupials that have been described are from the middle Upper Cretaceous Milk River Formation, which outcrops in the southernmost parts of the Province of Alberta, in western Canada. The rocks that contain the fossils are sombre-coloured, poorly lithified sandstones, siltstones, and claystones, the products of fluvial deposition across a broad, aggrading floodplain adjacent to the Western Interior Seaway (Williams and Stelck 1975). The marsupial fossils come from the upper member of the formation; this member is early Campanian in age (Russell 1975) and, hence, about 83 my old (Hartland *et al.* 1982). The Milk River marsupials, then, lived nearly 20 my before the end of the Mesozoic, the time of extinction of dinosaurs and, hence, nearly 20 my before the Age of Mammals began (Lerbekmo *et al.* 1979). Clemens *et al.* (1979) and Woodburne and Zinsmeister (1984), in influential papers concerning the early evolution of marsupials, indicate that the Milk River marsupials are only about 75 my old, which is incorrect; see, for example, Meyboom (1960), Obradovich and Cobban (1975), Williams and Stelck (1975) and Caldwell *et al.* (1978).

The Milk River marsupials were discovered by a field party from the University of Alberta, Edmonton, in 1968 (Fox 1968), and initial descriptions of them were published shortly thereafter (Fox 1971a). Field collecting has continued in the years since and we now know as a consequence that these early marsupials were accompanied by a diverse array of other mammals, including triconodontids (Fox 1969, 1976), multituberculates (Fox 1971b), acute- and obtuse-angled symmetrodonts (Fox 1972a, 1976, in press), tribotheres belonging to several different lineages (Fox 1972b, 1980, 1982) and generalized insectivorans (Fox 1970, 1984). The Milk River mammals are represented at present by 32 species, all of which are endemic to the Formation (several more species remain to be described); nine of the genera and six of the families are endemic, as well, with the assemblage in its entirety comprising a mix of archaic and progressive taxa not known from elsewhere in the Mesozoic. Fossils of other vertebrates (*Lepisosteus*, *Amia*, allocaudates (Fox and Naylor 1982), salamanders, anurans, turtles, champsosaurs, lizards, snakes, crocodilians and all of the major groups of Late Cretaceous dinosaurs) occurring in sedimentary association with the mammalian remains suggest freshwater and terrestrial communities living in a warm temperate or subtropical climate; their closest modern analogues are probably provided by the vertebrate communities inhabiting the humid Coastal Plain of the United States bordering the eastern Gulf of Mexico (Fox 1972b).

The Milk River marsupials are known primarily from teeth, and from occasional maxillary and dentary fragments containing teeth (with disarticulation and breakage having occurred mostly as a consequence of transportation by streams between death and final burial). In this respect, the fossil record of the Milk River species differs in no substantial way from the fossil record of all Cretaceous marsupials presently known from North America. Consequently, the material evidence on which

[1]Laboratory for Vertebrate Paleontology, Departments of Geology and Zoology, Biological Sciences Centre, The University of Alberta, Edmonton, Alberta, Canada T6G 2E9.

	Alberta	Saskatchewan	Montana	Wyoming	New Mexico	
64 my Lancian	Scollard Fm.	Frenchman Fm.	Hell Creek Fm.	Lance Fm.		Maestrichtian
Edmontonian	Horseshoe Canyon Fm.				Kirtland Fm.	
					Fruitland Fm.	
Judithian	Oldman Fm.		Judith River Fm.	Mesa Verde Fm.		Campanian
Aquilan **83 my**	Milk River Fm.					

Fig. 1. Stratigraphic distribution of Upper Cretaceous formations in the North American Western Interior that have yielded important collections of early marsupial fossils.

palaeontological interpretation of the early history of marsupials is based consists mostly of dental evidence — especially the configuration of the biting surfaces of the molar crowns. Although often not available, dental formulae and knowledge of replacement patterns are helpful as well, especially in determining what are and what are not early marsupials. In any case, the dental evidence and the little skeletal evidence known, indicate that these early marsupials were small animals, probably adaptively similar to modern New World opossums, an inference reflected in the informal name commonly given them, "opossum-like" marsupials.

The Milk River species mark the beginning of a North American record of early marsupials that is especially rich (Fig. 1): this record is documented by several thousand identifiable specimens from as many as seven different stratigraphic horizons spanning the second half of the Upper Cretaceous. Most of these specimens come from localities in the easternmost Rocky Mountain region and adjacent parts of the Great Plains, from Alberta to New Mexico, but a few specimens come from Baja California, Mexico: these are the only early marsupials

presently known from the Pacific Rim of North America (Lillegraven 1972).

In the Western Interior, the stratigraphic position of most Upper Cretaceous marsupials is well established: superposition of formations that contain marsupials, especially in Alberta (Fig. 2), supplemented by biostratigraphic and, increasingly, geochronologic and magneto-stratigraphic data enabling relatively precise correlation with localities elsewhere, reveals a succession of marsupial-containing local faunas that are of known temporal relationship to each other, thereby permitting determination of the direction and rate of evolutionary change in dental morphology in real time. From this sort of information, it has been possible to infer what appear to be realistic approximations of lineages of descent at the species level, especially within the genera *Alphadon* and *Pediomys*, and for the stagodontids *Eodelphis* and *Didelphodon* (Fox 1979a and b, 1981). Furthermore, a well corroborated stratigraphic framework for determining the temporal succession of local faunas that contain marsupials allows one to test concepts of relationship among early marsupial groups inferred initially from what were

	Didelphidae	"Pediomyidae"	Stagodontidae
64 my **Lancian**	Glasbius Alphadon	"Pediomys" Pediomys	Didelphodon
Edmontonian	Alphadon	"Pediomys"	Didelphodon Eodelphis
Judithian	Alphadon	"Pediomys"	Eodelphis
Aquilan **83 my**	Albertatherium Alphadon	"Pediomys" Aquiladelphis	Eodelphis

Fig. 2. Stratigraphic distribution of Upper Cretaceous marsupial genera in North America. *"Pediomys"* and "Pediomyidae" are taxa whose members are of uncertain relationship, as discussed in text.

essentially comparative studies lacking a time dimension. Both approaches, stratigraphic and comparative, have provided the palaeontological data for the focus of this paper.

Conventionally, North American Upper Cretaceous marsupials are classified in three families, Didelphidae, Pediomyidae and Stagodontidae; representatives of each of these are first seen in the Milk River Formation. This classification stems from the work of O. C. Marsh (1889) and G. G. Simpson (1927, 1935) and, more recently, from that of W. A. Clemens (1966). According to Clemens' (1966) interpretation of the dental evidence, the Didelphidae include the structurally most primitive and the phylogenetically ancestral marsupials; the Pediomyidae and the Stagodontidae are thought to have descended from primitive didelphids, as have all subsequent and more derived marsupials, including those that came to inhabit Australia and South America down to the present day. Both pediomyids and stagodontids are believed to have gone extinct at the end of the Cretaceous (although Slaughter (1978) has reported a possible pediomyid from the Eocene of Texas) and apparently were ancestral to no other groups.

Central to the concept of the Didelphidae as both primitive and ancestral is the genus *Alphadon* (Fig. 3). The geologically oldest species of *Alphadon* that have been described occur in the Milk River Formation (Fox 1971a) and *Alphadon* has been found at all important mammal localities that are younger in the North American Cretaceous. *Alphadon* is known only from isolated teeth and jaw fragments with teeth, including a specimen that consists of upper and lower dentitions found in association at a latest Cretaceous horizon in the Scollard Formation (= "upper Edmonton" Formation in part (Gibson 1977)) of Alberta (Lillegraven 1969). This specimen is very important in that it furnished the first direct evidence that upper and lower dentitions which earlier had been separately referred to *Alphadon* from their occlusal anatomy did, in fact, belong together.

Clemens (1966, 1968) argued that the molar structure of *Alphadon* provides the closest known

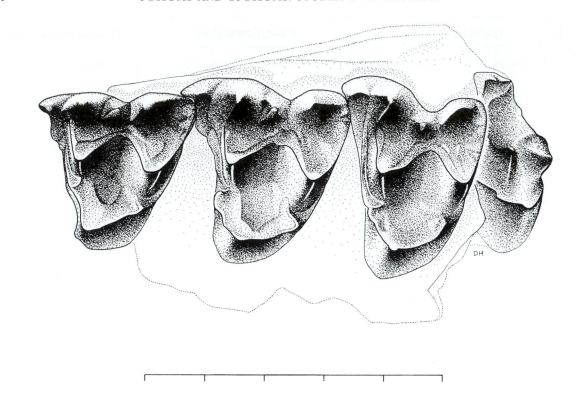

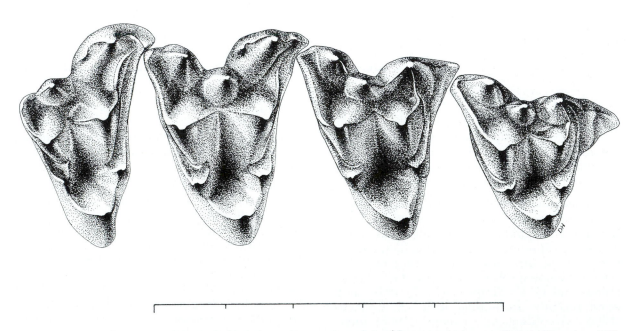

Fig. 3. *Alphadon marshi*, Scollard Formation, Upper Cretaceous, Alberta, left M$^{1-4}$, UA 2846 (above); *Albertatherium primus*, Milk River Formation, Upper Cretaceous, Alberta, reconstructed right M$^{1-4}$ (below). *Albertatherium* appears to be a close relative of *Alphadon*, but one perhaps more primitive in having a smaller protocone and less evident grinding adaptations of the molar crowns.

approximation of the ancestral molar structure for all marsupials, living and extinct. By reference to evidence from several sources, including that provided by the work of Patterson (1956) on therian molars from the Lower Cretaceous (Albian) of Texas, Clemens hypothesised that the ancestral marsupial upper molar was one in which there was a well developed stylar cusp B (homologous to the stylocone in more primitive therians), the stylar shelf was wide anteriorly and posteriorly, and four other stylar cusps (A, C, D and E), all lower than cusp B, were developed.

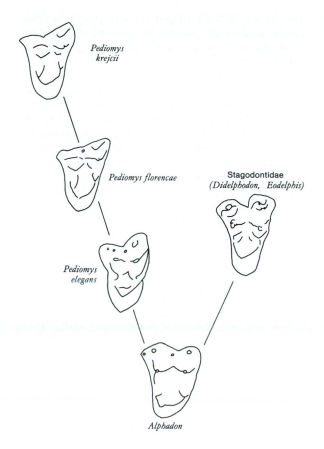

Fig. 4. Diagram showing the succession of structural stages in upper molar evolution from an *Alphadon*-like ancestor to species of *Pediomys*, in which the anterior part of the stylar shelf is progressively suppressed, and to the Stagodontidae, in which stylar cusp C is suppressed.

Any departures from this arrangement (which usually involved a decrease in the width of the stylar shelf (especially anteriorly), and suppression of the anterior stylar cusps, including cusp B), were interpreted as derivatives from the more primitive, *Alphadon*-like, condition. Accordingly, then, the Cretaceous Pediomyidae, in which cusps A and B were suppressed and often C as well, must have descended from species having *Alphadon*-like molars, as must have the Stagodontidae, in which the central stylar cusp, cusp C, is lacking (Fig. 4).

Clemens' interpretation of the polarity in evolution of the marsupial stylar shelf was opposite to that of Simpson (1929, 1948): Simpson had supposed that the pediomyid style of stylar shelf was primitive; widening of the shelf and addition of stylar cusps would then have led to the condition in *Alphadon* and other didelphines. His view seemed to be refuted, however, by the upper molars of the Albian therians of primitive tribosphenic grade, and by those of still earlier, pre-tribosphenic therians, which had been shown to have a well developed stylocone and wide anterior stylar shelf (Patterson 1956).

Other aspects, explicit or implicit in Clemens' (1966) model, were as follows: the ancestral marsupial upper molar possessed a protocone (was, then, tribosphenic) that occluded within a talonid basin on the lower molar; the paracone and metacone were well developed, separate from one another, and the metacone was subequal to or larger than the paracone and the more lingually extending at its base; conules were developed; and the lower molars had a differentiated trigonid and tricuspid talonid, on which the entoconid and hypoconulid were closely approximated or "twinned" (Slaughter (1971) was later to relate this twinning functionally to the enlarged metacone on the occluding upper molars in primitive marsupials, while Clemens (1966, 1979) has maintained that the "twinning" came about as a consequence of expansion of the talonid). Finally, on the lower molars of *Alphodon*, as illustrative of the ancestral didelphid, the cristid obliqua meets the back or postvallid wall of the trigonid beneath the notch between the protoconid and metaconid. A more labial position of the cristid obliqua, as seen for example in pediomyids, is derived.

Clemens' (1966, 1968) theory of dental evolution in early marsupials had a great many strengths and quite rightly became widely accepted: it accounted for the diversity of molar architecture in early marsupials known at the time by appealing to stratigraphic succession to define the direction of morphological change; it was consistent with inferences that could be made about the polarity of change in the stylar shelf from outgroup comparisons without reference to geologic time; and it clarified aspects of molar evolution in descendant, Tertiary marsupials, including evolution in lineages that apparently independently suppressed the anterior parts of the stylar shelf and the anterior stylar cusps (as in various of the so-called "dog-like" marsupials of Australia and South America (Clemens 1966; Fox 1979d)).

Recently, however, Clemens (1979) has speculated on an alternative view of early marsupial molar evolution: the molars of *Alphadon* are still conceived of as structurally ancestral to those of all Cainozoic marsupials and Late Cretaceous didelphids; however, the frequency distribution of stylar cusp C among Late Cretaceous marsupials in general raises the possibility that the ancestral marsupial upper molar may have lacked a cusp at the C position; this would have been added later, during the early evolution of the group. Fossils collected from the Milk River and Oldman formations (Alberta) since publication of Clemens' (1979) paper bear directly on his speculations about the ancestral pattern of the stylar shelf in marsupials; the fossils in question pertain to species that conventionally would have been classified as species of *Pediomys,* or as species closely related to them.

The name *"Pediomys"* was originally coined by O. C. Marsh (1889) for an opossum-like marsupial from the latest Cretaceous Lance Formation of Wyoming. Clemens, in his description (1966) of the Lance marsupial fauna, argued that Marsh's type species, *P. elegans,* was the most primitive of the several species of *Pediomys* then known from the Lance. This argument was based on the fact that whereas upper molars of *P. elegans* show a relatively narrow stylar shelf, especially anteriorly, and suppression of the anteriormost stylar cusps (features then thought to be diagnostic of *Pediomys* among Late Cretaceous marsupials generally), these features are normally better developed than in other Lancian species of *Pediomys.* It was as if *P. elegans* represented an early morphological stage in the suppression of the anterior parts of the stylar shelf and anterior stylar cusps from an ancestry that had originally a more *Alphadon*-like configuration of the area, including a prominent cusp A and C and a still larger cusp B or stylocone (Fig. 4).

In 1972, Sahni described the fossil vertebrates, including mammals, from the upper part of the Judith River Formation, Montana — coeval with the Oldman Formation of southeastern Alberta (Russell 1975), middle Campanian in age, and hence older than the Lance Formation by at least 12 my (see Obradovich and Cobban 1975; Harland *et al.* 1982). In this paper, Sahni provided the first review of a taxonomically diverse pre-Lancian Late Cretaceous mammal local fauna from North America. Sahni recognized two species of *Pediomys* in the Judith River, but in spite of the difference in age, the architecture of the stylar shelf in these is no more *Alphadon*-like than is that of their Lancian relatives, *P. cooki* and *P. krejcii* (Fox 1979b). My own study of the marsupials from the Oldman Formation (Fox 1979 a and b, 1981) confirmed Sahni's recognition of *Pediomys* from the Judith River. Owing to better preserved material, however, I was able to reconstruct what appears to have been a direct lineage of descent leading to the Lancian *P. krejcii* and to show that Sahni's (1972) *P. clemensi* was more closely related to *P. cooki* than to *P. elegans,* as Sahni had proposed. Other Oldman species of *Pediomys* are likely to have been near to the ancestry of one or both of the large Lancian *P. hatcheri* and *P. florencae* (Fox 1979b). As with *Pediomys* from the Judith River, however, none of the Oldman *Pediomys* are more *Alphadon*-like in the construction of their molars than are their Lancian successors.

It is not until Milk River time that specimens of *Pediomys*-like marsupials which exhibit a change in state of the stylar shelf towards a more primitive condition are encountered. These specimens include upper molars that resemble upper molars of species in each of the groups cited above *P. prokrejcii-krejcii, P. clemensi-cooki,* and the larger Lancian species *P. hatcheri* and *P. florencae* (Fig. 5). (These new Milk River taxa will be named and described elsewhere and do not include *"Pediomys" exiguus* Fox, 1971a, which, from newly collected material, proves not to be *Pediomys*-like at all). The resemblances, especially among the *P. prokrejcii-krejcii*-like molars, extend to details of the crowns generally and to the stylar areas specifically. On molars of each of the Milk River species, an enlarged, almost blade-like cusp is developed in the D position, labial to the metacone, and no definite C cusp is evident — but unlike the Judithian and Lancian molars, the Milk River specimens show a moderately prominent stylocone connected to the paracone along a preparacrista. I interpret the Milk River fossils as representing early members of the three pediomyid lineages in question which show two trends not documented before by stratigraphic sequences of early marsupials: the independent and parallel suppression of the anterior parts of the stylar shelf and of stylar cusps A and B, occurring in animals in which no C cusp is present primitively (Fig. 6). If, in fact, these patterns can be accepted at face value, one is led to conclude that these lineages of *Pediomys*-like marsupials never passed through a *Pediomys elegans* or *P. elegans*-like stage in upper molar evolution; and furthermore, never had a more remote ancestry from a species having five stylar cusps, that is, having an *Alphadon*-like construction of the upper molar crowns.

The morphologies of the Milk River molars lead to the formulation of several alternatives in respect to the evolution of early marsupials:

ALTERNATIVE 1. The Milk River forms are not, in fact, related to the Judithian-Lancian *Pediomys* species groups. This seems unlikely owing to the resemblances in molar structure between the older and younger groups, seen especially well among the *P. krejcii*-like fossils.

ALTERNATIVE 2. *Pediomys,* whether it includes all of the species it traditionally has contained or not, is not a genus (or genera) of marsupials, so the configuration of the stylar shelf in *Pediomys* is not relevant to reconstruction of the ancestral molar condition for marsupials. This alternative is probably not valid: the Judithian and Lancian species currently classified as *Pediomys* are known from specimens that exhibit marsupial features in addition to those of the upper molars: inflected angle of the dentary; three premolars and four molars, with the premolars being simple, premolariform teeth and the third premolars the only post-canine teeth replaced (the replacement pattern for the ante-premolar teeth is unknown for Cretaceous marsupials), and so forth. If the Milk River species, then, are

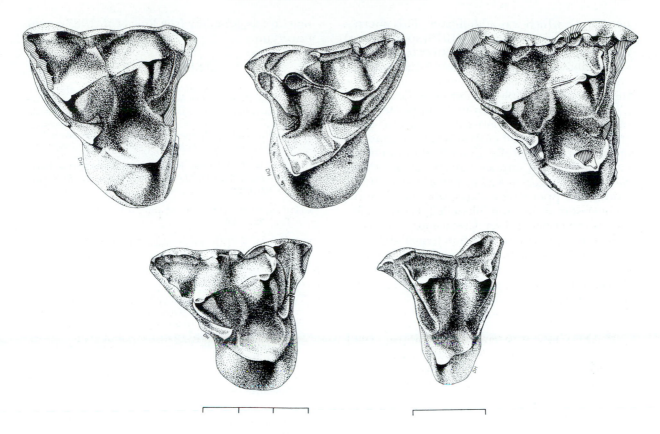

Fig. 5. Representative upper molars of Upper Cretaceous *Pediomys*-like marsupials. Top row, left to right: *Pediomys florencae*, M³; *Pediomys florencae*, M² (both from Lance Formation, Wyoming); M³, un-named species, resembling *P. florencae*, and bottom row, left: M², un-named species resembling *P. florencae*; right: M³, un-named species resembling *P. prokrejcii* (all from Milk River Formation, Alberta).

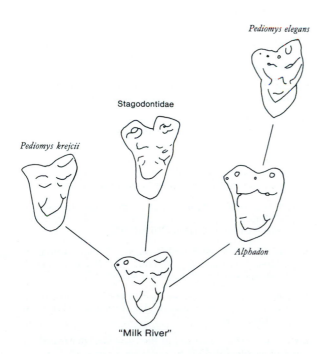

Fig. 6. Diagram showing alternative hypothesis of molar evolution in primitive, Upper Cretaceous marsupials, from an ancestral molar anatomy ("Milk River") in which cusp B and D were developed, but not cusp C, which is added later.

related to geologically younger *Pediomys*, all are marsupials, in so far as these can be recognized by dental and mandibular synapomorphies in Judithian and Lancian species and the resemblances of these to their homologues in Cenozoic didelphids.

ALTERNATIVE 3. Clemens (1966) is correct and the observed patterns are a consequence of cusp C having been suppressed in some lineages of early marsupials prior to the occurrence of the Milk River species. If this were true, the only revision required in current practice would be a taxonomic one; that is, one involving the systematics of *Pediomys* and, hence, the Pediomyidae as a valid family name. But this simply can not be the case: the upper molars of *P. elegans* imply that cusp B is suppressed before cusp C in ancestral pediomyids, whereas the Milk River forms do not show any indication of their ever having possessed a stylar cusp at the C position. Irrespective of the relationships of the groups concerned, the morphological condition illustrated by the upper molars of *P. elegans* can not be ancestral to species of *Pediomys* that lack cusp C.

ALTERNATIVE 4. The ancestral marsupials had a well developed stylocone and cusp D, but no cusp C, which was added later. This alternative (the subject of Clemens' (1979) speculations) is that exemplified by the Milk River fossils as discussed above; it receives support from outgroup comparison with other primitive tribosphenic therians. For example, *Paranyctoides* Fox, 1979c, a primitive insectivoran from the upper member of the Milk River Formation, has a small stylocone and a small cusp at what corresponds, at least topographically, to the D position in the early marsupial upper molar (Fox 1984). Similarly, *Kennalestes* Kielan-Jaworowska, 1969, a primitive Upper Cretaceous eutherian from the Gobi Desert of the Mongolian People's Republic, and undescribed Lower Cretaceous eutherians from the same region, also show a small stylocone and a small cusp in the D position, labial or buccal to the metacone, as does the North American Albian eutherian *Pappotherium* (Slaughter 1965, 1971, 1981).

If Alternative 4 is true, then Simpson (1929, 1948) and Clemens (1966) are each partly correct: Clemens, because the ancestral state of the upper molars was one in which the stylocone was well developed and the anterior stylar shelf wide; Simpson, in that additions to the stylar shelf, namely at the position of stylar cusp C, did take place after the origin of marsupials. If this is the case, its ramifications are intriguing and will cause our notions of evolution in early marsupials to be altered in major ways. For example, those Cretaceous marsupials in which cusp C is small could have been in the process of *gaining* the cusp, instead of losing it; *P. elegans* may be a derivative of an *Alphadon*-like ancestor, but probably other *Pediomys* are not; the Stagodontidae are primitive and not derived in having a large stylocone and cusp D on the upper molars and no cusp C; *Pediomys*, as presently conceived, is polyphyletic, containing species lineages that independently suppressed the stylocone and anterior stylar shelf and, except for *P. elegans*, never passed through a stage in which cusp C was present; and the origin of marsupials was not coincident with the development of stylar cusp C on the upper molars (Slaughter 1971), although it has yet to be shown that any Cretaceous therian possessing a cusp C is not a marsupial. In any case, Alternative 4 further removes the Pediomyidae from relevance in providing a morphotype for molar architecture of the "dog-like" marsupials (Marshall 1977; Fox 1979d) and makes use of the name "Pediomyidae" for stem marsupials (Szalay 1982), whatever their composition might be, inappropriate (Fox 1983). And finally, if Alternative 4 is valid, then the likelihood of close relationship between the Lancian *Glasbius* and *Pediomys* (Reig *et al.* 1987, this Volume) becomes more remote, and, indeed, the geometry of relationships of the living *Microbiotherium* and *Dromiciops* to Cretaceous marsupials generally (see Reig *et al.* 1987), quite uncertain.

REFERENCES

CALDWELL, W. G. E., NORTH, B. R., STELCK, C. R. AND WALL, J. H., 1978. A foraminiferal zonal scheme for the Cretaceous System in the interior plains of Canada. Pp. 496-575 *in* "Western and Arctic Canadian biostratigraphy" ed by C. R. Stelck and B. D. E. Chatterton. *Geol. Assoc. Canada, Spec. Pap.* 18.

CLEMENS, W. A., Jr., 1966. Fossil mammals of the type Lance Formation, Part II. Marsupialia. *Univ. Calif. Publ. Geol. Sci.* **62**: 1-222.

CLEMENS, W. A., 1968. Origin and evolution of marsupials. *Evol.* **22**: 1-18.

CLEMENS, W. A., 1979. Marsupialia. Pp. 192-220 *in* "Mesozoic mammals" ed by J. A. Lillegraven, Z. Kielan-Jaworowska and W. A. Clemens. University of California Press: Berkeley.

CLEMENS, W. A., LILLEGRAVEN, J. A., LINDSAY, E. H. AND SIMPSON, G. G., 1979. Where, when, and what — a survey of known Mesozoic mammal distribution. Pp. 7-58 *in* "Mesozoic mammals" ed by J. A. Lillegraven, Z. Kielan-Jaworowska and W. A. Clemens. University of California Press: Berkeley.

FOX, R. C., 1968. Early Campanian (Late Cretaceous) mammals from Alberta, Canada. *Nature* **220**: 1046.

FOX, R. C., 1969. Studies of Late Cretaceous vertebrates. III. A triconodont mammal from Alberta. *Can. J. Zool.* **47**: 1253-56.

FOX, R. C., 1970. Eutherian mammal from the early Campanian (Late Cretaceous) of Alberta, Canada. *Nature* **227**: 630-31.

FOX, R. C., 1971a. Marsupial mammals from the early Campanian Milk River Formation, Alberta, Canada. Pp. 145-54 *in* "Early mammals" ed by D. M. Kermack and K. A. Kermack. Academic Press: London.

FOX, R. C., 1971b. Early Campanian multituberculates (Mammalia: Allotheria) from the Upper Milk River Formation, Alberta. *Can. J. Earth Sci.* **8**: 916-38.

FOX, R. C., 1972a. An Upper Cretaceous symmetrodont mammal from Alberta, Canada. *Nature* **239**: 171.

FOX, R. C., 1972b. A primitive therian mammal from the Upper Cretaceous of Alberta. *Can. J. Earth Sci.* **9**: 1479-94.

FOX, R. C., 1976. Additions to the mammalian local fauna from the upper Milk River Formation (Upper Cretaceous), Alberta. *Can. J. Earth Sci.* **13**: 1105-18.

FOX, R. C., 1979a. Mammals from the Upper Cretaceous Oldman Formation, Alberta. I. *Alphadon* Simpson (Marsupialia). *Can. J. Earth Sci.* **16**: 91-102.

FOX, R. C., 1979b. Mammals from the Upper Cretaceous Oldman Formation, Alberta. II. *Pediomys* Marsh (Marsupialia). *Can. J. Earth Sci.* **16**: 103-13.

FOX, R. C., 1979c. Mammals from the Upper Cretaceous Oldman Formation, Alberta. III. Eutheria. *Can. J. Earth Sci.* **16**: 114-25.

FOX, R. C., 1979d. Ancestry of the "dog-like" marsupials. *J. Paleont.* **53**: 733-35.

FOX, R. C., 1980. *Picopsis pattersoni*, n. gen. and sp., an unusual therian from the Upper Cretaceous of Alberta, and the classification of primitive tribosphenic mammals. *Can. J. Earth Sci.* **17**: 1489-98.

Fox, R. C., 1981. Mammals from the Upper Cretaceous Oldman Formation, Alberta. V. *Eodelphis* Matthew, and the evolution of the Stagodontidae. *Can. J. Earth Sci.* **18**: 350-65.

Fox, R. C., 1982. Evidence of a new lineage of tribosphenic therians (Mammalia) from the Upper Cretaceous of Alberta, Canada. *Geobios, mem. spec.* **6**: 7-13.

Fox, R. C., 1983. Notes on the North American Tertiary marsupials *Herpetotherium* and *Peradectes. Can. J. Earth Sci.* **20**: 1565-78.

Fox, R. C., 1984. *Paranyctoides maleficus* (n. sp.), an early eutherian mammal from the Cretaceous of Alberta. *In* "Papers in vertebrate paleontology honoring Robert Warren Wilson" ed by R. M. Mengel. *Carnegie Mus. Nat. Hist., Spec. Pap.* 9.

Fox, R. C. (in press). A primitive, "obtuse-angled" symmetrodont (Mammalia) from the Upper Cretaceous of Alberta, Canada. *Can. J. Earth Sci.*

Fox, R. C. (in press). Upper molar structure in the Late Cretaceous symmetrodont *Symmetrodontoides* Fox, and a classification of the Symmetrodonta (Mammalia). *J. Paleont.*

Fox, R. C. and Naylor, B. G., 1982. A reconsideration of the relationships of the fossil amphibian *Albanerpeton. Can. J. Earth Sci.* **19**: 118-23.

Gibson, D. W., 1977. Upper Cretaceous and Tertiary coal-bearing strata in the Drumheller-Ardley region, Red Deer River Valley, Alberta. *Geol. Surv. Canada Paper,* **76-35**: 1-41.

Harland, W. B., Cox, A. V., Llewellyn, P. G., Pickton, C. A. G., Smith, A. G. and Walters, R., 1982. A geologic time scale. Cambridge University Press: Cambridge.

Kielan-Jaworowska, Z., 1969. Preliminary data on the Upper Cretaceous eutherian mammals from Bayn Dzak, Gobi Desert. Pp. 171-91 *in* "Results of the Polish-Mongolian palaeontological expeditions, Pt I" ed by Z. Kielan-Jaworowska. *Palaeontologia Polonica* **19**: Warsaw.

Lerbekmo, J. F., Evans, M. E. and Baadsgaard, H., 1979. Magnetostratigraphy, biostratigraphy and geochronology of Cretaceous-Tertiary boundary sediments, Red Deer Valley. *Nature* **279**: 26-30.

Lillegraven, J. A., 1969. Latest Cretaceous mammals of upper part of Edmonton Formation of Alberta, Canada, and review of marsupial-placental dichotomy. *University of Kansas Paleontological Contributions* **50** (Vertebrata 12): 1-122.

Lillegraven, J. A., 1972. Preliminary report on Late Cretaceous mammals from the El Gallo Formation, Baja California del Norte, Mexico. *Contribs Sci. Nat. Hist. Mus. Los Angeles Co.* **232**: 1-11.

Marsh, O. C., 1889. Discovery of Cretaceous **Mammalia.** *Amer. J. Sci.* ser. **3**: 81-92.

Marshall, L. G., 1977. Cladistic analysis of borhyaenoid, dasyuroid, didelphoid and thylacinid (Marsupialia: Mammalia) affinity. *Syst. Zool.* **26**: 410-25.

Meyboom, P., 1960. Geology and groundwater resources of the Milk River Sandstone in southern Alberta. *Res. Counc. Alberta, Geol. Div., Mem.* **2**: 1-89.

Obradovich, J. D. and Cobban, W. A., 1975. A time-scale for the Late Cretaceous of the Western Interior of North America. Pp. 31-54 *in* "The Cretaceous System in Western North America" ed by W. G. E. Caldwell. *Geol. Assoc. Canada, spec. Paper* 11.

Patterson, B., 1956. Early Cretaceous mammals and the evolution of mammalian molar teeth. *Fieldiana (Geology)* **13**: 1-105.

Reig, O. A., Kirsch, J. A. W. and Marshall, L. G., 1987. Systematic relationships of the living and Neocenozoic American "opossum-like" marsupials (Suborder Didelphimorphia), with comments on the classification of these and of the Cretaceous and Paleogene New World and European metatherians. Pp. 1-89 *in* "Possums and opossums: Studies in evolution" ed by M. Archer. Surrey Beatty & Sons and the Royal Zoological Society of New South Wales: Sydney.

Russell, L. S., 1975. Mammalian faunal succession in the Cretaceous System of western North America. Pp. 137-61 *in* "The Cretaceous System in Western North America" ed by W. G. E. Caldwell. *Geol. Assoc. Canada, Spec. Paper* 11.

Sahni, A., 1972. The vertebrate fauna of the Judith River Formation, Montana. *Bull. Amer. Mus. Nat. Hist.* **147**: 323-412.

Simpson, G. G., 1927. Mesozoic Mammalia. VIII: Genera of Lance mammals other than multituberculates. *Amer. J. Sci.,* ser. **5**: 121-30.

Simpson, G. G., 1929. "American Mesozoic Mammalia". Yale Univeristy Press: New Haven.

Simpson, G. G., 1935. Notes on the classification of recent and fossil opossums. *J. Mammal.* **16**: 134-37.

Simpson, G. G., 1948. The beginning of the Age of Mammals in South America. Part I. Introduction. Systematics: Marsupialia, Edentata, Condylarthra, Litopterna and Notioprogonia. *Bull. Amer. Mus. Nat. Hist.* **91**: 1-232.

Slaughter, B. H., 1971. Mid-Cretaceous (Albian) therians of the Butler Farm local fauna, Texas. Pp. 131-43 *in* "Early mammals" ed by D. M. Kermack and K. A. Kermack. Academic Press: London.

Slaughter, B. H., 1981. The Trinity therians (Albian, Mid-Cretaceous) as marsupials and placentals. *J. Paleont.* **55**: 682-83.

Szalay, F. S., 1982. A new appraisal of marsupial phylogeny and classification. Pp. 621-40 *in* "Carnivorous marsupials" ed by M. Archer. Royal Zoological Society of New South Wales: Sydney.

Williams, G. D. and Stelck, C. R., 1975. Speculations on the Cretaceous palaeography of North America. Pp. 2-20 *in* "The Cretaceous System in the Western Interior of North America" ed by W. G. E. Caldwell. *Geol. Assoc. Canada, Spec. Paper* 11.

Woodburne, M. O. and Zinsmeister, W. J., 1984. The first land mammal from Antarctica and its biogeographic implications. *J. Paleont.* **58**: 913-48.

SPERM STRUCTURE
AND MARSUPIAL PHYLOGENY

Sperm structure is usually regarded as a conservative character which reflects evolutionary affinities. In this review, sperm morphology has been used to explore further the relationships between Australian possums and American opossums. Four features of sperm structure — sperm maturation and pair formation, and sperm head and tail morphology — were considered and compared both within and between these two marsupial groups.

The process of sperm maturation in possums and opossums is generally similar to that described in placental mammals although some maturation changes, such as the covalent crosslinking of sperm nuclear chromatin, are not shared by marsupials. However, marsupial spermatozoa, including those of most possums and opossums, show a variety of unique structural changes during passage through the epididymis which are associated with sperm maturation. These include nuclear rotation, the development of a midpiece fibre network, and loss of the cytoplasmic droplet from the anterior midpiece.

Sperm pairing, also regarded as a feature of sperm maturation, is restricted to the American marsupial fauna and has been used to separate them from their Australian counterparts. The evidence provided suggests sperm pairing will occur in all those American species not yet examined, such as *Lestodelphys, Lutreolina, Caluromyciops, Glironia, Lestoros* and *Rhyncholestes*, with the exception perhaps of *Dromiciops*, which has sperm which more closely resemble those of Australian marsupials. Recent ultrastructural studies of possum and opossum spermatozoa enable a more comprehensive comparison of sperm morphology and its phylogenetic significance in these two groups. In the possums, with the exception of *Phascolarctos* and *Tarsipes*, sperm head and tail structure is relatively uniform. Tail structure of *Phascolarctos* is perhaps the most undifferentiated of the possums and its sperm head morphology confirms a close phylogenetic relationship with the Vombatidae. *Tarsipes* in contrast appears to have the most derived sperm morphology of the possums and its phylogenetic affinities remain controversial. Four distinct sperm head shapes (didelphid, caluromyid, caenolestid and microbiotheriid) are now recognized in the American opossums. The trident-shaped didelphid sperm head is the most common of these, occurring in all genera of the Didelphidae examined so far except *Caluromys*. The characteristic sperm head structure of *Caluromys* confirms its phylogenetic isolation from other didelphids but close similarities in sperm tail morphology, especially the identical asymmetry of the dense fibres in the midpiece, clearly ties *Caluromys* to a didelphid ancestry. Caenolestid sperm structure is distinctive and supports the taxonomic separation of the Caenolestidae from the Didelphidae. Sperm head and neck morphology show no close affinities with any of the other opossum families, although caenolestid sperm tail structure has a simple radial symmetry which is superficially similar to that of *Dromiciops* and reminiscent of that found in the possums. Phylogenetically, *Dromiciops* has perhaps the most interesting sperm morphology which is almost indistinguishable from that found in Australian marsupials, especially the possums. This is consistent with the conclusions of some other recent studies of *Dromiciops*, an extant microbiotheriid, which suggest a close evolutionary affinity between this species and the Australian marsupial fauna and indicates a closer relationship than other extant opossums to the progenitor(s) of the Australian marsupial radiation.

Finally, studies of sperm tail structure in possums and opossums suggest that simple radial symmetry, like that seen in many placentals, monotremes, reptiles and birds, is likely to be the ancestral condition in marsupial spermatozoa and that the more complex asymmetrical tail morphologies seen in *Tarsipes* and the didelphids are more probably a derived condition.

Key Words: Spermatozoa; Sperm pairing; Ultrastructure; Sperm maturation; Marsupial; Phylogeny.

Pages 171-93 *in* POSSUMS AND OPOSSUMS: STUDIES IN EVOLUTION ed by M. Archer. Surrey Beatty & Sons and the Royal Zoological Society of New South Wales: Sydney, 1987.

INTRODUCTION

FEW BIOLOGISTS would argue with the proposition that all of the diverse forms of marsupials which now inhabit Australasia and the Americas were derived from common ancestors (Ride 1962; Clemens 1968, 1977; Tyndale-Biscoe 1973; Sharman 1974; Kirsch 1977; Keast 1977; Simpson 1980; Rich 1982; Archer 1984a). Although interesting and plausible alternatives have recently been suggested to the contrary (Kirsch 1984), considerable support has been given to the theory that the Australian marsupial fauna was derived from an ancestral American stock which reached Australia via the Antarctic land mass at a time when these continents were still joined together as the southern continent of Gondwanaland (Clemens 1977; Keast 1977; Martin 1977; Archer 1984a). In the past many aspects of marsupial biology have been used to probe the continental and intercontinental phylogenetic affinities of the various marsupial fauna. In particular comparisons of dental and skeletal characteristics, karyotypes, serology and other biochemical markers, DNA hybridization and various anatomical features have been used to produce a profusion of cladistic relationships between and within the various Australian and American marsupial families (see Archer 1984a, b).

Sperm morphology has also been regarded as an important component of the marsupial evolutionary equation although, until recently, interest was mostly centred on the unusual phenomenon of sperm pairing which has been

[1]Department of Anatomy, Monash University, Clayton, Victoria, Australia 3168.

conveniently used to separate American marsupials from their Australian counterparts (Biggers and DeLamater 1965; Hughes 1965; Biggers 1966). In mammals, as in many vertebrate and invertebrate groups, sperm morphology is usually regarded as conservative and highly heritable (Beatty 1971), traits which have made it extremely useful in establishing or confirming phylogenetic affinities. For example in the urodeles (Picheral 1979), monotremes (Carrick and Hughes 1982), rodent groups (Friend 1936; Breed 1983) and primates (Bedford 1974), species often can be assigned to their respective taxonomic groupings simply by using comparative sperm morphology. However, perhaps the most striking example of sperm morphology reflecting taxonomic relationships which have been established by more traditional methods, is seen in the marsupials (Biggers and DeLamater 1965; Harding *et al.* 1979). Despite the general conservative nature of sperm morphology within mammalian groups, the marsupials appear to have generated an impressive diversity in sperm morphology. This is perhaps a result of their long evolutionary history and the enormous species diversity which evolved to occupy the wide variety of available ecological niches. The extent of this diversity in sperm structure is even more apparent now as a result of recent efforts to document the ultrastructural characteristics of American and Australian marsupial spermatozoa.

Besides sperm pairing, light microscope studies have revealed other interesting aspects of marsupial sperm morphology which have been shown to have some phylogenetic significance. For example, sperm head shape appears to confirm some taxonomic groupings of the American marsupial fauna (Biggers 1966). It has been used to place *Caluromys derbianus* into a separate evolutionary branch of the Didelphidae and the unique sperm head shape of *Caenolestes obscurus* (Biggers and DeLamater 1965; Biggers 1966; Rodger 1982; Temple-Smith and Tyndale-Biscoe unpublished observations), like many other features of this interesting family (Kirsch 1977; Marshall 1980), isolates the Caenolestidae from the extant Didelphidae as a separate branch of the American marsupial ancestry (Hayman *et al.* 1971). Sperm head morphology has also been used to confirm phylogenetic relationships of the Australian marsupials (Hughes 1965; Harding *et al.* 1979; Harding *et al.* 1982; Temple-Smith 1984c). The unusual hook-shaped sperm heads of the Koala and the wombats supports their close evolutionary ties. Structural specializations of dasyurid marsupial spermatozoa clearly identify members of this family and also suggest a phylogenetic affinity between the dasyurids and the highly specialized *Tarsipes rostratus* (Cummins *et al.* 1980; Harding *et al.* 1981, 1982).

More recently, extensive ultrastructural studies of Australian marsupial spermatozoa (Harding *et al.* 1975; Olson 1975; Cummins 1976; Harding *et al.* 1976; Temple-Smith and Bedford 1976; Harding 1977; Harding *et al.* 1979, 1982, 1984; Temple-Smith 1984c) and, to a lesser extent, those of a few American species (Holstein 1965; Phillips 1970; Rattner 1972; Olson and Hamilton 1976; Olson *et al.* 1977; Krause and Cutts 1979; Temple-Smith and Bedford 1980) have considerably expanded knowledge and understanding of marsupial sperm structure and diversity. This knowledge is now being used to assess the intra- and inter-continental affinities of various species and groups of marsupials, sometimes with quite surprising and interesting results.

This paper reviews the currently available information on sperm structure of the possums and opossums under three main headings (i) sperm pairing and epididymal maturation, (ii) sperm head and (iii) sperm tail structure. These will be discussed with particular reference to their use in establishing and confirming continental and intercontinental relationships between these groupings of the Australasian and American marsupial faunas.

(i) *Sperm Pairing and Epididymal Maturation*

One of the most fascinating discoveries concerning reproduction in male marsupials was the phenomenon of sperm pairing. Paired or "conjugated" sperm were first described in vaginal flushings from recently mated *Didelphis virginiana* (Selenka 1887). Later Retzius (1909) reported that opossum spermatozoa were released unpaired from the testis and that sperm pairing occurred during their passage through the epididymis. This observation was subsequently confirmed by others (Jordan 1911; Duesberg 1920; Biggers and Creed 1962; Biggers and DeLamater 1965). More recent observations show that pairing occurs in particular regions of the epididymis in various American marsupials (Temple-Smith and Bedford 1980; Rodger 1982) and is regarded as a feature of sperm maturation in these species.

Until the early 1960's the phenomenon of sperm pairing in marsupials was almost completely ignored. McCrady (1938) had observed that sperm pairs in *D. virginiana* separate in the female reproductive tract prior to fertilization and that sperm trapped in the albumen egg coat were always single. However, there was still no confirmation that pairing was a normal event in male opossums nor any indication of its prevalence in other American or Australasian marsupial species. The studies of Biggers and his associates (Biggers and Creed 1962; Biggers and DeLamater 1965; Biggers 1966) rekindled interest in sperm pair formation in marsupials and considerably extended our knowledge and understanding of the pairing process and its occurrence in marsupials. Sperm pairs were observed in species representing eight genera of American marsupials and pair formation occurred in the epididymis of all specimens examined. Biggers and DeLamater (1965) also examined a small range of Australian marsupials and concluded that pairing was restricted to the American species. Absence of sperm pairing in Australian marsupials was confirmed in a more detailed study by Hughes (1965) in which representatives from all the Australian marsupial families, except the Notoryctidae, were surveyed. These two studies provided the basis for a diagnostic morphological dichotomy which has been used subsequently by evolutionists to separate phylogenetically these geographically isolated marsupial groups.

Although all the important groups of Australian marsupials, especially the possums, have been thoroughly surveyed for sperm pairing, some interesting American species still await examination. Difficulty in obtaining suitable specimens of these rarer American forms has prevented a complete assessment of sperm pairing and sperm structure in the American marsupial fauna. Species which still need to be studied are the putative caluromyiids *Caluromysiops irrupta* and *Glironia venusta*, the Chilean caenolestid *Rhyncholestes raphanurus* and its Peruvian relative *Lestoros inca*, the Patagonian carnivorous didelphid *Lestodelphys halli* — which would all be expected to show sperm pairing — and, perhaps most importantly, the only extant representative of the Microbiotheriidae, the Chilean monito del monte, *Dromiciops australis*. The need to examine *Dromiciops* for sperm pairing is obvious in the light of recent evidence (Szalay 1982, Sharman 1982, Temple-Smith in preparation) which suggests a close evolutionary affinity between this American species and the Australian marsupial radiation. The similarities in sperm morphology between *Dromiciops* and the Australian marsupials suggest that sperm pairing probably does not occur in this species. If this is so, it would suggest that sperm pairing occurred after the Microbiotheriidae had diverged from the ancestral American marsupial stock or that it was present in the ancestral stock and was subsequently lost from the Microbiotheriidae.

The ultrastructure features of sperm pairing have been described in detail for only two American species (Phillips 1970; Olson and Hamilton 1976; Krause and Cutts 1979; Temple-Smith and Bedford 1980). However, in both species, a similar association of each sperm pair was described. Attachment of each sperm pair is always between the adjacent acrosomal surfaces of the sperm heads in a very precise orientation

(Figs 1 and 6). Each sperm pair is then held together by a narrow membrane junction which forms a peripheral seal outside the periacrosomal ring (Fig. 2). The periacrosomal ring, or dense plaque (Olson and Hamilton 1976), marks a point of fusion between the plasma and nuclear membranes around the periphery of the acrosome. The peripheral junction resembles a gap junction rather than a zone of membrane fusion or tight junction, since the trilaminar structure and continuity of each plasma membrane is retained. This junction in *Didelphis virginiana* is 500 to 700 Å wide and the associated plasma membranes are separated by a 70 to 100 Å gap (Temple-Smith and Bedford 1980). A similar arrangement occurs in *Caluromys philander* (Olson and Hamilton 1976). In *Didelphis virginiana* pairing also involves a close association between the adjacent plasma membranes over the entire acrosomal surface with the exception of a small region, about 300-500 Å wide, at the edge of each acrosome where the plasma membrane makes contact with the periacrosomal ring (Figs 1, 2 and 7). However, in *Caluromys* there is no consistent attachment between adjacent acrosomal plasma membranes (Phillips 1970; Olson and Hamilton 1976) and it appears that the peripheral junction alone is responsible for maintaining the integrity of each sperm pair in this species. In *Didelphis* the acrosomal plasma membrane "junction" appears to be functionally important. Observations by Rodger and Bedford (1982) of sperm pairs undergoing separation in the female opossum reproductive tract show that the sperm heads remain united across the acrosomal surface following separation of the peripheral junction.

In American marsupials, sperm pairing is regarded as one of the visual correlates of sperm maturation in the epididymis. Until recently, little was known about sperm maturation in marsupials, even though the observations of sperm pairing in *Didelphis virginiana* by Retzius (1909) were probably the first evidence in any mammal of the epididymis influencing the structure and function of spermatozoa. In placental mammals the epididymis plays an important role in post-testicular maturation of spermatozoa; a process which eventually results in the acquisition of the capacity for motility and the ability to fertilize (Gaddum 1968; Orgebin-Crist 1969; Glover and Nicander 1971; Bedford 1975; Moore 1983). Various minor morphological changes were also identified as concomitants of sperm maturation (Bloom and Nicander 1961, Fawcett and Hollenberg 1963; Bedford 1965; Calvin and Bedford 1971; Bedford and Calvin 1974; Bedford 1975; Bedford and Cooper 1978; Nicholson and Yamagimachi 1979).

From the few species which have been examined in detail, it appears that sperm maturation in marsupials is basically similar to that observed in placental species; for example, the

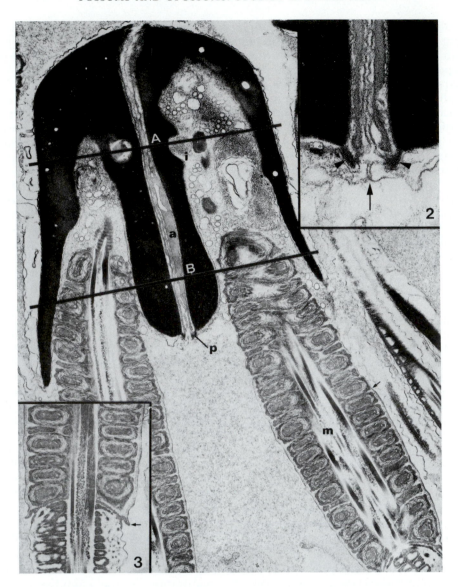

Fig. 1. Mature sperm pair from the cauda epididymidis of the Virginia Opossum, *Didelphis virginiana*, in longitudinal section showing the characteristic alignment of the two sperm heads across their acrosomal surfaces and the insertion and orientation of each sperm tail. Note the close apposition of the acrosomal plasma membrane between the attachment surfaces of the sperm pair and the peripheral junction. A and B correspond to the levels of transverse sections in Figure 7. Abbreviations: a, acrosome; m, midpiece; p, periacrosomal ring; i, implantation fossa; arrow, anterior edge of midpiece sub-membrane fibre network. x 17,300.

Fig. 2. (inset) Higher magnification of the peripheral junction region of a *Didelphis* sperm pair showing the periacrosomal rings (arrow heads) and the peripheral junction (arrow) forming a narrow outer junction between the two adjacent sperm plasma membranes. Close apposition of the acrosomal plasma membranes and the narrow membrane space between the peripheral and acrosomal membrane junctions can be clearly seen. x 55,800.

Fig. 3. (inset) Junction region between midpiece and principal piece of an opossum (*Didelphis virginiana*) sperm tail showing the distinctive flange-shaped annulus which extends outwards from the dense fibres as a thin plate of fibrous material and isolates midpiece and principal piece structures. At its exterior the annulus forms a characteristic flared rim (arrow) which is closely associated with the plasma membrane.

capacity for sperm motility is gradually acquired during passage through the epididymis (Temple-Smith and Bedford 1976, 1980; Cummins 1976), regional changes occur in the sperm plasma membrane (Temple-Smith and Bedford 1976, 1980; Olson *et al.* 1977; Temple-Smith 1981, 1984c), structural modification of the acrosome occurs in some species (Temple-Smith and Bedford 1976; Harding 1977; Harding *et al.* 1979), and the cytoplasmic droplet is shed during passage through the epididymis (Temple-Smith and Bedford 1976, 1980; Harding *et al.* 1979,

Temple-Smith 1984b, c; Jones *et al.* 1984; Cummins *et al.* 1985). Some features of placental sperm maturation are not shared by marsupials, for example the covalent crosslinking of nuclear chromatin, but, perhaps of more importance, marsupials show many interesting changes in sperm structure during passage through the epididymis which are unique. These peculiarities of marsupial sperm maturation include nuclear rotation (Hughes 1965; Temple-Smith and Bedford 1976, 1980; Cummins 1976; Harding 1977; Harding *et al.* 1979; Krause and Cutts 1979), loss of the cytoplasmic droplet, without distal migration, from the anterior region of the midpiece (Temple-Smith and Bedford 1976, 1980), development of a parallel fibre network beneath the plasma membrane (Olson 1975; Harding *et al.* 1975, 1979, 1981, 1982; Cummins 1976; Olson *et al.* 1977; Temple-Smith 1984c) and, in the American species, the formation of sperm pairs (Biggers 1966; Phillips 1970; Krause and Cutts 1979; Temple-Smith and Bedford 1980). In particular, nuclear rotation, development of the midpiece fibre network and loss of the cytoplasmic droplet from the sperm neck are maturation events which have been observed in most of the marsupial species examined. Exceptions to this generalization are the absence of nuclear rotation in the Koala (Hughes 1965; Harding *et al.* 1979), the wombats (Hughes 1965), and *Caenolestes obscurus* (Temple-Smith and Tyndale-Biscoe, unpubl. obs.), and the absence of a midpiece fibre network in the Koala, and probably also in the peramelids (Harding *et al.* 1979). Other structural changes which have been described in some species but are not generally associated with sperm maturation in marsupials include the formation of midpiece membrane invaginations and whorls in some Australian possums (Harding *et al.* 1975, 1979; Temple-Smith and Bedford 1976; Temple-Smith 1984c), extensive structural modification of the acrosome (Temple-Smith and Bedford 1976; Cummins 1976; Harding *et al.* 1976, 1979) and deposition of a prominent extracellular coating of dense material over the surface of the midpiece in *Caluromys philander* (Phillips 1970; Olson and Hamilton 1976).

(ii) *Sperm Head Structure*

Amongst vertebrates, the marsupial sperm head is unique in both its general shape and its diversity between various marsupial families (Retzius 1909; Duesberg 1920; Biggers and Creed 1962; Biggers and DeLamater 1965; Hughes 1965; Holstein 1965; Phillips 1970; Rattner 1972; Temple-Smith and Bedford 1976, 1980; Olson and Hamilton 1976; Cummins 1976; Harding 1977; Harding *et al.* 1979; Krause and Cutts 1979; Temple-Smith 1984c). In possums and opossums, sperm head structure has now been examined in detail for many species and recent ultrastructural observations on some key evolutionary species in these two groups (Cummins *et al.* 1980; Harding *et al.* 1984; Temple-Smith and Tyndale-Biscoe in prep.; Temple-Smith in prep.) enable a more comprehensive comparison of sperm head morphology in these groups.

Possums

Sperm head structure, and especially ultrastructure, has been examined now for most of the phylogenetically interesting representatives of the Australian possum families (Hughes 1965; Temple-Smith and Bedford 1976; Cummins 1976; Harding 1977; Harding *et al.* 1979; Temple-Smith 1984c). The reader is particularly referred to the comprehensive review by Harding *et al.* (1979) for a detailed discussion of the structural diversity of spermatozoa in the possums and other Australian marsupial groups and its phylogenetic implications. For purposes of comparison in this review, the following summary information on sperm head structure in possums and other Australian marsupials is included.

Within many Australian marsupial families, for example the Dasyuridae, the Macropodidae and the Peramelidae, sperm head morphology is remarkably uniform. It is apparently difficult using gross sperm morphology to distinguish between different genera within the Dasyuridae and Peramelidae, although various macropodid species are recognizable by minor variations in size and shape of their spermatozoa (Harding *et al.* 1979).

Despite some characteristic differences in sperm morphology between these Australian groups, the shape and ultrastructure of their sperm heads follow a general pattern of bilateral symmetry. This is common to all Australian marsupials, although family differences are apparent in sperm head size and shape, and in the organisation of perinuclear components such as the acrosome, perinuclear membranes and the shape of the implantation fossa, which accommodates the neck and anterior midpiece of the tail.

Of all the Australian marsupial families, the possums show the largest diversity in sperm head morphology, due mostly to the inclusion in this group of two controversial diprotodont families: the Phascolarctidae, containing the Koala, and the Tarsipedidae, with the Honey Possum as its sole representative (Harding *et al.* 1979; Temple-Smith 1984c). The evolutionary affinities of these two families are still not completely resolved, although sperm morphology suggests that they are not closely related to other possum families (Hughes 1965, 1977; Harding 1979; Cummins *et al.* 1980; Harding *et al.* 1979, 1981, 1982; Temple-Smith 1984c).

The strange hook-shaped appearance of Koala spermatozoa sets this species and the wombats (Vombatidae) apart from all other Australian marsupials (Hughes 1965) and seems clearly to confirm the phylogenetic relationship of these marsupial species which has been suggested from studies of dental, skeletal and serological characteristics (Archer 1976b, 1984b and c; Kirsch 1977). In fact, the close similarity in sperm morphology of these ecologically-diverse marsupial forms is an indication of the usefulness of sperm morphology in taxonomy.

Sperm head morphology of the Honey Possum, *Tarsipes rostratus*, shows closer similarities in size and shape to dasyurids than possums (Cummins *et al.* 1980; Harding *et al.* 1982, 1984; Temple-Smith 1984c). The Honey Possum displays an unusual mosaic of anatomical, reproductive and ecological features (Wooller *et al.* 1981; Renfree *et al.* 1984), and its highly specialized life style suggests that many of its anatomical characteristics are highly derived, reflecting its specialized existence rather than its phylogenetic affinities. Perhaps, under these circumstances, more weight should be given to sperm morphology as an indicator of evolutionary affinities than other anatomical and dental characteristics. Harding *et al.* (1981, 1982), have concluded that Honey Possum spermatozoa display a number of primitive, rather than highly derived, features. They suggest that *Tarsipes* and the peramelids and dasyurids represent the original stem radiation in the evolution of Australian marsupials and that the phalangerid and macropodid sperm morphology is derived from a *Tarsipes*-like sperm structure (Harding *et al.* 1981). However, more recent comparisons with sperm structure of American marsupials suggest that the reverse of this is more likely. This aspect will be discussed in more detail later in this review.

Apart from these two rather aberrant possum families, most possums share very similar sperm head shape and structure (Harding *et al.* 1979; Temple-Smith 1984c) which is characterized by a cuneiform-shaped sperm nucleus. The acrosome is situated over the anterior dorsal surface and a prominent ventral groove in the nucleus forms a tapering, semi-conical fossa into which the connecting piece and a portion of the anterior midpiece is located in the mature spermatozoon (Fig. 4). This relationship is a structural feature of all Australian marsupial spermatozoa, except those of the Koala and wombats.

In the possums the size of the acrosome varies between species and there is, as might be expected, an associated difference in the area of the dorsal nuclear surface covered by the acrosome. In the Phalangeridae and Burramyidae which have been examined the acrosome is relatively small and restricted to the anterior region

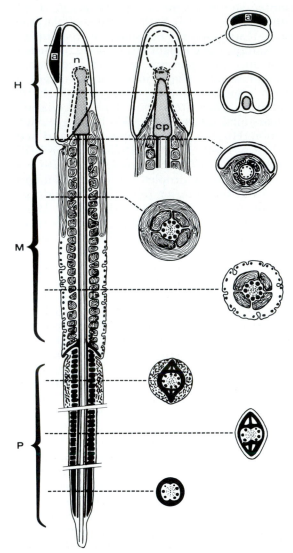

Fig. 4. Generalized diagram of an Australian possum spermatozoa in lateral, ventral (head, neck and anterior midpiece only) and cross-sectional views to show the distinctive structural features, in particular, bilateral symmetry, nuclear shape, structural specializations and neck insertion. This diagram is based on sperm structure of the Brush-tailed Possum, *Trichosurus vulpecula* (Temple-Smith and Bedford 1976). Sperm of the Honey Possum, *Tarsipes rostratus*, and the Koala, *Phascolarctos cinereus*, show marked differences from this generalized form (see text). Abbreviations: H, head; M, midpiece; P, principal piece; a, acrosome; n, nucleus; cp, connecting piece of the neck.

of the dorsal nuclear surface (Fig. 4). In contrast, the Petauridae, Pseudocheiridae and the Honey Possum have elongated acrosomes which cover most of the dorsal surface of the nucleus (Harding 1977; Harding *et al.* 1979, 1984; Cummins *et al.* 1980) and in the Koala (Hughes 1977) the acrosome occupies a long thin groove along most of the inner curvature of the hooked sperm head.

Opossums

Sperm head shape and structure in the American opossums show some fascinating and phylogenetically important differences both within and between the various opossum families. From light microscope studies of sperm structure in American marsupials, Biggers and DeLamater (1965) described three types of sperm head morphology which they designated as didelphid, caluromyid and caenolestid (Fig. 5). This extensive study examined sperm head and pair shapes in species from all the major genera of the American marsupial fauna with the exception of *Caluromysiops*, *Glironia*, *Lestodelphys*, *Lutreolina*, *Rhyncholestes*, *Lestoros* and *Dromiciops*.

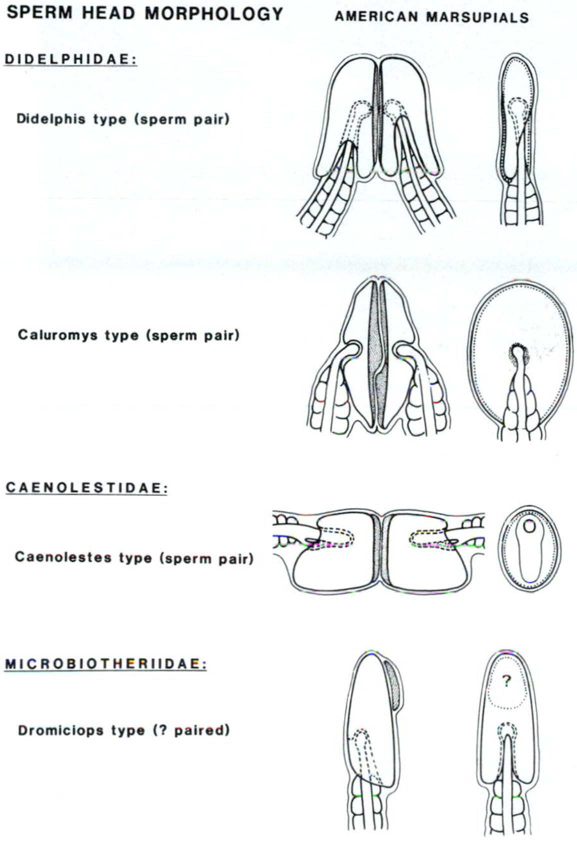

SPERM HEAD MORPHOLOGY   AMERICAN MARSUPIALS

DIDELPHIDAE:

Didelphis type (sperm pair)

Caluromys type (sperm pair)

CAENOLESTIDAE:

Caenolestes type (sperm pair)

MICROBIOTHERIIDAE:

Dromiciops type (? paired)

Fig. 5. Diagrammatic representation of the morphology and orientation of sperm pairs in the three extant families of American opossums (see text for description and explanation). This diagram shows the characteristic nuclear shape and neck insertions of the four sperm types which have been described in the American opossums, in lateral and abacrosomal view. Shaded portions represent the acrosomes and the dotted outline on each abacrosomal diagram corresponds to the peripheral limits of the acrosome.

Ultrastructural studies (Holstein 1965; Phillips 1970; Rattner 1972; Olson and Hamilton 1976; Krause and Cutts 1979; Temple-Smith and Bedford 1980) have confirmed the observations of Biggers and DeLamater (1965) and provided additional morphological detail which extends considerably their findings. From these published accounts and some unpublished ultrastructural observations on *Caenolestes obscurus* (Temple-Smith and Tyndale-Biscoe, unpubl. obs.) and *Dromiciops australis* (Temple-Smith, in prep.), a new perspective is presented on the significance of sperm head structure and marsupial relationships.

Didelphidae

Two distinct sperm head shapes have been described and confirmed in the Didelphidae (Biggers 1966; Phillips 1970; Olson and Hamilton 1976; Krause and Cutts 1979; Temple-Smith and Bedford 1980). The most common form, the didelphid-type (Fig. 5), is typical of all didelphid marsupials examined except *Caluromys*.

Didelphis-type

This sperm head shape is found in *Chironectes* and various species of *Didelphis*, *Marmosa*, *Monodelphis*, *Philander* and *Metachirus* and will probably also be found in *Lutreolina crassicaudata* and the Patagonian *Lestodelphys halli* from which reproductive material is not yet available for study. In all these species the mature sperm pairs have a characteristic trident shape (Figs 1, 5 and 6). Unlike the Australian possums, the *Didelphis*-type sperm head has no axis of bilateral symmetry. Instead it comprises a straight, major nuclear extension, along the outer surface of which the acrosome is attached and pairing occurs, and a more streamlined, recurved, minor nuclear extension which forms the outer prongs of each sperm pair trident (Fig. 6). This nuclear asymmetry is clearly shown in transverse sections taken at various levels through paired sperm heads (Fig. 7) and confirms that, although asymmetrical, the single sperm heads in each pair are not morphological mirror images of each other. Each sperm is produced with the same structural orientation in the testis and sperm pairs are formed by precise alignment of the two pairing spermatozoa along their opposing acrosomal surfaces (Figs 1, 5 and 6; Krause and Cutts 1979; Temple-Smith and Bedford 1980). Perhaps even more fascinating than the actual sperm pairs is the mechanism by which they form. Some specific recognition factors between the two pairing surfaces must presumably direct the precise orientation and alignment of the sperm heads during the pairing process. Such is the accuracy of these recognition factors that there have, as yet, been no recorded examples of misaligned or inverted sperm pairs; quite a remarkable phenomenon since pair formation seems to be a random and apparently undirected event in the epididymis.

Between the two nuclear extensions the curved connecting piece of the sperm tail gains access to the implantation fossa on the abacrosomal surface of the major extension (Figs 1, 5 and 7). In *Didelphis*-type sperm pairs, the sperm tail from each individual sperm head extends out from the opposite side of the sperm head to its counterpart sperm tail (Figs 6 and 7).

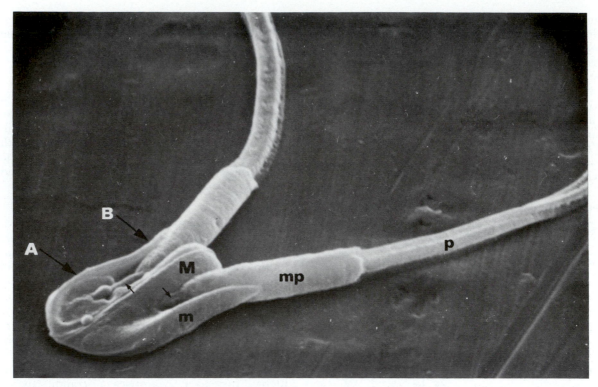

Fig. 6. Scanning electron micrograph showing paired spermatozoa from the Virginian Opossum, *Didelphis virginiana*. Note the characteristic trident-shape of the paired didelphis-type sperm heads, the major (M) and minor (m) nuclear extensions, connecting pieces (arrows), midpiece (mp) and principal piece (p) of the tail. The central ridge between the paired sperm heads represents the line of pairing and may correspond to the raised edge of the periacrosomal ring at the periphery of each acrosome. A and B correspond approximately to the levels of transverse section shown in Figure 7. (Courtesy of Dr W. Krause) x 5,000.

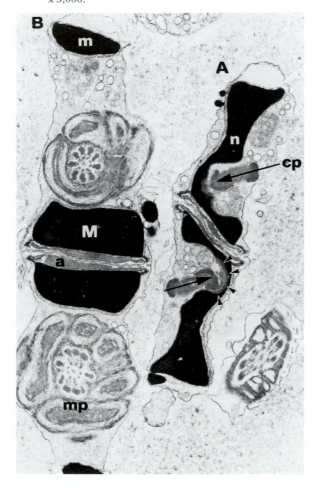

Fig. 7. Paired Opossum (*Didelphis virginiana*) sperm heads shown in transverse sections A and B taken at the levels shown in Figures 1 and 6. Section A clearly shows the structural asymmetry of the rostral portion of each sperm nucleus (n) in *Didelphis* and the opposing direction of insertion of each connecting piece (cp). Arrow along each connecting piece indicates its direction of insertion into the corresponding implantation fossa (arrow heads). Section B is caudal to Section A and shows the structure and organization of the major (M) and minor (m) nuclear extensions, midpiece (mp) and acrosome (a) of each sperm head. Note the close apposition of acrosomal plasma membranes, and the peripheral junctions and nuclear rings in each section. x 16,600.

Caluromys-type

Although included in the family Didelphidae, sperm head morphology of the genus *Caluromys*, from which this sperm head shape is named, is strikingly different to the other genera in this family (Fig. 5). *Caluromys* sperm heads are shaped like wide, flattened saucers with the sperm tail inserting into an open central implantation fossa on the convex ventral surface (Biggers and DeLamater 1965; Phillips 1970), and the acrosome covering the entire, slightly concave, dorsal surface of each sperm head (Phillips 1970; Olson and Hamilton 1976). Unlike the didelphid-type, *Caluromys* sperm heads have an axis of bilateral symmetry through the long axis of the sperm.

The genus *Caluromys* has been regarded for some time as a distinct group within the Didelphidae based on differences in dental, anatomical and serological characteristics (Reig 1955; Kirsch 1970, 1977). By comparison, the other didelphid genera form a relatively cohesive phylogenetic grouping. The differences in sperm head shape seem to confirm the phylogenetic isolation of *Caluromys* within the Didelphidae. Based on close similarities in dental and skeletal features, the two relatively unknown opossum genera *Caluromysiops* and *Glironia* have also been grouped with *Caluromys* rather than the other didelphids (Reig 1955; Kirsch 1970, 1977; Archer 1976a). Observations on sperm structure in these two species may also help to elucidate their phylogenetic affinities.

Caenolestidae

This family includes three closely related genera *Caenolestes*, *Lestoros* and *Rhyncholestes* which represent the only extant relatives of a once more widely distributed and ancient branch of the South American marsupial radiation (Marshall 1980). Until recently, little was known about sperm structure in this group. Sperm pairing in this family had only been described from light microscope observations of epididymal sperm from *Caenolestes obscurus* (Biggers and DeLamater 1965; Rodger 1982). These observations clearly demonstrated that sperm head morphology in this species was radically different from those of the Didelphidae and, although structural details were not easily discernible from these studies, it is clear that the sperm nucleus was more compacted forming a dome or rectilinear shape (Biggers and DeLamater 1965) and that pairing occurred across the flattened rostral surface of adjacent sperm. The lateral nuclear extensions which characterise didelphid-type sperm head morphology are not present in the caenolestid sperm (Figs 5 and 8). Recent ultrastructural observations of sperm from *Caenolestes obscurus* (Temple-Smith and Tyndale-Biscoe in prep.) provide a clearer indication of sperm head structure in this species, although the tissue was

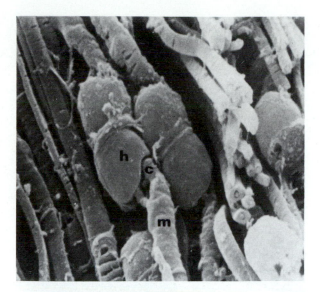

Fig. 8. Scanning electron micrograph showing sperm pairs in the caenolestid marsupial, *Caenolestes obscurus*. Note the characteristic shape of the sperm head (h), and the unusual opposing tail insertions of each sperm pair. Abbreviations: m, midpiece; c, connecting piece. x 6,600.

not initially fixed for ultrastructural studies. Towards its flattened rostral end, over which the acrosome is situated, the nucleus has an almost circular cross-section (Fig. 9). This becomes more ovoid towards the end of the implantation fossa (Fig. 5) with the caudal trailing edge of the sperm head recurving over the connecting piece and anterior midpiece (Figs 5 and 8). In sagittal section, the shape of the nucleus is almost rectilinear (Figs 5 and 10). An interesting aspect

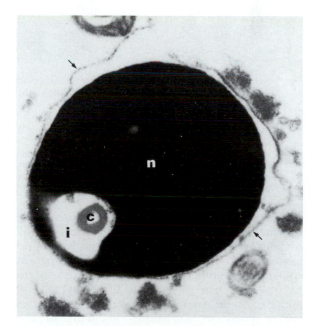

Fig. 9. Transverse section through a caenolestid (*Caenolestes obscurus*) sperm head showing the nucleus (n), implantation fossa (i), connecting piece (c) and plasma membrane (arrow). Note the characteristic rounded shape of the nucleus in this species and the acentric position of the implantation fossa. x 34,100.

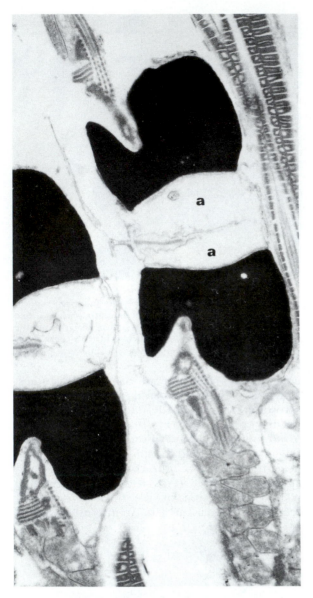

Fig. 10. Longitudinal section through paired caenolestid (*Caenolestes obscurus*) spermatozoa. Although the plasma membranes and acrosomal structure have been poorly preserved, the general morphology and organization of *Caenolestes* sperm pairs are clearly shown. Note the complex structure of the neck in this species, the deep implantation fossa and the rostral position of the acrosome (a). Sperm pairing occurs between the flattened rostral surface of each sperm. x 18,800.

of sperm head morphology in *Caenolestes* is the unusually deep caudal invagination in the nucleus which forms the point of insertion and implantation fossa for the tail. The apparent spatial constraints of this fossa direct the sperm tails of each pair in opposite directions (Figs 5 and 8); an unusual arrangement which causes sperm pairs in *Caenolestes* to align in the epididymis with the leading sperm in each pair moving tail-first along the duct. This spatial arrangement of sperm tails in each caenolestid pair makes it difficult to understand how forward, progressive motility is achieved, unless

there is an immediate post-ejaculatory separation of each pair or a structural reorientation of the tails. More thorough investigations of sperm structure and function in the Caenolestidae are clearly needed to provide both comparative information on sperm ultrastructure in representatives from each of the three extant genera and *in vitro* observations on motility and swimming behaviour of caenolestid sperm pairs.

Microbiotheriidae

All members of this family were thought to be extinct until the recent studies of Reig (1955) provided convincing evidence that *Dromiciops australis*, the Chilean monito del monte, was a living representative of the Microbiotheriidae. Because of its unique evolutionary position and some recent studies which have suggested a distant relationship with the extant American marsupial fauna (Kirsch 1977) and apparent affinities with the Australian marsupial radiation (Szalay 1982; Sharman 1982), sperm pairing and structure in *Dromiciops* have been regarded with some interest.

As indicated previously, there are still no observations on sperm pairing in *Dromiciops australis*, but recent observations on sperm structure, and in particular sperm head morphology, have cast more light on the evolutionary affinities of this species (Temple-Smith 1984a; Temple-Smith in prep.). Although these ultrastructural observations, like those of *Caenolestes obscurus*, have been made on inappropriately fixed, field material, sufficient detail is present in spermatozoa from the cauda epididymidis to indicate a remarkable similarity in sperm head structure between *Dromiciops* (Figs 5, 11, 12 and 13) and many Australian marsupial species (Figs 4 and 12 [inset]). In fact, with material appropriately prepared for electron microscopy, it would probably be difficult to readily distinguish the sperm head of *Dromiciops* from such species as *Trichosurus vulpecula*, *Cercartetus nanus* and some other related possum species (Fig. 12). The structural features which characterize these species (Fig. 4), such as the tapering cuneiform nucleus, the open, semiconical groove in the ventral surface of the nucleus and even, perhaps, a small acrosome restricted to the anterior, dorsal surface of the nucleus, are all features shared by *Dromiciops* spermatozoa (Figs 5, 11, 12 and 13), but not by those of any other extant American species (Fig. 5). This close similarity in sperm head morphology provides, with the conclusions from other studies (Szalay 1982; Sharman 1982; further compelling evidence to imply a close phylogenetic relationship between *Dromiciops australis* and the Australian marsupial radiation. Observations on sperm morphology suggest that, of all the Australian marsupial families, the Phalangeridae share the closest affinity with

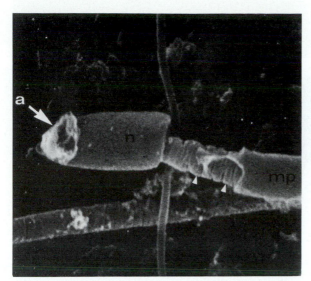

Fig. 11. Scanning electron micrograph of a microbiotheriid (*Dromiciops australis*) sperm. This shows the dorso-lateral surface of the sperm head and the remnants of acrosome (a) which appear to be restricted to the anterior third of the dorsal surface of the nucleus (n). The plasma membrane has not been preserved over the head or anterior midpiece (mp), where the mitochondrial gyrae are clearly visible (arrows). The cuneiform sperm head shape resembles that of Australian marsupials. x 8,500.

Dromiciops but more comprehensive studies of various other structural, biochemical and serological features are needed to confirm this association.

(iii) *Sperm Tail Structure*

The general structural organization of marsupial sperm tails is similar to that of prototherian and placental mammals: the sperm tail consists of the neck, containing striated connecting piece and centriolar complex, midpiece and principal piece. However, as indicated from various structural studies (Holstein 1965; Phillips 1970; Olson 1975; Harding *et al.* 1975; Cummins 1976; Temple-Smith and Bedford 1976, 1980; Olson and Hamilton 1976; Olson *et al.* 1977; Harding 1977; Harding *et al.* 1979, 1981, 1982, 1984; Temple-Smith 1984c), there are many structural specializations which are unique to marsupials and set them recognizably apart from all other vertebrate groups. In addition, the structure and organization of the sperm tail differs both within and between various marsupial families. As with sperm head morphology, the marsupial sperm tail provides a suite of characteristics which may also be important in indicating phylogenetic affinities of this group.

Before comparing the various structural characteristics and specializations of the possums and opossums it is important to draw attention to features of the marsupial sperm tail which are in general characteristic of this group of mammals

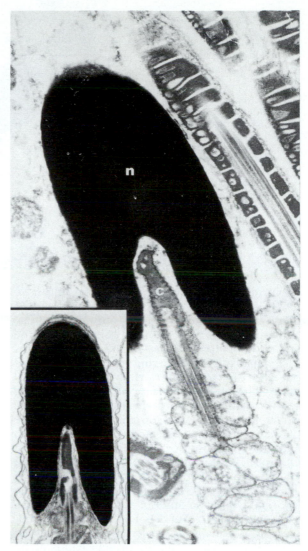

Fig. 12. Coronal section through the nucleus (n), implantation fossa, connecting piece (c) and anterior midpiece of a microbiotheriid sperm (*Dromiciops australis*). The plasma and inner mitochondrial membranes have not been preserved because of poor initial fixation. x 24,500. Inset: a similar coronal section through the sperm head of an Australian Brush-tailed Possum, *Trichosurus vulpecula*, shows the close structural similarities between sperm head morphology in *Dromiciops* and the Australian possums. x 13,700.

(Figs 3, 4 and 5). The neck consists of an elongated connecting piece which usually tapers towards the implantation fossa and terminates in a spherical expansion which provides the only point of contact between the tail and head, as a ball and socket joint. This arrangement presumably helps to facilitate the characteristic changes in orientation of the sperm head which occur during spermiogenesis and sperm maturation (Temple-Smith and Bedford 1976, 1980; Cummins 1976; Harding 1977; Harding *et al.* 1979; Temple-Smith 1984c). Like other mammals, the midpiece is characterized by the presence of mitochondria and terminates in the annulus which in most marsupials, although not in the

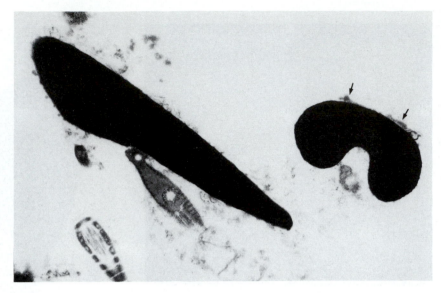

Fig. 13. Electron micrograph showing sagittal and transverse sections through the nucleus and connecting piece of two *Dromiciops* spermatozoa. Note the remnants of the acrosome (arrows), and similarities in the shape of the ventral nuclear implantation cleft between *Dromiciops* and the generalized Australian possum sperm (see transverse head sections in Figure 4). x 19,300.

dasyurids, *Tarsipes* or peramelids (Harding *et al.* 1979), consists of a characteristic broad flange of dense fibrous material curving outwards and distally from the periphery of the dense fibres at the end of the midpiece (Figs 3, 4 and 22). This forms a characteristic overhanging rim which is flared out and directed posteriorly over the principal piece. The posterior surface of the annulus is in contact with the fibrous sheath and loose fibres which are found between this sheath and the plasma membrane. The midpiece is divided into anterior and posterior segments (Phillips 1970; Olson 1975; Harding *et al.* 1975; Temple-Smith and Bedford 1976, 1980; Olson *et al.* 1977; Temple-Smith 1984c). The plasma membrane associated with the anterior midpiece is structurally unspecialized whereas in the posterior midpiece it is usually associated with a specialized parallel network of fibrous protein bands along the cytoplasmic surface of the plasma membrane (Olson 1975; Harding *et al.* 1975; Temple-Smith and Bedford 1976, 1980; Olson *et al.* 1977; Harding *et al.* 1979). This specialization has been found in all marsupial species examined with the exception of the Koala, *Phascolarctos cinereus* (Harding 1979) and peramelids (Harding *et al.* 1979). Freeze-fracture studies of the plasma membrane over the posterior midpiece in *Didelphis virginiana* (Olson *et al.* 1977) and *Trichosurus vulpecula* (Temple-Smith 1981, 1984c) clearly show intra-membranous particles aggregated into highly ordered parallel rows separated by particle-free interspaces. In *T. vulpecula* these particle rows are directly associated with the submembrane fibre networks rather than the interspaces between them (Temple-Smith 1981, 1984c). With the exception of the Dasyuridae and *Tarsipes rostratus*, marsupial sperm tails are circular in transverse sections through the midpiece although the principal piece is usually more ovoid in transverse section. Principal piece shape is related to the organization of the fibrous columns of the fibrous sheath. A standard 9 + 2 axonemal complex forms the axial core of the tail and this

is associated with nine peripherally arranged dense fibres. As will be discussed in more detail in subsequent sections, the disposition of these dense fibres may have phylogenetic significance in marsupials.

Possums

In the possums, sperm tail structure has been examined in detail for only a few species, for example *Trichosurus vulpecula* (Olson 1975; Harding *et al.* 1975, 1979; Cummins 1976; Temple-Smith and Bedford 1976; Temple-Smith 1981, 1984c), *Pseudocheirus peregrinus* (Harding *et al.* 1979) and *Tarsipes rostratus* (Cummins *et al.* 1981; Harding *et al.* 1981, 1982, 1984). However, many other species which have been studied in less detail provide sufficient information for a comprehensive appraisal of sperm tail structure in this group.

The generalized possum sperm tail structure shown in Figure 4 is found in representatives from all possum families examined so far with the exception of the Phascolarctidae, Pseudocheiridae and Tarsipedidae. Possum sperm tails are, in general, characterized by a straight connecting piece, a well-defined midpiece fibre network (Fig. 14) and, in some species, small plasma membrane invaginations which occur randomly in the interspace between these fibres (Fig. 4) (Harding *et al.* 1975, 1979; Temple-Smith and Bedford 1976; Temple- Smith 1984c). The midpiece is circular in cross-section with nine rounded dense fibres arranged symmetrically around the central axoneme (Figs 4 and 15).

Phascolarctos cinereus sperm tails have a similar structural organization to the generalized possum sperm tail but appear to be the least specialized of all the possums (Hughes 1977; Harding *et al.* 1979, 1987). Midpiece cross-sections are circular and radially symmetrical and the dense fibres are, like other possums, rounded and in close apposition to the axoneme. Little comparative information is available on the

Fig. 14. Oblique section through the sperm midpiece fibre network in *Trichosurus vulpecula* showing the parallel array of submembrane fibres winding around the distal midpiece. x 59,000.

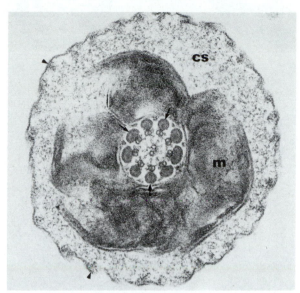

Fig. 15. Transverse section through the midpiece of a maturing Brush-tailed Possum sperm (*Trichosurus vulpecula*) taken from the proximal corpus epididymidis. Note the standard 9 + 2 axoneme, nine rounded dense fibres (arrows) closely apposed to each respective outer doublet pair of the axoneme, mitochondria (m), cytoplasmic sheath (cs) and plasma membrane, and the submembrane midpiece fibre network (arrow heads). Note also the rounded shape of the midpiece and the radial symmetry of its component structures. x 42,300.

structure of the connecting piece and implantation fossa of this species. However, *P. cinereus* lacks a well-defined midpiece fibre network and plasma membrane invaginations (Harding *et al.* 1979). The pseudocheirid sperm tail differs from the standard possum sperm tail structure in only two respects. It has a short, broad midpiece (about $7\mu \times 3\mu$ compared to approximately $12\mu \times 1.5\mu$ in other species), which reflects the extreme development of the mitochondria, a partially developed midpiece fibre network and lateral compression of the principal piece fibrous sheath (Harding *et al.* 1979). The Honey Possum, *Tarsipes rostratus*, differs most in sperm tail structure from the generalized possum sperm tail model. Honey Possum sperm tails appear to more closely resemble dasyurid and peramelid sperm tails in both size and structure than those of other possums (Cummins *et al.* 1980; Harding *et al.* 1981, 1982, 1984; Temple-Smith 1984c). They are longer than those of any other marsupial species (length $\simeq 350\mu$; midpiece length $\simeq 90\mu$). In both the midpiece and principal piece the dense fibres show an extreme and asymmetric

displacement from the axoneme (Fig. 16), except at their extreme ends (close to the neck and the end piece) where they fit more closely the traditional possum pattern. In *Tarsipes*, like the peramelids and dasyurids, the displaced dense fibres are connected to the other doublet microtubules of the axoneme by sinuous, diffuse, bands of fibrous material (Figs 16 and 25). With the exception of its proximal and distal ends, the flagellum is laterally compressed in the midpiece and principal piece such that in transverse section it appears ovoid in shape with the long axis always perpendicular to the plane which bisects the central pair of the axoneme (Harding *et al.* 1981). This feature is also shared by the dasyurids. The only obvious change in the structural organization of these dense fibres in *Tarsipes* sperm tails occurs in the principal piece where the two opposing dense fibres situated along the short axis of the tail section are gradually incorporated into the structure of the fibrous sheath. A distinct plasma membrane associated fibre network develops in the midpiece of *Tarsipes*, as in those of other possums, during passage through the epididymis. However, each helically arranged fibre in this network is uniquely associated with underlying pairs of microtubules (Figs 16 and 25), the functions of which are unknown. Clearly the structure of the sperm tail in this species is quite different from all of the other possum species examined. From the similarities in sperm structure to the peramelids

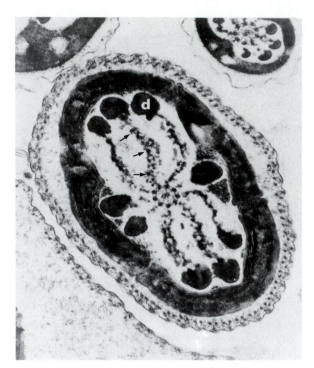

Fig. 16. Transverse section through the midpiece of a mature Honey Possum sperm (*Tarsipes rostratus*) showing extreme asymmetric displacement of the dense fibre (d) from the axoneme, and the diffuse, sinuous bands of fibrous material (arrows) which connect them to the outer doublet microtubules of the axoneme. Note the unusual arrangement of submembrane fibres and microtubules in this species. x 40,100.

and dasyurids and other structural and biochemical features, Harding *et al.* (1981, 1984) have suggested that sperm morphology in this species represents the ancestral sperm type from which the generalized possum sperm was derived. However, more recent observations on sperm tail structure in some of the American opossums suggest an alternative and, perhaps, more plausible evolutionary relationship.

Opossums

As with sperm head morphology, sperm tail structure has been examined in detail in only a few species of the American marsupial fauna. Fortunately, however, at least one species from each of the American opossum families have now been examined so that some meaningful intra- and inter-family comparisons can be presented.

Didelphidae

Although sperm tail morphology of only two didelphids *Didelphis virginiana* and *Caluromys philander*, has been described in detail (Phillips 1970; Olson and Hamilton 1976; Olson *et al.* 1977; Temple-Smith and Bedford 1980), these species represent the two different sperm head morphology subgroups of the Didelphidae. However, in contrast to the differences in sperm head morphology which have already been used to separate

Caluromys from the main stream didelphid species, the structure of caluromyid and didelphid sperm tails is almost identical. Sperm tails of both species have very short midpieces, about 5% of the length of the flagellum (compared with about 10% in most possums and about 25% in *Tarsipes rostratus*), which terminate in a distinct flange-shaped annulus with a posteriorly flared rim overlying the most anterior position of the principal piece (Fig. 3). The connecting piece in the Didelphidae is similar in structure to that seen in the possums except that its anterior end is curved away from the long axis of the flagellum at an angle of about 60°. This curved profile develops during spermatogenesis and can be clearly seen in immature caput sperm (Fig. 17) prior to the maturational reorientation of the nucleus (Phillips 1970; Temple-Smith and Bedford 1980). This structural specialization of the connecting piece provides an appropriate

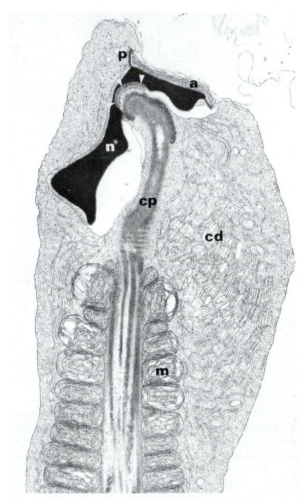

Fig. 17. Longitudinal section through an immature opossum sperm (*Didelphis virginiana*) showing the characteristically curved connecting piece (cp), and the basal plate (arrows) in the implantation fossa. The acrosomal surface of the nucleus in immature sperm from this species faces rostrally but during sperm maturation rotates to a lateral position. Abbreviations: a, acrosome; p, periacrosomal ring; m, mitochondria; cd, cytoplasmic droplet; n, nucleus. x 20,900.

orientation for the flagellum of each sperm after pairing has occurred. If the connecting piece was straight, as in the Australian species, the paired sperm tails would be set perpendicular, rather than almost parallel, to the long axis of the paired sperm heads (Figs 5 and 6). Presumably curved connecting pieces have evolved in these species to provide the most favourable position of the paired flagella for efficient progressive motility. Didelphid mitochondria are arranged in 16 to 18 gyrae (Fig. 1) around the axonemal complex and are dorsoventrally compressed giving a depth and width ratio of about 1:4, compared with approximately 1:1 in most possums except the Pseudocheiridae. In *Didelphis* and *Caluromys* the midpiece plasma membrane is separated from the mitochondrial sheath by a very narrow layer of granular cytoplasm (Figs 1 and 18) and, in the mature sperm, is associated with a very fine parallel network of discrete electron-dense ridges along its inner surface (Figs 18, 19 and 25) (Phillips 1970; Olson and Hamilton 1976; Olson *et al.* 1977, Temple-Smith and Bedford 1980). In *Didelphis*, these ridges of the midpiece fibre network have been shown to coincide with parallel, linear aggregations of intramembranous particles (Olson *et al.* 1977) which, if arranged in a similar manner to those in *Trichosurus vulpecula* (Temple-Smith 1981, 1984c), form a direct structural association. Although the midpiece is circular in transverse section in *Didelphis* (Temple-Smith and Bedford 1980), the organization of the

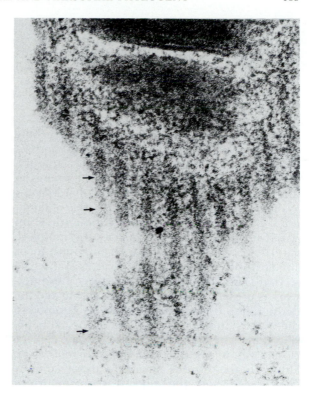

Fig. 19. Oblique, glancing section through the distal midpiece of a mature Virginian Opossum (*Didelphis virginiana*) sperm showing the fine, linear aggregations (arrows) of fibrous material which form the midpiece fibre network in this species. x 88,000.

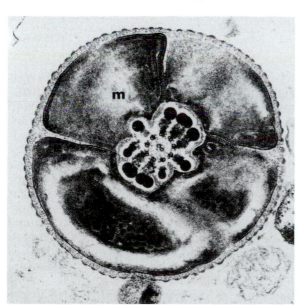

Fig. 18. Transverse section through the midpiece of a mature Virginian Opossum sperm showing the large mitochondria (m), the scalloped appearance of the plasma membrane over the midpiece fibre network (see Figure 19) and the asymmetrical displacement of dense fibres from their respective outer doublet pair in the axoneme. Dense fibres 1, 5 and 6 (see Fawcett 1975 for explanation of numbering system) are bilobed, the midpiece cross-section is circular and straight amorphous strands of fibrous material connect each dense fibre to its adjacent outer doublet. x 29,700.

dense fibres around the central axoneme is characteristically asymmetrical (Figs 18 and 25). An identical asymmetrical arrangement of dense fibres as seen in *Caluromys* (Phillips 1970; Olson and Hamilton 1976). The dense fibres, especially 1, 2, 5, 6 and 9, are displaced away from the outer doublet microtubules of the axoneme but connected to them by loosely arranged bands of fibrous material (Fig. 18). In addition there are distinct and characteristic differences in size and shape of individual dense fibres and in their displacement from the axoneme. For example dense fibres 1, 5 and 6 are large, bilobed and most distantly situated from the axoneme where as fibres 4 and 7 are small, rounded and more closely associated with their respective doublet microtubules (Figs 18 and 25). These relationships continue into the principal piece of the sperm tail in both species. Anteriorly, as in other marsupials (Harding 1977), the plasma membrane is separated from the fibrous sheath by cytoplasm containing diffusely distributed fibres however, this is absent further down the principal piece and the plasma membrane is in direct contact with the outer surface of the sheath. Other structural changes which occur in the principal piece include changes in the size, shape and position of the dense fibres (Phillips 1970; Harding 1977). The functional importance of these structural features of *dilelphid* sperm tails have not yet been resolved. There is some controversy as to whether sperm tail structure in these species has developed to provide specific tail movements which will ensure efficient and co-ordinated progressive motility for sperm pairs.

Caenolestidae

Sperm tail structure has been examined in only one caenolestid species, *Caenolestes obscurus* (Temple-Smith and Tyndale-Biscoe unpubl. obs.). Although, as with sperm head studies in this species, only inadequately fixed material was available for electron microscopy, the observations provide sufficient information for comparative purposes. In this species the midpiece contributes to about 10% of total tail length and is circular and radially symmetrical in transverse section (Fig. 20). A distinct midpiece fibre network is present in mature, paired sperm and the dense fibres are rounded and closely apposed to the outer doublet microtubules of the axoneme (Fig. 20). The structure of the principal piece closely resembles that described for *didelphids*, although the fibrous sheath is more laterally compressed and the fibrous segments from which it is composed are thinner. However, perhaps the most interesting feature of the sperm tail in this species is the structure of the neck, in particular the connecting piece and its insertion into the implantation fossa of the nucleus (Figs 5, 10 and 21). In *Caenolestes* the

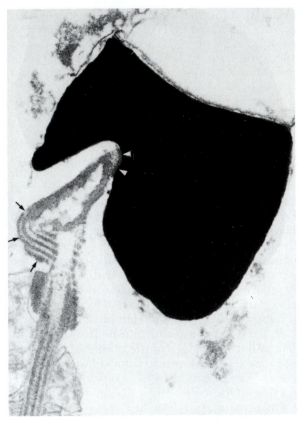

Fig. 21. Longitudinal section through a caenolestid (*Caenolestes obscurus*) sperm showing the unusual structure of the neck and its insertion into the nucleus. The connecting piece is straight and part of its structure appears to be derived from accessory fibres (arrowed) which create an unusual neck morphology. The basal plate (white arrows) is restricted to the medial wall of the implantation fossa. x 32,200.

connecting piece is straight and, unlike other marsupial species, additional accessory structures are present which create a unique and characteristic neck morphology in this species. During sperm maturation in this species there is no change in the sperm head-tail orientation; an unusual situation which may be caused by the deep insertion of the connecting piece into the apparently restricting confines of the nuclear implantation fossa in *Caenolestes*. This arrangement may also explain the rather unusual lateral placement of the sperm tails in paired *Caenolestes* sperm (Figs 5 and 8). Clearly many aspects of caenolestid sperm structure and function require further examination. In particular it will be important to compare sperm structure in the other caenolestid genera *Lestoros* and *Rhyncholestes* and to examine the swimming behavior of caenolestid sperm *in vitro* to gain an understanding of how sperm pairs in this family manage to achieve progressive motility.

Microbiotheriidae

Like *Caenolestes*, observations of sperm tail structure of *Dromiciops australis* have been

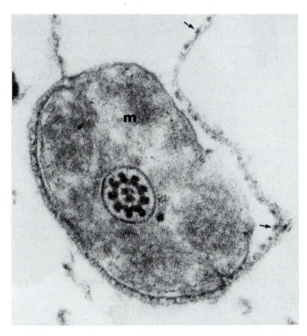

Fig. 20. Transverse section through the midpiece of a caenolestid (*Caenolestes obscurus*) sperm showing, despite poor fixation, the radially symmetrical arrangement of the small dense fibres around the axoneme. This section is through the distal midpiece and shows a distinct midpiece fibre network (arrows) associated with the plasma membrane. Abbreviations: m, mitochondria. x 50,400.

restricted to an examination of very poorly fixed material (Temple-Smith in prep.). However, sufficient details are available to make a general structural comparison with sperm tail structure of the other American opossum families. As with sperm head morphology, sperm tail structure resembles more closely that of the Australian possums, in particular the phalangerids, petaurids and burramyids, than the American opossums. The connecting piece is straight (Figs 5, 12 and 13) and in the mature sperm is aligned parallel to the long axis of the sperm head inside the ventral nuclear groove (Figs 5 and 13). The midpiece accounts for about 10% of the total length of the tail and the annulus, which forms a typical posteriorly directed flange with an outer rim overlapping the midpiece-principal piece junction (Fig. 22), provides a distinct posterior demarcation of the midpiece. A distinct midpiece fibre network is present in the posterior region of the midpiece (Fig. 23) which winds obliquely around the midpiece. In transverse section the midpiece is circular and, like the caenolestids and most Australian possums, radially symmetrical (Fig. 24). The dense fibres show no displacement from the axonemal complex and each is unilobed, rounded and in direct contact with its

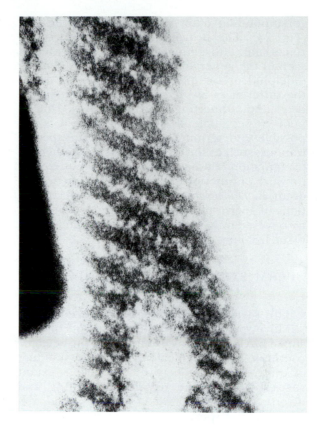

Fig. 23. High magnification, oblique section through the midpiece fibre network of a *Dromiciops* sperm tail. x 105,000.

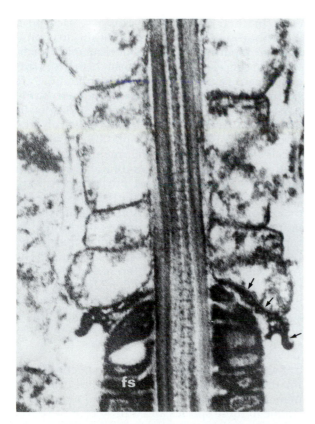

Fig. 22. Midpiece — principal piece junction of a *Dromiciops* spermatozoon showing the structure of the annulus (arrows) in this species. The irregular blocks and central cavities of the principal piece fibrous sheath (fs) can be seen and, for each mitochondrion, only the outer mitochondrial membrane has been retained by the fixation procedure. x 58,300.

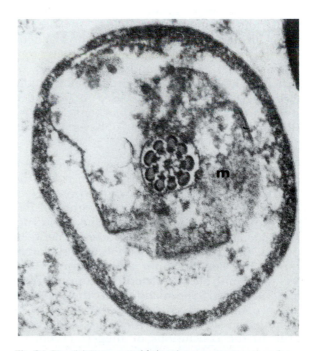

Fig. 24. *Dromiciops* sperm midpiece in transverse section showing nine, rounded dense fibres in direct contact with their respective axonemal doublet pairs. The midpiece is essentially circular in cross-section and its contents are distributed in a radially symmetrical fashion in relation to the long axis of the tail. Abbreviation: m, mitochondria. x 52,800.

corresponding outer doublet microtubule pair (Fig. 24). The principal piece structure is unremarkable and compares more closely with that seen in the generalized possum sperm than the didelphid or caenolestid sperm tails, although the differences are less marked than in sperm head and connecting piece structures. Due to poor preservation of the membranes it is not possible to determine if any membrane specializations, such as flask-shaped plasma membrane invagination between the fibres of the midpiece network or anterior midpiece membrane whorls occur in *Dromiciops*. These and other clarifications of sperm ultrastructure in this species will be resolved only when more appropriately fixed tissue becomes available for study.

SPERM STRUCTURE AND PHYLOGENY: CONCLUDING REMARKS

This review confirms the importance of comparative sperm structure in assessments of marsupial phylogeny. It also emphasizes the need for continuing studies to extend our understanding of the relationships between the various extant species which make up this interesting and diverse group of mammals. As suggested, the usefulness of sperm morphology in elucidating marsupial phylogeny has been enhanced by recent ultrastructural studies which have identified new suites of structural characters in marsupial spermatozoa. However it is important when drawing conclusions about the evolutionary affinities of groups of animals based on similarities in sperm morphology, to consider them in the wider context of phylogenetic relationships which have been suggested from other studies; for example, comparisons of various dental and skeletal characteristics and the more recent conclusions based on detailed serological (Kirsch 1977) and biochemical studies (Richardson and Sharman 1976; Baverstock *et al.* 1982; Sarich *et al.* 1982).

As indicated in this review, sperm pairing is still a phenomenon which occurs only in the American marsupial species and at present can still be regarded as a diagnostic feature of this marsupial subgroup. Except for the ultrastructural details of plasma interactions between paired sperm (Phillips 1970; Olson and Hamilton 1976; Temple-Smith and Bedford 1980; Temple-Smith and Tyndale-Biscoe unpubl. obs.) and informed speculation on the possible absence of sperm pairing in the extant microbiotheriid *Dromiciops australis*, this review adds little to the conclusions of Biggers (1966) on sperm pairing. The interesting American species which had been suggested for future studies on sperm pairing still remain to be examined, due mostly to the difficulties in obtaining specimens. However, with the exception of *Dromiciops australis*, the proposed phylogenetic affinities of

these species (*Caluromyciops, Glironia, Lestodelphys, Lestoros* and *Rhyncholestes*) suggest that each will have paired, mature spermatozoa which are structurally similar to other representatives of their major group affiliations.

Because of its proposed relationship with the Australian marsupial fauna (Szalay 1982; Sharman 1982; Temple-Smith 1984a and in prep. *Dromiciops* will be perhaps the most interesting species for future studies on sperm pairing. The striking similarities in sperm morphology between *Dromiciops* and some of the Australian phalangerids and macropodids suggest that it may also have unpaired mature spermatozoa. Confirmation of this prediction will be an exciting adjunct to information already gathered on sperm structure in this species (Temple-Smith in prep.) and will provide an important additional point of phylogenetic congruity between *Dromiciops* and the Australian marsupial radiation. Confirmation of sperm pairing in *Dromiciops*, however, is unlikely to significantly alter current thinking on its phylogenetic relationship with the Australian marsupials. It would imply either that the progenitors of the Australian marsupial radiation separated from the ancestral American stock before pairing had evolved or that sperm pairing disappeared very early in their evolutionary history. Despite a more thorough documentation of the structural aspects of sperm pairing in American marsupials, various aspects of the mechanism of sperm pairing remain unanswered. For example, how are sperm able to become so precisely aligned during passage through the epididymis? Do specific membrane recognition factors exist on the acrosomal surface of the plasma membrane which aid in sperm head alignment? What is the nature of the process of membrane attachment? How does the epididymis contribute to sperm pair formation (i.e., are specific proteins added to or removed from the acrosomal plasma membrane)? How do specific regions of the female reproductive tract separate sperm pairs and retain sperm viability? Does sperm pairing in the American species reflect differences in the structure of the sperm or the epididymal environment when compared with Australian species? Future studies may help to answer some of these questions and may also provide insights into why sperm pairing occurs in these species.

Two principal suggestions have been offered for the evolution of sperm pairing in American marsupials. One is that the acrosome is more fragile in these species and is protected by sperm pairing until required for fertilization in the oviduct (Biggers 1966), where sperm pair separation occurs in *Didelphis virginiana* (Rodger and Bedford 1982). However, there are no apparent structural reasons why American marsupial acrosomes should be more susceptible to degradation than those of their Australian counterparts. The other suggestion is that sperm pairing is required for normal, progressive sperm motility (Phillips 1970). However, *in vitro* observations show that mature, unpaired spermatozoa are able to swim in a directed and progressive fashion (Temple-Smith and Bedford 1980; Rodger and Bedford 1982) and a more extensive study of the motility characteristics of marsupial spermatozoa in the female reproductive tract is therefore needed to determine if paired spermatozoa have an advantage over unpaired, motile spermatozoa in reaching the site of fertilization. Future studies of the structure and function of paired caenolestid spermatozoa may also provide insights into the functional importance of sperm pairing. As indicated previously, paired caenolestid spermatozoa differ from those of *Didelphis* and *Caluromys* in the disposition of sperm tails. In the latter, each pair of sperm tails is directed outwards and posteriorly at an angle of 30°-45° to the longitudinal axis of pairing. In *Caenolestes obscurus*, however, the tails of each sperm pair appear to be set at 90° to this axis and 180° away from each other. This arrangement would seem to prevent rather than promote progressive sperm motility in *Caenolestes* and *in vitro* studies are needed to determine what functional and structural adaptations occur in caenolestid spermatozoa to enable effective, progressive motility.

Sperm structural comparisons appear to provide a sensitive indication of inter- and intra-family associations in marsupials. For example, sperm structure readily confirms the previous isolation of *Tarsipes* from other phalangerids (Kirsch 1977) and suggests phylogenetic links with the dasyurid and, perhaps, peramelid marsupials (Harding *et al.* 1981, 1982). Similarly the isolation of the Koala, *Phascolarctos cinereus*, from the phalangerids and its close phylogenetic relationship with the wombats, based on previous morphological (Archer 1976b, 1984b, c) and serological studies (Kirsch 1977), is supported by sperm morphology (Hughes 1965, 1977; Harding 1979; Harding *et al.* 1979; 1987).

The current taxonomic divisions of the American marsupials are also supported by sperm structural studies. Caenolestid spermatozoa clearly differ from those of the extant didelphids (Biggers and DeLamater 1965; Rodger 1982; Temple-Smith and Tyndale-Biscoe unpubl. obs.) and the structural studies of *Dromiciops australis* spermatozoa, although limited by available tissue and poor fixation, also demonstrate the clear phylogenetic divergence of this species both from the Didelphidae and Caenolestidae. Sperm morphology also appears to confirm the close relationship of *Caluromys* to the didelphid evolutionary lineage, by the close similarities in sperm tail morphology, but at the

same time substantial differences in sperm head morphology support the previous separation of this species into a separate subgroup within the Didelphidae (Kirsch 1977). Again, these findings support phylogenetic interrelationships which have already been suggested by previous paleontological, skeletal, dental and biochemical studies. It will be interesting to see if the structural features of spermatozoa from *Caluromyciops* and *Glironia* resemble those of *Caluromys*, thereby supporting the grouping by Reig (1955) of these two species with *Caluromys* on the basis of similarities in cranial and dental morphology.

The apparent fidelity of sperm structure as a predictor of evolutionary affinities in the marsupials, and indeed other vertebrates (Friend 1936; Furieri 1970; Bedford 1974; Picheral 1979; Carrick and Hughes 1982), encourages further discussion of recent attempts to define plesiomorphic characteristics in the sperm tail structure of marsupials (Harding *et al.* 1981). From their observations on sperm morphology in the enigmatic Honey Possum, *Tarsipes rostratus*, Harding *et al.* (1981) concluded that the structural organisation of its sperm tail represented the "primitive" evolutionary condition for Australian marsupials and that an early divergence from an ancient tarsipedoid line eventually gave rise to the radiation of phalangeroid marsupials in Australia. However, more recent observations on sperm morphology, especially in the American marsupials, appear to support an alternative suggestion, especially if the concept of a monophyletic derivation of the Australian marsupial fauna is correct. The information presented here supports the conclusion that simple radial symmetry, circular transverse section and rounded dense fibres immediately adjacent to the outer doublet microtubules together represent the plesiomorphic state of the marsupial sperm tail and that the more elaborate shape and asymmetrical arrangement of tail structures in *Tarsipes*, dasyurids and, to a lesser extent, peramelids is probably derived rather than ancestral for Australian marsupials. This would therefore argue against a tarsipedoid ancestry and suggest instead that sperm tail structure in *Tarsipes*, dasyurids and peramelids is a morphological specialization associated with specific functional requirements of the unusually large sperm which have evolved in these marsupial groups (Harding *et al.* 1981). This conclusion does not alter the suggestion by Harding *et al.* (1981) that the shared similarities in sperm structure may reflect phylogenetic affinities between these groups rather than convergent evolution, although serological studies (Kirsch 1977) suggest that, in an evolutionary sense, *Tarsipes* is as distant from the dasyurids and peramelids as it is from the phalangeroid marsupials.

Simple, circular symmetry is a feature of sperm tail structure in many species of placental mammals, including man, and all species of prototherian mammals which have been examined (Carrick and Hughes 1982). It is also a characteristic of some reptilian spermatozoa (Furieri 1970). This provides circumstantial evidence to suggest that simple radial symmetry was a plesiomorphic feature of the sperm tail in evolving mammalian groups and makes it more difficult to regard the sperm tail of *Tarsipes* as a "primitive" evolutionary form for Australian marsupials. Further evidence to support this conclusion can be found in the tail structure of the various American marsupial species which have now been described. In general the sperm tail of American marsupials has a relatively simple organization (Fig. 25). This is particularly so for *Caenolestes* and *Dromiciops* which represent two of the major evolutionary groups of the American marsupials. Sperm tails of the two didelphid species, *Didelphis virginiana* and *Caluromys derbianus*, which have been described in detail (Holstein 1965; Phillips 1970; Olson and Hamilton 1976; Olson *et al.* 1977; Temple-Smith and Bedford 1980), are circular in transverse midpiece sections but show minor radial displacement and structural asymmetry in the relationship of dense fibres to the axoneme (Fig. 25).

This, however, is not as exaggerated as that seen in *Tarsipes*, the dasyurids or the peramelids (Fig. 25). Asymmetry of intra-flagellar structures in didelphids may represent a morphological adaptation which was required to provide co-ordinated and controlled sperm tail movements during motility in paired spermatozoa from this group. Interestingly in the macropodoid *Potorous tridactylus* minor displacement of the dense fibres from the axoneme has also been reported (Harding *et al.* 1979, 1981). This is a more symmetrical radial displacement than in didelphids but, like them, the dense fibres appear heterogeneous in shape.

The paucity of information on sperm motility in marsupials at present allows only for speculation on the functional importance of the various sperm tail structures and organizations which have been described. Clearly there is now a need for a detailed comparative examination of marsupial sperm tail structure-function relationships. Functional aspects aside, however, recent structural studies suggest that a simple, rounded, radially symmetrical sperm tail with round dense fibres in close approximation to the outer microtubule doublets of the axoneme was most probably the archetypal form in marsupials. Since this format is found in some extant American

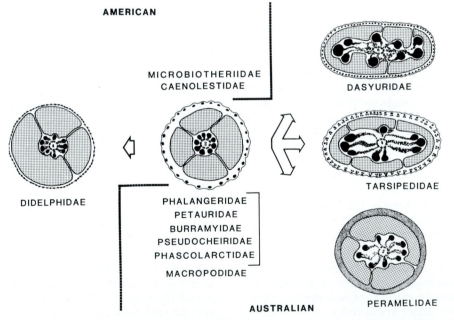

Fig. 25. Comparison of the morphology and distribution of midpiece organelles shown in transverse sections through the distal midpiece of spermatozoa from representatives of various families of possums and opossums and three other Australian marsupial families (Macropodidae, Dasyuridae and Peramelidae). This evolutionary scheme suggests that simple radial symmetry, circular midpiece section and rounded dense fibres immediately adjacent to the outer doublet microtubules of the axoneme, as seen in the Microbiotheriidae, Caenolestidae, the Australian possums (except the *Tarsipedidae*) and the Macropodidae, represent the primitive ancestral condition in marsupials. The minor radial displacement and structural asymmetry of dense fibres in the Didelphidae, and the more extreme displacement of dense fibres and asymmetry of tail structures in the Peramelidae, Dasyuridae and Tarsipedidae probably represent derived structural specializations superimposed on the symmetrical ancestral form.

marsupials, and in many Australian groups, and the most accepted theory at present is that Australian marsupials were derived from ancestral American stock, it seems unlikely that a tarsipedoid-like sperm was the ancestral type in Australian marsupials, or even the possum radiation. In fact the observations in this review support previous suggestions that *Tarsipes* is a highly specialized end product of a sterile branch of the Australian marsupial radiation and has only distant phylogenetic affinity with the possum radiation. The conclusion that a simple, radially symmetrical sperm tail morphology was probably the ancestral type for Australian marsupials further strengthens the association between *Dromiciops australis* and the Australian radiation. This association has already been alluded to by the remarkable similarities in sperm structure between *Dromiciops* and Australian marsupials, in particular the Phalangeridae and Macropodoidea. This strongly supports the conclusions from other studies (Szalay 1982; Sharman 1982), that *Dromiciops* is the closest living relative of the Australian marsupial radiation and that ancestral microbiotheriid stock, which may have evolved in either the American or Antarctic regions of Gondwanaland, was the progenitor stock from which both *Dromiciops* and the Australian marsupial fauna evolved.

Sperm structure is usually regarded as a conservative evolutionary characteristic which, protected by the more constant internal environment of the body, is less susceptible to the vagaries of changing conditions in the external environment. This is an important consideration, since changes in external environment provide strong selective forces for evolutionary change. This and other studies have demonstrated considerable stability of various sperm structural features both within and often between groups of species at various evolutionary levels. Yet, even within apparently closely related groups of species, large differences in sperm structure have sometimes been described, the adaptive significance of which are often as obscure as our understanding of the selective forces that created them. The experimental studies of Beatty (1971), Pant (1970) and others show clearly that sperm structure is under strong genetic control and those of Phillips (1970), Rattner (1972), and Fawcett *et al.* (1971) indicate the importance of a highly organised series of events in the testis during spermatogenesis which control the shape and structural organization of spermatozoa. Comparative studies of spermiogenesis may therefore provide important additional information not only about the genesis of these differences in sperm morphology but also perhaps about factors which influence their control.

ACKNOWLEDGEMENTS

We thank Drs Leon Hughes and Ronnie Harding for constructively reviewing the manuscript, Dr Bill Krause (University of Missouri), for supplying the SEM micrograph of paired *Didelphis* sperm, Shirley Goldsworthy, Shirley Wainer, Terry Martin, Lorraine Alter and Sue Simpson for their assistance in the preparation of this manuscript and, especially, Meredith Taylor for her helpful comments and inspiration and for carefully proof-reading the text. Dr Tom and Gina Grant assisted with proof-reading the page proofs.

REFERENCES

ARCHER, M., 1976a. The basicranial region of marsupicarnivores (Marsupialia), interrelationships of carnivorous marsupials and affinities of the insectivorous marsupial peramelids. *Zool. J. Linn. Soc. Lond.* **59:** 217-322.

ARCHER, M., 1976b. Phascolarctid origins and the potential of the selendont molar in the evolution of diprotodont marsupials. *Mem. Qd Mus.* **17:** 367-71.

ARCHER, M., 1984a. Origins and early radiations of marsupials. Pp. 585-626 *in* "Vertebrate zoogeography and evolution in Australasia" ed by M. Archer and G. Clayton. Hesperian Press: Perth.

ARCHER, M., 1984b. The Australian marsupial radiation. Pp. 633-808 *in* "Vertebrate zoogeography and evolution in Australasia" ed by M. Archer and G. Clayton. Hesperian Press: Perth.

ARCHER, M., 1984c. On the importance of being a koala. Pp. 809-15 *in* "Vertebrate zoogeography and evolution in Australasia" ed by M. Archer and G. Clayton. Hesperian Press: Perth.

BAVERSTOCK, P. R., ARCHER, M., ADAMS, M. AND RICHARDSON, B. J., 1982. Genetic relationships among 32 species of Australian dasyurid marsupials. Pp. 461-650 *in* "Carnivorous marsupials" ed by M. Archer. Royal Zoological Society of New South Wales: Sydney.

BEATTY, R. A., 1971. The genetics of size and shape of spermatozoan organelles. Pp. 97-115 *in* "Edinburgh symposium on the genetics of the spermatozoon" ed by R. A. Beatty and S. Gluecksohn-Waelsch. Bogtrykkeriet Forum: Copenhagen.

BEDFORD, J. M., 1965. Changes in fine structure of the rabbit sperm during passage through the epididymis. *J. Anat.* **99:** 891-906.

BEDFORD, J. M., 1974. Biology of primate spermatozoa. *Contrib. Primat.* **3:** 97-139.

BEDFORD, J. M., 1975. Maturation, transport and fate of spermatozoa in the epididymis. Pp. 303-17 *in* "Handbook of physiology", Section III Endocrinology, Vol. V., Male Reproductive System ed by D. W. Hamilton and R. O. Greep. American Physiological Society: Bethesda.

BEDFORD, J. M. AND CALVIN, H. I., 1974. Changes in the -S-S-linked structures of the sperm tail during epididymal maturation with comparative observations in submammalian species. *J. Exp. Zool.* **187:** 181-204.

BEDFORD, J. M. AND COOPER, G. W., 1978. Membrane fusion events in the fertilization of vertebrate eggs. Pp. 65-125 *in* "Cell surface reviews: membrane fusion, Vol. V" ed by G. Poste and G. L. Nicholson. Elsevier/North Holland: Amsterdam.

BIGGERS, J. D., 1966. Reproduction in male marsupials. *In* "Comparative biology of reproduction in mammals" ed by I. W. Rowlands. *Symp. Zool. Soc. Lond.* **15:** 251-80.

BIGGERS, J. D. AND CREED, R. F. S., 1962. Conjugate spermatozoa of the North American Opossum. *Nature Lond.* **196:** 1112-13.

BIGGERS, J. D. AND DeLAMATER, E. D., 1965. Marsupial spermatozoa pairing in the epididymis of American forms. *Nature Lond.* **208:** 402-4.

BLOOM, G. AND NICANDER, L., 1961. On the ultrastructure and development of the protoplasmic droplet of spermatozoa. *Z. Zellforsch. Mikrosk. Anat.* **55:** 833-44.

BREED, W. G., 1983. Variation in sperm morphology in the Australian rodent genus, *Pseudomys* (Muridae). *Cell Tissue Res.* **229:** 611-25.

CALVIN, H. I. AND BEDFORD, J. M., 1971. Formation of disulphide bonds in the nucleus and accessory structures of mammalian spermatozoa during maturation in the epididymis. *J. Reprod. Fert. Suppl.* **13:** 65-75.

CARRICK, F. N. AND HUGHES, R. L., 1982. Aspects of the structure and development of monotreme spermatozoa and their relevance to the evolution of mammalian sperm morphology. *Cell Tissue Res.* **222:** 127-47.

CLEMENS, W. A., 1968. Origin and early evolution of marsupials. *Evolution* **22:** 1-18.

CLEMENS, W. A., 1977. Phylogeny of marsupials. Pp. 51-68 *in* "The biology of marsupials" ed by B. Stonehouse and D. Gilmore. MacMillan Press: London.

CUMMINS, J. M., 1976. Epididymal maturation of spermatozoa in the marsupial *Trichosurus vulpecula*: changes in motility and gross morphology. *Aust. J. Zool.* **24:** 499-511.

CUMMINS, J. M., RENFREE, M. B. AND TEMPLE-SMITH, P. D., 1980. Sperm maturation and ultrastructure in the Honey Possum *(Tarsipes spencerae). Proc. Aust. Mammal Soc. Univ. of NSW:* 30-31.

CUMMINS, J. M., TEMPLE-SMITH, P. D. AND RENFREE, M. B., 1985. Reproduction in the male Honey Possum (*Tarsipes rostratus*: Marsupialia): the epididymis. *Am. J. Anat.* (submitted for publication).

DUESBERG, J., 1920. Cytoplasmic structures in the seminal epithelium of the Opossum. *Contr. Embryol.* **9:** 47-84.

FAWCETT, D. W., 1975. The mammalian spermatozoon. *Dev. Biol.* **44:** 394-436.

FAWCETT, D. W., ANDERSON, W. A. AND PHILLIPS, D. M., 1971. Morphogenetic factors influencing the shape of the sperm head. *Dev. Biol.* **26:** 220-51.

FAWCETT, D. W. AND HOLLENBERG, R. C., 1963. Changes in the acrosome of the guinea pig spermatozoa during passage through the epididymis. *Z. Zellforsch* **60:** 276-92.

FRIEND, G. F., 1936. Sperms of British Muridae. *Quart. J. Micr. Sci.* **78:** 419-43.

FURIERI, P., 1970. Sperm morphology in some reptiles: Squamata and Chelonia. Pp. 115-31 *in* "Comparative spermatology" ed by B. Baccetti. Academic Press: New York.

GADDUM, P., 1968. Sperm maturation in the male reproductive tract: development of motility. *Anat. Rec.* **161:** 471-82.

GLOVER, T. D. AND NICANDER, L., 1971. Some aspects of structure and function in the mammalian epididymis. *J. Reprod. Fertil.* **13:** 39-70.

HARDING, H. R., 1977. "Reproduction in male marsupials: a critique, with additional observations on sperm development and structure".Ph. D. Thesis: University of New South Wales, Sydney.

HARDING, H. R., 1979. Affinities of the Koala on the basis of sperm ultrastructure. *Bull. Aust. Mammal Soc.* **6:** 22-23.

HARDING, H. R., CARRICK, F. N. AND SHOREY, C. D., 1975. Ultrastructural changes in spermatozoa of the Brush-tailed Possum, *Trichosurus vulpecula* (Marsupialia) during epididymal transit Part I: The flagellum. *Cell Tissue Res.* **164:** 121-32.

HARDING, H. R., CARRICK, F. N. AND SHOREY, C. D., 1976. Spermiogenesis in the Brush-tailed Possum, *Trichosurus vulpecula* (Marsupialia). The development of the acrosome. *Cell Tissue Res.* **171:** 75-90.

HARDING, H. R., CARRICK, F. N. AND SHOREY, C. D., 1979. Special features of sperm structure and function in marsupials. Pp. 289-303 *in* "The spermatozoon" ed by D. W. Fawcett and J. M. Bedford. Urban and Schwarzenberg Inc.: Baltimore-Munich.

HARDING, H. R., CARRICK, F. N. AND SHOREY, C. D., 1981. Marsupial phylogeny: new indications from sperm ultrastructure and development in *Tarsipes spenserae? Search* **12:** 45-47.

HARDING, H. R., WOOLLEY, P. A., SHOREY, C. D. AND CARRICK, F. N., 1982. Sperm ultrastructure, spermiogenesis and epididymal sperm maturation in dasyurid marsupials: phylogenetic implications. Pp. 659-73 *in* "Carnivorous marsupials" ed by M. Archer. Royal Zoological Society of New South Wales: Sydney.

HARDING, H. R., CARRICK, F. N. AND SHOREY, C. D., 1984. Sperm ultrastructure and development in the Honey Possum, *Tarsipes rostratus*. Pp. 451-61 *in* "Possums and gliders" ed by A. P. Smith and I. D. Hume. Aust. Mammal Soc.: Sydney.

HARDING, H. R., CARRICK, F. N. AND SHOREY, C. D., 1987. The affinities of the Koala *Phascolarctos cinereus* (Marsupialia: Phascolarctidae) on the basis of sperm ultrastructure and development. Pp. 353-64 *in* "Possums and opossums: studies in evolution" ed by M. Archer. Surrey Beatty & Sons and the Royal Zoological Society of New South Wales: Sydney.

HAYMAN, D. L., KIRSCH, J. A. W., MARTIN, P. G. AND WALLER, P. F., 1971. Chromosomal and serological studies of the Caenolestidae and their implications for marsupial evolution. *Nature Lond.* **231:** 194-95.

HOLSTEIN, A. F., 1965. Electronen mikroskopische Untersuchungen am Spermatozoon des Opossum (*Didelphys virginiana* Kerr). *Z. Zellforsch.* **65:** 904-14.

HUGHES, R. L., 1965. Comparative morphology of spermatozoa from five marsupial families. *Aust. J. Zool.* **13:** 533-43.

HUGHES, R. L., 1977. Light and electron microscope studies on the spermatozoa of the Koala, *Phascolarctos cinereus* (Marsupialia). *J. Anat.* **124:** 513.

JONES, R. C., HINDS, L. A. AND TYNDALE-BISCOE, C. H., 1984. Ultrastructure of the epididymis of the Tammar, *Macropus eugenii*, and its relationship to sperm maturation. *Cell Tissue Res.* **237:** 525-35.

JORDAN, H. E., 1911. The spermatogenesis of the Opossum (*Didelphis virginiana*) with special reference to the accessory chromosome and the chondrisomes. *Arch. Zellforsch.* **7:** 41-86.

KEAST, A., 1977. Historical biogeography of the marsupials. Pp. 69-96 *in* "The biology of marsupials" ed by B. Stonehouse and D. Gilmore. MacMillan Press: London.

KIRSCH, J. A. W., 1970. Serological affinities of six genera of Didelphidae: significance for the phylogeny of American and Australian marsupials. *Acta IV Congreso Latino-Americano de Zoologia, Caracas (1968):* 117-28.

KIRSCH, J. A. W., 1977. The comparative serology of Marsupialia and a classification of marsupials. *Aust. J. Zool. Suppl. Ser.* **52:** 1-152.

KIRSCH, J. A. W., 1984. Marsupial origins: taxonomic and biological considerations. Pp. 627-32 in "Vertebrate zoogeography and evolution in Australasia" ed by M. Archer and G. Clayton. Hesperian Press: Perth.

KRAUSE, W. J. AND CUTTS, J. H., 1979. Pairing of spermatozoa in the epididymis of the Opossum (*Didelphis virginiana*): A scanning electron microscopic study. *Arch. histol. Jap.* **42**: 181-90.

MARSHALL, L. G., 1980. Systematics of the South American marsupial family Caenolestidae. *Fieldiana: Geology* n.s. **5**: 1-145.

MARTIN, P. G., 1977. Marsupial biogeography and plate tectonics. Pp. 97-116 in "The biology of marsupials" ed by B. Stonehouse and D. Gilmore. MacMillan Press: London.

McCRADY, E., 1938. The embryology of the Opossum. *Am. Anat. Mem.* **16**: 1-233.

MOORE, H. D. M., 1983. Physiological and *in vitro* models of sperm maturation. Pp. 9-38 in "In vitro fertilization and embryo transfer" ed by P. G. Crosignani and B. L. Rubin. Academic Press: London.

NICHOLSON, G. L. AND YANAGIMACHI, R., 1979. Cell surface changes associated with the epididymal maturation of mammalian spermatozoa. Pp. 187-94 in "The spermatozoon" ed by D. W. Fawcett and J. M. Bedford. Urban and Schwarzenberg: Baltimore.

OLSON, G. E., 1975. Observations on the ultrastructure of a fiber network in the flagellum of sperm of the Bush Tailed Phalanger, *Trichosurus vulpecula. J. Ultrastruct. Res.* **50**: 193-98.

OLSON, G. E. AND HAMILTON, D. W., 1976. Morphological changes in the midpiece of Wooly Opossum spermatozoa during epididymal transit. *Anat. Rec.* **186**: 387-404.

OLSON, G. E., LIFSICS, M., FAWCETT, D. W. AND HAMILTON, D. W., 1977. Structural specializations in the flagellar plasma membrane of Opossum spermatozoa. *J. Ultrastruct. Res.* **59**: 207-21.

ORGEBIN-CRIST, M.-C., 1969. Studies on the function of the epididymis. *Biol. Reprod.* **1**: 155-75.

PANT, K. P., 1970. Patterns of inheritance in the midpiece length of mouse spermatozoa. Pp. 116-20 in "Edinburgh symposium on the genetics of the spermatozoon" ed by R. A. Beatty and S. Glueckson-Waelsh.

PHILLIPS, D. M., 1970. Ultrastructure of spermatozoa of the Wooly Opossum *Caluromys philander. J. Ultrastruct. Res.* **33**: 381-97.

PICHERAL, B., 1979. Structural, comparative and functional aspects of spermatozoa in urodeles. Pp. 267-87 in "The spermatozoon" ed by D. W. Fawcett and J. M. Bedford. Urban and Schwarzenberg Inc.: Baltimore-Munich.

RATTNER, J. B., 1972. Nuclear shaping in marsupial spermatids. *J. Ultrastruct. Res.* **40**: 498-512.

REIG, O., 1955. Noticia preliminar sobre la presencia de microbiotherinos vivientes en la fauna sudamericana. *Invest. Zool. Chilenas* **2**: 121-30.

RENFREE, M. B., RUSSELL, E. M. AND WOOLLER, R. D., 1984. Reproduction and life history of the Honey Possum, *Tarsipes rostratus*. Pp. 427-37 in "Possums and gliders" ed by A. P. Smith and I. D. Hume. Australian Mammal Society: Sydney.

RETZIUS, G., 1909. Die spermien von *Didelphys. Biol. Untersuch.* **14**: 123-26.

RICH, T. H., 1982. Monotremes, placentals and marsupials: their record in Australia and its biases. Pp. 386-477 in "The fossil vertebrate record of Australasia" ed by P. V. Rich and E. M. Thompson. Monash University Printing Unit: Melbourne.

RICHARDSON, B. J. AND SHARMAN, G. B., 1976. Biochemical and morphological observations on wallaroos (Macropodidae: Marsupialia) with a suggested new taxonomy. *J. Zool. Lond.* **179**: 499-513.

RIDE, W. D. L., 1962. On the evolution of Australian marsupials. Pp. 281-306 in "The evolution of living organisms" ed by G. E. Leeper. Melbourne University Press: Melbourne.

RODGER, J. C., 1982. The testis and its excurrent ducts in American caenolestid and didelphid marsupials. *Amer. J. Anat.* **163**: 269-82.

RODGER, J. C. AND BEDFORD, J. M., 1982. Separation of sperm pairs and sperm-egg interaction in the Opossum, *Didelphis virginiana. J. Reprod. Fert.* **64**: 171-79.

SARICH, V., LOWENSTEIN, J. M. AND RICHARDSON, B. J., 1982. Phylogenetic relationships of the Thylacine (*Thylacinus cynocephalus*, Marsupialia) as reflected in comparative serology. Pp. 707-09 in "Carnivorous marsupials" ed by M. Archer. Royal Zoological Society of New South Wales: Sydney.

SELENKA, E., 1887. Studien uber entwickelungsgeschichte der there. Vietes heft das Opossum (*Didelphys virginiana*). C. W. Kriedel's Verlag: Wiesbaden.

SHARMAN, G. B., 1974. Marsupial taxonomy and phylogeny. *Aust. Mammal* **1**: 137-54.

SHARMAN, G. B., 1982. Karyotypic similarities between *Dromiciops australis* (Microbiotheriidae: Marsupialia) and some Australian marsupials. Pp. 711-14 in "Carnivorous marsupials" ed by M. Archer. Royal Zoological Society of New South Wales: Sydney.

SIMPSON, G. G., 1980. "Splendid isolation. The Curious History of South American Mammals". Yale University Press: New Haven.

SZALAY, F., 1982. A new appraisal of marsupial phylogeny and classification. Pp. 621-40 in "Carnivorous marsupials" ed by M. Archer. Royal Zoological Society of New South Wales: Sydney.

TEMPLE-SMITH, P. D., 1981. Redistribution of intramembranous particles in the midpiece plasma membrane of the Brush-tailed Possum spermatozoon during epididymal transit. *J. Anat.* **132**: 301.

TEMPLE-SMITH, P. D., 1984a. *Dromiciops australis* — a South American connection. *Bull. Aust. Mammal Soc.* **8**(2): 77.

TEMPLE-SMITH, P. D., 1984b. Phagocytosis of sperm cytoplasmic droplets by a specialized region in the epididymis of the Brushtailed Possum, *Trichosurus vulpecula. Biol. Reprod.* **30**: 707-20.

TEMPLE-SMITH, P. D., 1984c. Reproductive structures and strategies in male possums and gliders. Pp. 89-106 in "Possums and gliders" ed by A. P. Smith and I. D. Hume. Australian Mammal Society: Sydney.

TEMPLE-SMITH, P. D. AND BEDFORD, J. M., 1976. The features of sperm maturation in the epididymis of a marsupial, the Brush-tailed Possum, *Trichosurus vulpecula. Am. J. Anat.* **147**: 471-500.

TEMPLE-SMITH, P. D. AND BEDFORD, J. M., 1980. Sperm maturation and the formation of sperm pairs in the epididymis of the Opossum, *Didelphis virginiana. J. Exp. Zool.* **214**: 161-71.

TYNDALE-BISCOE, C. H., 1973. "Life of marsupials". Edward Arnold: London.

WOOLLER, R. D., RENFREE, M. B., RUSSELL, E. M., DUNNING, A., GREEN, S. W. AND DUNCAN, P., 1981. Seasonal changes in populations of the nectar-feeding marsupial *Tarsipes spencerae* (Marsupialia: Tarsipedidae). *J. Zool. Lond.* **195**: 267-79.

INTERRELATIONSHIPS OF THE FAMILIES OF THE DIPROTODONTA — A VIEW BASED ON SPERMATOZOAN ULTRASTRUCTURE

H. R. HARDING[1]

Sperm ultrastructure is compared for 21 species representing all currently recognised diprotodontan families. Systematic phylogenetic analysis has not been carried out on this data, but preliminary comments are given.

Spermatozoa from *Tarsipes rostratus* and the vombatiforms each have features which set these groups apart from other diprotodontans. Whether these features represent plesiomorphies or apomorphies has not been determined, but it is concluded that *present* available evidence suggests that convergence (with dasyurids and peramelids) provides the most economic explanation for the unusual flagellar features of the tarsipedid spermatozoon. This contrasts with our previous conclusion, based on evidence which pre-dated information on microbiotheriid sperm structure and indications of a microbiotheriid role in Australian marsupial evolution.

The remaining diprotodonts share sufficient similar sperm features for their suprafamilial grouping to be plausible. However, marked differences in sperm form and dimensions are evident even at the intrafamily and lower taxonomic levels, and within most genera the species can be separated by sperm form. In the case of *Pseudocheirus,* the differences are very marked at the intrageneric level. In this respect the diprotodontan families differ from the Dasyuridae and Peramelidae which are characterised by intrafamilial consistency in sperm form.

Current phylogenies of diprotodontans are discussed in the light of the comparative sperm data. Perhaps most interesting of the *very tentative* conclusions reached, is that sperm studies do not necessarily disagree with Baverstock's (see Fig. 32) monophyletic grouping of pseudocheirids, *Tarsipes, Distoechurus/Acrobates,* and petaurids and more particularly his conclusion that *Distoechurus* is not a burramyid and *may* be related to *Tarsipes.*

Key Words: Diprotodonta; Marsupialia; Possums; Spermatozoa; Phylogeny; Ultrastructure.

Pages 195-216 *in* POSSUMS AND OPOSSUMS: STUDIES IN EVOLUTION ed by M. Archer. Surrey Beatty & Sons and the Royal Zoological Society of New South Wales: Sydney, 1987.

INTRODUCTION

OUR KNOWLEDGE of spermatozoan structure in diprotodont marsupials is extremely uneven. Epididymal sperm maturation and mature sperm ultrastructure in the Common Brushtail Possum, *Trichosurus vulpecula,* have been described in detail by a number of authors (Harding *et al.* 1975, 1976; Olson 1975; Cummins 1976; Temple-Smith and Bedford 1976) and recently the unique spermatozoan form in *Tarsipes rostratus* has attracted much attention (Cummins *et al.* 1980, 1986; Harding *et al.* 1980, 1981, 1984). For the remainder of the group, a reasonably detailed description of spermatozoan structure has been published only for the Koala, *Phascolarctos cinereus* (Harding *et al.* 1987), whilst sperm ultrastructure in *Potorous tridactylus, Petaurus norfolcensis, Petaurus breviceps* and *Pseudocheirus peregrinus* is covered in some detail in an unpublished thesis (Harding 1977), and briefly in comparative papers (Harding *et al.* 1977, 1979, 1983). The remaining literature on diprotodont spermatozoa comprises light microscope reports on a number of macropodoids (Retzius 1906; Hughes 1964, 1965; Rodger 1978) and species of *Petaurus, Pseudocheirus* and *Vombatus* (Hughes 1965) and brief abstracts describing the ultrastructure of *Phascolarctos* spermatozoa (Hughes 1977; Harding 1979).

Even on the basis of this limited information it is clear that sperm form varies considerably within the Diprotodonta. The vombatiforms[2] and *Tarsipes* show most departure from the well-described *Trichosurus* pattern, whilst the remaining species display variations of the *Trichosurus* theme.

The purpose of this paper is to review the currently available information on diprotodontan spermatozoan structure and provide preliminary observations on additional species, in order to assess whether comparative sperm structure can contribute at the inter- and intrafamily levels to our interpretations of diprotodontan phylogeny. Each of the currently recognised (Strahan 1983) diprotodontan families is represented in this survey.

MATERIALS AND METHODS

Epididymal samples have been examined from the species shown in Table 1. The information for *Phalanger gymnotis, P. vestitus, Distoechurus pennatus, Cercartetus caudatus, Dactylopsila trivirgata, Pseudocheirus corinnae* and *P. cupreus* is based on preliminary observations. The material from these species has been examined only briefly and some is of poor quality as a result of sampling in far from ideal conditions. A detailed description

[1]Department of General Studies, University of New South Wales, P.O. Box 1, Kensington, New South Wales, Australia 2033.
[2] Following Woodburne (1984) who has placed the Koala in the superfamily Phascolarctoidea, the wombats in the superfamily Vombatoidea, and grouped these superfamilies in the suborder Vombatiformes.

Table 1. Species investigated for this study and occurrence of certain sperm characteristics.

Family *Species* (source)[1]	Midpiece fibre network	Midpiece plasma membrane invaginations	Approximate extent of acrosomal cover of nucleus(%)[2]
PHASCOLARCTIDAE			
Phascolarctos cinereus (A)	−[3]	−	? 90
POTOROIDAE			
Potorous tridactylus (A)	x	−	25
MACROPODIDAE			
Macropus parma (A)	x	−	45
Macropus robustus (A)	x	−	40-45
Macropus rufus (A)	x	−	45-50
Setonix brachyurus (A)	x	−	30
PHALANGERIDAE			
Phalanger gymnotis (N.G.)	?	?	40
Phalanger vestitus (N.G.)	?	?	40
Trichosurus vulpecula (A)	x	x	35
BURRAMYIDAE			
Cercartetus caudatus	x	?	?
Cercartetus concinnus	x	x	35
Cercartetus lepidus	x	x	30
ACROBATIDAE			
Distoechurus pennatus (N.G.)	x	x	55-60
PETAURIDAE			
Dactylopsila trivirgata (N.G.)	x	x	35
Petaurus breviceps (A)	x	x	65
Petaurus norfolcensis (A)	x	x	65
PSEUDOCHEIRIDAE			
Pseudocheirus archeri (A)	x	x	90
Pseudocheirus corinnae (N.G.)	?	x	?
Pseudocheirus cupreus (N.G.)	?	?	?
Pseudocheirus herbertensis (A)	?	?	80
Pseudocheirus peregrinus (A)	x[4]	x	75
TARSIPEDIDAE			
Tarsipes rostratus (A)	x	−	60

x = structure present; − = structure absent; ? = not determined.

[1] The family divisions in this table conform to those given by Strahan (1983) except that: following Smith (1984) the Petauridae is divided into the two families, Petauridae and Pseudocheiridae; and following Aplin and Archer (1987) *Distoechurus* (and *Acrobates*) are placed in the Acrobatidae. The source of animals is (A) for Australia and (N.G.) for New Guinea.

[2] Approximate extent of acrosomal cover of the dorsal nuclear surface in the anterior-posterior dimension expressed as a percentage of the length of the nucleus in this dimension.

[3] Probably absent, but see comments in the observations section.

[4] Network is incomplete.

of spermatozoan structure in these New Guinean species will follow further collections in 1985 (Harding, Shorey, Archer, Baverstock and Aplin). Likewise only preliminary observations have been made for *Cercartetus concinnus* and *C. lepidus* spermatozoa and a detailed description will follow (Harding, Shorey and Aplin).

Epididymal samples were obtained from the proximal, central and distal regions of the epididymis and are referred to as caput, corpus and cauda epididymidal samples respectively. In some species intermediate samples were taken from between the caput and corpus regions (caput-corpus samples) and corpus and caudal regions (corpus-cauda samples). Removal and processing of the epididymal samples was as described previously (Harding *et al.* 1982b). Epididymides from the New Guinea species

Psuedocheirus corinnae, *P. cupreus*, *Phalanger vestitus* and *Cercartetus caudatus* were stored in liquid nitrogen before thawing and processing as described above.

Material from *Trichosurus vulpecula* epididymides was also prepared for scanning electron microscopy using the methods described previously (Harding *et al.* 1975).

Specimens were examined using a Philips EM 200, EM 201, EM 300, or a JEOL 100 CX for transmission electron microscopy, and JEOL JSM-U3 for scanning electron microscopy.

OBSERVATIONS

A comprehensive description of spermatozoan ultrastructure in each of the diprotodont families is beyond the scope of this paper. Instead, two aspects will be emphasized: flagellar structure in the midpiece region; and, sperm head form, including nuclear shape and placement and proportion of the acrosome relative to the nucleus. The form of these structures in *Trichosurus vulpecula* will be used as the comparative reference for the group, since among marsupials, spermatozoa of this species have been most comprehensively examined.

PHALANGERIDAE

Sperm ultrastructure and epididymal maturation in *Trichosurus vulpecula* have been well-described (Harding *et al.* 1975, 1976; Olson 1975; Cummins 1976; Temple-Smith and Bedford 1976). The sperm head, as in other marsupials, undergoes dorsoventral[1] flattening as it develops during spermiogenesis. In *Trichosurus* this results in a mature sperm head which is elongated in the anterior-posterior dimension (Fig. 1) and slightly dorsoventrally flattened (the dorsoventral: lateral dimensions have approximately a 2:3 relationship) (Figs 2 and 3). Viewed dorsally or ventrally the head tapers anteriorly to a rounded tip. The lateral margins are slightly convex, such that the nucleus is widest just posterior to its mid-point and then narrows slightly to form a blunt posterior extremity (Fig. 1). A deep groove runs posteriorly from approximately the mid-point of the long axis of the ventral nuclear surface, where the neck is inserted (Fig. 3). It is narrow anteriorly and widens posteriorly such that it can accommodate the neck structures and proximal mitochondria of the midpiece (Fig. 4). This is necessary since in mature spermatozoa the long axis of the nucleus lies in a plane parallel with the long axis of the spermatozoon and thus the posterior half of the sperm head overlies the anterior extremity of the flagellum (Fig. 5).

In cauda epididymidal spermatozoa the acrosome covers approximately the anterior one-third of the dorsal nuclear surface. During epididymal maturation the acrosome is transformed from a projecting cup-like appendage sitting on this area to its compact mature form (Harding *et al.* 1976, 1983). In longitudinal sections of the mature sperm head the acrosome is seen to rest on a flattened or slightly concave region of the nuclear surface and its convex dorsal surface gives a stream-lined profile to the sperm head (Fig. 5).

Figure 7 illustrates a number of features of the midpiece of mature *Trichosurus vulpecula* spermatozoa. The midpiece is circular in transverse sections with the axoneme — dense outer fibe complex occupying approximately 25% of its total diameter. The dense outer fibres have an evenly rounded abaxial contour and a less well defined and somewhat flattened adaxial surface. Each closely overlies its corresponding axoneme doublet. Surrounding the mitochondria and underlying the plasma membrane in the posterior two-thirds of the midpiece is a helically wound fibre network (Harding *et al.* 1975; Olson 1975; Cummins 1976; Temple-Smith and Bedford 1976). In transverse sections (Fig. 7) this network appears as an array of circular electron dense profiles. Underlying the network is a layer of granular material and irregularly arranged between the fibres are numerous flask-like invaginations of the plasma membrane (Fig. 7). These structures develop as part of the epididymal sperm maturation process (Harding *et al.* 1975; Cummins 1976; Temple-Smith and Bedford 1976).

For the genus *Phalanger*, only brief observations have been made on poorly preserved material from *P. gymnotis* and *P. vestitus*. The sperm heads closely resemble those of *Trichosurus* except that in *Phalanger* the acrosome appears to be a slightly more prominent structure covering approximately the anterior 40% of the dorsal surface rather than around 35% as in *Trichosurus* (see Table 1). There are indications that an acrosomal maturation process occurs during epididymidal transit as in *Trichosurus*. In terms of relationship of the dense outer fibres to the axoneme and relative proportions of the axoneme-dense outer fibre complex to the mitochondria, the midpiece in transverse section closely resembles that in *Trichosurus*. However, it has not been possible to determine whether the fibre network and plasma membrane invaginations develop in *Phalanger*.

PETAURIDAE[2]

Spermatozoa of *Petaurus breviceps* and *P. norfolcensis* have been shown previously (Harding 1977; Harding *et al.* 1979) to closely resemble those of *T. vulpecula* in flagellar structure. A fibre network and invaginations of the plasma membrane are found in the midpiece of mature spermatozoa, whilst the relative proportions of mitochondria to the axoneme-dense outer fibre complex, together with the close relationship between each dense outer fibre and its associated axoneme doublet result in midpiece transverse sections which are almost identical to those in *Trichosurus* (compare Figs 7 and 11). However, there are marked differences between *Petaurus* and *Trichosurus* in sperm head form. The most obvious concerns the acrosome, which in *Petaurus* is not a prominent feature of the sperm head. It forms a very thin layer covering about three quarters of the long axis of the dorsal

---

[1] By convention the surface of the nucleus into which the neck is inserted in marsupial spermatozoa is described as ventral, whilst the opposite surface of the flattened nucleus is called the dorsal surface.

[2] I have followed Archer (1984) and Smith (1984) in separating the pseudocheirines and the petaurines into two families rather than including both in the family Petauridae (as in Strahan 1983).

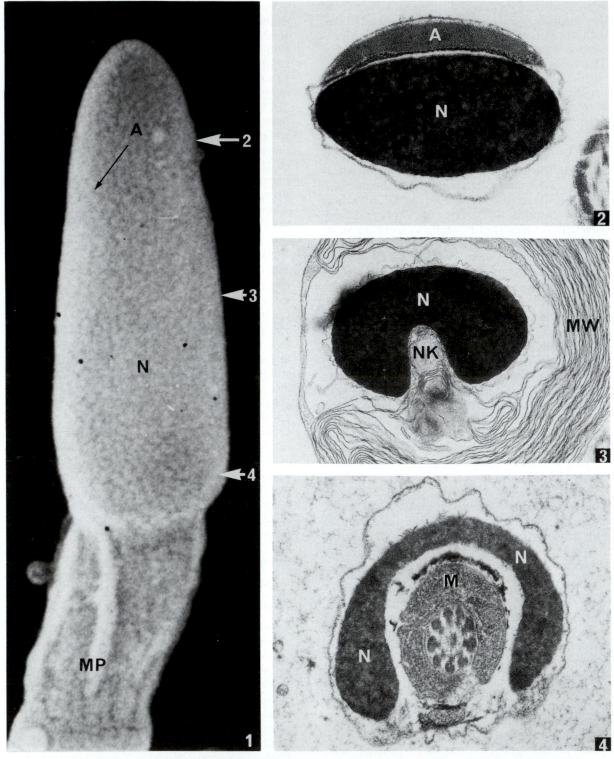

Fig. 1. Scanning electron micrograph of a cauda epididymidal spermatozoon of *Trichosurus vulpecula,* viewed dorsally. The posterior boundary of the acrosome (A) is just visible (arrowed). MP, midpiece; N, nucleus. The arrows labelled 2, 3 and 4 approximately mark the positions of the transverse sections shown in Figs 2, 3 and 4. X28500.

Fig. 2. Transverse section through the anterior region of the head of a cauda epididymidal spermatozoon of *T. vulpecula.* The acrosome (A) covers the convex dorsal surface of the anterior region of the nucleus (N). X46000.

Fig. 3. Transverse section through the head of a cauda epididymidal spermatozoon of *T. vulpecula* in the region of insertion of the neck (NK) into a narrow groove in the ventral surface of the nucleus (N). MW, membranous whorls surrounding the head. X35000.

Fig. 4. Transverse section through the posterior region of the head of a cauda epididymidal spermatozoon of *T. vulpecula.* The deep and wide ventral groove in the nucleus (N) enables the mitochondria (M) and other components of the anterior midpiece to be accommodated in the region where the nucleus overlies the flagellum. X43000.

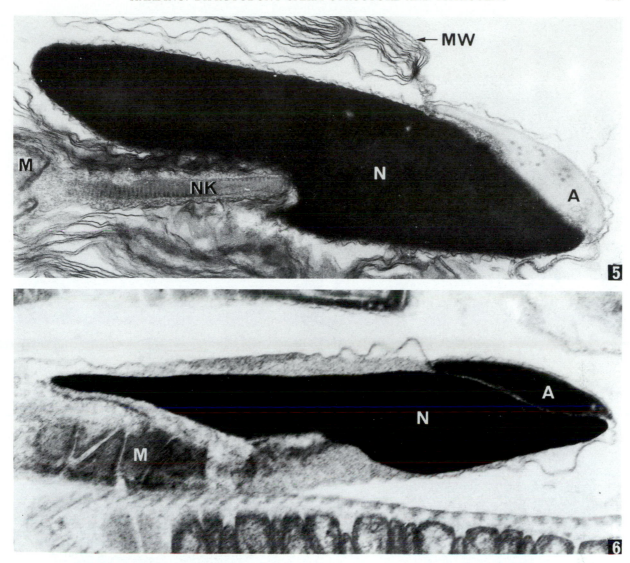

Fig. 5. Longitudinal section, in a dorsoventral plane, through the head of a cauda epididymidal spermatozoon of *T. vulpecula*. A, acrosome; M, mitochondrion of midpiece; MW, membranous whorls; N, nucleus; NK, neck. X37000.

Fig. 6. Longitudinal section, in a dorsoventral plane, through the head of a cauda epididymidal spermatozoon of *Cercartetus lepidus*. A, acrosome; M, mitochondria of midpiece; N, nucleus. X29000.

nuclear surface, although unlike most other Australian marsupial species it does not extend to the anterior tip of the nucleus. In the lateral dimension it is centrally located and covers between one-half and two-thirds of the nuclear surface (Fig. 15). Sperm heads in Petaurus also show overall shape differences to those in *Trichosurus*. They are more flattened dorsoventrally (the dorsoventral: lateral dimensions have a 1:2 relationship in contrast to the 2:3 found in *Trichosurus*) and the ventral groove is far less pronounced than in *Trichosurus* (compare Figs 3 and 15). *P. norfolcensis* and *P. breviceps* spermatozoa differ slightly in sperm dimensions and form (particularly head shape), but further analysis is necessary to confirm and fully identify these distinguishing features.

Preliminary observations on *Dactylopsila trivirgata* spermatozoa show a close resemblance to *T.*

vulpecula in both flagellar and head form. The acrosome definitely resembles that of *Trichosurus* rather than *Petaurus*. It seems likely that more detailed observations will enable *Trichosurus* and *Dactylopsila* spermatozoa to be differentiated on the basis of minor size and shape differences.

PSEUDOCHEIRIDAE

Further observations and analysis are required for each of the five pseudocheirids examined. This is particularly necessary for the New Guinean species *Pseudocheirus cupreus* and *P. corinnae* for which the available material is poorly preserved.

Previous observations (Harding 1977; Harding *et al.* 1979) have shown that *P. peregrinus* spermatozoa are readily distinguished from those of non-pseudocheirid marsupials. The

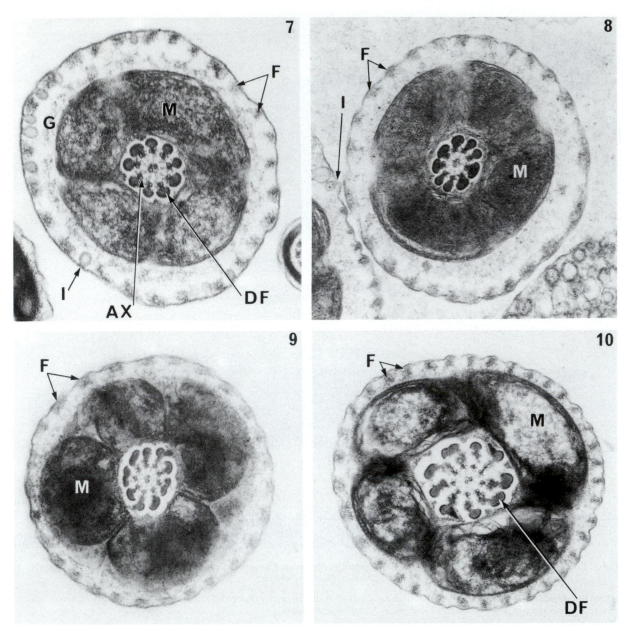

Fig. 7. Transverse section through the midpiece of a cauda epididymidal spermatozoon of *T. vulpecula.* AX, doublet of the axoneme complex; DF, dense outer fibre; F, fibre network; G, granular layer; I, invagination of the plasma membrane; M, mitochondria. X54500.

Fig. 8. Transverse section through the midpiece of a corpus epididymidal spermatozoon of *Cercartetus concinnus.* Invaginations of the plasma membrane (I) are evident in the transverse section through an adjacent midpiece. F, fibre network; M, mitochondria. X38000.

Fig. 9. Transverse section through the midpiece of a cauda epididymidal spermatozoon of *Macropus parma.* F, fibre network; M, mitochondria. X39500.

Fig. 10. Transverse section through the midpiece of a cauda epididymidal spermatozoon of *Potorous tridactylus.* Dense outer fibres (DF) are radially displaced from their corresponding axoneme doublets. F, fibre network; M, mitochondria. X40000.

most distinctive feature is the midpiece which is extremely wide relative to its length and to the diameter of the principal piece (Fig. 18). Transverse sections show the major contributor to the midpiece diameter to be the unusually large mitochondria (Fig. 13). Consequently, the axoneme-dense outer fibre complex constitutes only about 12% of the midpiece diameter compared with about 25% in *Trichosurus* and *Petaurus*.

The dense outer fibres closely overlie the axoneme doublets as in these groups but in contrast to *Trichosurus* and *Petaurus* they are relatively poorly developed (Fig. 13). A fibre network and plasma membrane invaginations develop in the midpiece during epididymal transit but the network is unusual in that it is incomplete. In transverse sections it is seen to cover only one-quarter of the circumference of the midpiece (Fig. 13) whilst in

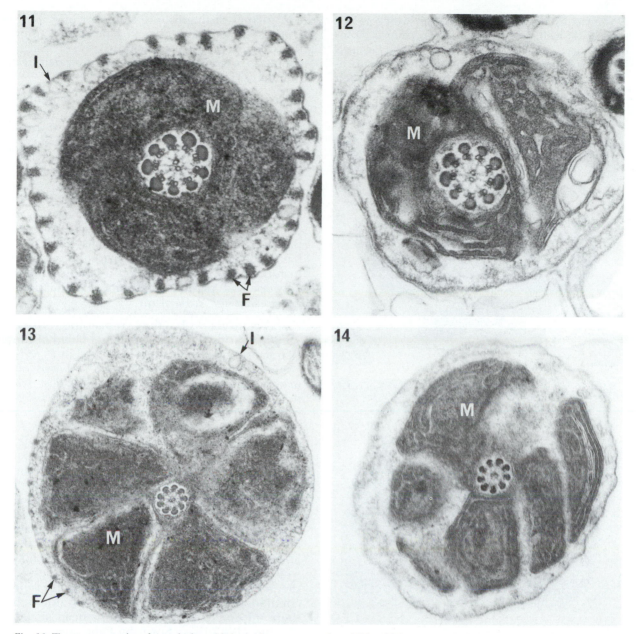

Fig. 11. Transverse section through the midpiece of a corpus-cauda epididymidal spermatozoon of *Petaurus norfolcensis.* F, fibre network; I, invagination of the plasma membrane; M, mitochondria. X60000.

Fig. 12. Transverse section through the midpiece of a caput epididymidal spermatozoon of *Pseudocheirus archeri.* Note that the fibre network and plasma membrane invaginations are not yet developed in this immature spermatozoon. M, mitochondria. X49500.

Fig. 13. Transverse section through the midpiece of a cauda epididymidal spermatozoon of *Pseudocheirus peregrinus.* Note that the fibre network (F) covers only one-quarter of the sectioned midpiece circumference. The mitochondria (M) are very well developed in the radial dimension relative to the centrally located axoneme-dense outer fibre complex. I, invaginations of the plasma membrane. X25500.

Fig. 14. Transverse section through the midpiece of a cauda epididymidal spermatozoon of *Pseudocheirus herbertensis.* M, mitochondria. X37000.

longitudinal sections the fibres are obvious only in the posterior region close to the annulus and extending anteriorly up one side of the longitudinally sectioned midpiece (Fig. 18 and see Harding *et al.* 1979 for discussion).

Using light microscopy Hughes (1965) showed the sperm head of *P. peregrinus* to be ovoid in shape and rather chunky relative to other marsupials, and the acrosome to cover all but a marginal annular zone of the dorsal nuclear surface. Electron microscopy reveals that a further distinctive feature of the sperm head is a centrally located depression along the long axis of the dorsal nuclear surface (Fig. 16). The acrosomal cover of this surface is similar to that in

Petaurus as is the ventral groove, which is not nearly as well developed as in *Trichosurus* (compare Figs 3, 15 and 16). However, unlike *Petaurus* the acrosome undergoes an epididymal maturation similar to *Trichosurus* (Fig. 16) and when mature is much thicker dorsoventrally than in *Petaurus* and consequently is a far more pronounced feature of the sperm head.

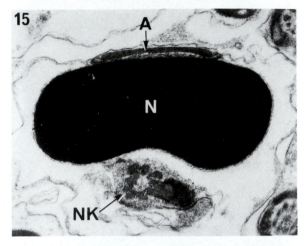

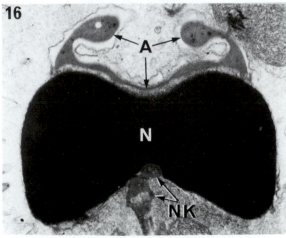

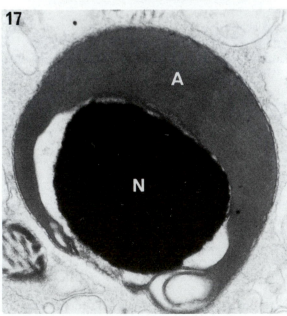

A further distinguishing character of *P. peregrinus* spermatozoa is the extreme flattening of the principal piece viewed in transverse section in comparison with other non-pseudocheirid diprotodonts (Figs 29 and 30).

Of the other four pseudocheirids examined, spermatozoa of *P. herbertensis* most closely resemble those of *P. peregrinus*. The relative proportions of the various regions of the spermatazoon are similar to *P. peregrinus* (Fig. 19), although very approximate measurements suggest that spermatozoa of *P. herbertensis* are smaller. Sperm head shape and acrosomal coverage closely resemble *P. peregrinus* spermatozoa as does the general appearance of the mid- and principal pieces in transverse sections (Fig. 14). It has not been possible with currently available material to determine whether a midpiece fibre network and plasma membrane invaginations are present in *P. herbertensis*.

Spermatozoa of the remaining three pseudocheirids show an interesting blend of the characteristics seen in *Trichosurus* and in *P. peregrinus*. The following description is based on material from *P. archeri* but preliminary observations on poorly preserved material from *P. corinnae* and *P. cupreus* indicate that the spermatozoa of these species closely resemble those of *P. archeri*. Similarities to *Trichosurus* sperm include: the general appearance of the mid- and principal pieces in transverse section (Fig. 12); the presence of a complete midpiece fibre network and numerous midpiece plasma membrane invaginations (Fig. 20); and the proportional relationship of the mid- and principal pieces seen in longitudinal section and the form of the annulus which separates them (Fig. 20). Links with *P. peregrinus* are evident in the shortness of the midpiece relative to head length and in the chunkiness of the sperm head (Fig. 20), although there appear to be differences in head shape. Hughes (1965) has shown by light microscopy that the sperm head in *P. cupreus* is more pointed anteriorly than in *P. peregrinus* and ultrastructural examination of sections of the head in *P. archeri* show no sign of the median dorsal concavity seen in *P. peregrinus*. The most distinctive

Fig. 15. Transverse section through the head, in the region of the neck components (NK), of a cauda epididymidal spermatozoon of *Petaurus breviceps*. A, acrosome; N, nucleus. X30000.

Fig. 16. Transverse section through the head, in the region of the neck (NK) of a caput epididymidal spermatozoon of *Pseudocheirus peregrinus*. The acrosome (A) displays the form typical of immature spermatozoa. Note the depression in the dorsal surface of the nucleus (N). X24500.

Fig. 17. Transverse section through the head of a corpus epididymidal spermatozoon of *Pseudocheirus archeri*. Note that the acrosome (A) extends down the lateral margins of the nucleus (N) and under part of the ventral nuclear surface. X32500.

feature of the sperm head in these three species is the extreme development of the acrosome. Its form is variable, but generally the entire dorsal nuclear surface is covered by the acrosome and in some transverse sections it extends laterally down the sides of the head to the ventral surface (Fig. 17). The presence of such variability in acrosomal form in cauda epididymidal spermatozoa is puzzling; possibly various stages of acrosomal maturation are represented.

MACROPODIDAE AND POTOROIDAE

The macropodids and *Potorus* show general similarity in sperm form to *Trichosurus*. However, there are clear differences both between the species examined in the Macropodoidea and between this group and *Trichosurus*.

The sperm head form, acrosomal size, location and epididymal maturation changes are similar to those in *Trichosurus* (compare Figs 5 and 22) and the closest resemblance is found in *Macropus robustus*. However, differences are evident. Although the form of the acrosome resembles that in *Trichosurus*, the extent of its coverage of the dorsal nuclear surface in the anterior-posterior dimension is greater in the macropodids (approximately 45% compared with about 33% in *Trichosurus*). This is not true of the very elongated sperm head of *P. tridactylus*, for which the corresponding figure is about 25%. The sperm head is relatively short and chunky in *M. rufus*, moderately elongated in *M. parma* and *Setonix brachyurus* and extremely elongated and streamlined in *P. tridactylus* (see Fig. 2a in Harding *et al.* 1979). Unlike *Trichosurus*, in which the posterior section of the sperm head forms a curved arch overlying the anterior mitochondria of the midpiece, only a fairly flat dorsal plate is found in this area in macropodoids (Fig. 24). That this is universally characteristic of this group is clear from Cleland's drawings of sperm heads from 20 macropodoid species (Rodger 1978). *Macropus parma*, *S. brachyurus* and *P. tridactylus* are similar in that the acrosome lies in a deep concavity of the anterior dorsal nuclear surface (Figs 22 and 23). In contrast, this region of the nucleus viewed in transverse section in *M. robustus* and *M. rufus* is flattened or saddle-shaped and, in *Trichosurus*, slightly convex (Fig. 2).

The macropodoids all resemble *Trichosurus* in the development of a midpiece fibre network during epididymal maturation but, unlike *Trichosurus*, invaginations of the midpiece plasma membrane do not develop. Although the macropodoid midpiece in transverse section is similar to *Trichosurus* in terms of the relative proportions of the axoneme-dense outer fibre complex and the mitochondria, slight differences are evident (Figs 9 and 10). The dense outer fibres do not so closely overlie the axoneme doublets as they do in *Trichosurus*, *Petaurus* and *Pseudocheirus*. The separation is slight in *M. rufus*, *M. robustus*, *S. brachyurus* and *M. parma* (Fig. 9) and a distinctive separation is evident in *P. tridactylus* (Fig. 10).

Although there are spermatozoan characteristics enabling the macropodoids examined in this study to be separated as a group from the other diprotodontans, there is sufficient disparity in sperm structure within this group to allow the individual macropodoid species to be distinguished from one another. The closest similarity is seen between *M. parma* and *S. brachyurus* whilst certain features, such as the elongated nucleus and wider radial displacement of the dense outer fibres, set *Potorous* apart. Cleland's light microscope study of spermatozoa from 20 macropodoid species showed variation between all species in 8 sperm characters, but no apparent grouping of species corresponding to recognised taxonomic divisions. Only *Lagorchestes* (*conspicillatus* and *hirsutus*) stood apart from the group on the basis of sperm form (Cleland, cited by Rodger 1978).

BURRAMYIDAE[1]

Only preliminary observations have been made for the burramyid species. Spermatozoa from *Cercartetus concinnus* and *C. lepidus* (Harding, Shorey and Aplin, unpublished observations) are almost identical to one another in both head and midpiece form and proportions and their overall sperm form closely resembles that in *Trichosurus* (Fig. 21). The head (Fig. 6) has a general similarity to *Trichosurus* and the macropodids although subtle shape differences will probably enable *Cercartetus* sperm heads to be distinguished from those of these phalangerid and macropodid species. The acrosomal maturation process and the form, siting and nuclear surface coverage of the resulting acrosome are very similar to *Trichosurus*. Likewise, in transverse sections the midpiece resembles that of *Trichosurus* in terms of the proportion of the dense outer fibre-axoneme complex to the mitochondria and presence in mature spermatozoa of a fibre network and invaginations of the midpiece plasma membrane (Fig. 8). Although the dense outer fibres lie close to their corresponding axoneme doublets, the relationship is not as tight as in *Trichosurus* and *Petaurus* but closer than in *M. parma*, *M. rufus* and *M. robustus* (compare Figs 7, 8, 9 and 11). Spermatozoa of the New Guinea species *C. caudatus* are too poorly preserved to determine other than a similarity in sperm head shape and internal midpiece proportions to *C. concinnus* and *C. lepidus* and the presence of a midpiece fibre network.

---

[1] As discussed below, recent evidence presented in this Volume (Aplin and Archer 1987) supports the separation of *Distoechurus* and *Acrobates* from the Burramyidae as the Acrobatidae.

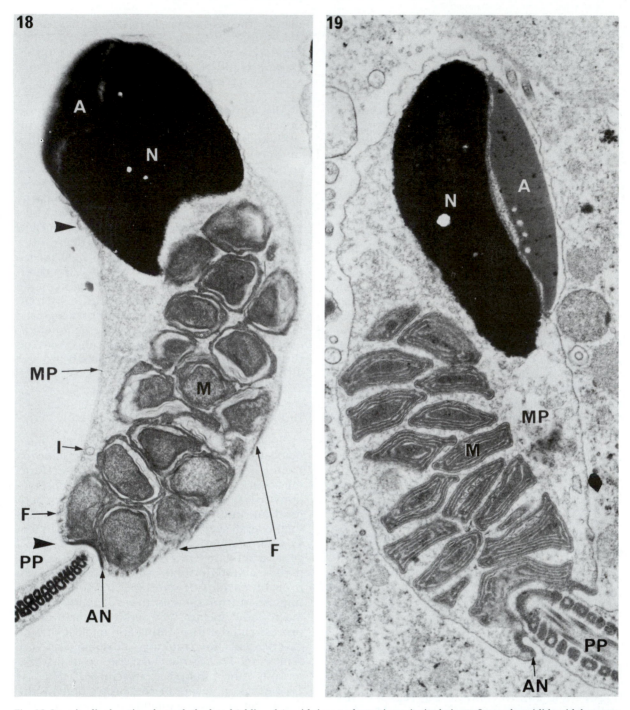

Fig. 18. Longitudinal section through the head (obliquely), midpiece and anterior principal piece of a cauda epididymidal spermatozoon of *Pseudocheirus peregrinus*. Note the proportional differences in these regions compared to typical diprotodontan proportions shown in Fig. 21. A, acrosome; AN, annulus; F, midpiece fibre network; I, invaginations of the midpiece plasma membrane; M, mitochondria; MP, midpiece (anterior and posterior extent marked by arrow heads); N, nucleus; PP, principal piece. X18000.

Fig. 19. Longitudinal section through the head, midpiece and anterior principal piece of a corpus-cauda epididymidal spermatozoon of *Pseudocheirus herbertensis*. A, acrosome; AN, annulus; M, mitochondria; MP, midpiece; N, nucleus; PP, principal piece. X28000.

ACROBATIDAE

Spermatozoa of *Distoechurus pennatus* resemble those of *Cercartetus* and *Trichosurus* in general head shape, midpiece proportions, acrosomal maturation and presence of a fibre network and plasma membrane invaginations. However,

Distoechurus spermatozoa have some very distinctive features. Except for the giant spermatozoa of *Tarsipes* (see Cummins and Woodall 1985), *Distoechurus* spermatozoa with a total length of approximately 149 μm are the largest reported so far among diprotodontans. The acrosome is larger than in *Cercartetus* and *Trichosurus* and

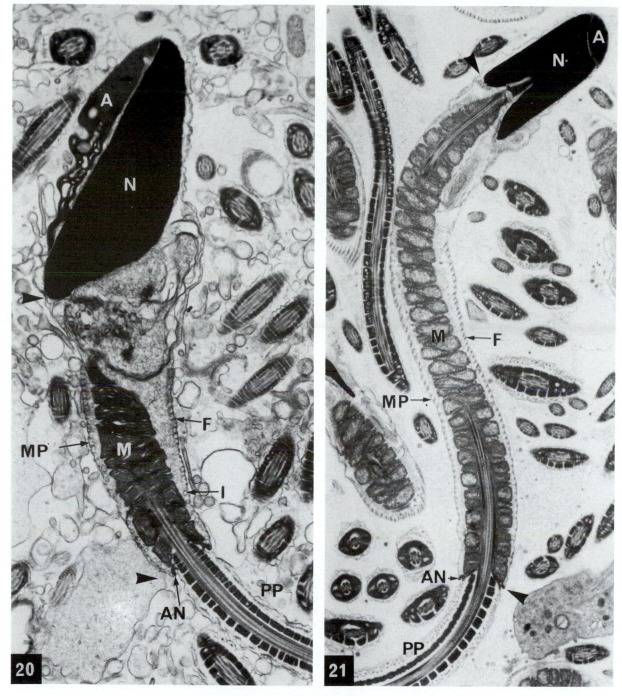

Fig. 20. Longitudinal section through the head, midpiece and anterior principal piece of a cauda epididymidal spermatozoon of *Pseudocheirus archeri*. A, acrosome; AN, annulus; F, midpiece fibre network; I, invaginations of the midpiece plasma membrane; M, mitochondria; MP, midpiece (anterior and posterior extremities marked by arrow heads); N, nucleus; PP, principal piece. X13500.

Fig. 21. Longitudinal section through the head (obliquely), midpiece and anterior principal piece of a cauda epididymidal spermatozoon of *Cercartetus lepidus* to show the regional proportions typical of the burramyids, phalangerids, petaurids and macropodoids examined for this study. A, acrosome; AN, annulus; F, midpiece fibre network; M, mitochondria; MP, midpiece (anterior and posterior extremities marked by arrow heads); N, nucleus; PP, principal piece. X10500.

covers the anterior half of the dorsal nuclear surface (Fig. 26). Preliminary observations show some variability in acrosomal form, but typically in longitudinal sections the anterior two-thirds of the acrosome forms a very prominent structure whilst the posterior one-third is reduced dorso-ventrally (Fig. 26). A further common feature is

that the anterior tip of the acrosome forms a 'lip' which projects rostrally over the leading edge of the sperm head. A distinct but narrow gap separates the acrosome from the nuclear surface in the region of the 'lip'. The posterior dorso-ventral reduction in acrosomal thickness and the anterior 'lip' are reminiscent of the acrosome in

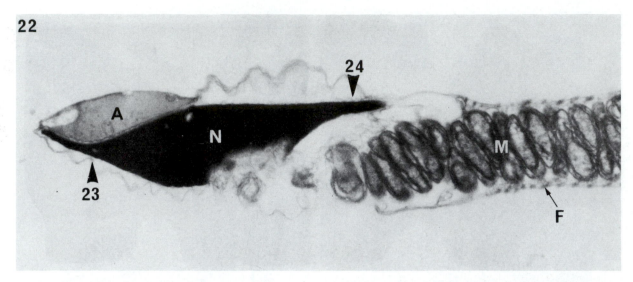

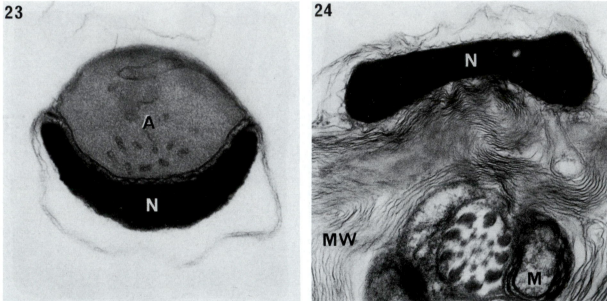

Fig. 22. Longitudinal section through the head (in a dorsoventral plane) and anterior midpiece of a cauda epididymidal sperm-
    atozoon of *Macropus parma*. The acrosome (A) sits in a concavity in the anterior dorsal surface of the nucleus (N). F, midpiece
    fibre network; M, mitochondria. The arrow heads labelled 23 and 24 mark the approximate positions of the sections shown
    in Figs 23 and 24 respectively. X17000.

Fig. 23. Transverse section through the anterior sperm head of a cauda epididymidal spermatozoon of *M. parma*. A, acrosome;
    N, nucleus. X44000.

Fig. 24. Transverse section through the posterior region of the head of a cauda epididymidal spermatozoon of *M. parma*. M,
    mitochondria; MW, membranous whorls; N, nucleus. X43000.

Tarsipes rostratus (Harding *et al.* 1984 and compare Figs 25 and 26). The midpiece in *Distoechurus* is characterised by the pronounced
development of the fibre network (Fig. 27). Very
approximate measurements show the fibres to
have a diameter around 70 nm compared with
around 30 nm in *Tarsipes* (Harding *et al.* 1984)
and 40-60 nm in the other diprotodonts. The
dense outer fibres are more widely separated
from the underlying axoneme doublets than in
Cercartetus and approximately match the condition seen in *M. parma* and *S. brachyurus* (compare Figs 8, 9 and 27).

TARSIPEDIDAE

Spermatozoa of *Tarsipes rostratus* have been
described in some detail elsewhere (Harding *et al.*
1984). As the largest mammalian spermatozoon
reported to date (356 μm total length) they are
most notable for their size (Cummins and Woodall
1985). However, they have a number of other
unusual features which make their sperm form
quite unique. The sperm head resembles that of
the dasyurids in its length (approximately 12.4 μm)
and very elongated and stream-lined form (Fig.
25), but in terms of acrosomal maturation it is

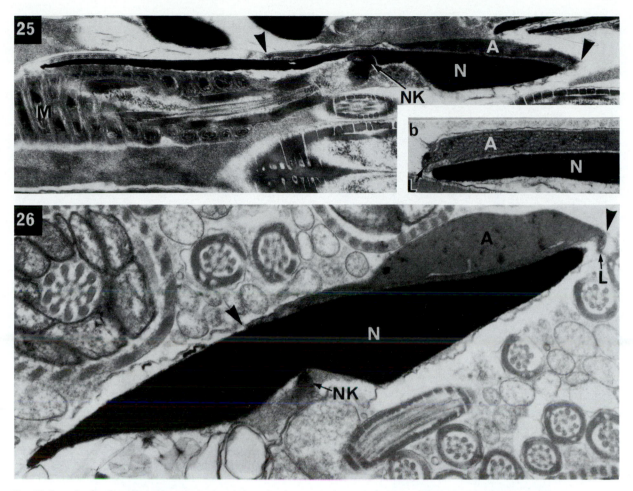

Fig. 25. Longitudinal section through the head (in a dorsoventral plane) and anterior midpiece of a cauda epididymidal spermatozoon of *Tarsipes rostratus.* Note the extensive overlying of the mitochondria (M) of the anterior midpiece by the elongated nucleus (N). The anterior and posterior extremities of the acrosome (A) are marked by arrow heads. NK, neck. X14500. The inset (b) gives an enlargement of a longitudinal section (in the dorsoventral plane) of the anterior tip of the head showing that the acrosome (A) forms a 'lip' (L) over the tip of the nucleus (N). X25000.

Fig. 26. Longitudinal section through the head (in a dorsoventral plane) of a cauda epididymidal spermatozoon of *Distoechurus pennatus.* Note that the anterior tip of the acrosome (A) forms a 'lip' (L) over the tip of the nucleus (N) and that the posterior third of the acrosome is reduced in the dorsoventral dimension. Arrow heads mark the anterior and posterior extent of the acrosome. NK, neck. X28500.

typically diprotodontan and preliminary observations indicate similarity with *Distoechurus* in mature acrosomal form. However, as discussed previously (Harding *et al.* 1984), there are also unique acrosomal features in *Tarsipes,* particularly those concerning intra-acrosomal structure. The midpiece in *Tarsipes* spermatozoa is striking in its similarity to that in dasyurids. This is due primarily to its flattened shape, evident from the oval profile in transverse sections, and extreme radial displacement of the dense outer fibres from the underlying axoneme doublets (Fig. 28). There are also unique features in the *Tarsipes* midpiece, particularly the additional microtubule network (Fig. 28) which underlies the fibre network (Harding *et al.* 1984).

The spermatozoon of *Tarsipes* does not conform to the generalised pattern seen in the diprotodontan species discussed above. Its form is quite unique but the strong similarities with

dasyurid sperm flagella are striking and lesser similarities with peramelid (in the flagellum) and diprotodont (in the spermatozoan head) spermatozoa are also evident.

PHASCOLARCTIDAE

Phascolarctid spermatozoa like those of *Tarsipes* do not sit comfortably with the generalised diprotodontan sperm form as outlined above. Since a detailed description of the Koala spermatazoon forms a separate chapter in this Volume (Harding *et al.* 1987), it is not necessary to repeat the details here. As shown previously by Hughes (1965) using light microscopy, phascolarctid (and vombatid) spermatozoa have, for marsupials, a most unusual and *superficially* placental-like head form. The midpiece appears relatively unspecialised and transverse and longitudinal sections are remarkably similar to those seen in the monotremes (Harding *et al.* 1979, 1987). A

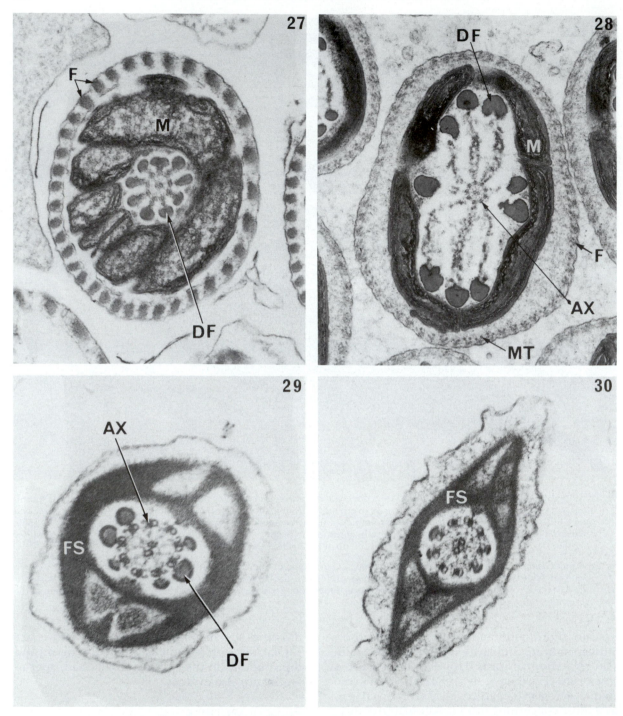

Fig. 27. Transverse section through the midpiece of a cauda epididymidal spermatozoon of *Distoechurus pennatus*. DF, dense outer fibre; F, fibre network; M, mitochondria. X38500.

Fig. 28. Transverse section through the midpiece of a caput-corpus epididymidal spermatozoon of *Tarsipes rostratus*. AX, axoneme; DF, dense outer fibre; F, fibre network; M, mitochondria; MT, microtubule component of the fibre network. X29000.

Fig. 29. Transverse section through the anterior principal piece of a cauda epididymidal spermatozoon of *Macropus robustus* showing the principal piece form typical of macropodoids, phalangerids, petaurids and burramyids. AX, axoneme; DF, dense outer fibre; FS, fibrous sheath. X111500.

Fig. 30. Transverse section through the anterior principal piece of a corpus epididymidal spermatozoon of *Pseudocheirus peregrinus*. Note the flattening of the fibrous sheath (FS) compared with that in Fig. 29. X68500.

well developed midpiece fibre network is certainly not present but occasional superficial longitudinal sections indicate that a rudimentary network may be present (Harding 1977; Harding

et al. 1979). There are also some unusual features of spermiogenesis and the Sertoli cell which more closely resemble certain placentals and/or monotremes than other marsupials (Harding *et al.*

1982a, 1987). Although wombat spermatozoa have not been described at the ultrastructural level, light microscopy (Hughes 1965) and ultrastructure of spermiogenesis in *Lasiorhinus latifrons* (Harding, Carrick and Shorey unpublished observations) indicate close similarity in sperm form in the Phascolarctidae and Vombatidae.

DISCUSSION

Compared with the Dasyuridae or Peramelidae, the Diprotodonta do not form a well defined group on the basis of sperm ultrastructure. If *Tarsipes* and the vombatiforms are left aside, then, although there are considerable differences in sperm structure among the remaining diprotodontans, they share sufficient similar spermatozoan features compared with other marsupials for their grouping to be plausible.

However, this view is based on a *phenetic* (stressing apparent similarities) rather than a *phylogenetic* approach to classification since no rigorous attempt has been made to determine whether the apparently linking or differentiating features represent apomorphic or plesiomorphic characters in the sense of Hennig (1979). Such analysis is underway and should be completed following the collection of further data during 1985 (Harding, Shorey, Archer, Baverstock and Aplin work in progress). In the meantime, the discussion below must be introduced with this cautionary remark.

We will consider first the problems *apparently* posed by sperm structure in *Tarsipes* and the vombatiforms for the unity of the Diprotodonta.

In previous papers (Harding *et al.* 1981, 1982b, 1984) we showed *Tarsipes* spermatozoa to possess unique features but also to display a blend of the sperm characteristics of dasyurids, peramelids and some phalangeroids. We drew particular attention to the unusual nature of the flagellum and its likeness to that of the dasyurids and to a lesser extent the peramelids, and discussed whether this similarity should be considered to result from phylogenetic links or convergence. On the basis of the available evidence we concluded that phylogenetic links provided the most economic explanation for the similarities in flagellar structure and further suggested that the common features of *Tarsipes*, dasyurid and peramelid spermatozoa may have been present in the ancestral sperm form for the major Australian marsupial radiation (Harding *et al.* 1981, 1982b). However, we noted that our suggested 'ancestral sperm form' is ". . . counter-intuitive in flagellar structure at least, in that it appears highly derived rather than generalised" (Harding *et al.* 1982b:671).

These suggestions now require reconsiderations in the light of new information relating to American-Australian marsupial phylogeny.

Following an extensive survey of tarsal morphology in living and fossil marsupials, Szalay (1982) suggested the monophyly of the South American microbiotheriid, *Dromiciops australis*, and all Australian marsupials. Further supporting evidence has been suggested from karyotype analysis (Sharman 1982) and sperm ultrastructure (Temple-Smith 1984, 1987). Although only poorly preserved material has been examined, Temple-Smith (1987) has shown *Dromiciops* spermatozoa to resemble those of Australian phalangerids, petaurids and burramyids in terms of sperm head and flagellar form. He also makes a case that simple radial symmetry, circular transverse section and rounded dense fibres immediately adjacent to the axoneme doublets (as found in these groups) most likely comprise the plesiomorphic condition of the marsupial sperm tail, whilst the arrangement found in *Tarsipes*, the dasyurids and to a lesser extent the peramelids is more likely to be derived. He suggests that this conclusion strengthens the association, indicated by sperm structure, between *Dromiciops* and the phalangeroid segment of the Australian radiation and lends support to the conclusions from other character systems (Szalay 1982; Sharman 1982) that ancestral microbiotheriid stock gave rise to *Dromiciops* and the Australian marsupial radiation. It should be noted, however, that close phylogenetic links between the microbiotheriids and Australian marsupials have been questioned by others on serological and anatomical grounds and the pedal similarities seen as convergent (Kirsch and Reig 1984).

Some recent phylogenies have incorporated the emerging evidence of links between *Dromiciops* (and extinct microbiotheriids) and the Australian marsupials (e.g. Szalay 1982, Archer 1983). Assuming for the moment that these phylogenies (see Fig. 31) represent reality, we will briefly explore the implications from the perspective of evolution of sperm form.

If the plesiomorphic condition for the marsupial sperm tail included, as suggested by Temple-Smith (1987), simple radial symmetry, circular transverse section, and rounded dense outer fibres immediately adjacent to the axoneme doublets, then in conformity with the tree in Fig. 31 we must assume either: (a) independent derivation of the radial displacement of the dense outer fibres in the dasyurid, peramelid and tarsipedid branches of the line leading to the phalangeroid radiation; or (b) that the shared flagellar features of these three groups were present at the base of the Australian radiation after divergence from the microbiotheriid line, and were lost before derivation of the phalangeroids and replaced by the microbiotheriid-like features. Neither of these alternatives is entirely satisfactory in terms of providing an economic explanation for the derivation of the

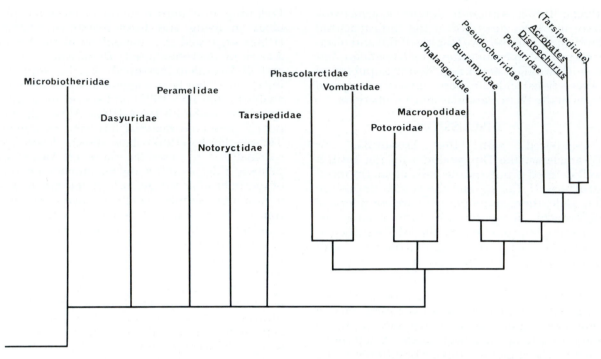

Fig. 31. Phylogenetic tree of Australian marsupial groups emphasising microbiotheriid links. Modified from Archer's (1984) figures 306 and 310. Tarsipedidae (unbracketed) represents the position for the tarsipedids discussed in the text with reference to Fig. 31. Tarsipedidae (bracketed) represents the recent suggested links for tarsipedids discussed in the text with reference to Fig. 32.

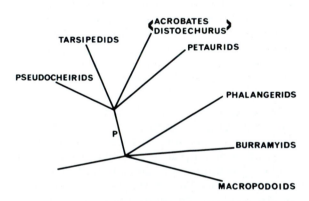

Fig. 32. A tentative phylogenetic interpretation by Baverstock of the results of comparative albumin studies using the technique of microcomplement fixation (MC'F) (Baverstock pers. comm. work in progress).

present distribution of Australian marsupial spermatozoan types. Hypothesis (b) is excessively convoluted and (a) seems unlikely in view of the most unusual features of the sperm flagellum that must be independently derived three times, and in the case of the dasyurids and *Tarsipes,* result in such closely similar flattened sperm tails. However, we must further consider the likelihood of convergence in the production of these flagellar characteristics.

Previously, in discussing this possibility (Harding *et al.* 1981, 1982b) we concluded that for convergence to have produced such an unusual (in terms of other mammals), but similar (particularly in the dasyurids and *Tarsipes*) flagellar structure, two conditions would be required: (1) a

strong functional rationale; and (2) that the prototype sperm form on which this functional requirement was acting, possessed constraints which allowed for a very restricted range of structural 'solutions'. The most likely functional rationale concerns motility, particularly since these groups possess the largest spermatozoa found among marsupials, whilst the basis for the second of the above conditions is that large placental spermatozoa apparently display a range of structural 'solutions' to a suggested motility 'problem' (Fawcett 1970).

On the basis of present evidence it seems reasonable to conclude that if the shared features of dasyurid, tarsipedid and peramalid spermatozoa were produced by convergence then these two conditions must have applied. However, since these conditions are postulated to rest on motility requirements related to large sperm size, we are still left with the need to explain why large spermatozoa were independently derived in these three branches of the Australian marsupial radiation. Alternatively, if we reject the hypothesised phylogeny shown in Fig. 31, other evolutionary sequences for sperm form become possible. Further information and analysis are necessary before we should explore such sequences.

The other group which, on the basis of sperm structure, does not appear to rest easily within the Diprotodonta, is the Vombatiformes. We have argued in detail elsewhere (Harding *et al.* 1987) against a derived position for the vombatiform lineage within the Diprotodonta on sperm-based criteria, and suggested that the unique features and great similarity of sperm form in the Phascolarctidae and Vombatidae support close phylogenetic links between these groups and indicate conservatism in sperm structure within the Vombatiformes. Whether the special features of vombatiform spermatozoa should be regarded as plesiomorphic or apomorphic with respect to other marsupial groups is unclear at present and although we have drawn attention to some similarities with placentals and monotremes in aspects of sperm form and development these may be convergent rather than plesiomorphic features.

Assuming once again the phylogenetic tree shown in Fig. 31 represents reality, and that shared features of *Dromiciops* and phalangeroid spermatozoa were present in the lineage leading from the microbiotheriids to the phalangeroid radiation, then numerous gains and losses in sperm structural features must be assumed in the early vombatiform lineage in order to achieve modern vombatiform sperm structure. These include: acquisition of the unique head form and siting of the neck insertion; loss of a well-developed midpiece fibre network; acquisition of crystalloids in the Sertoli cell; development of a discrete pro-acrosomal granule in early spermiogenesis; development of an uneven pattern of nuclear chromatin condensation during spermiogenesis (see Harding *et al.* 1987 for detailed discussion).

At present, the most fitting conclusion from sperm-based criteria seems to be that we should keep an open mind regarding placement of the vombatiforms in Australian marsupial phylogenies.

Leaving *Tarsipes* and the vombatiforms aside, the remaining diprotodontans share sufficient similar sperm features to make their monophyletic grouping plausible. These features include: a '*Trichosurus*-like' proportionality of the axoneme-dense outer fibre complex and mitochondria in the midpiece; close association between the dense outer fibres and associated axoneme doublets; presence of a well developed midpiece fibre network and invaginations of the midpiece plasma membrane in mature spermatozoa; and an obvious acrosomal maturation process. These remaining diprotodonts do not all match *Trichosurus vulpecula* in displaying each of these characters, but there is sufficient similarity within the group to consider the differences as variations on a common theme.

We will turn now to these variations and their possible significance for diprotodontan phylogeny and classification. Once again, it must be stressed that the following comments should be regarded as 'work-in-progress'. Insufficient

data are available at present to undertake the rigorous analysis, cladistic or otherwise, necessary to produce a sperm-based phlogeny.

At the intra/interfamily level the most significant matter requiring discussion is the suggestion based on basicranial and other anatomy (Aplin and Archer 1983, 1984; Archer 1983; Aplin and Archer 1987) and on microcomplement fixation (MC'F) of albumin (Baverstock 1984 and Baverstock *et al.* 1987), that *Acrobates* and *Distoechurus* stand far enough apart from *Cercartetus* and *Burramys* to warrant their removal from the Family Burramyidae. Indeed, the MC'F studies show *Distoechurus/Acrobates* to be closer to *Petaurus* and *Tarsipes* than to *Cercartetus* (Baverstock 1984; Baverstock *et al.* 1987). On the basis of this evidence, Archer (1984:780) in a recently constructed phylogenetic tree, has removed *Acrobates/Distoechurus* from the Burramyidae and placed them as the sister group to *Tarsipes*, and these three genera combined as the sister group to the Petauridae (see Fig. 31). The Burramyidae is placed as the sister group to the Phalangeridae. More recently Aplin and Archer (1984), primarily on evidence from basicranial anatomy, have stressed the separateness of *Acrobates/Distoechurus* from the phalangeroids (phalangerids, petaurids, pseudocheirids and burramyids) but, although they note some shared basicranial features of this duo with *Tarsipes*, they do not attempt to link the two groups phylogenetically as Archer (1983:780) has done (see Fig. 31). Finally, Aplin and Archer (1987) have formally established the Acrobatidae to include just the genera *Acrobates* and *Distoechurus*.

The relevant evidence from comparative sperm structure is based on *Cercartetus* and *Distoechurus* because *Acrobates* and *Burramys* have yet to be examined. *Cercartetus* spermatozoa closely resemble those of the phalangerid *Trichosurus* although most of the shared characters are probably plesiomorphic within the phalangeroid radiation. Although *Distoechurus pennatus* spermatozoa likewise share many similarities with those of *Trichosurus* and *Cercartetus*, they also display distinctive features, some of which are reminiscent of *Tarsipes*. In terms of their large size they more closely approach the sperm dimensions in *Tarsipes* than do spermatozoa of other diprotodontans. Even so the size difference between *Tarsipes* and *Distoechurus* spermatozoa is considerable. The acrosome in *Distoechurus* is also unusual among diprotodontans in its extent and form and shows most resemblance to that in *Tarsipes*. A further distinctive feature is the very pronounced nature of the midpiece fibre network. Thus, although *Distoechurus* spermatozoa are overall more like those of the phalangeroids than any other marsupial group, there are sufficient distinctive features to set this genus apart from *Cercartetus*, and the slight similarities to *Tarsipes* spermatozoa are interesting in view of Baverstock's (1984 and Baverstock *et al.* 1987) MC'F evidence of links between these groups.

Is it possible to reconcile the data from comparative sperm studies with Baverstock's phylogenetic reconstruction based on his MC'F data (see Fig. 32)? The following is an attempt to fit a plausible sequence of spermatozoan evolution to this phylogeny.

At the base stem of the tree (Fig. 32) we could postulate a spermatozoon with the features shared by the phalangerids/burramyids and macropodids/potoroids; namely, *Trichosurus*-like head shape and acrosomal form, a *Trichosurus*-like acrosomal maturation process, midpiece fibre network, and *Trichosurus*-like proportions of the intra-flagellar structures. Additionally, invaginations of the midpiece plasma membrane, which on present evidence are present in all other groups shown except *Tarsipes* and the macropodoids, would probably be developed at this point of the tree (see Table 1). Spermatozoa in the branches leading to the phalangerids, burramyids and macropodoids retain these features and develop few, if any, apomorphies. As the tree stands it would be necessary to assume loss of the midpiece plasma membrane invaginations in the macropodoid branch. Clearly, if instead of the unresolved quadrichotomy shown in the tree, the macropodoid lineage diverged earlier at a point preceding the development of these invaginations, such loss would not be required. A likely development in spermatozoa at the base of the branch labelled P would be a more extensive cover of the dorsal nuclear surface by the acrosome, a feature common to all the species in this segment examined so far (see Table 1). In the pseudocheirid branch the development of the various apomorphic features described earlier and discussed below would be necessary, whilst the major modification in the petaurid lineage would be in sperm head shape. Assuming for the present that unexamined spermatozoa of *Acrobates* are similar to those of *Distoechurus*, likely developments in spermatozoa of organisms in the *Tarsipes*/*Acrobates*-*Distoechurus* branches include increased sperm size and development of the similar features of the acrosome in *Tarsipes* and *Distoechurus*, described above. Comparatively rapid sperm evolution would be necessary in the *Tarsipes* branch involving a spectacular size increase and associated structural modifications to produce the distinctive flagellum. These include: flattening of the midpiece, extreme radial displacement of the dense outer fibres, changes in mitochondrial shape to accommodate the spatial requirements of the dense outer fibres; and development of the paired microtubule component of the midpiece fibre network. Also necessary would be loss of the midpiece plasma membrane invaginations. Clearly rapid

evolution of sperm form would be required to produce these features, but if as suggested previously, the changes are functionally interconnected and follow from an increase in sperm size, then such developments are plausible.

Considering the remaining interfamily relationships within the Phalangeroidea and Macropodoidea, the sperm data appear not to be discordant with the sister group relationships suggested by Archer (1984:780) and shown in Fig. 31. Once again, it should be stressed that this interpretation of the sperm data is *primarily* a phenetic one since the phylogenetic analysis of the data is still very rudimentary. The discussion above has demonstrated that some features (chiefly extent of acrosomal coverage) are shared within the pseudocheirid-petaurid-acrobatid group, whilst the phalangerids and burramyids (at least as represented by *Trichosurus* and *Cercartetus*) share very similar sperm types. In turn the macropodoids are readily distinguished from the phalangeroids since they lack the invaginations of the midpiece plasma membrane which are typical of the latter group. Since Kirsch's (1968, 1977) serological studies, the pseudocheirids have been linked most closely with the petaurids and the inclusion of both groups in the family Petauridae has been accepted practice. As mentioned earlier, I have followed the recent views of Archer (1983) and Smith (1984) in separating these groups into two families, and such separation is certainly supported by the sperm data. Whether the sister grouping of the petaurids/acrobatids with the pseudocheirids (see Fig. 31) is supported by sperm criteria, requires further investigation. However, the discussion above of Fig. 32 has shown such grouping to be plausible.

Sperm form in *T. vulpecula* has been used as a comparative reference point throughout this paper since all of the sperm features which recur consistently throughout the Diprotodonta are found in this species. Considering both the head and flagellum, preliminary observations show *Trichosurus* spermatozoa to be most similar to those of *Dactylopsila trivirgata* and *Cercartetus*. Although there is also close overall resemblance to macropod spermatozoa, these groups are readily distinguished by the absence of invaginations of the midpiece plasma membrane in the macropodoids. Further character analysis is required in order to determine the phlyogenetic significance of these similarities. The shared features in spermatozoa of *Trichosurus*, *Dactylopsila*, *Cercartetus* and the macropods *may* simply represent plesiomorphies for the Diprotodonta.

At the intrafamilial level, little can be said at present regarding intraphalangerid relationships on the basis of sperm structure, since apart from *Trichosurus vulpecula* only very poorly preserved material from two *Phalanger* species has been examined.

Likewise, insufficient sperm data are available at present to comment on intra-macropodid and potoroid relationships. It is clear from our observations on sperm ultrastructure in five macropodoid species and Cleland's (cited by Rodger 1978) light microscopy observations on twenty species, that sufficient variation exists in sperm dimensions, head form, and details of intra-flagellar structure, to separate these species. As mentioned above, Cleland concluded that only *Lagorchestes* spermatozoa are sufficiently different to stand apart from the group, and that variations in sperm form in the remaining species do not coincide with currently recognised lineages within the Macropodoidea. Further analysis using ultrastructural data from more macropodoid species, and character analysis based on cladistic methodology is required in order to assess Cleland's conclusions.

Judged by comparative sperm studies involving two of the three petaurid genera (*Petaurus* and *Dactylopsila*) the Petauridae does not appear as a close family group. Preliminary observations show *D. trivirgata* spermatozoa to closely resemble those of the phalangerid *Trichosurus vulpecula*. Although *Petaurus* sperm flagella are likewise very similar to those in *Trichosurus* and hence to *Dactylopsila,* the same is not true of their sperm heads. These have unique features which on available evidence are probably best regarded as apomorphic. Sperm heads of *Petaurus* share a few similarities with those of *Pseudocheirus peregrinus,* namely, the extent of acrosomal coverage of the dorsal nuclear surface and the reduced ventral groove relative to that in other diprotodonts but overall their sperm heads are not very alike. Whether these few similarities are indicative of phylogenetic links is not clear at present but if we regard the distinctive head features in each of these groups as apomorphic and if *Dactylopsila* is correctly placed within the Petauridae then convergence in these features seems more likely.

With regard to *Dactylopsila's* placement, Archer (1983) points out that the monophyly of the Petauridae is in doubt and that the doubt concerns *Dactylopsila,* whilst Baverstock (1984) found MC'F studies of albumin indicated *D. palpator* to be a petaurid, but not especially close to *Petaurus.* Given these doubts regarding the placement of *Dactylopsila* and on the basis of admittedly incomplete sperm data, it seems reasonable to suggest that if *Dactylopsila* is to be grouped with *Petaurus* then it should occupy a more plesiomorphic position in the branching sequence than *Petaurus.* The sperm data providing the basis for this suggestion include: the possession of more features tentatively judged as apomorphic in *Petaurus* spermatozoa (primarily sperm head) than in *Dactylopsila;* and, the likeness of *Dactylopsila* spermatozoa to those of *Trichosurus* and numerous other diprotodontan species, a

likeness resting or shared features which are suggested to be plesiomorphic with respect to the major diprotodont radiation (i.e. excluding vombatiforms and *Tarsipes*). Indeed, this view matches the phylogenetic hypothesis of relationships within the Petauridae, recently proposed by Archer (1983:721).

Within the Pseudocheiridae examined so far, *Pseudocheirus peregrinus* and *P. herbertensis* stand apart from the group and indeed from all phalangeroids by virtue of their distinctive midpieces. These are short and fat relative to overall sperm dimensions with the width of the midpiece resulting from extreme radial development of the mitochondria relative to the dense outer fibre-axoneme complex. The very flattened principal pieces in these species likewise serve to distinguish them from the other diprotodontans in which there is a fair degree of uniformity between species in principal piece structure. *Pseudocheirus peregrinus,* at least, is also distinguished by the incomplete nature of the midpiece fibre network. It seems most likely that these features are synapomorphies for spermatozoa of these species.

Spermatozoa of *P. archeri, P. corinnae* and *P. cupreus* appear very similar to one another and, as discussed above, show a blend of *Trichosurus*-like features and those of *P. peregrinus.*

These observations from sperm structure are certainly not discordant with recent phylogenetic studies of the pseudocheirids based on other character systems. Using microcomplement fixation (MC'F) of albumin, Baverstock (1984) found the genus *Pseudocheirus* to be twice as diverse as the entire dasyurid family. A similar conclusion can be drawn from sperm studies (Harding *et al.* 1982b). Archer (1983:712) considered information from many character systems in producing a cladogram for the Pseudocheiridae. His monophyletic grouping of *P. herbertensis, P. peregrinus, Hemibelidus* and *Petauroides* is supported for the former two species by sperm ultrastructure, but no information is yet available for the latter two species. Likewise his monophyletic grouping of *P. archeri, P. corinnae* and *P. cupreus* is (admittedly on the basis of poorly preserved material) supported by sperm structure. It should be noted, however, that on present evidence, these species appear to have less synapomorphic sperm characters than *P. peregrinus* and *P. herbertensis.*

CONCLUDING REMARKS

Insufficient data are currently available to attempt construction of a diprotodontan phylogeny based on sperm structure alone. Nevertheless, sufficient information is at hand to enable preliminary consideration of the sperm transformations which must be assumed if we accept the evolutionary events proposed in certain recently constructed phylogenetic trees involving Australian marsupials. Even though discussion in this paper has not attempted more than this latter format, any conclusions offered should be regarded as tentative.

At the intrafamilial[1] level the major conclusion is that most diprotodont families show great diversity in sperm form compared with the Dasyuridae and Peramelidae. This is well demonstrated by the Macropodidae which form a fairly cohesive group on the basis of sperm structure, yet Cleland (cited by Rodger 1978) found that the 17 species he examined could be separated on sperm form and dimensions. In contrast to the Macropodidae, the Petauridae does not appear to form a cohesive group in terms of sperm structure. In this case, the sperm data support information emerging from other character systems (see Archer 1983).

At the generic level, diversity in sperm form is most extreme in *Pseudocheirus* and this observation supports similar indications of diversity in this genus from serology (Baverstock 1984 and Baverstock *et al.* 1987) and karyotype analysis (McKay 1984).

At the interfamilial level the most notable observation is that separation of *Tarsipes* and of the phascolarctids and vombatids from the other diprotodonts, is indicated by comparative sperm structure. However, the *phylogenetic* implications of this separation are unclear at present. Perhaps the most fitting conclusion is simply that we should keep an open mind regarding the placement of these groups in marsupial phylogenies including, perhaps, reassessment of the assumed monophyly of the Australian radiation.

A critical question is whether the unusual similar features of dasyurid, *Tarsipes* and peramelid sperm flagella indicate phylogenetic links or convergence. Recent evidence concerning microbiotheriid sperm structure, together with a suggested role for this group in the Australian marsupial radiations, makes convergence the most economic explanation *if* we accept monophyly of the Australian radiation. Indeed, with regard to *Tarsipes* the discussion has shown that we cannot rule out a phylogenetic association between *Tarsipes* and acrobatids on the basis of sperm structure, even though sperm form in the modern species does not make this an obvious association. Clearly, convergence would be the required explanation for *Tarsipes'* unusual flagellar features in this case.

A further major conclusion at the interfamilial level is that, although the remaining diprotodontans (excluding Tarsipedidae and Vombatiformes) show considerable variation in sperm form, there are sufficient similarities for their higher ordinal grouping to be clearly supported by sperm studies. The sperm data also *appear* to be in agreement with the sister group arrangements shown in Archer's (1983:780) recent phylogeny (see Fig. 31) for the Phalangeridae, Potoroidae, Macropodidae, Burramyidae, Pseudocheiridae, Petauridae and *Distoechurus*. This latter conclusion must be regarded as very tentative and as with most aspects of this study its support or otherwise awaits the collection of more data and the application of systematic analysis to the comparative sperm information.

ACKNOWLEDGEMENTS

A number of the species discussed in this paper have been included in previous papers as part of a joint project with Cedric Shorey and Frank Carrick and I thank them for their roles in this work. I also thank Cedric Shorey for his help with the electron microscopy of further species not previously described, and Professor K. W. Cleland for generously allowing me to use the electron microscope and photographic darkroom facilities in the Department of Histology and Embryology at the University of Sydney. Leisa Manning of that Department has provided invaluable technical assistance.

For the species from Papua New Guinea described in a preliminary manner only in this paper, I am grateful to Michael Archer who organised the collecting expedition with funds from the National Geographic Society (grant number 2699-83). The expedition was also sponsored by the South Australian Museum Evolutionary Biology Unit and I thank Peter Baverstock from this unit for his role, and Ken Aplin, Tim Flannery and many others who collected the material in P.N.G. I also thank Ken Aplin for constructively commenting on a draft on the manuscript.

Many people/organisations have provided animals and/or taken samples from them. For this I thank Ken Aplin, Andrew Burbidge, Frank Carrick, Peter Johnston, Gerry Maynes, Meredith Smith, Rick Speare, the Lone Pine Koala Sanctuary and the Zoology Department of the University of Western Australia. Finally, I would like to thank Maureen Kelleher for her efficient typing of the manuscript.

The work which contributed to the initial stages of this study was supported by an A.R.G.C. grant (number D7715525).

REFERENCES

APLIN, K. AND ARCHER, M., 1984. Interrelationships of the diprotodontan families — the basicranial evidence and a synergistic view. *In* "possums and opossums: studies in evolution". *Abstracts* p.1.

---

[1] Using the family divisions outlined in Strahan (1983) except that, following Archer (1983) and Smith (1984), the Pseudocheirinae and Petaurinae are given separate family status.

APLIN, K. AND ARCHER, M., 1987. Recent advances in marsupial systematics with a new syncretic classification. Pp. xv-lxxii *in* "Possums and opossums: studies in evolution" ed by M. Archer. Surrey Beatty & Sons and the Royal Zoological Society of New South Wales: Sydney.

ARCHER, M., 1984. The Australian marsupial radiation. Pp. 633-808 *in* "Vertebrate zoogeography and evolution in Australasia" ed by M. Archer and G. Clayton. Hesperian Press: Perth.

BAVERSTOCK, P. R., 1984. The molecular relationships of Australasian possums and gliders. Pp. 1-8 *in* "Possums and gliders" ed by A. P. Smith and I. D. Hume. Surrey Beatty & Sons and the Australian Mammal Society: Sydney.

BAVERSTOCK, P. R., BIRRELL, J. AND KRIEG, M., 1987. Albumin immunologic relationships among Australian possums: a progress report. Pp. 229-34 *in* "Possums and opossums: studies in evolution" ed by M. Archer. Surrey Beatty & Sons and the Royal Zoological Society of New South Wales: Sydney.

CUMMINS, J. M., 1976. Epididymal maturation of spermatozoa in the marsupial, *Trichosurus vulpecula:* changes in motility and gross morphology. *Aust. J. Zool.* **24:** 499-512.

CUMMINS, J. M., RENFREE, M. B. AND TEMPLE-SMITH, P. D., 1980. Sperm maturation and ultrastructure in the Honey Possum, *Tarsipes spencerae. Aust. Mamm. Soc. Bull.* **6**(2): 30-31.

CUMMINS, J. M., TEMPLE-SMITH, P. D. AND RENFREE, M. B., 1986. Reproduction in the male Honey Possum *(Tarsipes rostratus:* Marsupialia): the epididymis. *Am. J. Anat.* **177:** 385-401.

CUMMINS, J. M. AND WODDALL, P. F., 1985. On mammalian sperm dimensions. *J. Reprod. Fert.* **75:** 153-175.

FAWCETT, D. W., 1970. A comparative view of sperm ultrastructure. *Biol. Reprod.* **2,** suppl. **2:** 90-127.

HARDING, H. R., 1977. "Reproduction in male marsupials: a critique, with additional observations on sperm development and structure". Ph.D. Thesis: University of New South Wales: Sydney.

HARDING, H. R., 1979. Affinities of the Koala on the basis of sperm ultrastructure. *Aust. Mamm. Soc. Bull.* **6:** 22-23.

HARDING, H. R., CARRICK, F. N. AND SHOREY, C. D., 1975. Ultrastructural changes in spermatozoa of the Brush-tailed Possum, *Trichosurus vulpecula* (Marsupialia), during epididymal transit. Part I: the flagellum. *Cell Tiss. Res.* **164:** 11-32.

HARDING, H. R., CARRICK, F. N. AND SHOREY, C. D., 1976. Ultrastructural changes in spermatozoa of the Brush-tailed Possum, *Trichosurus vulpecula* (Marsupialia), during epididymal transit. Part II: the acrosome. *Cell Tiss. Res.* **171:** 61-73.

HARDING, H. R., CARRICK, F. N. AND SHOREY, C. D., 1977. Spermatozoa of Australian marsupials: ultrastructure and epididymal development. Pp. 151-2 *in* "Reproduction and evolution: fourth international symposium on comparative biology of reproduction" ed by J. H. Calaby and C. H. Tyndale-Biscoe. Australian Academy of Science: Canberra.

HARDING, H. R., CARRICK, F. N. AND SHOREY, C. D., 1979. Special features of sperm structure and function in marsupials. Pp. 289-303 *in* "The spermatozoon" ed by D. W. Fawcett and J. M. Bedford. Urban and Schwarzenberg Inc.: Baltimore-Munich.

HARDING, H. R., CARRICK, F. N. AND SHOREY, C. D., 1980. Affinities of the Honey Possum, *Tarsipes spencerae,* on the basis of sperm ultrastructure: preliminary observations. *Aust. Mamm. Soc. Bull.* **6**(2): 37-38.

HARDING, H. R., CARRICK, F. N. AND SHOREY, C. D., 1981. Marsupial phylogeny: new indications from sperm ultrastructure and development in *Tarsipes spenserae? Search* **12:** 45-47.

HARDING, H. R., CARRICK, F. N. AND SHOREY, C. D., 1982a. Crystalloid inclusions in the Sertoli cell of the Koala, *Phascolarctos cinereus* (Marsupialia). *Cell Tiss. Res.* **221:** 633-42.

HARDING, H. R., CARRICK, F. N. AND SHOREY, C. D., 1983. Acrosome development during spermiogenesis and epididymal sperm maturation in Australian marsupials. Pp. 411-4 *in* "The sperm cell" ed by J. André. Procs. 4th Internatl. sympos. on spermatology, Seillac, France, 1982. Martinus Nijhoff Publishers: The Hague.

HARDING, H. R., CARRICK, F. N. AND SHOREY, C. D., 1984. Sperm ultrastructure and development in the Honey Possum, *Tarsipes rostratus.* Pp. 451-61 *in* "Possums and gliders" ed by A. P. Smith and I. D. Hume. Surrey Beatty & Sons and the Australian Mammal Society: Sydney.

HARDING, H. R., CARRICK, F. N. AND SHOREY, C. D., 1987. The affinities of the Koala, *Phascolarctos cinereus* (Marsupialia: Phascolarctidae), on the basis of sperm ultrastructure and development. Pp. 353-64 *in* "Possums and opossums: studies in evolution" ed by M. Archer. Surrey Beatty & Sons and the Royal Zoological Society of New South Wales: Sydney.

HARDING, H. R., WOOLLEY, P. A., SHOREY, C. D. AND CARRICK, F. N., 1982b. Sperm ultrastructure, spermiogenesis and epididymal sperm maturation in dasyurid marsupials: phylogenetic implications. Pp. 659-73 *in* "Carnivorous marsupials" ed by M. Archer. Roy. Zool. Soc. New South Wales: Sydney.

HENNIG, W., 1979. "Phylogenetic systematics". 2nd ed. University of Illinois Press: Urbana.

HUGHES, R. L., 1964. Sexual development and spermatozoon morphology in the male macropod marsupial *Potorous tridactylus* (Kerr). *Aust. J. Zool.* **12:** 42-51.

HUGHES, R. L., 1965. Comparative morphology of spermatozoa from five marsupial families. *Aust. J. Zool.* **13:** 533-43.

HUGHES, R. L., 1977. Light and electron microscope studies on the spermatozoa of the Koala, *Phascolarctos cinereus* (Marsupialia). *J. Anat.* **124:** 513.

KIRSCH, J. A. W., 1968. Prodromus of the comparative serology of Marsupialia. *Nature* **217:** 418-20.

KIRSCH, J. A. W., 1977. The comparative serology of Marsupialia, and a classification of marsupials. *Aust. J. Zool. Suppl. Ser.* No. **52:** 1-152.

KIRSCH, J. A. W. AND REIG, O., 1984. The ordinal-level relationships of marsupials: inconsistencies in the evidence for competing cladograms. *In* "Possums and opossums: studies in evolution". *Abstracts* p.5.

McKAY, G. M., 1984. Cytogenetic relationships of possums and gliders. Pp. 9-14 *in* "Possums and gliders" ed by A. P. Smith and I. D. Hume. Surrey Beatty & Sons and the Australian Mammal Society: Sydney.

OLSON, G., 1975. Observations on the ultrastructure of a fibre network in the flagellum of sperm of the Brush-tailed Phalanger, *Trichosurus vulpecula. J. Ultrastruct. Res.* **50:** 193-98.

RETZIUS, G., 1906. Die Spermien der Marsupialier. *Biol. Untersuch.* N.S. **13:** 75-86.

RODGER, J. C., 1978. Male reproduction: its usefulness in discussions of Macropodidae evolution. *Aust. Mammal.* **2:** 73-80.

SHARMAN, G. B., 1982. Karyotypic similarities between *Dromiciops australis* (Microbiotheriidae, Marsupialia) and some Australian marsupials. Pp. 711-14 *in* "Carnivorous marsupials" ed by M. Archer. Roy. Zool. Soc. New South Wales: Sydney.

SMITH, A., 1984. The species of living possums and gliders. Pp. xii-xv *in* "Possums and gliders" ed by A. P. Smith and I. D. Hume. Surrey Beatty & Sons and the Australian Mammal Society: Sydney.

STRAHAN, R. (ed), 1983. "The complete book of Australian mammals". Angus and Robertson: Sydney.

SZALAY, F. S., 1982. A new appraisal of marsupial phylogeny and classification. Pp. 621-40 *in* "Carnivorous marsupials" ed by M. Archer. Roy. Zool. Soc. New South Wales: Sydney.

TEMPLE-SMITH, P. D., 1984. *Dromiciops australis* — a South American connection. *Aust. Mammal. Soc. Bull.* **8**(2): 77.

TEMPLE-SMITH, P. D., 1987. Sperm structure and marsupial phylogeny. Pp. 171-93 *in* "Possums and opossums: studies in evolution" ed by M. Archer. Surrey Beatty & Sons and the Royal Zoological Society of New South Wales: Sydney.

TEMPLE-SMITH, P. AND BEDFORD, J. M., 1976. The features of sperm maturation in the epididymis of a marsupial, the Brushtailed Possum, *Trichosurus vulpecula. Amer. J. Anat.* **147**: 471-500.

WOODBURNE, M. O., 1984. Families of marsupials: relationships, evolution and biogeography. *In* "Mammals: notes for a short course" ed by T. W. Broadhead. Department of Geological Science, Studies in Geology 8: University of Tennessee.

THE SEMINIFEROUS TUBULES, RETE TESTIS AND EFFERENT DUCTS IN DIDELPHID, CAENOLESTID AND MICROBIOTHERIID MARSUPIALS

P. A. WOOLLEY[1]

The anatomy of the seminiferous tubules, rete and efferent ducts of the testis of American opossums of the families Didelphidae, Caenolestidae and Microbiotheriidae has been investigated and comparisons made with these structures in Australian marsupials of the families Dasyuridae, Thylacinidae, Myrmecobiidae, Peramelidae, Tarsipedidae, Macropodidae and Phalangeridae. In the general organisation of the testis, diameter of the seminiferous tubules and form of the rete the didelphids and caenolestids are most similar to the first five of the Australian families. The seminiferous tubules are larger than in all other mammals investigated, and the rete is a simple duct. The rete in the microbiothere *Dromiciops australis* is a more complex structure and is different from that of all other marsupials (American and Australian) previously studied. The urinogenital connection appears to occur outside the body of the epididymis in American opossums, in contrast to Australian marsupials belonging to the first five families listed above, in which it appears to occur within the epididymis.

Key Words: Anatomy; Seminiferous Tubules; Rete Testis; Efferent Ducts; American Marsupials.

Pages 217-27 in POSSUMS AND OPOSSUMS: STUDIES IN EVOLUTION ed by M. Archer. Surrey Beatty & Sons and the Royal Zoological Society of New South Wales: Sydney, 1987.

INTRODUCTION

IN EUTHERIAN mammals the seminiferous tubules open via the tubuli recti into a chamber or into an interconnecting tubular system, the rete testis, which is drained by efferent ducts that connect with the epididymal duct. The rete is lined with nonciliated squamous to cuboidal cells and the efferent ducts (vasa efferentia) contain ciliated columnar to pseudostratified columnar epithelial cells (Gier and Marion 1970). The rete, which may be surrounded by fibrous connective tissue (the mediastinum) is variously situated within the testis in different species. Setchell (1970) recognises three basic patterns exemplified by the rat (rete laterally placed, very little mediastinum), man (rete laterally placed, encased in a well developed mediastinum) and ram (rete centrally placed, very little mediastinum). In the rat the rete penetrates the tunica albuginea close to the anterior pole of the testis and forms a small cap-like extratesticular rete which connects with from three to nine efferent ducts about 2 to 3 mm long. In man and the ram there is no extratesticular rete and the efferent ducts, which are much shorter than in the rat, connect with the anterior end of the rete.

In marsupials, as in eutherians, the seminiferous tubules open via tubuli recti into a rete but the arrangement of the rete within the testis is different. Two patterns have been recognised. In one the rete is horseshoe-shaped, lying under the tunica albuginea beneath the major testicular blood vessels on the epididymal and free margins of the testis (Setchell 1970); in the other the rete consists of a simple, sometimes branched, duct system lying deep within the anterior pole of the testis and close to the major testicular artery (Woolley 1975; Rodger 1982). A horseshoe-shaped rete is found in members of the Australian families Phalangeridae and Macropodidae, and a simple duct system in the Australian families Dasyuridae, Thylacinidae, Myrmecobiidae, Peramelidae and Tarsipedidae and the American families Caenolestidae and Didelphidae. Multiple efferent ducts are associated with the horseshoe-shaped rete system. In the simple duct system the arrangement of the duct(s) connecting the testis to the epididymis is variable. The rete penetrates the tunica albuginea and emerges from the anterior end of the testis. It then either continues as a large diameter, thin walled duct, considered to be of rete origin, suspended in the mesorchium to enter the caput epididymis (in the Australian forms) or, so far as is known, it connects with one or more efferent ducts in a thickened, segmented structure connecting the testis with the epididymis (in the American forms). In both cases the length of the connecting duct(s) is considerably greater than in most eutherians and, as a consequence, the epididymis can be more widely separated from the testis. The marsupial species in which the arrangement of the rete and excurrent ducts have previously been investigated are listed in Table 1.

[1]Department of Zoology, La Trobe University, Bundoora, Victoria, Australia 3083.

Table 1. Studies on the rete and excurrent ducts of marsupials. (Species in parentheses studied by author in parentheses).

PHALANGERIDAE
    Trichosurus vulpecula   Fraser 1919, Setchell 1977

MACROPODIDAE
    Macropus eugenii   Setchell 1970
    M. rufogriseus (as *Halmaturus benetti*)   Setchell 1970, (de Burlet 1921)
    M. giganteus   Setchell 1970
    M. rufa (as *Megaleia rufa*)   Setchell 1970

DASYURIDAE
    Antechinomys spenceri   Woolley 1975
    Sminthopsis crassicaudata   Woolley 1975
    S. leucopus   Woolley 1975
    Antechinus stuartii   Woolley 1975
    Dasycercus cristicauda   Woolley 1975
    Dasyuroides byrnei   Woolley 1975
    Sarcophilus harrisii   Woolley 1975
    Satanellus hallucatus   Woolley unpublished
    Dasyurus viverrinus   Fraser 1919, Ullman 1984

THYLACINIDAE
    Thylacinus cynocephalus   Woolley unpublished

MYRMECOBIIDAE
    Myrmecobius fasciatus   Woolley unpublished

PERAMELIDAE
    Perameles sp.   Fraser 1919
    Isoodon obesulus (as *Perameles obesula*)   (de Burlet 1921)
    P. gunnii   Woolley unpublished

TARSIPEDIDAE
    Tarsipes rostratus   Woolley and Scarlett 1984

CAENOLESTIDAE
    Caenolestes obscurus   Rodger 1982

DIDELPHIDAE
    Didelphis virginiana   van den Broek 1906, de Burlet 1921, Burns
                              1941, Ladman 1967, Martan *et al.* 1967
    D. albiventris (as *D. azarae*)   Rodger 1982, (Nogueira *et al.* 1977)
    D. marsupialis (as *D. aurita*)   Rodger 1982, (Fraser 1919)
    Philander opossum (as *Metachirus opossum*) Rodger 1982, Martan *et al.* 1967, (Fraser
                              1919)
    Marmosa impavida (as *M. murina*)   (Rodger 1982)
    Marmosa sp.   Martan *et al.* 1967
    Caluromys derbianus   Martan *et al.* 1967

In the Australian families in which the rete is a simple duct the seminiferous tubules are larger in diameter than those of other mammals studied in this respect (see Woolley 1975 and Table 24.4 Setchell 1977) and, in dasyurids (Woolley 1975) and a peramelid (de Burlet 1921) they are known to be simple in form and few in number, with usually two openings into the rete. The paucity of comparable information for members of the American families Caenolestidae and Didelphidae, and for the Microbiotheriidae in which previously not even the form of the rete was known, led to the present study. Observations have been made on the diameter of the seminiferous tubules, the number of tubuli recti, the form of the rete and the arrangement of the excurrent ducts between the testis and epididymis in six species representing the three families. Comparisons are made between Australian and American marsupials.

MATERIALS AND METHODS

The species of American marsupials examined and the collection numbers of the individuals studied are as follows:- Didelphidae — *Didelphis albiventris* 575 (CM 12932) and 631 (CM 12933), *Philander opossum* 668 (CM 12950), *Monodelphis brevicaudata* 590, *Marmosa impavida* 632 (CM 12934); Caenolestidae — *Caenolestes obscurus* 521 (CM 12954); Microbiotheriidae — *Dromiciops australis* IEEUA 1092. The didelphids and caenolestid were collected in Colombia and Venezuela by Dr C. H. Tyndale-Biscoe and information relating to them can be found in Tyndale-Biscoe (1980). The specimens are lodged with the Australian National Mammal Collection in Canberra. Body weight and testis width and length was measured by the collector. One preserved testis and attached epididymis from each of these animals was examined. Serial 10μm sections stained with haematoxylin and eosin were prepared from the anterior half of each testis and all or a portion of the attached first segment of the duct(s) leading from it, and from the entire excurrent duct system, together with the anterior half of the epididymis of *D. albiventris* 631 and *P. opossum* 668. Material from some of the same animals was examined by Rodger (1982). The *Dromiciops* was collected at Tres Bocas, Valdivia

Province, Chile and the specimen is in the collection of the Instituto de Ecologia y Evolucion, Universidad Austral de Chile. Dr M. Gallardo fixed the scrotum entire in aqueous Bouin's solution. The right testis and attached epididymis and the anterior half of the left were serially sectioned at 8μm and stained as above.

A drawing mirror attached to a microscope was used to project the sections of each testis to make outline drawings of the rete and the connections of the seminiferous tubules to it. A plan diagram of the form of the rete was prepared from these drawings.

The diameter of the seminiferous tubules in histological sections of testes from five specimens in which spermatogenesis was taking place was measured using a microscope with a measuring eyepiece calibrated with a stage micrometer. Only tubules containing spermatozoa as well as other stages of the spermatogenic cycle were measured. The nature of the excurrent ducts lying between the testis and epididymis was analysed visually from the serial sections.

RESULTS

Testis and Epididymis

The form of the testis and epididymis and connecting ducts in the caenolestid and didelphids examined have been illustrated and described by Rodger (1982). Figure 1 shows the structures in *Dromiciops*. In all species examined the duct(s)

connecting the testis with epididymis lie in a thick band of tissue divisible into two segments. The testicular blood vessels were closely associated with the first segment which joined the rete testis. The second segment, which was generally broader than the first, joined the caput epididymis.

In *Dromiciops* there was no obvious pigmentation of the scrotal skin but both the parietal and visceral layers of the tunica vaginalis contained black pigment.

The internal organisation of the testis (tubules and intertubular tissue) of *Dromiciops* was found to be similar to that of the other species examined. That of the didelphids and caenolestid has previously been described (Rodger 1982). The seminiferous tubules of *Dromiciops* appeared to be highly coiled and the intertubular spaces filled largely with Leydig cells. No lobulation was apparent. The animal was judged to be sexually immature because most of the tubules were without lumina and spermatogenesis had not proceeded beyond the stage of spermatocytes. The *M. brevicaudata* male also was considered to be immature; most of the tubules were open but spermatogenesis was incomplete. The diameter of the seminiferous tubules in the other four species, in which all stages of spermatogenesis were present in the tubules, ranged from 0.17 mm in *C. obscurus* to 0.27 mm in *D. albiventris* (Table 2). Testis dimensions correlated roughly with body weight, the smaller species having smaller testes.

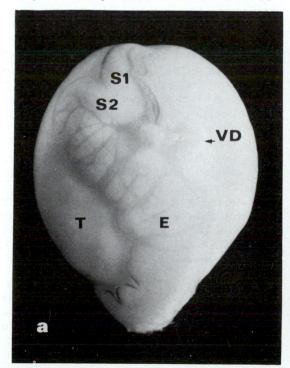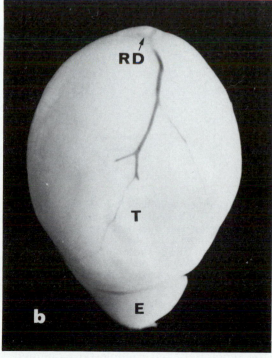

Fig. 1. Testis, efferent duct system and epididymis of *Dromiciops.* a) dorsal aspect and b) ventral aspect. T = testis; E = epididymis; S1 = first segment of efferent duct system; S2 = second segment of efferent duct system; RD = extratesticular rete duct; VD = vas deferens.

Table 2. Data on seminiferous tubules and rete testis in American marsupials. Asterisk denotes sexually immature specimen. Figures in parentheses are the numbers of tubuli recti entering each branch.

Species	No.	Body weight (g)	Testis (mm)	Diameter of tubules (mm)	Form of rete	Depth of penetration (mm)	Length of 'main' duct (mm)	Number of tubuli recti
DIDELPHIDAE								
Didelphis albiventris	575	1275	17.2 × 13.7	0.27	simple duct (branched)	4.30	2.50	32 (16, 16)
Didelphis albiventris	631	1450	16.5 × 14.0	0.25	simple duct	6.14	3.00	24
Philander opossum	668	675	13.6 × 11.2	0.24	simple duct	2.15	0.95	22
Monodelphis brevicaudata	590*	34	6.4 × 5.0	–	simple duct	0.50	0.45	6
Marmosa impavida	632	61	8.5 × 6.6	0.23	simple duct	1.70	1.20	6
CAENOLESTIDAE								
Caenolestes obscurus	521	34	7.4 × 6.0	0.18	simple duct	1.20	0.50	10
MICROBIOTHERIIDAE								
Dromiciops australis	1092*	45	R 8.0 × 6.0	–	network of inter-connecting	R 0.64	0.21	36 (17, 19)
			L 8.2 × 6.0	–	channels (branched)	L 0.58	0.26	41 (18, 23)

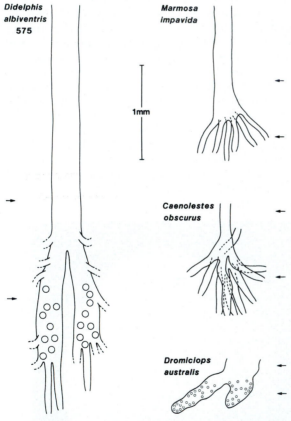

Fig. 2. Plan diagrams of the form of the testicular rete in *D. albiventris* 575, *M. impavida*, *C. obscurus* and *Dromiciops*. In *D. albiventris* and *Dromiciops* some or all of the positions of entry of the tubuli recti into the rete are shown by open circles. The arrangement of the rete and tubuli recti in *D. albiventris* 631 and *P. opossum* was like that in *D. albiventris* 575, except that there was no branching of the rete duct, and in *M. brevicaudata*, like *M. impavida*.

Form of the Rete and Number of Tubuli Recti

In the didelphids and caenolestid examined the observation, by Rodger (1982), that the rete is a simple duct system lying within the testicular tissue in the anterior pole of the testis was confirmed. A single duct was found in all except one *D. albiventris* (see Table 2 and Fig. 2), in which the main duct bifurcated. No distinct mediastinum was found in any species. Close to the hilus of the testis a layer of loose connective tissue surrounded the duct but deeper within the testis the layer was less distinct. In *C. obscurus* (contrary to the finding of Rodger 1982) and all other species testicular blood vessels entered the testis at the hilus and a large artery followed the course of the main rete duct, giving off branches in its passage through the intertubular tissues. Ductules, the tubuli recti, connected the seminiferous tubules to the duct(s); in *C. obscurus* (Fig. 2) the tubuli recti were relatively broad. The number of tubuli recti ranged from six to 32, with the larger species having the greatest numbers (Table 2). The form of the rete in various species is shown in Figure 2. The cells of the simple epithelium lining the rete had characteristically dark staining nuclei and they varied between species from squamous to low columnar cells in the tubuli recti to cuboidal to columnar cells in the duct(s) (Figs 3, 4 and 5). Some increase in the height of the epithelium towards the hilus was observed. The epithelium, in *D. albiventris*, had a folded appearance and in *P. opossum* there were villus-like projections of the wall into the lumen of the duct. A single rete duct penetrated the tunica albuginea at the anterior end of the testis in each species and entered the first segment of the duct system connecting the testis with the epididymis (see below).

In the microbiotheriid, *Dromiciops*, the rete (Fig. 6) consisted of a network of interconnected channels lying among the seminiferous tubules in the anterior pole of the testis. It was encased in a thin but distinct mediastinum formed as an inturning of the fibrous tissue of the tunica albuginea. After penetrating the testicular tissue for a short distance the rete branched into two, one branch being a little shorter than the other.

The testicular blood vessels entered the testis at the hilus alongside the rete and the major artery followed the course of the undivided rete and the shorter of the two branches. The tubuli recti (36 in the right testis, 41 in the left) opened into the branches of the rete, with slightly fewer opening into the short branch in each testis (Table 2). Low columnar cells lined the tubuli recti and rete channels.

The depth of penetration of the rete into the testis (from hilus to union of most distal tubulus rectus with a seminiferous tubule) and the length of the 'main' rete duct or channel (from hilus to point of entry of most proximal tubulus rectus or formation of branches) in each testis sectioned can be found in Table 2.

Duct System Between Testis and Epididymis

After leaving the testis the rete duct shortly entered the first segment of the excurrent duct system and a change in the epithelium lining the duct was seen in most specimens. In *D. albiventris* (Fig. 7a), the duct followed a straight course for approximately one-third of the length of the first segment of the system before branching and/or coiling. In one specimen it appeared to form three branches and in the other five, one of which later anastomosed. Throughout the remainder of the first segment in both specimens the ducts were coiled and in individual sections from four to eight sections through ducts were seen. The ducts were highly convoluted in their course through the second segment of the duct system. In both segments the ducts were lined with a columnar epithelium of irregular height and some of the cells were ciliated.

In *P. opossum* (Fig. 7b) the duct followed a straight course for more than half the length of the first segment. It then bifurcated to form ducts of unequal diameter which also followed a straight course to the second segment of the duct system where they became highly convoluted. The smaller of the two ducts was without a lumen in the second segment. Throughout the first and second segments the ducts were lined with ciliated low columnar epithelium.

No change in the rete duct of *M. brevicaudata* was seen in the very short extratesticular portion (less than 0.5 mm) sectioned.

The duct in *M. impavida* became coiled about halfway along the length of the first segment. It was lined with a ciliated columnar epithelium and the general appearance of the duct was similar to that seen in *D. albiventris*.

In *C. obscurus* (Fig. 8) the rete penetrated the tunica albuginea as a flattened duct which almost immediately coiled and there was a rapid transition from a dark staining low columnar epithelium to a paler staining ciliated cuboidal

epithelium lining the duct in the portion (about 0.5 mm) of the first segment sectioned.

The network of channels in the rete of *Dromiciops* (Fig. 9) gradually coalesced in the passage of the rete through the tunic to form a single duct which, immediately upon emergence, joined with one or more ducts lined with ciliated columnar cells. The duct(s) became increasingly convoluted, and appeared to branch in their passage through the first segment of the duct system. There was no obvious difference between the terminal portion of the first segment and the second segment, which was demarcated only by the flexure in the system.

DISCUSSION

The gross morphology of the seminiferous tubules could not be directly investigated because fresh material was not available for dissection. Whether the tubules are in the form of simple loops mostly with only two openings into the rete, as in dasyurid marsupials (Woolley 1975) or are branched and interconnected with more than two openings into the rete, as in the mouse (Hirota 1952) is not known. However, the low number of tubuli recti found in *M. brevicaudata*, *M. impavida* and *C. obscurus* suggests that, at least in these species, the form is simple.

The diameter of the seminiferous tubules in the sexually mature didelphids and the caenolestid examined ranged from 0.17 to 0.27 mm. This range is similar to that found in a large number of eutherians (0.1834 to 0.2654 mm) and to phalangerid (0.226 to 0.288 mm) and macropodid (0.18 to 0.32 mm) marsupials (see Woolley 1975), but lower than that found in dasyurids (0.36 to 0.52 mm, Woolley 1975), peramelids (0.39 mm, Benda 1906; 0.355 mm, Setchell 1977), the thylacinid *Thylacinus cynocephalus* (0.43 mm, Woolley unpublished) and the tarsipedid, *Tarsipes rostratus* (0.35 mm, Woolley and Scarlett 1984).

In didelphids the number of tubuli recti ranged from six to 32 with the smaller species having the lowest numbers. A similar relationship was found in dasyurids (Woolley 1975) in which the number of tubuli recti ranged from two to 10 in species with body weights ranging from 17 to 163 g. The higher number of 10 was also found in a much larger species, *Sarcophilus harrisii* (body weight 6850 g). More recently *Satanellus hallucatus*, an intermediate sized dasyurid (body weight 809 g), has been found to have 22 tubuli recti (Woolley unpublished), providing further support for the general relationship between body size and number of tubuli recti in didelphid and dasyurid marsupials.

In the microbiotheriid, *Dromiciops*, the number of tubuli recti (36, 41) is greater than the number found in didelphid (six to 32), caenolestid (10),

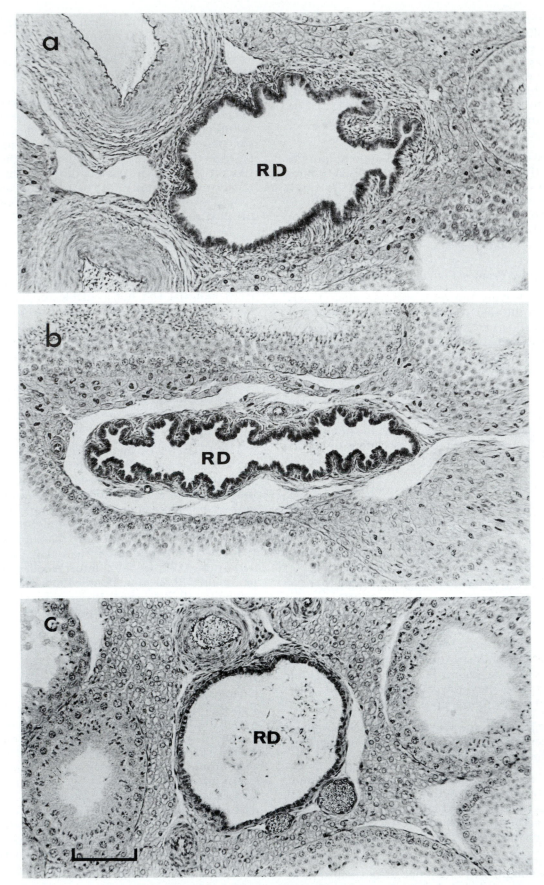

Fig. 3. Sections through the rete duct of a) *D. albiventris*, b) *P. opossum* and c) *M. impavida* (level of sections for *D. albiventris* and *M. impavida* shown in Fig. 2). RD = rete duct; scale line = 0.1 mm (a, b and c at same magnification).

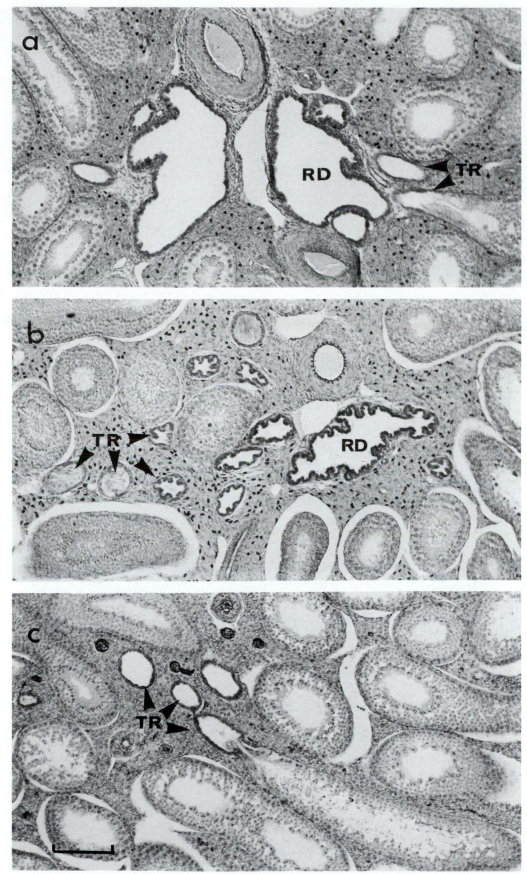

Fig. 4. Sections through the rete and tubuli recti of a) *D. albiventris*, b) *P. opossum* and c) *M. impavida* (level of sections for *D. albiventris* and *M. impavida* shown in Fig. 2). RD = rete duct; TR = tubuli recti; scale line = 0.2 mm (a, b and c at same magnification).

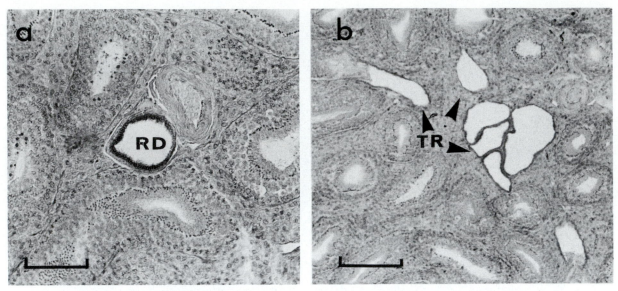

Fig. 5. Sections through a) the rete duct and b) the tubuli recti of *C. obscurus* (level of sections shown in Fig. 2). RD = rete duct; scale
     line a) = 0.1 mm b) = 0.2 mm.

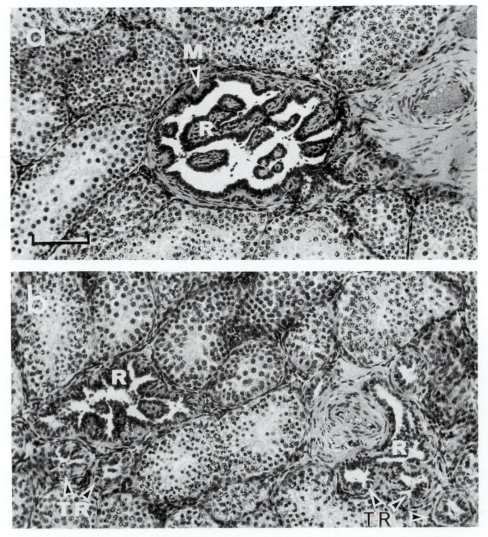

Fig. 6. Sections through a) the main channel and b) the two branches of the rete in *Dromiciops*
     (level of sections shown in Fig. 2). R = rete; M = mediastinum; TR = tubuli recti; scale
     line = 0.1 mm (a and b at same magnification).

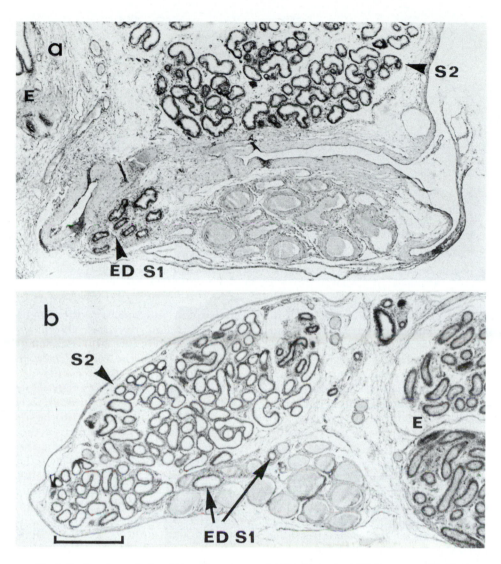

Fig. 7. Section through the efferent duct system of a) *D. albiventris* and b) *P. opossum*. ED S1 = efferent ducts in segment 1; S2 = segment 2; E = epididymis; scale line = 0.5 mm. (a and b at same magnification).

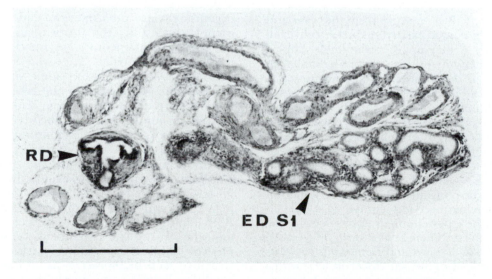

Fig. 8. Section through the extratesticular rete duct and part of first segment of the efferent duct system of *C. obscurus*. RD = extratesticular rete duct; ED S1 = efferent ducts in segment 1; scale line = 0.5 mm.

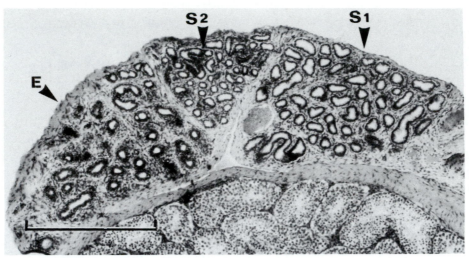

Fig. 9. Section through the first and second segments of the efferent duct system of *Dromiciops.* S1 = segment 1; S2 = segment 2; E = epididymis; scale line = 0.5 mm.

dasyurid (two to 22), tarsipedid (eight to 10 in *Tarsipes*, Woolley and Scarlett 1984) and peramelid (15 in *Isoodon obesulus*, de Burlet 1921) marsupials. No information is available on the number of tubuli recti in other groups of marsupials but in the macropodid, *Macropus rufogriseus* it seems likely that it is much higher because it has been estimated that there are between 200 and 300 seminiferous tubules in the testis (de Burlet 1921).

The rete testis of *Dromiciops* differs from that of all other marsupials. In most families it is a simple duct, the known exceptions being the Macropodidae and Phalangeridae. Although the location within the testis in *Dromiciops* is the same as in didelphid, caenolestid, dasyurid, tarsipedid, peramelid, thylacinid and myrmecobiid marsupials its structure is different, and in addition it is encased in a mediastinum.

A single rete duct emerges from the testis and enters the segmented excurrent duct system in all the American marsupials examined. Superficially the anatomy of this system was similar in all species but differences were seen internally. The rete duct joined with a duct or ducts in which the epithelium was different shortly after entering the system. The excurrent duct(s) were single or branched, straight or coiled in their passage through the first segment but in all species they were highly convoluted in the second segment. In *P. opossum* one of the two ducts found appeared to be nonfunctional, because it did not have a lumen in the second segment. The ciliated nature of the epithelium in the ducts in the excurrent duct system in *D. albiventris*, *P. opossum*, *M. impavida*, *C. obscurus* and *D. australis* suggests that in these species the rete duct connects with a duct or ducts of mesonephric origin (efferent ducts). If so, the urinogenital connection occurs outside the epididymis in these American

marsupials, in contrast to the situation found in the Australian dasyurid (Woolley 1975) and tarsipedid (Woolley and Scarlett 1984) marsupials in which the rete duct enters the caput epididymis. (The single excurrent duct of *T. rostratus* was referred to as an efferent duct by Woolley and Scarlett but it was in fact considered, on histological appearance, to be a duct of rete origin). Although the nature of the epithelium lining the single duct between the testis and epididymis in peramelid, thylacinid and myrmecobiid marsupials has not been investigated it seems likely, on the basis of gross anatomy, that the urinogenital connection occurs within the epididymis in these groups also. Ontogenetic studies are required to verify the conclusions drawn on the site of the urinogenital connection in all groups.

It can be seen from the above that the groups with the greatest similarity in testis and rete organisation are the American didelphids and caenolestids and the Australian dasyurid, peramelid, tarsipedid, myrmecobiid and thylacinid marsupials. The American microbiotheriid *Dromiciops* shows some similarity (in the location of the rete in the testis) to these families but is distinct from them, and from other Australian families, in the structure of the rete and in the possession of a mediastinum.

The distinctness of *Dromiciops* from other American opossums is also evident from studies on serology (Kirsch 1977), chromosomes (Sharman 1982) and pedal morphology (Szalay 1982). Both Sharman and Szalay consider *Dromiciops* and the Australian marsupials to be monophyletic. While it is possible to envisage the rete structure seen in *Dromiciops* as an intermediate condition between the simple duct system and the horseshoe-shaped system too little is known of the structure of the rete in Australian marsupials, other than those referred to above, to warrant further speculation.

ACKNOWLEDGEMENTS

I am indebted to Dr C. H. Tyndale-Biscoe and Dr M. Gallardo who collected and made available the material upon which this paper is based. Assistance from Jennifer Cheney, especially in the tedious drawing of serial sections, and from David Walsh for photomicrography is gratefully acknowledged.

REFERENCES

BENDA, C., 1906. Die spermiogenese der marsupialier. *Denkschr. Med. Naturw. Ges. Jena.* **6:** 441-58.

BURNS, R. K., 1941. The origin of the rete apparatus in the opossum. *Science* **94:** 142-44.

DE BURLET, H. M., 1921. Zur Entwicklung und Morphologie des Saugerhodens. II. Marsupialer. *Z. Anat. EntwGesch.* **61:** 19-31.

FRASER, E. A., 1919. The development of the urogenital system in the Marsupialia, with special reference to *Trichosurus vulpecula*. Part II. *J. Anat.* **53:** 97-129.

GIER, H. T. AND MARION, G. B., 1970. Development of the mammalian testis. Pp. 1-45 *in* "The Testis" ed by A. D. Johnson, W. R. Gomes and N. L. VanDemark. Academic Press: New York.

HIROTA, S., 1952. The morphology of the seminiferous tubules. 1. The seminiferous tubules of the mouse. *Kyushu Mem. med. Sci.* **3:** 121-28.

KIRSCH, J. A. W., 1977. The comparative serology of Marsupialia, and a classification of marsupials. *Aust. J. Zool.* Suppl. Ser., No. **52:** 1-152.

LADMAN, A. J., 1967. Fine structure of the ductuli efferentes in the opossum. *Anat. Rec.* **157:** 559-76.

MARTAN, J., HRUBAN, Z. AND SLESERS, A., 1967. Cytological studies on the ductuli efferentes of the opossum. *J. Morph.* **121:** 81-102.

NOGUEIRA, J. C., GODINHO, H. P. AND CARDOSO, F. M., 1977. Microscopic anatomy of the scrotum, testis with its excurrent duct system and spermatic cord of *Didelphis azarae*. *Acta Anat.* **99:** 209-19.

RODGER, J. C., 1982. The testis and its excurrent ducts in American caenolestid and didelphid marsupials. *Am. J. Anat.* **163:** 269-82.

SETCHELL, B. P., 1970. Testicular blood supply, lymphatic drainage, and secretion of fluid. Pp. 101-239 *in* "The Testis" ed by A. D. Johnson, W. R. Gomes and N. L. VanDemark. Academic Press: New York.

SETCHELL, B. P., 1977. Reproduction in male marsupials. Pp. 411-57 *in* "The Biology of Marsupials" ed by B. Stonehouse and D. Gilmore. The Macmillan Press Ltd. London and Basingstoke.

SHARMAN, G. B., 1982. Karyotypic similarities between *Dromiciops australis* (Microbiotheriidae, Marsupialia) and some Australian marsupials. Pp. 711-14 *in* "Carnivorous marsupials" ed by M. Archer. Roy. Zool. Soc. New South Wales: Sydney.

SZALAY, F. S., 1982. A new appraisal of marsupial phylogeny and classification. Pp. 621-40 *in* "Carnivorous marsupials" ed by M. Archer. Roy. Zool. Soc. New South Wales: Sydney.

TYNDALE-BISCOE, C., 1980. Observations on the biology of marsupials in Colombia and Venezuela. National Geographic Society Research Reports **12:** 711-20.

ULLMAN, S. L., 1984. Early differentiation of the testis in the native cat, *Dasyurus viverrinus* (Marsupialia). *J. Anat.* **138:** 675-88.

VAN DEN BROEK, A. J. P., 1906. On the relation of the genital ducts to the genital gland in marsupials. *Proc. K. ned. Akad. Wet.* 396-402.

WOOLLEY, P., 1975. The seminiferous tubules in dasyurid marsupials. *J. Reprod. Fert.* **45:** 255-61.

WOOLLEY, P. A. AND SCARLETT, G., 1984. Observations on the reproductive anatomy of male *Tarsipes rostratus* (Marsupialia: Tarsipedidae). Pp. 445-50 *in* "Possums and gliders" ed by A. Smith and I. Hume, Australian Mammal Society, Sydney.

ALBUMIN IMMUNOLOGIC RELATIONSHIPS AMONG AUSTRALIAN POSSUMS: A PROGRESS REPORT

P. R. BAVERSTOCK[1], J. BIRRELL[1] and M. KRIEG[1]

Microcomplement fixation of albumin was used to test some of the more unexpected results from an earlier albumin immunologic assessment of relationships among the possums (Baverstock 1984). The present analysis strengthens the conclusion that *Acrobates* is not in the Burramyidae but is allied to the Petauridae and Pseudocheiridae. It also confirms the close similarities of the albumin of *Tarsipes* with the albumins of the Petauridae and Pseudocheiridae. The anomolous cross-reactions involving *Lasiorhinus* and *Phascolarctos* are also clarified in the present study. Finally, the results confirm that the genus *Pseudocheirus,* as currently recognised, is both highly divergent at the molecular level and paraphyletic.

Key Words: Albumin immunology; Possums; Acrobatids; Tarsipedids; Petaurids; Pseudocheirids; Vombatids; Phascolarctids.

Pages 229-34 *in* POSSUMS AND OPOSSUMS: STUDIES IN EVOLUTION ed by M. Archer. Surrey Beatty & Sons and the Royal Zoological Society of New South Wales: Sydney, 1987.

INTRODUCTION

IN A PAPER on the molecular relationships of Australian possums and gliders, Baverstock (1984) presented a preliminary report on their albumin immunologic relationships based on microcomplement fixation. Briefly, MC'F gives a measure of the degree of amino acid sequence divergence between taxa for their albumins. Albumins from reference taxa are purified and injected into rabbits which raise antisera. The antisera are then cross-reacted to albumins of other taxa to yield an albumin immunologic distance (AID). The immunologic distance is highly correlated with the degree of amino acid sequence divergence (Wilson *et al.* 1977). The paper by Baverstock (1984) should be consulted for details of methods used and the rationale of the MC'F technique.

The most unexpected results from the antisera available in the Baverstock (1984) study were:

1, *Tarsipes* appeared to be allied to the families Petauridae and Pseudocheiridae (it was placed in its own superfamily by Kirsch and Calaby 1977).
2, *Acrobates* and *Distoechurus* were not members of the Burramyidae (as indicated by Kirsch and Calaby 1977) but appeared to be allied to the Petauridae and Pseudocheiridae.
3, There was no *especially* close relationship between *Lasiorhinus* and *Phascolarctos*(placed in a single superfamily by Kirsch and Calaby 1977).
4, The genus *Pseudocheirus* was highly divergent and possibly paraphyletic with *Petauroides*.
5, The genus *Phalanger* was highly divergent and possibly paraphyletic with *Trichosurus* and *Wyulda*.

Since that time a number of additional antisera have been prepared in order to test further some of the above results. Antibodies to additional members of the genus *Phalanger* have only recently become available and cross-reactions involving these are incomplete. Therefore we have no additional information on point 5 above. However, cross-reactions have been completed involving antisera to *Acrobates* and *Gymnobelideus* (to test the *Acrobates*/*Tarsipes*/Petauridae/Pseudocheiridae relationships), *Vombatus* and a repeat *Lasiorhinus* (to test the *Lasiorhinus*/*Phascolarctos* relationship), and *Pseudocheirus cupreus*, *Ps. archeri*, and *Ps. herbertensis* (to examine further relationships among *Pseudocheirus* species and *Petauroides*).

We do stress that our efforts since the 1984 report have centred specifically on points 1 to 4 above rather than a total overview of marsupial phylogeny. Such an overview will require raising additional antisera and conducting more cross-reactions. The purpose of this paper is therefore to specifically explore further points 1 to 4 above.

We follow the familial boundaries used in the present volume and by Smith (1984) in recognising the separate families for Pseudocheiridae and Petauridae.

RESULTS AND DISCUSSION

The Relationships of Acrobates *and* Tarsipes

Table 1 shows the albumin immunologic distances (AIDs) among represenative members of the Australian possum families. *Caluromys, Dasyurus, Isoodon* and *Macrotis* were included as outgroups.

[1] Evolutionary Biology Unit, South Australian Museum, North Terrace, Adelaide, South Australia, Australia 5000.

Table 1. Corrected AIDs among representatives of the families of possums using *Caluromys, Dasyurus, Isoodon,* and *Macrotis* as outgroups, with special reference to *Acrobates* and *Tarsipes.*

Antigen		Antibody										
		A	B	C	D	E	F	G	H	I	J	K
Dasyurus viverrinus	A	0	56	67	87	93	88	59	80	74	116	76
Isoodon macrourus	B	59	0	49	66	79	113	71	78	65	108	88
Macrotis lagotis	C	62	55	0	92	82	106	74	79	91	107	105
Vombatus ursinus	D	76	68	75	0	58	72	66	51	69	62	84
Phascolarctos cinereus	E	102	87	99	54	0	76	72	57	62	62	66
Trichosurus vulpecula	F	97	103	128	71	85	0	49	66	74	80	75
Cercartetus nanus	G	62	71	69	60	83	59	0	47	68	73	53
Acrobates pygmaeus	H	87	87	76	51	63	78	47	0	31	60	38
Petaurus breviceps	I	96	89	94	60	69	81	65	38	0	55	47
Pseudocheirus peregrinus	J	108	112	93	67	66	81	68	66	47	0	73
Tarsipes rostratus	K	82	73	91	83	68	74	63	40	43	69	0
Caluromys sp.	L	60	73	75	67	82	87	57	78	103	104	87
CF*		1.05	0.86	1.11	1.11	0.78	0.91	0.86	1.12	1.05	1.05	1.05

*CF = Correction factor for reciprocity.
% Standard deviation for reciprocity = 6.6%.

These data were used to construct a tree by the Fitch-Margoliash criterion for goodness-of-fit (Fitch and Margoliash 1967), using the PHYLIP programme (Phylogeny Inference Package) of J. Felsenstein. The sequence of taxon input was varied using the SHUFFLE programme of PHYLIP. The tree shown in Figure 1 is the best-fitting tree obtained of over 2000 trees tried, and was the best tree irrespective of the ordering of taxa. Branch lengths of 4 units or less in length were collapsed to form unresolved nodes.

Other than the placement of *Tarsipes* and *Acrobates,* the topology of the tree is about as expected from other criteria. Petauridae *(Petaurus)* and Pseudocheiridae *(Pseudocheirus)* are monophyletic. There is a single unresolved node at the separation of this group, Phalangeridae *(Trichosurus),* Burramyidae *(Cercartetus),* Phascolarctidae *(Phascolarctos)* and Vombatidae *(Vombatus).* Because this node is unresolved, we are unable to either support or deny the concept of monophyly between the Phascolarctidae and Vombatidae, or the possible superfamilial ranking of such a group.

On the tree, *Petaurus* and *Pseudocheirus* form a monophyletic unit, although the branch in common is very short. Thus whether Petauridae and Pseudocheiridae should be considered as confamilial (e.g., Kirsch and Calaby 1977) or as separate families (Smith 1984) depends upon where one wishes to draw the family line.

Our preliminary analysis using albumin (Baverstock 1984) suggested affinities among *Acrobates, Tarsipes* the Petauridae and the Pseudocheiridae. The present more complete analysis serves to strengthen this suggestion. The tree in Figure 1 shows *Tarsipes* and *Acrobates* to be monophyletic with Petauridae *(Petaurus)* to the exclusion of Pseudocheiridae. We would not push such a monophyly too far, since the shared branch is only 4.5 units long. However, the length of the branch shared by *Tarsipes, Acrobates, Petaurus* and *Pseudocheirus* is 9.1 units long. The

albumin molecule is therefore clearly indicating a relationship of *Acrobates* and *Tarsipes* to the Petauridae/Pseuodocheiridae group.

A feature of the branch lengths shown in Figure 1 is their relative uniformity. The average branch length from the common node of the five possum families is 30.3 units. The largest deviations from this mean are -11.7 to *Cercartetus nanus* and +9.2 to *Pseudocheirus peregrinus.* This is compelling evidence that the rate of albumin evolution in these taxa is relatively constant (range is ± about 30%). These data show that here, as elsewhere, albumin is evolving in a somewhat clock-like fashion (Maxson *et al.* 1975; Wilson *et al.* 1977). The rate of albumin evolution in marsupials has been calibrated at about 0.8 units per million years (Maxson *et al.* 1975). This gives a time for the radiation of the Diprotodonta into the present-day families in the late Eocene/early Oligocene.

To explore further the relationships of *Tarsipes* and *Acrobates* to the Petauridae and Pseudocheiridae, antisera to albumins of *Gymnobelideus* and *Petauroides* were included in cross-reactions with other members of the Petauridae and Pseudocheiridae, along with *Tarsipes, Acrobates,* and *Cercartetus; Trichosurus* and *Phalanger* were used as outgroups. The results of these cross-reactions are shown in Table 2. These data were subjected to analysis by the Fitch-Margoliash method as above, and the best-fitting of over 400 trees is shown in Figure 2. Again branch lengths of less than or equal to 4.0 units were collapsed into single nodes.

Once again *Acrobates* stands well apart from *Cercartetus,* and falls within the Petauridae/Pseudocheiridae group. *Tarsipes* too falls well within such a group. Here the branch length uniting *Acrobates* and *Tarsipes* within the Petauridae/Pseudocheiridae group is a total of 17.2 units. Thus the inclusion of antisera to *Gymnobelideus* and *Petauroides* strengthens still further the associations of *Tarsipes* and *Acrobates* with the Petauridae and Pseudocheiridae.

The inclusion of *Gymnobelideus* and *Petauroides* also strengthens the concept of monophyly of the Petauridae and Pseudocheiridae, with a common branch length of 12.8 units.

The Pseudocheirus *Radiation*

In our earlier preliminary analysis (Baverstock 1984), we found evidence for extensive molecular diversity among members of the genus *Pseudocheirus* at the albumin level. We have since raised antisera to albumins of *P. cupreus, P. archeri* and *P. herbertensis* to test this suggestion. Cross-reactions involving these additional antisera along with *P. peregrinus* and *Petauroides volans* are shown in Table 3. *Gymnobelideus* and *Acrobates* were used as outgroups. *Pseudocheirus lemuroides* was available as an antigen only.

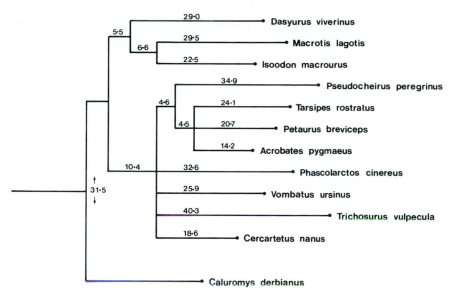

Fig. 1. The best-fit Fitch-Margoliash tree for the albumin immunologic relationships among the major branches of the Australian possums. The tree was rooted using *Caluromys*, *Dasyurus*, *Macrotis* and *Isoodon* as outgroups. Branch lengths shown are proportional to the proposed amount of albumin evolution along each branch. Branch lengths of less than 4 units in length were collapsed into unresolved nodes.

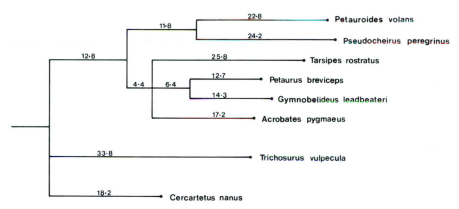

Fig. 2. The best-fit Fitch-Margoliash tree for the albumin immunologic relationships among the Petauridae, *Tarsipes*, *Acrobates* and the Burramyidae using *Trichosurus* as an outgroup. Branch lengths of less than 4 units in length were collapsed into unresolved nodes.

Table 2. Corrected AIDs among representatives of Petauridae (*Petaurus* and *Gymnobelideus*), *Pseudocheirus*, *Tarsipes*, *Acrobates*, and *Cercartetus* using *Trichosurus* and *Phalanger* as outgroups.

Antigen		Antibody							
		A	B	C	D	E	F	G	H
Trichosurus vulpecula	A	0	47	80	70	79	86	65	74
Cercartetus nanus	B	56	0	75	50	72	64	47	52
Petaurus breviceps	C	78	63	0	30	54	52	36	47
Gymnobelideus leadbeateri	D	69	61	24	0	56	57	40	51
Pseudocheirus peregrinus	E	78	66	52	68	0	46	65	72
Phalanger vestitus	F	74	69	52	63	47	0	66	63
Acrobates pygmaeus	G	75	46	33	34	59	71	0	38
Tarsipes rostratus	H	71	61	47	45	69	61	39	0
	CF	0.86	0.83	1.15	1.08	1.04	0.93	1.08	1.04

% Standard deviation for reciprocity = 5.6%.

The most striking feature of this table is the extent of divergence at the albumin level for the genus *Pseudocheirus*. Typically, we find that genera of mammals show AIDs of 0 to 10 (unpublished data), yet here all values are well above this.

The best-fit tree (of about 400 tried) by the Fitch-Margoliash criterion is shown in Figure 3. There are two points to note about this tree — the topology of the tree and the rates of albumin evolution among *Pseudocheirus*.

Tree topology. *Pseudocheirus lemuroides* appears to be monophyletic with *Petauroides volans*; they are united by a branch length of 11.5 units. Such an association is also supported by other data (McQuade 1984). Of the remaining *Pseudocheirus* species, *P. archeri* and *P. cupreus* form a group with a common branch length of 15.8 units. This group, *P. peregrinus,* and *P. herbertensis* in turn form a monophyletic assemblage with a common branch length of 7.0 units.

Rates of albumin evolution. Of the lineages in Figure 1, the branch leading to *P. peregrinus* has had a relatively fast rate of albumin evolution, being 9.2 units faster than average (Fig. 1). Yet in Figure 3, all other members of the genus included

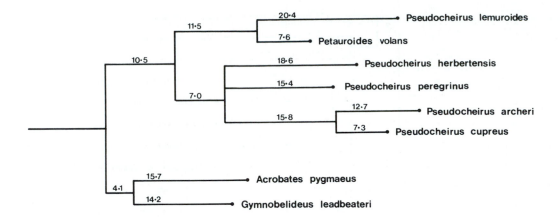

Fig. 3. The best-fit Margoliash tree for the Pseudocheiridae, using *Acrobates* and *Gymno-belideus* as outgroups. Branch lengths of less than 4 units in length were collapsed into unresolved nodes.

Table 3. Corrected AIDs among six Pseudocheirids (*Pseudocheirus* and *Petauroides*), using *Gymnobelideus* and *Acrobates* as outgroups.

Antigen		Antibody						
		A	B	C	D	E	F	G
Gymnobelideus leadbeateri	A	0	32	47	69	54	51	46
Acrobates pygmaeus	B	28	0	55	59	49	59	51
Pseudocheirus cupreus	C	57	58	0	17	49	35	38
Pseudocheirus archeri	D	60	58	22	0	62	45	32
Petauroides volans	E	50	52	56	54	0	42	28
Pseudocheirus peregrinus	F	55	51	40	42	43	0	18
Pseudocheirus herbertensis	G	52	52	28	39	32	17	0
Pseudocheirus lemuroides	H	56	60	69	78	28	50	49
CF		0.88	0.86	1.30	1.26	0.88	0.95	0.88

% Standard deviation for reciprocity = 7.0%.

Table 4. AIDs of *Pseudocheirus herbertensis* and *P. peregrinus* to other *Pseudocheirus*..

Antigen	Antibody	
	P.h.	P.p.
P. herbertensis	0	17
P. cinereus	2	25
P. forbesi	6	12
P. canescens	11	23
P. peregrinus	18	0
P. archeri	36	45
P. cupreus	33	35
P. lemuroides	49	50

have still faster rates of albumin evolution. Thus, it is apparent that at least some of the albumin diversity of the genus can be ascribed to faster than average rates of albumin evolution. However, the following considerations show that even after taking account of the faster rates, the divergence of the genus is still seen to be quite old.

The average amount of albumin evolution in the *Pseudocheirus/Petauroides* group since the Petauridae/Pseudocheiridae separation is 26 units of albumin (Fig. 3). Based on Figure 1, this divergence occurred about 37 MY before present (BP). Thus the rate of albumin evolution in the *Pseudocheirus/Petauroides* group has been about 1.4 units per MY which is about twice the average. Using 1.4 units/MY as an average for the group, the radiation into *P. herbertensis*, *P. peregrinus* and the *P. archeri/P. cupreus* group occurred about 14 MY BP, while the *P. lemuroides* separation occurred earlier at about 18 MY BP. Thus even after taking account of the faster rates of albumin evolution in *Pseudocheirus*, the radiation of the genus is still seen to be quite old.

Of the *Pseudocheirus* other than *P. lemuroides*, *P. herbertensis*, *P. peregrinus*, *P. archeri* and *P. cupreus*, we also have material from *P. forbesi* and *P.*

canenscens. Antisera have not yet been raised to albumins of these species, but cross-reactions involving antisera to the other *Pseudocheirus* have included them as antigens. These cross-reactions show that *P. forbesi* is very close to *P. herbertensis*, while the affinities of *P. canescens* appear to lie with *P. herbertensis* (Table 4).

The Burramyidae

Of the Burramyidae (which does not include *Acrobates*) antisera were available only to albumin of *Cercartetus nanus*. Cross reactions of this antiserum with *C. lepidus* and *C. concinnus* yield AIDs of 0-3, as expected for a genus.

The relationships between *C. nanus* and *Burramys parvus* were explored using *Trichosurus* and *Petaurus* as outgroups. The best-fit Fitch-Margoliash tree is shown in Figure 4. The rate of albumin evolution in the branch leading to *Petaurus breviceps* is about average for the possums (Fig. 1). Based on Figure 4, the rate of albumin evolution in the Burramyidae has been about half the average i.e., 0.4 units per MY. This gives a divergence time for *Burramys* and *Cercartetus* in the early Oligocene.

The Relationships of the Koala and Wombat

An anomolous result in our earlier study (Baverstock 1984) was a failure to demonstrate a

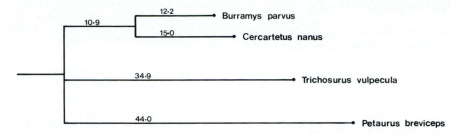

Fig. 4. The best-fit Fitch-Margoliash tree for the Burramyidae (excluding *Acrobates*) using *Trichosurus* and *Petaurus* as outgroups. Branch lengths of less than 4 units in length were collapsed into unresolved nodes.

close relationship between the Koala, *Phascolarctos*, and the Wombat, *Lasiorhinus*. While anti-*Phascolarctos* recognised *Lasiorhinus* as being closer than any other Diprotodonta, anti-*Lasiorhinus* did not recognise *Phascolarctos* as being especially close. In view of the compelling evidence for a close relationship of the Koala and the Wombats (see Kirsch 1977), it seemed that the anti-*Lasiorhinus* results were anomolous. We therefore purified a new batch of albumin from *Lasiorhinus* and raised new antisera in an additional three rabbits. We also raised antisera to *Vombatus* albumin. The results are shown in Table 5.

The new anti-*Lasiorhinus* albumin gave essentially similar results to the old antisera, although the *Phascolarctos* distance was now slightly less (56 versus 63) and the *Pseudocheirus* distance was slightly greater (72 versus 59). Indeed, these data indicate the extent of variability in MC'F results that can be ascribed to the immune recognition system of the rabbits. In contrast, anti-*Vombatus* does recognise *Phascolarctos* as the most similar diprotodont taxon other than *Lasiorhinus*.

CONCLUSIONS

The contrasting results for *Vombatus* and *Lasiorhinus* emphasize several important points about the technique of MC'F.

Firstly, wherever possible, relationships of a group should be explored using as many cross-reactions as possible involving members of that group, rather than a single representative. Indeed this is the very problem that plagues the placement of taxa such as *Tarsipes,* where only one taxon is available representing the entire superfamily Tarsipedoidea.

Secondly, MC'F is not the be all and end all of taxonomy. Even where actual amino acid sequence data are available, quite anomolous results often appear (e.g., Romero-Herrara *et al.* 1978), while different proteins often yield different trees (e.g., Penny *et al.* 1982). Distance data of the sort exemplified by MC'F obviously will be plagued by the same anomalies inherent in sequence data, but will have two additional

Table 5. Uncorrected AIDs involving antisera to *Vombatus ursinus* (Vu), *Phascolarctos cinereus* (Pc), and two separate antisera to *Lasiorhinus latifrons* (L1(1) and L1(2)).

Antigen	Antibody			
	Vu	Pc	L1(1)	L1(2)
Vombatus ursinus	0	75	9	8
Phascolarctos cinereus	48	0	63	56
Lasiorhinus latifrons	10	78	0	0
Trichosurus vulpecula	63	110	59	59
Cercartetus nanus	54	107	44	50
Petaurus breviceps	54	90	57	55
Pseudocheirus peregrinus	61	86	59	72

problems: firstly, information on ancestral sequences is lost, and secondly the correlation between MC'F distance and amino acid substitutions is not perfect (Wilson *et al.* 1977).

We suggest that MC'F data should test conventional hypotheses, supporting them in many cases, but challenging them in others. An excellent example concerns the placement of *Acrobates*. No one seems to have questioned its placement in the Burramyidae until MC'F data suggested that it was unrelated to that group. Since that time, the availability of *Distoechurus* material has supported our earlier contention. More significantly, Aplin (work in progress) has assessed basicranial data on *Acrobates* and *Distoechurus*, and he too has concluded that there is little basis for their inclusion in the Burramyidae.

It is in this light that we view the albumin data on *Tarsipes*, particularly in relation to *Acrobates,* the Petauridae and the Pseudocheiridae. Previous assessments on its relationships should be reviewed, and most importantly any other data should be assessed in a cladistic background. *Tarsipes* is a highly derived animal so it will possess many autapomorphies. It may well be that it is so modified morphologically that few synapomorphies remain to tie it with any family. The only clues to its true relationships may require other molecular data, using other proteins or DNA analysis.

The present analysis has also highlighted the extent of molecular diversity in the genus *Pseudocheirus*. These data suggest that the genus could validly be split into several genera, a view supported by other evidence (McKay, pers. comm.).

ACKNOWLEDGEMENTS

We are grateful to the many persons who contributed the material that made this study possible (see Appendix 1). We thank B. Young, S. Merrett and P. Altman for assistance with the care of rabbits, and P. Kidd for typing the manuscript. R. Andrews, S. Donnellan and K. Aplin kindly read and criticized the manuscript. We are most grateful to A. Gunjko for assistance with the computer analysis. Collection of some of the specimens by Aplin and Flannery was made possible by a National Geographic Society Research Grant to M. Archer.

REFERENCES

BAVERSTOCK, P. R., 1984. The molecular relationships of Australasian possums and gliders. Pp. 1-8 in "Possums and gliders" ed by A. P. Smith and I. D. Hume. Australian Mammal Society: Sydney.

FITCH, W. M. AND MARGOLIASH, E., 1967. Construction of phylogenetic trees. Science 155: 279-84.

KIRSCH, J. A. W. AND CALABY, J. H., 1977. The species of living marsupials: an annotated list. Pp 9-26 in "The biology of marsupials" ed. by B. Stonehouse and D. Gilmore. MacMillan: London.

MAXSON, L. R., SARICH, V. M. AND WILSON, A. C., 1975. Continental drift and the use of albumin as an evolutionary clock. Nature 255: 397-400.

McQUADE, L. R., 1984. Taxonomic relationship of the Greater Glider, Petauroides volans, and the Lemur-like Possum, Hemibelideus lemuroides. Pp 303-10 in "Possums and gliders" ed by A. P. Smith and I. D. Hume. Australian Mammal Society: Sydney.

PENNY, D., FOULDS, L. R. AND HENDY, M. D., 1982. Testing the theory of evolution by comparing phylogenetic trees constructed from five different protein sequences. Nature 297: 197-200.

ROMERO-HERRERA, A. E., LEHMANN, H., JOYSEY, K. A. AND FRIDAY, A. E., 1978. On the evolution of myoglobin. Proc. R. Soc. B 283: 61-163.

SMITH, A., 1984. The species of living possums and gliders. Pp xiii-xv in "Possums and gliders" ed by A. P. Smith and I. D. Hume. Australian Mammal Society: Sydney.

WILSON, A. C., CARLSON, S. S. AND WHITE, T. J., 1977. Biochemical evolution. Ann. Rev. Biochem. 46: 573-639.

APPENDIX 1

Sources of material: *Acrobates pygmaeus*, A. Smith; *Burramys parvus*, G. McKay (Mt. Kosciusko, NSW); *Caluromys* sp., D. Hayman; *Cercartetus nanus*, S.A. National Parks (Bray Junction, S.A.); *C. lepidus*, P. Baverstock (Kingston, S.A.); *C. concinnus*, P. Bird (Yumbarra Cons. Pk, S.A.); *Dasyurus viverrinus*, Adelaide Zoo; *Distoechurus pennatus*, K. Aplin (Naru Village, Papua New Guinea); *Gymnobelideus leadbeateri*, A. Smith; *Isoodon macrourus*, P. Baverstock (Nourlangie, N.T.); *Lasiorhinus latifrons*, P. Baverstock (Sedan, S.A.); *Macrotis lagotis*, P. Baverstock (Papunya, N.T.); *Petauroides volans*, D. Spratt (Bondo State Forest); *Petaurus breviceps*, P. Baverstock (Laboratory colony); *Phascolarctos cinereus*, T. Schwaner (Adelaide Zoo); *Phalanger vestitus*, J. Nelson (Papua New Guinea); *Pseudocheirus archeri*, G. McKay; *Pseudocheirus canescens*, J. Nelson (Papua New Guinea); *Pseudocheirus cupreus*, K. Aplin and T. Flannery (Papua New Guinea); *Pseudocheirus forbesi*, J. Nelson (Papua New Guinea); *Pseudocheirus herbertensis*, G. McKay (Mareeba, Qld); *Pseudocheirus lemuroides*, G. McKay; *Pseudocheirus peregrinus*, I. Beveridge (Marysville, Qld); *Tarsipes rostratus*, D. King (Captive Colony, Murdoch University); *Trichosurus vulpecula*, D. Hayman (Adelaide, S.A.); *Wyulda squamicaudata*, A. Bradley (Mitchell Plateau, W.A.).

PARASITES AND AUSTRALIAN MARSUPIAL PHYLOGENY

IAN HUMPHERY-SMITH[1]

An hypothesis is presented about how and when 'parasitic' groups can be of value in clarifying host phylogenetic relationships. The literature, pertaining to parasites exhibiting varying degrees of phylogenetic specificity to Australian marsupials, is reviewed. Parasites are demonstrated to be useful tools for retracing Australian marsupial phylogeny, both intra- and intercontinentally. The evidence strongly suggests a South American origin for the Australian marsupials and their dispersal into Australia via Antarctica when the continents were connected as part of Gondwanaland. Some parasites of dasyurids show marked affinities with those of didelphids. Within Australia, generalised trends include clear-cut divisions between the parasites of diprotodont and polyprotodont marsupials, with those of the peramelids aligning with the latter group. The present inadequacies in the parasitological record of Australian marsupials, in most cases, do not permit comment on relationships between host groups at the generic and specific level.

Key Words: Host; Parasite; Evolution; Marsupials; Phylogenetic specificity.

Pages 235-43 in POSSUMS AND OPOSSUMS: STUDIES IN EVOLUTION ed by M. Archer. Surrey Beatty & Sons and the Royal Zoological Society of New South Wales: Sydney, 1987.

INTRODUCTION

THE NOTION that a knowledge of parasitic fauna can be of value in helping to clarify host phylogeny was first put forward by Kellogg (1896) in a work dealing with the Mallophaga (biting-lice) of birds. Later (1913:158) he wrote: "I do indeed believe that it is a commonness of the genealogy rather than a commonness of adaption that is the chief explanation of this restriction of certain parasite groups to certain host groups". Along with Fuhrmann (1908), who worked on the cestodes of birds, Kellogg had laid the foundation to the study of parasitogenesis; that is, the evolution of the host-parasite relationship.

Although this concept has been given much credence by parasitologists, Patterson (1957:15) commented on the apparent lack of attention paid by mammalogists ". . . to the light which may be shed by parasites on problems of phylogeny and toxonomy within the class" a situation which, at least with respect to marsupials, has improved little in the last 30 years.

There have been several attempts to make generalisations for the phylogenetic rapport between host and parasite. These include the so-called 'parasitophyletic rules' of Fahrenholz, Szidat, Eichler and Manter, and the implied rule of Inglis and have been reviewed by Noble and Noble (1964), Dogiel (1964), Inglis (1971) and Brooks (1979). The parasites have not always heeded these rules and have left many exceptions in their evolutionary paths. It is therefore not the exercise of this review to place further constraints on the host-parasite relationship, but rather to postulate a contingency table for these relationships. Then, in turn, the value of each category to the study of host phylogeny, as told by the parasites, will be examined. Once a probable

explanation has been provided of how and when parasites can be of value in this respect, examples will be drawn, both intra- and intercontinentally, which appear to highlight phylogenetic associations relevant to Australia's marsupial fauna. In so doing, 'possums and opossums' will come under scrutiny.

The possible relevance of parasites to the phylogeny of Australian marsupials and in particular, their possible South American origin, was first examined by Zschokke (1898). Analyses of the topic were later carried out by Harrison (1928) and Sandars (1956). Unfortunately the now-antiquated taxonomies utilised have lessened the impact of this work, but each of these studies pre-empted the current acceptance of plate tectonic theories.

Mayr (1957:7) stated: "If, as seems to be true in some cases, the parasite is more conservative than the host (that is, its evolution is retarded as compared to that of the host) then the parasite will show phylogenetic relationships where they have become obscured in vertebrate hosts". Undoubtedly, a certain level of evolutionary conservatism must be attributed to all parasites with respect to their hosts, in that, in order to adapt a parasitic mode of life, the prior existence of a host is a prerequisite; the host constitutes the parasite's environment. Sprent (1963) has reviewed the range of biological associations into which 'parasites' may fall and these will not be discussed further here.

The following hypothesis on host-parasite evolution and its signficance to phylogenetic specificity is discussed in relation to continua of host specificity and pathogenicity. Examples at the extremes of each of these continua are given in Figure 1.

[1]Department of Parasitology, University of Queensland, St. Lucia, Queensland, Australia 4067.

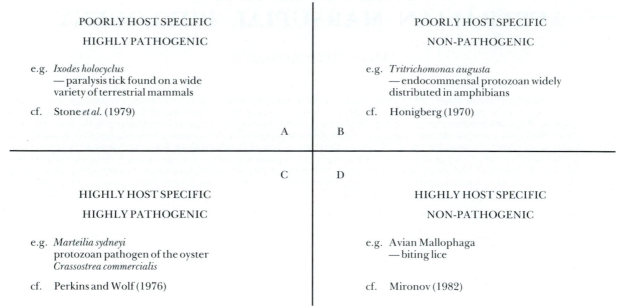

Fig. 1. Contingency table for host pathogenicity: host specificity.

Once formed, the host-parasite relationship can be one of considerable dynamism (cf. Mauel 1982; May and Anderson 1983). None better exemplifies this than *Trypanosoma brucei* (causative agent of African trypanosomiasis) which, according to Cross (1978), evades the host immune response by manifesting a range of antigens caused by variation of the glycoprotein covering the cell surface. However, if the host-parasite relationship could be viewed as an evolutionary continuum, one should see at one extreme a relationship dominated by the host. Here the 'parasite' is subservient to the various facets of host biology, for example, members of the Opalinata (endocommensal protozoa) which inhabit the hind-gut of cold blooded vertebrates. They inflict no known harm to the host and provide no benefit; they merely occupy a previously unoccupied niche with the host providing shelter and nutrients which would otherwise be lost (Wessenberg 1978).

At the other extreme the parasite can be expected to dominate the relationship. This role is filled by highly pathogenic organisms which select against non-resistant or weaker hosts, for example, the myxoma virus has been recorded as accelerating lagomorph evolution (Fenner and Ratcliffe 1965), or the *Melampsora* rust of *Linum* flax, which is used as a model for the gene-for-gene concept in the evolution of hosts and their parasites (Person *et al.* 1962; Flor 1971). Anderson and May (1982) and Ewald (1983) have examined the theoretical development of an equilibrium in the degree of parasite virulence and reproductivity and host recovery rate which is evident in the host-parasite relationship of severely pathogenic organisms. However, this work does not challenge the dominance of the parasite in the relationship. Other examples of

modified host biology in response to parasite infection have been cited by Price (1980).

Most assuredly, the bulk of 'parasitic' associations, with regard to pathogenicity, will fall between these two extremes and, in an evolutionary sense, involve some give and take between host and parasite.

The level of pathogenicity is not a static character associated with an organism. It may vary according to environmental conditions, host physiology and strain differences of both host and parasite, as seen in infections of *Entamoeba histolytica* (Albach and Booden 1978) or allergy to infection (Jarrett and Miller 1982). Nonetheless, in an evolutionary context, the potential for pathogenicity should equate to pathogenicity as it would have been manifested during the time involved.

A further theoretical complication with respect to pathogenicity is seen amongst parasitoid development in insects (cf. Knutson and Berg 1966) by organisms such as mermithid nematodes and ichneumonoid, vespoid, chalcidoid, and prostotrupoid wasps. In the above instances, the parasite would appear to be dominant in any individual host-parasite association, although it is questionable whether any evolutionary pressures on the host population have been forthcoming because of the host's large reproductive capacity in what are probably quite ancient associations. Populations of host and parasite species probably occur in equilibrium, even though the parasite exhibits extreme pathogenicity. Noteworthy is the fact that the gene-for-gene concept has yet to be demonstrated in nature, as opposed to 'man-guided' evolution. Although conducted *in vitro*, the work of Ron and Tal (1985) has demonstrated concomitant evolution in both host (L-cells) and parasite (minute virus of mice) to produce persistent infection, and may therefore be a step closer to proving the existence of this phenomenon.

Of relevance to the present discussion is the degree of host specificity exhibited by a parasite. Host specificity, including anatomical, physiological and behavioural specificity, can be expected to manifest itself as a continuum. On the one hand we find examples where parasites are known to occur in association with only one host species or a closely related group of hosts, for example *Wuchereria bancrofti* (nematode), the adults of which are found only in man. (One must acknowledge that in many instances these restricted host distributions may be the product of a poor parasitological record). On the other hand, many examples are known where parasites show limited or negligible host specificity. This is often the case where a trophic level in a food chain represents the degree of host specificity, for example *Contracaecum* spp (nematode) can be found in a wide variety of fish-eating birds and some mammals. Parasite species cannot always be assigned a position on this continuum and may confuse the issue by being highly host specific towards one or other intermediate hosts, but encompassing a wide range of definitive hosts, for example heterophyid trematodes.

In Fig. 1, the extremes in the continua of host specificity and host pathogenicity have been grouped together to form a contingency table. If each of the above categories is examined in turn, what implications can be drawn with respect to host phylogeny? Immediately groups A and B can be discounted because a lack of host specificity excludes these parasites from highlighting phylogenetic relationships between unrelated groups. Parasites in group C should not trace phylogenetic trends in the host simply because, as already stated, they tend to produce trends in host phylogeny. In addition, the 'classical contention' extolled by Allison (1982) and Holmes (1982) that a pathogenic parasite is poorly adapted and therefore more recent, would also support this view. Thus, only in group D can we expect to find parasites which fulfil our prerequisites for following host evolution. At this point, the task falls upon the relevant group specialist to detail trends in parasite evolution.

Furthermore, it is necessary for a parasite group to be widely distributed within the host group to be examined before meaningful conclusions can be forthcoming on host phylogeny. A final requirement must still be placed upon parasites in order that they might afford phylogenetic specificity (cf. Sprent 1969) and that is for host specificity to be maintained through time. Although at first glance this requirement would appear difficult to ascertain, as will be

demonstrated, this is not the case. Parasitic groups whose phylogeny clearly follows that of the host must have undergone lengthy co-evolution with the host, as opposed to colonisation subsequent to host speciation. In the former case, once co-evolution has been substantiated by comparison with accepted host phylogeny, the parasite's worth then becomes apparent. Questionable areas in the host's phylogeny can be clarified through use of parasite phylogeny.

Although there is difficulty in ascribing significance to distinct or bizarre forms because of possible convergence, several such instances may also establish a meaningful rapport between hosts.

As with any attempted synthesis of a biological phenomenon, there are bound to be examples which do not fit well the proposed schema. These inevitable inadequacies should not prevent us from trying to explain these phenomena as analysis is likely to stimulate further thought and thus improve our ideas.

The present discussion will examine the Australian marsupial parasite fauna under works concerned with the Protozoa, Helmintha and Arthropoda which have commented on host phylogeny as perceived by parasitic associations.

PROTOZOA

Cameron (1952:3) commented in an overview of world parasitism that ". . . the entozoa of the opossums and the Australian marsupials are quite different" and suggested that the Australian and American marsupials are not closely related. However, no detailed examination of the Australian protozoan fauna of marsupials has been conducted with a view to understanding phylogeny of either host or parasite groups.

HELMINTHA
Nematoda

Humphery-Smith (1983) found members of the Herpetostrongylinae (Trichostrongyloidea) to have embraced 18 genera of Australian marsupials and ramified into three distinct phylogenetic lineages, each derived from a form resembling existing species found in Dasyuridae and Thylacinidae:

(i) a short lineage restricted to the Dasyuridae;

(ii) a more extensive radiation involving not only Dasyuridae but also Peramelidae and Myrmecobiidae; and

(iii) a mainly diprotodont radiation derived from dasyurid parasites and involving Macropodidae followed by Potoroidae, Petauridae and Phalangeridae.

This phylogeny was clearly divided between the diprotodont and polyprotodont marsupials and seemed to support a diprotodont radiation akin to that proposed by Ride (1971). However, the concept of ecological partitioning as propounded by Archer (1981) was used to explain the parasite distribution in Australian possums and potoroids (the bunodont forms); their herpetostrongylin fauna having been derived from lophodont kangaroos via a *Hypsiprymnodon*-like ancestor. The parasite phylogeny was supportive of the Myrmecobiidae and Peramelidae having been derived from dasyurid stock. For the discussion relevant to the affinities of these host families, refer to Humphery-Smith (loc. cit.). The basal stock which gave rise to this radiation was considered to have occurred in a Gondwanaland didelphoid. This was substantiated by the numerous similarities which exist between the Australian Herpetostrongylinae and the South American Viannaidae, each group parasitic in the Marsupialia.

Durette-Desset and Beveridge (1981) have clarified a homogeneous evolutionary line distinct from other Australian Trichostrongyloidea within the Dromaestrongylidae. The more primitive parasites are found in the Dasyuridae, Peramelidae and Potoroidae, followed by examples in the Phalangeridae and subsequently in the Macropodidae.

Further evidence of the apparent affinity between the Peramelidae and the Dasyuridae, as demonstrated by parasites, can be found amongst the Mackerrastrongylinae (Trichostrongyloidea). This subfamily is deemed to have originated from monotremes and undergone its radiation entirely within these two host groups (Durette-Desset and Cassone 1980; Durette-Desset and Chabaud 1981).

Although the strongyloid nematodes have undergone considerable radiation within Australian marsupials, they are restricted to the Macropodidae, Potoroidae and Vombatidae (Beveridge 1982a). Again we see a major division between the parasite fauna of the bunodont and lophodont marsupials. The 'ancestral' genera were present in the large intestine of wombats and the potoroid *Hypsiprymnodon moschatus*. The stomach of these animals is argued to have been unsuitable for colonisation by strongyloid nematodes because of low pH. It was therefore not until the appearance of the theoretically-derived, grazing, Macropodidae with their highly succulated fermentative stomachs at a high pH, that the bulk of speciation occurred. Presumably the Potoroidae took no further part in the parasite phylogeny until a recent re-invasion. Because of this parasite group's restricted distribution and supposed origins in ratite birds, further phylogenetic conclusions are not possible.

Of note are the Metastrongyloidea parasites, which are extremely well-distributed in Australian marsupials and do occur in some South American marsupials. However, Spratt (1984) is of the opinion that those in Australian marsupials do not show any substantial degree of definitive host specificity.

Cestoda

An apparent intercontinental rapport between parasites of dasyurid and didelphid marsupials was demonstrated by Beveridge (1977) within the linstowid genus *Oochoristica*. It is quite possible, however, that these forms arose from reptilian cestodes on both continents rather than the *Oochoristica* of marsupials being related. The presence of the acraspedote, reptilian type, *O. eremophila*, in a mammal would support this view (Beveridge, pers. comm.). Beveridge (1982b and 1983) found further trans-Pacific links within another linstowid genus, *Linstowia*, which occurs in monotremes, peramelids and didelphids. These supposed affinities are, however, weakened by the similarities shared with other more widely distributed cestode genera. While these findings were somewhat antedated by Zschokke (1898), Beddard (1914), Johnston (1914a) and Sandars (1956), Nybelin (1917) was critical of these ostensible rapports.

Anoplocephalid cestodes are well distributed in herbivorous diprotodont marsupials and in hosts in Palaearctica, but do not occur in South American marsupials. Within the genus *Bertiella*, Beveridge (1982b) found that host distribution more strongly reflected feeding behaviour than host taxonomic relationships. *Progamotaenia* species were supportive of the sub-division within the Macropodidae as proposed by Kirsch (1977), which grouped together *Lagorchestes, Thylogale, Onychogalea* and *Petrogale* as opposed to *Macropus* and *Wallabia*. In addition, the more recent parasite genera were restricted to the Macropodidae.

Beveridge (1976, in press) has demonstrated species of the genus *Bertiella* as being highly host specific, with each species (except *B. trichosuri*) occurring in one species of host. This fauna has undergone considerable radiation within Australian and New Guinean possums, namely the Pseudocheiridae and the Phalangeridae, and is also present in koalas and rodents (*Uromys* and *Rattus*). A cladistic analysis of the 24 species involved showed four of the major clades to be restricted to particular host families and although some parallel evolution seems to have taken place between hosts and parasites, host switching was also apparent.

Trematoda

Examples of phylogenetic specificity are unlikely to be found amongst the trematode parasites of marsupials because of extensive definitive host range and the greatest host specificity being evident at the level of the molluscan, first intermediate host. The intercontinental rapport alluded to by Sandars (1958) for two *Zonorchis* species is no longer considered valid. The species were supposedly indistinguishable from one another. *Zonorchis australiensis* has now been synonymised into the dicrocoeliid genus *Platynosumum* by Angel and Pearson (1977) and the family is notoriously non-specific in its choice of hosts.

Johnston (1914a, b) was of the opinion that two species of *Brachylaema* (syn. *Harmostomum*, Brachylaemidae) from the Australian marsupials, *Dasyurus* and *Perameles*, were most closely related to congeneric parasites from American *Didelphis*, even though numerous *Brachylaema* species were known from birds. He stated (1914a:243): "They are so closely related that they must be considered as derived from common ancestors, and they thus present strong circumstantial evidence of the phylogenetic connection of their hosts". Unfortunately, it is difficult to attribute credibility to these findings owing to the extensive host range of the genus in snail-eating birds, mammals and, rarely, reptiles.

ARTHROPODA

In a monumental work Traub (1980) examined the zoogeography and phylogeny of the fleas and lice and devoted considerable attention to the question of parasite phylogenetic specificity towards host groups, in particular the marsupials. He justified the use of fleas and lice in this context and went on to conclude that marsupials arose in South America and dispersed via Antarctica into Australia. In addition, his evidence directly refuted an Asian/Indonesian entry of marsupials into Australia. The Siphonaptera generally parasitised hosts with which they had evolved or else those which developed later, rather than hosts lower on the evolutionary scale. Primitive hosts and their fleas and lice tended to be conservative and change very slowly at the generic or even specific level.

Siphonaptera

The above conclusions, at least with respect to the fleas of marsupials, are based on the work of Traub (1972) and Traub and Dunnet (1973). The family Stephanocircidae (or helmet fleas) is limited to Australia (not present in New Guinea) and South America, and is intimately associated with marsupials in each area, although more recently they have encompassed murids and cricetids. The recent nature of the rodent immigration to Australia is supported by all Siphonaptera-bearing Australian marsupials possessing the more primitive fleas. Similar conclusions were reached with regard to the pygiopsyllids of murids and this group's radiation into New Guinea was deemed as recent. The pygiopsyllid

fleas originated in Gondwanaland and are primarily parasites of Australian marsupials but are represented by one genus in South America and two in Africa. A third family of relevance, the Hystrichopsyllidae, is cosmopolitan. However, within the subfamily Doratopsyllinae are two endemic tribes found in Australia and one in South America, which are primarily parasitic on marsupials but also parasitize insectivores. Asian and North American examples of this subfamily are known, and the level of differentiation between the tribes is in accord with the concept of a common ancestry but long separation. Unfortunately, no discussion was possible pertaining to intracontinental associations within the Marsupialia. Traub (1980:124) concluded, "There are far too many fundamental similarities throughout the body of these fleas for one to espouse convergence".

Anoplura

Traub (1980) interpreted the absence of the Anoplura (or sucking lice) from the Australian marsupials as evidence that marsupials could not have dispersed to Australia via North America and Asia. Hopkins (1949, 1957) estimated the Anoplura to have originated in the Jurassic or Cretaceous, thus marsupials should have had contact with these parasites during a northern migration to Australia. Traub's case is strengthened by the absence of Anoplura on South American marsupials today. He contends that marsupials must have arisen where and/or when sucking-lice were absent.

Mallophaga

Harrison (1916), when arguing monophyly for the rhea and ostrich, predicted the occurrence of data he presented in 1922. In this latter article, he described for the first time a species of biting louse from a South American marsupial in a group previously known only from Australian marsupials and South American rodents, and conjectured (Harrison 1922:154): ". . . that such Amblyceran Mallophaga as occur on South American rodents have been migrants in the past from the marsupial stock". He went on to discuss the similarities between the Boopiidae (Australian) and the Trimenoponidae (South American) lice and was so much assured of the closeness of their relationship as to forecast that ". . . the discovery of further connecting forms will make it advisable to unite these three (also the South American Gyropidae) anomalous groups under one family name". These families have remained distinct, but subsequent authors who have reaffirmed this dramatic intercontinental association, which is believed demonstrative of a common ancestry, include Harrison (1924, 1928), Hopkins (1949, 1957), Vanzolini and Guimaraes (1955) and Traub (1980). Within

Australia, species of *Boopia* occur on Dasyuridae, Peramelidae, Vombatidae and Macropodidae and some genera seem to demonstrate affinities within the Macropodidae (Murray and Calaby 1971). Such host distributions may strengthen these authors' disagreement with the view outlined above, although von Keler (1971), whose work they revised posthumously, referred to the Trimenoponidae as being the family most closely related to the Boopiidae.

Acarina

The tick fauna of Australian marsupials is relatively impoverished and thus excludes phylogenetic consideration of relevance to their hosts. On this paucity of tick species specific for contemporary Australian marsupials, Hoogstraal and Aeschlimann (1982) concluded it may have resulted from a combination of factors, including harsh environment and low population densities of many marsupial species, together with extensive home ranges and solitary habits.

Some mites found on marsupials are very catholic in their choice of host, for example, the Trombiculidae, but some demonstrate a high level of host specificity. Examples of host specificity can be seen within the Laelapidae and Atopomelidae which have undergone extensive radiation on Australian marsupials. A recent northern origin would appear likely for the laelapids (Domrow pers. comm.). This is evident from the disorderly fashion in which the marsupial genera have been colonised and the Old World origins of the host taxa responsible for their entry into Australia. Phylogenetic considerations arising from this group indicate indistinct affinities, as was demonstrated by Domrow (1966).

The same does not apply for the Atopolmelidae, which, although their more recent distribution is cosmopolitan (e.g. *Listrophoroides* on *Rattus*), seem to be derived from Gondwanaland stock (Fain 1972). Here *Austrochirus* parasitic on the Peramelidae share numerous characters with *Dasyurochirus* parasitic on Dasyuridae. Within the genus *Cytostethum*, Fain and Domrow (1974a) interpreted macropodid parasites as having been derived from those found on potoroids. Fain (1971) found more primitive parasites in this lineage (*Procytostethum*) parasitic on *Hypsiprymnodon*. Fain and Domrow (1974b) further refined the use of this group as a phylogenetic indicator by producing evidence supporting conspecific status for continental and Tasmanian potoroos, *Potorous tridactylus*.

Although apparently in conflict with a view expressed in 1972, Fain (1979, 1982a) drew attention to the similarity between *Dasyurochirus*, which is well represented on dasyurids, and *Didelphoecius*, which is widespread amongst the didelphids. Fain (1979) also pointed out the more

derived characteristics seen amongst the Australian genera, for example, *Austrochirus* and *Petrogalochirus*.

A further example of Gondwanaland relationships can be seen within the Listropsoralginae (Psoroptidae) (cf. Fain 1982b). This group encompasses a monotypic genus on *Petaurus* (Petauridae) in Australia, a monotypic genus on *Proechimys* (Muridae) and a small genus on Didelphidae, the latter two host genera from South America. This represents the only example of a direct relationship between parasites of phalangeroids and didelphids. According to Fain (loc. cit.), the closest relatives of the Listropsoralginae occur amongst the Makialginae and Cheirogalalginae (Psoroptidae) on Madagascar lemurids and the monotypic Galogalgidae parasitic on African lorisids.

Affinities between Australian and South American marsupials are also borne out by the myobiid genera *Australomyobia* and *Archemyobia* which are found on Dasyuridae and Didelphidae respectively (Fain 1982a). The possibility of related forms occurring in the two continents was first raised by Jameson (1955), when he described the latter genus.

Domrow (1971:3) stated that the relationships of Australian mites ". . . of higher vertebrates are clearly with the Old World, those of flying forms are less restricted, while those of marsupials and lower vertebrates show the clearest New World links".

CONCLUSIONS

In summary, the parasite groups which were of most use to phylogenetic considerations relating to the Australian marsupials, were deemed to display a high level of host specificity, low pathogenicity, a wide distribution in the host fauna and appeared to have maintained host specificity through time (phylogenetic specificity). Of paramount importance amongst these were herpetostrongylin nematodes, atopomelid mites, stephanocircid fleas and amblyceran lice.

Although a certain amount of coincidence of form can be attributed to convergence, the dramatic number of similarities shared by parasites of Marsupialia in Australia and South America is strongly supportive of there being common ancestry for both the host and the parasite groups. The present geographical isolation is readily explained by continental drift and its applicability to Australian marsupial evolution (Fooden 1972; Jardine and MacKenzie 1972; and Smith 1974). This southern mode of dispersal into Australia of the marsupials, as opposed to one via Asia, is supported by parasitological evidence.

The more primitive parasite stock appears to have originated in South American marsupials and entered Australia via Gondwanaland. The subsequent radiation amongst the Australian marsupials has occurred during the 37 to 50 million years of isolation from the southern continents (Vogt and Connolly 1971; Jardine and McKenzie 1972; Talent 1984).

The parasite radiation can be divided into lineages found in association with the Diprotodonta and Polyprotodonta. Peramelid parasites clearly align with the latter group, while parasites primitive within the Australian radiation are found on, or in, the Dasyuridae and Thylacinidae.

Not all parasites provide clues to host evolution. When setting out to establish an hypothesis, one is prone to select supportive 'evidence' and reject other material on the grounds of, for example, limited host specificity or supposed recent northern immigration into Australia. The author wishes to make the reader aware that these short-comings are possible, no matter how impartial the approach.

At present, subtleties within the Australian marsupial radiation at the specific and generic levels are largely beyond the scope of parasitological consideration. This is due to the considerable gap in our knowledge concerning parasites of Australian fauna. Of the 180 species of Australian marsupials there are no parasites recorded from 88, approximately 50% (Spratt pers. comm.). Furthermore, the host species for which the parasite fauna has been well-documented, comprises only a miniscule proportion. The last checklist of Australian marsupial parasites, prepared by Mackerras in 1958, is currently being updated. This should provide an insight into additional areas of interest to the present discussion, as well as highlighting remaining areas of ignorance.

Parasites have been demonstrated as useful adjuncts to serology, palaeontology, physiology, cytology and karyology in the elucidation of host phylogeny. It would be presumptuous, however, to expect to produce a host phylogeny based entirely on parasite phylogeny. Once it has been demonstrated that the radiation of a parasite group follows that of the host, then it can be utilised to clarify uncertain affinities. For this reason it is hoped to examine parasites of groups such as the Notoryctidae in the near future. Cressey et al. (1983:227) concluded, in relation to the host specificity of copepods parasitic on scombrid fishes: "When a parasite group is taxonomically well understood, it can be treated as a host character with as much validity as host morphology, serology and ecology".

What then of 'possums and opossums'? The parasitological evidence is supportive of a common ancestry, but the fauna associated with the opossums shows more marked affinities with those of dasyurid marsupials, from which the possum fauna appear to be more recently derived.

ACKNOWLEDGEMENTS

I wish to extend my appreciation to M. Archer, I. Beveridge, R. Domrow, G. M. Dunnet, J. C. Pearson, P. J. A. Presidente, D. F. Sandars, D. M. Spratt, J. F. A. Sprent and R. Traub for their input to and comments on this article. In particular I wish to thank Professor J. F. A. Sprent whose stimulation was directly responsible for this article coming to fruition. Through numerous discussions over a number of years he helped develop my ideas on phylogenetic specificity. Discussions with departmental colleagues are also gratefully acknowledged. Finally, I wish to extend my gratitude to W. Gardiner for her editorial assistance.

REFERENCES

ALBACH, R. A. AND BOODEN, T., 1978. Amoebae. Pp. 455-506 in "Parasitic protozoa". Vol. 2 ed by J. P. Kreier. Academic Press: New York.

ALLISON, A. C., 1982. Co-evolution between hosts and infectious disease agents and its effects on virulence. Pp. 245-67 in "Population biology of infectious diseases" ed by R. M. Anderson and R. M. May. Springer-Verlag: Berlin and New York.

ANDERSON, R. M. AND MAY, R. M., 1982. Co-evolution of hosts and parasites. Parasitology 85: 411-26.

ANGEL, L. M. AND PEARSON, J. C., 1977. Dicrocoellid trematodes from Australian hosts. Trans. R. Soc. S. Aust. 101: 115-32.

ARCHER, M., 1981. A review of the origins and radiations of Australian mammals. Pp. 1436-88 in "Ecological biogeography of Australia". Vol. 3 ed by A. Keast. W. Junk: The Hague.

BEDDARD, F. E., 1914. Contributions to the anatomy and systematic arrangement of the Cestoidea. XIII. On two new species belonging to the genera Oochoristica and Linstowia, with remarks upon those genera. Proc. Zool. Soc. (Lond.): 263-83.

BEVERIDGE, I., 1976. A taxonomic revision of the Anoplocephalidae (Cestoda: Cyclophyllidea) of Australian marsupials. Aust. J. Zool., Suppl. 44: 1-110.

BEVERIDGE, I., 1977. On two new linstowiid cestodes from Australian dasyurid marsupials. J. Helminthol. 51: 31-40.

BEVERIDGE, I., 1982a. Evolution of the strongyloid nematodes of Australian marsupials. Mem. Mus. natn. Hist. nat. (Paris), Ser. A Zool. 123: 87-92.

BEVERIDGE, I., 1982b. Specificity and evolution of the anoplocephalate cestodes of marsupials. Mem. Mus. natn. Hist. nat. (Paris), Ser. A Zool. 123: 103-9.

BEVERIDGE, I., 1983. The genus Linstowia Zschokke, 1899 (Cestoda: Anoplocephalidae) in Australian mammals with the description of a new species, L. macrouri. Syst. Parasitol. 5: 291-304.

BEVERIDGE, I. (in press). The genus Bertiella (Cestoda: Anoplocephalidae) from Australasian mammals: new species, new records and redescriptions. Syst. Parasitol.

BROOKS, D. R., 1979. Testing the context and extent of host-parasite co-evolution. Syst. Zool. 28: 299-307.

CAMERON, T. W. M., 1952. Parasitism, evolution and phylogeny. *Endeavour* **11**(44): 1-4.

CRESSEY, R. F., COLLETTE, B. B. AND RUSSO, J. L., 1983. Copepods and scombrid fishes: a study in host-parasite relationships. *Fishery Bull.* **81**: 227-65.

CROSS, G. A. M., 1978. Antigenic variation in trypanosomes. *Proc. R. Soc. Lond. B.* **202**: 55-72.

DOGIEL, V. A., 1964. "General Parasitology". Revised by Y. I. Polyanski and E. M. Kheisin. Translated by Z. Kabota. Oliver and Boyd: Edinburgh and London.

DOMROW, R., 1966. Some laelapid mites of syndactylous marsupials. *Proc. Linn. Soc. NSW* **90**: 164-75.

DOMROW, R., 1971. "Mite parasites of Australian and oriental vertebrates: their taxonomy, host specificity, and medical importance". D.Sc. Thesis, University of Queensland, St. Lucia.

DURETTE-DESSET, M-C. AND BEVERIDGE, I., 1981. *Peramelistrongylus* Mawson, 1960 et *Profilarinema* n.gen. Nématodes Trichostrongyloides paraissant faire transition entre les Strongyloidea et le genre atypique *Filarinema* Mönnig, 1929. *Ann. Parasitol. (Paris)* **56**: 183-91.

DURETTE-DESSET, M-C. AND CASSONE, J., 1980. *Mackerrastrongylus* Mawson, 1960, et *Sprattellus* n.gen. (Nematoda: Trichostrongyloidea) parasites de Perameloidea et de Dasyuroidea (Marsupiaux australiens). *Bull. Mus. natn. Hist. nat. (Paris), 4 ser.* **2**: 943-54.

DURETTE-DESSET, M-C. AND CHABAUD, A. G., 1981. Nouvel essai de classification des Nématode Trichostrongyloidea. *Ann. Parasitol. (Paris)* **56**: 297-312.

EWALD, P. W., 1983. Host-parasite relations, vectors, and the evolution of disease severity. *Ann. Rev. Ecol. Syst.* **14**: 465-85.

FAIN, A., 1971. Notes sur quelques Atopomelidae de la région australienne. *Rev. Zool. Bot. Afr.* **83**: 238-42.

FAIN, A., 1972. Les listrophorides d'Australie et Nouvelle-Guinee (Acarina: Sarcoptiformes). *Bull. Inst. r. Sci. nat. Belg.* **48**: 1-196.

FAIN, A., 1979. Les listrophorides d'Amerique Neotropicale (Acarina: Astigmates). II. Famille Atopomelidae. *Bull. Inst. r. Sci. nat. Belg.* **51**: 1-158.

FAIN, A., 1982a. Notes et discussion sur les marsupiaux et leurs parasites. 2. Acariens parasites de quelques mammiféres d'origine ou d'affinités incertaines. *Mem. Mus. natn. Hist. nat. (Paris), Ser. A, Zool.* **123**: 112-4.

FAIN, A., 1982b. Notes et discussion sur les marsupiaux et leurs parasites. 1. Acariens de la sous-famille Listropsoralginae (Psoroptidae, Astigmates) représentés chez les hystricomorphes et les marsupiaux sud-americains et chez les marsupiaux australiens. *Mem. Mus. natn. Hist. nat. (Paris), Ser. A, Zool.* **123**: 111-2.

FAIN, A. AND DOMROW, R., 1974a. The subgenus *Cytostethum* Domrow (Acari: Atopomelidae): Multiple speciation on the marsupial *Potorous tridactylus* (Kerr). *Aust. J. Zool.* **22**: 549-72.

FAIN, A. AND DOMROW, R., 1974b. The subgenus *Metacytostethum* Fain (Acari: Atopomelidae): parasites of macropodid marsupials. *Acarologia* **16**: 719-38.

FENNER, F. AND RATCLIFFE, F. N., 1965. "Myxomatosis". Cambridge University Press: London.

FLOR, H. H., 1971. Current status of the gene-for-gene concept. *Ann. Rev. Phytophysiol.* **9**: 275-96.

FOODEN, J., 1972. Breakup of Pangea and isolation of relic mammals in Australia, South America and Madagascar. *Science (Wash., D.C.)* **175**: 894-98.

FUHRMANN, O., 1908. Die Cestoden der Vogel. *Zool. Jahrb., Suppl.* **10**: 1-232.

HARRISON, L., 1916. Bird-parasites and bird-phylogeny. *Ibis* **4**: 254-63.

HARRISON, L., 1922. On the mallophagan family Trimenoponidae; with a description of a new genus and species from an American marsupial. *Aust. Zool.* **2**: 154-8.

HARRISON, L., 1924. The migration route of the Australian marsupial fauna. *Aust. Zool.* **3**: 247-63.

HARRISON, L., 1928. Presidential address: Host and parasite. *Proc. Linn. Soc. NSW* **53**: ix-xxxi (foreword).

HOLMES, J. C., 1982. Impact of infectious disease agents on the population growth and geographical distribution of animals. Pp. 37-51 *in* "Population biology of infectious diseases" ed by R. M. Anderson and R. M. May. Springer-Verlag: Berlin and New York.

HONIGBERG, B. M., 1970. Trichomonads. Pp. 469-550 *in* "Immunity to parasitic animals" Vol 2 ed by G. J. Jackson, R. Herman and I. Singer. Appleton-Century-Crofts: New York.

HOPKINS, G. H. E., 1949. The host-associations of the lice of mammals. *Proc. Zool. Soc. Lond.* **119**: 387-604.

HOPKINS, G. H. E., 1957. The distribution of Phthiraptera on mammals. Pp. 88-117 *in* "First symposium on host specificity among parasites of vertebrates". Attinger: Neuchatel.

HOOGSTRAAL, H. AND AESCHLIMANN, A., 1982. Tick-host specificity. *Mem. Mus. natn. Hist. nat. (Paris), Ser. A. Zool.* **123**: 157-71.

HUMPHERY-SMITH, I., 1983. An hypothesis on the evolution of Herpetostrongylinae (Trichostrongyloidea: Nematoda) in Australian marsupials, and their relationship with Viannaidae, parasites of South American marsupials. *Aust. J. Zool.* **31**: 931-42.

INGLIS, W. G., 1971. Speciation in parasitic nematodes. *Adv. in Parasitol.* **9**: 185-223.

JAMESON, E. W., Jr., 1955. A summary of the genera of Myobiidae (Acarina). *J. Parasitol.* **41**: 407-16.

JARDINE, N. AND MCKENZIE, D., 1972. Continental drift and the dispersal and evolution of organisms. *Nature (Lond.)* **235**: 20-4.

JARRETT, E. E. E. AND MILLER, H. R. P., 1982. Production and activities of IgE in helminth infections. *Progress in Allergy* **31**: 178-233.

JOHNSTON, S. J., 1914a. A preliminary contribution to the zoogeography of Australian trematodes and cestodes. *Med. J. Aust.* **1**: 243-4.

JOHNSTON, S. J., 1914b. Trematode parasites and the relationships and distribution of their hosts. *Rep. Australasian Assoc. Adv. Sci.* **14**: 272-8.

KELER, VON S., 1971. A revision of the Australian Boopiidae (Insecta: Phthiraptera) with notes on the Trimenoponidae. *Aust. J. Zool., Suppl.* **6**: 1-126.

KELLOGG, V. L., 1896. New Mallophaga, 1, — with special reference to a collection made from maritime birds of the Bay of Monteray, California. *Proc. Cal. Acad. Sci., 2nd Ser.* **6**: 31-168, pls 2-15.

KELLOGG, V. L., 1913. Distribution and species-forming of ectoparasites. *Amer. Nat.* **47**: 129-58.

KIRSCH, J. A. W., 1977. The comparative serology of Marsupialia and a classification of marsupials. *Aust. J. Zool., Suppl.* **52**: 1-132.

MACKERRAS, M. J., 1958. Catalogue of Australian mammals and their recorded internal parasites. I-IV. *Proc. Linn. Soc. NSW* **83**: 101-60.

MAUËL, J., 1982. Effector and escape mechanisms in host-parasite relationships. *Progress in Allergy* **31**: 1-75.

MAY, R. M. AND ANDERSON, R. M., 1983. Epidemiology and genetics in coevolution of parasites and hosts. *Proc. R. Soc. Lond. B.* **219**: 281-313.

MAYR, E., 1957. Evolutionary aspects of host specificity among parasites of vertebrates. Pp. 7-14 *in* "First symposium on host specificity among parasites of vertebrates". Attinger: Neuchatel.

MIRONOV, S. V., 1982. A distribution of feather mites on recent birds: *Parazitologiia* **16**: 404-11.

MURRAY, M. D. AND CALABY, J. H., 1971. The host relations of the Boopiidae. *Aust. J. Zool., Suppl.* **6**: 81-2.

NOBLE, E. R. AND NOBLE, G. A., 1964. "Parasitology. The biology of animal parasites". 2nd edition. Kimpton: London.

NYBELIN, O., 1917. Australische Cestoden. *Kungl. Svenska Vetenskapsakademiens Handlingar* **52**: 1-48.

PATTERSON, B., 1957. Mammalian phylogeny. Pp. 15-49 *in* "First symposium on host specificity among parasites of vertebrates". Attinger: Neuchatel.

PERKINS, F. O. AND WOLF, P. H., 1976. Fine structure of *Marteilia sydneyi* sp.n. — Haplosporidan pathogen of Australian oysters. *J. Parasitol.* **62**: 528-38.

PERSON, C., SAMBORSKI, D. J. AND ROHRINGER, R., 1962. The gene-for-gene concept. *Nature (Lond.)* **194**: 561-2.

PRICE, P. W., 1980. "Evolutionary biology of parasites". Princeton University Press: Princeton.

RON, D. AND TAL, J., 1985. Coevolution of cells and virus as a mechanism for the persistence of lymphotropic minute virus of mice in L-cells. *J. Virol.* **55**: 424-30.

RIDE, W. D. L., 1971. On the fossil evidence of the evolution of the Macropodidae. *Aust. Zool.* **16**: 6-16.

SANDARS, D. F., 1956. "A comparative study of helminth parasites of Australian and American marsupials". Ph.D. Thesis: University of London.

SANDARS, D. F., 1958. A pancreatic fluke, *Zonorchis australiensis* sp. nov. (Trematoda) from Australian marsupials. *Ann. trop. Med. Parasitol.* **52**: 129-38.

SMITH, J. M. B., 1974. Southern biogeography on the basis of continental drift: a review. *Aust. Mammal.* **1**: 213-29.

SPRATT, D. M., 1984. Further studies of lung parasites (Nematoda) from Australian marsupials. *Aust. J. Zool.* **32**: 283-310.

SPRENT, J. F. A., 1963. "Parasitism". University of Queensland Press: St. Lucia.

SPRENT, J. F. A., 1969. Evolutionary aspects of immunity in zooparasitic infections. Pp. 3-62 *in* "Immunity to parasitic animals" ed by G. J. Jackson, R. Herman and I. Singer. Appleton-Century-Crofts: New York.

STONE, B. F., DOUBE, B. M., BINNINGTON, K. C. AND GOODGER, B. V., 1979. Toxins of the Australian paralysis tick *Ixodes holocyclus*. Pp. 347-56 *in* "Recent advances in Acarology", Vol. 1 ed by J. G. Rodriguez. Academic Press: London and New York.

TALENT, J. A., 1984. Australian biogeography past and present: determinants and implications. Pp. 57-93 *in* "Phanerozoic early history of Australia" ed by J. J. Veevers. Clarendon Press: Oxford.

TRAUB, R., 1972. The zoogeography of fleas (Siphonaptera) as supporting the theory of continental drift. *J. med. Ent.* **9**: 584-9.

TRAUB, R., 1980. The zoogeography and evolution of some fleas, lice and mammals. Pp. 93-172 *in* "Proceedings of the International Conference on Fleas, Ashton Wold/Peterborough/UK/ 21-25 June 1977" ed by R. Traub and H. Starcke. A. A. Balkema: Rotterdam.

TRAUB, R. AND DUNNET, G. M., 1973. Revision of the siphonapteran genus *Stephanocircus* Skuse, 1893 (Stephanocircidae). *Aust. J. Zool., Suppl.* **20**: 1-128.

VANZOLINI, P. E. AND GUIMARÃES, L. R., 1955. South American land mammals and their lice. *Evolution* **9**: 345-7.

VOGT, P. R. AND CONNOLLY, J. R., 1971. Tasmantid guyots, the age of the Tasman Basin, and motion between the Australia plate and the mantle. *Bull. Geol. Soc. Am.* **82**: 2577-84.

WESSENBERG, H. S., 1978. Opalinata. Pp. 552-81 *in* "Parasitic Protozoa", Vol. 2 ed by J. P. Kreier. Academic Press: London and New York.

ZSCHOKKE, F., 1898. Die Cestoden der Marsupialia und Monotremata. *Denkschr. med-naturiv. Gesellsch. Jena.* **8**: 357-80.

NOTE ADDED IN PRESS

Pertinent to a comment I made in the conclusion, Beveridge and Durette-Desset (1985, *Bull. Mus. natn. Hist. nat. (Paris), 4 ser.* **7**: 341-47) have described an herpetostrongylin nematode from the marsupial mole, *Notoryctes typhlops*. Although they considered *Austrostrongylus notoryctis* to represent a 'recent capture' because of some derived characters, its similarities with forms found in macropodids and its inclusion in the genus *Austrostrongylus* suggests diprotodontan affinities for the Notoryctidae because this parasite group has already been shown to demonstrate phylogenetic specificity.

The cytological study of Calaby *et al.* (1974, *Aust. J. Biol. Sci.* **27**: 529-32) supported the separation of the Notoryctidae from the Dasyuridae and Peramelidae and their association with the phalangeroids. Apomorphic status amongst the *Austrostrongylus* nematodes would also group notoryctid parasites near to phalangeroid parasites (Humphery-Smith 1981, *Bull. Mus. natn. Hist. nat. (Paris), 4 ser.* **3**: 509-13). The views of Archer (1976, *Zool. J. Linn. Soc.* **59**: 217-322; 1984, pp 633-808, *in* "Vertebrate zoogeography and evolution in Australasia" ed by M. Archer and G. Clayton, Hesperian Press, Perth) and Szalay (1982, pp 621-40, *in* "Carnivorous marsupials" ed by M. Archer, Royal Zoological Society of New South Wales, Sydney) were supportive of notoryctid affinities not including the Dasyuroidea. Because they also grouped the Perameloidea with the Diprotodonta in the Syndactyla instead of with the Dasyuroidea their opinions differ from that of Humphery-Smith (1983) and the parasitic evidence presented in this article.

Sweet (1904, *Proc. R. Soc. Victoria* **17**: 76-108) and Klaauw (1931, *Bull. Am. Mus. Nat. Hist.* **62**: 1-352), on the basis of the anatomy of Jacobson's organ and the auditory region, grouped the Notoryctidae with the diprotodonts.

SYNDACTYLY IN MARSUPIALS — PROBLEMS AND PROPHECIES

LESLIE S. HALL[1]

Syndactyly of the 2nd and 3rd toes occurs in all Australian diprotodont marsupials and in the polyprotodont family Peramelidae. Despite the diversity in foot morphology shown in these groups, the syndactylous toes are remarkably consistent in their manifestation. There is always a reduction in their size but the two toes show no loss of osseous or muscular components. A number of theories have been advanced to explain the origin and function of marsupial syndactylous toes. They include a form of degeneration, from an ancestral mutant, for arboreal life, and as a grooming organ. Syndactyly is almost non-existant in other mammals but a large number of substances are known to induce syndactyly. The hypothesis that syndactyly arose only once in marsupials is reiterated and it is suggested that this occurred in the Eocene prior to Australia separating from Antarctica. The origin of the marsupial syndactylous toes remains unresolved but their functional retention as a grooming organ is without doubt.

Key Words: Marsupials; Syndactyly; Origins; Evolution.

Pages 245-55 *in* POSSUMS AND OPOSSUMS: STUDIES IN EVOLUTION ed by M. Archer. Surrey Beatty & Sons Pty Ltd and the Royal Zoological Society of New South Wales: Sydney, 1987.

INTRODUCTION

ONE of the first things noticed by the early European explorers about the Australian marsupials was their "very ill-sorted toes". They were of course referring to the syndactylous condition of the 2nd and 3rd toes which are characterised by their reduction to a common size, retention of claws (even when other claws are lost), by the elaboration of specialised musculature and by almost complete containment in a common integument which results in the two toes being manipulated as a single structure. These toes are found in all the diprotodont marsupials and in the polyprotodont perameloids.

A combination of dentition and syndactyly is frequently used in marsupial classification but poses problems. Which condition arose first and which, therefore, should be taken as the key character in the original differentiation of the taxon is considered to be a vexed question by Kirsch (1977). To explain the syndactylous condition in the polyprotodont perameloids several authors (e.g., Abbie 1937, Kirsch 1977) have suggested that syndactyly has arisen twice and represents a convergent state but others, such as Ride (1962), consider that syndactyly arose only once and that this occurred prior to diprotodonty.

The vertebrate limb is a unique structure that has undergone an enormous amount of adaptive change but the precise nature of the mechanisms that control its developmental program are virtually unknown (Krey *et al.* 1984). Lewis (1980 a, b, c) and Szalay (1981) both consider that the evolving foot is replete with apomorphic characters and that these can be keyed to reasonable functional and adaptive roles, thus making the foot an important phylogenetic indicator. Szalay (1982) proposed a classification for marsupials based on a detailed character analysis of their tarsal morphology. In this paper Szalay (1982) implies that syndactyly evolved only once.

Despite its obvious importance, few studies exist on syndactyly *per se* in marsupials. Müller (1930) described the anatomy of the syndactylous macropodid foot in detail and Marshall (1972) looked at the functional evolution of the syndactylous peramelid foot. The functional use of the syndactylous toes has often been commented upon, perhaps most extensively by Boas (1918), Jones (1926) and Müller (1930).

This paper describes the diverse morphology of marsupial feet and reviews theories about the origin and possible function of their syndactylous toes. The occurrence of syndactylous toes in other mammals is compared with the marsupial condition and the use of syndactyly in marsupial phylogenetics is discussed.

PEDAL MORPHOLOGY OF MARSUPIALS

The most significant paper dealing with the anatomy of marsupial feet is that of Müller (1930). Although this work was based on macropodids, observations were also made on the feet of a wide range of both syndactylous and diadactylous marsupials. Studies by Lewis (1964), Marshall (1972) and Szalay (1982) have also contributed to the understanding of marsupial foot structure and function. The osteology and myology of the feet of a number of marsupials has been described by Coues (1872), Macalister (1872), Cunningham (1881, 1882), Owen (1874), Sidebotham (1885), MacCormick (1887), Parsons (1896, 1903), Alezias (1901), McMurrich (1907),

[1]Department of Anatomy, University of Queensland, St. Lucia, Queensland, Australia 4067.

Thompson and Hiller (1905), Glaesmer (1908, 1910), Carlsson (1914), Osgood (1921), Sonntag (1922), Forster (1926), Heighway (1939) and Lansmeer (1979). Most of these papers have been reviewed in the monograph on the musculature of *T. vulpecula* by Barbour (1963).

There is much variation in the external pedal morphology of marsupials (see Winge 1893, Dollo 1899, Bensley 1903, Jones 1925 and Müller 1930). Figure 1 shows the morphological variation of marsupial feet from a primitive form, to the least and most highly modified forms of both syndactylous and diadactylous marsupials. Modifications to the primitive foot form have occurred as a result of specialisations in life style and or mode of locomotion.

The 1st toe, when present, is either at right angles to the other toes, forming an opposing hallux (most arboreal marsupials), or is directed along the side of the foot (most terrestrial marsupials). Except for *Hypsiprymnodon*, the first toe is absent in macropodoids and in some of the terrestrial dasyurids (Fig. 1f, m). This toe is always clawless.

The 2nd and 3rd toes are usually reduced and equal in length, particularly when they form the syndactylous toes. These toes are generally closely appressed and always possess terminal claws. In the case of *Tarsipes*, these are the only claws on the foot (Fig. 1d). A number of didelphids have subequal 2nd and 3rd toes which are smaller than the other toes (Fig. 1n).

The 4th toe is always the longest and contains the largest elements of the foot. This is most pronounced in terrestrial groups such as the macropodoids and perameloids (Fig. 1e,l,m). The 5th toe is usually smaller than the 4th, particularly in syndactylous marsupials, and reflects the elements of the 4th toe.

A detailed description and diagrams of the foot musculature of T. *vulpecula* can be found in Barbour (1963) and for a number of marsupials in Müller (1930). Despite a complexity, there is a close similarity in the anatomy of the syndactylous digits. A brief description of the general pattern of the muscles relative to the syndactylous 2nd and 3rd toes follows.

The two syndactylous toes are always reduced in size but there is no reduction in the number of bones or muscles in these toes. In fact, the intrinsic muscles of the 2nd and 3rd toes are well developed (Müller 1930, Barbour 1963).

The basis of movement of the syndactylous toes is the fixed position of metatarsophalangeal joints 2 and 3. The syndactylous toes 2 and 3 can be abducted by the action of the extensor digitorum longus. They can be adducted by the action of the short flexors of the toes and of the medial bands of the deep flexors of the toes. The short flexors of the 2nd and 3rd toes act in adduction of the foot while the deep flexor tendons and the tibialis posterior act through their insertions of the distal phalanx of the toes to adduct them so that the closed position is put at the best angle against the fur. This movement is completely independent of the rest of the movements of the leg and foot, suggesting that these two toes are specialised (Müller 1930).

It is generally agreed that the primitive marsupial foot would have been the diadactylous foot of a small arboreal and insectivorous mammal showing four equal toes and an opposing hallux (Fig. 1s). This theory was first put forward by Huxley (1880) and further supported by Dollo (1899), Bensley (1903) and Mathew (1904). The least specialised marsupial hind feet (i.e., most primitive) are now found in dasyurids and didelphids (Fig. 1). The most primitive form of a syndactylous foot is seen in some diprotodontans (e.g., *Petaurus* and *Cercartetus*) where the combined 2nd and 3rd syndactylous toes form a single digit equal in diameter to, but shorter in length than digits 4 and 5), and the 1st toe forms an opposing hallux (Fig. 1p). One could assume from this that the arboreal lifestyle of these diprotodontans resembles that of ancestral marsupials.

By these criteria the most specialised syndactylous feet are those of *Chaeropus ecaudatus* (Peramelidae) (Fig. 1e), where the 4th toe is greatly enlarged and elongated, the 1st toe is short and the 2nd, 3rd and 5th toes are greatly reduced. *Tarsipes rostratus* (Tarsipedidae) shows an interesting modification in having toes 2 and 3 joined over their entire length and the "single" toe capped with two non-opposing claws (Fig. 1d). None of the other toes possesses a claw. The 5th toe of *Tarsipes* is angled laterally and away from the line of the foot. The opposing 1st toe is not directed 90° medially but is inclined forward at 45°. These modifications presumably reflect its arboreal lifestyle but the unusual nature of them suggest that a detailed study of the *Tarsipes* foot may prove interesting in relation to the development of syndactyly as a response to arboreal habits.

The macropodoids have feet that have been modified for saltatory locomotion. With the exception of the syndactylous toes there has been a lateral fixation of all the joints from the knee to the end of the toes. The 4th toe has been greatly elongated to form a median toe which is balanced on either side by the smaller 5th and combined 2nd and 3rd toes (Fig. 1m). A similar situation occurs in perameloids (Figs 1e, l and Fig. 2).

Intermediate between the arboreal diprotodontans and terrestrial macropodoids and perameloids is *Hypsiprymnodon* which shows elongation of the pes and 4th toe, a medium

VARIATIONS IN THE PEDAL MORPHOLOGY OF MARSUPIALS

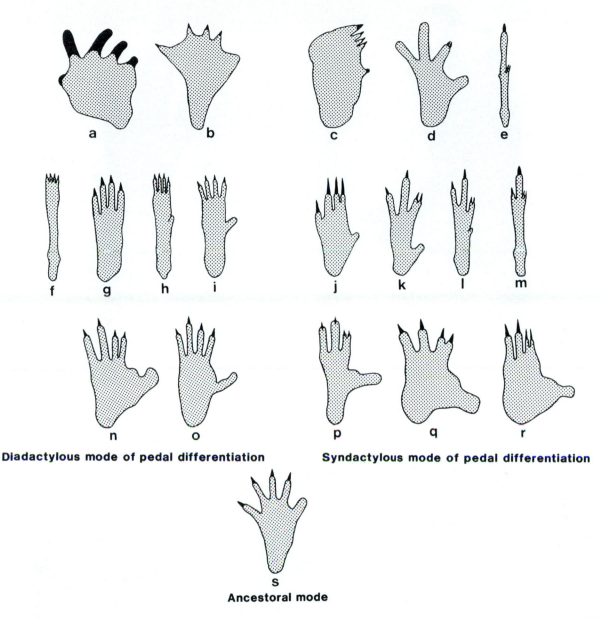

Diadactylous mode of pedal differentiation   **Syndactylous mode of pedal differentiation**

Ancestoral mode

Fig. 1. Outlines of the pedal morphology of marsupials drawn in proportion but at various scales to facilitate comparisons. Top line (a to e) are highly divergent feet of both diadactylous and syndactylous marsupials, second line (f to m) are terrestrial marsupials, and third line (n to r) are arboreal marsupials. Species and source of information are: a, *Notoryctes typhlops*, Jones 1925; b, *Chironectes minimus*, Winge 1893; c, Diprotodont sp., Bensley 1903; d, *Tarsipes rostratus*, Dollo 1899; e, *Chaeropus ecaudatus*, Jones 1925; f, *Antechinomys laniger*, Jones 1925; g, *Myrmecobius fasciatus*, Jones 1925; h, *Sminthopsis crassicaudata*, Jones 1925; i, *Sminthopsis murina*, Dollo 1899; j, *Vombatus ursinus*, Jones 1925; k, *Hypsiprymnodon moschatus*, Jones 1925; l, *Perameles bougainville*, Jones 1925; m, *Macropus giganteus*, Müller 1930; n, *Didelphis microtarsus*, Winge 1893; o, *Didelphis elegans*, Dollo 1899; p, *Cercartetus concinnus*, Jones 1925; q, *Phalanger celebensis*, Dollo 1899; r, *Trichosurus vulpecula*, Jones 1925; s, *Didelphis crassicaudata*, Winge 1893.

reduction of toe 5 and combined 2 and 3 and a prominent lst toe (Fig. 1k). It is the only macropodoid with 5 toes and on this basis is regarded to exhibit an ancestral macropodoid condition (Archer 1984). The pedal morphology of *Hypsiprymnodon* also reflects its quadrupedal gait and the complex habitat in which it is found (Johnson and Strahan 1982).

It is interesting to note that no terrestrial diadactylous marsupials have elongated toes. However, *Antechinomys*, which is known to gallop (Ride 1965), does have an elongated pes (Fig. 1f).

Syndactyly of digits 3 and 4 has been recorded on the front feet of *Thylogale billardierii* (Lyne 1953) but this was an obvious abnormality and will be discussed later in the text.

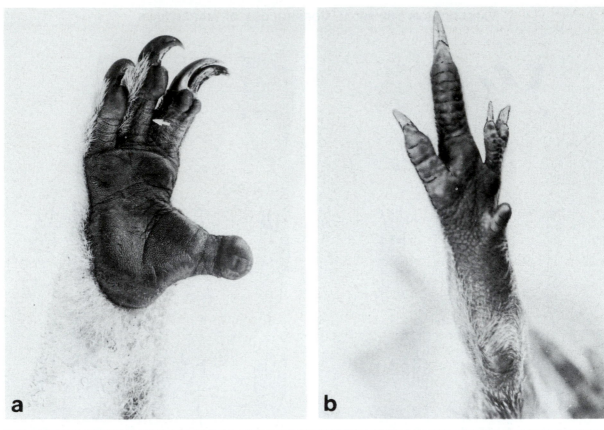

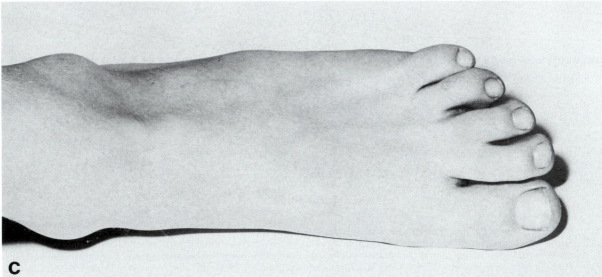

Fig. 2. Syndactyly in the feet of mammals: a, *Phascolarctos cinereus* (note additional webbing between toes 3 and 4; arrow). b, *Isoodon macrourus*; first toe reduced to a stump and the 4th toe is elongated. The nails on the syndactylous 2nd and 3rd toes oppose each other forming cathaerodactylous toes used for grooming. c, Human foot showing soft tissue syndactyly between 2nd and 3rd toes to the level of the lst phalangeal joint.

Although syndactyly is regarded to be restricted to marsupials found in the Australian region, incipient syndactyly has been recorded in *Marmosa* and *Caluromys* by Bensley (1903) and in American didelphids by Tate (1933). Goodrich (1935) investigated the foot of *Marmosa pusilla* and reported that interdigital webbing occurred between the subequal 2nd and 3rd toes and extended half the length of the 1st phalanges. He did not consider that it represented syndactyly.

Kirsch (1977) reports sending a photograph of the foot of *Caluromys derbianus* to G. G. Simpson which showed toes 2 and 3 nearly equal and less divergent than the other toes but apparently without a common integument beyond the metatarsals. It was not regarded as a syndactylous foot but it was thought that it could well represent a primitive basis for the evolution of syndactyly. It would appear that "incipient syndactyly" occurs in a number of other South American marsupials (Fig. 1n). Winge (1893) and Müller (1930) depicted a number of didelphids whose feet had sub-equal 2nd and 3rd toes which were reduced in relation to the 4th and 5th toes.

Chironectes minimus, a diadactylus South American marsupial, exhibits complete webbing between all five toes (Fig. 1b). This is due to its aquatic lifestyle and is termed zydactyly. It also exhibits subequal and reduced 2nd and 3rd toes.

Additional webbing of the integument between the 3rd and 4th toes is a feature in a number of the possums and also occurs in *Phascolarctos* and *Vombatus* (Fig. 1j and Fig. 2) and has been illustrated by Jones (1920, 1922, 1923, 1924) and Boardman (1943). In *Trichosurus vulpecula* the webbing reaches the level of the first phalangeal joint and is almost as extensive as the syndactylous webbing between the 2nd and 3rd toes. The significance of this "double" syndactyly has not been commented upon and requires further work.

A comparative study of the embryological development of diadactylous and syndactylous digits in marsupials has not yet been attempted. Such a study could prove rewarding in terms of the ontogeny of syndactyly in marsupials.

SYNDACTYLY IN OTHER MAMMALS

Natural Occurrence

Only two other mammals naturally exhibit a similar type of syndactyly that occurs in marsupials. One is the small African Otter Shrew, *Micropotamogale lamottei* (Insectivora; Potamogalidae) and, although the syndactylous digits are on the hind feet of this mammal, they involve the 3rd and 4th toes (Fig. 3a). It is known that *Micropotamogale* uses these toes for grooming (Walker 1968). The other mammal exhibiting syndactyly is the Great Gibbon, with the appropriate scientific name *Symphalangus syndactylus* (Primates; Pongidae). This primate from Borneo has a similar webbing between toes 2 and 3 and it is suspected that it uses its syndactylous toes as a grooming organ (Walker 1968). Further work on the osteology and myology of the syndactylous digits of these two mammals is necessary however, before an opinion can be reached as to whether the condition is convergent with that seen in marsupials.

Abnormal Occurrence

Genetic abnormalities. A large number of inherited abnormalities affecting the feet have been recorded and their genetics investigated (Grüneberg 1963, Zwilling 1969, Wolpert 1976). However, the precise mechanisms whereby the genotype controls this development remains, as in normal development, unknown. In general terms syndactyly results from a lack of differentiation between two digits. Normally, the mesenchyme in the periphery of the hand and foot plates condenses to form the primordia of the fingers and toes and the thinner tissue between them breaks down via necrosis and cell death. Webbing between digits results from failure of this tissue breakdown to occur.

Epidermal syndactylism occurs in a recessive laboratory mutant in the Vole *Microtus arvalis*. It is usually manifested in digits 3 and 4 and hind limbs are more affected than fore limbs (Frank 1959). In all cases of what Frank termed "primary" syndactyly, the digits were joined by the skin complete with all its layers and there were no osseous abnormalities. Unfortunately there are no data on the associated digital musculature.

Bovine syndactylism is manifested more frequently and is more complete in the fore limbs of Cattle (Eldridge *et al.* 1951). Syndactylous (or "mule-footed") swine which were known to Aristotle, are the result of a simple dominant gene (Detlefsen and Carmichael 1921, Kosswig and Ossent 1936). However, due to the modifications already present on the feet of Cattle and Pigs (Artiodactyla) the two above cases serve merely to show that syndactyly occurs as a genetic abnormality in other mammalian orders.

Chondrodystrophy in the Rabbit is usually lethal and Grüneberg (1963) records that Rabbits suffering from this occasionally had syndactylous toes on the hind feet. Grüneberg regards syndactyly to be caused by a recessive gene (sm/sm), a double recessive producing the syndactylous condition. The primary fusion is always between the digits of the hind feet, usually digits 3 and 4, but digit 2 is frequently involved.

Kadam (1962) studied the anatomy of syndactylous (sm/sm) Mice and showed that any osseous fusion only involved the phalanges and not the metacarpals or metatarsals which remain separate. Muscular anomalies are more frequent in the hind limb and the tendons seem to be more affected than the muscles.

Brachydactylia in the Rabbit is due to a recessive gene with complete penetrance but very viable manifestation. All homozygotes show defects on one or more of the feet ranging from minor involvement of the digit down to complete absence of hands or feet (acheiropodia). Syndactyly is frequent and invariably involves digits 2, 3 and 4 or the hind feet (Grüneberg 1963).

Grüneberg (1956) showed that syndactyly was a spontaneous genetic mutation in the A/Fa inbred strain of Mice.

Syndactyly in Humans is inherited as a simple dominant or simple recessive trait and is most frequently observed between the 3rd and 4th fingers and between the 2nd and 3rd toes (Moore

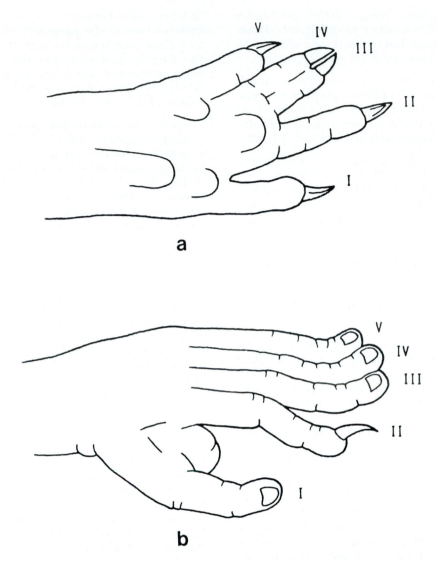

Fig. 3. Pedal morphology of: a, *Micropotamogle lamottei* showing syndactyly in the 3rd and 4th
toes; and b, *Perodicticus potto* with the specialised grooming nail on the 2nd toe.

1982) (Fig. 2c). It is the most common limb malformation in Humans.

Induced abnormalities. A large number of substances affect development of the limbs, though usually they are not specific to limbs alone. Krey *et al.* (1984) have pointed out that many of these substances affect the developing cardiovascular system, cell proliferation and differentiation, cell death and carbohydrate metabolism. Unfortunately, in no case is the mechanism clearly understood and even a description of the development of the abnormalities, particularly at the cellular level, is lacking. However, it is known that damage to the vascular system does cause abnormalities to the surrounding tissue, as occurs with thalidomide (Jurand 1966).

The compound 2,2'-depyridyl (DIP), a chelator for ferrous iron, produces limb defects. When pregnant female Rats were injected with DIP, a high proportion of their young were born with defects to the limbs. 70% of the defects, which ranged from fusion to missing toes, involved digits 3 and 4 (Oohira and Nogami 1978). Kochhar *et al.* (1978) showed that cytoseine arabionoside caused polydactyly in 5% of forelimbs and 20% of hindlimbs of Rats. Kochhar (1977) has shown that when retinoic acid (Vitamin A) is administered to pregnant Rats on the 12th and 13th day of gestation, it caused abnormalities which were confined to the limbs and palate. Most authors attribute the teratogenic effect to an inhibition of DNA synthesis causing a lack of collagen biosynthesis.

POSSIBLE ORIGINS AND FUNCTIONAL USE OF SYNDACTYLY IN MARSUPIALS

Syndactyly Represents a Form of Degeneration

Owen (1874) considered that syndactyly was a condition of degeneration but this has been refuted by a number of authors including Jones

(1925) and Müller (1930). Against the concept that syndactylous toes are involved in the process of degeneration is the fact that the muscles connected to these toes although reduced in size, are well developed (see Müller 1930, Barbour 1963). If the syndactylous toes were in the process of degeneration the muscles involved would have shown some sign of reduction. Also, a positional change of the body of individual muscles would be expected to accompany degeneration of digits.

In relation to the other toes, the skin and underlying connective tissue of the syndactylous toes are well supplied with nerve endings (Müller 1930). This would not be the case if degeneration was occurring. Also the claws on the syndactylous toes are frequently a different shape to the others and in the case of *Tarsipes* are the only claws on the feet (Fig. 1d). This differentiation suggests specialisation rather than degeneration.

Syndactyly Resulted from an Ancestral Mutant

Since only one other species of mammal has developed the same form of syndactyly in its hind limb digits, Hall (1981) considers that it is unlikely that syndactyly has arisen twice in marsupials.

Hall (1981) believes that the syndactylous condition arose in the late Cretaceous or Paleocene which closely agrees with the thoughts of Archer (1984) who considers that syndactyly probably arose at least as early as the Eocene prior to the separation of Australia from Antarctica. Abbie (1937) also believed that the separation of the polyprotodont and diprotodont condition was already well established before marsupials reached Australia. Marsupial footprints, exhibiting syndactyly of digits 2 and 3 of the pes, have been described from Cetaceous sediments in British Columbia, Canada, by Sarjeant and Thulborn (ms. in preparation). It is also possible that other American Mesozoic marsupials could have been syndactylous but very few specimens are complete with postcranial skeleton details.

As for the cause of syndactyly, Hall (1981) speculates a link between a possible teratogenic reaction causing genetic change due to phytochemicals new to the diet when some of the marsupials changed from an insectivorous to a herbivorous diet. This occurred in the late Cretaceous and early Tertiary and corresponds to the appearance and radiation of the angiosperms and the arrival of marsupials in Australia.

Many plants are known to cause abnormalities to fetuses when eaten by pregnant animals. Angiosperms are the most frequently cited group (Kingsbury 1964), particularly those which contain high concentrations of alkaloids (Keeler 1975) and those which accumulate high levels of minerals (Kingsbury 1964). Although a wide variety of mutations can be induced it is not known if any plants specifically cause syndactyly.

The main problem with this argument is the genetic fixing of a mutation caused by a teratogenetic agent. There is some evidence that this can occur (Landauer 1948) but differences between animals prevents extrapolation. Further detailed studies on the anatomy of induced syndactyly are also necessary before meaningful comparisons can be made with Müller's (1930) work on syndactyly in marsupials.

An alternative explanation for the genetic cause of syndactyly is that it resides in marsupials as a recessive gene (e.g. Lyne 1953) and, under a constant selective pressure such as the need to groom, could be manifested and maintained in the phenotype. If this were the case, syndactyly would have appeared many times and would still be occurring in marsupials. Given the obligatory nature of grooming, this argument also fails to explain why syndactyly has not appeared in extant South American marsupials and the Australian dasyurids.

Syndactyly is an Adaptation for an Arboreal Lifestyle

Dollo (1899) followed Huxley's (1880) theory that marsupials descended from arboreal ancestors and further speculated that syndactyly, in that it combines two digits into one allowing for a better grip, was an adaptation for climbing in trees. The hypothesis of an arboreal ancestry of marsupials has been further supported by Bensley (1903) and Mathew (1904) and recently by Szalay (1982) and Lewis (1983) who considered that the basic foot structure of early therians proved to be highly adaptable and allowed marsupials to become committed arborealists. Lewis (1983) also considered that the feet clearly possessed considerable evolutionary plasticity and permitted modifications that also catered for the invasion of a variety of terrestrial niches by both marsupials and placentals. This is well supported by Figure 1.

The combining of two digits into a structure which functions as a single digit as an aid to climbing amongst small branches is well displayed in *Tarsipes*, *Cercartetus* and *Petaurus breviceps* where the 2nd and 3rd toes are combined to form a "single" toe (Fig. 1d, p). In *Tarsipes*, claws are absent except on the 2nd and 3rd "single" toe and in *Cercartetus* the small claws are contained in a terminal groove on toes 2 to 5 (Jones 1925).

While this hypothesis holds true for a number of arboreal marsupials and possibly for ancestral marsupials, there are major exceptions. The terrestrial perameloids and predominantly terrestrial macropodoids are all syndactylous and, while this may merely reflect their ancestry, the arboreal dasyurids and didelphids are diadactylous. Also, the lack of syndactyly in virtually all other arboreal mammals fails to support Dollo's hypothesis.

The Syndactylous Toes are a Grooming Organ

Jones (1925) put forward a strong argument supporting Owen's (1874) statement that syndactyly consists essentially of the "conversion of the two inner toes into unguiculate appendages, useful only in cleansing the fur". Although Owen regarded degeneration as the basic cause of syndactyly, Jones described syndactyly as a persisting specialisation whilst undergoing degeneration. Jones (1925, 1926) considered that the stage of development of the syndactylous toes was related to the nature and texture of the coat of the animal. His observation was that the more dense and coarser the fur, the more pronounced was the syndactylous "comb". He also put forward the theory that the syndactylous marsupials were infested with more ectoparasites than the diadactylous marsupials. Apparently Jones was unaware of a similar argument put forward earlier by Brandes (1906) but published in German.

The statements by Jones (1925) relating fur density with the degree of syndactyly contain many exceptions and have been refuted by several authors (see Müller 1930). Müller (1930) also disagreed with Jones' ectoparasite concept and showed that almost all marsupials are infected equally with *Mallophaga* and other ectoparasites such as *Ixodes* and *Siphonoptera*. Müller pointed out that in many cases the same species of parasites infect both syndactylous and diadactylous marsupials. Müller concluded that the syndactylous toes were definitely a toilet organ and were a highly specialised development in marsupials.

The use of the syndactylous toes as a toilet organ was also supported by the observations of Boas (1918). He described a modification of the claws of toes 2 and 3 as "cathaerodactyly". He found that in these toes the slightly concave "sole" of these claws, which on other digits is symmetrical, is asymmetrically developed. The concavity of the claw of the 2nd toe extends on its postaxial side and that of the claw of the 3rd toe to its preaxial side. The concavities face each other and the claws are thus mirror images of each other. This would give the claws a gripping area between them and would make them more than just scratching implements (Fig. 2b).

Boas (1918) also describes a similar asymmetry of the 2nd claw in most didelphids and to a slight extent of claw 3. But in this family claw 3 is asymmetrical in the same direction as claw 2. Boas believes that all marsupials originally possessed such modified "Putzkrallen" but in some groups it was subsequently lost.

Relatively speaking, the syndactylous toes have a greater specialisation of their muscles than the rest of the toes which have very strong muscles to produce uncomplicated movement. This prompted Müller (1930) to suggest that the specialisation of the syndactylous toes evolved simultaneously with the fixation of the hind legs in the terrestrial macropodids. The arboreal marsupials which have syndactylous toes, especially *Phascolarctos cinereus*, can move their hind limb in a parality of directions (Müller 1930). This means they can move the whole foot with syndactylous toes against the body so that they can scratch and clean themselves with the whole foot. This capacity is reduced in perameloids and completely lost in macropodoid who can only groom (with syndactylous toes) in a zone parallel to the body. Although this sounds restrictive, anyone who has watched a macropodoid groom cannot help being amazed by the dexterity and use of their syndactylous digits.

While not exhibiting syndactyly, there are some similarities in the toes involved as grooming organs in other mammals.

Almost all the members of the primate Family Lorisidae have a claw on their 2nd toe but only a flat nail on the remaining toes including the opposing 1st toe. In the Potto (*Perodicticus potto*; Lorisidae), the 2nd toe which has an extended claw is markedly reduced (Fig. 3b). Other primates in the Family Lemuridae, such as the Hairy-eared Dwarf Lemur (*Cheirogaleus trichotis*) and the Fork-marked Lemur (*Phaner furcifer*), have their 2nd toe reduced in size and tipped with a claw instead of a nail as on their other digits. None of the above mentioned primates have dense fur but all use the clawed 2nd toe for grooming (Walker 1968).

Grooming is accomplished via specialised digits in several genera of rodents. The Spiny Pocket Mouse (*Liomys pictus*) and *Pedetes capensis* both have specialised claws on their 2nd toes for grooming (Walker 1968). The Beaver (*Castor canadensis*) has a further modification, having both the claws on the 2nd and 3rd toes split. The Beaver draws its coarse fur between these split claws while pushing its foot against its body to enhance the grip (Bailey 1923). The South American Plains Viscacha (*Lagostomus maximus*) grooms its coarse fur between its "normal" toes which are supplied with a tuft of bristles Walker (1968). Müller (1930) interpreted similar tufts of bristles on the feet of Kangaroo Mice (*Microdipodops* sp.) as a grooming device but they have since been shown to function as an aid for locomotion over loose sandy soil (Walker 1968).

It would seem then that in mammals the hind foot is definitely favoured over the front foot for the location of a grooming organ and that toes 2 and 3 and to a lesser extent 4 are the ones modified to form the grooming organ. There is no doubt that the syndactylous digits are an important toilet organ in marsupials but was this the inductive stimulus for its presence?

A major problem with the toilet organ theory is why extant South American marsupials do not possess syndactylous toes, or perhaps a specialised claw, if grooming is so important in marsupials. Also, one would expect to see a greater variation in the form of the two syndactylous toes due to fur characteristics and the various modes of locomotion as reflected in marsupial feet (Fig. 1).

The use of polyprotodont and diprotodont dentition in grooming requires closer observation. It may be that syndactyly developed as a result of diprotodont teeth being ineffective in the removal of ectoparasites from fur.

CONCLUSIONS

Most recent authors accept a South American origin for marsupials (see Archer 1984) except Kirsch (1984) who predicts that future fossil finds will establish Australia as the marsupial cradle.

Syndactyly predates diprotodonty (Ride 1962) and both were present before marsupials arrived in Australia. That stem marsupials had different diets in South America is indicated by there being both diprotodont and polyprotodont groups before they reached Australia. Polyprotodonty is regarded as more primitive than diprotodonty and reflects a change by early marsupials from an insectivorous to a herbivorous diet. This corresponds in time to the radiation of angiosperms and the possible arrival of marsupials in Australia (Archer 1984). Implications arising from Goodrich's (1935) observations are that *Marmosa* and the Australian syndactylous marsupials shared a common ancestor. Secondly, because there are both syndactylous and diadactylus marsupials in Australia, ancestral marsupials on entering Australia had either incipient syndactyly or were diadactylous.

There has been much speculation on the cause and origin of syndactyly in marsupials but the problem remains unresolved. The incipient syndactyly that occurs in the 2nd and 3rd toes of a number of South American marsupials (Fig. 1n) may represent an incomplete expression of syndactyly. Webbing is known to exist between the subequal 2nd and 3rd digits of *Marmosa pusilla* (Bensley 1903) but it is not regarded as a syndactylous condition (Goodrich 1935). Szalay's (1982) work on apomorphy in marsupial tarsal elements indicates that more studies in the pedal morphology of South American marsupials are warranted. In particular, the pedal morphology of *Dromiciops australis* needs close scrutiny as this animal has been shown to have striking similarities with Australian marsupials in tarsal morphology (Szalay 1982) and with its chromosomes (Sharman 1982).

That when pedal modifications occurred in marsupials, they did so in a short period of time, is supported by Maderson (1975) who considers that morphological changes in structures have occurred suddenly and have had evolutionary significance. Such could be the case for the appearance of syndactylous digits in marsupials. A teratogenic syndactylous condition as suggested by Hall (1981) could have been followed by an intense selective pressure for its retention, possibly for grooming or climbing.

Since it usually involves the 2nd and 3rd toes, teratogenic syndactyly in other mammals exhibits close similarities to the syndactylous condition in marsupials. However, further detailed studies on the anatomy of teratogenic syndactyly are needed before the symplesiomorphy of all syndactylous feet can be established.

With most of the stages of toe formation and separation occurring after the young is born and located in a pouch, the marsupial neonate foot would make an excellent paradigm for studies on the development of the foot. In particular, marsupials with syndactylous toes provide a unique opportunity to study the mechanisms of cell death and digit separation side by side.

ACKNOWLEDGEMENTS

During the writing of this article a number of people provided helpful comments and stimulus. In particular I would like to thank Drs M. Archer, M. M. Bryden, R. L. Hughes and F. S. Szalay and an anonymous referee.

REFERENCES

ABBIE, A. A., 1937. Some observations on the major subdivisions of the Marsupialia with special reference to the position of the Peramelidae and Caenolestidae. *J. Anat.* **71**: 429-35.

ALEZAIS, H., 1901. Les muscles du membre postérieur du Kangouru. *C.R. Soc. Biol. (Paris)* **53**: 971.

ARCHER, M., 1984. The Australian marsupial radiation. *In* "Vertebrate zoogeography and evolution in Australasia" ed by M. Archer and G. Clayton. Hesperian Press: Perth.

BAILEY, V., 1923. The combing claws of the beaver. *J. Mammal.* **4**: 77-79.

BARBOUR, R. A., 1963. The musculature and limb plexuses of *Trichosurus vulpecula*. *Aust. J. Zool.* **11**: 488-610.

BENSLEY, B. A., 1903. On the evolution of the Australian Marsupialia; with remarks on the relationships of the marsupials in general. *Trans. Linn. Soc. Lond. Zool. (2nd ser.)* **9**: 83-217.

BOARDMAN, W., 1943. On the external characters of the pouch young of some Australian marsupials. *Aust. Zool.* **10**: 138-60.

BOAS, J. E. V., 1918. Zur Kenntnis des Hinterfusses der Marsupialier. *Kgl. Danske Vidensk. Selsk. Biol. Medd.* **1**(8): 1-26.

BRANDES, G., 1906. Von den Känguruhs. *Mitt. Zoo Halle, H.G.*

CARLSSON, A., 1914. Uber *Dendrolagus dorianus*. *Zool. Jb.* **36:** 547-617.

COUES, E., 1872. On the osteology and myology of *Didelphis virginiana*. *Mem. Boston Soc. Nat. Hist.* **2:** 41-149.

CUNNINGHAM, D. J., 1981. The nerves of hind limb of *Thylacinus* and *Cuscus*. *J. Anat. Physiol.* **15:** 265-77.

CUNNINGHAM, D. J., 1882. Some points in the anatomy of the thylacine (*Thylacinus cynocephalus*), collected by H.M.S. Challanger, during the years 1873-1876; with an account of the comparative anatomy of the intrinsic muscles and nerves of the mammalian pes. *In* Report on the scientific results of the voyage of H.M.S. Challenger during the years 1873-1876. (Zool.) Vol. **5:** 1-192.

DETIEFSEN, J. A. AND CARMICHAEL, W. J., 1921. Inheritance of syndactylism, black and dilution in swine. *J. Agric. Res.* **20:** 595-604.

DOLLO, L., 1899. Les ancestres des marsupiaux, étaient ils arboricoles? *Trav. Sta. Zool. Wimereux* **7:** 188-203.

ELDRIDGE, F. E., SMITH, W. H. AND McLEOD, W. H., 1951. Syndactylism in Holstein-Fresian cattle. Its inheritance, description and occurrence. *J. Hered.* **42:** 241-50.

FORSTER, A., 1926. Considérations sur la statique due pied chez le kangourou (*Macropus rufus, Petrogale pencillata*). *Arch. Anat. Hist. Embryol.* **5:** 141-59.

FRANK, F., 1959. 'Geheftete Zehen' — eine neve Mutation bei der Feldmaus (*Microtus arvalis* Pallas). *Z. Säugertier K.* **24:** 89-91.

GLAESMER, E., 1908. Untersuchung über die Flexorengruppe am Unterschenkel und Fuss der Saugetiere. *Morph. Jb.* **38:** 36-90.

GLAESMER, E., 1910. Die Beugemuskeln am Unterschenkel und Fuss bei den Marsupialia, Insectivora, Edentata, Prosimiae und Simiae. *Morph. Jb.* **41:** 149-336.

GOODRICH, E. S., 1935. Syndactyly in marsupials. *Proc. Zool. Soc. Lond.* **1935:** 175-78.

GRUNEBERG, H., 1956. Genetical studies on the skeleton of the mouse. XVIII. Three genes for syndactylism. *J. Genet.* **54:** 113-45.

GRUNEBERG, H., 1963. "The pathology of development. A study of inherited skeletal disorders in animals." Blackwell: Oxford.

HALL, L. S., 1981. The problem with syndactyly in marsupials. *Bull. Aust. Mammal. Soc.* **7:** 36-37.

HEIGHWAY, F. R., 1939. "The anatomy of *Hypsiprymnodon moschatus*". M.D. Thesis: University of Sydney.

HUXLEY, T. H., 1880. On the application of the laws of evolution to arrangement of the vertebra, and more particularly of the Mammalia. *Proc. Zool. Soc. Lond.* **1880:** 649-62.

JOHNSON, P. M. AND STRAHAN, R., 1982. A further description of the Musky Rat-kangaroo, *Hypsiprymnodon moschatus* Ramsay, 1876 (Marsupialia, Potoridae), with notes on its biology. *Aust. Zool.* **21:** 27-46.

JONES, F. W., 1920. The external characters of pouch embryos of marsupials. 1. *Trichosurus vulpecula*, var. *typicus*. *Trans. Proc. Roy. Soc. S. Aust.* **44:** 360-73.

JONES, F. W., 1922. The external characters of pouch embryos of marsupials. 4. *Pseudochirops dahli*. *Trans. Proc. Roy. Soc. S. Aust.* **46:** 119-30.

JONES, F. W., 1923. The external characters of pouch embryos of marsupials. 5. *Phascolarctos cinereus*. *Trans. Proc. Roy. Soc. S. Aust.* **47:** 129-35.

JONES, F. W., 1924. The external characters of pouch embryos of marsupials. 9. *Phascolomys tasmaniensis*. *Trans. Proc. Roy. Soc. S. Aust.* **48:** 145-48.

JONES, F. W., 1925. "The mammals of South Australia". Parts 1-3. South Aust. Gov. Printer: Adelaide.

JONES, F. W., 1926. The R. M. Johnston memorial lecture, 1925. The mammalian toilet and some considerations arising from it. *Proc. Roy. Soc. Tasmania* **1926:** 14-62.

JURAND, A., 1966. Early changes in limb buds of chick embryos after thalidomide treatment. *J. Embryol. Exp. Morphol.* **16:** 289-300.

KADAM, K. M., 1962. Genetical studies on the skeleton of the mouse. XXXI. The muscular anatomy of syndactylism and oligosyndactylism. *Genet. Res. Camb.* **3:** 139-56.

KEELER, R. F., 1975. Toxins and teratogens of higher plants. *Lloydia* **38:** 56-86.

KINGSBURY, J. M., 1964. "Poisonous plants of the United States and Canada". Prentice-Hall: New Jersey.

KIRSCH, J. A. W., 1977. The classification of marsupials. *In* "The biology of marsupials" ed by D. Hunsaker. Academic Press: New York.

KIRSCH, J. A. W., 1984. Marsupial origins: taxonomic and biological considerations. *In* "Vertebrate zoogeography and evolution in Australia" ed by M. Archer and G. Clayton. Hesperian Press: Perth.

KOCHHAR, D. M., 1977. Limb development in mouse embryos. I. Analysis of teratogenic effects of retinoic acid. *Teratology* **7(3):** 238-98.

KOCHHAR, D. M., PENNER, J. D. AND McDAY, J. A., 1978. Limb development in mouse embryos. II. Reduction defects, cytotoxicity and inhibition of DNA synthesis produced by cytosine aribinoside. *Teratology* **18:** 71-92.

KOSSWIG, C. AND OSSENT, H. P., 1936. Syndaktylie einhugiger Schweine und weiterc Ergänzungen zur Vererburg der Schweinefarben. *Zuchter* **8:** 324-29.

KREY, A. D., DAYTON, D. H. AND GOETINCK, P. F., 1984. NICHD Research Workshop: normal and abnormal development of the limb. *Teratology* **29:** 315-23.

LANDAUER, W., 1948. The phenotypic modification of heredity polydactylism of fowl by selection and by insulin. *Genetics* **33:** 133-57.

LANSMEER, J. M. F., 1979. The extensor assembly in two species of opossum, *Philander opossum* and *Didelphis marsupialis*. *J. Morph.* **161:** 337-46.

LEWIS, O. J., 1964. The homologies of the mammalian tarsal bones. *J. Anat.* **98:** 195-208.

LEWIS, O. J., 1980a. The joints of the evolving foot. Part I. The ankle joint. *J. Anat.* **130:** 527-43.

LEWIS, O. J., 1980b. The joints of the evolving foot. Part II. The intrinsic joints. *J. Anat.* **130:** 833-57.

LEWIS, O. J., 1980c. The joints of the evolving foot. Part III. The fossil evidence. *J. Anat.* **131:** 275-98.

LEWIS, O. J., 1983. The evolutionary emergence and refinement of the mammalian pattern of foot architecture. *J. Anat.* **137:** 21-46.

LYNE, A. G., 1953. Syndactyly in the manus of a marsupial. *Proc. Roy. Soc. Tasmania* **87:** 31-32.

MACALISTER, A., 1872. The muscular anatomy of the koala (*Phascolarctos cinereus*). *Ann. Mag. Nat. Hist.* (4) **10:** 127-34.

MACCORMICK, A., 1887. The myology of the limbs of *Dasyurus viverrinus*. *J. Anat.* **21:** 103-37, 199-226.

MADERSON, P. F. A., 1975. Embryonic tissue interactions as the basis for morphological change in evolution. *Amer. Zool.* **15**: 315-27.

MARSHALL, L. G., 1972. Evolution of the peramelid tarsus. *Proc. Roy. Soc. Vict.* **85**: 51-60.

MATHEW, W. D., 1904. The arboreal ancestry of the Mammalia. *Amer. Nat.* **38**: 811-18.

MCMURRICH, J. P., 1907. The phylogeny of the plantar musculature. *Amer. J. Anat.* **6**: 407-37.

MOORE, K. L., 1982. "The developing human", 3rd ed. Saunders: Philadelphia.

MÜLLER, R. J., 1930. Die Mechanik der syndactylen Zehen von *Macropus* und auderen Beuteltieren und ihre Verwendung als Putzorgan. *Z. Morph. Ökol. Tierc Berlin* **17**: 154-218.

OOHIRA, A. AND NOGAMI, H., 1978. Limb abnormalities produced by 2,2'-Dipyridyl in rats. *Teratology* **18**: 63-70.

OSGOOD, W. H., 1921. A monographic study of the American marsupial, *Caenolestes*. *Field Mus. Nat. Hist. Chicago (Zool.)* **14**: 3-156.

OWEN, R., 1874. On the osteology of the Marsupialia. *Trans. Zool. Soc. Lond.* **8**(1874): 483-500.

PARSONS, F. G., 1896. On the anatomy of *Petrogale xanthopus*, compared with that of other kangaroos. *Proc. Zool. Soc. Lond.* **1896**: 683-714.

PARSONS, F. G., 1903. On the anatomy of the pig-footed bandicoot *(Chaeropus castanotis)*. *J. Linn. Soc. (Zool.)* **29**: 64-80.

RIDE, W. D. L., 1962. On the evolution of Australian marsupials. *In* "The evolution of living organisms" ed by G. Leeper. Melbourne Uni. Press: Melbourne.

RIDE, W. D. L., 1965. Locomotion in the Australian marsupial *Antechinomys*. *Nature* **205**: 199.

SARJEANT, W. A. S. AND THULBORN, R. A. (in prep.). Marsupial footprints from the Cretaceous sediments of British Columbia.

SHARMAN, G. B., 1982. Karyotypic similarities between *Dromiciops australis* (Microbiotheriidae, Marsupialia) and some Australian marsupials. *In* "Carnivorous marsupials" ed by M. Archer. Royal Zoological Society of New South Wales: Sydney.

SIDEBOTHAM, E. J., 1885. On the myology of the water-opossum. *Proc. Zool. Soc. Lond.* **1885**: 6-22.

SONNTAG, C. F., 1922. On the myology and classification of the wombat, koala and phalangers. *Proc. Zool. Soc. Lond.* **1922**: 863-69.

SZALAY, F. S., 1981. Functional analysis and the practice of the phylogenetic method as reflected by some mammalian studies. *Amer. Zool.* **21**: 37-45.

SZALAY, F. S., 1982. A new appraisal of marsupial phylogeny and classification. *In* "Carnivorous marsupials" ed by M. Archer. Royal Zoological Society of New South Wales: Sydney.

TATE, G. H. H., 1933. A systematic revision of the marsupial genus *Marmosa*, with a discussion of the adaptive radiation of the murine opossums. *Bull. Amer. Mus. Nat. Hist.* **66**: 1-125.

THOMPSON, P. AND HILLER, W. T., 1905. The myology of the hind limbs of the marsupial mole *Notoryctes typhlops*. *J. Anat. Physiol.* **39**: 308-31.

WALKER, E. P., 1968. "Mammals of the World". John Hopkins: Baltimore.

WINGE, H., 1893. Jordfunde og nulevende Pungdyr fra Lagoa Santa, Minas Geraes, Brasilien, med Udsigt over Pungdyrenen Slaegskab. *Aftryk af Musco Lundii, Copenhagen:* **2**(1).

WOLPERT, L., 1976. Mechanisms of limb development and malformation. *Brit. Med. Bull.* **32**: 65-70.

ZWILLING, E., 1969. Abnormal morphogenesis in limb development. *In* "Limb development and deformity: problems of evaluation and rehabilitation" ed by C. A. Swinyard. Thomas Springfield: Illinois.

AN EVOLUTIONARY PERSPECTIVE OF STRUCTURAL ADAPTATIONS FOR ENVIRONMENTAL PERCEPTION AND UTILIZATION BY THE NEONATAL MARSUPIALS *Trichosurus vulpecula* (PHALANGERIDAE) AND *Didelphis virginiana* (DIDELPHIDAE)

LESLIE S. HALL[1] and R. L. HUGHES[1]

Using known age embryos and pouch young of *T. vulpecula* comparisons were made with the published accounts for similar stages of development in *D. virginiana*. Emphasis was placed on tissues and organs associated with environmental perception by these two marsupials at birth. Apart from a few minor differences, *D. virginiana* and *T. vulpecula* are at a similar stage of development at birth. Like all marsupials, both species are faced with a very short period of foetal development in the terminal states of gestation. The early state of development in the caudal end of the foetus at birth, results in the characteristic marsupial cloaca. A more advanced state of the cranial end of the newborn marsupial coincides with the need for teat attachment, sucking, co-ordinated crawling movements, and possibly olfaction. Ossification is already occurring in the mandible and in several bones of the skull and while this is probably necessary for firm teat attachment and jaw/middle ear development, it could be responsible for the reduced size of the marsupial brain. Similarly, the precociously developed forearms which enable the newborn marsupial to climb to the pouch, may have precluded further developmental specialisations of the forearm. Although it can be shown morphologically that breathing and sucking can occur simultaneously in the newborn marsupial, confirmation via appropriate experiments is required. Likewise, the concept that newborn marsupials locate the pouch and teat via olfactory clues requires both ultrastructural and experimental studies. There is a grading in the state of development of marsupials at birth which shows the Macropodidae as the most advanced and the Dasyuridae as the least advanced while the Phalangeridae and Didelphidae lie in between.

Key Words: Embryology; *Trichosurus; Didelphis;* Perception; Foetal development.

Pages 257-71 *in* POSSUMS AND OPOSSUMS: STUDIES IN EVOLUTION ed by M. Archer. Surrey Beatty & Sons and the Royal Zoological Society of New South Wales: Sydney, 1987.

INTRODUCTION

THE EMBRYOLOGY and early development of pouch young of marsupials has been the subject of an extensive literature spanning more than a century. This literature has been reviewed by Müller (1967, 1969a, 1969b), Tyndale-Biscoe (1973), Lillegraven (1975) and Kirsch (1978) with little reference to original material.

The detailed observations of McCrady (1938) on *Didelphis virginiana* overshadow the small number of contributions describing the development and embryology of other didelphids. The postnatal development of *D. virginiana* has been studied by Krause and his colleagues who have produced an extensive literature (see references). With regard to Australian phalangerids, the major studies have been done on *Trichosurus vulpecula* and consist of descriptions of the development of a diversity of organ systems and extends back into the 19th century. On the basis of examining histological material from the personal collection of known age embryos and pouch young of *T. vulpecula* from which an overview has already been published (Hughes and Hall 1984), we now compare additional data with the published observations on *D. virginiana* principally contained in the opus of McCrady

(1938). In this paper we do not propose to give a full account of the state of development at birth in *D. virginiana* and *T. vulpecula*. The main purpose is to consider adaptations for survival and environmental perception at birth in these species. The perceptive faculties considered are taste, touch, sight, olfaction, hearing and balance, along with selected life support systems such as respiration, digestion, excretion, locomotion and endocrine. We do not consider in any detail the cardio-vascular, skeletal, or central nervous system, except in terms for their support for perceptive faculties and the listed life support systems.

In presenting the data we use a comparative functional interpretation of histogenesis between *D. virginiana* and *T. vulpecula* at birth. This leads us to make some comments on the evolutionary significance of marsupial development during the neonatal period.

MATERIALS AND METHODS

The source and age of the material has been described in Hughes and Hall (1984). The bulk of the observations were made on two foetuses of 16 days, a one-day-old pouch young, and one three-day-old pouch young *T. vulpecula*. A 13-day-

[1]Department of Anatomy, University of Queensland, St. Lucia, Queensland, Australia 4067.

old foetus was also used for developmental perspective. The specimens were wax embedded, serially sectioned at 5μ, and stained with haematoxylin and eosin.

Terminology for skeletal muscle development follows Bridge and Allbrook (1970) and references to generalised placental embryogenesis and development are from Moore (1982), unless otherwise stated.

PERCEPTIVE FACULTIES

The external appearance of new-born *D. virginiana* and *T. vulpecula* is similar and reflects the early state of development characteristic of marsupials at birth (Fig. 1). A distinguishing feature is the didactylous hind foot of *D. virginiana* and the syndactylous 2nd and 3rd toes of *T. vulpecula*.

Taste

In *T. vulpecula* the tongue is deeply grooved on its upper surface as an adaptation for teat accommodation several days prior to birth. At 16 days gestation its smooth upper surface has small developing papillae and is covered by stratified squamous epithelium (Fig. 2a). One day after birth a number of taste buds are found scattered over the surface of the tongue with several near the tip. These taste buds were innervated, suggesting that they could be a mechanism for the newborn to sample teats. The presence of the glossopharyngeal nerve (CN, IX) supports the contention that the taste buds could be functional at this stage. At the same time a number of papillae have appeared (Fig. 2b) and by three days taste buds showed internal differentiation (Fig. 2c), and a number showed distinct innervation. These findings in *T. vulpecula* are in contrast with those of McCrady (1938) who found no taste buds in *D. virginiana* at birth.

Touch

No touch receptors could be found in the lips of *T. vulpecula* at birth. The obicularis oris muscle for lip closure, was only in an early stage of differentiation and consisted of myoblasts. This suggests that the teat is not located by the lips, and that the lips are not responsible for attachment to, or extracting milk from, the teat at birth.

Vibrissal primordia of the mystical region were found in *T. vulpecula* at 16 days gestation. They were innervated by fibres from the trigeminal nerve (CN, V) but had not produced hair shafts. Vibrissae first appear externally between 16 and 21 days after birth (head length 12 mm-14 mm; Broom, 1898; age by head length regression from Lyne and Verhagen 1957). No Meisner or Pacinian corpuscles were found on the forearms or nose area of *T. vulpecula* at birth.

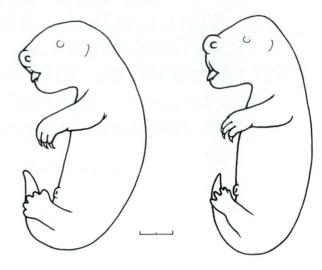

Fig. 1. The external appearance of *D. virginiana* (left) at birth and *T. vulpecula* (right) one day after birth. Note pronated and clawed forelimbs and lesser developed hindlimbs. Relatively minor differences include shape of head, size and position of nostril and mandible, and digital development on hindlimbs. Data from McCrady (1938), and Hughes and Hall (1984). Scale bar = 1 mm.

Selenka (1887) claimed that no organs of touch are present in *D. virginiana* at birth but McCrady (1938) considered that the crawling movements and clasping of maternal hair by the neonate could not be accomplished without some sort of exteroceptive apparatus of the skin. However, McCrady (1938) failed to identify any such apparatus and Brenowitz *et al.* (1980) reported that no identifiable receptors could be found in the dermis of the forepaw of *D. virginiana* until 25 days after birth. Primordia of Pacinian corpuscles are not seen until between 21-30 days and Meisner corpuscles do not commence to develop until 42 days after birth in *D. virginiana* (Brenowitz *et al.* 1980).

Eye

In *T. vulpecula* the eyes are superficial and uncovered at 13½ days gestation and lack eyelids. At 16 days gestation the pigmentation of the retina makes the location of the eye visible (Fig. 2d). One day after birth the lens is still differentiating and the cornea is a thick layer of undifferentiated cells. The retina has separated into a pigmented and a thick nervous layer. This suggests that the eye at birth could be slightly photosensitive, but the lack of differentiated retinal ganglion cells and fused eyelids preclude visual acuity. The lack of a ciliary body and an undifferentiated lens indicates a non-functional focussing system at birth.

Three days after birth some neural connections from the eye to the diencephalon are evident and the optic nerve fibres are visible. The

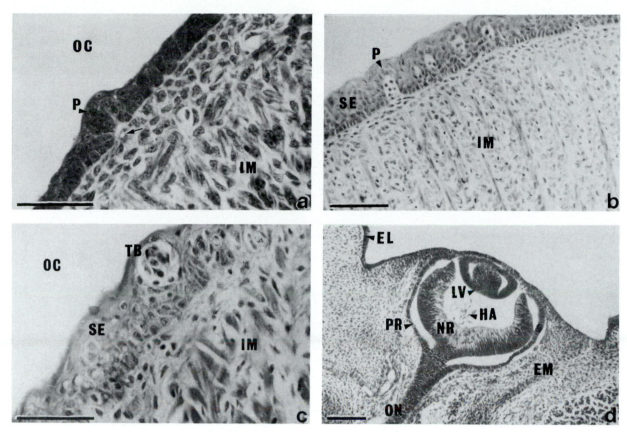

Fig. 2. (a) Surface of tongue of *T. vulpecula* at 16 days gestation. P = developing papilla with associated capillary (arrow); IM = intrinsic tongue muscles showing primary fibres; OC = oral cavity. Scale bar = 50 μm. (b). Tongue surface of *T. vulpecula* one day after birth. P = papillae; SE = stratified squamous epithelium; IM = intrinsic tongue muscles showing primary muscle fibres with striations and peripheral nuclei. Scale bar = 50 μm. (c) Surface of tongue of *T. vulpecula* three days after birth. OC = oral cavity; TB = taste bud; SE = stratified squamous epithelium with some surface cornification; IM = intrinsic tongue muscles. Scale bar = 50 μm. (d) Developing eye of *T. vulpecula* at 16 days gestation. Note that the eye is superficial and not invested by the fusion of eyelids. LV = lens vesicle; ON = optic nerve; NR = nervous layer of retina; PR = pigmented layer of retina; HA = hyaloid artery remnants; EM = developing extrinsic eye muscles; EL = developing eyelid. Scale bar = 100 μm.

eyelids are still fused and the lacrimal glands are small condensations of mesenchyme without lacrimal ducts.

McCrady (1938) found that the eyes of *D. virginiana* at birth were completely non-functional and his description indicates that the eye of *D. virginiana* and *T. vulpecula* are at a similar stage of development at birth.

Olfaction

There is an indication of differentiation in the olfactory epithelium of *T. vulpecula* at 16 days gestation. Although it is incompletely developed, there is a thickening of the olfactory epithelium, numerous cells are undergoing mitosis, and occasional ciliated cells are present (Fig. 3a). However, very few nerve fibres and only the early rudiments of Bowman's glands are present.

One day after birth in *T. vulpecula* the olfactory epithelium is stratified, numerous cells are ciliated, the tissue is richly innervated, and Bowman's glands, which are responsible for flushing the olfactory epithelium surface, have

differentiated into early tubular glands. These features are prominent three days after birth (Fig. 3b).

McCrady (1938) believed that the olfactory organ in *D. virginiana* was in a functional condition at birth. McCrady (1938) describes the epithlium as ciliated and having nerve fibres entering the olfactory lobe of the forebrain. This agrees with the earlier work on *D. virginiana* by Selenka (1885) who, on the basis of morphology, considered that the olfactory organ was functional at birth.

A secondary olfactory component, the vomeronasal organ, is present in *T. vulpecula* as a prominent pocketing on the nasal septum adjacent to the external nares, at 16 days gestation (Fig. 3c). One day after birth the epithelium lining the vomeronasal organ is continuous with the surface epithelium of the nasal chamber and contains ciliated cells. Some of the non-ciliated cells show cytoplasmic secretory domes and others have clear cytoplasm and resemble phagocytes. There is a layer of basally located nuclei. The lamina propria is well developed with

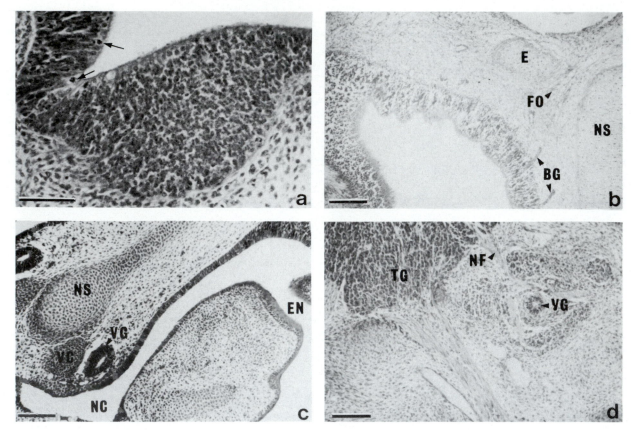

Fig. 3. (a) Olfactory epithelium of *T. vulpecula* at 16 days gestation showing thickened stratified epithelium with cells undergoing mitoses (arrowed). Scale bar = 50 μm. *(b)* Olfactory epithelium of *T. vulpecula* three days after birth. BG = developing Bowman's glands; FO = filum olfactorum; NS = nasal septum; E = ethmoid cartilage. Scale bar = 100 μm. *(c)* Frontal section through nasal region of *T. vulpecula* at 16 days gestation. NS = nasal septum; EN = external nares; VC = vomeronasal cartilage; VG = vomeronasal organ; NC = nasal cavity. Scale bar = 100 μm. *(d)* The innervation of the vomeronasal organ in *T. vulpecula* at 16 days gestation. TG = trigeminal ganglion; NF = vomeronasal component of trigeminal nerve fibres; VG = vomeronasal gland. Scale bar = 100 μm.

prominent venous sinuses and bundles of nerve fibres were located in close proximity to the vomeronasal tissue at 16 days (Fig. 3d). This would suggest that this organ may be functional at birth as a supplementary olfactory receptor, and could be used in pouch and teat location by the neonate.

One day after birth an extensive vomeronasal gland has developed caudal to the vomeronasal organ. The vomeronasal gland is lined by a stratified columnar epithelium and at the most caudal end the cells possess microvilli. Leucocytes are found in the surrounding blood vessels and large nerve fibre tracts are located in adjacent tissue. Vomeronasal cartilage is present at 1 day in *T. vulpecula*. Both McCrady (1938) and Russell (1944) recorded a similar stage of vomeronasal gland development in *D. virginiana* at birth but neither commented on its possible function.

From the above observations it is obvious that the role of the vomeronasal organ at birth in marsupials needs further studies.

Hearing and Balance

The inner ear of *T. vulpecula* at 16 days gestation is a multichambered vessel with some

walls lined with differentiated and innervated epithelium varying from simple cuboidal to a stratified layer of up to five cells thick. The eustachian tube is prominent and leads to the pharynx. The vestibular cochlea nerve (CN, VIII) lies adjacent to the inner chamber and afferent fibres can be observed immediately adjacent to the epithelium of the main part of the chamber. The external pinna is partly formed and is covered with epitrichium. The epitrichium completely fills the external auditory canal.

At one day after birth the semicircular canals have appeared and there are several adjacent areas of thickened epithelium which will eventually form maculae and cristae. The cochlear duct is a simple tube partly coiled and lined by a dense stratified epithelium but with no evidence of the organ of Corti. This would suggest that at birth *T. vulpecula* has no sensation of balance or gravity and lacks the ability to hear.

Extensive observations and experiments on pouch young *D. virginiana* conducted by Larsell and McCrady (1935) and Larsell *et al.* (1935) showed that nerve terminations are present in relation to both vestibular and cochlear hair cells long before there is any indication of functional

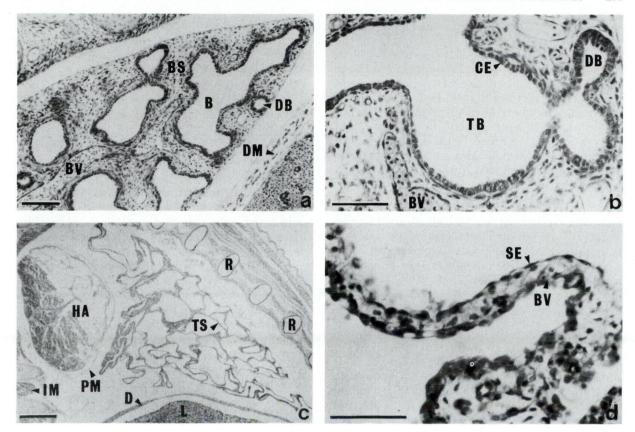

Fig. 4. (a) Longitudinal section of the lung of *T. vulpecula* at 16 days gestation. BV = pulmonary blood vessel; BS = thick inter-bronchial septa; DB = developing bronchiole; B = bronchus. Note poorly developed muscles of diaphragm (DM). Scale bar = 100 μm. (b) Longitudinal section of the lung of *T. vulpecula* at 16 days gestation. TB = lumen of terminal bronchus; DB = developing bronchiole; BV = pulmonary blood vessel; CE = simple cuboidal epithelium. Note relatively wide separation of pulmonary blood vessels from lumen of bronchial tree. Scale bar = 50 μm. (c) Longitudinal section of thoracic cavity of *T. vulpecula* one day after birth. HA = atria of heart; PM = pleuro-pericardial membrane; D = diaphragm; L = liver; R = ossifying ribs; IM = intercostal muscles; TS = thin walled septa of lung. Presumptive bronchioles are no longer distinguishable. Scale bar = 750 μm. (d) Longitudinal section of respiratory septa of *T. vulpecula* three days after birth. The squamous epithelium (SE) has been displaced by a dense network of underlying capillaries containing a mixture of foetal and adult erythrocytes (BV). Scale bar = 50 μm.

response to either type of stimulus. However, the first reflexes to hearing and balance did not appear until 43 days after birth.

Larsell *et al.* (1935) consider that it is the shape of the newborn *D. virginiana* that makes it respond to gravity. The lesser developed hindquarters dangle beneath the large grasping forearms and perhaps prevent the neonate from rolling over.

Following a series of experiments with Quokkas *(Setonix brachyurus)*, Cannon *et al.* (1976) concluded that gravity was the sole navigational aid used by the newborn to reach the pouch. Their observations suggest that the newborn Quokka has a righting reflex present at birth but then lost for the next 65-70 days of pouch life. They consider that the transient righting reflex could be mediated by muscle stretch receptors, particularly in the neck.

Lung and Breathing

At 16 days gestation the lung of *T. vulpecula* has the typical appearance of a mammalian lung at the canicular stage with large blood vessels separated by mesoderm from the air spaces which are lined by cuboidal epithelium (Fig. 4a, b). One day after birth, bronchi are divided into many enlarged terminal sacs and the mesoderm is greatly reduced and contains many thin walled capillaries. These capillaries are separated from the surface epithelium of the air passages by a layer of at least three cells, often more. This constitutes quite a considerable blood-air barrier. At the same time there is an increased proportion of airways occupying the lung with an associated reduction of the septal thickness (Fig. 4c). Three days after birth a large proportion of the air passages contain numerous capillaries bulging into their lumens. These capillaries were filled with a mixture of adult and nucleated foetal erythrocytes and were lined with endothelial cells. The septae were highly vascularised (Fig. 4d).

The considerable observations on the lung of *D. virginiana* at birth (Selenka 1887; Bremer 1904; McCrady 1938; Sorokin 1962; Krause and Leeson 1973) indicate a similar stage of morphological development as seen in *T. vulpecula*. Initially it was considered that the lungs of *D. virginiana* at birth were of reptilian form (Selenka 1887) but subsequent studies have shown that the lung consists of a primitive system of branching airways that ends in a number of terminal chambers or sacs which further differentiate during pouch life (Krause and Leeson 1973). Krause and Leeson (1973) consider that the air chambers are adequate to supply the metabolic requirements of the pouch young of *D. virginiana*.

One day after birth the diaphragm in *T. vulpecula* consists of a relatively thin muscular sheet (Fig. 4c) but the intercostal muscles are well developed and innervated, suggesting that the diaphragm could be of subservient functional importance in respiration.

Cutaneous Respiration

One day after birth in *T. vulpecula* the nearest capillary is 80-100μ from the surface of the skin and separated by at least 5-6 cell layers. This indicates that cutaneous respiration should be precluded as a mechanism for oxygen exchange at birth. A similar finding has been reported in *Macropus eugenii* by Randall *et al.* (1984). The skin of the newborn marsupial is highly transparent and has probably led to the spurious cutaneous respiration concept which is frequently quoted.

Tongue

The extrinsic musculature of the tongue one day after birth in *T. vulpecula* exhibits one of the most advanced stages of muscle development seen in the neonate, being composed of primary muscle fibres with striations and peripheral nuclei (Fig. 5b).

At 13 days of gestation fibres from the hypoglossal and glossopharyngeal nerves are already prominent in the tongue of *T. vulpecula*. At birth the tongue protrudes from the oral cavity and is deeply grooved (Fig. 1). It has been suggested by Broom (1898) that the teat is initially grasped by the tongue and the upper lip. Only after several days, by which time the tongue is entirely contained in the mouth, are both the upper and lower lips used to grasp the teat.

Oral Cavity

At 16 days gestation in *T. vulpecula* the roof of the mouth is lined with stratified squamous epithelium with several poorly developed palatine ridges (Fig. 5a, c). The main body of the maxilla and mandible commence to ossify at 16 days gestation and are still ossifying one day after birth. Gingival lamellae are present at 16 days gestation and dental lamellae are present one day after birth. One day after birth, masseter muscles are relatively poorly developed and appear to be incapable of locking the jaw shut. Except for the area surrounding the teat, the mouth is kept shut by the epitrichium and fused lips. The lips remain fused until at least 47 days (head length 23.5 mm; Broom, 1898; age from head length regression in Lyne and Verhagen 1957). The secondary palate has fused, but the medium junction of the two shelves is still evident at 16 days gestation in *T. vulpecula*.

Salivary Glands

One day after birth in *T. vulpecula* the ducts of the submandibular and parotid glands are patent and lined with cuboidal epithelium. The acini have commenced to differentiate from surrounding mesenchymal tissue but are not advanced enough to secrete any products. Three days after birth the acini of the parotid have branched and secretory material is found in the lumen of developing intralobular ducts (Fig. 5d). At birth in *D. virginiana* the submandibular gland consists of primitive tubular elements that terminate in end pieces composed of proacinar cells and it takes a further 31 days to appear structurally mature (Leeson *et al.* 1978).

Epiglottis

In *T. vulpecula* the epiglottis projects into the nasopharynx well above the posterior end of the soft palate at 16 days gestation (Fig. 5a). At birth milk is diverted around each lateral surface of the epiglottis to enter the esophagus. The digestive and respiratory tracts are separated at the level of the epiglottis so that air from the nasal cavity passes medially and separately into the trachea (Fig. 5e, f). A similar situation exists at birth in *D. virginiana* (McCrady 1938) and in *Dasyurus viverrinus* (Hill and Hill 1955). This arrangement persists into the adult *D. virginiana* (McCrady 1938) and is also present in adult cetaceans and some bats. Although the tissue of the epiglottis one day after birth is mainly undifferentiated and consists of a dense mesenchymal core covered with a stratified epithelium, it is probable that the neonate *T. vulpecula* has some control over the epiglottis. This is evident by the nerve fibres and extrinsic musculature.

The movement of the tongue and associated movements at its base have the potential for valvular action. The contours of the surfaces of the epiglottis, soft palate, pharynx and glottis exhibit near parallel contours. This could also involve laryngeal muscles. This relationship requires the appropriate experimental demonstration of the simultaneous patentcy of the digestive and respiratory passages.

Esophagus

One day after birth the esophagus is a patent duct lined with a low stratified epithelium in

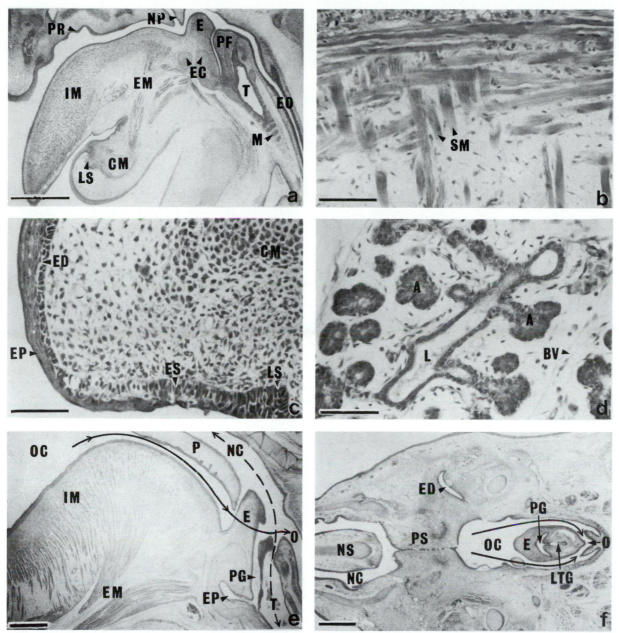

Fig. 5. *(a)* Sagittal section of oro-pharangeal region of *T. vulpecula* at 16 days gestation. LS = lip sulcus; CM = cartilaginous mandible; IM = intrinsic tongue muscles; EM = extrinsic tongue muscles; PR = palatine ridges; EO = esophagus; NP = nasopharangeal fold; E = proximal portion of epiglottis; PF = post epiglottal fold; T = lumen of trachea; M = models of cartilage rings; EC = differentiating epiglottal cartilage. Scale bar = 500 μm. *(b)* Longitudinal section of tongue of *T. vulpecula* one day after birth. The intrinsic tongue muscles exhibit the most advanced stage of differentiation of the entire muscular system at this stage. SM = striated skeletal muscle. Scale bar = 50 μm. *(c)* Junctional area of the lip and oral cavity of *T. vulpecula* at 16 days of gestation. CM = maxillary cartilage model; LS = lip sulcus; EP = superficial cornified epitrichial layer; ED = deep nucleated layers of epidermis; ES = stratified squamous epithelium of oral cavity. Scale bar = 50 μm. *(d)* Branched tubular parotid gland of three-day-old pouch young *T. vulpecula.* A = branched acini; BV = blood vessel of capillary bed; L = presumptive secretory material in lumen of intralobular collecting duct. Scale bar = 50 μm. *(e)* Sagittal section of oro-pharangeal area of three-day-old *T. vulpecula.* IM and EM = intrinsic and extrinsic tongue muscles; OC = oral cavity; P = palate; NC = nasal cavity; T = trachea; E = epiglottis; O = esophagus; PG = post epiglottal groove; EP = basal epiglottal pocket; arrow = presumptive food passage; broken arrow = air passage. Scale bar = 750 μm. *(f)* Coronal section of epiglottis and oro-nasal region of *T. vulpecula* at 16 days. NS = nasal septum; NC = nasal cavity; PS = fused shelves of 2nd palate; ED = endolymphatic duct; OC = oral cavity; E = epiglottis; PG = post-epiglottal groove; O = entrance to esophagus; LTG = laryngo-tracheal groove (partly fused); arrow = presumptive food passage. Scale bar = 750 μm.

T. vulpecula. There is an undifferentiated lamina propria and a ring of myoblasts where the muscularis externa will eventuate. Although the vagus nerve can be traced down the full length of

the esophagus, the lack of development in the muscularis mucosa and externa, make it unlikely that any peristaltic movement of milk occurs at birth. A study by Krause *et al.* (1976a) on the

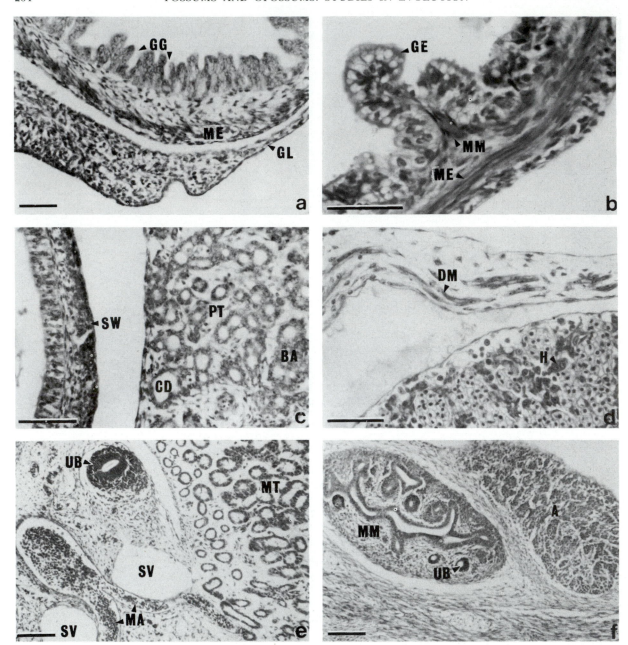

Fig. 6. (a) Longitudinal section of stomach of *T. vulpecula* one day after birth. GG = differentiating gastric glands; ME = muscularis externa; GL = gastro-splenic ligament. Scale bar = 100 μm. *(b)* Longitudinal section of stomach of *T. vulpecula* three days after birth. MM = muscularis mucosa; GE = gastric epithelium (metabolically active); ME = muscularis externa. Scale bar = 50 μm. *(c)* Longitudinal section of pancreatic area in *T. vulpecula* one day after birth. PT = highly vascularised pancreatic tissue; BA = branching acini; CD = collecting ducts; SW = stomach wall. Scale bar = 50 μm. *(d)* Longitudinal section of hepatic regions of *T. vulpecula* at 16 days gestation. Note haemopoietic liver containing extensive vascular network. H = scattered hepatocytes; DM = tubular muscle fibres of diaphragm. Scale bar = 50 μm. *(e)* Longitudinal section of excretory system of *T. vulpecula* at 16 days gestation. UB = ureteric bud; MA = mesonephric artery running from dorsal aorta; MT = vascularised mesonephric collecting tubules; SV = subcardinal veins. Scale bar = 100 μm. *(f)* Longitudinal section of metanephros and adrenal gland of *T. vulpecula* one day after birth. MM = metanephrogenic mesoderm; UB = secondary branch of ureteric bud. Note lack of zonation in adrenal gland (A). Scale bar = 100 μm.

esophagus of *D. virginiana* at birth shows similar immature features.

Stomach

At 16 days gestation in *T. vulpecula* the stomach is a simple holding chamber of limited capacity which lacks rugae and is without cardiac or pyloric sphincters. The gastric mucosa is an unspecialised columnar epithelium. One day

after birth gastric glands are differentiating (Fig. 6a) and gastric epithelium appears to be metabolically active three days after birth (Fig. 6b). The entire gastrointestinal tract of *D. virginiana* has been described as being embryonic at birth (Krause *et al.* 1976a, 1976b, 1976c). However it appears to be slightly more advanced at birth in *D. virginiana* than in *T. vulpecula* (Hughes and Hall 1984) as clefts, which may represent the initial formation of gastric glands, and parietal cells, can be identified (Krause *et al.* 1976b) and elements of the myenteric plexus are present (Cutts *et al.* 1978). Krause *et al.* (1976b) consider that the stomach does not act only as a storage vehicle for ingested milk, but morphologically appears active in the absorption of lipid during the first three weeks after birth.

Duodenum

This is a prominent organ one day after birth in *T. vulpecula* and its appearance is characteristic with a large pale undifferentiated lamina propria and muscularis mucosa, circled by developing myoblasts of the muscularis externa. The luminal epithelium is simple columnar with a few goblet cells. There is no evidence of a myentric plexus. In *D. virginiana* at birth the duodenum contains villi and the intestinal epithelium covering both the floor and villi shows an extensive apical endocytic complex, numerous dense aggregates of material in the supranuclear regions and large lipid droplets (Krause *et al.* 1976c).

Pancreas

This organ is prominent in neonatal *T. vulpecula* and consists of developing acini located in a meshwork of fine connective tissue (Fig. 6c). It is vascularised but the endocrine islets are not evident. The pancreas is connected to the duodenum via a patent pancreatic duct which is lined by a layer of low cuboidal cells. A similar situation occurs in the pancreas of *D. virginiana* and King *et al.* (1978) have suggested that the immature nature of the exocrine pancreas, combined with the delayed appearance of chief cells in the gastric mucosa, is related to the mechanism of transfer of immunoglobulins. The late development of pancreatic elements would permit direct absorption of intact macromolecules from the milk by the intestinal epithelium of the neonate marsupial (King *et al.* 1978). It is known that other marsupials (e.g. Tammar, *Macropus eugenii*; Quokka, *Setonix brachyurus*; and *T. vulpecula*) acquire maternal immunoglobulins by absorption from maternal milk and colostrum (Yadav 1971). However, Renfree (1973) observed immunoglobulins in embryonic blood of *M. eugenii* and a further discussion on the origins of immunoglobulins is found in Hines and Mizell (1976).

Liver

In perinatal *T. vulpecula* the liver is one of the largest abdominal structures occupying approximately 30% of the abdominal cavity. It is so densely packed with nucleated red blood cells that it is difficult to interpret any internal structures. The classical liver components seen in the adult marsupial are absent and structurally the liver appears as a series of large sinusoids joined by thin septae of developing hepatocytes (Fig. 6d).

In newborn *D. virginiana* the liver does not appear to be well enough established to assume normal hepatic function and is primarily a haematopoietic organ for several weeks of the postnatal period (Cutts *et al.* 1973; Krause *et al.* 1975). It has a similar appearance as described for *T. vulpecula* with an additional observation that some hepatocytes contain vesicular nuclei with distinct nucleoli and the cytoplasm is finely granular and appears vacuolated because of the presence of discrete lipid droplets (Cutts *et al.* 1973). In investigating the haematopoietic properties of the liver in neonatal *D. virginiana* Block (1964) found that half of the liver was occupied by haematopoietic cells.

The presence of a well developed gall bladder and common bile duct at birth in *T. vulpecula* indicate the possibility of early production of bile.

Excretion

At birth all marsupials have the mesonephros as the main functional excretory organ. In *T. vulpecula* and *D. virginiana* the mesonephros is highly vascular, supplied by branches from the dorsal aorta, and drained medially by the subcardinal veins (Fig. 6e). There are approximately 45 glomeruli which appear to be similar in structure to glomeruli described for foetal eutherian mesonephroi. The collecting ducts from the glomeruli empty into the mesonephric duct which in turn empties into a urogenital system at a point just lateral and cephalic to the entry points of the metanephric duct in the dorsal lateral wall (Krause *et al.* 1979; Hughes and Hall 1984).

In *D. virginiana* and *T. vulpecula* the metanephros is present at birth as a condensation of metanephrogenic mesoderm containing the ureteric bud which has undergone several branches (Krause *et al.* 1979; Hughes and Hall 1984; Fig. 6f). Ultrastructural observations by Krause *et al.* (1979) on *D. virginiana* suggest that metanephric nephrons are not functional during the first four days of postnatal life, while the mesonephros reaches the height of its development during this period. Krause *et al.* (1979) also think that there may be some functional overlap between meso- and metanephros during the later part of the first week of postnatal life in *D. virginiana*.

Locomotion

At birth marsupial forelimbs are pronated functional pentadactyl limbs which are capable of digito-palmar prehension which enables the newborn to climb unaided to its mother's pouch. Each digit has a pronounced claw which in *T. vulpecula* included a cornified epitrichial cap (Fig. 7a).

The skeletal muscle in the forelimb of *T. vulpecula* at birth is of the secondary type with

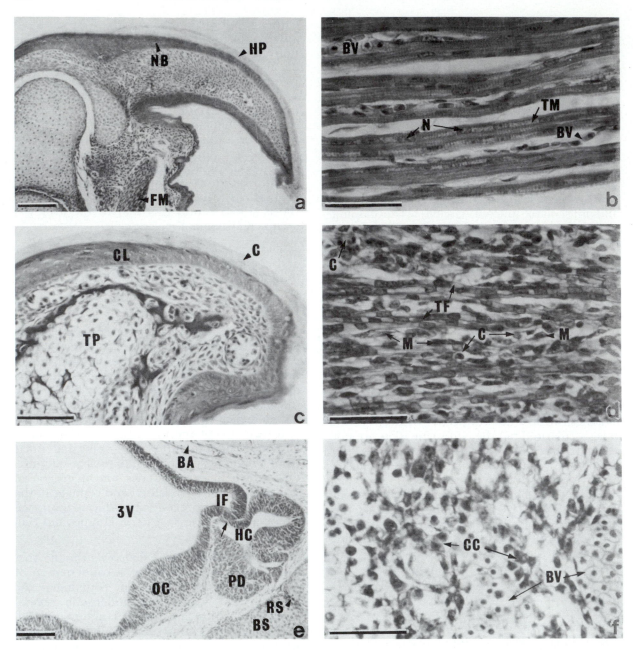

Fig. 7. (a) Oblique section of manus of T. vulpecula one day after birth. HP = horny epitrichial plate of claw; FM = flexor muscle; NB = nail bed. Note functional synovial joint between distal and second phalange. Scale bar = 100 μm. (b) Longitudinal section of forearm skeletal muscle of T. vulpecula one day after birth. TM = tubular striated skeletal muscle fibres; N = central nucleus. Note prominent vascularization by intra-fibrilar capillaries (BV). Scale bar = 50 μm. (c) Oblique section of manus of T. vulpecula one day after birth. TP = ossifying terminal phalange; cornified (C) and cellular layer (CL) of epidermis. Scale bar = 50 μm. (d) Longitudinal section of hindlimb skeletal muscle of T. vulpecula one day after birth. Note early stage of differentiation of tubular fibres (TF) from myoblasts (M); C = capillaries. Scale bar = 50 μm. (e) Sagittal section through pituitary gland of T. vulpecula at 16 days gestation. OC = optic chiasma; IF = infundibulum; PD = pars distalis; arrow = presumptive pars intermedia; HC = hypophyseal cleft; BS = basisphenoid cartilage; RS = remnants of Rathke's and Seessel's pouches; BA = basilar artery; 3V = 3rd ventricle. Scale bar = 100 μm. (f) Adrenal gland of T. vulpecula at 16 days gestation. CC = differentiating cortical cells; BV = vascular network. Scale bar = 50 μm.

external nuclei and long fibres. Occasional striations are seen, particularly in the flexor muscles (Fig. 7b). The synovial joint of the humerus/scapula is composed of articulating cartilage surfaces in a synovial capsule. The clavicle is ossified at birth in T. vulpecula (Broom 1898) and forms a firm anchor point for the forearm and scapula. At one day the distal phalange is ossifying and is highly vascular (Fig.

7c). Cheng (1955) reported for D. virginiana a detailed evolutionary study of the embryogenesis of the shoulder region with particular reference to musculature. The cervical and thoracic spinal ganglions are particularly well developed in T. vulpecula by 16 days gestation and exhibit prominent fibre tracts leading to the forearm area. The surface skin is thinly keratinised, stratified squamous, but lacks any pads, hair or vibrissae.

The external appearance of the forelimb of *D. virginiana* and *T. vulpecula* at birth is similar (Fig. 1). Comments on the forelimb of *D. virginiana* at birth by McCrady (1938) indicate a similar stage of development as seen in *T. vulpecula*. At birth the hindlimb of *T. vulpecula* is paddle shaped but the outline of primary phalanges can be seen (Fig. 1) and regions of tarsal, metatarsal, and phalangeal differentiation are apparent. Areas of developing muscle showed tubular fibres (Fig. 7d).

Observations on the hindlimb of *D. virginiana* show that it is non-motile but innervated at birth (Blinco 1962). However, most nerves are non-myelinated up to three weeks after birth (Mizell and Isaacs 1970) and are considered non-functional.

Endocrine

Just prior to birth the anterior pituitary of *T. vulpecula* consists of a dense area of undifferentiated tissue containing a large irregular lumen. This tissue lies adjacent to a downpocketing of the diencephalon (future pars nervosa) and sits in the sella tursica of the cartilage plate of the developing basisphenoid (Fig. 7e). Studies on the development of the pituitary gland of *D. virginiana* by Wheeler (1943) showed that at birth the cells of the anterior lobe, intermediate lobe, and pars tuberalis are entirely undifferentiated. The structural development at birth in *D. virginiana* appears to be similar to that of *T. vulpecula* where Shorey (1968) recorded an undifferentiated pituitary for the first four days of pouch life. A similar stage of development of the pituitary at birth has been recorded for *Isoodon macrourus* by Hall and Hughes (1980) who consider that the pituitary of this marsupial is non-functional at birth. The early development of the pituitary gland in marsupials has recently been reviewed by Hall and Hughes (1984).

At birth in *T. vulpecula* the adrenal is relatively large in comparison to other organs. There is no structural arrangement and the gland lacks a true capsule and medulla (Fig. 6f). The cells appear undifferentiated and are arranged into strands and cords surrounded by sinusoidal-like vessels similar to its appearance at 16 days (Fig. 7f). In *D. virginiana* at birth the medulla of the adrenal gland is in the process of formation from chromaffin cells (McCrady 1938), and is thus more advanced than in *T. vulpecula*.

DISCUSSION

Kirsch (1978) considers that therians must have evolved from an egg-laying form and it is still possible to regard the common features of marsupial and monotreme neonates as being of some phylogenetic significance. Since monotremes do not have well-formed nipples, Kirsch (1978) argues that it is vital that their young be able to cling to the fur and suck the exposed milk. Thus because of their egg-laying heritage, early marsupials were preadapted for the journey to the nipples by the extreme precociousness of the forequarters.

The precociously developed pentadactyl forelimb with separate and clawed digits seen in *D. virginiana* and *T. vulpecula* presumably represents the optimal adaptation for the newborn marsupial to crawl to the pouch, attach to a teat, and possibly stimulate milk flow. At the same time, however, it also confers a developmental adaptation that could preclude a number of specialisations of the forelimb. Müller (1967-69) and reviewed in Lillegraven (1975) suggests that the exigencies of the journey to the pouch have placed such limitations on the adaptive diversity of marsupial forelimbs that no equivalent to the Chiroptera, Cetacea, Sirenia or Pinnipedia have evolved in marsupials.

Kirsch (1976) thinks that the importance of the advanced state of the marsupial forelimb in preventing subsequent adaptive development has been over-emphasised and cites the forelimbs of *Chaeropus* and *Notoryctes* (reduced digits), and *Chironectes* (possess nails rather than claws) as being counter to the argument. It appears that *Notoryctes* may well not possess deciduous claws at birth as the description of the "pouch embryo" by Jones (1921) is of a specimen which is close to newborn and exhibits two enlarged spade-like and three greatly reduced digits. To this trio could be added *Dactylopsila* which possesses an elongated 4th digit, but obviously an inspection of newborn and known age pouch young (particularly *Notoryctes*) are essential before this matter can be resolved.

As a consequence of marsupial birth and monotreme hatching occurring a short interval after the onset of organogenesis, the caudal end of the newborn exhibits retarded development which gives rise to an undifferentiated anatomical arrangement as well as prematurely functional organs. Although the slower rate of development in the caudal end is characteristic of all amniote vertebrates, it usually occurs well before birth or hatching. These undifferentiated and prematurely functional organs remain as definitive characteristics of the adult pattern of both the marsupial and monotreme grade particularly in regard to the presence of a cloaca and the arrangement of the ureters in monotremes.

Not only does the slower development of the caudal end influence adult anatomy, it also influences the morphology of embryonic membranes. Since the cloacal membrane fails to rupture during gestation, the closely adhering amniotic membrane is not distended. The size of the allantois is related to the functional development of the mesonephros.

As shown by Hughes and Hall (1984) in *T. vulpecula* and McCrady (1938) in *D. virginiana*, an increase in size of the allantois only occurs in the terminal stages of gestation and corresponds to the onset of embryonic excretion.

The precocious development of the head in marsupial newborn, as reflected by the advanced state of the trigeminal nerve (CN, V), early ossification centres in maxilla and mandible, and the well developed tongue, can also be regarded as having phylogenetic implications. It has been pointed out by Hopson (1966) that marsupials are born with a typical reptilian articular-quadrate jaw and it is essential that the jaw be securely attached to the teat and "locked" shut to allow for the subsequent development of the mammalian dentary-squamosal type jaw. Closure of the lips by fusion and by the epitrichial layer are also mechanisms ensuring jaw closure (Müller 1968a, b). The fragility of the undifferentiated tissues in the head region (e.g., eye, ear) is complimented by the well developed protective layer of epitrichium (Fig. 2a).

The early ossification of parts of the cranial vault in neonatal marsupials could possibly have subsequent adverse results. Jerison (1973) considers that the brains of South American marsupials (including extinct forms) are only half the size of their placental counterparts, and that this factor is reflected in the dominance of placentals over marsupials in South America. Jerison (1973) believes that South American carnivorous marsupials, owing to unspecified causes, were unable to increase their relative brain size to keep pace with the increasing brain size in their placental prey and carnivorous competitors. Should this be substantiated, it may be that the unspecified causes stem from the early ossification of parts of the cranial vault in neonatal marsupials as microencephaly in humans is known to be caused by premature ossification of the cranial vault, resulting in reduced brain size (Moore 1982). Despite the validity of Jerison's (1973) conclusions it is obvious that the dimensions of the cranial vault can be causally related to the timing of ossification during encephalization.

The advanced state of development of the adrenal gland at birth in *D. virginiana* (McCrady 1938) and its size in *T. vulpecula* (Hughes and Hall 1984) suggest that it may be producing glucocorticoids as occurs in the bandicoot *Isoodon macrourus* (Gemmell *et al.* 1982) and the wallabies *Macropus eugenii* (Catling and Vinson 1976; Call *et al.* 1980), and *M. rufogriseus* (Walker and Gemmell 1983) which is critical for glycogen deposition in the liver and voluntary muscle, as well as for surfactant synthesis in the lung and possibly myelin synthesis (Smith *et al.* 1982).

It has been shown that short gestational placentals (e.g., rat) are susceptible to neonatal hypoglycaemia because they have no functional glucogenic pathway available at birth and rely on glycogen accumulations in foetal voluntary muscle at birth (Shelley 1961). Longer gestation periods seen in some placentals enable different metabolic processes to occur — i.e. fatty acids can be oxidised and may provide an important source of metabolic fuel (Jones and Rolph 1981). Although the energy requirements of neonatal marsupials must be low, an immediate and constant supply is required by the tongue and forearm muscles following birth. Studies are required on glucose storage sites and metabolism in newborn marsupials. In their extensive survey of brown fat deposits in newborn mammals Rowlatt *et al.* (1971) found no deposits in a number of marsupials. Measurements given for their specimens, however, suggest that the marsupials could not be classified as newborn.

Discussions relating to the length of the gestation period in various marsupials often fail to consider the actual period of organogenesis. For example, gestation in *I. macrourus* lasts 12½ days while that of *M. rufogriseus* lasts 26 days (Lyne and Hollis 1977; Merchant and Calaby 1981; Walker and Rose 1981). Organogenesis (post primitive streak), however, occurs in the last 5½ days of gestation in *I. macrourus* and last 11 days in *M. rufogriseus*. This is a considerable difference, considering the short period of organogenesis. The short period of organogenesis, restricted to the latter half of gestation, is a characteristic of all marsupials so far investigated.

The concept that newborn marsupials are capable of simultaneous breathing and sucking was introduced by Hill and Hill (1955) on the basis of morphological investigations of *D. viverrinus*. Our studies suggest that there is sufficient anatomical evidence to show valvular separation of the digestive and respiratory passageways at the epiglottis in newborn *T. vulpecula*. However, we suggest that the concept of simultaneous breathing and swallowing requires further confirmation by studies based on physiological experimentation.

Selenka (1887) as quoted by Bremer (1904) regarded the lung of *D. virginiana* at birth to have "exactly the form of a reptilian lung". This was due to the symmetry of the bronchial branches, general appearance, and character of the lining of the epithelium. However, as Bremer (1904) pointed out, Selenka was wrong in referring to alveoli which along with interalveolar septae are absent from neonatal marsupial lungs. Even though Hughes and Hall (1984) have shown that at the respiratory surface, the level of development at birth in *T. vulpecula* more approximates the canalicular stage of placental lung development occurring in early uterine life, there is still no placental parallel to lung development in marsupials.

The importance of the pouch to marsupial neonates is well exemplified by the observations of Block (1964) who showed that neonatal *D. virginiana* is so deficient in white blood cells and antibodies, that even a minor wound resulted in acute, massive and fatal bacterial invasion without inciting an inflammatory response. Since Block (1964) regards the marsupial pouch as an essential shield for the newborn young against noxious influences, it would be interesting to know how this problem is overcome in pouchless marsupials. *Marmosa* is a pouchless genus with some species having up to 19 teats distributed over the pectoral as well as the abdominal trunk (Kirsch 1978), yet the newborn are regarded as exhibiting the minimum anatomical development allowable at birth (Müller 1967).

Although *D. virginiana* and *T. vulpecula* appear to exhibit minimal development for environmental perception at birth, they are slightly more advanced than *Dasyurus viverrinus* and *Marmosa* (Hill and Hill 1955; Müller 1967) but less developed than *M. eugenii* and *M. rufogriseus* (Renfree 1972; Walker and Rose 1981). This is reflected mainly in the state of organogenesis of the kidney complex, urogenital organs and gastrointestinal tract.

CONCLUSION

The birth mode has had severe limitations on the evolutionary diversification within the Marsupialia, particularly in regard to the adaptation of the forelimbs and possibly brain size. The early stage of organogenesis at birth has resulted in the retention of embryonic structures in the adult. This is most evident in the urogenital system which is a distinguishing feature of the Marsupialia.

We have shown that diversity exists at the superfamily level between the stage of development in newborn marsupials. Macropodidae exhibit the most advanced developmental stages at birth and could be justifiably regarded as a separate group from the Phalangeridae. This concept is further supported by another paper in this Volume (Gemmell *et al.* 1987). Our observations combined with those of Hill and Hill (1955), Walker and Rose (1981) and Hall (1984) indicate that the superfamily Dasyuroidea shows a high level of homogeneity in being the least advanced newborn marsupials, and possessing unique neonatal features, such as the cervical pad. *D. virginiana* and *T. vulpecula* are both at a similar stage of development at birth in regard to their ability to perceive their environment.

We do not know how the newborn marsupial sufficiently perceives its environment to successfully achieve attachment to the maternal teat.

What we have shown is that it is easy to confirm the anatomical basis for the mechanism of reflex crawling, attachment to and sucking of the maternal teat.

At the light microscope level we cannot document the functional morphology for either a sense of touch (vibrissae primordia or skin) or for smell (vomeronasal organ or olfactory organ). It has been long known that the trigeminal nerve (CN, V) appears at an exceptionally early stage of development in both monotremes and marsupials (Wilson and Hill 1908; Hughes 1984) and it would seem that ultrastructural studies on the preoptic area of developing marsupials, particularly in relation to olfaction, would be rewarding.

ACKNOWLEDGEMENTS

The preparation of histological material was performed by L. Bell and R. Collins to whom we are most thankful.

REFERENCES

BLINCO, H., 1962. The structure of the non motile hindlimb of the pouch young Opossum. *Anat. Rec.* **142:** 89-93.

BLOCK, M., 1964. The blood forming tissues and blood of the newborn Opossum (*Didelphis virginiana*). I. Normal development through about the one hundredth day of life. *Ergeb. Anat. Entwickl.* **37:** 237-366.

BREMER, J. L., 1904. On the lung of the Opossum. *Amer. J. Anat.* **3:** 67-73.

BRENOWITZ, G. L., TWEEDLE, C. D. AND JOHNSON, J. I., 1980. The development of receptors in the glabrous forepaw skin of pouch young Opossums. *Neuroscience* **5:** 1303-10.

BRIDGE, D. T. AND ALLBROOK, D., 1970. Growth of striated muscle in an Australian marsupial (*Setonix brachyurus*). *J. Anat.* **106:** 285-95.

BROOM, R., 1898. A contribution to the development of the common phalanger. *Proc. Linn. Soc. NSW* **23:** 705-29.

CANNON, J. R., BAKKER, H. R., BRADSHAW, S. D. AND McDONALD, I. R., 1976. Gravity as the sole navigational aid to the newborn Quokka. *Nature* **259:** 42.

CALL, R. N., CATLING, P. C. AND JANSSENS, P. A., 1980. Development of the adrenal gland in the Tammar wallaby, *Macropus eugenii* (Desmarest) (Marsupialia: Macropodidae). *Aust. J. Zool.* **28:** 249-59.

CATLING, P. C. AND VINSON, G. P., 1976. Adrenocortical hormones in the neonate and pouch young of the Tammar wallaby *Macropus eugenii. J. Endocrinol.* **69:** 447-48.

CHENG, C. C., 1955. The development of the shoulder region of the Opossum, *Didelphis virginiana*, with special reference to the musculature. *J. Morph.* **97:** 415-71.

CUTTS, J. H., KRAUSE, W. J. AND LEESON, C. R., 1978. Development of the external muscle coats in the digestive tract of the Opossum, *Didelphis virginiana. Acta Anat.* **102:** 333-40.

CUTTS, J. H., LEESON, C. R. AND KRAUSE, W. J., 1973. The postnatal development of the liver in a marsupial, *Didelphis virginiana*. I. Light microscopy. *J. Anat.* **115:** 327-46.

GEMMELL, R. T., HUGHES, R. L. AND JENKIN, G., 1987. Comparative studies on the hormonal profiles of progesterone and prostaglandin F metabolite in the Common Brushtail possum, *Trichosurus vulpecula*. Pp. 279-91 *in* "Possums and opossums: studies in evolution" ed by M. Archer. Surrey Beatty & Sons and the Royal Zoological Society of New South Wales: Sydney.

GEMMELL, R. T., SINGH-ASA, P., JENKIN, G. AND THORBURN, G. D., 1982. Ultrastructural evidence for steroid-hormone production in the adrenal of the marsupial, *Isoodon macrourus*, at birth. *Anat. Rec.* **203**: 505-12.

HALL, L. S., 1984. Some comparisons of developmental stages of different marsupials at birth. *Bull. Aust. Mammal Soc.* **8**(2): 82.

HALL, L. S. AND HUGHES, R. L., 1980. Early cytodifferentiation of the pituitary gland in the Brindled Bandicoot *(Isoodon macrourus)*. *Bull. Aust. Mamm. Soc.* **6**(2): 36.

HALL, L. S. AND HUGHES, R. L., 1985. The embryological development and cytodifferentiation of the anterior pituitary in the marsupial, *Isoodon macrourus*. *Anat. Embryol.* **172**: 353-63.

HILL, J. P. AND HILL, W. C. O., 1955. The growth stages of the pouch-young of the Native Cat *(Dasyurus viverrinus)*, together with observations on the anatomy of the new-born young. *Trans. Zool. Soc. Lond.* **28**: 349-452.

HINDES, R. D. AND MIZELL, M., 1976. The origin of immunoglobulins in Opossum embryos. *Dev. Biol.* **53**: 49-61.

HOPSON, J. A., 1966. The origin of the mammalian middle ear. *Amer. Zool.* **6**: 437-50.

HUGHES, R. L., 1984. Structural adaptations of the eggs and foetal membranes of monotremes and marsupials for respiratory and metabolic exchange. Pp. 389-421 *in* "Respiration and metabolism of embryonic vertebrates" ed by R. S. Seymour. Junk: Dordrecht.

HUGHES, R. L. AND HALL, L. S., 1984. Embryonic development in the Common Brushtail Possum *Trichosurus vulpecula*. Pp. 197-212 *in* "Possums and gliders" ed by A. P. Smith and I. D. Hume. Australian Mammal Society: Sydney.

JERISON, H. J., 1973. "Evolution of the brain and intelligence". Academic Press: New York.

JONES, C. T. AND ROLPH, T. P., 1981. Metabolic events associated with the preparation of the foetus for independent life. Pp. 214-33 *in* "The foetus and independent life". Pitman: London (Ciba Foundation Symposium 86).

JONES, F. W., 1921. The external characters of pouch embryos of marsupials. No. 2 *Notoryctes typhlops*. *Trans. R. Soc. S. Aust.* **45**: 36-9.

KING, F. C., KRAUSE, W. J. AND CUTTS, J. H., 1978. Post natal development of the pancreas in the Opossum: Light microscopy. *Acta anat.* **101**(3): 259-74.

KIRSCH, J. A. W., 1977. Biological aspects of the marsupial-placental dichotomy: a reply to Lillegraven. *Evolution.* **31**: 898-900.

KIRSCH, J. A. W., 1978. The six-percent solution: second thoughts on the adaptations of the Marsupialia. *Amer. Scientist.* **65**: 276-88.

KRAUSE, W. J., CUTTS, J. H. AND LEESON, C. R., 1975. The postnatal development of the liver in a marsupial *Didelphis virginiana*. II. Electron microscopy. *J. Anat.* **120**: 191-206.

KRAUSE, W. J., CUTTS, J. H. AND LEESON, C. R., 1976a. The postnatal development of the alimentary canal in the Opossum. I. Oesophagus. *J. Anat.* **122**: 293-314.

KRAUSE, W. J., CUTTS, J. H. AND LEESON, C. R., 1976b. The postnatal development of the alimentary canal in the Opossum. II. Stomach. *J. Anat.* **122**: 499-19.

KRAUSE, W. J., CUTTS, J. H. AND LEESON, C. R., 1976c. The postnatal development of the alimentary canal in the Opossum. III. Small intestine and colon. *J. Anat.* **123**: 21-45.

KRAUSE, W. J., CUTTS, J. H. AND LEESON, C. R., 1979. Morphological observations on the mesonedelsphros in the post-natal Opossum *Didelphis virginiana*. *J. Anat.* **129**(2): 377-97.

KRAUSE, W. J. AND LEESON, C. R., 1973. The postnatal development of the respiratory system of the Opossum. I. Light and scanning electron microscopy. *Amer. J. Anat.* **137**: 337-56.

LARSELL, O., McCRADY, E. AND ZIMMERMANN, A. A., 1935. Morphological and functional development of the membranous labyrinth in the Opossum. *J. Comp. Neurol.* **63**: 95-118.

LEESON, C. R., CUTTS, J. H. AND KRAUSE, W. J., 1978. Post natal development and differentiation of the Opossum submandibular gland. *J. Anat.* **126**(2): 329-52.

LILLEGRAVEN, J. A., 1975. Biological considerations of the marsupial-placental dichotomy. *Evolution.* **29**: 707-22.

LYNE, A. G. AND HOLLIS, D. E., 1977. The early development of marsupials, with special reference to bandicoots. Pp. 293-302 *in* "Reproduction and evolution" ed by J. H. Calaby and C. H. Tyndale-Biscoe. Aust. Acad. Sci.: Canberra.

LYNE, A. G. AND VERHAGEN, A. M. W., 1957. Growth of the marsupial *Trichosurus vulpecula* and a comparison with some higher mammals. *Growth* **21**: 167-95.

McCRADY, E., 1938. The embryology of the Opossum. *Amer. Anat. Mem.* **16**: 1-233.

MERCHANT, J. C. AND CALABY, J. H., 1981. Reproductive biology of the Red-necked Wallaby *(Macropus rufogriseus banksianus)* and Bennett's Wallaby *(M. r. rufogriseus)* in captivity. *J. Zool.* (London) **194**: 203-17.

MIZELL, M. AND ISAACS, J. J., 1970. Induced regeneration of hindlimbs n the newborn Opossum. *Amer. Zool.* **10**: 141-55.

MOORE, K. L., 1982. "The developing Human. Clinically oriented embryology". 3rd Edition. Saunders: Philadelphia.

MÜLLER, F., 1967. Zum Vergleich der Ontogenesen von *Didephis virginiana* und *Mesocricetus auratus*. *Rev. Suisse Zool.* **74**: 607-13.

MÜLLER, F., 1968a. Die transitorischen Verschlüsse in der postnatalen Entwicklung der Marsupialia. *Acta Anat.* **71**: 581-624.

MÜLLER, F., 1968b. Zur Phylogenese des sekundären Kiefergelenks. *Rev. Suisse Zool.* **75**: 373-414.

MÜLLER, F., 1969a. Verhaltnis von Körperentwicklung und Cerebralisation in Ontogenese und Phylogenese der Säuger. Versuch einer Übersicht des Problems. *Verh. naturf. Ges. Basel.* **80**: 1-31.

MÜLLER, F., 1969b. Zur frühen Evolution der Säuger-Ontogenesetypen. Versuch einer Rekonstruktion aufgrund der Ontogenese-Verhaltnisse bei den Marsupialia. *Acta Anat.* **74**: 297-404.

RANDALL, D., GANNON, B., RUNCIMAN, S. AND BAUDINETTE, R. V., 1984. Gas transfer by the neonate in the pouch of the Tammar wallaby *Macropus eugenii*. Pp. 423-36 *in* "Respiration and metabolism of embryonic vertebrates" ed by R. S. Seymour. Junk: Dordrecht.

RENFREE, M. B., 1972. "Embryo-maternal relationships in the Tammar wallaby, *Macropus eugenii* (Desmarest)". Ph.D. Thesis, Dept. of Zoology, Australian National University: Canberra.

RENFREE, M. B., 1973. The composition of foetal fluids of the marsupial *Macropus eugenii*. *Devel. Biol.* **33**: 62-79.

ROWLATT, U., MROSOVSKY, N. AND ENGLISH, A., 1971. A comparative survey of brown fat in the neck and axilla of mammals at birth. *Biol. Neonate.* **17:** 53-83.

RUSSELL, A. M., 1944. The embryological development of Jacobson's organ in *Didephis virginiana* (Kerr). *Anat. Rec.* **90:** 149-53.

SELENKA, E., 1887. "Studien über Entwichelungsgeschichte. Viertes Heft. Das Opossum". C. W. Kreidels: Verlag, Wiesbaden.

SHELLEY, H. J., 1961. Glycogen reserves and their changes at birth and in anoxia. *Brit. Med. Bull.* **17:** 137-43.

SHOREY, C. D., 1968. "An electron microscope study of pituitary-ovarian changes during organogenesis, and embryonic development in *Trichosurus vulpecula*". M.Sc. Thesis: University of New South Wales: Sydney.

SMITH, B. T., TANSWELL, A. K., MINSHALL, D., BOGUES, W. N. AND VREEKEN, N., 1982. Influence of corticosteroids on glycogen content and steroid II-reductase activity in lung and liver of the foetal and newborn rat. *Biol. Neonates.* **42:** 201-7.

SOROKIN, S., 1962. A note on the histochemistry of the Opossum's lung. *Acta. Anat.* **50:** 13-21.

TYNDALE-BISCOE, H., 1973. "Life of marsupials". Edward Arnold: London.

WALKER, M. T. AND GEMMELL, R. T., 1983. Organogenesis of the pituitary, adrenal, and lung at birth in the wallaby, *Macropus rufogriseus. Amer. J. Anat.* **168:** 331-44.

WALKER, M. T. AND ROSE, R., 1981. Prenatal development after diapause in the marsupial *Macropus rufogriseus. Aust. J. Zool.* **29:** 167-87.

WHEELER, R. S., 1943. Normal development of the pituitary in the Opossum and its responses to hormonal treatments. *J. Morph.* **73:** 43-87.

WILSON, J. T. AND HILL, J. P., 1908. Observations on the development of *Ornithorhynchus. Phil. Trans. R. Soc. Lond.* Ser. B **199:** 31-168.

YADAV, M., 1971. The transmission of antibodies across the gut of pouch-young marsupials. *Immunology* **21:** 839-51.

OBSERVATIONS ON TORPOR IN THE SMALL MARSUPIAL *Dromiciops australis* (MARSUPIALIA: MICROBIOTHERIIDAE) FROM SOUTHERN CHILE

T. R. GRANT[1] and P. D. TEMPLE-SMITH[2]

Five specimens of *Dromiciops australis* were kept in captivity in the field in southern Chile for periods of up to three weeks. During that time all maintained body weight and entered torpor on most days independently of food availability and T_a. A mean arousal rate from torpor of 0.40 ± 0.02 (S.E.M.)°C/minute was recorded and animals were found to be active with surface temperatures (T_s, skin surface ventral abdomen) of 31.6 ± 0.6°C. The mean T_b of three animals measured was 34.4 ± 0.7°C at T_a 15°C. Mean T_s of three individuals in torpor at T_a 4.0°C was 8.1 ± 0.6°C. These responses are compared to those of other marsupial species known to enter torpor.

Key Words: Torpor; *Dromiciops*; Microbiotheriidae.

Pages 273-77 *in* POSSUMS AND OPOSSUMS: STUDIES IN EVOLUTION ed by M. Archer. Surrey Beatty & Sons and the Royal Zoological Society of New South Wales: Sydney, 1987.

INTRODUCTION

DROMICIOPS AUSTRALIS is a small scansorial marsupial which inhabits the beech forests (*Nothofagus* sp.) of southern Chile and the Lago Nahuel Huapi area of Argentina (Marshall 1978; Pearson 1983). It is known to exhibit winter torpor and to be capable of storing fat reserves in the base of its tail prior to such periods of seasonal torpor (Mann 1955; Greer 1965; Pearson 1983). The observations presented here were made incidentally during studies on the reproductive anatomy and sperm morphology in the species (Temple-Smith 1987).

MATERIALS AND METHODS

Five individuals of *D. australis* (1 male and 4 females) were captured by local people near the Parque Nacional de Puyehue in southern Chile. Animals were fed each day with an excess of apple, banana, sugar and bread, and were given 2-5 large gryllid crickets per day when these were available. Water was given, but was not observed to be used by any of the animals.

Animals were weighed each morning. This normally initiated arousal if they were in torpor at the time, and the process of arousal was monitored for each individual over at least one ambient temperature range. Surface temperatures (T_s) were taken by pushing a thermistor probe (YSI, Telethermometer) onto the ventral surface of the curled and torpid animal. This probe had a response time of only a few seconds and was temperature sensitive only at the tip. Respiration rate (respirations per minute, R.P.M.) was observed by eye at regular intervals, and timed with an alarm stopwatch (Seiko Sports 1000). Notes of any movement were recorded.

In one instance spontaneous arousal was monitored by inserting the thermistor probe into the sleeping chamber to touch the animal inside.

Body temperatures (T_b) were obtained from three animals, immediately on the induction of anaesthesia (Sodium pentobarbitone, Nembutal), by opening the body cavity and inserting the tip of the thermistor probe. As no constant temperature facilities were available, induced arousals were monitored over a range of ambient temperatures (T_a) which varied over 1-4°C within an experiment. Minimum T_a was recorded next to the animal cages using a maximum/minimum thermometer.

RESULTS

Figure 1 shows the weights of the five individual *D. australis* during the investigation. All maintained their weights, and all became torpid on most days of their captivity, including three days of transport in a camper-van to Santiago at the end of the study. Two individuals remained torpid over 36 hours, not being induced to arouse even by handling to be weighed and have their T_s taken (T_a 4.0°C).

During torpor all individuals were curled on their sides with the tail passed over the head and the fore feet and nose applied closely to the curled ventral side of the body. The rear feet were often exposed and trembled during arousal. At some time during arousal all animals shifted their positions from this lying, curled posture to a sitting position. In all instances of torpor the eyes were closed and the ears folded down until T_s was close to that of an active animal (28-34°C).

Arousal rates recorded for all individuals had a mean (± S.E.M.) of 0.40 ± 0.02°C/min (Table 1),

[1]School of Zoology, University of New South Wales, Kensington, New South Wales, Australia 2033.
[2]Department of Anatomy, Monash University, Clayton, Victoria, Australia 3168.

Table 1. Details of induced arousals from torpor in the five *D. australis* studied. Initial air temperature and surface temperatures (T_a init and T_s init), final T_s at activity (T_s fin), and duration and rates of arousal are shown. The predicted arousal rates were calculated from the equation: °C/min = 3.22 Body Weight (g)$^{-0.51}$ (Heindrich and Bartholomew 1971).

T_a init	T_s init	T_s fin	Time	°C/min	Predicted °C/min	T_a Range
ANIMAL 1. (Male)						
13.6	13.0	30.0	46	0.37	0.67	13.6-14.2
12.3	14.0	30.0	36	0.44	0.64	12.3-14.2
10.8	9.0	31.8	54	0.42	0.59	10.8-14.4
4.0	7.1	28.8	78	0.27	0.61	4.0-6.8
ANIMAL 2. (Female)						
11.2	10.0	34.0	58	0.41	0.64	11.2-11.8
ANIMAL 3. (Female)						
12.4	11.5	31.0	50	0.40	0.57	12.4-12.5
ANIMAL 4. (Female)						
13.6	13.6	34.0	42	0.49	0.53	13.6-13.7
13.6	13.0	32.0	42	0.45	0.53	13.6-15.0
ANIMAL 5. (Female)						
7.5	8.0	31.0	62	0.37	0.59	7.5-10.0
		31.6 ± 0.64 (S.E.M.)		0.40 ± 0.02 (S.E.M.)		

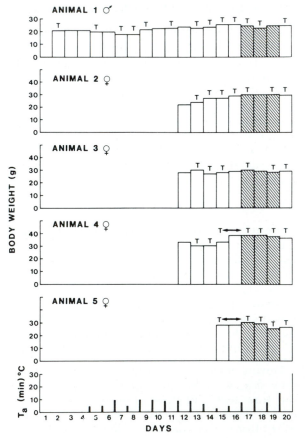

Fig. 1. Body weights of the five *D. australis* in captivity. T, torpid; ▨, en route to Santiago; T_a, minimum air temperature each day.

although a maximum rate of arousal over a 10 minute period in one individual was close to 0.8°C/min, and one animal aroused much more slowly from a T_s of 7.0°C (0.27°C/min) at a T_a of 4.0-6.0°C (Fig. 1D). In all except this last arousal, rates were between 60 and 85% of those

predicted for a placental species of the same body weight (Heindrich and Bartholomew 1971).

During torpor, periods of apnea were observed lasting from two to 30 minutes, followed by several deep breaths, or by a series of up to 30 rapid respirations. During arousal, respiration rate increased from below 100 R.P.M. to over 200 R.P.M. in all individuals. Figure 2 shows T_s, T_a and R.P.M. of animal No. 1 (male) over four different temperature ranges. Arousal for the other four animals (females) are illustrated in Figure 3. These data are summarised in Table 1.

Surface temperatures and T_a prior to, and during a spontaneous arousal are shown in Figure 4. In this instance the rate of arousal was much slower than that observed in induced arousal; i.e. 0.18°C/min compared to 0.4 ± 0.02°C/min.

The T_b of three anaesthetised *D. australis* were 33.0, 34.9 and 35.2°C respectively (Mean 34.4 ± 0.7°C). Animals were found to be active with T_s as low as 28°C. When animals became active it was no longer possible to obtain T_s measurements. The highest T_s taken was 34.0°C.

DISCUSSION

The T_b of 34.4 ± 0.7° for *D. australis* is similar to two measurements made by Pearson (1983) on a sleeping, but not torpid, specimen in Patagonia. On two separate occasions he recorded deep rectal temperatures of 33.0 and 33.5°C for the same individual which became active as soon as it was disturbed from sleep. These observations indicate that the T_s measured in the present study were probably quite close to the actual T_b of the animals.

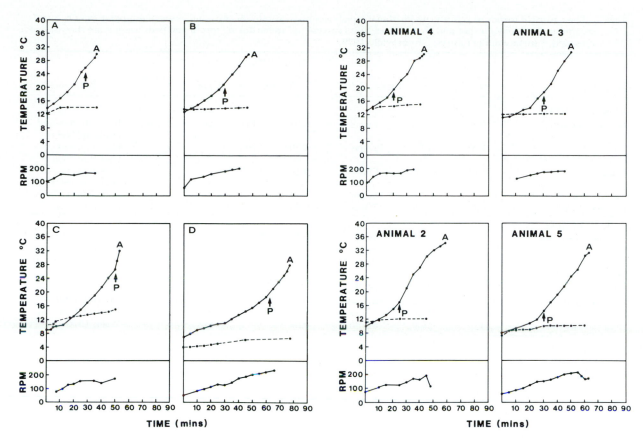

Fig. 2. Induced arousals in the male *D. australis* ●——●, T_s; ●--●, T_a; RPM, respirations per minute; P, change to sitting position; A, active.

Fig. 3. Induced arousals in four female *D. australis*. Symbols as for Figure 2.

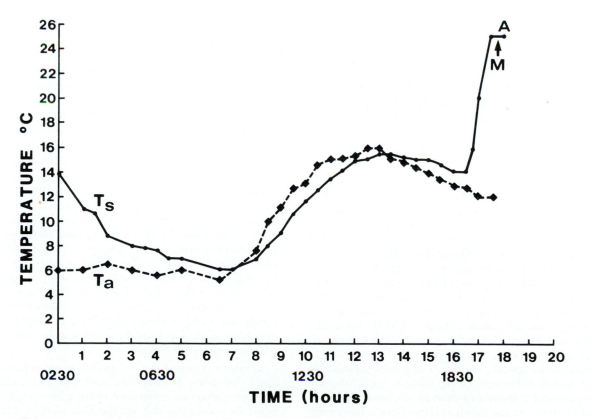

Fig. 4. Spontaneous arousals in the male *D. australis*. A, activity; M, first body movement; ◆--◆, T_a; ●——●, T_s.

Table 2. The nature of torpor in various species of marsupial mammals. Minimum body and ambient temperatures (T_b and T_a), maximum duration of torpor and arousal rates, the occurrence of apnea during torpor and body weights are shown. — not recorded; * surface temperatures; [+] from Wallis 1979.

Species	Min T_b °C	Min T_a °C	Duration Max	Arousal Rate Max °C/min	Overall °C/min	Apnea	Weight g	Reference
BURRAMYIDAE								
Cercartetus nanus	10.8	5.0	2 weeks	0.33	0.14	yes	70	Bartholomew and Hudson 1962
	12.5*	12.5	12 days	0.70	0.10	—	—	Hickman and Hickman 1960
C. lepidus	15.0*	15.0	6 days	0.30	0.07	—	—	Hickman and Hickman 1960
C. concinnus	—	—	11 days	—	—	—	14	Wakefield 1970[+]
Burramys parvus	6.0	5.5	7 days	0.44	0.17	—	45	Fleming 1985a
ACROBATIDAE								
Acrobates pygmaeus	6.9	6.0	44 hours	0.58	0.31	yes	14	Fleming 1982 Fleming 1985b
DASYURIDAE								
Sminthopsis crassicaudata	17.0*	14.0	10 hours	0.45	—	—	14	Godfrey 1968
S. macroura	18.0*	14.0	8 hours	0.29	—	—	20	Godfrey 1968
Antechinus stuartii	21.0	10.0	8 hours	0.22	—	yes	25	Wallis 1976
PETAURIDAE								
Petaurus breviceps	15.0	8.0	12 hours	0.30	0.22	yes	132	Fleming 1980
DIDELPHIDAE								
Marmosa sp.	16.0	11.0	6 hours	0.63	—	—	13	Morrison and McNab 1962
MICROBIOTHERIIDAE								
Dromiciops australis	7.1*	4.0	>36 hours	0.80	0.40	yes	30	This study

A number of marsupial species undergo torpor. The details of these torpid conditions are outlined in Table 2 where it can be seen that *D. australis* is similar to most of the members of the Australian Burramyidae in its thermoregulatory characteristics. Species in this family undergo torpor which lasts several days. In the Feathertail Glider, *Acrobates pygmaeus*, torpor has been found to last up to 44 hours, but more usually this species is like the other marsupials shown in Table 2, which exhibit torpor of less than 24 hours in duration.

At a T_a of 4°C, three *D. australis* had T_s of 7.1, 8.2 and 9.0°C respectively (mean 8.1 ± 0.8°C). Hibernation has been defined in terms of animals which allow their T_b to fall below 10°C for several days or weeks, with reduced metabolism and heart rate, while retaining the ability to rewarm without the need for exogenus heat (Lyman *et al.* 1982). Although metabolic rate was not measured in this study, Rosenmann (pers. comm.) has indicated reduced metabolism during torpor, and Mann (1955) recorded reduced heart rate in *D. australis*. The thermo-regulatory strategy adopted by the various burramyid species and by *D. australis*, whether it be called hibernation or torpor, undoubtedly represents a means of energy saving in habitats where winters are harsh and food availability may be reduced.

Animals in this study became torpid in spite of being fed each day. Both Hickman and Hickman (1960) and Bartholomew and Hudson (1962) made similar observations in *Cercartetus nanus* which became torpid independently of T_a, food availability, time of day and season. However, all of these studies have been carried out on captive animals and behaviour in wild individuals of the various species may be quite different.

The pygmy possums studied by Bartholomew and Hudson (1962) were considerably over-weight. Wild-caught *C. nanus* weigh between 15 and 42 g. Like *D. australis*, the Eastern Pygmy Possum has the ability to store fat in the base of its tail before entering torpor, it also builds nests of leaves (although these are considerably smaller than those built by *D. australis*) and undergoes periods of apnea during torpor (Turner 1983). *Dromiciops australis* had a more rapid rate of arousal than *C. nanus*. However, as was found for *D. australis* in this study, Hickman and Hickman (1960) reported a slower rate of arousal in *C. nanus* arousing without disturbance. Morrison and McNab (1962) found the same to be true for *Marmosa* sp. from Brazil.

While it is tempting to speculate on relation-ships between *D. australis* and the Australian Burramyiidae in terms of their physiology, it must be said that the similarities observed are probably more the result of convergent evolution into similar niches on the two continents than to close relatedness of the species.

ACKNOWLEDGEMENTS

Luis Pena and Checho are gratefully acknowledged for their invaluable help and good company in the field. Dr Mario Rosenmann, who is carrying out thorough investigation of thermoregulation in *D. australis* at the University of Chile in Santiago, is thanked for his personal communication on metabolic rate in the species. Dr Merv Griffiths is also acknowledged for his financial support of the trip to Chile, as are the wives of the authors, and Monash University.

REFERENCES

BARTHOLOMEW, G. A. AND HUDSON, J. W., 1962. Hibernation, estivation, temperature regulation, evaporative water loss and heart rate in the pygmy possum, *Cercartetus nanus*. *Physiol. Zool.* **35:** 94-107.

FLEMING, M. R., 1980. Thermoregulation and torpor in the Sugar Glider, *Petaurus breviceps* (Marsupialia: Petauridae). *Aust. J. Zool.* **28:** 521-34.

FLEMING, M. R., 1982. "Thermal strategies of three small possums from southeastern Australia". Ph.D. Thesis: Department of Zoology, Monash University, Clayton.

FLEMING, M. R., 1985a. The thermal physiology of the Mountain Pygmy Possum, *Burramys parvus* (Marsupialia: Burramyidae). *Aust. Mammal.* **8:** 79-90.

FLEMING, M. R., 1985b. The thermal physiology of the Feathertail Glider, *Acrobates pygmaeus* (Marsupialia: Burramyidae). *Aust. J. Zool.* **33:** 667-81.

GODFREY, G. K., 1968. Body temperature and torpor in *Sminthopsis crassicaudata* and *S. larapinta* (Marsupialia: Dasyuridae). *J. Zool. Lond.* **156:** 499-511.

GREER, J. K., 1965. Mammals of Malleco Province, Chile. *Publ. Mus. Michigan State Univ., Biol. Ser.* **3:** 49-152.

HEINDRICH, B. AND BARTHOLOMEW, G. A., 1971. An analysis of preflight warm-up in the sphynx moth, *Manduca sexta*. *J. exp. Biol.* **55:** 223-39.

HICKMAN, V. V. AND HICKMAN, J. L., 1960. Notes on the habits of the Tasmanian dormouse phalangers *Cercartetus nanus* (Desmarest) and *Eudromicia lepida* (Thomas). *Proc. Zool. Soc. Lond.* **135:** 365-74.

LYMAN, C. P., WILLIS, S. J., MALAN, A. AND WANG, L. C. H., 1982. "Hibernation and torpor in mammals and birds". Academic Press: New York.

MANN, G. F., 1955. Monito del Monte, *Dromiciops australis* Philippi. *Invest. Zool. Chilienas* **4:** 159-66.

MARSHALL, L. G., 1978. *Dromiciops australis. Mammal species* (No. 99, pp. 1-5). American Society of Mammalogists.

MORRISON, P. AND McNAB, B. K., 1962. Daily torpor in a Brazilian murine opossum *(Marmosa). Comp. Biochem. Physiol.* **6:** 57-68.

PEARSON, O. P., 1983. Characteristics of a mammalian fauna from forests in Patagonia, southern Argentina. *J. Mammal.* **64:** 476-92.

TEMPLE-SMITH, P. D., 1987. Sperm structure and marsupial phylogeny. Pp. 171-93 *in* "Possums and opossums: studies in evolution" ed by M. Archer. Surrey Beatty & Sons and the Royal Zoological Society of New South Wales: Sydney.

TURNER, V., 1983. Eastern Pygmy Possum, *Cercartetus nanus*. Pp. 160-61 *in* "The complete book of Australian mammals" ed by R. Strahan. Angus and Robertson: Sydney.

WALLIS, R., 1976. Responses to low temperature in small marsupial mammals. *J. Thermal Biol.* **4:** 105-11.

COMPARATIVE STUDIES ON THE HORMONAL PROFILES OF PROGESTERONE AND PROSTAGLANDIN F METABOLITE IN THE COMMON BRUSHTAIL POSSUM *Trichosurus vulpecula*

R. T. GEMMELL[1], R. L. HUGHES[1] and G. JENKIN[2]

The American opossum, *Didelphis virginiana* and the Australian possum *Trichosurus vulpecula* are similar in that the plasma progesterone profile during pregnancy is not significantly different from that obtained during an oestrous cycle.

In pregnant possums, the mean concentration of progesterone at oestrus was 1.10 ± 0.29 ng/ml (SEM, n=10). This hormonal concentration increased to 3.49 ± 0.88 ng/ml (SEM, n=10) 10 days before birth and reached a peak of 10.49 ± 1.64 ng/ml (SEM, n=10) five days before birth. Plasma progesterone concentrations had commenced to decline or had declined to basal levels when possums gave birth.

The large increase in prostaglandin F metabolite observed in *I. macrourus* at birth was not observed with *T. vulpecula*.

D. virginiana and *T. vulpecula* exhibit a similar progesterone profile contrasting with at least two other diverse marsupial patterns evidenced within the families Macropodidae and Peramelidae. Comparison of the plasma progesterone profiles in the seven marsupial species studied during pregnancy would suggest that the Peramelidae have modified the post luteal phase of the basic *D. virginiana* -type profile whereas the other marsupial families have extended the pre-luteal phase.

Key Words: Marsupial; Progesterone; PGFM; Pregnancy; Oestrous cycle.

Pages 279-91 *in* POSSUMS AND OPOSSUMS: STUDIES IN EVOLUTION ed by M. Archer. Surrey Beatty & Sons and the Royal Zoological Society of New South Wales: Sydney, 1987.

INTRODUCTION

MONOTREMES and marsupials have long been recognised as providing important models in comparative vertebrate reproductive biology (Owen 1834). The turn of the present century marked a period of accelerated understanding of comparative monotreme and marsupial reproductive biology. Natural history data and gross anatomical observations were complemented by the microscopical examination of preserved wax sectioned material. The basis for the present understanding of comparative marsupial reproductive biology had been established by the end of the 1950s and was reviewed by Sharman (1959). This publication and subsequent reviews during the next decade not only reported diversity between marsupial species but also recurrently featured the theme of "the marsupial pattern" versus "the placental pattern" (Waring *et al.* 1955; Sharman 1963, 1970). Marsupial-placental comparisons continue as one of the dominant themes in many contemporary publications. However, there is currently an increasing awareness of the taxonomic and other significance of comparisons of reproductive parameters between marsupial families that frequently exhibit separated affinity groups.

Differing aspects of reproduction have been used to form these various affinity groups.

Certain reproductive features such as morphology of the reproductive tract (Hill 1900; Pearson 1945, 1946, 1950a, b), sperm morphology (Hughes 1965) and embryonic diapause (Renfree 1981) allow for the formation of definite groups. A measure of the specialisation in reproduction reached by the Macropodidae is illustrated by embryonic diapause. This reproductive strategy has been mainly associated with the Macropodidae, having been observed in all but one of the 20 or so kangaroos and wallabies examined (Renfree 1981). The different members of the families Macropodidae, Dasyuridae, Phascolarctidae and Peramelidae can be readily identified from the distinctive morphology of their spermatozoa. Distinctive morphological characteristics occur within families, while spermatozoa structure varies between marsupial families (Hughes 1965).

The previous reproductive features allow the formation of definite groups; other features have not been as helpful. Pearson (1949) stated that the grouping of the marsupials on the basis of placentation produces groups which have no relation to "phylogenetic affinities". Hughes (1974, 1984) has shown that marsupial extra-embryonic membranes are similar in some related marsupial groups. An unusual feature of placentation in the marsupials is the presence of

[1]Department of Anatomy, University of Queensland, St. Lucia, Brisbane, Australia 4067.
[2]Department of Physiology, Monash University, Melbourne, Australia 3052.

an invasive chorio-allantoic placenta within the Peramelidae (Hill 1895, 1897; Hughes 1974; Padykula and Taylor 1976). In most marsupials, yolk-sac placentation is the sole mechanism for intra uterine transfer between the foetus and the mother, however placentation is variable (Pearson 1949; Hughes 1974, 1984).

There are an estimated 250 species of living marsupials (Kirsch and Calaby 1977) and information concerning reproducton is known for about 29 species (reviews by Sharman, Calaby and Poole 1966; Tyndale-Biscoe 1973, 1984; Renfree 1981). A widely held but unsubstantiated view is that in general, marsupials with a short gestation period produce smaller young. This correlation has contributed to the postulate that evolution in marsupials has been accompanied by a lengthening of the gestation period (Sharman 1959). Although the gestation periods are known for some 29 marsupial species the evolutionary adaptations required to allow for an extension of the gestation period have not as yet become apparent. A comparison of the hormonal profiles obtained from those marsupials with a short gestation, for example the possum and opossum, with those with a longer pregnancy, such as the macropodids, may allow an understanding of the hormonal and other changes required for a lengthening of the gestation period.

The major secretory hormone of the corpus luteum is progesterone. This steroid hormone is required for the initiation of the luteal phase within the uterus. The concentration of progesterone in peripheral plasma has been determined throughout pregnancy in four species of marsupials, the opossum *Didelphis virginiana* (Harder and Fleming 1981), the bandicoot *Isoodon macrourus* (Gemmell 1981), the macropodid *Macropus eugenii* (Hinds and Tyndale-Biscoe 1982), and the dasyurid *Dasyuroides byrnei* (Fletcher 1983a). The profile of progesterone has been determined in three species of macropodids during delayed pregnancy, the Quokka *Setonix brachyurus* (Cake, Owen and Bradshaw 1980), the Tammar *M. eugenii* (Hinds and Tyndale-Biscoe 1982) and The Red-necked Wallaby *Macropus rufogriseus* (Walker and Gemmell 1983a).

Further investigation of progesterone profiles in other marsupials would assist in obtaining an understanding of the role of progesterone and the corpus luteum in the maintenance of pregnancy in marsupials. The concentration of progesterone postoestrus in the peripheral plasma of both pregnant and nonpregnant *Trichosurus vulpecula* (a phalangerid) has been previously reported. However, the small number of plasma samples assayed did not enable the authors to fully describe the plasma progesterone profile throughout pregnancy (Thorburn, Cox and Shorey 1971; Shorey and Hughes 1973). In this study we have determined the concentration of progesterone in the peripheral plasma of *T. vulpecula* in greater detail during pregnancy as well as during the oestrous cycle.

Although the actual role of prostaglandin F2α (PGF2α) during parturition remains to be elucidated, it is now clear that prostaglandin production is a prerequisite for normal birth in a wide range of placental mammals. Although an increased concentration of 13,14-dihydro-15-keto prostaglandin F2α (PGFM), the circulating metabolite of PGF2α, has already been observed in the plasma of the bandicoot at birth (Gemmell, Jenkin and Thorburn 1980), there was no comparable increase in the concentration at birth in *M eugenii* (Tyndale-Biscoe, Hinds, Horn and Jenkin 1983; Shaw 1983), or in *M. rufogriseus* (Walker and Gemmell 1983a). It remains to be established whether PGF2α is involved in parturition in marsupials other than the bandicoot. The concentration of PGFM in the peripheral plasma of *T. vulpecula* was, therefore, also measured during pregnancy.

In this study, the differing profiles of plasma progesterone throughout pregnancy in the marsupial will be examined. The placement of the Macropodidae and the Peramelidae at opposite grades of marsupial reproduction (Hill 1895) will be confirmed and possible relationships within and between marsupial groups will be suggested after examination of the plasma progesterone and prostaglandin F metabolite profiles during pregnancy.

MATERIALS AND METHODS

Animals

The breeding colony was established from *T. vulpecula* trapped in the environs of Brisbane. Two female and two male *T. vulpecula* were housed in one of four fully enclosed yards, each 20 square metres in area. A total of eight males and eight females were held simultaneously. The date of birth of captured and captive pouch young was determined either by observing the day of birth or by estimating the age of the pouch young from the length of the head and the total body weight according to the method described by Lyne and Verhagen (1957).

Blood samples were obtained every second or third day during the months March to July from nine females. The day of pregnancy was calculated from the date of birth. The pouch young were removed (RPY) from a further ten females during the months March to October. Blood samples were collected daily from five days after RPY until eight days after birth. The possums were monitored for the presence of sperm around the cloaca or a newborn young. To obtain the plasma progesterone profile in nine possums during the oestrous cycle, pouch young were removed and the females housed without males.

Blood samples were collected by cardiac puncture daily from five days to 40 days after RPY. The possums were lightly anaesthetized with a halothane (Fluothane-I.C.I.) and oxygen mixture. The blood was placed into heparinized tubes, centrifuged within 15 mins of collection and the plasma was frozen until assayed.

Assays

Progesterone

Plasma progesterone concentrations were determined by radioimmunoassay using the method described previously for the bandicoot (Gemmell 1979), using sheep anti-progesterone-11-hemisuccinate-bovine serum albumin (anti-serum No. 334) kindly donated by Dr R. I. Cox, CSIRO, Prospect, New South Wales, Australia. The anti-serum is highly specific for progesterone, the only significant cross-reaction being that with 11β-hydroxyprogesterone (11.9%). The limit of the sensitivity of the assay was 25 pg/tube and the intra- and inter-assay coefficients of variation were 12.5% (n = 5) and 14.0% (n = 10) respectively. The assay buffer blank was less than 0.1 ng/ml and the efficiency of extraction was 87.0% (n = 10).

PGFM. 13,14-dihydro-15-keto-prostaglandin F2α was determined in plasma samples by a modification of the method of Mitchell, Flint and Turnbull (1976), described previously for the measurement of PGFM in bandicoot plasma (Gemmell, Jenkin and Thorburn 1980). PGFM antiserum was kindly donated by Dr R. Fairclough and authentic PGFM was kindly donated by Dr K. T. Kirkton (Upjohn Company). The assay was specific for PGFM. The only prostaglandins which significantly cross-reacted in the assay were PGFM (100%), 15-keto-PGF2α (12.0%), PGEM (2.8%) and 13,14-dihydro PGF2α (0.75%). 6-15-dihydro-PGF1αα, 13,14-dihydro-6-keto PGF1αα, PGD1, PGD2, TXB2, PGE1, PGE2α, 6-keto PGF1αα, PGF1α and PGF2α all cross reacted less than 0.3%. The correlation coefficient of PGFM added to plasma and subsequently assayed was 0.99, the regression line being y = 5.91-0.92X. The intra- and inter assay coefficients of variation were 12% and 14%. The non-specific binding was 0.75%, assay buffer blank not less than 0.1 ng/ml and the efficiency of extraction of [3H] PGFM from possum plasma was 86%.

RESULTS

In Queensland, *T. vulpecula* births are first observed during March. The peak breeding period is during April and the breeding season continues through to November (Fig. 1). During 1984, all the female possums (n = 7) in the breeding colony gave birth between 11 and 30 March. All seven possums had given birth to young during the previous breeding season of

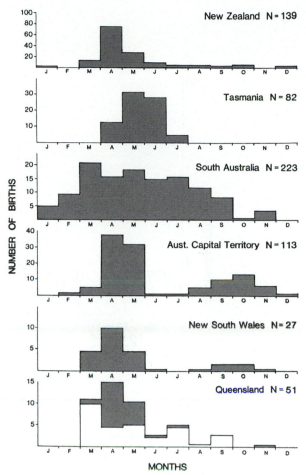

Fig. 1. The breeding season for the Brushtail Possum *T. vulpecula* in Christchurch, New Zealand (Tyndale-Biscoe 1955) in Tasmania (Lyne and Verhagen 1957), in South Australia (Pilton and Sharman 1962), in the Australian Capital Territory (Dunnet 1964), in New South Wales (Lyne and Verhagen 1957) and in Queensland. Animals trapped in the wild are represented in open boxes and animals giving birth in captivity are in shaded boxes.

1983. In four of these females, no rise in plasma progesterone concentrations or births were observed following RPY between 4 October and 11 October, 1983.

Possums normally mated eight to 11 days following RPY, although sperm was not seen around the cloaca of all possums which gave birth. The mean period between RPY and the presence of sperm was 9.57 ± 0.48 days (SEM, n = 7). Birth occurred 25 to 28 days after RPY, the mean period from RPY to birth was 26.11 ± 0.26 days (SEM, n = 9). RPY day 1 was taken as the day young were removed from the mother. The circulating level of progesterone increased from basal concentrations of 0.45 ± 0.15 ng/ml (mean, SEM, n = 8) 19 days prior to birth (day of mating) to peak concentrations 10.49 ± 1.64 ng/ml (mean, SEM, n = 10) 6 days prior to birth. The concentration of progesterone was 2.68 ± 0.78 ng/ml (mean, SEM, n = 16) on the day of birth (Fig. 2). Basal concentrations were not reached until at least 7 days after birth.

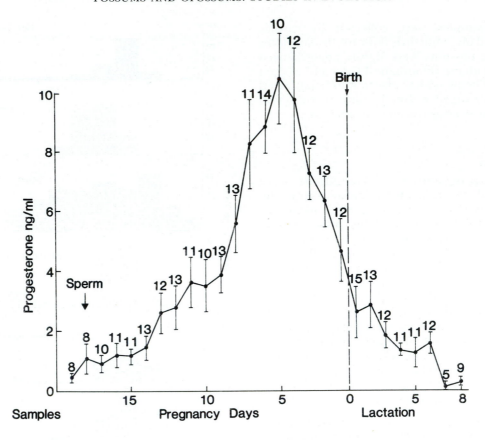

Fig. 2. The plasma progesterone profile during pregnancy for *T. vulpecula*. This graph of the mean plasma progesterone concentration was constructed from samples (numbers of samples denoted above standard error of mean) taken from 19 pregnant possums.

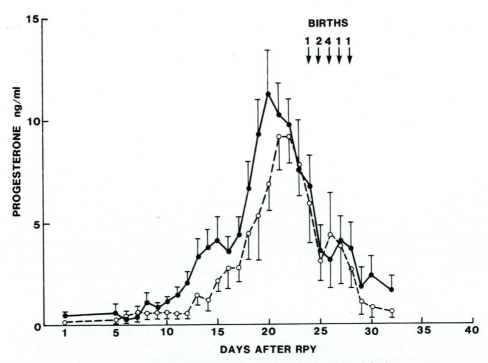

Fig. 3. The plasma progesterone profile during pregnancy (●—●) and during an oestrous cycle (○– –○). These graphs were constructed from daily samples taken from nine pregnant possums and nine possums during an oestrous cycle. Sampling started immediately after removal of the pouch young and the days of birth of the nine pregnant possums are indicated.

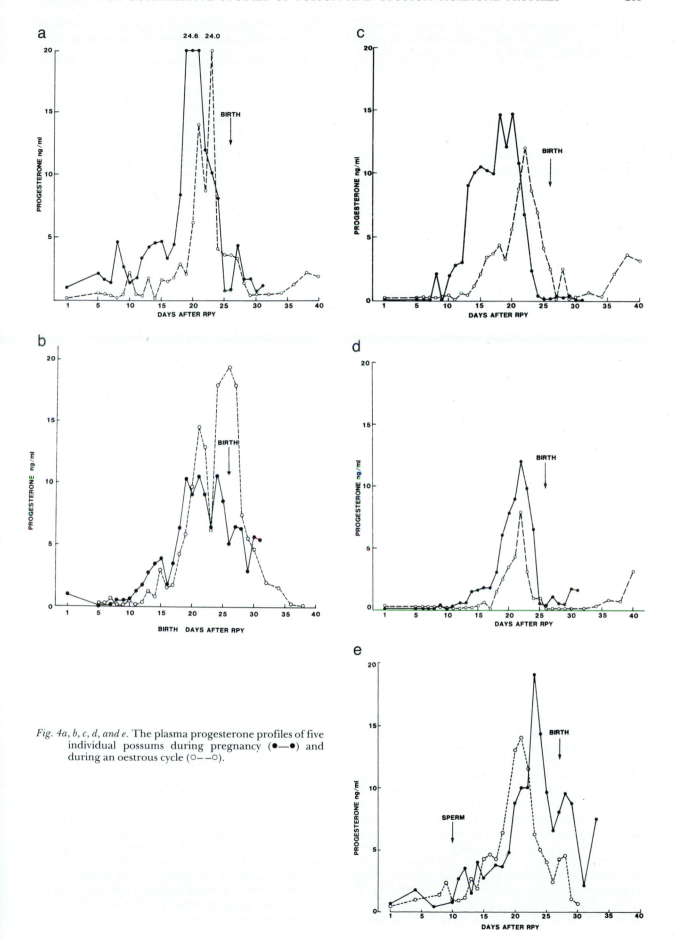

Fig. 4a, b, c, d, and e. The plasma progesterone profiles of five individual possums during pregnancy (●—●) and during an oestrous cycle (○– –○).

There was no significant difference between the plasma progesterone profile of the nine pregnant possums and that of the nine possums during an oestrous cycle (Fig. 3). When plasma progesterone profiles from individual possums were compared, a remarkable similarity was apparent between the profile during pregnancy and that of the oestrous cycle (Figs 4a, b, c, d, e). In the profiles shown in Fig. 4a, the maximum progesterone concentration observed was 24.6 ng/ml during pregnancy and 24.0 ng/ml during an oestrous cycle. In the profiles for the possum shown in Fig. 4d, the maximum concentration during pregnancy was 12.0 ng/ml and 7.8 ng/ml during an oestrous cycle.

Although the mean results from 19 possums indicate that plasma progesterone was declining at birth in the possum, (Fig. 2), the change in progesterone concentration in individual possums at birth is masked. Figures 4b, c and e, demonstrate that plasma progesterone was declining at birth in some possums, whereas in others (Figs 4a and d), the plasma progesterone had reached basal levels at the time of parturition.

The concentration of PGFM did not alter significantly throughout pregnancy or during the first eight days of lactation except in two possums close to parturition. The range of PGFM in samples from ten pregnant possums was 0.14 to 4.04 ng/ml, with a mean concentration of 0.52 ± 0.06 ng/ml (mean, SEM, n = 67) during the 17 days prior to birth and 0.44 ± 0.05 ng/ml (mean, SEM, n = 35) during the eight days following the birth.

Although the plasma PGFM concentration was not elevated in four possums one day before birth, and the mean PGFM concentration was 0.53 ± 0.09 ng/ml (SEM, n = 4), in one possum a concentration of 4.04 ng/ml was obtained. Similarly the mean PGFM concentration was 0.37 ± 0.07 ng/ml (SEM, n = 6) in six possums on the day of birth. However, in one possum a concentration of 1.88 ng/ml was measured.

DISCUSSION

The start of the breeding season for the *T. vulpecula* appears to vary from January in Adelaide through to April in Tasmania (Fig. 1). The main breeding period is from March through to June. However, births have been observed in all months of the year. It is unknown whether the possum employs environmental cues, such as daylength, rainfall, nutrition or temperature, to regulate the onset and duration of the breeding season. The occurrence of seven births all within 19 days would suggest some degree of synchrony so that a parameter such as daylength might be involved. Nevertheless, when the breeding records from other studies are examined (Fig. 1) it is apparent that the commencement of breeding varies greatly so that it is unlikely that day length is the only parameter governing the start and duration of the breeding season.

The profile of progesterone observed in this study is in reasonable accord with previous reports of the plasma profile in possums during pregnancy (Thorburn *et al.* 1971; Shorey and Hughes 1973). Nevertheless, this extended study has shown that the mean peak plasma progesterone concentrations of 10.49 ng/ml occurred five days before birth and progesterone concentrations had started to decline at least three days prior to birth. The maximum plasma progesterone concentrations in the possum are therefore similar in magnitude to those of the didelphid *D. virginiana* (maximum concentration 14.0 ng/ml; Harder and Fleming 1981) the bandicoot *I. macrourus* (maximum concentration 12.6 ng/ml; Gemmell 1981) (Fig. 5) and *D. byrnei* (maximum concentration 10.32 ng/ml; Fletcher 1983a) during pregnancy.

The three species of macropodids examined, which all exhibit embryonic diapause, have appreciably lower maximum plasma progesterone concentrations than those reported for on-macropodid marsupials. In the Quokka *S. brachyurus*, a peak concentration of 2.24 ng/ml was reported five days prior to birth (Cake *et al.* 1980); in the Tammar *M. eugenii*, the peak concentration of 500 pg/ml was observed just one day prior to birth (Hinds and Tyndale-Biscoe 1982) and similarly in the Red-necked Wallaby, *M. rufogriseus*, a peak concentration of only 300 pg/ml was reported two days prior to birth (Walker and Gemmell 1983a) (Fig. 6). Further study will reveal whether this association of greatly reduced peak progesterone plasma concentration during pregnancy within the macropodids, in contrast with non-macropodid marsupials, is a valid characteristic of the family Maropodidae.

There is no measurable difference in the plasma progesterone profiles when pregnancy values are compared with equivalent days post oestrus throughout the oestrous cycle in *D. virginiana* (Harder and Fleming 1981) and in *T. vulpecula*. This differs from the pattern reported for *D. byrnei* (Fletcher 1983a) and *S. brachyurus* (Cake *et al.* 1980). Both Fletcher (1983) and Cake *et al.* (1980) observed a transient peak of progesterone at day seven of pregnancy and at day three of delayed pregnancy respectively. This transient peak of progesterone was not observed during the oestrous cycle. With *M. eugenii* this transient peak of progesterone was observed during pregnancy and during the oestrous cycle (Hinds and Tyndale-Biscoe 1982). Further study is required to ascertain the functional significance of this transient peak of progesterone in *D. byrnei* and the three macropodids before a full understanding of the progesterone profile in these marsupials can be obtained.

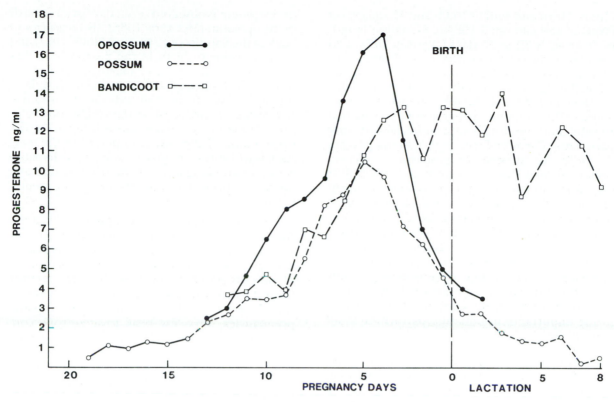

Fig. 5. A comparison of the plasma progesterone profiles of the opossum *D. virginiana* (Harder and Fleming 1981), the bandicoot *I. macrourus* (Gemmell 1981) and the possum *T. vulpecula* during pregnancy.

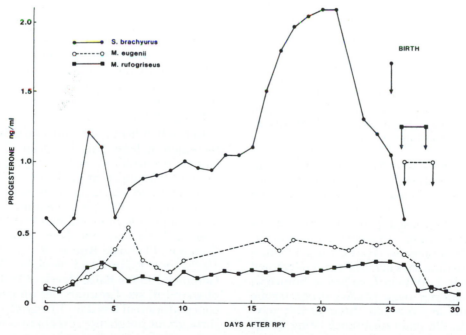

Fig. 6. A comparison of the plasma progesterone profiles of *S. brachyurus* (Cake *et al.* 1980), *M. eugenii* (Hinds and Tyndale-Biscoe 1982) and *M. rufogriseus* (Walker and Gemmell 1983a).

Variations are apparent in the progesterone profiles of all seven marsupial species during pregnancy. As has been shown for placental mammals, a decline in progesterone is not always observed immediately prior to parturition and there is a large variation in the pattern of plasma progesterone just before birth in the marsupials. Plasma progesterone declines in *D. virginiana* at least four days prior to birth (Harder and Fleming 1981), at least two days prior to birth in *S. brachyurus* (Cake *et al.* 1980), and at least 24 hours prior to birth in *M. eugenii* (Tyndale-Biscoe,

Hinds, Horn and Jenkin 1983) and *M. rufogriseus* (Walker and Gemmell 1983a). Gestation length in *D. byrnei* is 30 to 35 days (Woolley 1971) and the plasma progesterone decreases from a concentration of 10.32 ng/ml on day 21, to 7.91 ng/ml on day 28 and eventually to 3.54 ng/ml on day 35 (Fletcher 1983a). In the bandicoot *I. macrourus*, the plasma progesterone remained elevated and did not change appreciably during parturition (Gemmell *et al.* 1980). In the possum *T. vulpecula*, plasma progesterone starts to decline at least three days prior to birth, and is therefore similar in pattern to the American opossum. With some *T. vulpecula*, plasma progesterone concentrations had declined to basal concentrations before birth and with others progesterone concentrations were still declining at parturition. These differences in progesterone profiles close to parturition in individual possums and with other species of marsupial suggest that a drop in progesterone concentration is not of prime importance in initiating parturition in all marsupials.

Recent studies in *M. eugenii*, *I. macrousus* and *M. rufogriseus* have indicated that the adrenal gland of the newborn marsupial is capable of steroid production at birth. A cortisol concentration of 0.9 and 1.5 μg/100 ml was obtained in fetal samples, from the Tammar 1 or 2 days before birth (Catling and Vinson 1976). A lower concentration of 0.094 ng of cortisol per adrenal was determined in *I. macrousus* (Gemmell, Singh-Asa, Jenkin and Thorburn 1982) and 0.58 ng per adrenal in *M. rufogriseus* at birth (Walker and Gemmell 1983b). This evidence of cortisol synthesis within the marsupial adrenal at birth indicates that, as in many placentals, the marsupial adrenal could play an important role in determining the length of gestation.

When the plasma progesterone profile during pregnancy of the three marsupials which do not exhibit embryonic diapause are examined, some obvious similarities are observed (Fig. 5). In all three species, the plasma progesterone concentration increases eight to 12 days prior to birth; 12 days in *D. virginiana* (Harder and Fleming 1981), eight to nine days in *I. macrourus* (Gemmell 1981) and 10 to 11 days in *T. vulpecula*. Similarly, in all three species peak plasma progesterone concentrations are reached three to five days prior to birth, the main difference being in the bandicoot *I. macrourus* where elevated progesterone concentrations persist into lactation. Those of *D. virginiana* and *T. vulpecula* decline prior to birth. The plasma progesterone profiles of all three marsupials are similar during the last 10 days of pregnancy. Consequently, *T. vulpecula* has a longer gestation period than *D. virginiana* due to a longer pre-luteal phase. There is a similarity in the timetable of foetal development in all three marsupials. The primitive streak stage of

development is observed seven days prior to birth in the opossum (McCrady 1938), six days prior to birth in the bandicoot (Lyne and Hollis 1977) and 7.5 days prior to birth in the possum (Hughes 1974). With the dasyurid marsupial *Antechinus stuartii* which has a gestation period of 27.2 days, organogenesis from the advanced primitive streak stage to birth took six days (Selwood 1980). From these meagre examples it is apparent the lengthening of the gestation period in the marsupials which do not possess embryonic diapause is due to an elongation of the pre-luteal phase, the period prior to the significant rise in progesterone.

The macropodid marsupials, a more specialised group of marsupials than *D. virginiana* and *T. vulpecula*, have a longer gestation and give birth to larger young. Macropodid plasma progesterone profiles also differ greatly from those of the opossum, possum and bandicoot. However, in the latter third of macropodid pregnancies, very low peak progesterone concentrations are observed approximately two days prior to birth in the Tammar (Hinds and Tyndale-Biscoe 1982) and in the Red-necked Wallaby (Walker and Gemmell 1983a) and approximately four days prior to birth in the Quokka (Cake *et al.* 1980) (Fig. 6). In all three macropodids, the plasma progesterone declines prior to birth, as observed with the opossum and the possum. The primitive streak stage of development in the macropodids is observed at 11 and 14 days before birth in *M. rufogriseus* (Walker and Rose 1981) and *M. eugenii* (Renfree and Tyndale-Biscoe 1973) respectively. Thus, this longer period for embryogenesis, as observed in these two macropodids, may be a distinctive feature of this marsupial group.

In the bandicoot, the concentration of PGFM increased at least one day prior to birth from basal concentrations of 0.29 ng/ml to maximal levels of 2.53 ng/ml immediately after birth with these elevated concentrations persisting for at least two days after birth (Gemmell *et al.* 1980). This magnitude of increase in PGFM concentration at birth has not been observed in the Tammar (Tyndale-Biscoe *et al.* 1983; Shaw 1983), nor Red-necked Wallaby (Walker and Gemmell 1983a). Raised levels of PGFM were not observed in six Tammars before parturition. A concentration of less than 1.0 ng/ml was observed throughout pregnancy, except in two animals. A concentration of 3.5 ng/ml was measured in one sample taken when the newborn young was moving from the uterus to the pouch and in another Tammar a concentration of 1.4 ng/ml was obtained when the young was first observed and already attached to a teat in the pouch (Tyndale-Biscoe *et al.* 1983). Shaw (1983) did not observe elevated concentrations of plasma PGFM throughout parturition in four Tammar wallabies. The concentration of PGFM ranged from 0.2 to 0.8 ng/ml with a modal value of 0.2 to 0.3 ng/ml. There was no significant increase in PGFM at birth in the Red-necked Wallaby (Walker and Gemmell 1983a). The mean PGFM concentrations in the peripheral plasma of this wallaby were less than 1.0 ng/ml throughout parturition. Similarly, in the possum, there was no signifcant increase in PGFM at birth. However, in two out of a total of ten possums elevated PGFM concentrations were observed close to parturition, similar in magnitude to that observed in the Tammar (Tyndale-Biscoe *et al.* 1983). It is possible that due to infrequent sampling, peaks of PGFM have been missed. Nevertheless, similar sampling procedures in the bandicoot have revealed a large increase in PGFM prior to birth, during parturition and into lactation. Studies so far suggest that peramelids may be unique amongst marsupials in displaying a large increase in PGFM at birth.

From the scant information now available concerning plasma progesterone profiles in pregnant marsupials, a tentative grouping of the seven marsupials can be made (Fig. 7). The plasma progesterone profile of *D. virginiana*, which possesses one peak of progesterone during pregnancy is probably the basic marsupial progesterone profile during pregnancy. *T. vulpecula* has modified this basic profile by extending the pre-luteal phase. *I. macrourus* has also modified the basic profile by extending the post-luteal phase into lactation. This latter modification might be viewed as a pre-adaptation that could provide the evolutionary basis for extending the

length of the gestation period within the Peramelidae. The macropodids, *M. eugenii*, *S. brachyurus* and *M. rufogriseus* which all use embryonic diapause in their reproductive strategy, represent a specialised group of marsupials. All three macropodids display a biphasic pattern for plasma progesterone during pregnancy. *D. byrnei* appears to be a somewhat intermediate between the opossum, possum and bandicoot and the macropodids. This marsupial has a biphasic progesterone profile during pregnancy. However, the peak plasma progesterone concentrations are more like those of the opossum. *D. byrnei* would appear to have the endocrinological features which would allow for embryonic diapause but the marsupial does not show this form of reproductive control. Although *D. byrnei* does not possess embryonic diapause, it has been suggested on the basis of the long gestation period, 30 to 35 days (Fletcher 1983b), that *D. byrnei* and *A. stuartii* may have a period of slow development or developmental arrest (Sharman 1963; Selwood 1980).

If Sharman (1959) is correct with his postulation that evolution in marsupials has been accompanied by a lengthening of the gestation period, a possible heirarchy can be suggested from the basic marsupial profile of progesterone of *D. virginiana*, to the more complicated biphasic version of the macropodids. As stated previously, *T. vulpecula* has extended gestation by extending the pre-luteal phase. *D. byrnei* may be displaying a further evolutionary step with still greater

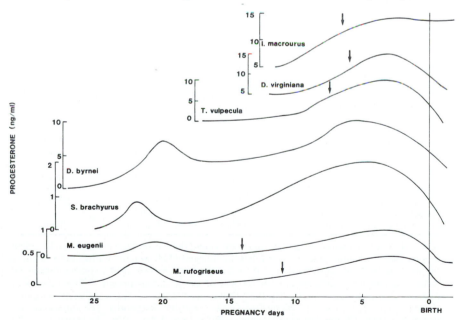

Fig. 7. Plasma progesterone profiles through pregnancy for seven marsupials. Progesterone data was obtained from Gemmell (1981) for *I. macrourus*, Harder and Fleming (1981) for *D. virginiana*, this study for *T. vulpecula*, Fletcher (1983a) for *D. byrnei*, Cake *et al.* (1980) for *M. eugenii* and Walker and Gemmell (1983a) for *M. rufogriseus*. The stage of pregnancy at which the primitive streak blastocyst stage is observed is denoted with an arrow. This data was obtained from Lyne and Hollis (1977) for *I. macrourus*, McCrady (1938) for *D. virginiana*, Hughes (1974) for *T. vulpecula*, Renfree and Tyndale-Biscoe (1973) for *M. eugenii* and Walker and Rose (1981) for *M. rufogriseus*.

extension with the addition of the biphasic pattern of progesterone during pregnancy. The second peak of progesterone is believed to correlate with the monophasic peak observed in *D. virginiana* and *T. vulpecula*. The majority of the macropodids have proceeded beyond the pattern exhibited by *D. byrnei* with evolution and use embryonic diapause in their reproduction.

Whereas *D. byrnei* and the macropodids modified the pre-luteal phase, the bandicoots altered the post luteal phase. The bandicoots are unique amongst the marsupials in two important aspects. The first is the possession of chorio-allantoic placentae (Hill 1897; Padykula and Taylor 1976) and the second is the corpora lutea of pregnancy which persist for approximately 45 days into lactation (Hughes 1962a; Gemmell 1979; Lyne and Hollis 1979; Hollis and Lyne 1980). These corpora lutea continue to secrete progesterone throughout pregnancy and into lactation (Gemmell, Jenkin and Thorburn 1980; Gemmell 1981). The bandicoots therefore have

the structural and endocrinological features required for prolonging uterine foetal development (Lillegraven 1975, 1979; Taylor and Padykula 1978). However, although bandicoots seem to have crossed the marsupial-placental grade of specialization with a change that prolonged the life span of the corpora lutea, despite this potential evolutionary advance the bandicoots have retained the marsupial mode of reproduction and as yet have not capitalized on their apparent evolutionary option for extending the length of their gestation period.

The large peak of PGFM observed in the bandicoot at birth is not apparent in the other three marsupials, *M. eugenii*, *M. rufogriseus* and *T. vulpecula*. This supports the concept of a separate peramelid mechanism for the initiation of parturition from other marsupials.

In recent publications, the marsupials have been divided into three groups, based on their pattern of reproduction (Tyndale-Biscoe 1984;

Table 1. Gestation length and birth weight of various marsupials.

Family and Species	Gestation length (days)	Birth[1] weight (mgm)	Mother's[2] weight (gm)
PERAMELIDAE			
Perameles nasuta (1)	12.5	240	859
Isoodon macrourus (1)	12.5	240	1130
DIDELPHIDAE			
Didelphis virginiana (2)	13.0	130	1500
"POSSUMS"[3]			
Cercartetus concinnus (3)	–	125	–
Petaurus breviceps (4)	about 17.0	194	130
Trichosurus vulpecula (5)	17.5	200	2330
Pseudocheirus peregrinus (6)	–	300	700
DASYURIDAE			
Dasyurus viverrinus (7)	24	13	818
Antechinus stuartii (8)	27.2	16	30
Sarcophilus harrisii (9)	about 21 to 28	70	6700
MACROPODOIDEA			
Bettongia penicillata (10)	21.0	290	–
Bettongia gaimardi (11)	21.1	300	–
Bettongia lesueur (12)	21.3	320	1100
Macropus rufogriseus (13)	26.8	560	12000
Setonix brachyurus (14)	27.0	310	2750
Macropus eugenii (15)	29.3	500	5000
Macropus agilis (16)	29.4	630	12000
Macropus rufus (17)	33.0	750	27300
Wallabia bicolor (18)	35.0	600	–
Potorous tridactylus (19,20)	38.0	333-380	1000

Note: [1] In relation to marsupial maternal body weight the Family Dasyuridae has a low birth weight and the Family Macropodidae has an elevated birth weight. Within both of these families the birth weight increases with increasing maternal size of the species concerned rather than being correlated with the length of the gestation period. [2] The weights of mothers were obtained from Russell (1982). [3] "Possums" here refers to the following families: Burramyidae (*C. concinnus*); Petauridae (*P. breviceps*);Phalangeridae (*T. vulpecula*) and Pseudocheiridae (*P. peregrinus*).(1) Hughes 1962a. Lyne 1964, 1974. (2) Hartman 1928; McCrady 1938. (3) Clark 1967. (4) Clark 1967; Schultze-Westrum 1965. (5) Pilton and Sharman 1962. (6) Thomson and Owen 1964. (7) Hill and Hill 1955; Fletcher 1983b. (8) Marlow 1961; Woolley 1966; Selwood 1980. (9) Hughes 1982; Hughes, unpublished data. (10) Parker 1977; Rose 1978. (11) Rose 1978. (12) Tyndale-Biscoe 1968. (13) Walker and Rose 1981; Walker and Gemmell 1983. (14) Sharman 1955; Tyndale-Biscoe 1963. (15) Tyndale-Biscoe 1973; Merchant 1979. (16) Merchant 1976. (17) Sharman and Pilton 1964. (18) Sharman, Calaby and Poole 1966; Tyndale-Biscoe 1973. (19) Rose 1978; Shaw and Rose 1979. (20) Hughes 1962b.

Renfree 1981). Group one contains species in which the gestation period is considerably shorter than the oestrous cycle. This group includes most of the Phalangeridae, including *T. vulpecula*, the Didelphidae, the Dasyuridae, the Peramelidae and the four macropodids, *Macropus parma, Macropus parryi, Macropus giganteus* and *Macropus fuliginosus*. The second group contains the marsupials in which the gestation period is almost the same length as the oestrous cycle. The majority of the macropodid species belong to this group. The members of Group 3 are species with long gestation periods in which blastocysts are totally delayed, or slowly growing during lactation. Some members of the family Burramyidae and *Tarsipes* belong to this group. These groupings appear to have been formed without consideration of all the reproductive information available for marsupials. The Peramelidae, as stated by Hill (1899), is a distinct marsupial group and cannot be closeted with the Phalangeridae. The marsupial grouping obtained from sperm morphology (Hughes 1965), from observations of the marsupial placenta (Hill 1895, 1897; Padykula and Taylor 1976) and from the hormonal profiles obtained in this study, all suggest that the Peramelidae are a distinct group of marsupials. Similar arguments can be made for the separation of the Didelphidae and the Dasyuridae. Renfree (1981) has suggested that Group 2 and Group 3 marsupials have evolved separately from Group 1 animals. There would appear to be confusion about the taxonomic membership of each group. We believe that further investigation of reproductive relationships within valid affinity groups will often comprise whole marsupial families, and that this will be a fruitful area for biological study. We would urge caution in erecting affinity groups on the basis of reproductive parameters that lead to evolutionary speculations that confound well established marsupial taxonomic groupings particularly at the family level.

In conclusion, although some marsupials have extended their gestation length, especially the macropodids, when compared with *D. virginiana* and *T. vulpecula*, it has not always meant an increase in birth weight (Table 1). However, when the birth weight is considered with respect to maternal body weight and compared with gestation length within the Family Dasyuridae and the Family Macropididae, the birth weight increases with maternal size rather than being correlated with the length of the gestation period. It is therefore possible that the primary effect of the increase in gestation length in macropodid marsupials was to permit the evolution of embryonic diapause as a strategy of reproduction and only secondarily contributing to the production of young with an increased birth weight by increasing the interval between the primitive streak stage and birth.

ACKNOWLEDGEMENTS

The authors would like to thank the Australian Research Grants Scheme for financial assistance.

REFERENCES

CAKE, M. H., OWEN, F. J. AND BRADSHAW, S. D., 1980. Difference in concentration of progesterone in plasma between pregnant and non-pregnant quokkas *(Setonix brachyurus)*. *J. Endocr.* **84:** 153-58.

CATLING, P. C. AND VINSON, G. P., 1976. Adrenocortical hormones in the neonate and pouch young of the tammar wallaby *Macropus eugenii. J. Endocr.* **69:** 447-48.

CLARK, M. J., 1967. Pregnancy in the lactating pygmy possum, *Cercartetus concinnus. Aust. J. Zool.* **15:** 673-83.

DUNNET, G. M., 1964. A field study of local populations of the brush-tailed possum *Trichosurus vulpecula* in eastern Australia. *Proc. Zool. Soc. Lond.* **14:** 665-95.

FLETCHER, T. P., 1983a. Progesterone and luteinizing hormone concentration in peripheral plasma of *Dasyuroides byrnei* during the oestrous cycle and pregnancy. *Procs of Aust. Soc. Reprod. Biol.* Abst. 44.

FLETCHER, T. P., 1983b. "Endocrinology of reproduction in *Dasyuroides byrnei*". Ph.D. Thesis: La Trobe University, Melbourne.

GEMMELL, R. T., 1979. The fine structure of the luteal cells in relation to the concentration of progesterone in the plasma of the lactating bandicoot *Isoodon macrourus* (Marsupialia:Peramelidae). *Aust. J. Zool.* **27:** 501-10.

GEMMELL, R. T., 1981. The role of the corpus luteum of lactation in the bandicoot, *Isoodon macrourus* (Marsupialia: Peramelidae). *Gen. Comp. Endocr.* **44:** 13-19.

GEMMELL, R. T., 1982. Breeding bandicoots in Brisbane *(Isoodon macrourus;* Marsupialia, Peramelidae). *Aust. Mammal.* **5:** 187-93.

GEMMELL, R. T., JENKIN, G. AND THORBURN, G. D., 1980. Plasma concentrations of progesterone and 13,14-dihydro-16-keto- prostaglandin F2α at parturition in the bandicoot, *Isoodon macrourus. J. Reprod. Fert.* **60:** 253-56.

GEMMELL, R. T., SINGH-ASA, P., JENKIN, G. AND THORBURN, G. D., 1982. Ultrastructural evidence for steroid-hormone production in the adrenal of the marsupial, *Isoodon macrourus* at birth. *Anat. Rec.* **203:** 505-12.

HARDER, J. D. AND FLEMING, M. W., 1981. Estradiol and progesterone profiles indicate a lack of endocrine recognition of pregnancy in the opossum. *Science* **212:** 1400-02.

HARTMAN, G. G., 1928. The breeding season of the opossum *(Didelphys virginiana)* and the rate of intra uterine and post natal development. *J. Morph.* **46:** 143-215.

HILL, J. P., 1895. Preliminary note on the occurrence of a placental connection in *Perameles obesula*, and on the foetal membranes of certain macropods. *Proc. Linn. Soc. N.S.W.* **10:** 578-81.

HILL, J. P., 1897. The placentation of *Perameles. Quart. J. Micr. Sci.* **40:** 385-446.

HILL, J. P., 1900. On the fetal membranes, placentation and parturition of the native cat *(Dasyurus viverrinus). Anat. Anz.* **18:** 364-73.

HILL, J. P. AND HILL, W. C. O., 1955. The growth stages of the pouch young of the native cat *(Dasyurus viverrinus)* together with observations on the anatomy of the newborn young. *Trans. Zool. Soc. Lond.* **28:** 349-453.

HINDS, L. A. AND TYNDALE-BISCOE, C. H., 1982. Plasma progesterone levels in the pregnant and non-pregnant tammar, *Macropus eugenii. J. Endocr.* **93:** 99-107.

HOLLIS, D. E. AND LYNE, A. G., 1980. Ultrastructure of luteal cells in fully formed and regressing corpora lutea during pregnancy and lactation in the marsupials *Isoodon macrourus* and *Perameles nasuta. Aust. J. Zool.* **28:** 195-211.

HUGHES, R. L., 1962a. Role of the corpus luteum in marsupial reproduction. *Nature Lond.* **194:** 890-91.

HUGHES, R. L., 1962b. Reproduction in the macropodid marsupial *Potorous tridactylus* (Kerr). *Aust. J. Zool.* **10:** 193-224.

HUGHES, R. L., 1965. Comparative morphology of spermatozoa from five marsupial families. *Aust. J. Zool.* **13:** 533-43.

HUGHES, R. L., 1974. Morphological studies on implantation in marsupials. *J. Reprod. Fert.* **39:** 173-86.

HUGHES, R. L., 1982. Reproduction in the Tasmanian Devil *Sarcophilus harrisii* (Dasyuridae, Marsupialia). Pp.49-63 *in* "Carnivorous marsupials" ed by M. Archer. Royal Zoological Society of New South Wales: Sydney.

HUGHES, R. L., 1984. Structural adaptations of the eggs and fetal membranes of monotremes and marsupials for respiration and metabolic exchange. Pp.389-421 *in* "Respiration and metabolism of embryonic vertebrates" ed by R. S. Seymour. Dr. W. Junk, Dordrecht: The Netherlands.

KIRSCH, J. A. W. AND CALABY, J. H., 1977. The species of living marsupials: an annotated list. Pp. 9-25 *in* "The biology of marsupials" ed by B. Stonehouse and D. Gilmore. Macmillan: London.

LILLEGRAVEN, J. A., 1975. Biological considerations of the marsupial placental dichotomy. *Evolution* **29:** 707-22.

LILLEGRAVEN, J. A., 1979. Pp. 259 *in* "The Mesozoic mammals: the first two-thirds of mammalian history" ed by J. A. Lillegraven, A. Kielan-Jaworowska and W. A. Clemens. University of California Press: Berkley.

LYNE, A. G., 1964. Observations on the breeding and growth of the marsupial *Perameles nasuta* Geoffroy, with notes on other bandicoots. *Aust. J. Zool.* **12:** 322-39.

LYNE, A. G., 1974. Gestation period and birth in the marsupial *Isoodon macrourus*. *Aust. J. Zool.* **22:** 303-09.

LYNE, A. G. AND HOLLIS, D. E., 1977. The early development of marsupials, with special reference to bandicoots. Pp. 293-302 *in* "Reproduction and evolution" ed by J. H. Calaby and C. H. Tyndale-Biscoe. Australian Academy of Science: Canberra.

LYNE, A. G. AND HOLLIS, D. E., 1979. Observations on the corpus luteum during pregnancy and lactation in the marsupials, *Isoodon macrourus* and *Perameles nasuta*. *Aust. J. Zool.* **27:** 881-99.

LYNE, A. G. AND VERHAGEN, A. M. W., 1957. Growth of the marsupial *Trichosurus vulpecula* and a comparison with some higher mammals. *Growth* **21:** 167-95.

MARLOW, B. J., 1961. Reproductive behaviour of the marsupial mouse *Antechinus flavipes* (Waterhouse) (Marsupialia) and the development of the pouch young. *Aust. J. Zool.* **9:** 203-18.

McCRADY, E., 1938. The embryology of the opossum. *Am. Anat. Mem.* **16:** 1-233.

MERCHANT, J. C., 1976. Breeding biology of the agile wallaby *Macropus agilis* (Gould) (Marsupialia:Macropodidae) in captivity. *Aust. Wildl. Res.* **3:** 93-103.

MERCHANT, J. C., 1979. The effect of pregnancy on the interval between one oestrus and the next in the tammar wallaby, *Macropus eugenii*. *J. Reprod. Fert.* **56:** 459-63.

MITCHELL, M. D., FLINT, A. P. F. AND TURNBULL, A. C., 1976. Plasma concentrations of 13,14-dihydro-15-keto-prostaglandin F during pregnancy in the sheep. *Prostaglandins* **11:** 319-29.

OWEN, R., 1834. On the generation of the marsupial animals with a description of the impregnated uterus of the kangaroo. *Phil. Trans. R. Soc. Lond.* **124:** 333-64.

PADYKULA, H. A. AND TAYLOR, M. J., 1976. Ultrastructural evidence for loss of the trophoblastic layer in the chorioallantoic placenta of Australian bandicoots (Marsupialia: Peramelidae). *Anat. Rec.* **186:** 357-86.

PARKER, P., 1977. An ecological comparison of marsupial and placental patterns of reproduction. Pp. 273-86 *in* "The biology of marsupials" ed by B. Stonehouse and D. Gilmore. Macmillan: London.

PEARSON, J., 1945. The female urogenital system of the marsupialia with special reference to the vaginal complex. *Pap. Roy. Soc. Tas.* 1944, pp. 71-78.

PEARSON, J., 1946. The affinities of the rat-kangaroos (Marsupialia) as revealed by a comparative study of the female urogenital system. *Pap. Roy. Soc. Tas.* 1945, pp. 13-25.

PEARSON, J., 1949. Placentation of the Marsupialia. *Proc. Linn. Soc. Lond.* **161:** 1-9.

PEARSON, J., 1950a. A further note on the female urogenital system of *Hypsiprymnodon moschatus* (Marsupialia). *Pap. Roy. Soc. Tas.* 1949, pp. 203-10.

PEARSON, J., 1950b. The relationships of the Potoroidae to the Macropodidae (Marsupialia). *Pap. Roy. Soc. Tas.* 1949, pp. 211-29.

PILTON, P. E. AND SHARMAN, G. B., 1962. Reproduction in the marsupial *Trichosurus vulpecula*. *J. Endocr.* **25:** 119-36.

RENFREE, M. B., 1981. Embryonic diapause in marsupials. *J. Reprod. Fert. Suppl.* **29:** 67-78.

RENFREE, M. B. AND TYNDALE-BISCOE, C. H., 1973. Intrauterine development after diapause in the marsupial *Macropus eugenii*. *Devel. Biol.* **32:** 28-40.

ROSE, R. W., 1978. Reproduction and evolution in female Macropodidae. *Aust. Mammal.* **2:** 65-72.

RUSSELL, E. M., 1982. Patterns of paternal care and parental investment in marsupials. *Biol. Rev.* **57:** 423-86.

SHULTZE-WESTRUM, T., 1965. Innerartliche verstandigung durch Dufte beim Gleitbeutler *Petaurus breviceps papuanus* Thomas (Marsupialia, Phalangeridae). *Z. vergl. Physiol.* **50:** 151-220.

SELWOOD, L., 1980. A timetable of embryonic development of the dasyurid marsupial *Antechinus stuartii* (Macleay). *Aust. J. Zool.* **28:** 649-68.

SHARMAN, G. B., 1955. Studies on marsupial reproduction. 3. Normal and delayed pregnancy in *Setonix brachyurus*. *Aust. J. Zool.* **3:** 56-70.

SHARMAN, G. B., 1959. Marsupial reproduction. *Monogr. Biol* **8:** 332-68.

SHARMAN, G. B., 1963. Delayed implantation in marsupials. Pp. 3-14 *in* "Delayed Implantation" ed by A. C. Enders. University of Chicago Press: Chicago.

SHARMAN, G. B., 1970. Reproductive physiology of marsupials. *Science* **167:** 1221-28.

SHARMAN, G. B., CALABY, J. H. AND POOLE, W. E., 1966. Patterns of reproduction in female diprotodont marsupials. *Symp. Zool. Soc. Lond.* **15:** 205-32.

SHARMAN, G. B. AND PILTON, P., 1964. The life history and reproduction of the red kangaroo (*Megaleia rufa*). *Proc. Zool. Soc. Lond.* **1432:** 29-48.

SHAW, G., 1983. Effect of PGF2α on uterine activity, and concentrations of 13,14-dihydro-15-Keto-PGF2-PGF2α in peripheral plasma during parturition in the tammar wallaby (*Macropus eugenii*). *J. Reprod. Fert.* **69:** 429-36.

SHAW, G. AND ROSE, R. W., 1979. Delayed gestation in the potoroo, *Potorous tridactylus* (Kerr). *Aust. J. Zool.* **27:** 901-12.

SHOREY, C. D. AND HUGHES, R. L., 1973. Development, function and regression of the corpus luteum in the marsupial *Trichosurus vulpecula. Aust. J. Zool.* **21:** 477-89.

TAYLOR, J. M. AND PADYKULA, H. A., 1978. Marsupial trophoblast and mammalian evolution. *Nature Lond.* **271:** 588.

THOMSON, J. A. AND OWEN, W. H., 1964. A field study of the Australian ringtail possum, *Pseudocheirus peregrinus* (Marsupialia: Phalangeridae). *Ecol. Monogr.* **34:** 37-52.

THORBURN, G. D., COX, R. I. AND SHOREY, C. D., 1971. Ovarian steroid secretion rates in the marsupial *Trichosurus vulpecula. J. Reprod. Fert.* **24:** 139.

TYNDALE-BISCOE, C. H., 1963. Effect of ovariectomy in the marsupial, *Setonix brachyurus. J. Reprod. Fert.* **6:** 25-40.

TYNDALE-BISCOE, C. H., 1968. Reproduction and post-natal development in the marsupial *Bettongia lesueur* (Quoy and Gaimard). *Aust. J. Zool.* **16:** 577-602.

TYNDALE-BISCOE, C. H., 1973. "Life of marsupials". Edward Arnold: London.

TYNDALE-BISCOE, C. H., 1984. Mammalia-Marsupialia. Pp. 386-454 *in* "Marshall's physiology of reproduction" (4th Ed.) ed by G. E. Lamming. Churchill Livingstone: Edinburgh.

TYNDALE-BISCOE, C. H., HINDS, L. A., HORN, C. A. AND JENKIN, G., 1983. Hormonal changes at oestrus, parturition and post-partum oestrus in the tammar wallaby *(Macropus eugenii). J. Endocr.* **96:** 155-61.

WALKER, M. T. AND GEMMELL, R. T., 1983a. Plasma concentrations of progesterone, oestradiol-17β, and 13,14-dihydro-15-oxo- prostaglandin F2α in the pregnant wallaby *(Macropus rufogriseus rufogriseus). J. Endocr.* **97:** 369-77.

WALKER, M. T. AND GEMMELL, R. T., 1983b. Organogenesis of the pituitary adrenal and lung at birth in the wallaby, *Macropus rufogriseus. Am. J. Anat.* **168:** 331-44.

WALKER, M. T. AND ROSE, R., 1981. Prenatal development after diapause in the marsupial *Macropus rufogriseus. Aust. J. Zool.* **29:** 167-87.

WARING, H., SHARMAN, G. B., LOVAT, D. AND KAHAN, M., 1955. Studies on marsupial reproduction. 1. General features and techniques. *Aust. J. Zool.* **3:** 34-43.

WOOLLEY, P., 1966. Reproduction in *Antechinus* spp. and other dasyurid marsupials. *Symp. Zool. Soc. Lond.* **15:** 281-94.

WOOLLEY, P., 1971. Maintenance and breeding of laboratory colonies of *Dasyuroides byrnei* and *Dasycercus cristicauda. Int. Zoo. Yb* **11:** 351-54.

Section 3

AUSTRALIAN FAMILIES

PLATE 2: The living phascolarctid *Phascolarctos cinereus*, the Koala.
Photograph by M. Seyfort courtesy the Australian National Photographic Index.

Madakoala, A NEW GENUS AND TWO SPECIES OF MIOCENE KOALAS (MARSUPIALIA: PHASCOLARCTIDAE) FROM SOUTH AUSTRALIA, AND A NEW SPECIES OF Perikoala

MICHAEL O. WOODBURNE[1], RICHARD H. TEDFORD[2], MICHAEL ARCHER[3]
and NEVILLE S. PLEDGE[4]

A new genus of phascolarctid, *Madakoala* n. gen., is described from strata of Miocene age in South Australia. One species, *Madakoala devisi* n. sp., from the Pinpa, Ericmas and Tarkarooloo faunas of the Tarkarooloo Basin, possibly was ancestral to *Madakoala wellsi* n. sp., from the Ericmas Fauna, Tarkarooloo Basin. Specimens from the Ditjimanka Local Fauna, Lake Palankarinna, Lake Eyre Basin, probably are referable to *M. wellsi,* although at present this conclusion cannot be tested. Another species of *Madakoala* n. gen., probably is represented in the Wadikali Local Fauna of the Tarkarooloo Basin. *Madakoala* n. gen., does not appear to have been closely related to either *Perikoala robustus* n. sp., from the South Palankarinna Local Fauna, *P. palankarinnica* from the Ditjimanka Local Fauna or *Litokoala kanunkaensis* (Kanunka North Local Fauna; Springer, 1987) or *L. kutjamarpensis* from the Kutjamarpu Local Fauna, Lake Ngapakaldi, Lake Eyre Basin, and probably is even more distinct from the Pliocene-aged species of *Koobor,* known from the Bluff Downs and Chinchilla Local Faunas of Queensland.

Key Words: *Madakoala; Perikoala;* Miocene; Koala evolution; Pinpa Fauna; Ericmas Fauna; Wadikali Local Fauna; Ditjimanka Local Fauna; Tarkarooloo Local Fauna; Kanunka North Local Fauna; South Palankarinna Local Fauna.

Pages 293-317 *in* POSSUMS AND OPOSSUMS: STUDIES IN EVOLUTION ed by M. Archer. Surrey Beatty & Sons and the Royal Zoological Society of New South Wales: Sydney, 1987.

INTRODUCTION

In 1957, R. A. Stirton described *Perikoala palankarinnica*, the first known fossil koala of Miocene age, from the Etadunna Formation, Ditjimanka Local Fauna of present terminology, Lake Palankarinna, South Australia. Stirton (1957) noted features in which *Perikoala palankarinnica* resembles, and might have been allied with, the living *Phascolarctos cinereus*. This study was followed by that of Stirton, Tedford and Woodburne (1967), in which the single upper molar of *Litokoala kutjamarpensis* was described from the geologically younger Kutjamarpu Local Fauna (Wipajiri Formation, Lake Ngapakaldi, South Australia). One other phascolarctid of Miocene age is described in this Volume (see Springer 1987). Archer (in Archer and Wade 1976) described *Koobor jimbarratti*) from the Bluff Downs Local Fauna, of early Pliocene age, in northeastern Queensland. Archer (1977) referred *Pseudocheirus notabilis* DeVis (1899) to *Koobor,* as another Pliocene species from the Chinchilla Local Fauna of southeastern Queensland. Pledge (1987, this Volume) describes a new Pliocene species of *Phascolarctos* from South Australia. Finally, *Phascolarctos stirtoni*, a giant Pleistocene species, also known from Queensland, was described by Bartholomai (1968).

The materials reported upon in the present study were recovered from the Miocene Namba Formation that yields the Pinpa, Ericmas, Wadikali and Tarkarooloo faunas of the Tarkarooloo Basin, Lake Frome district of South Australia, and from the Etadunna Formation that yields the Ditjimanka Local Fauna, South Palankarinna Local Fauna of Lake Palankarinna, and the Kanunka North Local Fauna, Lake Kanunka, east of Lake Eyre in South Australia (Fig. 1). The specimens from the Tarkarooloo Basin were obtained by R. H. Tedford and colleagues in 1971 and 1973 and by T. H. Rich in 1977, 1980 and 1983, and by N. S. Pledge in 1983, whereas the Ditjimanka material was collected by M. O. Woodburne and M. Archer in 1972, and by M. Archer in 1977, and the other Lake Kanunka and Lake Palankarinna localities by M. O. Woodburne and colleagues, 1984, 1985. Woodburne prepared a first draft of the manuscript, with Tedford, Archer and Pledge contributing data, revisions, and refinements that resulted in the present version.

[1]Department of Earth Sciences, University of California, Riverside, California, USA 92521.
[2]Department of Vertebrate Paleontology, American Museum of Natural History, New York, USA 10024.
[3]School of Zoology, University of New South Wales, P.O. Box 1, Kensington, New South Wales, Australia 2033.
[4]South Australian Museum, North Terrace, Adelaide, South Australia 5000.

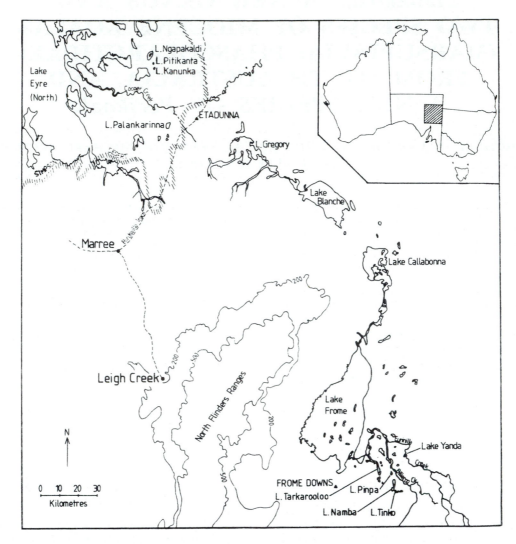

Fig. 1. Map of South Australia showing the location of sites that yielded the fossil koalas discussed in the text.

DEFINITIONS, TERMINOLOGY AND ABBREVIATIONS

All measurements are metric. The distinction between Faunas and Local Faunas follows Tedford (1970). Local Faunas (initial letters upper case) are assemblages — sometimes a single taxon — derived from one or a few closely spaced geographically and temporally contemporaneous sites. Faunas (initial letters upper case) are a collection of virtually contemporaneous local faunas that share a majority of taxa and occur over a wider geographic area. The stratigraphic and geochronologic relationships of these faunas and local faunas follow that of Woodburne *et al.* (1985).

Ma refers to year-ages in the radioisotopic time scale; m.y. to temporal duration. All radioisotopic ages have been calculated with respect to the constants established by Steiger and Jaeger (1977).

Dental terminology is either standard or explained in the text; numerical assignment of teeth in the dentition follows Archer (1978). In addition to accepted homologies for the cusp(id)s of the diprotodontan molar crowns, we identify the posterolingual cusp on the upper molars as a metaconule and the neomorphic cuspule that appears in phascolarctids at the lingual base of the metacone as the neometaconule (see Tedford and Woodburne 1987, this Volume, for extended discussion).

Higher level systematic nomenclature follows that of Aplin and Archer (1987).

RV- is a fossil locality of the Department of Earth Sciences, University of California, Riverside; and Riverside specimen are numbers preceded by UCR. V- refers to a fossil locality of the Department of Paleontology, University of California, Berkeley; and Berkeley specimen numbers are preceded by UCB. AMNH refers to both localities and specimen numbers of the

American Museum of Natural History, New York. CPC refers to specimens of the Commonwealth Palaeontological Collections, Bureau of Mineral Resources, Canberra. NMV refers to localities of the Museum of Victoria, Melbourne; specimen numbers of this institution are preceded by NMV P. UNSW refers to localities of the School of Zoology, University of New South Wales, Kensington; specimen numbers from the Australian Museum, Sydney are listed as AM F. QM refers to localities of the Queensland Museum, Brisbane; Queensland Museum specimen numbers preceded by QM F. SAM refers to localities of the South Australian Museum, Adelaide; specimen numbers of this institution are preceded by SAM P.

SYSTEMATICS

Order DIPROTODONTIA

Suborder Vombatiformes

Superfamily Phascolarctoidea

Family Phascolarctidae

Madakoala, n. gen.

Genotypic species: Madakoala devisi n. sp.

Additional species: Madakoala wellsi n. sp., *Madakoala* sp., aff. *M. wellsi, Madakoala* sp.

Distribution: Pinpa, Ericmas, Wadikali and Tarkarooloo Faunas, Namba Formation, Tarkarooloo Basin, South Australia; Ditjimanka and Palankarinna South local faunas, Lake Palankarinna, South Australia; Medial Miocene.

Etymology: Mada means "new" in the Dieri language (Austin 1981), in reference to the status of the genus so-named. Gender is considered to be masculine.

Diagnosis: Species of *Madakoala* differ from species of *Perikoala* in being slightly to much larger, in having proportionately more elongate molars, in lacking the strong lingual shelf basal to the metaconid and in having a weaker lingual shelf basal to the entoconid of the lower molars. *Madakoala wellsi*, has a much better-developed protostylid ridge on M_3 (not as well developed in *M. devisi*) and both species of *Madakoala* have weaker surface crenulations overall (based on comparison with the paratype and all molars referred to species of *Perikoala*). Adult specimens of *M. wellsi* have a completely non-ankylosed dentary symphysis, in contrast with the apparently strongly fused symphysis of *P. palankarinnica, P. robustus* and (in juveniles and adults) *Phascolarctos cinereus*.

Species of *Madakoala* differ from *Litokoala kutjamarpensis* in being larger and in having a smaller neometaconule, in lacking the protostyle, and in lacking the strongly bifurcate pair of ribs that connect the protocone and metaconule in a manner which results in the development of a deep pocket between them in the Kutjamarpu species. Species of *Madakoala* are larger, lower crowned, less selenodont and lack an entostylid ridge in M_3, as compared to *L. kanunkaensis*.

Species of *Madakoala* differ from species of *Koobor* in that the latter are smaller, have a distinctly bicusped P^3 (known only in *K. notabilis*), stronger buccal stylar cusps on the upper molars that effectively close off buccal basins of each tooth, lack a paraconule in M^3 and M^4, lack a neometaconule in M^2-M^4 (M^5 unknown), have unobstructed lingual exits of the transverse valley of the upper molars and upper molars that are much longer than wide.

Species of *Madakoala* differ from species of *Phascolarctos* in being somewhat larger (than *P. cinereus*) to very much smaller (than *P. stirtoni*, known only from upper molars), in having a scalpriform lower incisor, a more cuspate and less trenchant P_3, a less strongly-developed and less buccally situated protostylid on M_2, a more distinct and widely separated paraconid and metaconid on M_2, a better-developed metastylid on M_2 (i.e., a more triangular arrangement of the primary trigonid cuspids), a more complexly constructed talonid on M_2 (e.g., a hypoconid-entoconid crest is present), in lacking the metastylid fold in which the pre-entocristid is continuous with the postmetastylid cristid, in lacking the columnar stylids on the lingual faces of the metastylid and entoconid, in having a generally more robust and less trenchant P^3 which also lacks a cingular shelf, greater separation between the paracone and metacone across the transverse valleys of the upper molars, more vertically standing paracone and metacone, generally weaker paraconules, no protostyle (which in species of *Phascolarctos* is a longitudinally oriented posterior spur from the preprotocrista just lingual to the paraconule), a much smaller or absent neometaconule, a much less distinct hypocone-metacone crest on M^5, a horizontal ramus which lacks the prominent flange-like digastric process that lies below the masseteric fossa in *P. cinereus* (unknown in *P. stirtoni*), and at least in *M. wellsi* the dentaries of adult specimens are completely unankylosed (in striking contrast to the fused dentaries of even juvenile *P. cinereus*).

Madakoala devisi, n. sp.
(Figs 2 and 3; Tables 1, 2 and 7)

Holotype: SAM P24792 (AMNH 102242), right dentary with broken incisor, P_3, M_{2-5}, lacking most of the horizontal and ascending ramus, collected by R. H. Tedford *et al.* in 1971.

Type locality: AMNH Lake Pinpa Site D, Namba Formation, unit 3, section 8 of Callen and Tedford (1976); west side of Lake Pinpa, South

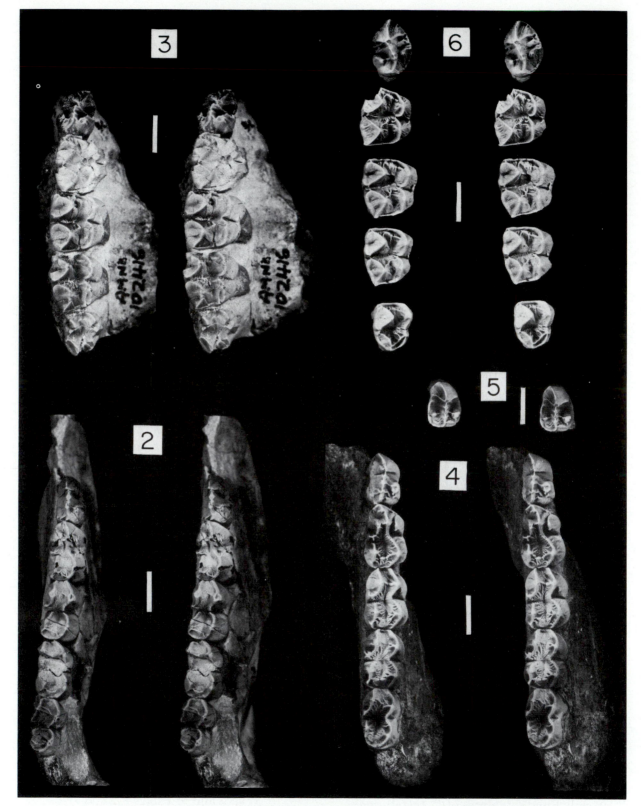

Fig. 2. Madakoala devisi, holotype, stereophotograph of SAM P24792, RP$_3$-M$_5$. Here and in all subsequent figures, white bar = 5 mm in length.

Fig. 3. Madakoala devisi, stereophotograph of AMNH 102415, RP^3, and AMNH 102416, RM^2-RM^5.

Fig. 4. Madakoala wellsi, stereophotograph of SAM P24793, RP$_3$-M$_5$.

Fig. 5. Madakoala wellsi, NMV P178115, RP$_3$.

Fig. 6. Madakoala sp., aff. *M. wellsi*, stereophotograph of UCR 16990, RP^3; UCR 15827, RM^2; UCR 16991, RM^3; UCR 16992, RM^4; UCR 16993, RM^5.

Australia, grid zone 6, reference 318148, Curnamona topographic series, SH 54-14; 1:250,000, 1965 ed.

Referred material: A. From the upper part of the lower member of the Namba Formation. From Lake Pinpa, Site C; AMNH 102399, left M_2; AMNH 102413, fragment of a left dentary with posterior half of P_3, complete M_2 and anterior part of M_3 alveolus; AMNH 102416, fragment of right maxillary with M^2 lacking the lingual enamel of the protocone and metaconule, complete M^3, M^4 lacking the buccal enamel of the paracone, M^5 lacking the buccal enamel of the paracone and part of the lingual enamel of the protocone; AMNH 102542, fragment of a right dentary with an isolated lower incisor and P_3, M_{2-3}. From Lake Pinpa, Site D, AMNH 102240, right M^3. From Billeroo Creek, Site 3: AMNH 102412, fragment of a right dentary with P_3 lacking the lingual enamel of the principal cusp, M_2 lacking the lingual enamel of the entoconid and part of the buccal enamel of the hypoconid; AMNH 102415, a right P^3 (mounted with AMNH 102416); AMNH 102573, badly crushed anterior part of a skull with incisor alveoli, partial left P^3 and roots of M^{2-3}.

B. From the base of the upper member of the Namba Formation. From Lake Pinpa, NMV locality North Pinpa A, in clay in unnamed upper member of the Namba Formation; NMV P180871, RM^5. From Ericmas Quarry, Lake Namba, reference 320140; SAM P24229, right M^2 and M^3. From Tom O's Quarry, Lake Tarkarooloo, reference 307147; NMV P178101, anterior half of right $M^{?1}$ and NMV P178096, rear half of a right lower molar.

Age: Pinpa, Ericmas and Tarkarooloo faunas; Medial Miocene.

Etymology: This species has been named in honour of Rev. C. W. DeVis, who, from 1882 to 1905, as curator and later Director of the Queensland Museum, was a productive researcher on fossil marsupials, including koalas.

Diagnosis: M. devisi differs from *Madakoala wellsi* in having a less robust lower dentition overall, including a much weaker posterobuccal cuspid on P_3, relatively narrower lower molar proportions, a less strongly-developed protostylid on M_2 and a weaker protostylid ridge on M_3.

This species differs from *Madakoala* sp., aff *M. wellsi* in having a much weaker paraconule on the upper molars, a weaker metaconule-metacone ridge on M^5, a less bilobed P^3 with a better-developed deep pit anterior to the posterolingual cusp, a less well-developed set of buccal ridges on P^3 and a much weaker anterior lingual ridge on that tooth, in having generally weaker buccal stylar structures in the upper molars, and thus more weakly-developed pockets buccal to the

paracones and metacones, and the generally less bilobed occlusal outline of P^3.

Description: The right dentary, SAM P24792, was chosen as the holotype because the lower dentition is the most widely comparable element among Miocene phascolarctids. The lower dentition and part of the mandibular symphysis are represented in the holotype and in referred material from the Pinpa Fauna. What can be seen of the dentary in the holotype resembles the better-preserved dentary in *M. wellsi*, including the short diastema (ca. 9.5 mm), the steeply rising symphysis and angle of the lower incisor (ca. 40 degrees relative to the alveolar row in lateral view).

The lower incisor is only partly preserved in the holotype but it and the perfectly preserved tooth associated with M_{2-3} in AMNH 102542 resemble that of *M. wellsi*. Both species exhibit a scalpriform tooth with a well-developed lateral flange on the crown, in contrast to the more gliriform morphology developed in I_1 of the living Koala (the only other phascolarctid in which this tooth is known).

The P_3 of the holotype is slightly fractured and is missing the enamel surface from its posterobuccal quadrant. It is essentially unworn in AMNH 102542 which (Table 1) also is distinctly shorter than in the holotype. The basic morphology of the two teeth is similar, however, and M_2 and M_3 in AMNH 102542 are essentially identical to those teeth in the holotype.

The P_3 is a semi-rectangular tooth in occlusal view (Fig. 2), with a prominent, longitudinal crest occupying the lingual third of the occlusal surface. The crest curves slightly anterolingually from its major apex, located above the rear of the anterior root of the tooth (Fig. 2). A major buccal and a less strongly developed lingual crest descend, respectively, from that apex to the base of the crown. There are three cusps on the longitudinal crest posterior to the main apex; the cusps diminish in size posteriorly. A buccal and very much weaker lingual cusp are situated near the base of the crown opposite the last cusp of the longitudinal crest. These basal cusps are not as strongly connected to the crest by ridges as in *M. wellsi* (in fact, the posterolingual cusp is isolated). A posterior continuation of the longitudinal crest descends from the last cusp toward the base of the crown and bifurcates into a week posterobuccal and a stronger posterolingual ridge. On the lingual side, another ridge lies anterior to its posterior counterpart but is much more weakly expressed in AMNH 102542 than it is in the holotype.

The P_3 of *M. devisi* is comparable to that of *M. wellsi* except that it is narrower and has one fewer cuspid on the longitudinal crest, a less strongly developed posterobuccal cuspid and weaker lingually directed crests on the posterolingual quadrant of the tooth.

Table 1. Measurements (in mm) of the dentary and lower dentition of species of *Madakoala*. Abbreviations: Lgth, length; Wdth, width; Ant, anterior; Pst, posterior; Dentary depth at M_2, vertical depth at anterior alveolus; Dentary depth at M_4, vertical depth at posterior alveolus.

| | M. devisi | | | | | | M. wellsi | |
	SAM P24792 type	102399	AMNH 102412	102413	102542	NMV P178096	SAM P24793	NMV P178115
Lgth P_3	6.4	–	5.7	–	6.0	–	7.1	6.1
Wdth P_3	3.6	–	3.5	3.5	3.4	–	4.2	4.1
Lgth M_2	7.8	7.8	7.9	8.0	7.8	–	8.3	–
Ant. wdth M_2	4.8	4.5	4.7	4.5	4.5	–	4.9	–
Pst wdth M_2	5.0	4.9	4.8	5.0	4.8	4.8	5.5	–
Lgth M_3	7.8	–	–	–	8.1	–	8.1	–
Ant. wdth M_3	4.9	–	–	–	4.7	–	5.5	–
Pst wdth M_3	5.0	–	–	–	5.1	–	5.7	–
Lgth M_4	–	–	–	–	–	–	8.0	–
Ant. wdth M_4	–	–	–	–	–	–	5.3	–
Pst wdth M_4	4.7	–	–	–	–	–	5.5	–
Lgth M_5	7.1	–	–	–	–	–	7.9	–
Ant. wdth M_5	4.5	–	–	–	–	–	4.8	–
Pst wdth M_5	4.0	–	–	–	–	–	4.6	–
Lgth P_3-M_5	38.0	–	–	–	–	–	38.5	–
Dentary depth at M_2	22a	–	–	–	–	–	19.2	–
Dentary depth at M_4	–	–	–	–	–	–	20.5	–

The M_2 is a semi-rectangular, subselenodont tooth in occlusal view. It is essentially identical to the holotype, AMNH 102542 and AMNH 102399 (Table 1; Fig. 2). The trigonid is narrower than the talonid (Fig. 2; Table 1) and has a protoconid, paraconid and metaconid, as well as a protostylid that is located buccal and slightly posterior to the protoconid. The protostylid is less well-developed in *M. devisi* than in *M. wellsi*. The three primary trigonid cuspids are connected by crests with the preprotocristid extending slightly anterolingually, past the paraconid, to a slight swelling just posterior to the end of the main longitudinal crest of P_3. The development of this anterolingual crest is comparable to, but less pronounced than, the parastylid developed in the same position in AMNH 102399 and (even more strongly) in *M. wellsi*. A short postprotocristid which extends lingually and slightly posteriorly from the protoconid to the metaconid is best preserved in AMNH 102542.

The metaconid is connected by a very weak posterolingually directed crest to the presumptive metastylid which is located on the lingual margin of the tooth near its midlength. This crest and the metastylid are much more strongly developed in *M. wellsi*. The presumptive metastylid is separated from the structures of the talonid by a sharp cleft at the lingual end of the transverse valley.

On the buccal side of M_2, a very weak rib descends anterolingually from the protoconid, past the anterior surface of the protostylid, to terminate at the base of the crown. There is a small cuspule at this point in *M. wellsi*. The protostylid is connected by a very weak strut (which is stronger in *M. wellsi*) to the buccal surface of the protoconid. A crevice which separates the

posterior base of the protostylid from the talonid is more linear and more distinct than the broadly defined and anteriorly curvilinear structure seen in *M. wellsi*.

The talonid of M_2 is composed on the entoconid and, directly opposite it, the hypoconid. An entostylid may be represented by a very slight swelling at the end of the postentocristid (the entostylid is more strongly developed in *M. wellsi*). The pre-entocristid is straight and extends anteriorly to the cleft forming the transverse valley between the talonid and trigonid. The postentocristid extends buccally around the posterior margin of the tooth to connect to the posthypocristid which extends posterolingually from the hypoconid. The cristid obliqua, the major crest of the talonid, extends anterolingually, toward the metaconid but, at the rear of the trigonid, turns anteriorly to connect to the rear of the protoconid. The hypoconid and entoconid are weakly connected by a crest that diverges lingually from the posthypocristid at a point opposite the posterior half of the entoconid. In *M. wellsi*, the crest is stronger, anteriorly concave, crosses the floor of the talonid basin and connects the hypoconid apex to that of the entoconid.

Surface ornamentation is poorly developed on molars of *Madakoala devisi*. It is somewhat more strongly developed on the talonid anterior to the hypoconid-entoconid crest of those teeth of *M. wellsi*.

The chief differences between M_2 and M_3 are found in the trigonid. In M_3, the generally triangular arrangement of the protoconid, paraconid and metaconid has been modified. A protostylid is not present. The protoconid and metaconid are connected by a weak transverse crest that descends from each cuspid to the longitudinal midline of the tooth (this is stronger in *M. wellsi*). The preprotocristid extends anteriorly and slightly lingually to the anterior margin of the tooth, opposite the point where the posthypocristid of M_2 joins the postentocristid. A postprotocristid is directed posterolingually to the base of the trigonid where it meets the cristid obliqua.

The premetacristid is directed anteriorly, through a very slight swelling (marginally stronger in *M. wellsi*) that probably is the remnant of the paraconid, to the anterior margin of the tooth. A parastylid, comparable to that found in M_2, is not developed here. The postmetacristid curves posterolingually to the metastylid which, as in M_2, is separated from the talonid by a cleft at the lingual end of the transverse valley. The talonid of M_3 is essentially as in M_2, including the weakly developed hypoconid-entoconid crest.

Except for size and proportions (Table 1), M_4 and M_5 (certainly preserved in the holotype only, although NMV P178096 could be M_4) resemble each other and M_3. The metastylid (not preserved in M_4) is weaker in M_5 than in M_3, the protostylid ridge is absent, a paraconid swelling is not developed in the two posterior teeth, and the hypoconid-entoconid crest is still present as is (in the fragmentary M_4 as well as in the better preserved M_5) a protoconid-metaconid crest.

Thus, except as noted above, the lower molars of *M. devisi* are generally similar to those of *M. wellsi*. In contrast, however, they have narrower proportions, a weaker protostylid and associated struts and crevices on M_2, where the protostylid varies from poorly developed (AMNH 102399) to only slightly weaker than in *M. wellsi* (AMNH 102413 or the holotype), weaker protostylid ridge on M_3, metastylids on all molars weaker and more posteriorly located, more weakly-developed entoconid-hypoconid crests, and weaker coronal surface ornamentation.

The upper dentition is represented by the following referred specimens: From Lake Pinpa, Site C; AMNH 102416, RM^{2-5}. From Billeroo Creek, Site 3; AMNH 102415, RP^3, AMNH 102573, LP^3. From Ericmas Quarry; SAM P24229, RM^{2-3} and NMV P178101. From NMV locality North Pinpa A: NMV P180871, RM^5.

The P^3 is a robust, semi-bilobed tooth that, in occlusal view, is wider posteriorly than anteriorly (Fig. 3). A nearly straight longitudinal crest extends from the anterior to the posterior margin of the tooth. A buccal and lingual crest intersect the longitudinal crest at a point over the rear of the anterior root and delineate the anterior apex of the coronal surface. This is followed posteriorly by three or more cusps, the last of which

occurs over the anterior part of the posterior root and defines the tallest point of the tooth crown. A major cusp is located on the lingual slope of this apex cusp and forms the posterior border of a narrow but deep pocket delimited anteriorly by a crest that descends posteroventrally (in occlusal aspect) from the anteriormost cusp. A smaller but still prominent cusp occurs on the posterobuccal quadrant of the crown. Breakage in AMNH 102415 has obscured what probably was the posterior continuation of the longitudinal crest but in AMNH 102573 this crest can be seen to terminate in a cuspule at the posterior cingulum.

The M^2 is broken at the lingual surfaces of the protocone, paracone and metacone, but is essentially complete in SAM P24229. A paraconule is present between the protocone and metacone. An anterobuccal ridge from the paraconule connects to a weak anterolingual crest at the base of the paracone. A comparable cusp is not present at the metaconule position as utilized in placental mammals, lending credence to the interpretation that the so-called hypocone in this taxon actually is an enlarged metaconule. The preparacrista bifurcates near the anterobuccal margin of the tooth into an anterolingually curved crest that is continuous with the anterior cingulum, and a posterobuccally directed crest. The former bears a swelling at its junction with the anterior cingulum that may represent the parastyle (stylar cusp A). The latter may represent stylar cusp B. A short, transversely directed crest is present anterior to stylar cusp B in SAM P24229 and NMV P178101 and the preserved portion of the broken crown of the holotype in this area is consistent with the possible presence of such a structure in the NMV specimen, as well. Stylar cusp B meets, at the buccal base of the paracone, a short ridge that similarly but oppositely extends anterobuccally from the postparacrista. This apparently represents another stylar cusp, possibly stylar cusp C. A short rib extends posteriorly from this junction to meet a similarly but oppositely directed crest from the premetacrista that is developed just lingual to what appears to be stylar cusp D. These short ridges, then, bridge the transverse valley that otherwise separates stylar cusps C from D and the paracone from the metacone. The W-shaped ectoloph, otherwise ubiquitous in phascolarctids, is thus also in evidence in *Madakoala*. The pre- and postmetacristae diverge from the metacone, with the postmetacrista terminating in what may be a remnant of stylar cusp E. A short, weak anterolingual and a weaker posterolingual strut from the apex of the metacone extend toward its base, and the anterior crest meets a small cuspule here regarded as a neometaconule.

The M^3 is essentially like M^2 as far as can be seen but preserves some structures not represented in M^2. These are a pair of short crests that extend antero- and posterolingually, respectively, along the base of the paracone and join the anterior and posterior wings, respectively, of the paraconule. The paraconule also bears a short lingual strut that extends toward, but does not reach, the preprotocrista. A protostyle, a new cusp that extends posteriorly as a narrow spur from the preprotocrista just lingual to the paraconule in species of *Phascolarctos* and *Litokoala*, is not seen in any upper molar of *M. devisi* (or *Madakoala* sp. aff. *M. wellsi*, or *Perikoala palankarinnca*). Stylar cusps B, C, D, and possibly E, are represented in M³ of *M. devisi*, and the selenodont construction of the protocone and metaconule is well shown.

The M⁴ and M⁵ are like M³ but in M⁵ the metaconule and metacone are very reduced, so that the tooth is posteriorly attenuated. NMV P180871 is better preserved than, but appears to be essentially identical to, M⁵ of AMNH 102416. Differences appear to be the somewhat more pronouced surface crenulation, a somewhat stronger metaconule-metacone crest anterior to the posterior cingulum (the broken enamel of the holotype in this region is suggestive of a weak metaconule-metacone crest in that specimen as well) and a slightly more sloping lingual surface of the crown. This tooth also is similar to that of *Madakoala* sp., aff. *M. wellsi* but differs in having a weaker metaconule-metacone crest, the greater lingual slope of the coronal surface and slightly broader occlusal proportions (Table 2).

Comparisons: The upper dentition of *M. devisi* is generally similar to that of *Madakoala* sp., aff. *M. wellsi*, as represented by material from the Ditjimanka Local Fauna. The P₃ is slightly shorter and upper molars slightly larger in *M. devisi*. The posterobuccal and possibly the posterolingual

cusps of P³ are better developed in the Ditjimanka tooth and there is no trace of the deep pit developed anterior to the posterolingual ridge in the Pinpa form. On the other hand, the Ditjimanka taxon has a more strongly-developed rectangular lingual space, floored (in occlusal view) with a strong basal cingulum, a feature not comparably developed in *M. devisi*.

In comparison to Ditjimanka specimens of *Madakoala* sp., aff. *M. wellsi*, the upper molars of *M. devisi* are somewhat larger (Table 2), have less strongly-developed stylar structures and a weaker lingual basal cingulum between the protocone and metaconule in M²⁻⁴. The M⁵ of the Ditjimanka form has an apparently more strongly-developed metacone-metaconule crest and a posterolingually more attenuated metaconule base than found in *M. devisi*.

Koobor notabilis (and *K. jimbarratti* to the extent that comparisons can be made) generally resembles species of *Madakoala*, but differs in having a distinctly bicusped P³, better-developed buccal stylar cusps on the upper molars that effectively close off the buccal side of each tooth, in lacking a paraconule on M³ and M⁴, in lacking a neometaconule on M²-M⁴ and in having proportionately more elongate upper molars. *Koobor* species share many plesiomorphic features with those of *Madakoala* (simpler dentition, strong separation of stylar cusps C and D, with longitudinal struts between them, absence of a protostyle, absence or only poor development of a neometaconule, linear rather than crescentic protoconule) but, as discussed below, these two genera apparently are not closely related.

The badly crushed anterior part of the skull of *Madakoala devisi* from Billeroo Creek (AMNH

Table 2. Measurements (in mm) of upper dentition of species of *Madakoala*. Abbreviations as in Table 1.

		Madakoala devisi				
	102416	AMNH 102415	103340	NMV 180871	SAM P24229	*Madakoala* sp. aff. *M. wellsi*
Lgth P³	–	7.4	–	–	–	7.9 (UCR 16990)
Wdth P³	–	5.3	–	–	–	5.1 (UCR 16990)
Lgth M²	7.9	–	–	–	8.2	7.4 (UCR 15827)
Lgth M²						7.3 (QM F00003)
Ant. wdth M²	7.3a	–	–	–	7.0	6.5a(UCR 15827)
Ant. wdth M²						6.8 (QM F00003)
Ant. wdth M²						7.1 (NMV P178101)
Post. wdth M²	7.4a	–	–	–	7.5	6.7 (UCR 15827)
Lgth M³	7.7	–	8.0	–	7.6	7.45(UCR 16991)
Ant. wdth M³	7.7	–	7.6	–	7.9a	7.2 (UCR 16991)
Post. wdth M³	7.3	–	7.2	–	7.2	6.2 (UCR 16991)
Lgth M⁴	7.6	–	–	–	–	7.1 (UCR 16992)
Ant. wdth M⁴	7.1	–	–	–	–	6.4 (UCR 16992)
Post. wdth M⁴	6.5	–	–	–	–	5.7 (UCR 16992)
Lgth M⁵	5.9	–	–	5.9	–	6.0 (UCR 16993)
Ant. wdth M⁵	5.3a	–	–	5.8	–	5.3 (UCR 16993)
Post. wdth M⁵	4.5	–	–	4.7	–	4.6 (UCR 16993)

102573) offers only a few points for comparison with *Phascolarctos cinereus*. As far as can be determined, the skull proportions are as in *P. cinereus*, with a short, deep antorbital region. The construction of the root of the zygoma is much like *P. cinereus* except that the maxillary process of the arch lies opposite M^3 rather than M^2 as seen in *P. cinereus*. Species of *Madakoala* and *Perikoala palankarinnica* seem to agree in this regard. The incisor alveoli imply an enlarged central pair and smaller, subequal I^{2-3}. The canine root is separated by a short diastema from the anterior root of P^3, this distance being about half that in the similar-sized *P. cinereus*. The diastema and the evidence from the proportions of the mandible of *M. wellsi* (partially corroborated by material of *M. devisi*) indicate that species of *Madakoala* have skulls with a proportionately shorter antorbital region than *P. cinereus*. At the same time, the nasal bones of *M. devisi*, although relatively and actually shorter than those of *P. cinereus* of comparable size, have widths of only half those of their living counterparts. The narial passage of *M. devisi* was thus not inflated to the extent seen in *P. cinereus*.

Thus, in comparison with *Phascolarctos cinereus*, *Madakoala devisi* has a short-faced skull with an uninflated narial region, a generally more robust, less trenchant P^3 that lacks an internal cingulum, greater separation between the paracone and metacone across the transervse valley, less vertical paracone and metacone, generally weaker paraconule, no protostyle (the longitudinally directed posterior spur from the preprotocrista just lingual to the paraconule seen, e.g., in *P. cinereus*), a much smaller or absent neo-metaconule and a much less distinct metaconule-metacone crest on M^5.

The fragmentary M^3 of the paratype of *Perikoala palankarinnica* shows little of significance except its smaller size, apparently shorter occlusal proportions and its much more pronounced surface crenulation compared to that tooth in species of *Madakoala*. Material herein referred to *P. robustus* confirms that species of *Perikoala* have proportionately shorter upper (and lower) molars than found in species of *Madakoala*. *Perikoala* species probably had a somewhat different P^3 morphology than seen in species of *Madakoala*, although species of both genera share a level of plesiomorphic characters not seen in more derived taxa. At the same time and consistent with the morphology of the lower dentition of the holotype and of the lower teeth referred below to *P. palankarinnica*, the morphology shown by the upper tooth (paratype) and by upper teeth referred to *Perikoala robustus* seem to ally species of this genus with species of *Phascolarctos*. A few features of the dentition, however, suggest that any affinity between species of *P. palankarinnica* (and *P. robustus*) and those of

Phascolarctos may be remote. These features are the apparent lack of a protostyle on the upper molar of the paratype of *P. palankarinnica* and referred material of *P. robustus*, the presence in the lower dentition of the type and referred material of both *Perikoala* species of distinct crosscrests between the protoconid and metaconid, and hypoconid and entoconid, respectively and the incomplete metastylid-entoconid crest. These same features resemble those seen in species of *Madakoala* and serve to show the more plesiomorphic condition of *P. palankarinnica* relative to species of *Phascolarctos*.

Litokoala kutjamarpensis differs from *Madakoala devisi* in having a buccal extension of the paraconule that connects to the buccal margin of the tooth, a much stronger development of the neometaconule, the crescentic rather than linear configuration of these structures, the presence of a protostyle and the strongly bifurcate pair of ribs that connect the protocone and metaconule and enclose between them a very deep pocket. In the second through fourth of the above features, *L. kutjamarpensis* closely resembles *Phascolarctos cinereus* but differs from it in the buccal extension of the paraconule, the less crenulated enamel surface and the stronger development of the bifurcate pair of ribs between the protocone and metaconule and the pocket enclosed between them. *Litokoala kanunkensis* (Springer 1987) differs from species of *Madakoala* (and *Perikoala*) in the much more selenodont and hyposodont form of its lower dentition, and the presence of a well-developed entocostylid ridge in M_3.

Stratigraphic comments: As herein described, *Madakoala devisi* occurs in the Pinpa, Ericmas and Tarkarooloo faunas. The Pinpa fauna is older than the Ericmas based on the evidence of stratigraphic superposition (Tedford *et al.* 1977) and the taxa of the Ericmas fauna generally seem to be evolutionarily more derived than comparable Pinpa forms (Tedford *et al.* 1977 and comments below). The sediments that yield the Tarkarooloo Local Fauna (lower arenaceous siltstone about 18-20 m above the local base of the Namba Formation outcrops at Lake Tarkarooloo; Callen and Tedford 1976: Fig. 14, p.132) have not been closely correlated with other outcrops or subsurface sections of the Namba Formation. Callen and Tedford (1976, p.130) suggest that the palygorskite-dolomite association of the Lake Tarkarooloo sequence favours correlation with the upper member of the Namba Formation that elsewhere yields — at its base — elements of the Ericmas Fauna but fewer appraisals of the data by Tedford, R. Callen and T. H. V. Rich (in preparation) suggest caution in utilizing such mineralogical data for purposes of fine-scale correlation. At the same time, these workers have developed reconstructions that suggest the Tarkarooloo Local Fauna occurs in Namba Formation sediments several metres stratigraphically above those that yield the Ericmas Fauna. Fossil marsupials of the Tarkarooloo Local Fauna generally support a correlation that is younger than the Ericmas Fauna or its purported correlate, the Ditjimanka Local Fauna (e.g., Woodburne and Clemens 1986; Pledge 1986, based on studies of the Ektopodontidae).

If the above-discussed age relationships, and the identification of specimens as *Madakoala devisi*, are correct, *M. devisi* appears to have been a relatively chronologically long-ranging species. SAM P24229 from Ericmas Quarry is virtually identical to the holotype of the Pinpa Fauna. NMV P178101, also from Ericmas Quarry, is very similar to M^1 of SAM P24229 and NMV P178096 clearly is comparable to *M. devisi* in size (Table 1), degree of enamel crenulation (very little) and in the fact that the pattern formed by the posthypocristid and cristid obliqua in lower teeth of *M. devisi* is a more arcuate and less angulate, selene than that of the holotype of *M. wellsi*. On present evidence, it appears that *Madakoala devisi* pre-dates, is contemporaneous with, and post-dates, *M. wellsi*.

Madakoala wellsi n. sp.
(Figs 4 and 5; Tables 1 and 7)

Holotype: SAM P24793 (AMNH 102174), a right dentary which has P_3 and M_{2-5} but lacks the incisor and ascending ramus.

Type locality: AMNH Ericmas Quarry, Lake Namba, Namba Formation, unit 4 of Callen and Tedford (1976); northwest side of Lake Namba, South Australia, grid zone 6, reference 320140, Curnamona topographic series, SH 54-14; 1:250,000, 1965.

Referred material: AMNH 102164, the posterior half of a left P_3; AMNH 102169, a fragment of a right dentary with roots of the M_2, part of a crown of M_3 and M_4; AMNH 102171, fragment of a nearly edentulous right dentary with the root of I_1 and alveoli of P_3, M_2 and the anterior root of M_3; AMNH 102172, a fragment of a left dentary with the trigonid of M_5; AMNH 102178, a right $I^{3(?)}$; AMNH 102177, a right I_1 lacking most of its root; AMNH 102179, a left I_1 with a portion of the root missing; NMV P178115, a RP_3.

Age: Ericmas Fauna; Medial Miocene.

Etymology: This species is named in honour of R. T. Wells who collected the holotype and contributed so much to the success of the field work in South Australia and Queensland in 1971 and 1973.

Diagnosis: This species differs from *Madakoala devisi* in having a more robust dentition,

including a much stronger posterobuccal cuspid on P_3, stronger metastylids, relatively wider lower molars, M_2 with a stronger protostylid and a more strongly-developed protostylid ridge on M_3.

Description: The right dentary, SAM P24793, was chosen as the holotype because the lower dentition is the most widely comparable element among Miocene phascolarctids. The dentary is relatively robust in lateral aspect and relatively deep (Table 1). The incisor-P_3 diastema is relatively shorter than in *Phascolarctos cinereus* and comparable to *M. devisi*. The symphysis extended upward at an angle of about 45 degrees relative to the alveolar border and projected postero-ventrally to a point below the mid-length of M_2. This is comparable to the condition in *Perikoala palankarinnica*, shown by QM F14693 and to *P. cinereus*. The mental foramen is higher on the lateral surface of the dentary and closer to the anterior root of P_3 than it is in *P. cinereus*. There is no trace of the prominent flange-like digastric process that lies below the masseteric fossa in *P. cinereus*. The digastric insertion on the dentary of *M. wellsi* seems to be represented by a slight bulge below the anterior part of M_4 in the otherwise smooth curve of the inferior border of the dentary. As a further expression of the relatively short jaw of *M. wellsi* compared to *P. cinereus*, the masseteric fossa extends forward beneath M_5 so that the root of the coronoid process barely clears the rear of M_5 in lateral view.

The referred lower incisor (AMNH 102179) agrees closely with those associated with cheek-teeth of *M. devisi* (e.g., AMNH 102542) and differs from *P. cinereus* in its scalpriform crown. The crown is short and bears a lateral enamel flange on the large, strongly laterally compressed root. The overall structure is of a more primitive diprotodontan form than the rather gliriform crown of the lower incisor of *P. cinereus*.

The cheekteeth are very similar to those of *M. devisi*. Only points of difference will be noted here. P_3 of *M. wellsi* has four (rather than three) cusps posterior to the main anterior cusp, each of which diminishes in prominence posteriorly. The second of these cusps is connected by a weak ridge to a cusp located above the posterior root in the posterobuccal quadrant of the tooth. On the lingual side, the first and third of these cusps bear weak ridges, each of which apparently connected to a cingular cusp at the base of the crown. The last cusp also bears a posterobuccally and a posterolingually extending ridge. On the buccal side, the ridge overlaps *en echelon* a similarly oriented ridge that is continuous with the posterobuccal base of the crown and which is faintly connected to the last (smallest) cusp of the longitudinal crest. The posterolingually directed ridge from cusp 3 is continuous with the lingual coronal base.

The P_3 of NMV P178115 (Fig. 5) is somewhat shorter (Table 1) than in the holotype, has a very weakly-developed last of four cusps behind the main cusp of the blade and has generally weaker accessory cusps and ridges. The large postero-buccal cusp is, however, developed comparably to that of the holotype, as is its connection to the second of the series of cusps behind the main cusp of the blade. The degree of differences versus similarities shown here is comparable to Pinpa Fauna specimens of *M. devisi* described above and supports the referral of NMV P178115, in this case, to *M. wellsi*.

The M_2 of *M. wellsi* differs from that of *M. devisi* chiefly in having a stronger protostylid, para-stylid, metastylid, entoconid-hypoconid crest and in the more broadly-floored valley between the protoconid and protostylid.

The chief differences between M_2-M_5 of *M. wellsi* and *M. devisi* are the more strongly-developed metastylids and cross-crests between the protoconid and metaconid, and hypoconid and entoconid, respectively, plus the apparently more strongly developed surface crenulations in *M. wellsi*.

Comparisons: Madakoala wellsi is similar to *Madakoala devisi* in overall dental morphology. *M. wellsi* is overall larger than *M. devisi*, but at the moment it is not clear to what extent these differences are due to intrapopulational or sexual variation. In comparison to *M. devisi, M. wellsi* is more robust and has proportionately wider lower cheekteeth. The protostylid of M_2 is larger relative to the size of that tooth in *M. wellsi* than in even the otherwise most similar specimen of *M. devisi* (AMNH 102413). The molars have more robust metastylids and better-developed cross-crests between the protoconid and metaconid, and the hypoconid and entoconid.

In comparison to *Perikoala palankarinnica, Madakoala wellsi* is much larger, lacks the strong lingual shelf basal to the metaconid and the weaker lingual shelf basal to the entoconid seen in M_3 and M_4 of *P. palankarinnica*, has a much better developed protostylid ridge in M_3 and much weaker expression of surface crenulation overall. *M. wellsi* differs from both species of *Perikoala* in the degree of surface enamel crenulation of the upper and lower cheek teeth, and in lacking on its lower molars any suggestion of the intermediate postprotocristid seen in species of *Perikoala*.

In comparison to *Phascolarctos cinereus, P. wellsi* is generally larger, has a more cuspate and less trenchant P_3, a less strongly-developed and buccally set off protostylid in M_2, a more centrally positioned protoconid in the trigonid of M_2, a more distinct and widely separated paraconid and metaconid and a better developed metastylid

on M_2 (i.e., a more triangular arrangement of the primary trigonid cuspids), a more complexly constructed talonid on M_2 (e.g., a hypoconid-entoconid crest is present) and lacks the meta-stylid fold by which the pre-entocristid is continuous with the postmetastylid cristid. It also lacks the columnar stylids on the lingual faces of the metaconid and entoconid that characterize the M_{3-5} of *P. cinereus*.

Madakoala sp. aff. *M. wellsi*
(Fig. 6; Table 2)

Locality: Lake Palankarinna, Etadunna Formation, unit 6 of Stirton *et al.* (1961), from RV-7230 (= V-5375), Tedford Locality, west side of Lake Palankarinna, South Australia, grid zone 5, reference 657431, Kopperamanna topographic series, SH 54-1; 1:250,000, 1965. From QM Locality SAM Quarry North. See locality data for *Perikoala* for the stratigraphic position of V-5375 and SAM Quarry North.

Material: From V-5375: UCR 16990, RP^3; UCR 15827, RM_2; UCR 16991, RM^3; UCR 16992, RM^4; UCR 16993, RM^5; NMV P178101, RM^2. From SAM Quarry North: QM F14694, RM^2.

Age: Ditjimanka Local Fauna; Medial Miocene.

Description: Madakoala sp., aff. *M. wellsi* differs from *M. devisi* in having a much stronger paraconule on the upper molars, stronger buccal stylar structures on the upper molars, a stronger metaconule-metacone ridge on M^5, more bilobed P^3 with a broadly rectangular space between the two major lingual ridges of P^3 (thus no deep pit anterior to the posterolingual cusp), a more well developed set of buccal ridges on P^3 and a much stronger anterior lingual ridge on that tooth.

The upper cheekteeth of *Madakoala* sp. aff. *M. wellsi* are very similar to those of *M. devisi*. Differences include the slightly smaller overall size, the somewhat more bilobed occlusal outline of P^3, the deep and strongly-floored lingual excavation between the two lingual ribs of P^3 (rather than having a deep, narrow, pocket in that location), upper molars with more strongly-developed stylar structures and stronger paraconules, M^5 with a stronger ridge between the metaconule and metacone, anterior to the posterior cin-gulum, the more posteriorly attenuated M^5, and less well-developed buccal ribs on P^3 than seen in *M. devisi*.

Comparisons: Differences between *Madakoala* sp., aff. *M. wellsi* and with respect to other Cainozoic and Recent koalas are comparable to those noted above for the other species of *Madakoala*, except that *Madakoala* sp., aff. *M. wellsi* appears to be morphologically derived with respect to *M. devisi* in having stronger buccal stylar structures in the upper dentition and thus stronger incipient development of enclosed pockets buccal to the paracones and metacones. The lower dentition *(M. wellsi)* is in a similarly more derived state in comparison to the upper dentition *(Madakoala* sp. aff. *M. wellsi),* and the degree of departure of both dental elements from *M. devisi* is interpreted here to represent a comparable evolutionary state relative to that of *M. devisi.* Again, the upper as well as the lower dentition of the Ericmas and Ditjimanka specimens under discussion appear to be similarly derived relative to *M. devisi,* and probably represent the same taxon.

Remarks: Madakoala is represented by at least three species of Miocene age. To the extent that *Madakoala* sp., aff. *M. wellsi* from the Ditjimanka Local Fauna can be regarded as being more derived than *M. devisi* of the Pinpa, Ericmas, and Tarkarooloo faunas, it resembles the comparably more derived state of the lower dentition of *M. wellsi* relative to that of *M. devisi.* This is consistent with the geologically younger age of strata that bear the Ericmas Fauna in comparison to the Pinpa based on physical stratigraphy (Callen and Tedford 1976), and the evolutionarily more derived aspect of other elements of the Ericmas Fauna as compared to those of the Pinpa (e.g., Tedford *et al.* 1977 — various taxa; Woodburne, Tedford and Archer 1987 — pseudocheirids; Woodburne and Clemens 1986 — ektopodontids).

Madakoala sp.

Material: From an NMV locality on the west side of un-named lake located about 4.3 km east of Lake Tinko, reference 323132, Curnamona topographic series, SH 54-14, 1:250,000, 1965 ed.; NMV P170091, RP^3; NMV P180872, fragment of left M$^{?3}$; NMV P178112, posterior half of LP$_3$; and NMV P178113, left lower molar fragment.

Age: Wadikali Local Fauna, Medial Miocene, probably close in age to the Tarkarooloo Local Fauna based on stratigraphic assessments by Tedford, R. Callen, and T. H. V. Rich (in preparation).

Description: Only NMV P170091, the RP^3, is complete enough to warrant description. It is about as long as (length, 7.1) but wider (width 5.6) than that of *M. devisi* and also differs as follows: The posterobuccal cusp is more discrete and prominent, being represented in *M. devisi* instead by a crest with associated crenulations; there is a deep fissure that isolates this crest from the posterior end of the main longitudinal crest of the tooth in NMV P170091; the anterobuccal face is vertically crenulated and exhibits an anterobuccal basal inflection (or concavity); the anterolingual face exhibits a pronounced trough

which is partially the result of the hypertrophy of the anterolingual crest into a prominent buttress; this buttress also has a large base that gradually narrows toward the occlusal edge into a very narrow crest.

NMV P170091 differs from P^3 of *Madakoala* sp., aff. *M. wellsi* in being somewhat shorter, but wider; in having a somewhat less well-developed mid-lingual excavation; in having one less cuspule on the main shearing crest, in having a less pronounced anterolingual buttress; in being transversely wider; and in having a more strongly-developed posterobuccal cusp. In contrast to *M. devisi* the Wadikali species partially resembles that of the Ditjimanka form in having a nearly as well-developed lingual excavation that is bounded basally by a nearly as well-developed basal shelf anterior to the posterolingual cuspule and in having what appears to be a generally better-developed P$_3$ blade than in the Pinpa taxon. The Wadikali species appears to be more derived than that of the Ditjimanka Local Fauna in the strength of the posterobuccal cusp and almost as derived as that species in the development of the lingual trough anterior to the posterolingual cusp of P$_3$.

Discussion: The other koala remains from the Wadikali Local Fauna are too incomplete to yield significant information. The upper premolar differs from both that of *M. devisi* and *Madakoala* sp. aff. *M. devisi.* It is not named as a new species because of lack of information on the upper dentition of *M. wellsi.* At the moment the other koala material of the Wadikali Local Fauna gives no information as to its age relative to other parts of the Namba Formation of the Tarkarooloo Basin. The ways in which the P^3 of the Wadikali form, NMV P170091, differ from *M. devisi* and *Madakoala* sp., aff. *M. wellsi* could reflect its derived state relative to both of those species. It appears on stratigraphic grounds that this site may be comparable to that which yields the Tarkarooloo Local Fauna at Lake Tarkarooloo (Tedford, R. Callen and T. H. V. Rich, in preparation).

Perikoala Stirton 1957

Genotypic species: Perikoala palankarinnica Stirton 1957.

Additional species: Perikoala robustus n. sp.

Distribution: Ditjimanka and Palankarinna South local faunas, Lake Palankarinna, South Australia; Medial Miocene.

Diagnosis: Species of *Perikoala* differ from species of *Madakoala* in P^3 having a anterobuccally elongate posterobuccal cusp, upper molars have a much more crenulate surface enamel pattern, shorter molar proportions, a more continuously bounded stylar rim and M^2 of

Perikoala robustus has an additional stylar cusp anterior to stylar cusp E. The lower dentition of species of *Perikoala* differs from those of *Madakoala* species in having a much more crenulate surface enamel pattern, shorter molar proportions and in having lingual flange-like extensions associated with the metaconid and entoconid of M_3 and M_4.

Species of *Perikoala* differ from those of *Phascolarctos* in being lower crowned, in having symplesiomorphic cross-crests on the lower molars, a strong posterobuccal cusp on P^3 and in having *(P. robustus)* a continuously closed-off stylar border on the upper molars and (in M^2) a stylar cusp anterior to stylar cusp E. A protostyle is absent in M^2 of *P. robustus*.

Species of *Perikoala* differ from those of *Litokoala* in being lower-crowned, less selenodont and in the features cited above as separating *Perikoala* species from those of *Phascolarctos*.

Species of *Perikoala* differ from those of *Koobor* in having weaker stylar cusps, a crested P^3 and generally greater crenulation of the enamel surface of the molars.

Perikoala palankarinnica Stirton 1957
(Fig. 7, Tables 3, 4 and 7)

Material: SAM P10893, holotype, a left dentary fragment with the talonid of P_3, M_2 lacking the protostylid region and missing enamel from the base of the metaconid, a nearly complete M_3 except enamel missing from the lingual surface of the metaconid and with an anterior portion of trigonid of M_4; UCMP 45343, paratype, a right maxilla fragment with the rear of the alveolus for the posterior root of P^3, the roots of M^2, M^3 with much of the enamel missing and having the anterior roots of M^4; QM F14693, a left dentary with part of the symphysis, the base of the ascending ramus, alveoli for P_3, M_2 lacking the protostylid and having M_3, M_4 and the alveoli for M_5; QM F14695, a RM^2 crown; NMV P180083, the rear two-thirds of a left upper molar and NMV P157582, a very worn right upper molar missing the ectoloph.

Distribution: Specimens representing *P. palankarinnica* are known only from the Ditjimanka Local Fauna, Lake Palankarinna, South Australia. The holotype is from the SAM Quarry North locality, Unit 6 of the Etadunna Formation (Stirton, Tedford and Miller 1961), but probably from the grey claystone or possibly immediately underlying pale grey argillaceous fine-grained sandstone that lies about 1.5 m stratigraphically above the lower main quarry level (which is a white to pale-green argillaceous sandstone), west side of Lake Palankarinna, Tirari Desert, South Australia, reference grid zone 5, 6565-4315, Kopperamanna topographic series SH-54-1,

1:250,000, 1965 ed. UCMP 45343 is from Tedford Locality, V-5375, in Unit 6 of the Etadunna Formation, in grey mudstone with arenaceous stringers and locally underlying white argillaceous sandstone, located on the north side of Tedford Locality Basin about 265 m north and 1.5 m stratigraphically above SAM Quarry North. UCMP 45343 thus appears to have been recovered from (or very close to) the same stratigraphic level as the holotype. QM F14693 is from QM Tedford Locality East, from the white sandstone located about 30 m east and 1.5 m stratigraphically below V-5375. QM F14695 is from SAM Quarry North, from the white sandstone of the main quarry level (stratigraphically equivalent to, but 265 m south of, QM F14693). NMV P180083 and NMV P157582 are from White Sands Basin locality, about 80 m south and stratigraphically equivalent to SAM Quarry North. Note that Stirton (1957: 72) records both the type and paratype of *P. palankarinnica* as deriving from V-5375. The above statements that the holotype probably came from sediments slightly above the SAM Quarry North white sandstone level at that locality stem from discussions with Paul F. Lawson who collected the specimen in 1954.

Age: Ditjimanka Local Fauna; Medial Miocene.

Revised diagnosis: (based on the holotype, paratype and new material herein referred to this species): *Perikoala palankarinnica* differs from species of *Madakoala* in being much smaller, in having broader and relatively shorter molars, a strong lingual shelf basal to the metaconid, a weaker lingual shelf basal to the entoconid of the lower molars, more strongly-crenulated molars,

Table 3. Measurements (in mm) of the dentary and lower dentition of *Perikoala palankarinnica*. Abbreviations as for Table 1.

	SAM P10893 type	QM F14693
Lgth P_3	–	–
Wdth P_3	2.9	–
Lgth M_2	6.1	–
Ant. wdth M_2	–	–
Pst wdth M_2	3.8a	4.5
Lgth M_3	6.4	6.3
Ant. wdth M_3	4.3a	4.5
Pst wdth M_3	4.6a	4.6
Lgth M_4	–	6.4
Ant. wdth M_4	–	4.35
Pst wdth M_4	–	4.4
Lgth M_5	–	–
Ant. wdth M_5	–	–
Pst wdth M_5	–	–
Lgth P_3-M_5	–	–
Dentary depth at M_2	–	17.2
Dentary depth at M_4	–	16.2

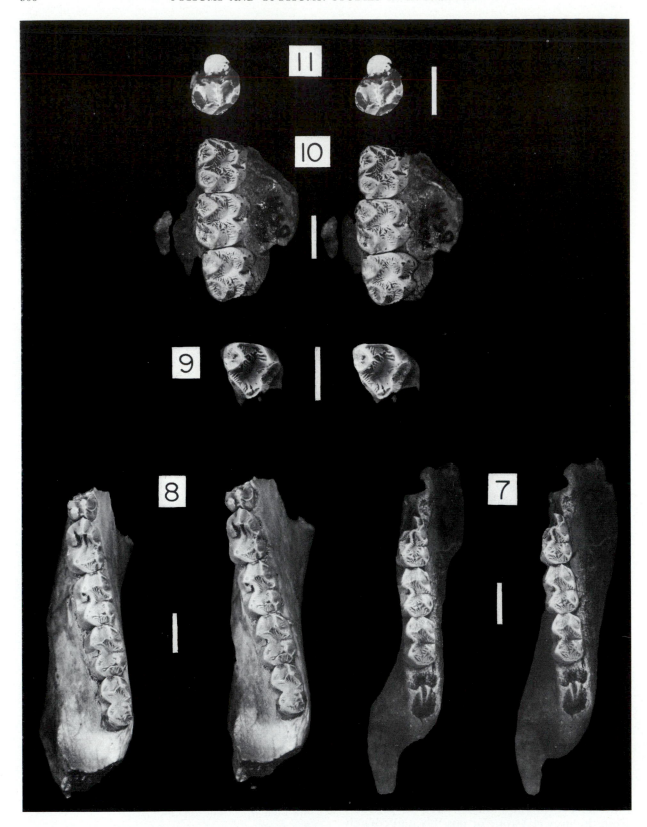

Fig. 7. *Perikoala palankarinnica*, stereophotograph of QM F14693, LM$_2$-M$_4$.

Fig. 8. *Perikoala robustus*, stereophotograph of SAM P26552, LM$_2$-M$_5$.

Fig. 9. *Perikoala robustus*, stereophotograph of UCR 21910, RM^5.

Fig. 10. *Perikoala robustus*, stereophotograph of UCR 21881, RM^2-M^4.

Fig. 11. *Perikoala robustus*, stereophotograph of UCR 21919, rear half, RP^3.

Table 4. Measurements (in mm) of the upper dentition of *Perikoala palankarinnica*. Abbreviations as in Table 1.

	QM F14695	UCM P45343 (paratype)	NMV P180083
Lgth P^3	–	–	–
Wdth P^3	–	–	–
Lgth M^2	7.2	–	–
Ant. wdth M^2	6.6	–	–
Post. wdth M^2	6.6	–	–
Lgth M^3	–	6.1a	–
Ant. wdth M^3	–	–	–
Post. wdth M^3	–	–	–
Lgth M^4	–	–	5.45
Ant. wdth M^4	–	–	5.4
Post. wdth M^4	–	–	–
Lgth M^5	–	–	–
Ant. wdth M^5	–	–	–
Post. wdth M^5	–	–	–

in the upper molars (paratype and referred material) being smaller and in being proportionately shorter, in having weaker buccal stylar structures, in being lower crowned, in having a more sloping rather than more vertical buccal ectoloph, in having lingually sloping rather than more vertical lingual surfaces of the protocones and metaconules and in having a small but distinct pair of short crests at the lingual base of the transverse valley of M^2 that rise, respectively, toward the protocone and metaconule, thereby blocking the lingual edge of the transverse valley (a less well-developed version of this structure also can be seen on M^3).

Perikoala palankarinnica differs from *Litokoala kutjamarpensis* in being somewhat larger, in having a much smaller neometaconule (or none at all), in having a linear rather than strongly crescentic paraconule, in lacking the protostyle and in lacking the strongly bifurcate pair of ribs that connect the protocone and metaconule and enclose a deep pocket between them in *L. kutjamarpensis*. *P. palankarinnica* differs from *Litokoala kanunkaensis* in being generally lower crowned, less selenodont, and in lacking a entostylid ridge in M$_3$.

Perikoala palankarinnica differs from species of *Koobor* in that the latter are slightly smaller, have much better-developed stylar cusps on the upper molars that effectively close off the buccal side of each tooth, lack paraconules in M^3 and M^4 and have molars that are much longer than wide.

Perikoala palankarinnica differs from species of *Phascolarctos* in being somewhat (*P. cinereus*) to greatly (*P. stirtoni*) smaller, in having a distinct basal shelf below the metaconid and smaller shelf below the entoconid, in having a more complexly constructed talonid of M$_2$ (e.g., a hypoconid-entoconid crest is present), in lacking the metastylid fold in which the pre-entocristid is continuous with the postmetastylid cristid, in having less

well-developed columnar stylids on the lingual faces of the metaconid and entoconid, in having a generally more linear and less crescentic paraconule, no protostyle (the longitudinally orientated posterior spur from the preprotocrista just lingual to the paraconule), a much smaller (or absent) neometaconule in the upper molars, a shorter diastema of the dentary, and a horizontal ramus that lacks the prominent flange-like digastric process that lies below the masseteric fossa in *P. cinereus*.

Description: The original description of the holotype (SAM P10893; Stirton 1957) can now be augmented by QM F14693 and QM F14696. In QM F14693 it can be seen that the depth of the dentary in *Perikoala palankarinnica* is relatively even from the rear of the symphysis to below the masseteric fossa (Table 3). As in species of *Madakoala*, the symphysis is steep, and the pre-premolar diastema must have been similarly short. In *P. palankarinnica* the snout must have been somewhat shorter proportionately than in *P. cinereus*. Based on QM F14693, it appears that the symphysis in *P. palankarinnica* may have been more firmly ankylosed than in species of *Madakoala* and similar to the strongly fused symphysis in *P. cinereus*. In that the ventral border of the mandible is somewhat inturned lingually in *P. palankarinnica*, it apparently had an inflected digastric process similar to that seen in *P. cinereus*. As for species of *Madakoala*, the ascending ramus would have barely cleared the rear of M^5 in *P. palankarinnica*.

The broken M$_2$ in QM F14693 resembles that of the holotype in most features. QM F14693 shows that a crest descends the anterolingual surface of the metaconid and extends forward as a cingular shelf to a strongly cuspate paraconid (also suggested in the holotype). Similarly, QM F14693 shows a distinct, short, basal ridge that extends anteriorly and slightly ventrally on the lingual surface of the crown from the posterior end of the postentocristid, a structure incompletely preserved in the holotype. This is important inasmuch as similar features are present in M$_3$ and M$_4$, although more strongly developed, in QM F14693, whereas they are not preserved in the holotype. Likewise, a basal cingular flange is present at the anterolingual corner of M$_3$ and M$_4$ in QM F14693, a flange also not preserved in the holotype, but duplicated (along with the flange lingual to the entoconid) in QM F14696. The latter is overall slightly larger and more robust than M$_3$ of QM F14693, and additionally differs from it in having a strong protostylid ridge that descends posterolingually from the rear of the protoconid to terminate in the valley buccal to the cristid obliqua. Although QM F14696 is referred to *Perikoala palankarinnica*, these differences, combined with the fact that QM F14696 occurs stratigraphically 1.5 m below the otherwise lowest stratigraphic occurrence of *P. palankarinnica* at Lake Palankarinna, suggest that this referral should be regarded as tentative.

The lower teeth described here nevertheless otherwise duplicate the essential features of *P. palankarinnica*, including the relatively short molars, high degree of surface crenulation, the distinct vertical ribs on the lingual surfaces of the metaconid and entoconid, the presence of distinct cross-crests between the protoconid and metaconid, and hypoconid and entoconid, respectively. The newly discovered specimens additionally show that a distinct basal flange at the anterolingual and posterolingual corners of the lower molars represent formerly unappreciated hallmarks of this species.

Similarly, the newly discovered upper teeth add to the definition of the species. The M^2 (QM F14695) is somewhat abraded and the once prominent surface crenulations have been subdued by wear. The tooth is about as wide as it is long (Table 4) and relatively low-crowned in that the buccal surfaces of the ectoloph cusps and lingual surfaces of the protocone and metaconule are not as steep nor are the cusps as high-standing as in species of *Madakoala* (especially *Madakoala* sp., aff. *M. wellsi.* Stylar cusps are present at the buccal ends of the ectoloph "V" 's but are not as strongly developed as they are in species of *Madakoala*. The paraconule is essentially linear (slightly posterolingually oriented) and connects by a short buccal crest to the anterolingual surface of the paracone and, lingually, by another short crest to the preprotocrista. A protostyle is not present in contrast to the condition in *P. cinereus* where a narrow but distinct spur extends posteriorly from the preprotocrista just lingual to the paraconule. A neometaconule is represented by a weak ridge at the anterolingual base of the metacone and a short but distinct crest blocks the lingual exit of the transverse valley, this crest being contributed from the opposing bases of the protocone and metaconule.

QM F14695 resembles what can be seen of the morphology of the paratype (UCMP 45343), especially in its lack of a protostyle and its only slightly-developed neometaconule. The strongly crenulated enamel surface of the paratype is interpreted to have been similar to the now more abraded surface of QM F14695.

NMV P180083 and NMV P157582 are well worn. In combination, however, they show the following morphology: upper molars that are relatively low-crowned, have distinct but relatively small stylar cusps, the paraconule is indistinct and the protostyle is absent; the lingual exit of the transverse valley is essentially blocked by a short, low crest at the opposing bases of the protocone and metaconule.

Remarks: As indicated in Fig. 12, *P. palankarinnica* appears to be more closely related to species of *Phascolarctos* than to any other phascolarctid, except for *P. robustus*, n. sp. (see below). This is especially shown by the form of the talonid of P_3 and the enhanced degree of crenulation of the enamel surface of the molars. *Perikoala palankarinnica* still retains some apparently plesiomorphic features, including the absence of a protostyle, the presence of cross-crests in the lower molars and the lack of a connection between the metastylid and entoconid. The short (especially in the lower dentition) molar proportions, the basal flange-like crests at the lingual surface of the metaconid and entoconid of M_{3-4} and the slight basal shelf at the terminus of the paraconid and at the entoconid of M_2 appear to be unique to this species (molar proportions and flange-like crests on M_3 apparently foreshadowed in *P. robustus*, n. sp., see below).

Perikoala palankarinnica shares with species of *Koobor* symplesiomorphies including stylar cusps B and C well separated and connected buccally by longitudinal crests, a linear rather than strongly crescentic paraconule, and the lack (or with little development of) the neometaconule on the upper molars.

Perikoala robustus, n. sp.
(Figs 8, 9, 10, 11; Tables 5, 6 and 7)

Holotype: SAM P26552, left dentary with most of cheek tooth dentition, lacking most of symphysis, ascending ramus and angular region, collected by M. O. Woodburne and party in 1985.

Type locality: UCR RV-8504, "The Mother Load", = V-5762, Turtle Quarry, Etadunna Formation, member 3 of Stirton, Tedford and Miller (1961); west side of Lake Palankarinna, South Australia, grid zone 5, reference 656431, Kopperamanna topographic series, SH 54-1; 1:250,000, 1965 ed.

Referred material: From the type locality; UCR 21878, LM_4; UCR 21879, RM_4; UCR 21919, rear of RP^3; UCR 21912, premolar fragment; UCR 21918, RM^2; UCR 21911, RM^3; UCR 21910, RM^5. From RV-8505, "Perikoala Max. Loc.," Etadunna Formation, member 3 of Stirton, Tedford and Miller (1961); west side of Lake Palankarinna, South Australia, grid zone 5, reference 656431, Kopperamanna topographic series, SH 54-1; 1:250,000, 1965 ed.; SAM P26553, right maxillary fragment with M^{2-4}. From UNSW "Croc. Pot 8" Locality, Etadunna Formation, member 3 of Stirton, Tedford and Miller (1961); west side of Lake Palankarinna, South Australia, grid zone 5, reference 656431, Kopperamanna topographic series, SH 54-1; 1:250,000, 1965 ed; CPC 25338, LM^4 lacking its anterior root; CPC 25339, LM_3. All specimens are from nominally

Table 5. Measurements (in mm) of lower dentition and dentary of *Perikoala robustus*. Abbreviations as in Table 1.

Parameter		Specimens			
		SAM P26552	UCR 21879	CPC 25338	UCR 21878
P_3	Lgth	–	–	–	–
	Wdth	3.8	–	–	–
M_2	Lgth	7.0	–	–	–
	Ant. wdth	4.3	–	–	–
	Post. wdth	4.7	–	–	–
M_3	Lgth	6.8	–	6.8	–
	Ant. wdth	4.8	–	4.7	–
	Post. wdth	4.8	–	4.7	–
M_4	Lgth	6.5	6.9	–	6.8
	Ant. wdth	4.7	4.7	–	4.7
	Post. wdth	4.5	4.9	–	4.9
M_5	Lgth	6.4	–	–	–
	Ant. wdth	3.9	–	–	–
	Post. wdth	3.8	–	–	–
Dentary depth at M_2		20.1			
Dentary depth at M_4		19.7			

Table 6. Measurements (in mm) of upper dentition of *Perikoala robustus*. Abbreviations as in Table 1.

Parameter		Specimens					
		UCR 21919	SAM P26553	UCR 21918	CPC 25339	UCR 21911	UCR 21910
P^3	Lgth	–	–	–	–	–	–
	Ant. wdth	–	–	–	–	–	–
	Post. wdth	4.8	–	–	–	–	–
M^2	Lgth	–	6.8	6.8	–	–	–
	Ant. wdth	–	6.1	5.6+	–	–	–
	Post. wdth	–	6.1	6.4	–	–	–
M^3	Lgth	–	6.6	–	–	6.3	–
	Ant. wdth	–	6.9	–	–	6.7	–
	Post. wdth	–	–	–	–	5.9	–
M^4	Lgth	–	6.4	–	6.3	–	–
	Ant. wdth	–	6.5	–	6.8	–	–
	Post. wdth	–	5.4	–	6.2	–	–
M^5	Lgth	–	–	–	–	–	5.7
	Ant. wdth	–	–	–	–	–	5.2

Table 7. Dimensions (in mm) and proportional relationships of M_2 trigonid cuspids in some fossil and living Koalas.

Species	Parameters						
	A	B	C	D	E	F	G
	Prst-Prd Dist.	Prst-Buccal Margin	Trigonid Width	Mid-longit. axis	Prd-Buccal Margin	Prst-Mid axis	Prd-Mid axis
M. devisi	1.0	1.2	5.0	2.50	2.35	+1.3	−0.15
M. wellsi	1.1	1.3	4.9	2.45	2.75	+1.15	−0.30
P. palankarinnica	1.0a	0.7a	3.45a	1.73	2.0a	+1.03	−0.27
P. robustus	1.1	1.45	4.3	2.15	2.4	+0.70	−0.25
P. cinereus	1.8	1.1	5.0	2.50	3.2	+1.40	−0.70

Percent A of C (X = 100A/C)

M. devisi	20%
M. wellsi	22.4%
P. palankarinnica	28%
P. robustus	25.6%
P. cinereus	36%

Explanation: A indicates distance between the protostylid and protoconid. B indicates the distance from the protostylid to the buccal margin of the tooth. C is the width of the trigonid. D is the location of the mid-longitudinal axis of the tooth (C/2). E is the distance from the protoconid to the buccal margin of the tooth. F is the distance between the protostylid and the mid-longitudinal axis of the tooth (+ indicates a buccalward displacement of the protostylid). G is the distance between the protoconid and the mid-longitudinal axis of the tooth (− indicates a lingual-ward displacement of the protoconid).

Percent A of C expresses the proportional separation of the protostylid and protoconid in terms of the width of the trigonid. Note that these dimensions in *Phascolarctos cinereus* show that the separation of the protostylid and protoconid, the separation of the protostylid from the mid axis of the trigonid, and the separation of the protoconid from the mid axis of the trigonid are significantly greater in the living Koala than in species of *Perikoala* or *Madakoala*.

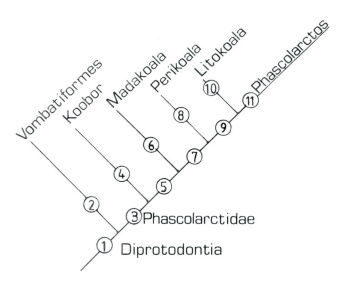

Fig. 12. Cladogram of intergeneric relationships of phasco-larctids. Characters at numbered points are: 1. Diprotodonta; epitympanic sinus developed relative to Syndactyla of Szalay (1982), dental formula 3/1, 1/1, 3/3, 5/5**; I[3] blade-like**; stylar cusps stronger than in Syndactyla; M_{2-5} lose hypoconulid; mandibular foramen still constricted**. 2. Vombatiformes; only cervical thymus gland retained; cardio-vascular gland patch developed; unique method of sperm neck insertion; loss of C_1; P3 bicuspid, tendency to be bladed; P[3] with posterobuccal cusp; incipient (at least) hypolophid developed; protolophid developed in lower molars. 3. Phascolarctoidea and Phascolarctidae; metastylid and protostylid developed in M_2, metastylid developed in posterior lower molars; P1-2 lost. 4. *Koobor*; additional hypertrophy of stylar cusps; more elongate molar proportions; suppression of para-conule in M[3] and M[4]; well developed molar enamel crenulations. 5. P[3] bladed; tendency to develop crest that blocks lingual exit of transverse valley of upper molars; neometaconule developed; posterobuccal cusp developed on P[3]; stylar cusps enlarged but not to degree in *Koobor*. 6. *Madakoala;* molars tend to be relatively hypsodont; strong posterobuccal cusp on P[3]; crest present between M[5] metaconule and metacone. 7. development of strong surface crenulations; reduction of stylar cusps; protoconid of M_2 in lingual third of trigonid; ectoloph cusps more lingually sloping; lingual surfaces of protocone and metaconule slope more lingually; symphysis of dentary ankylosed. 8. *Perikoala;* molar proportions short; distinct lingual shelves on the basal flanks of the metaconid and the entoconid. 9. suppression of stylar cusps; protostyle developed; neo-metaconule well developed; strong posterolingual crest from paracone is aligned with pre-paracrista; deep lingual pocket between opposing bases of protocone and metaconule. 10. *Litokoala;* buccal extension of paraconule connects to buccal tooth margin; neometaconule elongate posteriorly. 11. *Phascolarctos;* enlarged buccal crests on upper molars; large size; snout relatively longer (*than *Madakoala*); P3 trenchant, elongate*; M_2 protostylid very large*; lower molars relatively hypsodont*; cross crests of lower molars suppressed*; metastylid-entoconid crest as continuous fold*; I_1 gliriform*.

Characters marked by * are presumed to have been present in the next most closely related sister taxon, but are not present in the material available. Those marked by ** are symplesiomorphic but show the basic condition relative to more derived states.

the same stratigraphic position, in a white sandstone and interbedded grey-green aren-aceous claystone about 1.4 m stratigraphically above the lower dolomitic marker unit [unit 2, of Stirton, Tedford and Miller (1961)], and thus about 3 m stratigraphically below the levels of the Etadunna Formation that yielded specimens of *Perikoala palankarinnica* (see above) and other elements of the Ditjimanka Local Fauna.

Age: Palankarinna South Local Fauna; Member 3 of the Etadunna Formation of Stirton, Tedford and Miller (1961), about 3 m strati-graphically below intervals that bear elements of the Ditjimanka Local Fauna; Medial Miocene.

Etymology: In reference to the more robust morphology of specimens of this species in com-parison with those of *Perikoala palankarinnica*.

Diagnosis: P. robustus differs from *Perikoala palankarinnica* in having a more robust lower den-tition overall, including a much stronger lingual cingular ridge on the posterior portion of P_3, relatively broader lower molar proportions, a less strongly-developed paraconid on P_3, in possibly having the protostylid relatively nearer to the mid-axis of M_2, an apparently smaller paraconid on M_2, a weaker shelf on the lingual surface of the entoconid, especially of M_3, in having a more strongly developed intermediate postprotoconid crest in M_3 and M_4, and in the greater develop-ment of lingual metaconid and entoconid ribs on those teeth.

Description: The holotype dentary, SAM P26552, contains all of the cheek tooth dentition, save of the anterior half of P_3. The dentary is robust and deep (Table 5), with the lingual surface being nearly flat and the buccal surface bowed outward anterior to the masseteric fossa. The inturned ventral margin of the dentary is similar to that of *P. palankarinnica,* suggesting the presence of an inflected digastric process as, in *Phascolarctos cinereus* and in contrast to species of *Madakoala*. A small genial pit is present at the anteroventral margin of the lingual surface, but the symphysis is broken dorsal and anterior to that. This is persuasive but not necessarily com-pelling evidence for the symphysis having been fused in *P. robustus*, like the condition in *Phasco-larctos cinereus*, rather than having been loosely ankylosed as in species of *Madakoala*. As in koalas in general, the ascending ramus rises past the rear of M_5, but the ramus is broken a short dis-tance above the occlusal plane. Also as in koalas generally, the dentary diverges prosterobuccally at an angle of about 10 degrees relative to the cheektooth row.

The preserved dentary of *P. robustus* is rela-tively and actually deeper than in QM F14693, herein referred to *P. palankarinnica*, about as deep as in *M. devisi,* and relatively deeper (in comparison to the larger size of its dentition, e.g., Tables 1 and 5) than in *M. wellsi* and *P. cinereus*. Alternatively, dental dimensions of *P. robustus* are somewhat larger than those of *P. palankarinnica* (which also has a shallower dentary), and smaller than those of species of *Madakoala* and *P. cinereus*, which have a similarly deep dentary as compared with *P. robustus*. The possibly fused symphysis of *P. robustus* would be similar to the condition in *P. palankarinnica* and in *Phascolarctos cinereus*, and different from the less well ankylosed condition seen in at least *M. wellsi* of species of *Madakoala*.

As shown in Table 5 and Figure 8, the lower cheektooth dentition of *P. robustus* is essentially linear from front to rear, with the molar size gradient increasing from P$_3$-M$_3$ and decreasing from M$_3$-M$_5$. This is comparable to the general koala condition within reasonable limits of variation.

Only the rear portion of P$_3$ is preserved. Available features include a cuspate (two distinct cusps and some less distinct ones posteriorly) longitudinal crest located slightly lingual of the midline of the tooth and directed toward the paralophid of M$_2$. A worn but still well-developed, nearly circular, cusp is located in the posterobuccal quadrant of the tooth and is weakly connected by a short anterolingually-directed crest to the adjacent wall of the longitudinal crest. A buccal cingulum seems to have been absent, as preserved. Each cusp of the longitudinal crest bears relatively weakly-developed nearly vertical ridges that descend on either side of the cusps of the crest. A stronger ridge descends the lingual side of the second (last) major cusp of the longitudinal crest, curves anteriorly as a cingular ridge, and extends anteriorly to a point below the first of the preserved cusps of the longitudinal ridge.

The preserved portion of P$_3$ of *P. robustus* is wider than that of *Perikoala palankarinnica* and lacks a posterobuccally directed crest that passes from the rear of the longitudinal crest to the posterobuccal corner of the crown in *P. palankarinnica*. Also, the lingual crest of *P. robustus* is absent in *P. palankarinnica* (QM F14693). To judge from AMNH 102242, P$_3$ in *Madakoala devisi* is more gracile than in *P. robustus*, and lacks the lingual cingular crest and associated connecting ridge to the longitudinal crest seen in *P. robustus*. In comparison with *M. wellsi* (AMNH 102174), P$_3$ of *P. robustus* is smaller, has a more posteriorly positioned posterobuccal cusp and, apparently, a still better-developed cingular crest and associated connecting ridge to the longitudinal crest. Even though the enamel is broken at the relevant position of the lingual base of P$_3$ in AMNH 102174, the configuration of adjacent structures suggests that a cingular development was not present in this part of the crown of P$_3$ in *M. wellsi*. Neither does a cingular crest at this part

of P$_3$ appear to have been present in *P. cinereus*. On the other hand, P$_3$ of *P. cinereus* contrasts with that of all Miocene koalas yet known in having a derived, selene-like development of the posterobuccal cusp.

M$_2$ is represented by the holotype only in *P. robustus*. This tooth is generally similar to that of other Miocene koalas in having a five-cusped trigonid and a two-cusped talonid. The trigonid is narrower than the talonid and tapers anteriorly. The trigonid is composed of a weakly-expressed paraconid at the anterior end of the paralophid (the anterior tip of the tooth). In addition to the paraconid, the trigonid is composed of the protoconid, which is the tallest cusp, located near the mid-longitudinal axis of the tooth and connected by a strong, slightly antero-lingually-directed, paralophid to the paraconid; the metaconid, located a short distance (0.8 mm) posterolingual to the protoconid and connected to it by a weak crest; a metastylid, a small cusp situated on the lingual margin of the crown just anterior to the tranverse crevice that separates the trigonid from the talonid, and connected to the metaconid by a weak but distinct arcuate crest; a protostylid, situated on the posterobuccal quadrant of the trigonid, below the protoconid and nearly opposite the metaconid. A long posterior ridge from the protostylid terminates at the crevice between the trigonid and talonid. This crevice is posterobuccally oriented in its buccal half, but curves anteriorly to separate the protostylid from the protoconid. A short continuation of this crevice extends slightly anterobuccally and is bounded buccally by an equally short, anteriorly-directed, crest from the protocristid. A weakly developed ridge (essentially part of the crenulation pattern of the enamel surface) extends from the protoconid posterobuccally a short distance and terminates in the crevice that separates the protoconid and protostylid. This crest is better developed in QM F14693 (*P. palankarinnica*) and also can be seen in the surface enamel crenulation pattern on M$_2$ in *P. cinereus*. Similar structures are absent in species of *Madakoala* which have virtually no surface enamel crenulation.

The apex of the protostylid is situated about 1.1 mm posterobuccal to the protoconid, and about 0.70 mm buccal to the midline of the tooth. As indicated in Table 7, the locations of the trigonid cuspids in *P. robustus* are comparable to those for Miocene koalas and unlike proportional locations seen in *P. cinereus*, wherein the protoconid is set well lingual to the mid-axis of the tooth and the protostylid is well buccal to that axis, as well.

In species of *Perikoala* (Table 7), the relative separation of the protostylid from the mid-axis of the tooth (25%-28%) appears to be somewhat greater than for species of *Madakoala* (20%-22%), and thus perhaps somewhat more like the proportions found in *P. cinereus* (36%). The small number of specimens studied here promotes caution when interpreting these data, however.

The trigonid of M_2 of *P. robustus* differs from that of *P. palankarinnica* in having a weaker metastylid, in lacking the well-developed premetaconid crista at the lingual margin of the tooth, in apparently having a weaker paraconid (this seems to have been quite strong in both SAM P10893 and QM F14693, type and referred specimen of *P. palankarinnica*), in having a weaker crest buccal to the cristid obliqua posterobuccal to the protoconid, and, possibly, the protoconid being proportionately nearer to the protostylid (Table 7, column A; e.g., having a similar separation of the protostylid and protoconid, but having an actually wider trigonid — Column C).

The talonid of M_2 in *P. robustus* is generally comparable to that of *P. palankarinnica*. The similarities include tooth proportions, degree of enamel crenulation, degree of transverse (anteriorly concave) linking by low crests of the hypoconid and entoconid, and passage of the cristid obliqua from the hypoconid to the rear of the protoconid. M_2 of SAM P26552 differs from QM F14693 (*P. palankarinnica*) in having a smaller shelf-like development lingual to the anterior base of the entoconid.

The lingual entoconid rib and posterolingual entoconid shelf of M_2 are absent in species of *Madakoala* (not preserved in M_2 of *M. devisi*, but apparently absent to judge from more posterior molars of that species). The posterolingual entoconid shelf and the lingual entoconid rib are elaborated in *P. cinereus*, additionally suggesting that species of *Perikoala* are more closely related to the modern koala than to species of *Madakoala*.

M_3-M_5 of *P. robustus* resemble each other and differ from M_2 chiefly in details of the trigonid. As shown in Figure 8 the paralophid is present in M_3-M_5 of *P. robustus*, but the paraconid is not well developed. The tooth is broader anteriorly with a short cingulum present between the paraconid and premetacristid that extends anteriorly and slightly anterolingually from the metaconid. A slight, but distinct, lingual shelf is developed at the point where the premetacristid changes direction. The shelf is directed posterolingually and ventrally along the anterolingual base of the metaconid. This shelf is most strongly developed on M_4 and next strongly developed on M_5 of SAM P26552. The lingual metaconid shelf is quite robust in CPC 25338, here interpreted to be M_3. That shelf is but slightly developed in UCR 21879, here interpreted to be M_4 (the enamel is missing in the relevant location in UCR 21878, also interpreted to be M_4).

The specimens mentioned above also indicate that the metastylid of M_3 is stronger than that of M_2, M_4, or M_5. Finally, an apparent hallmark of this species is the presence — especially in M_3 and M_4 — of a pair of ribs on the posterior surface of the protoconid. The most buccal of these ribs is in a position homologous to that of the posterior protostylid crest of M_2 (assuming that a crest associated with a neomorph protostylid would occur functionally in association with the protoconid of more posterior molars). Especially in M_3, this crest extends posteriorly from the rear of the protoconid apex toward (but does not reach) the transverse crevice that buccally separates the trigonid and talonid. In fact this crest appears to terminate in a raised eminence that is additionally connected by another more lingually situated crest between the postprotocristid and the cristid obliqua (although less well differentiated in M_4 of SAM P26552 a strong wear facet developed on this eminence displays its formerly strong development on that tooth). CPC 25338 shows a strong intermediate ridge in this location, but a weaker postprotocristid than M_3 of SAM P26552. These structures are diminishingly well portrayed in M_4 and M_5 to judge from the available material. One apparent corollary of this is that the protoconid of M_4 and M_5 is situated relatively nearer the buccal margin of the tooth than seen in M_2 and M_3. Comparable features are very faintly developed in M_3 and M_4 of *P. palankarinnica* (holotype and QM F14693). Neither the postprotocristid nor the intermediate crest appears to have been present in M_3 (or M_4 or M_5) of *Madakoala devisi;* the postprotocristid alone is present on M_3 (not M_4 or M_5) of *M. wellsi*. Comparable features also appear to be absent in *P. cinereus* which also differs from these Miocene taxa in having developed a broad, cuspate, buccal cingulum between the trigonid and talonid.

The talonid of M_3-M_5 is nominally similar to that of M_2 in *P. robustus*, except that a lingual shelf at the rear half of the entoconid is better developed in M_3 than in M_2.

In summary, the lower cheektooth dentition of *P. robustus* most clearly resembles that of *P. palankarinnica* in having generally short molar proportions, in having a similar degree of enamel complexity and degree of transverse connection of the buccal to lingual molar cusps, in having relatively well developed anterolingual and posterolingual shelves on the metaconid and entoconid of M_3-M_4, respectively, and in having a distinct crest posterior to the protoconid that is intermediate between the postprotocristid and the cristid obliqua.

The lower dentition of *P. robustus* seems to be less derived than that of *P. palankarinnica* in the possibly closer approximation of the protostylid and the mid-axis of M_2, the seemingly smaller paraconid of M_2, the less well developed lingual entoconid shelf in M_2-M_4 and less well developed lingual metaconid shelf in M_3. *P. robustus* differs from (and seems to be more derived relative to) *P. palankarinnica* in the stronger development of the intermediate postprotoconid crest on M_3-M_4 and in the greater development of lingual metaconid and entoconid ribs (?prophetic of the condition seen in *P. cinereus*).

P. robustus appears to be unique relative to all known koalas in the development of a lingual cingular shelf in the posterior portion of P_3.

Referral of the upper dental elements described here to *Perikoala robustus* is based on their similar provenance and the fact that their analysis points to the same conclusion as derived from study of the lower dentition: that a new species of *Perikoala* is represented; the taxon is not referable to species of *Madakoala*, nor is it a new species of that genus; the morphology of these upper dental elements supports the concept that species of *Perikoala* are in some ways more derived than those of *Madakoala*, but not as morphologically advanced as *Litokoala kutjamarpensis* or *Phascolarctos cinereus*.

P^3 is represented by UCR 21919, which preserves only the rear portion of the crown (Fig. 11). The tooth is relatively broad (Table 6) and is generally similar to that of species of *Madakoala*, with some important differences. As preserved, P^3 of *P. robustus* is composed of a cuspate longitudinal crest that is situated slightly buccal of the midline of the tooth. The ridge is ornamented by four cusps, the second and third of which each bear a short vertical rib on either side of the tooth. The crest curves posterobuccally as it extends from the apex of the fourth cusp toward the coronal base. This is similar to the situation found in species of *Madakoala*, but exactly opposite from the condition seen in *Phascolarctos cinereus*, where the crest under discussion is oriented posterolingually.

The first of the preserved cusps of the longitudinal crest is broken anteriorly, but does show an anterolingually directed crest that extends to a basal cingulum that apparently was cuspate. A similarly oriented ridge extends from the fourth cusp of the longitudinal crest and becomes part of a series of low conules that form a cingular swelling along the lingual margin of the tooth. These ridges and associated lingual cingulum enclose a well-marked, anterolingually-directed pocket. The portion of the crown that shows this pocket and the ridges that bound it is supported by the anterior portion of the posterior root and adjacent interradicular area. In *Madakoala devisi* the ridge that forms the anterior border of the lingual pocket in P^3 is directed posterolingually so that the pocket is "V"-shaped in both occlusal

and lingual view (Fig. 11). In *Madakoala* sp., aff. *M. wellsi*, on the other hand, the ridge that forms the anterior border of the lingual pocket is developed at a coronal position that is supported by the anterior root, so that the pocket, itself, is much larger than in *M. devisi* or *P. robustus*. The lingual pocket of P^3 is not as strongly developed in *Phascolarctos cinereus* as in species of *Perikoala* or *Madakoala*. The ridge that forms the anterior ridge of the pocket in *Phascolarctos cinereus* connects to a cusp on the longitudinal crest that is located posterior to the anterior root of P^3. The lingual pocket of P^3 in *Phascolarctos cinereus* is thus less extensive than in *Madakoala* sp., aff *M. wellsi*, but more extensive than in *M. devisi* or *P. robustus*.

The posterobuccal side of UCR 21919 is ornamented by a low cusp that is elongate anterobuccally and connects weakly at its posterior end to the buccal flank of the fourth cusp of the longitudinal crest. The cusp in the comparable postion in species of *Madakoala* is essentially conical. In *Phascolarctos cinereus* the posterobuccal cusp of P^3 is essentially absent, and the posterior ridge from the last cusp of the longitudinal crest is oriented posterolingually, rather than posterobuccally as in the Miocene forms under discussion (see above). P^3 of *Phascolarctos cinereus* is thus distinctly derived over the condition shown in the Miocene forms.

The molar dentition in *Perikoala robustus* is best represented in SAM P26553, in which M^2-M^4 are associated. M^2 is also represented by UCR 21918. M^2 of *P. robustus* is of the general koala configuration, with a W-shaped ectoloph in which the paracone and metacone each are connected to a pair of stylar cusps; a paraconule is present as an anterobuccally oriented crest at the anterior base of the paracone; a short, slender, crest extends from the paraconule to the lingual apex of the protocone; the protocone and metaconule are subequally-sized selenodont cusps at the lingual margin of the crown; a neometaconule is not developed at the anterior base of the metacone; a short cingulum is present at the lingual mouth of the transverse valley; a parastyle is present anterior to Stylar cusp B. As commonly seen in koala M^2's, the buccal margin opposite the paracone is relatively straight, and oriented slightly anterolingually in contrast to the buccal margin opposite the metacone, which curves posterolingually.

Crenulations of the enamel surface are well developed in upper molars of *Perikoala robustus*, similar to the development seen in *P. palankarinnica* and *Phascolarctos cinereus*, but not in the essentially smooth enamel surface of species of *Madakoala*. As shown in Table 6, upper molars of *P. robustus* are relatively shorter than in species *Madakoala*, and actually larger than those of *P. palankarinnica*.

The stylar region of M^2 in *Perikoala robustus* is composed of the usual component of cusps but an additional cusp-like swelling is present anterior to Stylar cusp E (metastyle), and is connected by a short crest to the postmetacrista. Such an additional cusp is not present in more posterior molars of *P. robustus*, nor in any upper molar of species of *Madakoala*, *Litokoala kutjamarpensis* (only M^2 preserved) or *Phascolarctos cinereus*. The continuously bounded buccal border of the crown opposite the paracone and metacone, respectively, in M^2 of *P. robustus* is however, similar to the condition seen in *Litokoala kutjamarpensis* and *Phascolarctos cinereus*.

The more posterior upper molars of *Perikoala robustus* are generally similar to M^2 but display the usual progressively posterior alteration of form and character emphasis: e.g., diminution of the paraconule, decreased degree of development of the lingual cingulum, diminished degree to which the stylar border structures are continuously developed, and reduction of the metaconule (extremely reduced in M^5). The cusp anterior to Stylar cusp E in M^2 is not present in any of the more posterior molars of *P. robustus*.

In summary, the morphology of the upper dental elements of *Perikoala robustus* supports the concept that species of *Perikoala* are in some ways more derived than those of *Madakoala*, but not as morphologically advanced as *Litokoala kutjamarpensis* or *Phascolarctos cinereus*. Such points include: the continuously bounded stylar borders of M^2 (cf. *L. kutjamarpensis* and *Phascolarctos cinereus*), and the stronger degree of surface enamel complexity in contrast to species of *Madakoala*.

Features that show the more plesiomorphic state of *P. robustus* in comparison to *L. kutjamarpensis* and *Phascolarctos cinereus* include: the posterobuccally curved crest at the rear of the longitudinal crest of P^3; the presence of a posterobuccal cusp in P^3; absence of a protostyle in M^2.

Features that appear to be unique to *Perikoala robustus* include: the anteroposteriorly narrow, but anterolingually trapezoidal pocket on the lingual surface of P^3 that is located in the posterior portion of that tooth; relatively short molar proportions (the former cannot be determined in *P. palankarinnica*. The latter is comparable in both species of that genus).

Features that show that *Perikoala robustus* is more derived than species of *Madakoala* include: the anterobuccally oriented and elongate cusp on the posterobuccal corner of P^3; the presence in M^2 of a cusp anterior to Stylar cusp E and posterior to Stylar cusp D; increased crenulation of the enamel surface; a continuously developed stylar border opposite the paracone and metacone of M^2.

It is difficult to assess the relationships of *Perikoala robustus* relative to *P. palankarinnica* on the basis of upper teeth, inasmuch as the upper dentition of the latter species is represented by a partially preserved M^2 (Stirton, Tedford and Woodburne 1967). What can be seen of that tooth, however, is compatible with the general interpretation of the affinity of *Perikoala* as exemplified by *P. robustus* (see above).

Based essentially on an analysis of its lower dentition, *Perikoala robustus* appears to be a species of *Perikoala* that is overall less derived than *P. palankarinnica*, but possibly (increased development of lingual ribs of lower molar lingual cusps) more prophetic of *Phascolarctos cinereus* than is *P. palankarinnica*.

DISCUSSION

At the present time, there are five genera of living and fossil phascolarctids. Three (*Madakoala*, *Perikoala* and *Litokoala*) are of Miocene age, one is known from Pliocene deposits (*Koobor*), with *Phascolarctos* known to range from the Pliocene to the Recent (Pledge 1987).

As suggested in Fig. 12, species of *Phascolarctos* appear to be the most derived and those of *Koobor* among the most plesiomorphic. *Litokoala kutjamarpensis* appears to be phyletically closer to species of *Phascolarctos* than is *Perikoala palankarinnica*, but *L. kutjamarpensis* is known from only a single upper molar. Species of *Madakoala* are more plesiomorphic than species of *Perikoala*, *Litokoala* and *Phascolarctos*, but members of all four of these genera still appear to be more closely related than any is to species of *Koobor*. Species of *Koobor* are, however, known only from elements of the upper dentition.

In addition to features just mentioned above with respect to *P. palankarinnica*, *Koobor* species demonstrate their plesiomorphic sister-group status relative to all other koalas in having a bicusped P^3 (and likely a similarly bicusped P_3), absence of a neometaconule in upper molars, free exit of transverse valley of upper molars lingually, relatively weak molar enamel crenulations. Species of *Koobor* are derived over their conceptually most primitive status in having more strongly-developed stylar cusps in the upper molars, more elongate upper molar proportions and having suppressed the paraconule in M^3 and M^4.

Species of *Madakoala* are derived versus those of *Koobor* in having a more bladed P^3 (and $P_{3?}$), more hypsodont molars, a stronger posterobuccal cusp on P^3, and the development of a crest on M^5 between the metaconule and metacone.

Within *Madakoala*, *M. devisi* appears to be more plesiomorphic than *M. wellsi* in having a less robust lower dentition overall, including a much weaker posterobuccal cuspid on P_3, relatively narrower lower molar proportions, a less strongly-developed protostylid on M_2 and a weaker protostylid ridge on M_3.

Madakoala sp., aff *M. wellsi* is derived relative to *Madakoala devisi* in having a much stronger paraconule on the upper molars, a stronger metaconule-metacone ridge on M^5, a more bilobed P^3 in which the deep pit anterior to the posterolingual cusp has been lost, a more well-developed set of buccal ridges on P^3 and a much stronger anterior lingual ridge on that tooth, in having generally stronger buccal stylar structures in the upper molars, and thus more strongly-developed pockets buccal to the paracones and metacones.

Madakoala sp. from the Wadikali Local Fauna differs from both *M. devisi* and *Madakoala* sp., aff. *M. wellsi*. The robust P^3 of the Wadikali form and the presence of a strong posterobuccal cusp in addition to the main crest on P^3 may reflect its derived status versus that tooth in the other two species.

The form of the talonid of P_3 and trigonid of M_2 and the well-developed pattern of surface ornamentation of *Perikoala palankarinnica* from the Ditjimanka Local Fauna suggests affinity with species of *Phascolarctos* (Stirton 1957) and, compared to that, species of *Madakoala* appear to have retained more plesiomorphic traits than occur in *P. palankarinnica*.

Perikoala robustus from the South Palankarinna Local Fauna appears to be more plesiomorphic than *P. palankarinnica* in the possibly closer approximation of the protostylid and the mid-axis of M_2, in the seemingly smaller paraconid of M_2, the less well developed lingual entoconid shelf in M_2-M_4 and less well developed lingual metaconid shelf in M_3. On the other hand, the stronger development of the intermediate post-protoconid crest in M_3-M_4 and the greater development of the lingual metaconid and entoconid ribs on lower teeth of *P. robustus* may be prophetic of the condition in *Phascolarctos cinereus*, and suggest a closer relation of these two species than between the living koala and *P. palankarinnica*.

Apparently symplesiomorphic features shared between the species of *Madakoala* and species of *Perikoala* include the presence of relatively well-separated stylar cusps B and C, with longitudinal crests connecting them, in lacking (or having only a slightly developed) neometaconule, in having a relatively linear rather than crescentic proto-conule and in lacking a protostyle on the upper molars, the presence in the lower molars of a hypoconid-entoconid crest, a discontinuous metastylid-entoconid crest and the buccal selenes (postprotocristid-cristid obliqua) which are less lingually extensive.

As discussed above, species of *Koobor* probably are not closely related to any of the Pinpa, South Palankarinna, Ericmas, Ditjimanka, Tarkarooloo or Wadikali phascolarctids. In spite of the features cited above that appear to support a degree of relationship between *Perikoala palan-karinnica* and species of *Phascolarctos*, the latter appears to have shared an even closer relationship with *Litokoala kutjamarpensis* of the Kutjamarpu Local Fauna, based on the shared presence of a protostyle in M^2, a distinct development of a neometaconule, and a strong postero-lingual crest from which the paracone is aligned with the preparacrista in the upper molars (see also Springer 1987). For these reasons, and the presence of a deep lingual pocket between opposing bases of the protocone and metaconule, *L. kutjamarpensis* is interpreted here to be the sister-group of *Phascolarctos* species rather than the plesiomorphic sister-group to all other phascolarctids (Archer 1976).

Inasmuch as at least three separate lineages of phascolarctids appear to have been already present in Miocene-aged strata, their common ancestor must have been older than that. This is supportive of the concept of the antiquity of the Phascolarctidae within the Diprotodontia as previously suggested on the basis of dental criteria (Winge 1941; Archer 1976), by sharing with the Vombatidae the derived presence of a cervical thymus gland and loss of the thoracic thymus gland, the presence of a unique cardio-vascular gland patch, serological relationships (e.g. Kirsch 1977) and attachment style of sperm neck to head (e.g. Hughes 1965; Harding *et al.* 1987). At the same time, however, phascolarctids are more plesiomorphic than vombatimorphians in having retained an alisphenoid bulla in contrast to the derived squamosal bulla of the vombatimorphians.

Species of *Perikoala* and *Madakoala* appear to demonstrate potential for biostratigraphic-biochronologic correlation. *M. devisi* is chronologically older (although apparently ranging later in time than) *M. wellsi*. Features in which *M. wellsi* are derived relative to *M. devisi* are cited in the diagnosis for the Ericmas species. The apparent similarity in age of the Ericmas and Ditjimanka faunas, based on shared phascolarctids of a post-Pinpa evolutionary state, follows other evidence of shared taxa, such as the monotreme, *Obdurodon insignis* (Woodburne and Tedford 1975), a species of vombatiform (type B of Tedford *et al.* 1977) that appears to be more derived than a Pinpa Fauna form (Tedford and Wood-burne 1987) and a pseudocheirid that is more derived than a congener from the Pinpa Fauna (Woodburne, Tedford and Archer 1987). *Chunia* cf. *C. illuminata* from the Pinpa Local Fauna is close to but probably less derived than *Chunia illuminata* from the Ditjimanka Local Fauna (Woodburne and Clemens 1986).

Fossil phascolarctids are as old as about 15 Ma according to palynofloral correlations between the Namba and Etadunna formations and marine strata of coastal southern Australia (Callen and Tedford 1976). On the other hand, serological and other soft-anatomical evidence from living forms suggests that the Phascolarctidae has an ancient relationship with the vombatimorphians that is significantly older than the known fossil record (Miocene) for either group.

ACKNOWLEDGEMENTS

Elizabeth Archer, then of the Queensland Museum, and Winston and Brenda Head, of the South Australian Museum, contributed considerably to the successes of the 1972 field work at Lake Palankarinna. Tedford was aided by R. T. Wells, T. H. and P. V. Rich, M. Archer, M. Plane, N. Pledge, A. Bartholomai and R. Brown in the Tarkarooloo Basin expeditions. Archer was aided on various of his field trips to central Australia by M. Plane and R. Brown (Bureau of Mineral Resources), T. H. V. Rich (Museum of Victoria), by S. Van Dyck and H. Plowman (Queensland Museum), and by H. Godthelp, T. Flannery, S. Hand, G. Clayton, J. A. Case, P. Bridge and third year students (all associated with the University of New South Wales). T. H. V. Rich, Museum of Victoria, Melbourne, graciously allowed us to report upon specimens collected by him. Ms Jenny Taylor of the School of Zoology, University of New South Wales, Mrs Linda Bobbit, Department of Earth Sciences and Mrs Marcia Kooda-Cisco, Department of Biology, University of California, Riverside, helped produce the illustrations used in this report. Finally, we wish to recognize the hospitality of Bryan and Cath Oldfield of Etadunna Station, upon which Lake Palankarinna is located, and Mr Buddy Napier of Frome Downs Station, for allowing access to the Tarkarooloo Basin sites.

Work reported upon here was supported by National Science Foundation Grant GB-35488, EAR 840-3448 (Woodburne), GB-18273X1 (Tedford) and Australian Research Grants Scheme grants to Archer, T. H. V. Rich and P. V. Rich.

REFERENCES

APLIN, K. AND ARCHER, M., 1987. Recent advances in marsupial systematics with a new syncretic classification. Pp. xv-lxxii *in* "Possums and opossums: studies in evolution" ed by M. Archer. Surrey Beatty & Sons and The Royal Zoological Society of New South Wales: Sydney.

ARCHER, M., 1976. Phascolarctid origins and the potential of the selenodont molar in the evolution of the diprotodont marsupials. *Mem. Qd Mus.* **17:** 367-71.

ARCHER, M., 1977. *Koobor notabilis* (DeVis), an unusual koala from the Pliocene Chinchilla Sand. *Mem. Qd Mus.* **18:** 31-35.

ARCHER, M., 1978. The nature of the molar-premolar boundary in marsupials and a reinterpretation of the homology of marsupial cheek-teeth. *Mem. Qd Mus.* **18:** 157-64.

ARCHER, M. AND WADE, M., 1976. Results of the Ray E. Lemley Expeditions, part 1. The Allingham Formation and a new Pliocene vertebrate fauna from northern Queensland. *Mem. Qd Mus.* **17:** 379-97.

AUSTIN, P., 1981. "A grammar of the Diyari, South Australia". Cambridge Univ. Press: Melbourne.

BARTHOLOMAI, A., 1968. A new fossil koala from Queensland and a reassessment of the taxonomic postion of the problematical species, *Koalemus ingens* DeVis. *Mem. Qd Mus.* **15:** 65-73.

CALLEN, R. J. AND TEDFORD, R. H., 1976. New late Cainozoic rock units and depositional environments, Lake Frome area, South Australia. *Trans. Roy. Soc. S. Aust.* **100:** 125-68.

DEVIS, C. W., 1889. On the Phalangistidae of the Post-Tertiary period in Queensland. *Proc. R. Soc. Qd* **6:** 105-14.

HARDING, H. R., CARRICK, F. N. AND SHOREY, C. D., 1987. The affinities of the Koala, *Phascolarctos cinereus* (Marsupialia: Phascolarctidae), on the basis of sperm ultrastructure and development. Pp. 353-64 *in* "Possums and opossums: studies in evolution" ed by M. Archer. Surrey Beatty & Sons and the Royal Zoological Society of New South Wales: Sydney.

HUGHES, R. L., 1965. Comparative morphology of spermatozoa from five marsupial families. *Aust. J. Zool.* **13:** 533-43.

KIRSCH, J. A. W., 1977. The comparative serology of the Marsupialia, and a classification of marsupials. *Aust. Jour. Zool. Suppl. Ser.* **52:** 1-152.

PLEDGE, N. S., 1986. A new species of *Ektopodon* from the Miocene of South Australia. *In* Revision of the Ektopodontidae (Mammalia; Marsupialia; Phalangeroidea) of the Australian Neogene" ed by M. O. Woodburne and W. A. Clemens. *Univ. Calif. Publ. Geol. Sci.* **31:** 43-67.

PLEDGE, N. S., 1987. *Phascolarctos maris*, a new species of Koala (Marsupialia: Phascolarctidae) from the early Pliocene of South Australia. Pp. 327-30 *in* "Possums and opossums: studies in evolution" ed by M. Archer. Surrey Beatty & Sons and the Royal Zoological Society of New South Wales: Sydney.

SPRINGER, M. S., 1987. Lower molars of *Litokoala* (Marsupialia: Phascolarctidae) and their bearing on phascolarctid evolution. Pp. 319-25 *in* "Possums and opossums: studies in evolution" ed by M. Archer. Surrey Beatty & Sons and the Royal Zoological Society of New South Wales: Sydney.

STEIGER, R. H. AND JAEGER, E., 1977. Subcommission on geochronology: convention on the use of decay constants in geo- and cosmochronology. *Earth and Planet. Sci. Let.* **36:** 359-62.

STIRTON, R. A., 1957. A new koala from the Pliocene Palankarinna Fauna of South Australia. *Rec. S. Aust. Mus.* **13:** 71-81.

STIRTON, R. A., TEDFORD, R. H. AND MILLER, A. H., 1961. Cenozoic stratigraphy and vertebrate paleontology of the Tirari Desert, South Australia. *Rec. S. Aust. Mus.* **14:** 19-61.

STIRTON, R. A., TEDFORD, R. H. AND WOODBURNE, M. O., 1967. A new Tertiary formation and fauna from the Tirari Desert, South Australia. *Rec. S. Aust. Mus.* **15:** 427-62.

SZALAY, F. S., 1982. A new appraisal of marsupial phylogeny and classification. Pp. 621-40 *in* "Carnivorous marsupials" ed by M. Archer. Roy. Zool. Soc. New South Wales: Sydney.

TEDFORD, R. H., 1970. Principles and practices of mammalian geochronology in North America. *Proc. N. Amer. Paleo. Conven.* **F:** 666-703.

TEDFORD, R. H., ARCHER, M., BARTHOLOMAI, A., PLANE, M., PLEDGE, N. S., RICH, T., RICH, P. AND WELLS, R. T., 1977. The discovery of Miocene vertebrates, Lake Frome area, South Australia. *Bur. Min. Res. Jour. Aust. Geol. and Geophys.* **2:** 53-57.

TEDFORD, R. H. AND WOODBURNE, M. O., 1987. The Illariidae, a new family of vombatiform marsupials from Miocene strata of South Australia and an evaluation of the homology of molar cusps in the Diprotodontia. Pp. 401-18 *in* "Possums and opossums: studies in evolution" ed by M. Archer. Surrey Beatty & Sons and the Royal Zoological Society of New South Wales: Sydney.

WINGE, H., 1941. "The interrelationships of mammalian genera". Vol. 1, 413pp. C. A. Retizels Forlag: Copenhagen.

WOODBURNE, M. O. AND CLEMENS, W. A., 1986. A new genus of Ektopodontidae and additional comments on *Ektopodon serratus*. Pp. 10-42 *in* "Revision of the Ektopodontidae (Mammalia; Marsupialia; Phalangerioidea) of the Australian Neogene" ed by M. O. Woodburne and W. A. Clemens. *Univ. Calif. Publ. Geol. Sci.* **131**.

WOODBURNE, M. O. AND TEDFORD, R. H., 1975. The first Tertiary monotreme from Australia. *Novitates* 2588.

WOODBURNE, M. O., TEDFORD, R. H. AND ARCHER, M., 1987. New Miocene ringtail possums (Marsupialia: Pseudocheiridae) from South Australia. Pp. 639-79 *in* "Possums and opossums: studies in evolution" ed by M. Archer. Surrey Beatty & Sons and the Royal Zoological Society of New South Wales: Sydney.

WOODBURNE, M. O., TEDFORD, R. H., ARCHER, M., TURNBULL, W. D., PLANE, M. AND LUNDELIUS, E. L., 1985. Biochronology of the continental mammal record of Australia and New Guinea. *S. Aust. Dept. Mines and Energy Spec. Publ.* **5:** 347-64.

LOWER MOLARS OF *Litokoala* (MARSUPIALIA: PHASCOLARCTIDAE) AND THEIR BEARING ON PHASCOLARCTID EVOLUTION

MARK S. SPRINGER[1]

Middle Miocene phascolarctid teeth from Lake Kanunka, Etadunna Station, South Australia, have been described as *Litokoala kanunkaensis* n.sp. and as such represent the second known species of the genus. The lower molars of *L. kanunkaensis* are the first known for a species of the genus and, in their structure, reveal closer affinities to the species of *Phascolarctos* than to any other phascolarctid genus.

Key Words: Phascolarctidae; *Litokoala kanunkaensis*; Etadunna Formation; Kanunka North Local Fauna; Miocene.

Pages 319-25 *in* POSSUMS AND OPOSSUMS: STUDIES IN EVOLUTION ed by M. Archer. Surrey Beatty & Sons and the Royal Zoological Society of New South Wales: Sydney, 1987.

INTRODUCTION

In 1967, Stirton, Tedford and Woodburne described *Litokoala kutjamarpensis* from the Kutjamarpu Local Fauna, Wipajiri Formation, Lake Ngapakaldi, South Australia, based on a single upper molar. Since that time no material has been referred to the genus.

Specimens forming the basis of the present study were obtained from the Etadunna Formation, Lake Kanunka, South Australia, by M. O. Woodburne and colleagues in 1985.

Numerical assignment of teeth within the molar arcade follows Archer (1978); cusp homology and topographic dental terminology follow Archer (1984a:715) with the exception of modifications discussed by Tedford and Woodburne (1987).

RV- refers to a fossil locality at the Department of Earth Sciences, University of California, Riverside; Riverside specimen numbers are preceded by UCR. F refers to specimens at the School of Zoology, University of New South Wales.

All measurements are in millimetres.

SYSTEMATICS

Order Diprotodontia Owen, 1866
Suborder Vombatiformes Woodburne, 1984
Superfamily Phascolarctoidea Owen, 1839
Family Phascolarctidae Owen, 1839

Litokoala Stirton, 1967

Genotypic species: Litokoala kutjamarpensis.

Additional species: Litokoala kanunkaensis n. sp.

Distribution: Litokoala kutjamarpensis is represented by a single upper molar from the Kutjamarpu Local Fauna, Lake Ngapakaldi, South Australia. *Litokoala kanunkaensis* is represented by two isolated lower molars and two upper molar fragments, all from the Etadunna Formation, Lake Kanunka, South Australia.

Revised diagnosis of Litokoala (based on new material herein referred to *L. kanunkaensis*): The M_3 and M_5 of *Litokoala kanunkaensis* differ from other known phascolarctids in having an anterior bifurcation of the preentocristid. The M_3 of *Litokoala kanunkaensis* differs from other known phascolarctids in having a well developed entostylid ridge and a prominent posterobuccally directed crest extending from the apex of the metaconid. The M_5 of *L. kanunkaensis* differs from other known phascolarctids in having a reduced talonid, a shorter, ridge-like entoconid, an anteriorly displaced entoconid (relative to the hypoconid), a posteriorly displaced protoconid (relative to the metaconid), a strong lingual protocristid directed toward the metastylid, a distinct entostylid ridge and a parallel arrangement of the postprotocristid and cristid obliqua in which these cristids do not connect end to end.

Litokoala kanunkaensis differs from *Phascolarctos cinereus* in being smaller, in having proportionately less elongate lower molars, less buccal expansion at the base of the protoconid, a less crescentic premetacristid and free exit of the transverse valley at the buccal tooth margin. In addition, the cristid obliqua and postprotocristid of the M_3 unite at a more lingual position in *P. cinereus* and the transverse distances between the

[1]Department of Biology, University of California, Riverside, Califonia, USA 92521.

protoconid and metaconid and between the hypoconid and entoconid are greater in *Litokoala* relative to the width of the tooth. Also, the posterolingual corner of the M_5 is much more rounded in *Litokoala*.

Species belonging to *Litokoala* differ from *Madakoala* and *Perikoala* in being smaller, in having well developed selenodont molars and in lacking the incipient to moderately developed transverse lophodonty present in the latter genera. In addition, the M_3 cristid obliqua and postprotocristid unite at a more lingual position in *Litokoala* than in *Madakoala* or *Perikoala*. The M_3 protostylid ridge is present in *Perikoala* and variably present in *Madakoala* but not in *Litokoala*. The known lower molars of *Litokoala* differ from those of *Perikoala* in being proportionately more elongate and in lacking the lingual shelves around the entoconid and metaconid.

At this point, some justification needs to be provided as to why material in the present study is referred to the Phascolarctidae rather than to the Pseudocheiridae. In pseudocheirids, the cristid obliqua in the M_2 is directed toward and connects to the metaconid. This feature is not found in any other diprotodontian family and appears to represent a pseudocheirid synapomorphy (Woodburne *et al.* 1987). In phascolarctids, the M_2 cristid obliqua connects to the protoconid. Features in the M_3 and M_5 of *L. kanunkaensis* that appear to be convergent in their development on pseudocheirids will be discussed below.

Litokoala kanunkaensis n. sp.

Holotype: UCR 21926, a right M_3.

Type locality: UCR Locality RV-8453, Etadunna Formation, west side of Lake Kanunka, Lake Eyre Basin, South Australia, grid zone (5), reference (647482), Kopperamanna topographic series SH #54-1, 1:250,000, 1965 ed. The faunal assemblage from this locality is designated the Kanunka North Local Fauna.

Referred material: UCR 21945, a right M_5 from RV-8453. From RV-8452 (approximately the same grid coordinates as for holotype): UCR 21980, a metacone of a RM^4; UCR 21979, a metacone of a LM^2.

Age of the Kanunka North Local Fauna: Medial Miocene.

Placement within Litokoala: The genus *Litokoala* was previously known only from the Wipajiri Formation (Kutjamarpu Local Fauna) at Lake Ngapakaldi in South Australia and is represented by a single upper molar (Stirton, Tedford and Woodburne 1967). The placement of *Litokoala kanunkaensis* within that genus is based on the following criteria.

First, the length of the upper molar (M^2) in *L. kutjamarpensis* (5.7 mm) is comparable to the length of the M_3 (5.15 mm) and M_5 (4.50 mm) in *L. kanunkaensis*. Thus, the lower teeth being referred to *Litokoala* are of the proper size to occlude with the known upper molar of *L. kutjamarpensis*. Upper molar fragments being referred to *L. kanunkaensis* are also representative of teeth comparable in size to the upper molar of *L. kutjamarpensis*. In addition, molar teeth from other known phascolarctid genera are all significantly larger than in *Litokoala*. The upper molar fragments being referred to *L. kanunkaensis* show further resemblance to *L. kutjamarpensis* and differ from other phascolarctid genera in having a well developed neometaconule (seen in UCR 21980) and in exhibiting a similar degree of selenodonty. Overall, the style of the upper and lower molars in *L. kutjamarpensis* and *L. kanunkaensis* is similar, both being selenodont and exhibiting prominent intermediate cuspules (e.g., a neometaconule on upper molars of both species and an entostylid ridge on lower molars of *L. kanunkaensis*). The association of lower molars and upper molar fragments from the same lithostratigraphic level at Lake Kanunka and their similarity in size and style suggest that they are representative of the same species. Comparisons of these teeth with *Litokoala kutjamarpensis* suggests that all of these teeth belong to *Litokoala*.

Second, both species are found within the same geographic province. Lake Kanunka and Lake Ngapakaldi are only several kilometres apart.

Third, in addition to inferences regarding stage of evolution, lithostratigraphic and biostratigraphic correlation indicate that the remains from RV-8453 and RV-8452 are geologically younger than other previously known fossils from the Etadunna Formation and are much closer in age to the Kutjamarpu Local Fauna than any other Etadunna-derived specimens (Woodburne *et al.* 1986).

Specific diagnosis: Although the lower molars of *L. kutjamarpensis* are unknown, *L. kanunkaensis* is given its own specific name. This is based on differences which are apparent in direct comparison of the M^2 of *L. kutjamarpensis* and the M^2 metacone fragment of *L. kanunkaensis*. Most notably, the posterolingual metacrista is much more distinct in *L. kutjamarpensis*. Also, the buccal margin of the tooth is not as convex in *L. kanunkaensis* as in *L. kutjamarpensis*.

However, at this time the direct comparison that can be made between *L. kanunkaensis* and *L. kutjamarpensis* is limited and the amount of intraspecific variation is unknown. Species level differences consistently occur within most other lineages shared between the Kutjamarpu and Kanunka North Local Faunas (Woodburne *et al.* 1986) suggesting that the same pattern probably occurred within species of *Litokoala*.

Description: UCR 21926, an isolated tooth, may be either an M_2, M_3 or M_4. However, comparison with as yet undescribed *Litokoala* material from the Henk's Hollow Local Fauna from Riversleigh (Archer, pers. comm.) suggests it is an M_3. The M_3 of *L. kanunkaensis* (see Fig. 1A), based on UCR 21926, is a double-rooted, selenodont, subrectangular tooth that tapers slightly anteriorly. The protoconid shows significantly more wear than any of the other major cusps (metaconid, entoconid or hypoconid). The metaconid and protoconid are approximately equal in size. A well developed, somewhat crescentic paracristid is directed anterolingually from the apex of the protoconid and extends to an indistinct paraconid at the anterobuccal corner of the tooth. The metaconid is located opposite but slightly anterior to the protoconid and is sub-

pyramidal in shape. A well developed posterobuccally directed crest extends from the apex of the metaconid to the posterior border of the trigonid basin. The premetacristid is a strongly ridged crest that is parallel to the lingual border of the tooth and extends anteriorly from the metaconid to the anterior margin of the tooth. At this point it is deflected sharply in a buccal direction to form a short anterior cingulum that connects to the paraconid. A very small intermediate cuspule is found between the metaconid and protoconid. The trigonid basin is very distinct, especially where it is bounded by the premetacristid, anterior cingulum and pre-protocristid. The postmetacristid extends posterolingually from the apex of the metaconid to a small metastylid. The preentocristid extends anteriorly from the apex of the entoconid and

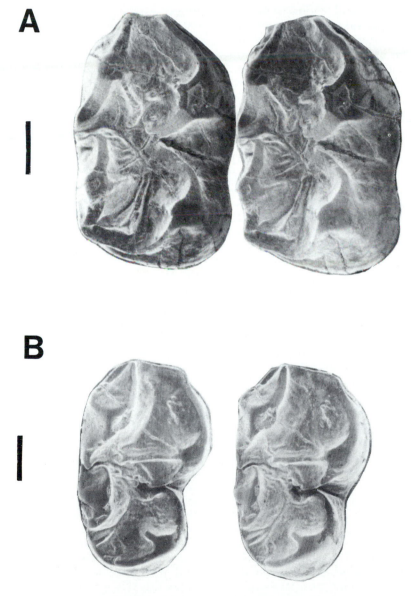

A

B

Fig. 1. A, SEM stereopair of the occlusal view of *Litokoala kanunkaensis*, holotype, UCR 21926, a right M_3. B, SEM stereopair of the occlusal view of *Litokoala kanunkaensis*, UCR 21945, a right M_5. All scale bars = 1 mm.

bifurcates before reaching the trigonid. The main branch of the preentocristid makes a sharp lingual bend from the unbranched segment of the preentocristid and connects to the metastylid. The preentocristid and postmetacristid are continuous but are not aligned en echelon as in some pseudocheirids. The weaker branch of the preentocristid extends buccally to a crevice at the anterobuccal margin of the entoconid. The hypoconid is slightly larger than the entoconid and is located at a slightly more posterior position on the tooth. The cristid obliqua is an anterobuccally convex crest that extends from the apex of the hypoconid and connects to the postprotocristid just lingual to the longitudinal axis of the tooth. The postentocristid is shorter than the preentocristid and curves lingually to a small entostylid, which is higher than the metastylid. The posthypocristid forms the posteriorly convex margin of the tooth and connects to the entostylid. An entostylid ridge arises from the buccal margin of the postentocristid and extends buccally almost as far as the longitudinal axis of the tooth before turning sharply in an anterior direction. Anteriorly, the entostylid ridge fades into the talonid basin. The entostylid ridge is moderately cuspate at the position where the kink occurs.

UCR 21945 is referred to as an M_5 (see Fig. 1B) based on its smaller size in comparison with the M_3 and on the reduction exhibited in the talonid, which is reminiscent of certain pseudocheirids. UCR 21945 (RM_5) is neatly reniform in shape and exhibits a trigonid that is distinctly larger than the talonid. The metaconid is taller than and is situated opposite but slightly anterior to the protoconid. The premetacristid and postmetacristid are well developed although the prominent posterobuccal strut arising from the apex of the metaconid in M_3 is not developed here. The metaconid is ribbed lingually. The premetacristid extends anteriorly to the anterior margin of the tooth and turns sharply to form an anterior cingulum which connects to a small but distinct paraconid. The anterior cingulum is proportionately wider in M_5 than M_3. The paracristid is buccally convex and crescentic as in the more anterior M_3 but it is not directed as lingually as in that tooth. A prominent lingual protocristid arises from the protoconid and is directed toward the metastylid. It extends almost as far as the preentocristid and connects to a short, low-lying crest that extends from the posterior margin of the metaconid. A weakly developed ridge extends anterolingually from the protoconid into the trigonid basin. This may be homologous with a much less prominent ridge arising from the protoconid of the M_3. The postprotocristid extends posterolingually from the protoconid, curves lingually and parallels the cristid obliqua and finally bends anterolingually and extends to

the base of the lingual protocristid. The cristid obliqua extends anteriorly from the hypoconid before being deflected lingually and paralleling the postprotocristid. This parallel arrangement continues for the remaining length of these cristids. The cristid obliqua finally connects to a buccal branch of the preentocristid. Small struts that radiate at right angles from the cristid obliqua and postprotocristid provide the only connection between these two crests, which are otherwise separated by a long crevice. The rectangular arrangement of the primary cusps (protoconid, metaconid, hypoconid and entoconid) in more anterior molars is much less evident in the M_5. This is primarily the result of a significant anterior shift of the entoconid (relative to the hypoconid) and to a lesser extent a posterior shift of the protoconid (relative to the metaconid). The hypoconid is subconical and is a much broader cusp than the entoconid. It exhibits a broad, moderately developed anterolingual strut that extends from its apex into the talonid basin. The entoconid is a ridge-like cusp that parallels the lingual margin of the tooth. Anteriorly, the preentocristid is divided into a buccal branch leading to the cristid obliqua and a lingual branch leading to the metastylid. Both of these connections are notched. The entostylid is not developed although a distinct entostylid ridge is present. It is short and linear and extends anterobuccally from the posterolingual corner of the tooth into a cleft that separates the entoconid from the hypoconid. The posthypocristid forms a smooth and rounded posterior tooth margin. The rounded posterolingual corner of the tooth contrasts with the more squared off posterolingual corner in the M_3.

UCR 21979 and UCR 21980 appear to represent metacones of an anterior and posterior molar, respectively. UCR 21979 is referred to as the metacone of a LM^2 (possibly LM^3) and is shown in Fig. 2A. The obtuse angle at the posterobuccal corner of the tooth, formed by the posterior cingulum and the buccal edge of tooth, is similar to that for the M^2 of *L. kutjamarpensis* and is less obtuse than in UCR 21980. In other phascolarctids the obtuse angle at the posterobuccal corner of the tooth increases posteriorly along the tooth row. UCR 21979 also represents a proportionately wider tooth than UCR 21980, providing additional support for its more anterior position in the tooth row. The length of the remaining ectoloph in UCR 21980 is protionately greater than would be expected if this fragment was a paracone. While this cannot be determined for UCR 21979 because paracone and metacone lengths are subequal on phascolarctid M^2s, the cingulum on UCR 21979 is similar to the posterior cingulum on the *L. kutjamarpensis* molar and is much less complex than the anterior cingulum on that tooth.

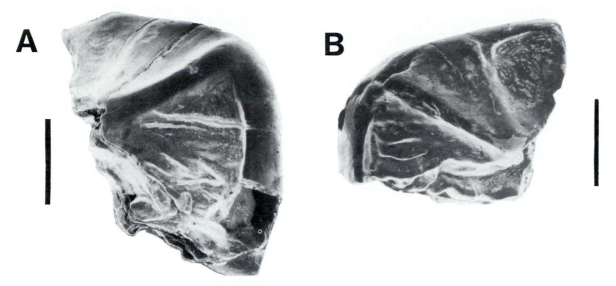

Fig. 2. A, SEM of *Litokoala kanunkaensis*, UCR 21979, a metacone of a LM^2. B, SEM of *Litokoala kanunkaensis*, UCR 21980, a metacone of a RM^4. All scale bars = 1 mm.

UCR 21979, the metacone of a LM^2 or LM^3, is broken along the anterior face of the cusp and along the longitudinal axis of the tooth. Because the anterior face of the metacone is largely missing, it is not possible to determine whether a prominent anterolingual metacrista was present. UCR 21979 is more bulbous than the metacone in UCR 21980. The posterolingual face of the cusp is highly crenulated although a distinct posterolingual metacrista is not well developed. A fine ridge extends anteriorly from the posterior cingulum but cannot be traced too far anteriorly because of breakage. It cannot be determined if this connects to fine ridges coming off of the lingual metacone face or to the neometaconule.

UCR 21980, the metacone of an RM^4 or RM^5 (see Fig. 2B), is broken along the transverse valley and along the longitudinal axis of the tooth. This metacone is proportionately more elongate than in UCR 21979. The premetacrista and postmetacrista are both well developed. There is a weakly developed buccal cingulum extending posteriorly from the mesostyle to a faint buccal rib on the metacone. Anterolingual and posterolingual metacristae are present although they are not as well developed as the pre- and postmetacristae. The premetacrista is crescentic and extends to the mesostyle, which has been broken. The postmetacrista is linear and extends to the posterobuccal corner of the tooth. The neometaconule is a finely ridged crest originating anterior to the posterior cingulum. Posteriorly, the neometaconule is linear, whereas anteriorly, it is deflected buccally before reaching the transverse valley. The anterolingual metacrista fades out just before joining the buccal deflection of the neometaconule. The posterolingual metacrista extends halfway down the metacone before

making a sharp kink in a more lingual direction and then subdividing into a series of smaller ridges. Two of these ridges extend to the longitudinal midline of the tooth while a third extends to the posterior cingulum and is paralleled by three other fine ridges originating on the posterolingual face of the metacone. Other fine ridges radiate at right angles from the neometaconule and these are paralleled by fine ridges on the lingual and anterolingual faces of the metacone. The extensive crenulation seen in this tooth fragment and in UCR 21979, contrast with its minimal development in the lower molars.

PHYLOGENETIC RELATIONSHIPS

Based on what is known of its lower dentition, species of *Litokoala* appear to be more closely related to those of *Phascolarctos* than to species of any other known phascolarctid genus. Possible synapomorphies uniting species of *Litokoala* and *Phascolarctos* include the emphasis on selenodonty, the development of an entostylid ridge (discussed below), the more lingual union of the postprotocristid and cristid obliqua (although *Litokoala kanunkaensis* shows additional modification in the M$_5$), and the connection between postmetacristid and preentocristid. *Litokoala kanunkaensis* also exhibits several autapomorphic features that serve to distinguish it from species of *Phascolarctos* and those of other phascolarctid genera and preclude it from being ancestral to these forms. These include the more prominent entostylid ridge (M$_3$ and M$_5$), the well developed posterobuccal metacristid (M$_3$), the parallel arrangement of the cristid obliqua and postprotocristid in the M$_5$ in which these cristids do not meet end to end, a reduced talonid (M$_5$), anteriorly bifurcated preentocristids, an anteriorly displaced entoconid (M_5), a posteriorly displaced protoconid (M_5) and a strong lingual protocristid (M_5). This suite of derived features may indicate that a significant period of time elapsed within the *Litokoala* clade between its divergence from a common ancestor with species of *Phascolarctos* and its first appearance in the fossil record. Thus, although species of *Phascolarctos* and *Litokoala* may be sister taxa, there still appears to be a considerable amount of divergence between them.

Features of the lower molars shared by species of *Litokoala*, *Madakoala* and *Perikoala* but not found in species of *Phascolarctos* include a relatively broad anterior cingulum, an essentially linear preentocristid, relatively greater transverse distances between the protoconid and metaconid and between the entoconid and hypoconid and the lack of a buccal cingulum blocking free exit of the transverse valley. All of these features are interpreted to be symplesiomorphic and indicate no special affinity between these genera. Derived character states for these characters are found in species of *Phascolarctos*.

Based on the known upper molar of *Litokoala kutjamarpensis*, Archer (1984b) grouped species of *Perikoala*, *Phascolarctos* and *Koobor* in a clade separate from *Litokoala kutjamarpensis* which was regarded as the most plesiomorphic of known phascolarctids. Based on the same upper molar, Woodburne *et al.* (1987) suggest that *Litokoala kutjamarpensis* is most closely related to species of *Phascolarctos*. The lower molars of *L. kanunkaensis* provide additional support for this latter hypothesis. Species of *Perikoala* and *Madakoala* may represent another clade in which lophodonty has been emphasised rather than selenodonty. Another potential synapomorphy uniting species of *Madakoala* and *Perikoala* is the presence of a protostylid ridge on M_3, although its variable presence in certain species of *Madakoala* may argue against this. According to this interpretation, both lophodonty and selenodonty are derived character states related to increased masticatory efficiency. Lophodonty and selendonty may both be derived from an ancestor with a more bandicoot-like dentition. However, Woodburne *et al.* (1987) consider incipient lophodonty to be primitive within the Phascolarctidae and suggest that species of *Perikoala*, *Litokoala* and *Phascolarctos* form a clade to the exclusion of *Madakoala* species. Several potential synapomorphies are suggested to unite this clade. In addition, Archer (1976, 1984b) considered selenodonty to be the primitive condition in the Phascolarctidae and suggested that the incipient transverse lophodonty seen in *Perikoala palankarinnica* was derived from this. Elucidation of the polarity of selenodonty and lophodonty and the ancestral state within

the Phascolarctidae will require additional material before a more conclusive phylogenetic interpretation can be made.

In the modern Koala, the entostylid ridge is most prominent on M_2. In this tooth, the entostylid ridge exhibits a fairly sharp bend so that it extends transversely from the postentocristid, paralleling the posterior tooth margin, and then turns in an anterior direction and runs along the longitudinal axis of the tooth. Whereas this entostylid ridge is not well developed (and less so on M_3), the approximately right-angle bend is similar to that found on the M_3 of *L. kanunkaensis*. Pseudocheirids, which may exhibit a prominent entostylid ridge buccal to the entoconid on the M_3 and M_4 (e.g., *P. cupreus*), are similar to *L. kanunkaensis* in this respect. However, when this ridge is present it tends to be linear or only slightly kinked, forming a much larger obtuse angle than the approximately right angle found in *Phascolarctos cinereus* or *L. kanunkaensis*. In addition, the orientation of the entostylid ridge in pseudocheirids tends to be more oblique, with respect to the longitudinal axis of the tooth, than in the phascolarctid species exhibiting an entostylid ridge (except M_5 in *L. kanunkaensis*).

In UCR 21926 (M_3 or possibly M_4), the postprotocristid extends posterolingually from the apex of the protoconid and meets the cristid obliqua to form a buccally open "V". This is similar to the arrangement of the cristid obliqua and postprotocristid for the M_3 and M_4 in species of *Phascolarctos*, *Perikoala* and *Madakoala*. In most fossil and extant pseudocheirids (e.g., *Pseudochirops albertisii*, *Pseudochirops cupreus*, *Petauroides volans*, *Pseudocheirus forbesi* and *Pseudocheirus peregrinus*), the postprotocristid is oriented transversely and extends from the apex of the protoconid toward the metastylid. In these forms the cristid obliqua extends anterolingually from the hypoconid to the transversely oriented postprotocristid.

The M_5 of *L. kanunkaensis* bears a striking resemblance to the M_5 of certain pseudocheirids. In fact, it probably shows more resemblance to *Petauroides volans* than to any known phascolarctid. The similarities include the reduced talonid, anteriorly shifted entoconid, posteriorly shifted protoconid and the arrangement of the cristids originating from the protoconid. In *P. volans*, there is a strong postprotocristid from the protoconid that is directed toward the metastylid. There is also a short posterior cristid from the protoconid apex that joins the cristid obliqua. Whereas this latter cristid is not present in more anterior molars (although it is present in *Pseudocheirus herbertensis*), the transversely orientated, metastylid-directed, postprotocristid is very prominent (as in other living pseudocheirids). Thus, the short cristid in the M_5 of *P. volans* that connects to the cristid obliqua, and which is not present in more anterior molars, may be regarded as a neomorph. In the two molars referred to *L. kanunkaensis*, assuming they represent the same taxon, the situation is reversed; the posterolingually directed post-protocristid in M_5 has a posterolingually directed counterpart in the M_3, whereas the strong, metastylid-directed protocristid (M_5) does not have a transversely oriented counterpart in the M_3. Thus, in this case the metastylid-directed protocristid may be the neomorph and the posterolingually directed postprotocristid may actually be homologous with the postprotocristid in more anterior molars.

CONCLUSION

Litokoala kanunkaensis is significant in that the teeth ascribed to this species represent the first lower molars known for the species of *Litokoala*. This species also extends the known temporal range of *Litokoala* and represents the first occurrence of *Litokoala* within the Etadunna Formation. *Litokoala kanunkaensis* represents the third phascolarctid genus known from the Etadunna Formation, the other two being represented by species of *Madakoala* and *Perikoala*, which are known from older stratigraphic horizons. The appearence of *L. kanunkaensis* in the Lake Eyre Basin during the medial Miocene may be related to the disappearance of the more lophodont phascolarctids but at this time this cannot be determined.

ACKNOWLEDGEMENTS

The specimens described herein were discovered in 1985 on a field expedition led by Dr M. O. Woodburne and supported by the National Science Foundation. I would also like to thank all members of the 1985 field party (J. A. Case, B. J. MacFadden, T. MacFadden, K. B. Johnson, J. D. Power and J. Woodburne) for their dedicated efforts. Dr M. O. Woodburne provided help, advice and criticism during the preparation of this manuscript. Dr M. Archer also provided help and advice involved in the analysis and in the preparation of this manuscript. The SEM work was done by Marcia Kooda-Cisco and J. A. Case, to whom I am very grateful. A Philips 515 Scanning Electron Microscope was used for all SEM photographs, which were shot at 5KV.

REFERENCES

ARCHER, M., 1976. Phascolarctid origins and the potential of the selenodont molar in the evolution of diprotodont marsupials. *Mem. Qd Mus.* **17**: 367-71.

ARCHER, M., 1978. The nature of the molar-premolar boundary in marsupials and a reinterpretation of the homology of marsupial cheekteeth. *Mem. Qd Mus.* **18**: 157-64.

ARCHER, M., 1984a. The Australian marsupial radiation. Pp. 633-808 *in* "Vertebrate zoogeography and evolution in Australasia" ed by M. Archer and G. Clayton. Hesperian Press: Western Australia.

ARCHER, M., 1984b. On the importance of being a koala. Pp. 809-15 *in* "Vertebrate zoogeography and evolution in Australasia" ed by M. Archer and G. Clayton. Hesperian Press: Western Australia.

STIRTON, R. A., TEDFORD, R. H. AND WOODBURNE, M. O., 1967. A new Tertiary formation and fauna from the Tirari Desert, South Australia. *Rec. S. Aust. Mus.* **15**: 427-62.

TEDFORD, R. H. AND WOODBURNE, M. O., 1987. The Ilariidae, a new family of vombatiform marsupials from Miocene strata of South Australia and an evaluation of the homology of molar cusps in the Diprotodontia. Pp. 401-18 *in* "Possums and opossums: studies in evolution" ed by M. Archer. Surrey Beatty & Sons and the Royal Zoological Society of New South Wales: Sydney.

WOODBURNE, M. O., TEDFORD, R. H., ARCHER, M. AND PLEDGE, N., 1987. *Madakoala*, a new genus and two species of Miocene Koalas (Marsupialia: Phascolarctidae) from South Australia and a new species of *Perikoala*. Pp. 293-317 *in* "Possums and opossums: studies in evolution" ed by M. Archer. Surrey Beatty & Sons and the Royal Zoological Society of New South Wales: Sydney.

WOODBURNE, M. O., MACFADDEN, B. J., CASE, J. A., SPRINGER, M. S., PLEDGE, N. S., POWER, J. D., WOODBURNE, J. M. AND JOHNSON, K. B., 1986. Land mammal biostratigraphy and magnetostratigraphy of the Etadunna Formation (medial Miocene) of South Australia: preliminary report. *Proc. Roy. Soc. S. Aust.* (in press)

Phascolarctos maris, A NEW SPECIES OF KOALA (MARSUPIALIA: PHASCOLARCTIDAE) FROM THE EARLY PLIOCENE OF SOUTH AUSTRALIA

NEVILLE S. PLEDGE[1]

A new species of Koala, *Phascolarctos maris,* is recorded in the predominantly marine Sunlands Faunal Assemblage of the Loxton Sands, of early Pliocene age, near Waikerie, South Australia. It is represented by a single lower molar tooth which is distinguished from the teeth of *Phascolarctos cinereus* mainly by its much greater size. It is the oldest known representative of the genus.

Key Words: Phascolarctidae; Pliocene; *Phascolarctos maris*; South Australia.

Pages 327-30 *in* POSSUMS AND OPOSSUMS: STUDIES IN EVOLUTION ed by M. Archer. Surrey Beatty & Sons and the Royal Zoological Society of New South Wales: Sydney, 1987.

INTRODUCTION

THE SUNLANDS Local Fauna is predominantly an assemblage of marine vertebrate fossil species, dominated by sharks (Pledge 1984), that accumulated on a near-shore oysterbank, now part of the Loxton Sands in the Murray Basin. A major stream apparently debouched nearby, supplying pebbles, grits of quartz and metamorphic rock and occasional marsupial remains to the reef.

The species described here is the third marsupial taxon to be recognised in the Sunlands Local Fauna. The previous two, however, are as yet too poorly known to be named.

SYSTEMATICS

Order Diprotodonta Owen, 1866

Family Phascolarctidae Owen, 1839

Genus *Phascolarctos* Blainville, 1816

Phascolarctos maris sp. nov.
(Figs 1-3).

Holotype: SAM P24801, a lower left molar, M_3 or $_4$.

Type Locality: 200 m upstream from the Sunlands Pumping Station, on the River Murray, 8 km west of Waikerie, South Australia.

Age: Early Pliocene (Kalimnan), Loxton Sands (Ludbrook 1961, 1973).

Etymology: Maris (Lat.) means "of the sea" and is in reference to its provenance and indirectly to its finder, Mr J. van der Meer.

Diagnosis: This species is about 15% larger than *Phascolarctos cinereus*. Otherwise it is indistinguishable from *P. cinereus* except for its relatively sharper cristids, stronger transverse apical ribs on principal cusps, weaker metastylid and presence of a linear entoconulid crest. It is

evidently smaller than *Phascolarctos stirtoni* (by about 11%) although it is not yet known from lower teeth.

Description: Tooth is large, rectangular and selenodont. It is about 15% larger than the equivalent M_3 or M_4 of *P. cinereus*. Pre- and post-cristids on the protoconid and hypoconid form a broad W. Those on the metaconid and entoconid

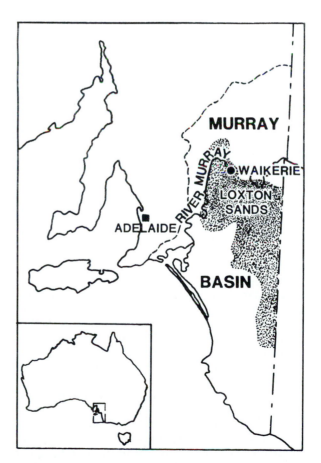

Fig. 1. Locality Map, Sunlands Local Fauna.

---

[1]South Australian Museum, Adelaide, South Australia, Australia 5000.

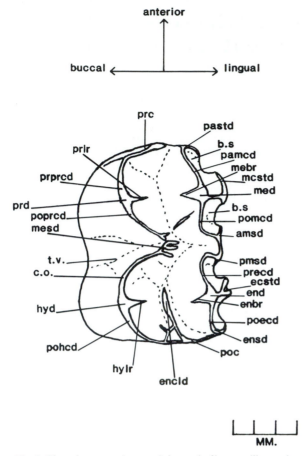

Fig. 2. Phascolarctos maris n. sp. Schematic diagram illustrating terminology used. Abbreviations as follows: amst, anterior branch of metastylid; b.s., basal shelf; c.o., cristid obliqua; ecstd, entoconid columnar stylid; enbr, buccal rib of entoconid; encld, entoconulid; end, entoconid; ensd, entostylid; hylr, lingual rib of hypoconid; hyd, hypoconid; mcstd, metaconid columnar stylid; mebr, buccal rib of metaconid; med, metaconid; mesd, mesostylid; pamcd, parametacristid; past, parastylid; pmsd, posterior branch of metastylid; poc, postcingulum; poecd, postentocristid; pohcd, posthypocristid; pomcd, postmetacristid; poprcd, postprotocristid; prlr, lingual rib of protoconid; prc, precingulum; prd, protoconid; precd, preentocristid; prprcd, preprotocristid; t.v., transverse valley.

are almost linear, with a slight lingual offset at the transverse valley. All cristids curve lingually towards the base of the crown, the premetacristid (or parametacristid) forming a small parastylid and the postentocristid forming an entostylid. Basal shelves exist at the base of the lingual faces of both meta- and entoconid, being most prominent just posterior to the parastylid. The postmetacristid and pre-entocristid curve lingually and meet at the metastylid, a rather complicated U-shaped structure with a short anterior spur and a smaller posterior rib. Both metaconid and entoconid are "double-cusped", the accessory cusp in each case being a columnar stylid slightly lower on the lingual side and joined to the main cusp by a short transverse crest. On the metaconid, this is slightly posterior to the apex

and joins the linear columnar stylid at its posterior end. There is a distinct sharp rib on the buccal side of the metaconid apex. A similar rib is present on the entoconid and continues directly across the apex as the linking rib to the bifid columnar stylid. There is a small centrally-directed rib at the base of the postmetacristid and two on the preentocristid, that converge with, but do not meet, a spread of short ribs on the mesostylid where the post-protocristid and cristid oblique meet.

A novel feature is a linear crest at the buccal base of the entoconid, interpreted as the entoconulid. It extends for about two thirds of the distance from the postcingulum, trending slighty buccally, to the mesostylid, and continues as the lingual end of the postcingulum. A short, subparallel cristid approaches it from the posterior end of the posthypocristid.

Protoconid and hypoconid are high, subequal and subpyramidal. Each has a strong lingual rib on the apex. The preprotocristid swings sharply lingually near the base of the cusp, then anteriorwards at the midline of the tooth, to meet the lingual precingulum. A lower, buccal precingulum joins at this point.

The buccal end of the transverse valley has fewer crenulations and cuspules than the living *P. cinereus*. There are two low subparallel transverse ridges that merge into the posterior base of the protoconid.

DISCUSSION

The most notable feature of this species is its size. Morphological differences from the M_4 of *P. cinereus* are few and minor, being mostly ones of degree. The exceptions are the development of the ridge-like entoconulid and the somewhat cuspate posterobuccal rib at the base of the postmetacristid. By comparison with *P. cinereus*, where equivalent upper and lower molars are almost the same length, it can be seen that *P. maris* was almost as big as *P. stirtoni* (Bartholomai 1968), a species known only from a partial maxilla (with P^3 to M^3) from a presumed Pleistocene cave deposit in southeastern Queensland. The length of M^3 of *P. stirtoni* is 10.7 mm, presumably also the length of M_3. This compares with 8.2 mm for *P. cinereus* and 9.5 mm for *P. maris*. Comparison with upper and lower dentitions of *P. cinereus* also indicates there are no distinctive characters between upper and lower teeth that could distinguish between two different species that do not preserve comparable elements. Thus we are left with age and size differences. It has been suggested to the author that because some Late Cainozoic macropodid species extended from the (late) Pliocene to the Pleistocene with negligible change, erecting a new species for the Sunlands specimen was not

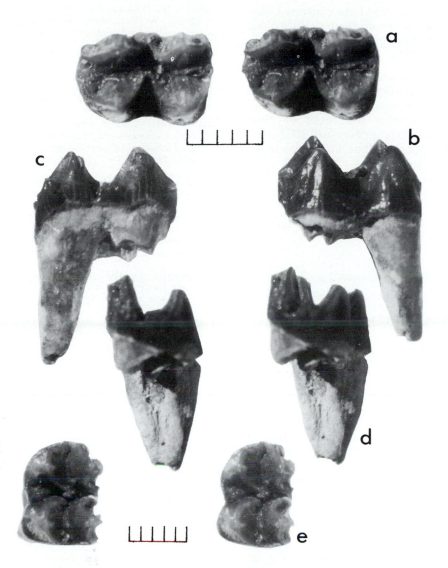

Fig. 3. Phascolarctos maris n. sp. Holotype, SAM P24801. L M$_3$ or $_4$. a, occlusal view, stereopair; b, buccal face; c, lingual face; d, anterior end, stereopair; e, occlusal view,
rotated 90° stereopair. Scale in millimetres.

justified and that it should be referred to *P.*
stirtoni. However, in view of the great age difference of some three million years between *P. maris*
and *P. stirtoni,* the fact that both are phascolarctids, a family that during the Late Miocene
proved to be quite diverse (Woodburne *et al.*
1987) and because there is a size difference
between the Sunlands phascolarctid and *P. stir-*
toni, I consider *Phascolarctos maris* to be a species
distinct from *P. stirtoni*.

Additional, circumstantial evidence for this conclusion has recently come to light in the form of a
left dentary of a Phascolarctid from a cave at Curramulka, South Australia. All four lower molars are
preserved and are, each and in total, more than
half as long again as their modern equivalents. It
is not possible to test possible occlusion with the
P. stirtoni maxilla, which is from the right side, but
the dentary is obviously larger. Because this may
be an example of Bergmann's Rule relating to

latitudinal clines, this dentary is tentatively
refered here to *P.* sp. cf. *P. stirtoni.*

Comparison of the M$_4$ of the Curramulka
specimen with the type of *P. maris* shows several
morphological differences, including the poor
development of the columnar stylids, the absence
of an entoconulid ridge, as well as a noticeably
greater size. The age of the Curramulka specimen is as yet unknown.

Phascolarctos maris is thus outstanding amongst
Phascolarctids, being much larger than any known
Pliocene or older species as well as *P. cinereus* and
the oldest known representative of *Phascolarctos*.
It should be noted that Koalas are not common in
Pliocene deposits: none is known in the Hamilton
Local Fauna of western Victoria (Turnbull and
Lundelius 1970), which is roughly contemporan-
eous with Sunlands and otherwise rich in possums.
Only species of the rather aberrant *Koobor* are

known, *K. notabilis* from the Late Pliocene Chinchilla Sands and *K. jimbarratti* from slightly older deposits in north Queensland (Archer 1977; Archer and Wade 1976).

It is noteworthy that the early species of *Phascolarctos* has been predicted by Archer (1984: 813). The relatively small size of the living *P. cinereus* is presumably the result of post-Pleistocene diminution (e.g., Marshall 1973:169) as larger variants occur in Late Pleistocene cave deposits (pers. obs.).

Table 1. Dimensions (mm) of Phascolarctid lower molars.

Species		Length	Anterior Width*	Posterior Width*
Phascolarctos maris n. sp.				
SAM P24801	$m_{3 \text{ or } 4}$	9.50	7.2	7.2
P. sp. cf. *P. stirtoni*	M_2	11.8	7.6	9.0
SAM P24904	M_3	12.9	8.5	8.8
	M_4	12.9	8.7	8.9
	M_5	12.7	8.6	7.8
P. cinereus				
	M_3 (N=9)	8.2	5.3	5.1
	M_4 (N=14)	8.2	5.1	4.9

* maximum width parallel to base of enamel.

ACKNOWLEDGEMENTS

Many thanks go to Mr Joe van der Meer of Waikerie for bringing the Sunlands Local Fauna and this specimen to my attention. Mrs Joan Murphy and Janine Flory typed the manuscript. Drawings by Kathy Bowshall and the author.

REFERENCES

Archer, M., 1977. *Koobor notabilis* (De Vis), an unusual koala from the Pliocene Chinchilla Sand. *Mem. Qd Mus.* **18:** 31-35.

Archer, M., 1984. On the importance of being a koala. Pp. 809-15 *in* "Vertebrate zoogeography and evolution in Australasia" ed by M. Archer and G. Clayton. Hesperian Press: Perth.

Archer, M. and Wade, M., 1976. The results of the Ray E. Lemley Expeditions, Part 1. The Allingham Formation and a new Pliocene vertebrate fauna from northern Queensland. *Mem. Qd Mus.* **17:** 379-97.

Bartholomai, A., 1968. A new fossil koala from Queensland and a reassessment of the taxonomic position of the problematical species, *Koalemus ingens* De Vis. *Mem. Qd Mus.* **15:** 65-73.

Ludbrook, N. H., 1961. Stratigraphy of the Murray Basin in South Australia. *Geol. Surv. S. Aust. Bull.* **36:** 1-96.

Ludbrook, N. H., 1973. Distribution and stratigraphic utility of Cenozoic molluscan faunas in southern Australia. *Tohoku Univ. Sci. Rep. 2nd Ser.* (Geol.) *spec. Vol. 6 (Hatai Memorial Volume):*241-61.

Marshall, L. G., 1973. Fossil vertebrate faunas from the Lake Victoria region, S.W. New South Wales, Australia. *Mem. Nat. Mus. Vict.* **34:** 151-71.

Pledge, N. S., 1985. An early Pliocene shark tooth assemblage in South Australia. *Spec. Publ., S. Aust. Dept. Mines and Energy* **5:** 287-99.

Rich, T. H., Archer, M., Plane, M. D., Flannery, T. F., Pledge, N. S., Hand, S. and Rich, P., 1982. Australian Tertiary mammal localities *in* "The fossil vertebrate record of Australasia" ed by P. V. Rich and E. M. Thompson. Monash Univ. Offset Print. Unit: Clayton.

Turnbull, W. D. and Lundelius, E. L. Jr, 1970. The Hamilton Fauna. A late Pliocene mammalian fauna from the Grange Burn, Victoria, Australia. *Fieldiana: Geology* **19:** 1-161.

Woodburne, M. O., Tedford, R. H., Archer, M. and Pledge, N. S., 1987. *Madakoala,* a new genus and two new species of Miocene koalas (Marsupialia: Phascolarctidae) from South Australia and a new species of *Perikoala*. Pp. 293-317 *in* "Possums and opossums: studies in evolution" ed by M. Archer. Surrey Beatty & Sons and the Royal Zoological Society of New South Wales: Sydney.

A BRAIN THAT DOESN'T FIT ITS SKULL:
A COMPARATIVE STUDY OF THE BRAIN
AND ENDOCRANIUM OF THE KOALA,
Phascolarctos cinereus
(MARSUPIALIA: PHASCOLARCTIDAE)

JOHN R. HAIGHT[1] and JOHN E. NELSON[2]

Endocranial volumes are good measures of brain size in mammals, especially those of small to medium size. The Koala provides an exception to this rule. With a mean brain volume ($\cong$ mass) of 16.7 ml and a mean endocranial volume of 27.4 ml the Koala's brain tissue occupies only 61% of its endocranial cavity though through ventricular enlargement and judicious shaping of its overall morphology the brain does attempt to fill most of its endocranial cavity. The gross morphology of the Koala's brain is, in consequence, unusual. The cerebral hemispheres are much reduced and the lateral ventricles much enlarged when compared with those of other, comparably sized, marsupials. The cerebellum is also smaller than would be expected. Microanatomical examination shows a reduction in the size of the parietofrontal or motor-somaesthetic cortex. Cortical lamination is unusual in much of this area with a reduced layer 5 and a virtually absent layer 6. The overall thickness of parietofrontal cortex is less than would be expected on allometric grounds. Other areas of the cerebral mantle are histologically normal. The striatum and hippocampus are reduced in size as are the deep cerebellar nuclei. In the thalamus the structure of the ventroposterior nucleus is unusual and is organised unlike that of other marsupials. The visual and auditory systems appear well-developed at both thalamic and cortical levels. Other motor centres such as the red nucleus and cranial nerve nuclei are normal in size and appearance. Overall, we conclude that the Koala displays a reduction of certain forebrain and cerebellar motor centres and their connecting pathways when these centres are compared with those found in other, related marsupials.

Key Words: Allometry; Brushtail Possum; Cerebellum; Comparative Neuroanatomy; Motor systems; Neocortex; Sensory systems; Thalamus; Wombat.

Pages 331-52 *in* POSSUMS AND OPOSSUMS: STUDIES IN EVOLUTION ed by M. Archer. Surrey Beatty & Sons and the Royal Zoological Society of New South Wales: Sydney, 1987.

" . . . . a bear of very little brain."
(A. A. Milne, The House at Pooh Corner)

INTRODUCTION

THE gross morphology of the Koala brain was first described by Ziehen (1897). He noted that cerebral surface features, aside from the rhinal sulcus, were not prominent, the few sulci present serving as shallow channels for surface blood vessels. Overall, he claimed, the Koala cerebrum resembled that of the much smaller Ringtail Possum, *Pseudocheirus peregrinus*, a marsupial with which the Koala was long thought to be linked at the subfamilial level (Jones 1924; Simpson 1945), primarily because of their similar selenodont dentitions. These dental similarities are now thought to be convergent, and the Koala is currently considered to be most closely related to the wombats though it is placed in a separate, monotypic family, the Phascolarctidae (see Kirsch 1977; Archer 1984 for a review of marsupial relationships).

Smith (1902a) also commented upon the "paucity of sulci in such a relatively large neopallium" while Jones and Porteus (1929), more correctly, commented on the "ridiculously small" size of the Koala's brain, especially the cerebral hemispheres. For when compared with other Australian marsupials of comparable size the Koala's brain stands in striking contrast (Figs 1-3). On allometric grounds the Koala's brain is considerably smaller than would be expected for a diprotodont marsupial of its size, and its endocranial cavity is much larger than the brain it contains. The cerebral hemispheres are small and shrunken in appearance and fail to meet over much of the midline, thus exposing not only the midbrain, but the entire diencephalon and other deep lying forebrain structures as well. Though essentially lissencephalic, the hemispheres present a knobly or warty appearance which is not seen in other marsupials, a number of whose brains have been illustrated recently by Johnson (1977) and by Haight and Murray (1981). The cerebellar cortex, though well fissured, also appears to be reduced in size. However, the brainstem presents no unusual gross anatomical features. To the authors' knowledge, the microanatomy of the Koala brain has not been previously described. The present paper addresses itself to some of the underlying microanatomical bases of this brain's unusual gross morphology. In particular we have examined specified structures and systems within

FOOTNOTE: Some of the findings presented in this study were given to the Lone Pine Koala Symposium held in Brisbane, Queensland, August 21-22, 1981 and sponsored by the Department of Veterinary Anatomy at the University of Queensland and the Lone Pine Koala Sanctuary.
[1]Department of Anatomy. The University of Tasmania, Hobart, Tasmania, Australia 7001.
[2]Department of Zoology, Monash University, Clayton, Victoria, Australia 3168.

the Koala's central nervous system in order to determine whether any peculiarities are general, affecting all of a major brain region more or less equally, or whether particular systems and areas were affected more than others. For clearest exposition this approach will involve comparisons with other, related, marsupial species. Thus, in addition to the Koala this paper will serve to introduce the Common Wombat (*Vombatus ursinus*), the Koala's closest living relative, and whose large and well-developed brain contrasts remarkably with that of the Koala. The results will also show that the overall histological and cytoarchitectural appearance of this brain is, for the most part, typically "marsupial". Nonetheless, there is a marked reduction of certain forebrain and cerebellar centres, and significant portions of the neocortical mantle and ventral thalamic tier are atypical in cytoarchitecture.

MATERIALS AND METHODS

Materials

Most of the marsupial crania and brains used in this study were collected in eastern Australia and on the island of Tasmania. The *Planigale* specimens came from the Northern Territory of Australia, and the cuscuses of the genus *Phalanger* came from the island of New Guinea. Approximately half the crania were obtained from the National Reference Collection held by the Division of Wildlife and Rangelands Research of the Commonwealth Scientific and Industrial Research Organisation (CSIRO) located in Canberra. Other crania were borrowed from the Tasmanian Museum and Art Gallery in Hobart and from the Institute of Anatomy in Canberra. Additional material was provided from the collections held by the Zoology Departments at the University of Tasmania and Monash University and from the Department of Prehistory in the Research School of Pacific Studies at the Australian National University.

The marsupial brain data had been collected over a number of years by the authors. However, some of the brain and body mass figures were supplied from Professor J. I. Johnson's collection in the Department of Anatomy at Michigan State University, East Lansing, Michigan, U.S.A.

For the Koala only data taken from the larger southern variety were used in this study though we did have occasion to examine two individuals of the smaller northern subspecies. All Koala brains and crania were originally collected from Victoria and from Kangaroo and French Islands off the South Australian and Victorian coasts.

Body Mass Measurements

All body mass figures were taken from animals whose brain mass *or* endocranial volume were also known. The sample consisted of mature animals; the few exceptions are noted below. Efforts were made to include both sexes in equal proportion, especially with the sexually dimorphic species. Some of the antechinuses (*Antechinus* spp.), Common Brushtail possums (*Trichosurus vulpecula*) and Tammar wallabies (*Macropus eugenii*) had been reared in captivity, but the sample also included wild-caught examples of these species. In the case of *Antechinus flavipes* the specimens which provided our brain mass measurements (as well as body mass measurements) came from a laboratory colony which had been breeding in captivity for several generations. The remaining animals were all captured from the wild and were weighed at or shortly after capture.

For the male representatives of our *Antechinus* sample we tried to include only mature specimens. These we defined as those captured between June and August prior to the annual male die off. Data from male antechinuses captured outside these months were not used unless these were the only specimens available as was the case for most of the *A. flavipes* specimens. Female antechinuses were accepted from any time of the year as long as their weights were within the normal adult range or, in the case of laboratory reared specimens, they were known to have produced a litter.

Endocranial Volume Measurements

As with body mass measurements, these data were taken, where possible, from mature animals. Both sexes are represented though unsexed material was also used where physical size and dental evidence indicated probable maturity. Prior to measurement each endocranium was cleared of all debris and the larger foramina were sealed with adhesive tape or plasticine. For larger skulls number eight lead shot (mean diameter = 2.3 mm; 6.66 g/ml) was poured into the foramen magnum until the cavity was filled. The skull was gently tapped to consolidate the shot and more was added until the cavity was completely filled. The shot was poured into a beaker, weighed and the figure obtained was converted into a volume. For smaller skulls the procedure was the same except that the glass beads (0.4-0.5 mm in diameter; 1.78 g/ml) were used in place of the lead shot.

Brain Mass Measurements

Brain mass data were obtained from animals which had been collected specifically for the purpose or which were destined for other experimental procedures. The animals were deeply anaesthetised and perfused through the heart with physiological saline followed by a mixture of 4% formaldehyde in physiological saline which latter solution also served as the preservation medium. The brains were removed,

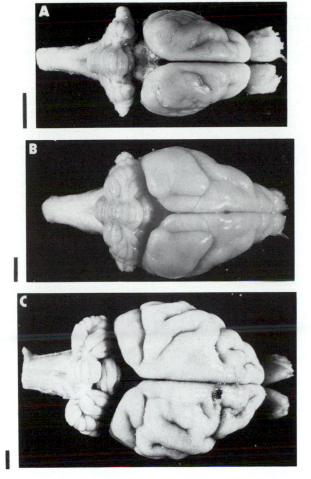

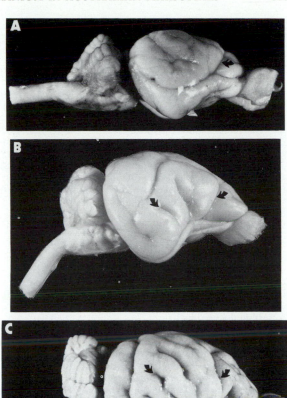

Fig. 1. Dorsal views of three marsupial brains: A. Tasmanian Devil, *Sarcophilus harrisii*; B. Tasmanian Pademelon, *Thylogale billardierii*; C. Common Wombat, *Vombatus ursinus*. Marker bars = 10 mm.

Fig. 2. Left lateral views of marsupial brains shown in Figure 1. The three rostrally located arrows on each cerebral hemisphere identify sulcus α, also known as the orbital sulcus. This sulcus delimits the rostral border of the motor-somaesthetic cortex. The caudal arrows in B and C indicate sulcus β, which forms the posterolateral boundary of the motor-somaesthetic region in diprotodont marsupials. This boundary is not known in *Sarcophilus*. Marker bars = 10 mm.

usually within a few days, and, after removal of the meninges, were weighed and photographed. Only brains which had been prepared in this way were used in this study. Other fixative and preservative materials such as glutaralderhyde and alcohols cause substantial changes in gross brain volume and are not suitable for purposes of comparative mass determinations. Even brains prepared in the above manner have been observed to change volume by up to 10% after a few months in storage. Brains, fixed in the manner described have densities very near 1.0 g/ml, hence we assume throughout this study that brain mass = brain volume.

Histological Procedures

Most of the comparative histological examples used in this study have been taken primarily from the Common Wombat, *Vombatus ursinus*, and the Common Brushtail Possum, *Trichosurus vulpecula*, and, of course, from the Koala (Figs 10-13 and 15-19). Illustrative examples from other marsupial species are used sparingly and only to focus attention upon specific regional anomalies (Fig. 13). These brains had been removed as

described above, dehydrated in graduated alcohols, embedded in celloidin and cut in the frontal plane at 30 μm. Alternate sections were stained for Nissl substance (thionin) or myelinated fibres (weil-haematoxylin). These procedures cause the brain to shrink approximately 20-25% in linear dimension. Nonetheless, *all the examples shown herein were identically processed and are directly comparable in terms of relative, if not absolute, dimension.*

RESULTS

Body mass, brain mass and endocranial volume data were collected for 33 Australasian marsupial species distributed within eight families and two orders. Body masses were obtained from 283 individuals, brain masses from 156 individuals and 328 endocranial volumes were measured. The body mass and sex were known for all but four of the 156 animals for

Table 1. Data summary of marsupials used in this study.

Order, species and family	BoM (N, SE) in gms	BrM (N, SE) in gms	ECV (N, SE) in ml	ECV/BrM × 100	E
Polyprotodonta (basal group)					
1. *Planigale maculata* (Dasyuridae)	9.45 (19, 0.59)	0.219 (4, —)	0.173 (15. 0.005)	79.0	61.9
2. *Sminthopsis murina* (Dasyuridae)	16.7 (9, 1.61)	0.434 (3, —)	0.398 (6, —)	91.7	88.3
3. *Sminthopsis leucopus* (Dasyuridae)	18.3 (2, —)	0.550 (2, —)	na	na	106.2
4. *Antechinus stuartii* (Dasyuridae)	24.0 (25, 1.52)	0.627 (6, —)	0.571 (19, 0.013)	91.1	109.7
5. *Antechinus flavipes* (Dasyuridae)	40.6 (8, 1.93)	0.746 (4, —)	0.633 (8, 0.011)	84.9	90.8
6. *Anthechinus swainsonii* (Dasyuridae)	41.7 (23, 1.22)	0.951 (1, —)	0.827 (26, 0.012)	87.0	114.0
7. *Dasyurus viverrinus* (Dasyuridae)	1217 (12, 368)	6.31 (12, 0.16)	6.19 (12, 0.13)	98.1	107.2
8. *Dasyurus maculatus* (Dasyuridae)	2768 (9, 307)	8.85 (2, —)	10.1 (18, 0.22)	114.1	93.3
9. *Sarcophilus harrisii* (Dasyuridae)	5143 (15, 350)	14.0 (70.5)	16.5 (38, 0.3)	117.8	103.3
10. *Perameles gunnii* (Peramelidae)	828 (3, —)	5.22 (3, —)	4.90 (6, —)	93.9	110.9
11. *Perameles nasuta* (Peramelidae)	1272 (6, —)	5.79 (2, —)	4.77 (7, 0.20)	82.4	95.9
12. *Isoodon obesulus* (Peramelidae)	1072 (8, 133)	4.96 (8, 0.23)	4.87 (10, 0.23)	98.2	90.8
Diprotodonta (possum-wombat-koala group)					
13. *Cercartetus lepidus* (Burramyidae)	6.4 (2, —)	0.376 (1, —)	0.32 (1, —)	85.1	133.4
14. *Gymnobelideus leadbeateri* (Petauridae)	104 (2, —)	2.39 (1, —)	2.43 (1, —)	101.7	168.7
15. *Petaurus breviceps* (Petauridae)	125 (6, —)	2.69 (5, —)	2.33 (6, —)	86.7	170.1
16. *Petaurus norfolcensis* (Petauridae)	237 (7, 12.8)	3.07 (2, —)	2.96 (7, 0.20)	96.4	134.3
17. *Pseudocheirus peregrinus* (Pseudocheiridae)	844 (6,—)	5.27 (3, —)	4.95 (11,0.12)	93.9	110.4
18. *Petauroides volans* (Pseudocheiridae)	1181 (5, —)	4.91 (1, —)	4.17 (4, —)	84.9	84.9
19. *Wyulda squamicaudata* (Phalangeridae)	1463 (2, —)	na	9.7 (2, —)	na	(148.3)[3]
20. *Trichosurus caninus* (Phalangeridae)	2899 (9, 11)	12.2 (2, —)	11.8 (7, 0.6)	96.7	125.3
21A. *Trichosurus vulpecula* (Phalangeridae)[1]	1630 (14, 74)	10.2 (14, 0.3)	9.4 (1, —)	92.2	146.6
21B. *Trichosurus vulpecula* (Phalangeridae)[2]	3231 (24, 194)	12.9 (22, 0.3)	12.7 (12, 0.3)	98.4	124.7
22. *Phalanger orientalis* (Phalangeridae)	2442 (3, —)	na	9.4 (3,—)	na	(107.2)[3]
23. *Phalanger maculatus* (Phalangeridae)	3025 (3,—)	na	15.9 (4, —)	na	(159.2)[3]
24. *Phascolarctos cinereus* (Phascolarctidae)	9614 (7, 928)	16.7 (11, 0.6)	27.4 (16, 0.9)	164.1	84.3
25. *Vombatus ursinus* (Vombatidae)	15979 (6, —)	55.7 (6, —)	59.6 (14, 2.35)	107.0	212.8
Diprotodonta (kangaroo group)					
26. *Potorous tridactylus* (Potoroidae)	1445 (11, 93)	12.1 (6, —)	11.7 (7, 0.5)	91.7	186.2
27. *Bettongia gaimardi* (Potoroidae)	1640 (12, 35)	12.5 (10, 0.3)	12.2 (14, 0.2)	97.6	178.7
28. *Setonix brachyurus* (Macropodidae)	2495 (4, —)	14.7 (4, —)	na	na	164.4
29. *Thylogale billardierii* (Macropodidae)	5570 (5, —)	23.3 (4, —)	25.3 (13, 0.5)	108.6	163.7
30. *Dendrolagus lumholtzi* (Macropodidae)	5883 (3, —)	22.6 (2, —)	26.0 (2, —)	115.0	153.9
31. *Macropus eugenii* (Macropodidae)	5655 (11, 448)	21.9 (7, 0.6)	22.8 (12, 0.9)	104.1	152.8
32. *Macropus rufogriseus* (Macropodidae)	12786 (1, —)	36.6 (1,—)	40.0 (22, 1.2)	109.3	159.2
33. *Macropus giganteus* (Macropodidae)	28650 (4, —)	59.5 (4, —)	64.3 (14, 2.8)	108.1	162.2

Notes: [1]Data derived from a South Australian population; [2]Data derived from a Tasmanian population; [3]Encephalisation derived from endocranial volume measurement rather than directly from brain mass measurement; na = Specimen not available.

which the brain mass data were available and for 127 of the 328 individuals which provided endocranial volume data. The object was to collect a sufficient number of specimens for each species in order to determine a range of normal values for these measures. This ideal was attained for most, but not all, of the species considered. The data are summarised in Table 1 where each species is listed, together with the mean values for body mass (BoM), brain mass (BrM) and endocranial volume (ECV). In brackets following these figures are the number of individual specimens measured for each species (N) and, where N ≥ 7, the standard error of the mean (SE). In the penultimate column the ratio (ECV/BrM × 100) provides a measure of the fit between brain and endocranial cavity. Values near 100 indicate that the brain mass, in grams, approximates the endocranial volume, in millilitres. Values in excess of 100 indicate an endocranial volume greater than the brain mass; values less than 100, the opposite. The final column lists an encephalisation value (E) which is a measure of the

relationship between brain and body size for each species. The encephalisation values were determined as follows.

An assemblage of Australasian polyprotodont marsupials (families Dasyuridae and Peramelidae) was taken as a basal reference group against which the remaining marsupials were to be compared. Following Nelson and Stephan (1982), we excluded the dasyurid genus *Planigale* from the basal group calculations as encephalisation values for these animals are atypically low. Unlike Nelson and Stephan, we have included the bandicoots (Peramelidae) with the basal group as the encephalisation values obtained for these marsupials differ little from those of the dasyurids. The composition of this basal group consists of relatively unspecialised carnivorous and insectivorous mammals displaying a 300-fold range of body mass. The remaining Australasian polyprotodont families are either monotypic or oligotypic and, frequently, rare. For these reasons

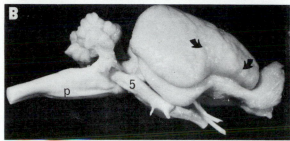

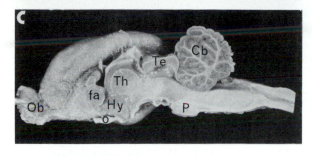

Fig. 3. Dorsal (A), left lateral (B) and mid-sagittal (C) views of the Koala brain. Note shrunken appearance of the cerebral hemispheres and the small size of the cerebellar hemispheres in comparison with the marsupials shown in Figs 1 and 2. The arrows in B indicate the extent of the motor-somaesthetic cortex. *Abbreviations*: Cb, cerebellum; fa, fasciculus aberrans-anterior commissure; Hy, hypothalamus; o, optic nerve; Ob, olfactory bulb; P, pons; p, medullary pyramid; Te, midbrain tectum (superior and inferior colliculi); Th, thalamus; 5, trigeminal nerve. Marker bar = 10 mm.

representatives of these families are not included in this study.

A least squares regression line was computed from the $\log_{10}$ values of the mean brain and body masses for the eleven basal species (Fig. 4). The resulting allometric equation yielded an expected value of brain mass for a species of given body mass:

$$\log_{10} \text{(expected brain mass)} =$$
$$0.581 \log_{10} \text{(body mass)} - 1.017$$

brain and body masses are given in grams. The basal group correlation coefficient was 0.998. The actual encephalisation (E) was then determined by dividing the measured brain mass (BrM) by the expected brain mass as determined from the above equation and multiplying by 100.

Thus, each encephalisation value can be represented as a vertical distance from the basal regression line. Values in excess of 100 indicate that the brain is larger than the expected value for a member of the basal group; values less than 100 that the brain is smaller than the expected value.

Basal Group Encephalisation

The basal group regression is plotted in Figure 4. The twelve numbered points correspond to the polyprotodont species listed in Table 1. However, as noted above, the *Planigale* (1) data were not used to compute the regression line even though these data are included in the graph. Standard error values, computed when N ≥ 7, can be found in Table 1 and are not plotted on this or succeeding graphs in order to avoid clutter. As can be seen most of the basal group species fall within ±10% of the expected value for that species. The mean value of all the computed encephalisations in this group (excluding *Planigale*) is 100.

Of the twelve species appearing on this plot most lie within ±10% of the basal group regression line. Of those which do not, *Antechinus swainsonii* (6) is represented by only one brain specimen, and its relatively high encephalisation value of 144.0 may not be typical. The data for *Planigale maculata* (1), the only major exception, are sufficient to show that the observed encephalisation of 61.9 is typical for this species and that it is abnormally low when compared with other members of the basal group. This finding agrees with that reported earlier by Nelson and Stephan (1982) on this and other *Planigale* species.

Encephalisation in the Possums, Wombat and Koala

Whereas the basal group marsupials, as defined above, all belong to the stem order Polyprotodonta, the marsupials in this and the next group to be considered belong to the advanced order, Diprotodonta. The major trend among these marsupials is toward an herbivorous diet though omnivorous and insectivorous species are found within this order. The grouping considered here comprises five families. They are placed together because in terms of gross and microscopic brain morphology they form a distinct group, clearly distinguished from the kangaroos, another diprotodont family (Haight and Murray 1981). Most of the marsupials known colloquially in Australia as possums are placed in five families: the Burramyidae or pigmy possums, the Petauridae which include most of the gliding forms, the Pseudocheiridae which include the ringtail possums, the Phalangeridae which contains the larger possums and the cuscuses and the Tarsipedidae, a monotypic group that is distantly

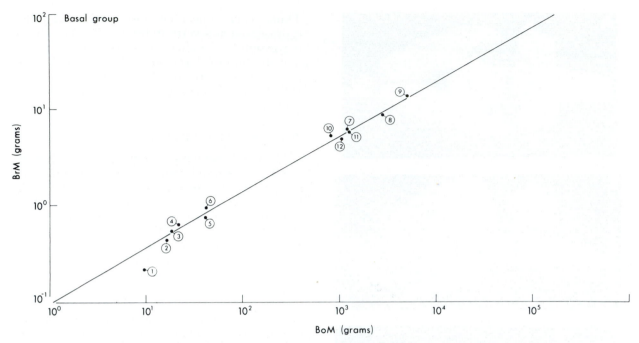

Fig. 4. Brain versus body mass plot for basal group polyprotodonts. Each numbered point represents the similarly numbered species listed in Table 1. The number of specimens used and the standard errors (where calculated) associated with each datum are listed in the Table and are, for reasons of clarity, not included on the graph. The solid line is the least squares regression line through points 2-12 and defines the basal level of encephalisation (E = 100) for the group, as described in text.

related to the other three. These animals occupy a variety of habitats from semi-arid to wet, tropical to alpine. Their dietary habits range from strict herbivory through to omnivory and insectivory. All are arboreal. Several petaurids, one pseudocheirid and one burramyid are specialised for gliding. The three species of wombat are burrowing grazers and the Koala is an arboreal browser. Further information about these and the other marsupials mentioned in this section can be found in Ride (1970 or Strahan 1983).

The brain and body mass regression line for this group is plotted in Fig. 5. As in Fig. 4, the numbers associated with each data point are those prefixing the indicated species in Table 1. The heavy solid line is the basal group regression curve and the dotted line above and parallel to it is not a regression line, but rather represents the mean encephalisation value (E = 136) obtained from averaging the encephalisation values obtained from members of this group. Though the mean encephalisation level for this group of marsupials is significantly higher than that of the basal group, there is considerable variation in the degree of encephalisation expressed by individual species within this group. For example, the Ringtail Possum, *Pseudocheirus peregrinus* (17), and the related Greater Glider, *Petauroides volans* (18), display values which do not exceed those of the basal group. *Petauroides*, the gliding form, displays, at E = 84.9, the second lowest encephalisation value for any diprotodont marsupial considered in this study. *Pseudocheirus*, a closely related, but non-gliding form, shows a

value of 110.7. In contrast, members of the genus *Petaurus* (15, 16), all of which glide, have relatively high values ranging from 134.3 to 170.1, as does the related but non-gliding form, *Gymnobelideus leadbeateri* (14) at 168.7. Our single burramyid, the pigmy possum, *Cercartetus lepidus* (13) at 133.4 falls near the mean value for the group. The larger possums and cuscuses of the family Phalangeridae also appear to occupy the midrange of observed encephalisation values for this group. Unfortunately, our data for *Wyulda* and *Phalanger* rely upon endocranial volume measurements rather than actual brain measurements as, for these species, brains were not available at the time of writing. An illustration of the endocranial cast of *Phalanger maculatus* has been published (Haight and Murray 1981) and from the detailing expressed upon that cast, it would seem that the endocranial volume is a reasonable measure of brain size in this species. The Scaly-Tailed Possum, *Wyulda squamicaudata* (19) and the Spotted Cuscus, *Phalanger maculatus* (23), seem to be relatively large-brained with encephalisation values of 148.3 and 159.2, respectively. Another cuscus, *P. orientalis* (22), would, on the basis of its endocranial measurement, appear to be rather small-brained with a value of 107.2.

The wombats are the Koala's closest living relatives and, according to Kirsch and Calaby (1977), form a "distinct group of Australian diprotodonts". The Common Wombat, *Vombatus ursinus* (24), at E = 212.8, is the most highly

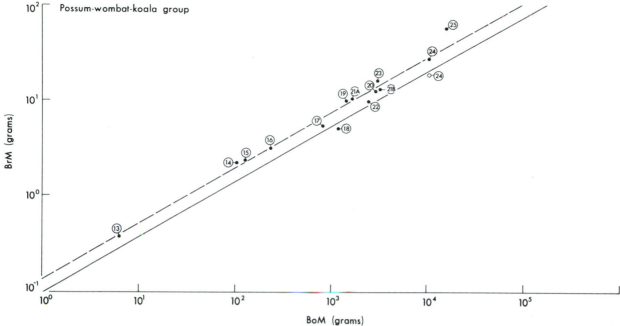

Fig. 5. Brain versus body mass plot for Possum-Wombat-Koala group. Each numbered point represents the similarly numbered species listed in Table 1. The number of specimens used and the standard errors (where calculated) associated with each datum are listed in the Table and are, for reasons of clarity, not included on the graph. The solid line represents the basal group mean encephalisation level (E = 100) as presented in Figure 4. The dotted line is the mean encephalisation value for the assemblage (E ≅ 135) and is not a regression line through the points on the graph. Values for the Koala appear twice. The lower point was calculated using the mean brain mass and yields an E of 84. The upper point was computed using the mean endocranial volume and yields an E of 134.

encephalised marsupial reported in this study. The Koala (25), E = 84.3, is the least.

In summary this group of marsupials displays a wide range of encephalisation values (E = 84-213). Some species fail to achieve a level of encephalisation higher than those of the basal group; others display some of the highest values recorded for living marsupials. Further, the range of observed values can vary widely among closely related species, even within the same family and genus. However, the mean encephalisation for the group, E = 136, is clearly greater than that shown by the basal group marsupials, reflecting a definite trend toward greater brain size in this assemblage of diprotodonts.

Encephalisation in Kangaroos

The Macropodoidea, which include the rat kangaroos (Potoroidae), wallabies and the familiar large kangaroos (Macropodidae), is a superfamily of diprotodonts which form a uniformly highly encephalised group, but whose brains are anatomically distinct from those in the Possum-Wombat-Koala group. The mean encephalisation for the eight species plotted in Fig. 6 is 165, ranging from 153 to 186. Again, the numbers on the graph correspond to the species in Table 1; the heavy line is the basal group regression line while the dotted line represents the mean encephalisation for the kangaroo group. As is easily seen, the macropodids are much more closely distributed around their mean encephalisation value of 165 than are the

rest of the diprotodont marsupials plotted in Fig. 5. The mean encephalisation value is also significantly higher, though it will be recalled that several members of the Possum-Wombat-Koala group ranged well into, and in one case, considerably beyond, the macropodid range.

Endocranial Volume and Brain Size Comparisons

If one examines the mean values for brain mass and endocranial volume shown in Table 1, it is apparent that the mean brain masses and mean endocranial volumes do not stand in fixed ratio, but vary in accordance with body size. The smaller marsupials in each of the three groups tend to have endocranial volumes (expressed in ml) which are somewhat less than the brain weights (expressed in gm). Note that we assume a brain density of approximately 1.0; hence brain mass can be taken as equivalent to the brain volume (see Jerison 1973, ch. 2). The situation reverses when the larger members of each group are considered. In these instances the brain comes to occupy less volume than its endocranial cavity. Expressed analytically, the slope of the regression line fitting *endocranial volume* and body mass is greater than that obtained for that fitting *brain mass* and body mass. Taking the basal group, for example, the slope of the $\log_{10}$ECV/BoM regresion is 0.618 as compared with the figure 0.581 obtained for the $\log_{10}$BrM/BoM curve. The two lines cross (i.e. BrM = ECV) where BoM = 1260 gm. The mean values of ECV

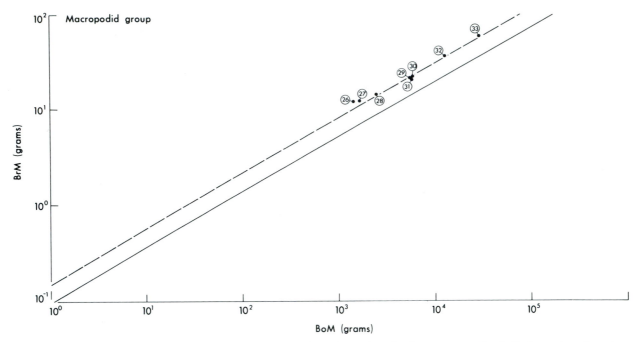

Fig. 6. Brain versus body mass plot for macropodoid marsupials. Each numbered point represents the similarly numbered species listed in Table 1. The number of specimens used and the standard errors (where calculated) associated with each datum are listed in the Table and are for reasons of clarity, not included on the graph. the solid line represents the basal group encephalisation (E = 100) as presented in Figure 4. The dotted line is the mean encephalisation value for this assemblage (E = 165).

and BrM are such that for animals with body masses below this value, the mean brain mass generally exceeds the mean endocranial volume whereas in cases where the body mass exceeds about 1200 gm the mean endocranial volume exceeds the mean measured brain mass. Thus, within the basal group the ratio of the mean values of ECV and BrM expressed as ECV/BrM × 100 (see Table 1) ranges from a low of 79 for *Planigale maculata*, the smallest member of the group, to nearly 118 for *Sarcophilus harrisii*, the Tasmanian Devil, which is the largest member of the basal group. Similar observations apply to the ECV/BrM ratios measured within the Possum-Wombat-Koala and macropodid groups.

If, on the other hand, the absolute ranges of brain mass and endocranial volume rather than their mean values are compared for each species, a different picture emerges. This is done in Figs 7-9. In these figures the left hand datum of each pair represents the range of brain mass measurements and the right hand datum, the range of endocranial volumes. The vertical axis is read in either grams or millilitres, whichever is appropriate. Mean or sole values are indicated by dots; the vertical bars represent the *total observed range* of measures where multiple specimens were available. The number of specimens obtained for each measure can be found in Table 1.

Among the basal group polyprotodonts (Fig. 7) the correspondence between endocranial volume and brain mass is generally very good.

The range of values obtained for both measures overlaps substantially in 8 of the 11 cases for which both ECV and BrM data were available. The degree of overlap between the two measures was nil in three species. In two of these, *Antechinus swainsonii* (6) and *Perameles nasuta* (11), the sample size for brains was small, N = 1 and 2, respectively. Hence, the data may not be representative. In addition, the skulls and brains for the *Perameles* data were derived from geographically distinct populations. For the third exception, *Antechinus flavipes* (5), the total non-overlap of the two measures is more difficult to explain. Six of the eight skulls used for endocranial measurements were from immature males. Nonetheless, two of the skulls and all four of the brains were taken from fully mature specimens, of both sexes. However, unlike the rest of our material, all the *A. flavipes* brains were obtained from a colony which had been maintained for several generations in a laboratory. The skulls were all taken from the wild. It should be appreciated, however, that the differences between the mean values for ECV and BrM in these three species, as expressed by ECV/BrM × 100 (Table 1), differed little from the values obtained for species in which the absolute ranges for the two measures did overlap substantially.

When ECV and BrM are compared within the diprotodont groups (Figs 8, 9), the results are, with the exception of the Koala (24), similar to those obtained for the basal group polyprotodonts. In all cases where the sample size was adequate, bar the Koala, the range of measured

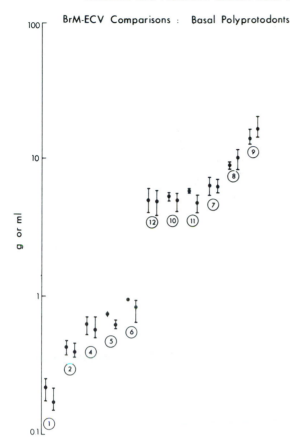

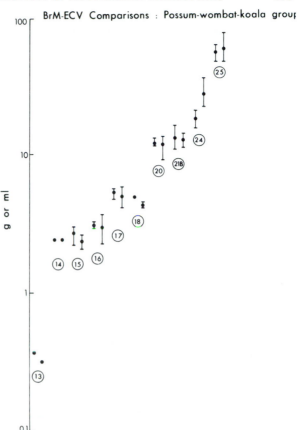

Fig. 7. Comparisons of brain mass and endocranial volume for basal group marsupials. Vertical axis is read in grams (brain mass) or millilitres (endocranial volume). Horizontal axis is meaningless. Left hand datum of each pair represents the brain mass measurements; the right hand datum, the endocranial volume measurements. Black dots represent the mean value or, where N = 1, the sole value for each group of measurements. The vertical lines represent the *total observed range* of values obtained for each measurement. The encircled numbers below each data pair correspond to similarly numbered species listed in Table 1.

Fig. 8. Comparisons of brain mass and endocranial volume for Possum-Wombat-Koala group marsupials. The data are presented as described in Fig. 7. Note that the absolute ranges of endocranial volume and brain mass for the Koala (#24) fail to overlap.

values for ECV and BrM largely overlap. Within the Diprotodonta, the Koala excepted, the only instances of non-overlap between the two measures occurred when the sample size for one or both measures was small (cf. *Petauroides volans* (18), Fig. 8; *Dendrolagus lumholtzi* (30), Fig. 9). Except for the Koala the ratio ECV/BrM × 100 varied ±15% (i.e. range = 85-115) or less around the central value of 100 for all of the diprotodont species measured. For the Koala the ratio was 164.1, and there was no overlap at all in the ranges of ECV and BrM. The smallest endocranial volume encountered in our Koala sample was 22.1 ml (mean value = 27.4 ml); the largest brain weighed 21.0 gm (mean value = 16.7 gm). As our Koala measurements were made on 11 brains and 16 skulls, we believe these observations are generally valid. Interestingly, if the koala's endocranial volume is used instead of the brain mass to compute an encephalisation value,

the result (E = 134) is very close to the expected value for the Possum-Wombat-Koala group.

Gross and Microscopic Anatomy of the Koala's Brain compared with other Marsupials

In Figs 1 and 2 dorsal and lateral views of three marsupial brains are shown. The Tasmanian Devil, *Sarcophilus harrisii* (Figs 1A, 2A) is the largest living Australian carnivorous marsupial (mean body mass = 5143 g, N = 15). As a member of the order Polyprotodonta, the Devil is distantly related to the other marsupials considered in this study, all of which belong to the order Diprotodonta. In Figs 1B and 2B the brain of a member of the kangaroo family (Macropodidae), the Pademelon or wallaby is pictured (mean body mass = 5570 g, N = 5). Finally, in Figs 1C and 2C is shown the brain of the Common Wombat (mean body mass = 15979 g, N = 6). Dorsal, lateral and mid-sagittal views of a Koala are seen in Fig. 3 (mean body mass = 9614 g, N = 7). As noted in the introduction, the Koala cerebrum is notably shrunken and "knobly" in appearance, and the lateral hemispheres of the cerebellum are much reduced in comparison with the marsupials illustrated in Figs 1 and 2. The Devil, Pademelon and Wombat's neocortices

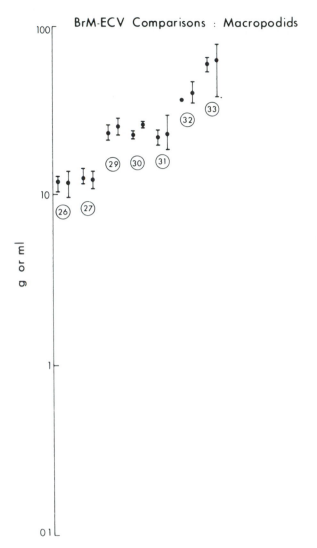

Fig. 9. Comparisons of brain mass and endocranial volume for macropodoids. The data are presented as described in Fig. 7.

are clearly gyrencephalic; the Wombat's is very much so. The Koala's brain is nearly lissencephalic, a feature unusual in any mammal, marsupial or placental, of this size.

Portions of the midbrain tectum can be seen deep within the transverse fissure between cerebrum and cerebellum in the Tasmanian Devil. The neocortical hemispheres cover this region much more completely in the Pademelon and the Wombat. In the Koala, however, the entire tectum is visible, along with the roof of the diencephalon and the large forebrain commissure known as the fasciculus aberrans (Smith 1902b); these latter structures being exposed by the failure of the cerebral hemispheres to meet over much of the midline. Though the neocortex is reduced in the Koala, the structures of the rhinencephalon appear to be in normal proportion. The mid-sagittal view, Fig. 3C, presents fewer exceptional features. The prominent fasciculus aberrans and subjacent anterior

commissure (fa) are clearly seen. The thalamus (Th), hypothalamus (Hy) and midbrain tectum (Te) are present in proper proportion. The pons (P) and medullary pyramids are clearly seen (p, Fig. 3B only). The lingula of the cerebellum is much reduced (see also Fig. 17) along with the lateral hemispheres. The trigeminal (5, Fig. 3B) and optic (o, Fig. 3C, unlabelled in Fig. 3B) nerves are prominent, the former especially so.

In the remainder of this section we will describe some major internal and surface features of the Koala's brain. The emphasis will be on differences between the Koala and other marsupials rather than upon similarities, and the treatment makes no pretence to completeness. The information will be presented primarily in comparison with the brains of the Common Wombat and the Common Brushtail Possum. The former is the Koala's closest living relative and the latter is the Australian marsupial for which the most neurobiological information is currently available. Also, within the order to which all three of the marsupials belong, the Koala's brain is, on allometric grounds, small for its size, that of the Possum is close to the norm for this group while that of the Wombat is quite large.

Neocortex

Much of the Koala neocortex is typically six layered and displays cytoarchitectural regionation similar to that observed in other marsupials. For example, the cytoarchitecture of prefrontal, temporal and posterior parietal areas in the Koala does not differ grossly from similar regions in the Brushtail Possum as illustrated by Haight *et al.* (1980) and by Neylon and Haight (1983). The Koala diverges from the general pattern in a very major way within the parietofrontal cortex which physiological mapping studies in other marsupials, including the Possum and Wombat, have been shown to be devoted to motor and somaesthetic functions (Abbie 940; Adey and Kerr 1954; Haight and Weller 1973; Johnson *et al.* 1973; Lende 1963a, 1963b, 1963c; Rees and Hore 1970; Weller and Haight 1973; Weller *et al.* 1976). In a number of marsupials the parietofrontal cortex and its regional subdivisions has been defined anatomically, on the basis of ascending and descending thalamic relationships (Donoghue and Ebner 1981; Haight and Neylon 1978, 1979, 1981). In medium to large diprotodont marsupials the parietofrontal region is delimited by constant sulci, particularly sulci α-α' which separate the prefrontal lobe from the parietofrontal region and sulcus β which separates the parietofrontal from the temporal and posterior parietal cortex. These limiting sulci are indicated, respectively, by the rostral and caudal sets of arrows in Fig. 2. The caudal and lateral limits of the parietofrontal cortex have not yet been identified in the Tasmanian Devil (Fig. 2A), thus the caudal arrow

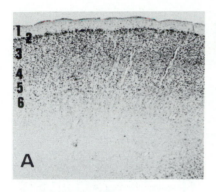
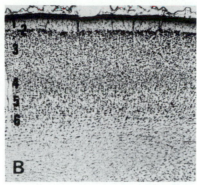
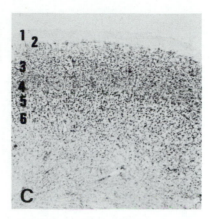

Fig. 10. Nissl (thionin) stained sections through occipital (visual) cortex in Koala (A), Brushtail Possum (B) and Wombat (C). Numbers to left of each illustration indicate cytoarchitectural laminae. Note the relative reduction in thickness of layer three in the Koala (A) and the near absence of layer two in the Wombat (C). Otherwise, except for allometric differences in cortical thickness, all three sections appear very similar. Marker bar = 1.0 mm.

is missing in this case. From these illustrations, it can be appreciated that in most diprotodont marsupials the motor-somaesthetic cortex occupies a large fraction of the total neocortex, a fraction which, for example, has been estimated at 35% in the Brushtail Possum (Haight and Neylon 1979). Traces of sulci α-α' and β are present in the Koala (Fig. 3B), but are very much reduced. However, these incipient sulci do mark regions of sharp cytoarchitectural transition in the Koala as in other diprotodont marsupials. The area thus defined in the Koala is proportionately much smaller than in the Wallaby (Fig. 2B), Wombat (Fig. 3C) and Brushtail Possum (see Haight and Neylon 1978, their Fig. 1). It appears, qualitatively at least, that a major, if not sole, cause of the reduction in size of the Koala's cerebral mantle is the diminution in area of the parietofrontal cortex when compared with other, related marsupials.

While, for example, the occipital (visual) area of cortex does display differences in laminar emphasis among the three marsupials (Fig. 10), in the Koala the cortical histology is essentially normal in appearance. Layer four is prominent in all three examples. In the Koala (Fig. 10A) layer three is diminished in thickness and layers five and six are not easily separated. In the Wombat (Fig. 10C) layers two and three are very much reduced in cell density when compared with the other two marsupials. In the Possum (Fig. 10B) all six layers are easily distinguished. While these differences do suggest that the organisational and connective properties of occipital cortex may differ among these three marsupials, the differences are probably not profound. The visual cortex is somewhat thinner in the Koala (1.0 mm) than it is in the other two marsupials (1.2 mm) with the major deficiency appearing, as noted above, in the thickness of layer three. As in the following illustration (Fig. 11), efforts were made to obtain the examples shown from presumably homologous regions of

cortex and to ensure that the sections were cut at approximately the same angle in order that cortical thicknesses might be accurately compared.

In contrast, portions of the parietofrontal cortex in the Koala differ quite radically from the other two marsupials (Fig. 11). The parientofrontal region receives input from a number of thalamic centres as has been reported in other marsupials (Donoghue and Ebner 1981; Haight and Neylon 1981; Neylon 1983) and, in the Possum, is among the thickest and architecturally complex regions of neocortex to be found (Haight *et al.* 1980, Fig. 1). In the Wombat (Fig. 11C) the example shown is approximately 2.1 mm thick and in the Possum (Fig. 11B), approximately 1.8 mm. In these two marsupials all six layers are prominent. Layers five and six contain the projection neurons whose axons extend to areas outside the neocortex. Layer six, in particular, projects strongly to the thalamus as has been shown in both marsupials (Mayner and Haight 1983a, b) and placental mammals (Jones and Wise 1977; Wise and Jones 1977). Neurons located in cortical layer five tend to project primarily to brainstem and spinal cord as is well documented in a large number of mammals (cf. Brodal 1981), though in placentals and in at least one marsupial there is also a small but significant thalamic projection from cortical layer 5 (Jones and Wise 1977; Mayner and Haight 1983a, b; Wise and Jones 1977). In both the Wombat and the Brushtail Possum cortical layers five and six are prominent and tightly packed with neurons. In the Koala (Fig. 11A) the overall cortical thickness is much less (1.3 mm). On allometric grounds one would have expected it to be intermediate in thickness to that of the Possum and the Wombat. Layer six is very much reduced, implying an impoverishment of descending corticothalamic projections. The neurons in layer five are less densely packed than in the other two marsupials though some of the neurons contained in this layer are very much larger

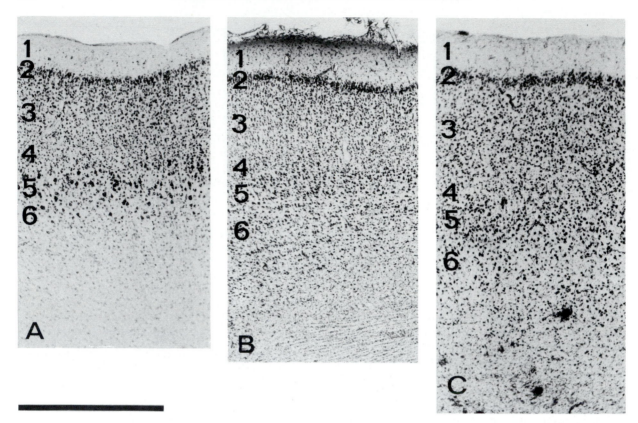

Fig. 11. Nissl (thionin) stained frontal sections through rostromedial part of the motor-somaesthetic cortex in Koala (A), Brushtail Possum (B) and Wombat (C). Numbers to left of each illustration indicate cytoarchitectural laminae. Note the very large cells in lamina 5 and the reduced lamina 6 in the Koala as compared with the other two examples. Marker bar = 1.0 mm.

(approaching 40-50μm in diameter) than those observed in the other two species (none larger than 25-30μm in diameter). These large pyramidal neurons are clearly seen in Fig. 11A. This region of the Koala's cortex is cytoarchitecturally very distinctive; we have not observed cortex of this type anywhere within the cortical mantle in any of some 27 different species of Australian and American marsupials (authors' observations).

Other Telencephalic Structures

Myelin-stained frontal sections through the maximum extent of the corpus striatum of the Koala, Possum and Wombat are shown in Fig. 12. All are printed to the same scale; consequently, only the left hand side of the Wombat example (Fig. 12C) is seen. The levels shown are just rostral to the point where the commissural fibres of the forebrain meet to form the fasciculus aberrans and the anterior commissure (Fig. 3C). The Koala example (Fig. 12A) is slightly rostral to the other two. The septal (s) region is prominent in all three marsupials. The corpus striatum, represented here by the caudatoputamen or neostriatum (ns) is especially prominent in the Wombat. A notable feature is the small size of the hippocampal formation (hc) in the Koala together with its efferent pathway, the fornix

(fx). That this complex is, in absolute size, very little different to that of the Possum can be seen by comparing Figs 12A and 12B.

Diencephalon

In fibre stained material (Fig. 13) the thalamus and hypothalamus appear quite similar in the Koala, Wombat and Possum. Barring size differences, those of the Wombat and Possum are nearly identical. The nuclei of the somaesthetic and visual thalamus are prominent in all diprotodont thalami. In three of the four Nissl-stained examples shown in Fig. 14 a clear differentiation within the somaesthetic thalamus or ventroposterior complex (vp) into medial (m) and lateral (l) divisions is apparent. The cells in the medial division are smaller and more densely packed. In the two possums, *Petaurus* (Fig. 14A) and *Trichosurus* (Fig. 14B), and in the Wombat (not illustrated in Nissl stain) the two divisions are roughly equal in size. In the Pademelon, *Thylogale* (Fig. 14C), the medial division is much larger than the lateral division, though the demarcation between the divisions remains distinct. In the Koala (Fig. 14D) this region presents an entirely different appearance than it does in the other diprotodonts. Vp is extremely irregular in appearance. In what we assume to be the medial

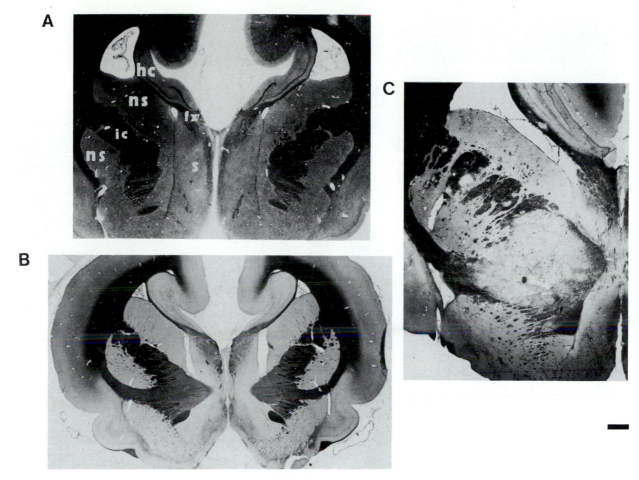

Fig. 12. Myelin (Weil) stained frontal sections through the corpus striatum of Koala (A), Brushtail Possum (B) and Wombat (C). These sections are located immediately rostral to the crossing of the forebrain commissures. Section A is slightly rostral to the other two. *Abbreviations*: fx, fornix; hc, hippocampal formation; ic, internal capsule; ns, neostriatum = caudatoputamen; s, septal region. Marker bar = 1.0 mm.

division clusters of densely packed cells can be seen arrayed across the dorsolateral edge of the nucleus. Though the Koala vp is clearly subdivided into different zones, the odd shape and other irregularities of these when compared with the other three examples prevents any firm establishment of corresponding regions in lieu of experimental findings.

The posterior thalamic nucleus (po, Fig. 14) is large in both Possums (and in the Wombat) and comes to occupy a considerable area dorsal to the ventroposterior complex (see Neylon and Haight 1983 for details of this thalamic area). Po is also prominent and clearly defined in the Koala, forming a distinct wedge of cells inserted between vp and the internal medullary lamina. In the wallaby (Fig. 14C) po is much less clearly delineated.

The visual thalamus, comprised of the dorsal lateral geniculate nucleus (1g in Fig. 13, d and v in Fig. 14) and the lateroposterior complex (lp) display considerable differences among the various marsupials. In the Possum the geniculate is highly laminated, reflecting in its cytoarchitecture, a pattern of alternating inputs from each

eye (Hayhow 1967; Sanderson *et al.* 1978). In the species of wombat used to illustrate this study the geniculate displays very indistinct lamination and is, in fact, rather small (Fig. 13C). However, recent information (K. J. Sanderson pers. comm.) concerning the experimental investigation of retinogeniculate projections in this species indicates that inputs from each eye are segregated within the nucleus. In another Wombat species, *Lasiorhinus latifrons* the Hairy-nosed Wombat, the geniculate is larger and displays a considerable degree of cytoarchitectural lamination (Sanderson and Pearson 1981). As is seen in Fig. 14, the two possums and the wallaby also exhibit regional differentiation within their respective geniculate nuclei. The Koala geniculate is also laminated, though not quite to the extent found in the possums and in some kangaroos (Sanderson *et al.* 1984). The laminated nature of the Koala geniculate is shown in Fig. 14D, where it can be seen that this geniculate is proportionately large in comparison to those of the possums and, especially, that of the wombat. One further feature (not discernable at the magnifications used in Fig. 14) distinguishes the Koala geniculate from the

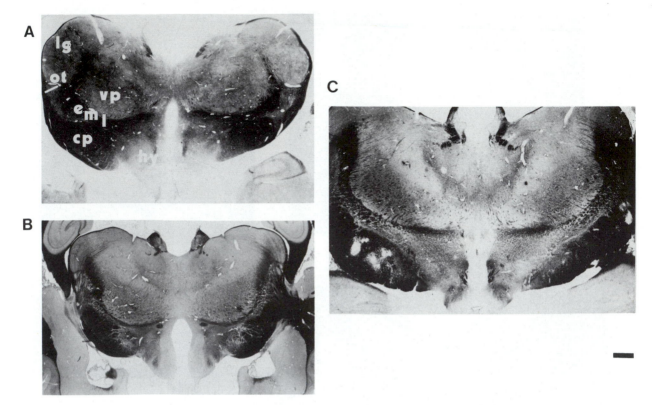

Fig. 13. Myelin (Weil) stained frontal sections through posterior half of thalamus in Koala (A), Brushtail Possum (B) and Wombat (C). *Abbreviations*: cp, fibres of cerebral peduncle = crus cerebri; eml, external medullary lamina of thalamus; hy, hypothalamus; lg, dorsal lateral geniculate nucleus; ot, optic tract; vp, ventroposterior nuclear complex. Marker bar = 1.0 mm.

others. In the Koala the cell layer immediately adjacent to the optic tract contains rather large cells while the next deepest layer is comprised of small, thinly scattered neurons. In many other diprotodont marsupials, notably the kangaroos (Sanderson *et al.* 1984), this pattern is reversed and the small cells are found next to the optic tract. Yet other diprotodont marsupial species do not appear to possess this small cell layer at all (Hayhow 1967; Johnson and Marsh 1969; Sanderson *et al.* 1984). Finally, in all these examples the lateroposterior complex (lp) is large; in the Koala (Fig. 14D) it appears proportionately larger than in the other species illustrated.

Midbrain

The essential features of the midbrain are shown in Fig. 15 which shows, in frontal section, the exit of the oculomotor nerve in all three marsupials. The Wombat midbrain (Fig. 15C) has been slightly reduced in scale in comparison to the other two. All three marsupials possess a prominent red nucleus (rn) and substantia nigra (sn), the latter being particularly large in the Wombat. The crura cerebri or descending fibres of the cerebral peduncle (cp) form a large and prominent tract in the Wombat. These fibre tracts are approximately the same size in the Possum and Koala, suggesting a major reduction of this system in the Koala. Unlike in the Possum and

Wombat, in the Koala these peduncular fibres fail to extend over the lateral margin of the substantia nigra.

In the Possum the superior colliculus is highly laminar in appearance. This is clearly seen in both Nissl (not shown) and fibre-stained material (Fig. 15B). This lamination is not as prominent in the Wombat and Koala. The inferior colliculus is similar in all three animals, as are the tegmentum and periaquaductal regions. The nucleus and tract of the mesencephalic portion of the trigeminal complex (5m) are a pronounced feature within the Koala's midbrain.

Pons

The frontal sections shown in Fig. 16 were chosen to illustrate the maximum extent of the middle cerebellar peduncle (mp) in each marsupial. As is readily seen, that of the Koala (Fig. 16A) is proportionately and absolutely smaller than the others. Similarly, the major ascending efferent cerebellar pathway, the superior cerebellar peduncle (sp) is small, especially in comparison with that of the Wombat. The corticospinal and corticobulbar tracts (cst) are slightly larger in the Koala than in the Possum, as would be expected on allometric grounds. The lateral lemniscus (l) and nucleus thereof (ln) form a major feature of the pons at this level, especially in the Possum (Fig. 16B).

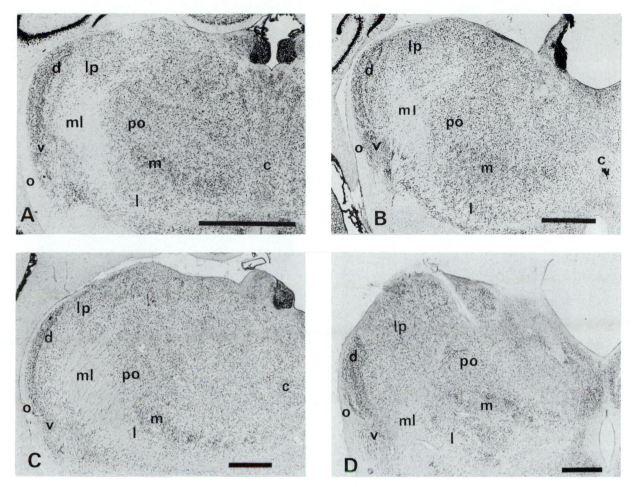

Fig. 14. Nissl (thionin) stained sections through left caudal thalamus in the Sugar Glider, *Petaurus breviceps* (A), Brushtail Possum, *Trichosurus vulpecula* (B), Pademelon, *Thylogale billardierii* (C) and Koala (D). Sections are chosen to display the extent of the visual and somaesthetic thalamic areas. *Abbreviations*: c, central nucleus, not present in D; d, dorsal lateral geniculate nucleus; v, ventral lateral geniculate nucleus; lp, lateroposterior nuclear complex; ml, medial lemniscus; o, optic tract; po, posterior nucleus; l, ventroposterior nucleus, lateral division; m, ventroposterior nucleus, medial division. Some of these abbreviations differ from those used elsewhere in this paper. This is done in order that the labels do not obscure the cytoarchitectural features being demonstrated. Marker bars = 1.0 mm.

Medullary-Pontine Junction

The sections shown in Fig. 17 were chosen to display the deep cerebellar nuclei (dcn). This level corresponds to the entry of the vestibulo-cochlear nerve (8). The three sections are printed to the same scale. In the Koala (Fig. 17A) the deep cerebellar nuclei are reduced, being approximately the same size as those of the Possum, a much smaller animal. The lingula of the cerebellum fails to extend in to the space of the fourth ventricle as well, reflecting the general poverty of cerebellar development in this animal. Similarly, the inferior cerebellar peduncle (ip) is relatively small and poorly developed. However, the motor nuclei of the brain stem are comparable in all three marsupials. The motor nuclei of the facial nerves (f) can be seen in all three marsupials.

Pyramidal Decussation and Caudal Medulla

In Fig. 18 are shown sections through the pyramidal decussation and the cuneate-gracile-trigeminal nuclear complex in the three species. The Wombat section (Fig. 18C) has been reduced in size in comparison with the other two. The spinal trigeminal tract and nucleus (5s) figure prominently in each animal. The cuneate (c) and gracile (g) nuclei are proportionately smaller in the Wombat and Koala than in the Possum, particularly so in the Wombat. The pyramidal decussation and tract is comparable in all three animals.

DISCUSSION

With E = 84.3 the Koala's brain is about 60% smaller by weight and volume than would be predicted for a typical diprotodont marsupial of the Possum-Wombat-Koala group (mean E ≅ 134). In this feature the koala is not alone as the possums of the *Pseudocheirus-Petauroides* group are also small-brained (E = 84.9-110.4), as is the Grey Cuscus, *Phalanger orientalis* (E = 107.2), though not, apparently, the spotted cuscus, *Phalanger maculatus* (E = 159.2). The

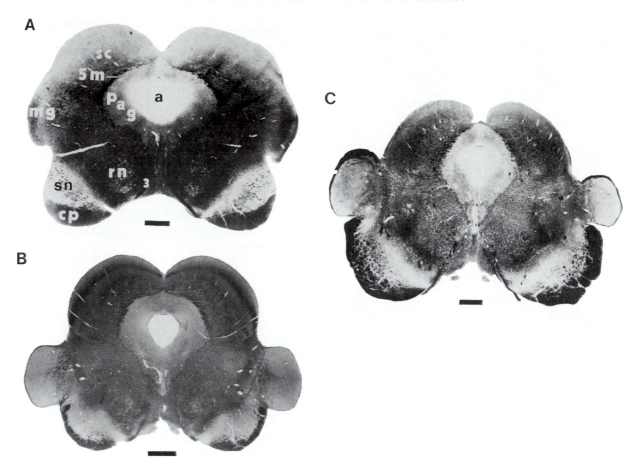

Fig. 15. Myelin (Weil) stained frontal sections through midbrain at the level of the exit of the third cranial nerve in Koala (A), Brushtail Possum (B) and Wombat (C). *Abbreviations*: a, cerebral aqueduct; cp, descending cortical fibres in cerebral peduncle; mg, caudal part of medial geniculate nucleus; pag, periaqueductal grey matter of midbrain; rn, red nucleus, sc, superior colliculus; sn, substantia nigra; 3, exiting oculomotor nerve; 5m, mesencephalic tract and nucleus of trigeminal nerve. Marker bars = 1.0 mm.

encephalisation value of 133 which we obtained from our single specimen of Pigmy Possum, *Cercartetus lepidus*, may be atypically high as Nelson and Stephan (1982) have reported values for this group of burramyids comparable to those of the basal polyprotodont group. It should also be noted here that our encephalisation value for the Wombat at 213 is higher than that reported by Nelson and Stephan (1982) who obtained values of 130-150 using a slightly different basal group regression as base line. Our wombat measurements do agree with those of Moeller (1973), who used endocranial measurements, and Pirlot (1981), who used actual brains. Both of these authors reported that the wombats were the most highly encephalised marsupials among the groups that they had examined.

What makes the Koala's brain exceptional, in addition to its strange appearance and small size, is the disparity between its mean brain mass of 16.7 g (N = 11) [cf. Ziehen's (1897) value of 17.0 g (N = 3)] and the mean endocranial volume of 27.4 ml (N = 16) [cf. Moeller's (1973) value of 23.2 (N = 5)]. Even in a very large-brained mammal such as ourselves, the discrepency

between brain mass and endocranial volume is only of the order 20% or less (Tobias 1971). A difference of more than 60% for a mammal the size of a Koala is most unusual (see also Jerison 1973, ch. 2).

As noted in the results, the endocranial volume for the Koala approximates closely the expected brain size for a diprotodont marsupial belonging to the possum group. In the other cases presented in this study, whether of high, low or moderate encephalisation, the disparity between brain mass and endocranial volume never approaches that observed in the Koala.

It is not that the Koala brain, in a proper Parkinsonian manner, does not attempt to fill the space available to it. Quite the contrary! Examination of the endocranial region shows that the surface features of the cerebrum and cerebellum, for example, are impressed clearly and accurately upon the inner surface of the cranial vault, indicating a fairly tight fit. The Koala brain, in its efforts to fit the available space, contains large amounts of non-neural "space". The large cavity overlying the midbrain and

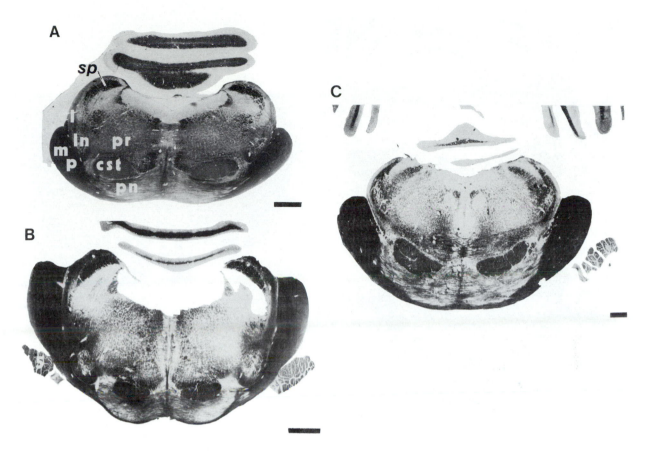

Fig. 16. Myelin (Weil) stained frontal sections through pons at level of middle cerebellar peduncle in Koala (A), Brushtail Possum (B) and Wombat (C). *Abbreviations*: cst, descending corticospinal and corticobulbar tracts; l, lateral lemniscus; ln, nucleus of lateral lemniscus; mp, middle cerebellar peduncle; pn, pontine nuclei; pr, pontine reticular formation; sp, superior cerebellar peduncle; 5, trigeminal nerve (present only in B and C). Marker bars = 1.0 mm.

diencephalon caused by the failure of the neocortical hemispheres to meet along the midline comprises the largest empty area encountered. However, the ventricular system is enlarged, especially the lateral ventricles of the telencephalon. These features are illustrated in Fig.19 The exaggeration of "emptiness" within the Koala's brain is obvious; the resemblance to the condition of hydrocephalus is suggestive.

The question is: why and how does this happen? Conventional understanding of skull growth in mammals has the brain case expanding in response to the growth of the brain contained within it. Suppression of bone growth following hypophysectomy in young rats, for example, does not result in cessation of brain growth, and the animal eventually dies from the resulting compression of the neuraxis (Asling *et al.* 1952). The factors which cause the Koala to produce a "normal-sized" brain case which contains a brain that is much too small present an intriguing, but as yet unanswerable developmental problem. We doubt, for example, that the explanation lies in a mechanistic interpretation involving, say, expansion of the cranial vault to accommodate this animal's massive masticatory apparatus. Other relatively small-brained marsupials such

as the Tasmanian Devil, *Sarcophilus harrisii*, possess massive jaw musculature, yet their brains are of normal configuration and the ECV and BrM values are comparable (see also Haight and Murray 1981, their Figs 3 and 6).

While we are left with only speculations as to why the brain-endocranial cavity relationship in the Koala is so at variance with that seen in other marsupials, we can make some useful comment on the contribution to this condition by structural peculiarities within the brain itself together with the implications these peculiarities may have for the Koala's behaviour.

The incoming sensory pathways involving vision and audition appear normal and are equally well developed when compared with other marsupials, the visual system especially so. The neocortical areas associated with vision appear unexceptional though the reduced thickness of layer three could suggest a reduction in inter and intracortical connections relative to other marsupials. As is now well known, the neocortical representation of somaesthesis and motor function is closely linked and, to varying degrees, overlapping, in marsupials (Donoghue and Ebner 1981; Haight and Neylon 1978, 1979,

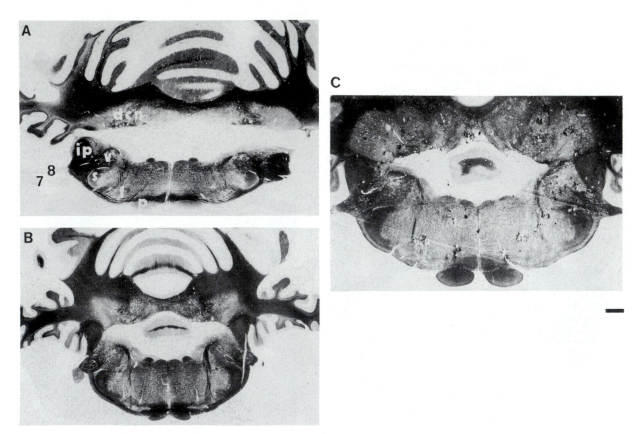

Fig. 17. Myelin (Weil) stained frontal sections through entry of the vestibulocochlear nerve in Koala (A), Brushtail Possum (B) and Wombat (C). *Abbreviations*: dcn, deep cerebellar nuclei; facial motor nucleus; ip, inferior cerebellar peduncle; p, pyramidal tract; v, vestibular nuclei; 5s nucleus and tract of spinal part of trigeminal complex; 7, facial nerve (not present in B and C); 8, vestibulocochlear nerve. Marker bar = 1.0 mm.

1981; Lende 1963a, b, c), it would appear that in the Koala this region of neocortex is both deficient and histologically distinctive. First, there is the disproportionately small area of cortex devoted to these functions. Second, there is the suggestion of abnormal functional organisation due to the thinness of cortex in this area plus the overall diminution of efferent connections to subcortical centres, as evidenced by the near absence of layer six and the reduction of layer five when compared with other marsupials. The presence of extremely large pyramidal cells within layer five is yet another puzzling feature. These very large cells resemble the giant Betz cells found in agranular cortex of many placental mammals and, as noted earlier, have not been seen in other marsupials.

Further evidence that the relationships with subcortical centres are impoverished is indicated by the small size of the crura cerebri in comparison with other diprotodonts (Fig. 13). If the somatotopic organisation within the marsupial crura is the same as has been reported for placental mammals, the reduction of the lateral crural fibres could suggest a deficiency in the corticopontine projection (Brodal 1981). This observation is substantiated by the reduced size of the pontine nuclei and their major efferent

pathway, the middle cerebellar peduncle (Fig. 14). The cerebellum itself is much reduced, particularly the neocerebellum (Fig. 3), and here is a concomitant decrease in the size of the deep cerebellar nuclei, particularly of the dentate nucleus. However, the corticospinal and corticobulbar pathways and their nuclear target areas appear to be in normal proportion in this animal (Figs 14, 15, 16). The structures of the corpus striatum, compared with those of the Wombat, are small, though their histological appearance is normal (Fig. 11). Another major forebrain structure, the hippocampus, is also very much smaller than would be expected in an animal of this size. The thalamus appears to be normal in size, especially with respect to the nuclei of the ventral tier. It was noted that the cytoarchitectural organisation of the ventroposterior or somaesthetic thalamic region was atypical. This is of interest because this nucleus is anatomically and functionally related to the cortical regions which we have observed to be unusual.

Though the cytoarchitectural organisation of the dorsal lateral geniculate nucleus differs from that reported in either kangaroos (Sanderson *et al.* 1984), possums (Hayhow 1967; Johnson and Marsh 1969; Pearson *et al.* 1976) or Hairy-nosed

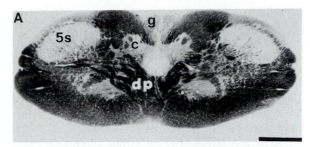

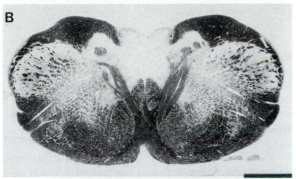

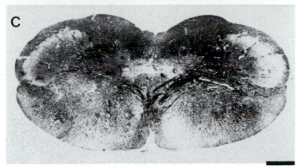

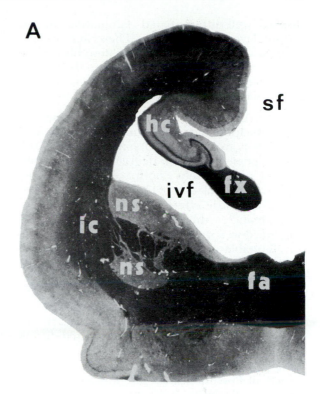

Fig. 18. Myelin (Weil) stained frontal sections through lower medulla at the level of the pyramidal decussation in Koala (A), Brushtail Possum (B) and Wombat (C). *Abbreviations*: c, cuneate nucleus; dp, decussation of pyramidal tract; g, gracile nucleus; 5s, nucleus and tract of spinal part of trigeminal complex. Marker bars = 1.0 mm.

Wombat (Sanderson and Pearson 1981), the difference cannot be considered particularly anomalous in view of the known variability in the anatomy and in the organisation of retinal inputs to this nucleus in all of these diprotodont marsupials (Sanderson *et al.* 1984 for discussion).

Within the brainstem the major structures such as the red nucleus, pontine and medullary reticular formations and the motor and sensory cranial nerve nuclei are of normal appearance and proportion.

All of this suggests that the Koala's behaviour should not be notable for agility or fine control over its motor activities. There is general (anecdotal) consensus that Koalas are slow and clumsy, though willing enough to bite when provoked. In fact the widely used term "native bear" might have better been "native sloth". Though the sloth, an edentate placental, does possess a reasonably gyrencephalic neocortex, it fails, according to Gerebtzoff and Goffart (1966), to

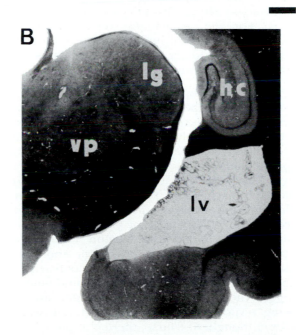

Fig. 19. Myelin (Weil) stained frontal sections through the Koala brain at the level of the major forebrain commissures (A) and the posterior part of the diencephalon and temporal lobe (B). The broad cavity of the sagittal fissure (sf), the enlarged interventricular foramen (ivf) and the temporal lobe portion of the lateral ventricle convey the extent of non-neural space within this brain. *Abbreviations*: fa, fasciculus aberrans (anterior commissure occupies the lower half of this fibre bundle); fx, fornix; hc, hippocampal formation; ic, internal capsule; ivf, interventricular foramen; lg, dorsal lateral geniculate nucleus of thalamus; lv, temporal lobe portion of lateral ventricle; ns, neostriatum = caudatoputamen; sf, sagittal fissure; vp, ventroposterior nuclear complex of thalamus. Marker bar = 1.0 mm.

display well-differentiated lamination in the output layers five and six. These authors also claimed that intracortical association fibres were rare in the sloth. Seemingly, a sedentary life style is correlated with neocortical anomalies in both the Koala and the sloth.

The Brushtail Possum's behaviour is somewhat better documented. Two studies (Megirian *et al.* 1977; Rees and Hore 1970) have shown that this animal possesses a high degree of manipulative skill. Unfortunately, both of these studies reported that ablation of the motor-somaesthetic cortex had little effect upon subsequent motor performance, though admittedly the tests employed were not sophisticated. The Possum is an adaptable and highly successful animal. Its introduction to New Zealand, where it has become a pest of major economic proportions (Pracy 1974), proves that this Possum has the ability to cope with changed circumstances in a way in which the Koala would clearly be incapable, if only because of its restricted dietary requirements (see Bergin 1978). If the behavioural flexibility and adaptability of the Brushtail Possum are accepted as a measure of that animal's "braininess", it becomes difficult to understand why the Wombat's brain is as large as it is. For the behavioural repertory of the Wombat is not, according to folklore, enough in advance of the Koala's, let alone the Possum's, to warrant the size brain it has.

The "hydrocephalic" appearance of the Koala's brain is, perhaps worth a few concluding comments. Clinically, two aetiological factors seem to be involved in hydrocephalus; one involving a compensatory increase in cerebro-spinal fluid (CSF) volume due to cerebral hypoplasia; the other, an increase in pressure due to imperfect CSF resorption or ventricular blockage and which also involves a concomitant reduction in the volume of brain tissue due to compression (Brain and Walton 1969). The presence of specific motor system anomalies rather than generalised CNS deficiencies observed within the Koala's CNS would tend to suggest the former rather than the latter cause as being responsible for the enlarged ventricular system though, of course, there is at present no real evidence to support either contention.

Mutations resulting in minor to gross neurological deficiencies are well documented among laboratory and domestic mammals. The various mutations which cause neurological anomalies in mice (Caviness and Rakic 1978) and the anatomical variations seen in the visual system of Siamese and albino cats (Guillery 1969) are but two well-known examples. Outside of our own and domestic species documentation of genetically stable gross CNS abnormalities have been rare; it usually being assumed that such abnormalities are incompatible with survival. In the case of the Koala it is difficult to imagine how (or why) slow selection for its particular brain morphology could proceed. Further, one would also assume that slow changes in brain structure would be accompanied by a more normal brain-endocranial cavity relationship. A possible explanation could be forthcoming in the idea of macromutations or "hopeful monsters" (Gold-schmidt 1940), a concept which has been resurrected a number of times over the years (Dover and Flavell 1982; Robertson 1981). An hypothesis could be considered that the Koala, having already adopted a sedentary, sloth-like, lifestyle, might not have been particularly disadvantaged by a meiotic malfunction which produced a sudden, major alteration of its neuraxis, profoundly affecting the animal's motor abilities. However, one is forced to assume that this condition tended to recur reproducibly (*i.e.* the appearance of a constant mutagen in the animal's environment) that the condition be maintained within the population and that it give a definite advantage when it came to repro-ductive success, for the hypothetical, "normally-brained" Koala is rare or non-existent today. Examination of fossil endocranial material would be of extreme interest in this regard, as such endocasts could indicate whether this unique relationship between brain and skull is one of long standing or whether it is a comparatively recent development in Koala phylogeny.

ACKNOWLEDGEMENTS

We gratefully acknowledge the kind assistance and cooperation of the staff of the Australian Institute of Anatomy, particularly Mr David Kaus, of Mr Phil Andrews, Curator of Vertebrate Zoology at the Tasmanian Museum and Art Gallery, or Randy Rose of the Zoology Depart-ment, The Univeristy of Tasmania, Dr Janette Hope of the Department of Prehistory, The Australian National University and especially of Dr John Calaby of the CSIRO Division of Wildlife and Rangelands Research in Canberra, all of whom kindly permitted access to the records and cranial material in their care.

This study was supported in part by a special grant from the Univeristy of Tasmania to JRH, by the Zoology Department at Monash Univer-sity, Clayton, Victoria and by the Research School of Biological Sciences at The Australian National University. Mr Garry Brown drew Figs 4-9 and Ms Barbara Piper typed an initial draft of the manuscript. We also wish to thank the photo-graphic section at the Research School of Biological Sciences in the Australian National University and Mr Michael Stranger of the Anatomy Department, the Univeristy of Tasmania for their photographic assistance.

Section 18C (INNER)

REFERENCES

ABBIE, A. A., 1940. The excitable cortex in *Peramales, Sarcophilus, Dasyurus, Trichosurus* and *Wallabia* (*Macropus*). *J. comp. Neurol.* **72**: 469-87.

ADEY, W. R. AND KERR, D. I. B., 1954. The cerebral representation of deep somatic sensibility in the marsupial phalanger and the rabbit; an evoked potential and histological study. *J. comp. Neurol.* **100**: 597-626.

ARCHER, M., 1984. The Australian marsupial radiation. Pp. 633-808 *in* "Vertebrate zoogeography and evolution in Australasia" ed by M. Archer and G. Clayton. Hesperian Press: Perth

ASLING, C. W., WALKER, D. G., SIMPSON, M. E., LI, C. H. AND EVANS, H. M., 1952. Death in rats submitted to hypophysectomy at an extremely early age and the survival effected by growth hormone. *Anat. Rect.* **114**: 49-65.

BERGIN, T. J., 1978. "The Koala, proceedings of the Taronga Symposium on Koala biology, management and medicine." Zoological Parks Board of New South Wales.

BRAIN, LORD AND WALTON, J. N., 1969. "Brain's diseases of the nervous system", 7th ed. Oxford University Press: London.

BRODAL, A., 1981. "Neurological anatomy in relation to Clinical medicine", 3rd ed. Oxford University Press: New York.

CAVINESS, V. S. AND RAKIC, P., 1978. Mechanisms of cortical development: a view from mutations in mice. *Ann. Rev. Neurosci.* **1**: 297-326.

DONOGHUE, J. P. AND EBNER, F. F., 1981. The organisation of thalamic projections to the parietal cortex of the Virginia opossum. *J. Comp. Neurol.* **198**: 365-88.

DOVER, G. A. AND FLAVELL, R. B., 1982. "Genome evolution". Academic Press: New York.

GEREBTZOFF, M. A. AND GOFFART, M., 1966. Cytoarchitectonic study of the isocortex in the sloth (*Choloepus hoffmanni* Peters). *J. Comp Neurol.* **126**: 523-34.

GOLDSCHMIDT, R., 1940. "The material basis of evolution". Yale University Press: New Haven.

GUILLERY, R. W., 1969. An abnormal retinogeniculate projection in Siamese cats. *Brain Res.* **14**: 739-41.

HAIGHT, J. R. AND MURRAY, P. F., 1981. The cranial endocast of the early Miocene marsupial, *Wynyardia bassiana*: an assessment of taxonomic relationships based upon comparisons with recent forms. *Brain, Behav. Evol.* **19**: 17-36.

HAIGHT, J. R. AND NEYLON, L., 1978. The organisation of neocortical projections from the ventroposterior thalamic complex in the marsupial brush-tailed possum, *Trichosurus vulpecula*: a horseradish peroxidase study. *J. Anat.* **126**: 459-85.

HAIGHT, J. R. AND NEYLON, L., 1979. The organisation of neocortical projections from the ventrolateral thalamic nucleus in the brush-tailed possum, *Trichosurus vulpecula*, and the problem of motor and somatic sensory convergence within the mammalian brain. *J. Anat.* **129**: 673-94.

HAIGHT, J. R. AND NEYLON, L., 1981. An analysis of some thalamic projections to parietofrontal neocortex in the marsupial native cat, *Dasyurus viverrinus* (Dasyuridae). *Brain, Behav. Evol.* **19**: 180-91.

HAIGHT, J. R. AND WELLER, W. L., 1973. Neocortical topography in the brush-tailed possum: variability and functional significance of sulci. *J. Anat.* **116**: 473-74.

HAIGHT, J. R., SANDERSON, K. J., NEYLON, L. AND PATTEN, G. S., 1980. Relationships of the visual cortex in the marsupial brush-tailed possum, *Trichosurus vulpecula*, a horseradish peroxidase and autoradiographic study. *J. Anat.* **131**: 387-412.

HAYHOW, W. R., 1967. The lateral geniculate nucleus of the marsupial phalanger, *Trichosurus vulpecula*, an experimental study of cytoarchitecture in relation to the intranuclear optic nerve projection fields. *J. Comp. Neurol.* **131**: 571-604.

JERISON, H. J., 1973. "Evolution of the brain and intelligence". Academic Press: New York.

JOHNSON, J. I., 1977. Central nervous system of marsupials. Pp. 157-278 *in* "Biology of marsupials" ed by D. Hunsaker. Academic Press: New York.

JOHNSON, J. I. AND MARSH, M. P., 1969. Laminated lateral geniculate in the nocturnal marsupial *Petaurus breviceps* (sugar glider). *Brain Res.* **15**: 250-54.

JOHNSON, J. I., HAIGHT, J. R. AND MEGIRIAN, D., 1973. Convolutions related to sensory projections in cerebral neocortex of marsupial wombats. *J. Anat.* **114**: 153.

JONES, E. G. AND WISE, S. P., 1977. Size, laminar and columnar distribution of efferent cells in the sensory-motor cortex of monkeys. *J. Comp. Neurol.* **175**: 391-438.

JONES, F. W., 1924. "The mammals of South Australia, Part II". Government Printer: Adelaide.

JONES, F. W. AND PORTEUS, S. D., 1929. "The matrix of the mind". Edward Arnold: London.

KIRSCH, J. A. W., 1977. The comparative serology of Marsupialia, and a classification of marsupials. *Aust. J. Zool. Suppl. Ser. No. 52.*

KIRSCH, J. A. W. AND CALABY, J. H., 1977. The species of living marsupials: an annotated list. Pp. 9-26 *in* "The biology of marsupials" ed by B. Stonehouse and D. Gilmore. Macmillan: London.

LENDE, R. A., 1963a. Sensory representation in the opossum (*Didelphis virginiana*). *J. Comp. Neurol.* **121**: 395-403.

LENDE, R. A., 1963b. Motor representation in the cerebral cortex of the opossum (*Didelphis virginiana*). *J. Comp. Neurol.* **121**: 405-15.

LENDE, R. A., 1963c. Cerebral cortex: a sensorimotor amalgam in the Marsupialia. *Science* **141**: 730-32.

MAYNER, L. AND HAIGHT, J. R., 1983. Neocortical efferents to thalamus in the Tammar wallaby, *Macropus eugenii*. *J. Anat.* **136**: 667.

MAYNER, L. AND HAIGHT, J. R., 1983. Efferent and afferent connections between neocortex and thalamus in a marsupial wallaby, *Macropus eugenii*. *Neurosci. Lett. Suppl.* **11**: S60.

MEGIRIAN, D., WELLER, W. L., MARTIN, G. F. AND WATSON, C. R. R., 1977. Aspects of laterality in the marsupial *Trichosurus vulpecula* (brush-tailed possum). *Ann. N. Y. Acad. Sci.* **299**: 197-212.

MOELLER, H., 1973. Zur Evolutionshöhe des Marsupialia-gehirns. *Zool. Jb. Anat.* **91**: 434-48.

NELSON, J. E. AND STEPHAN, H., 1982. Encephalisation in Australian marsupials. Pp. 699-706 *in* "Carnivorous marsupials" ed by M. Archer. Royal Zool. Soc. N.S.W.: Sydney.

NEYLON, L., 1983. "The thalamus and its cortical projection in the brush-tailed possum, *Trichosurus vulpecula*, and the native cat, *Dasyurus viverrinus*, with a comparative review of thalamic organisation in marsupial and placental mammals". Ph.D. Thesis: University of Tasmania, Hobart.

NEYLON, L. AND HAIGHT, J. R., 1983. Neocortical projections of the suprageniculate and posterior thalamic nuclei in the marsupial brush-tailed possum, *Trichosurus vulpecula* (Phalangeridae) with a comparative commentary on the organisation of the posterior thalamus in marsupial and placental mammals. *J. Comp. Neurol.* **227:** 357-75.

PEARSON, L. J., SANDERSON, K. J. AND WELLS, R. T., 1976. Retinal projections in the ringtailed possum, *Pseudocheirus peregrinus. J. Comp. Neurol.* **170:** 227-40.

PIRLOT, P., 1981. A quantitative approach to the marsupial brain in an eco-ethological perspective. *Rev. Can. Biol.* **40:**
229-50.

PRACY, L. T., 1974. Introduction and liberation of the opossum *(Trichosurus vulpecula)* into New Zealand. *New Zealand Forest Service, Information Series No. 45.*

REES, S. AND HORE, J., 1970. The motor cortex of the brush-tailed possum *(Trichosurus vulpecula)*: Motor representation, motor function and the pyramidal tract. *Brain Res.* **20:** 439-51.

RIDE, W. D. L., 1970. "A guide to the native mammals of Australia". Oxford University Press: Melbourne.

ROBERTSON, M., 1981. Gene families, hopeful monsters and the selfish genetics of DNA. *Nature* **293:** 333-34.

SANDERSON, K. J. AND PEARSON, L. J., 1981. Retinal projections in the hairy-nosed wombat, *Lasiorhinus latifrons. Aust. J. Zool.* **29:** 473-81.

SANDERSON, K. J., PEARSON, L. J. AND DIXON, P. G., 1978. Altered retinal projections in the brush-tailed possum, *Trichosurus vulpecula,* following the removal of one eye. *J. Comp. Neurol.* **180:** 841-68.

SANDERSON, K. J., HAIGHT, J. R. AND PETTIGREW, J. D., 1984. The dorsal lateral geniculate nucleus of macropodid marsupials: cytoarchitecture and retinal projections. *J. Comp. Neurol.* **224:** 85-106.

SIMPSON, G. G., 1945. The principles of classification and a classification of mammals. *Bull. Am. Mus. Nat. Hist.* **85:** 1-350.

SMITH, G. E., 1902a. "Descriptive and illustrated catalogue of the physiological series of comparative anatomy contained in the Museum of the Royal College of Surgeons of England", Vol. II. Roy. Coll. Surgeons: London.

SMITH, G. E., 1902b. On a peculiarity of the cerebral commissure in certain Marsupialia, not hitherto recognised as a distinctive feature of the Diprotodonta. *Proc. Roy. Soc. Lond.* **70:** 226-31.

STRAHAN, R. (ed), 1983. "The Australian Museum complete book of Australian Mammals". Angus and Robertson: Sydney.

TOBIAS, P. V., 1971. "The brain in hominid evolution". Columbia University Press: New York.

WELLER, W. L. AND HAIGHT, J. R., 1973. Barrels and somatotopy in SI neocortex of the brush-tailed possum. *J. Anat.* **116:** 474.

WELLER, W. L., HAIGHT, J. R., NEYLON, L. AND JOHNSON, J. I., 1976. Single representation of mystachial vibrissae in SI neocortex of rufus wallaby, *Thylogale billiardierii. Neurosci. Abstr.* **2:** 926.

WISE, S. P. AND JONES, E. G., 1977. Cells of origin and terminal distribution of descending projections of the rat somatic sensory cortex. *J. Comp. Neurol.* **175:** 129-158.

ZIEHEN, T., 1897. Das Centralnervensystem der Monotremen und Marsupialer. I. Theil: Makroskopische Anatomie. In "Zoologische Forschungsreisen in Australien und dem Malayischen Archipel", Bd. III von Denkschriften der Medicinisch-Naturwissenschaftlichen Gesellschaft zu Jena. Gustav Fischer Verlag: Jena.

THE AFFINITIES OF THE KOALA *Phascolarctos cinereus* (MARSUPIALIA: PHASCOLARCTIDAE) ON THE BASIS OF SPERM ULTRASTRUCTURE AND DEVELOPMENT

H. R. HARDING[1], F. N. CARRICK[2] and C. D. SHOREY[3]

In a light microscope study, Hughes (1965) described the mature sperm heads of the Koala *(Phascolarctos cinereus)* and Common Wombat *(Vombatus ursinus)* as having features unique among marsupials, and as showing a superficial resemblance to the sperm heads of certain placental species.

Electron microscopy reveals further unusual features of sperm structure and development and Sertoli cell structure in the Koala. These include: an uneven condensation of the chromatin during nuclear development, somewhat similar to that in the Platypus (Monotremata); a mode of nuclear flattening which, relative to the flagellum, differs somewhat from that of other marsupials, and results in the unusual and 'placental-like' neck insertion of the Koala spermatozoon; features of early acrosome development; the presence of crystalloid inclusions in the basal region of the Sertoli cell, which resemble the Charcot-Böttcher crystalloids of Human Sertoli cells; features of the organisation and internal structure of the midpiece mitochondria.

In this paper we review the taxonomic and phylogenetic placements of *Phascolarctos* by authors using skeletal and other criteria, and discuss the extent to which sperm-based criteria support these phylogenies. Interpreted phenetically, the sperm-based data suggest a wide separation of the Vombatiformes from other Australian marsupials. Preliminary phylogenetic analysis of the sperm information suggests we should keep an open mind regarding the placement of the Vombatiformes in marsupial phylogenies, but that an early origin of the lineage seems most likely.

Key Words: Phascolarctidae; *Phascolarctos cinereus* (Marsupialia); Koala; Spermatozoan ultrastructure; Phylogeny.

Pages 353-64 *in* POSSUMS AND OPOSSUMS: STUDIES IN EVOLUTION ed by M. Archer. Surrey Beatty & Sons and the Royal Zoological Society of New South Wales: Sydney, 1987.

INTRODUCTION

BENDA (1906) first described Koala sperm form and development, but his description was brief and based on inadequate material. It remained for Hughes (1965) in a comparative light microscope study sixty years later, to draw attention to the unusual features shared by Koala and Common Wombat *(Vombatus ursinus)* spermatozoa. Hughes showed the sperm heads of the Koala and Wombat to be very similar to one another but quite different from those of other Australian marsupials he examined, and to bear a superficial resemblance to the sperm heads of certain placental (rodent) species. On this basis he concluded that sperm structure supported the suggestion of Sonntag (1923), that the Koala should be united taxonomically with the wombats in contrast to the prevailing view based on skeletal and dental criteria, which linked *Phascolarctos* with the ringtail possums *Pseudocheirus* (e.g., Simpson 1945).

Published reports of Koala sperm structure and development at the ultrastructural level are limited to short abstracts (Hughes 1977; Harding, Carrick and Shorey 1978), and a brief description of mature sperm structure as part of a broad comparative study (Harding, Carrick and Shorey 1979), whilst more lengthy descriptions

are included in an unpublished thesis (Harding 1977). In addition, the presence of crystalloid structures in the Sertoli cells has been noted and described (Harding *et al.* 1978; Harding, Carrick and Shorey 1982).

In this paper we emphasise the unusual features of Koala sperm structure and development evident from our ultrastructural studies, and examine the implications from these sperm based features, for placement of *Phascolarctos* in classificatory and phylogenetic systems.

MATERIALS AND METHODS

Testicular and epididymal material was obtained from four adult male Koalas *(Phascolarctos cinereus)*. Three of these animals were from south-east Queensland and were sampled in July (two animals) and August, the fourth was from Victoria and was sampled in March. Testicular material was obtained in late November from a fifth animal, from south-east Queensland.

Four of the Koalas were sampled as follows: to obtain the testicular material, the tunica albuginea of the testis was cut to expose the seminiferous tubules, small pieces of the tubules were cut and placed for 30-60 minutes in Karnovsky's fixative with picric acid added and

[1]Department of General Studies, University of New South Wales, Kensington, New South Wales, Australia 2033.
[2]Department of Veterinary Anatomy, University of Queensland, Queensland, Australia 4067.
[3]Department of Histology and Embryology, University of Sydney, New South Wales, Australia 2000.

then washed (in several changes) and left overnight in cacodylate/sucrose buffer. Post fixation was carried out in the cold for 60 minutes using osmium tetroxide/cacodylate/glucose, and the samples were stained for 30 minutes in uranyl acetate prior to dehydration. Epididymal samples were obtained from five regions. Those taken from the proximal, central and distal regions of the epididymis are referred to as caput, corpus and cauda epididymidal samples respectively, whilst intermediate material from between the former two and latter two regions is referred to as caput-corpus and corpus-cauda respectively. The samples were cut from the appropriately located segments of the epididymis using fine scissors, and then transferred to a drop of fixative (Karnovsky's with added picric acid) on a piece of dental wax. Multiple parallel cuts (1 mm or less apart) were made in the epididymal sample with a new razor blade whilst the sample was under fixative, and the process was repeated making a second series of cuts at right angles to the first. The cut epididymal tissue was then transferred to a specimen tube containing a large volume of fixative for 30-60 minutes. Further processing was as for the testicular samples.

For the remaining animal (from south-east Queensland, sampled in August) the material was fixed as follows: the animal was anaesthetised with ketamine hydrochloride (Ketalar, Parke-Davis) and heparinised by injection of 5000 units of heparin (Heparin injection B.P., Commonwealth Serum Laboratories). A cannula was then inserted into the aorta via the left ventricle, tied in place, and the animal's cardio-vascular system flushed with 250 ml of 0.1M cacodylate buffer (pH7.3). Both atria were opened to facilitate drainage. Perfusion then continued using 1 l of 2% glutaraldehyde in 0.1M cacodylate buffer (pH7.3). Immediately after the fixative ceased flowing, small pieces of testicular parenchyma were excised. These were placed in Karnovsky's fixative (with picric acid added) to give a total fixation time of 60 minutes from the start of the perfusion. The specimens were then transferred to buffer (cacodylate/sucrose) and processed in the same manner as the immersion fixed samples.

Specimens were examined using a Philips EM201 or EM300 or a JEOL 100CX. The JEOL 100CX was fitted with a goniometer stage enabling the specimens to be rotated through 80°.

OBSERVATION AND DISCUSSION

Mature Epididymal Spermatozoa

As previously described by Hughes (1965) using light microscopy, the most unusual features of mature epididymal Koala spermatozoa relative to those of other marsupials, are the form of the head and the position of insertion of the neck into the nucleus. The neck is inserted distally in a depression to one side of the caudal pole of the nucleus (Fig. 9). The latter is elongated and, in mature sperm, markedly recurved such that the neck is inserted towards the convex side of the nucleus whilst the acrosome covers the opposite, or concave side, into which it is deeply inserted centrally (Fig. 10).

The distal insertion of the neck into the nucleus in Koala spermatozoa ensures that the long axes of the flagellum and sperm head (in its immature, uncurved form) lie in parallel planes (Fig. 8). As a result Koala spermatozoa show a superficial similarity to certain placental (rodent) spermatozoa and are quite unlike those of other marsupials, at least with respect to overall head form and its relationship with the flagellum. In other marsupials, the neck is typically inserted into approximately the central region of the ventral nuclear surface whilst the acrosome lies on the dorsal surface. The nucleus is dorso-ventrally flattened and thus the long axis of the nucleus lies at right angles to the long axis of the flagellum in immature sperm. During epididymal maturation the nucleus rotates about its axis with the flagellum so that in mature spermatozoa, its long axis lies in a plane parallel to the long axis of the flagellum and thus the distal nucleus overlies the proximal midpiece (see Harding, Carrick and Shorey 1976a, 1979).

As discussed previously (Harding et al. 1979), in other respects the Koala spermatozoon is relatively unspecialised. It lacks the well developed midpiece fibrous network found in mature sperm of dasyurids, Tarsipes rostratus (Harding, Carrick and Shorey 1981, 1984) and the phalangeroids, and the midpiece invaginations found in Trichosurus vulpecula and the petaurids (Figs 11, 12, 13). The arrangement of the mitochondria and dense outer fibres in transverse sections of the midpiece is likewise unspecialised (Fig. 13), and not unlike that of a monotreme, the Platypus (Fig. 14).

Although unspecialised, the shape of the mitochondria and their internal organisation are characteristic in the Koala. In longitudinal sections the long axes of the mitochondria are seen to be orientated parallel with the long axis of the flagellum (Figs 11, 12) rather than at right angles to it as in other marsupials (Harding 1977). In addition, the arrangement of the cristae within the mitochondria is distinctive (Figs 11, 12). The cristae are generally circumferentially orientated, but are not confined to the periphery of the mitochondria as in most other marsupials (Harding 1977). The matrix is filled with electron dense material and the intracristal spaces are somewhat distended. In these two respects (orientation and internal appearance) the mitochondria of the

Koala as seen in longitudinal section show a striking resemblance to those of a monotreme, the Echidna *Tachyglossus aculeatus* (see Fig. 21, Bedford and Rifkin 1979).

Spermiogenesis and the Sertoli cell

As for mature sperm structure, there are a number of features of spermiogenesis which serve to distinguish the Koala from other Australian marsupials.

Early spermatids in the Koala are notable for the presence of a large spherical acrosomal granule within the acrosomal vacuole (Figs 1b, 1c). As spermiogenesis progresses this granule loses its spherical shape and flattens over the region of the acrosomal vacuole which is apposed to the nucleus (Fig. 1d). However, even in its somewhat flattened form, the granule remains a prominent structure. In contrast, the granular component of the acrosome is not pronounced in

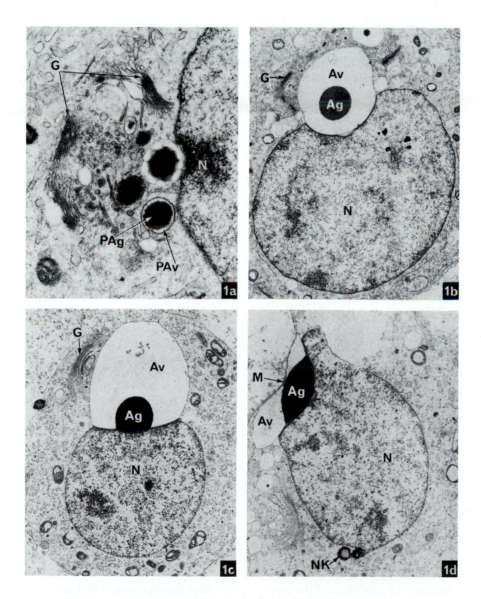

Fig. 1. Early stages of acrosome formation in spermatids of *Phascolarctos.* (a) Golgi-derived proacrosomal granules within proacrosomal vacuoles approach the surface of the nucleus. x 28500. (b) large spherical acrosomal granule lies within the acrosomal vacuole which is now clearly apposed to the nucleus. x 4700. (c) acrosomal granule has begun to flatten onto the region of the acrosomal vacuole apposed to the nucleus. x 6800. (d) acrosomal granule further flattened towards the nucleus. Acrosomal vacuole is now in contact with the spermatid plasma membrane and is in the process of collapsing. Nucleus slightly flattened and early development of the neck is evident. x 8300.

other Australian marsupial species, in which dispersed granular material within the acrosomal vacuole condenses to form a thin layer covering the acrosomal membrane adjacent to the nucleus (Harding, Carrick and Shorey 1976b). Thus, a Golgi-derived discrete proacrosomal granule housed within a proacrosomal vacuole (i.e., the granule is a discrete structure *before* apposition of the vacuole to the nuclear surface) is not a typical feature of marsupial spermiogenesis (Harding 1977) as it is in placentals (Burgos 1974).

Early acrosome development in *Phascolarctos* appears therefore to resemble the placental rather than the typical marsupial pattern with respect to the derivation of the acrosomal granule. This difference was first noticed by Benda (1906). One of the authors (Harding 1977) previously questioned Benda's finding and suggested instead that the prominent acrosomal granule in *Phascolarctos* develops rapidly after apposition of the vacuole to the nuclear surface. Further investigation suggests however, that a true proacrosomal granule may be present in the Koala (Fig. 1a). It has also been reported that a proacrosomal granule develops in the Golgi region of spermatids of the American Woolly Opossum, *Caluromys philander* (Phillips 1970).

A second unusual feature of Koala spermiogenesis concerns the manner of chromatin condensation during nuclear development. In the other marsupial species examined to date (see Harding 1977; Harding *et al.* 1976b) the fine granularity of the nucleus gradually increases in electron density as spermiogenesis progresses. However, it is typical of the Koala that the chromatin condensation is extremely uneven in the mid and later stages of spermiogenesis, with markedly electron dense areas scattered throughout the otherwise evenly granular nucleus (Fig. 4). Our preliminary observations of spermiogenesis in the Southern Hairy-nosed Wombat, *Lasiorhinus latifrons*, indicate that a similar pattern of uneven chromatin condensation occurs in this species also. As far as we know the only other mammal showing a similar pattern of chromatin condensation is *Ornithorhynchus anatinus*, but in this species the precociously electron dense regions are very organised (Fig. 5) in contrast to their apparently random nature in the Koala (Fig. 4).

A particularly interesting aspect of spermiogenesis in *Phascolarctos* is the mode of nuclear flattening relative to the long axis of the spermatid and to the position of insertion of the neck, since this determines the orientation of the head with respect to the flagellum, and thus may explain the unusual features of mature sperm head form and orientation described above for the Koala. Previously, we considered that this unusual head form, and orientation relative to the flagellum in *Phascolarctos*, may result from a flattening of the nucleus during spermiogenesis in a plane parallel to the long axis of the spermatid (as in placentals), rather than in a plane at right angles to this long axis, as is typical of marsupials (Harding 1977; Harding *et al.* 1979). Further observations have not, however, supported this suggestion. The evidence available to us now suggests the following: nuclear flattening in the Koala is similar to that in other marsupials, namely it occurs in a plane at right angles to the long axis of the flagellum; instead of insertion into the central region of the ventral nuclear surface, the neck in Koala spermatids is inserted toward the periphery of this ventral surface; as in other marsupials, the nucleus in Koala spermatids rotates around the neck insertion during late spermiogenesis so that the plane of the long axis of the nucleus approaches that of the long axis of the flagellum; due to the lateral insertion of the neck 'pre-rotation', the nucleus is distally articulated with the flagellum 'post-rotation', and this leads to its placental-like appearance (Figs 2, 3, 8).

The remaining feature to be discussed in this section concerns the Sertoli cell which is notable in *Phascolarctos* for the presence of crystalloid inclusions in its basal region (Fig. 6). As we have described these crystalloids in detail elsewhere (Harding *et al.* 1982) only a brief description is necessary here.

The crystalloids are generally located near the Sertoli cell nucleus (Fig. 6). They are non membrane-bound (Fig. 7), slender rectangular structures ranging in length from 1.8 to 15.3 μm and in width from 0.4 to 1.6 μm. They are composed of tubules (of about 15 nm diameter) which are orientated at right angles to the long axis of the crystalloid and regularly arranged in rows parallel to this long axis (Fig. 7). The tubules in adjacent rows are offset from one another at definite angles and extensively inter-connected by filaments (Fig. 7). Although the crystalloids display a range of 'patterns' in section (Fig. 7), rotation of the sections on a goniometer stage shows that these 'patterns' result from viewing crystalloids of the same basic structure from differing aspects (see Fig. 2 in Harding *et al.* 1982). We have not determined either the composition or function of the crystalloids but their association with tonofilaments and the presence of ribosomes in the vicinity suggests that they are proteinaceous (see Harding *et al.* 1982).

The Koala crystalloids are similar in appearance in many respects to the well known crystalloids of Charcot-Böttcher which are a characteristic feature of Human Sertoli cells. As far as we know Sertoli cell crystalloids are a normal and consistent constituent in only two mammalian species, the Koala and Man. Since the function of the crystalloids is unknown in both cases, and

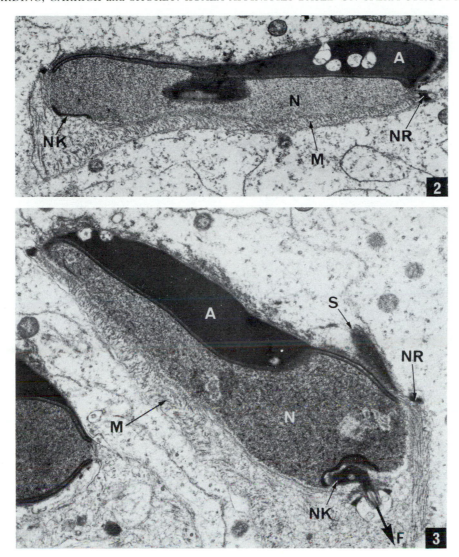

Fig. 2. Spermatid of *Phascolarctos* at the stage of nuclear flattening. The nucleus (N) is dorso-ventrally flattened with the acrosome (A) covering the dorsal surface and the neck (NK) inserted peripherally into the ventral surface. M, manchette microtubules; NR, nuclear ring. x 11000.

Fig. 3. Longitudinal section of a spermatid at a slightly later stage than that in Fig. 2. The uneven condensation of the nuclear chromatin is becoming evident. Note that the nucleus (N) has rotated such that its long axis is now approximately 30° from the plane of the long axis of the flagellum (indicated by orientation of arrow F). A, acrosome; M, microtubules of manchette; NK, neck; NR, nuclear ring; S, spurs of Sertoli cell reaction. x13500.

since the two species are widely separated phylo-genetically, we cannot assume either functional similarity or usefully draw any other inference from the presence of these structures at present.

Epididymal sperm maturation

Epididymal sperm maturation is a less obvious process in the Koala than in many other mar-supial species (see Harding *et al.* 1979). As previously reported (Hughes 1965, 1977), its chief feature in *Phascolarctos* is a marked increase in the recurvature of the sperm head (Figs 9, 10). Loss of the cytoplasmic droplet (Figs 8, 9, 10) and some rearrangement of the acrosomal material

(Fig. 8) also occur, but these are not as pro-nounced as in a number of other marsupials (Harding *et al.* 1976a, 1979). In addition, since midpiece invaginations and a well developed midpiece fibrous network are lacking in *Phas-colarctos* (Fig. 12), the development of these struc-tures does not form part of the sperm maturation process as it does in a number of other marsupial species. Similarly, since the head already lies in the same plane as the long axis of the flagellum at spermiation, rotation of the nucleus from a per-pendicular to a nearly parallel orientation with respect to the flagellar long axis, as occurs in other marsupials, is not a feature of maturation.

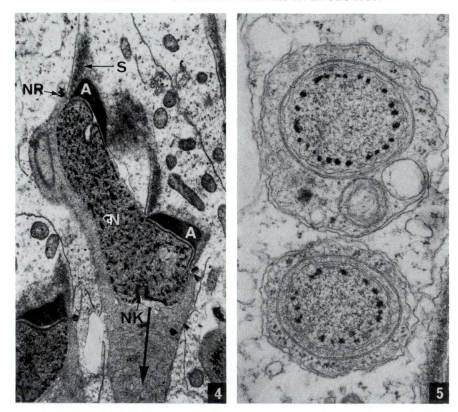

Fig. 4. Longitudinal section of a Koala spermatid at a later stage than that in Fig. 3. The long axis of the nucleus (N) is now approximately 20° from the plane of the long axis of the flagellum (arrowed). The uneven chromatin condensation is obvious in the nucleus. A, acrosome; NK, neck insertion; NR, nuclear ring; S, spurs of Sertoli cell reaction. x11500.

Fig. 5. Transverse sections through the nucleus of two spermatids of the Platypus (*Ornithorhynchus anatinus*) showing the regular arrangement of the uneven chomatin condensation pattern. x 38300.

Concluding Remarks and
Comments on Taxonomy and Phylogeny

Although Koala spermatozoa possess features which mark them undeniably as 'marsupial spermatozoa' (see Harding *et al.* 1979), they also display a number of unusual features in their mature form and in their development which set them apart from those of other marsupials (except the wombats). Hughes (1965) drew attention to the unusual features of the sperm head displayed at the light microscopy level, and noted the extreme similarity between Koala and Common Wombat *(Vombatus ursinus)* spermatozoa in this regard. The present study has confirmed Hughes' findings on the Koala and drawn attention to a number of other unusual features discernible at the ultra-structural level. Most significantly, these include: aspects of the early development of the acrosome; the mode of flattening of the nucleus and manner of condensation of its chromatin during spermiogenesis; the presence of crystalloid inclusions in the basal region of the Sertoli cell; and less significantly, the orientation of the midpiece mitochondria and their internal organisation.

Preliminary observations on spermiogenesis in the Southern Hairy-nosed Wombat *(Lasiorhinus latifrons)* (Harding, Carrick and Shorey, unpub. obs.) show that this species shares at least the unusual features of chromatin condensation and nuclear flattening with the Koala, whilst the situation regarding Sertoli cell crystalloids and early acrosomal development is not clear at present.

A relationship between *Phascolarctos* and the wombats is therefore strongly suggested by sperm studies, whilst a wide separation of the Vombatiformes (Koala and wombats) from other marsupial groups is also indicated. Thus sperm studies support the currently accepted grouping of the families Phascolarctidae and Vombatidae into the superfamily Vombatoidea (Kirsch 1977), or their more recent grouping in the suborder Vombatiformes (Woodburne *et al.* 1984), and raise a number of questions regarding the appropriate placement of the Vombatiformes in marsupial phylogenies. The remainder of this paper will discuss the latter point.

It is necessary to first outline the various placements of phascolarctids in marsupial phylogenies and then consider how well these placements fit

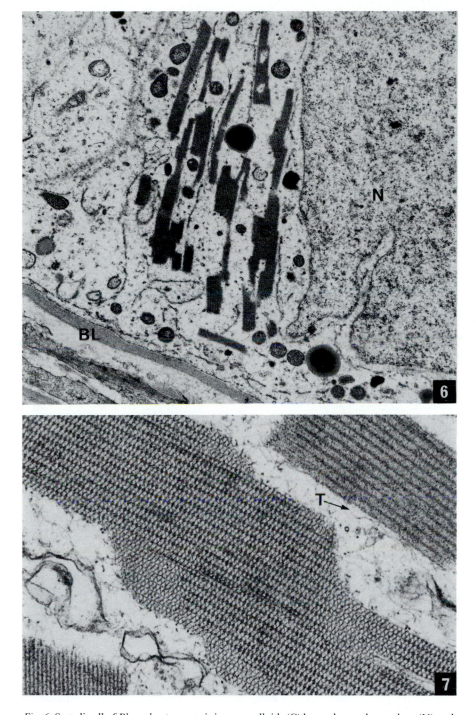

Fig. 6. Sertoli cell of *Phascolarctos* containing crystalloids (C) located near the nucleus (N) and basal lamina (BL). x 8700.

Fig. 7. Higher magnification of Sertoli cell crystalloids showing three typical 'patterns'. T, tonofilaments. x 62500.

with our knowledge of comparative sperm structure in marsupials. Since Strahan (1978) and Archer (1976, 1978) have discussed these earlier phylogenies in detail, it is necessary only to emphasise their chief features here. In most early phylogenies (e.g., Bensley 1903) the phascolarctids were placed at the end of the diprotodontan radiation as highly specialised phalangeroids. The exception was provided by Winge (1893/1941) who saw *Phascolarctos* and *Vombatus* as "specially adapted" survivors of the primitive group derived separately from didelphid stock, that gave rise to the line leading to all other modern diprotodontans (i.e., possums and kangaroos; see Ride 1978).

More recently, Archer (1976, 1978) suggested that on the basis of their proto-selenodont upper

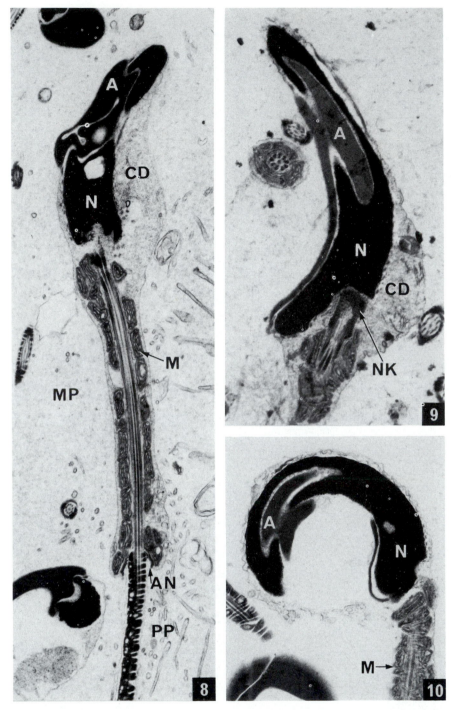

Fig. 8. Longitudinal section of a spermatozoon from the corpus-cauda epididymidis of *Phascolarctos*. The long axis of the nucleus (N) lies approximately parallel to the plane of the long axis of the midpiece (MP) and principal piece (PP) of the flagellum. Note that the acrosome (A) contains vacuities suggesting maturation is not yet complete. AN, annulus; CD, cytoplasmic droplet; M, midpiece mitochondria. x 11000.

Fig. 9. Longitudinal section through the head and proximal midpiece of a corpus epididymidal spermatozoon of *Phascolarctos*. The neck (NK) is inserted into a notch on the ventral side of the caudal pole of the nucleus (N). The nucleus is slightly recurved with the acrosome (A) covering the concave dorsal surface into which it is deeply inserted. CD, cytoplasmic droplet. x17000.

Fig. 10. Longitudinal section through the head and proximal midpiece of a distal cauda epididymidal spermatozoon of *Phascolarctos*. The nucleus (N) is now markedly recurved and the cytoplasmic droplet has been shed. A, acrosome; M, mitochondria of the midpiece. x 12000.

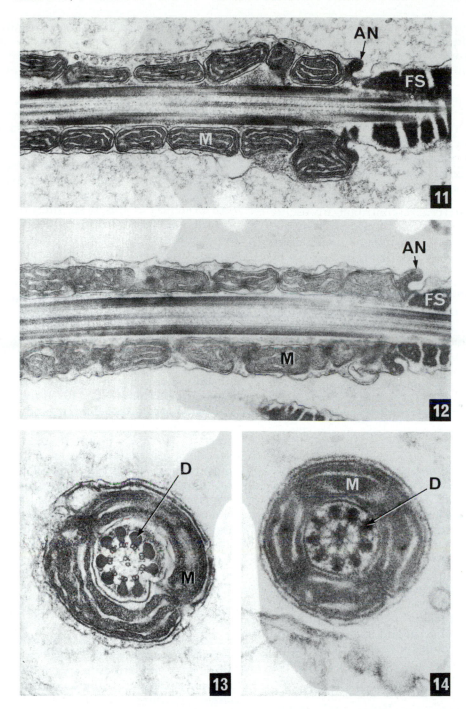

Fig. 11. Longitudinal section through the annular region of the flagellum of a caput-corpus epididymidal spermatozoon of *Phascolarctos*. Note that the long axes of the midpiece mitochondria (M) lie in a plane parallel to the long axis of the flagellum. AN, annulus; FS, fibrous sheath of principal piece. x 30500.

Fig. 12. Longitudinal section through the annular region of a cauda epididymidal spermatozoon of *Phascolarctos*. Note the lack of obvious maturation changes in the midpiece when compared with Fig. 11. AN, annulus; M, midpiece mitochondria; FS, fibrous sheath of principal piece. x 30500.

Fig. 13. Transverse section through the midpiece of a caput-corpus epididymidal spermatozoon of *Phascolarctos*. Note the unspecialised arrangement of the mitochondria (M) and dense outer fibres (D). x 49000.

Fig. 14. Transverse section through the midpiece of a corpus-cauda epididymidal spermatozoon of the Platypus (*Ornithorhynchus anatinus*). Note the general similarity to that of the Koala in Fig. 13. D, dense outer fibres; M, midpiece mitochondria. x 90000.

molar structure and syndactylous pes, the peramelids might have been ancestral to those selenodont possums which formed the base of the diprotodont radiation. In addition, he suggests that of these selenodonts, the phascolarctids' molar morphology most closely resembles that of the peramelids and therefore is in this regard the most primitive surviving selenodont family. Thus Archer sees the modern vombatiforms as retaining some characters of the group which emerged near the base of the diprotodontan radiation.

Can sperm-based criteria contribute any useful pointers on these versions of the place of the phascolarctids in Australian marsupial phylogenies?

Firstly, some comments on the idea of phascolarctids as highly derived phalangeroids (e.g., Bensley 1903). It is important to note that despite the predominance of this view among earlier authors, as outlined above, those most recently concerned with this topic (Kirsch 1968; Strahan 1978; Archer 1976, 1978) have followed the arrangement first proposed by Winge (1893/1941) in arguing (on the basis of serological, skeletal and other criteria) for derivation of the vombatiform line from the base of the diprotodontan tree.

As argued above, on sperm-based criteria we must assume a special relationship between the modern Koala and wombats relative to other marsupials. The many and extreme differences between their spermatozoa in form and development, compared to the phalangeriforms, means that if we are to accept the vombatiforms as lying at the end of the diprotodontan tree as Bensley saw them, then the rather recent acquisition of a number of characters of phascolarctid spermatozoa must be assumed. These include: the distinctive sperm head form and its manner of insertion with the flagellum; the characteristic mode of nuclear flattening; the uneven manner of chromatin condensation; an alteration in early acrosome development; the development of Sertoli cell crystalloids; the changes in the orientation and appearance of the midpiece mitochondria. In addition the loss of other features must also be assumed, namely, the midpiece fibrous network.

To bring about these changes would require either a very long evolutionary separation of the phalangeriform and vombatiform lines, or a very fast evolving sperm form in the vombatiforms. It is difficult to envisage the existence of selective pressures which might be necessary to bring about rapid evolution of sperm form. If vombatiforms are highly specialised phalangeriforms we would expect to see a greater similarity between vombatiform spermatazoa and some section of

the phalangeriforms that gave rise to them, than between all phalangeriforms to each other. As discussed above, this is not the case. Instead phalangeriform spermatozoa (based at least on examination of the genera *Petaurus, Pseudocheirus, Trichosurus, Macropus, Potorous, Setonix, Distoechurus Cercartetus, Dactylopsila*) form a recognisable group and differ markedly from vombatiform sperm.

Archer (1978) suggests that the phascolarctids and vombatids are themselves widely separate from each other. This is indicated by skeletal and fossil evidence. If this is so, then the extreme similarity of sperm form between these two groups must be due either to conservatism with respect to sperm form, or parallelism. The former is far more plausible and further suggests that on the basis of sperm criteria, the view of vombatiforms as a specialised group derived from phalangerforms is most unlikely.

What of the idea that the vombatiforms represent the end products of a lineage branching from near the base of the Australian marsupial radiation? There are at least two variants of this suggestion. Archer (1976, 1978) followed well-established precedent but argued from new evidence that the perameloids might have been ancestral to the diprotodonts, of which selenodont diprotodontans with cheek-teeth similar to those of ancestral Koalas provide likely progenitors of all other diprotodontan radiations. But Winge (1893/1941) saw the phascolarctids (within which he included the phascolarctids, vombatids, thylacoleonids and the diprotodontids) as derived directly from didelphid ancestors and he states ". . . the most primitive of the phascolarctids hardly differed in any essentials from the lowest didelphids . . .".

On the basis of sperm structure and development the following is relevant to the hypothesis that the perameloids should be considered an intermediate stage. We recently drew attention to a number of similarities in sperm form between the modern dasyurids, peramelids and *Tarsipes rostratus* (Harding *et al.* 1981, 1984). On the basis of these similarities and on the place of these groups in current phylogenies of Australian marsupials, we suggested the following. The ancestral sperm form for the major Australian marsupial radiation most likely displayed characteristics common to the sperm of these three groups: large sperm size; extreme radial displacement of the dense outer fibres from the axoneme in all but proximal and distal regions of the flagellum; a smooth profile of the flagellum at the annulus as a result of equal diameter of the mid and principal pieces in this region. In addition we suggested that a possible alternative to the controversial question of diprotodontan origins may be as follows: the dasyuroids, perameloids, thylacinids and tarsipedoids all arose from a common marsupial stock which had sperm possessing the 'primitive' features listed above. The first three groups all underwent relatively narrow radiations whilst the tarsipedoid line led to the modern *T. rostratus* and to the diprotodontan on. This explanation accommodates the anatomical links between the modern *Tarsipes* and the diprotodontans, whilst *Tarsipes* spermatozoa show some similarity to phalangeriform spermatozoa with respect to acrosome form and development, making it possible to postulate a sequence by which diprotodontan spermatozoa could have been derived from the postulated ancestral sperm type.

Where does this suggestion leave the vombatiform lineage as seen in its two modern representatives the Koala and wombats? Certainly, without an ancestral role with respect to phalangeriforms. Whilst it is undoubtedly simpler on sperm-based criteria to see the vombatiforms as an ancestral rather than a descendant group with respect to the phalangeriforms, a number of complex sperm transformations would still have to be assumed. Given the positions accorded the dasyuroids, perameloids and tarsipedoids in current phylogenies, it is simpler on sperm-based criteria to accept the postulated tarsipedoid role in phalangeriform evolution and leave the vombatiforms uninvolved in this sequence.

Winge's (1893/1941) idea of a more remote origin for the vombatiform line is more acceptable on sperm criteria. Indeed, in agreement with Winge it is tempting to suggest that modern vombatiforms are the remnants of a once much more species-rich group that was separately derived from primitive didelphoid stock. One could support such a suggestion by arguing that the unusual neck insertion and the developmental derivation of this suggested above (i.e., the relationships of neck insertion and plane of nuclear flattening during spermiogenesis), represents an early stage in a transfer from a typical monotreme/placental mode which may have existed in prototherians, and a true marsupial mode as seen in all marsupials examined so far except the vombatiforms. It could also be argued that the nuclear chromatin condensation pattern which shows similarity to that in the monotremes, is likewise a primitive feature, as is the mitochondrial form and arrangement in the midpiece, which also resembles that in the monotremes. If these monotreme features are truly primitive in marsupials they clearly imply that Koala and wombat spermatozoa have been extremely conservative in many respects. The considerable similarity of modern Koala and wombat spermatozoa, despite the long evolutionary separation suggested for the two groups (Archer 1978) supports this view of conservatism.

This interpretation of vombatiform origins and evolution clearly implies a polyphyletic origin for Australian marsupials and a wider representation of vombatiform species than those currently extant might be expected if this were so. That the phascolarctids, at least, once formed a very diverse family is now becoming clear from the fossil evidence, even though the palaentological data are still scant (Archer 1977, 1978; Archer and Bartholomai 1978; Woodburne *et al.* *1986*).

If on the other hand, we were to accept that the monotreme-like features in Koala and wombat spermatozoa are secondary, then a more conservative conclusion is possible. However, in our opinion a very early origin of the lineage is still necessary. That is, the sperm-based features support a derivation of the vombatiform line from near the base of Australian marsupial radiations in contrast to placing this group at the end of the diprotodontan tree. However, unless the features appeared after separation from the group of "early phascolarctids" they do not support the more specific derivation of the vombatiform line from the base of the diprotodontan tree. Rather, sperm criteria suggest that until the palaentological data increase, we should keep an open mind regarding the placement of the vombatiform line in marsupial phylogenies.

ACKNOWLEDGEMENTS

The authors would like to thank, Dr W. D. L. Ride for his constructive criticisms of an earlier draft of this paper; Mr D. Harbrow, Mr S. Brown and Ms B. Canty for technical assistance; the Queensland National Parks and Wildlife Service and the Victorian Ministry of Conservation (Fisheries and Wildlife Division) for providing permits to work on this protected species; the staff of the Lone Pine Koala Sanctuary, Brisbane, for their assistance in obtaining animals. This work was supported by a grant from the Australian Research Grants Committee, Number DI-77/15525.

REFERENCES

ARCHER, M., 1976. Phascolarctid origins and the potential of the selenodont molar in the evolution of diprotodont marsupials. *Mem. Qd Mus.* **17**: 367-71.

ARCHER, M., 1977. *Koobor notabilis* (De Vis), an unusual koala from the Pliocene Chinchilla Sand. *Mem. Qd Mus.* **18**: 31-35.

ARCHER, M., 1978. Koalas (phascolarctids) and their significance in marsupial evolution. Pp. 20-28 *in* "The Koala" ed by T. J. Bergin.Proceedings of the Taronga Symposium, Sydney, 1976. Zoological Parks Board of NSW: Sydney.

ARCHER, M. AND BARTHOLOMAI, A., 1978. Tertiary mammals of Australia: a synoptic review. *Alcheringa* **2**: 1-19.

BEDFORD, J. M. AND RIFKIN, J. M., 1979. An evolutionary view of the male reproductive tract and sperm maturation in a monotreme mammal — the Echidna, *Tachyglossus aculeatus. Am. J. Anat.* **156**: 207-30.

BENDA, C., 1906. Die Spermiogenese der Marsupialier. *Denkschr. med-naturw. Ges. Jena* **6:** 441-58.

BENSLEY, A. B., 1903. On the evolution of the Australian Marsupialia; with remarks on the relationship of the marsupials in general. *Trans. Linn. Soc. Lond. Zool.* **9:** 83-217.

BURGOS, M. H., 1974. Ultrastructure of the mammalian sperm head during differentiation and maturation. Pp. 209-25 *in* "Physiology and Genetics of Reproduction". Part A ed by E. M. Couthino and F. Fuchs. Plenum Press: New York.

HARDING, H. R., 1977. "Reproduction in male marsupials: a critique, with additional observations on sperm development and structure". Ph.D. Thesis: University of New South Wales, Sydney.

HARDING, H. R., CARRICK, F. N. AND SHOREY, C. D., 1976a. Ultrastructural changes in spermatozoa of the Brush-tailed Possum, *Trichosurus vulpecula* (Marsupialia), during epididymal transit. Part II: The acrosome. *Cell Tiss. Res.* **171:** 61-73.

HARDING, H. R., CARRICK, F. N. AND SHOREY, C. D., 1976b. Spermiogenesis in the Brush-tailed Possum, *Trichosurus vulpecula* (Marsupialia). The development of the acrosome. *Cell Tiss. Res.* **171:** 75-90.

HARDING, H. R., CARRICK, F. N. AND SHOREY, C. D., 1978. Unusual features of spermiogenesis and the Sertoli cell in the Koala *(Phascolarctos cinereus). Procs. 10th Ann. Conf. Aust. Soc. Reprod. Biol.* Pp. 53.

HARDING, H. R., CARRICK, F. N. AND SHOREY, C. D., 1979. Special features of sperm structure and function in marsupials. Pp. 289-303 *in* "The spermatozoon" ed by D. W. Fawcett and J. M. Bedford. Urban and Schwarzenberg: Baltimore-Munich.

HARDING, H. R., CARRICK, F. N. AND SHOREY, C. D., 1981. Marsupial phylogeny: new indications from sperm ultrastructure and development in *Tarsipes spenserae? Search* **12:** 45-47.

HARDING, H. R., CARRICK, F. N. AND SHOREY, C. D., 1982. Crystalloid inclusions in the Sertoli cell of the Koala, *Phascolarctos cinereus* (Marsupialia). *Cell Tiss. Res.* **221:** 633-42.

HARDING, H. R., CARRICK, F. N. AND SHOREY, C. D., 1984. Sperm ultrastructure and development in the Honey Possum *Tarsipes rostratus.* Pp. 451-61 *in* "Possums and gliders" ed by A. P. Smith and I. D. Hume. Australian Mammal Society: Sydney.

HUGHES, R. L., 1965. Comparative morphology of spermatozoa from five marsupial families. *Aust. J. Zool.* **13:** 533-43.

HUGHES, R. L., 1977. Light and electron microscope studies on the spermatozoa of the koala, *Phascolarctos cinereus* (Marsupialia). *J. Anat., Lond.* **124:** 513.

KIRSCH, J. A. W., 1968. Prodromus of the comparative serology of Marsupialia. *Nature, Lond.* **217:** 418-20.

KIRSCH, J. A. W., 1977. The comparative serology of Marsupialia, and a classification of marsupials. *Aust. J. Zool. Suppl. Ser. No.* **52:** 1-152.

PHILLIPS, D. M., 1970. Development of spermatozoa in the Wooly Opossum with special reference to the shaping of the sperm head. *J. Ultrastruct. Res* **33:** 369-80.

RIDE, W. D. L., 1978. An historical introduction to studies on the evolution and phylogeny of the Macropodidae. *Aust. Mammal.* **2:** 1-14.

SIMPSON, G. G., 1945. The principles of classification and a classification of mammals. *Bull. Amer. Mus. Nat. Hist.* **85:** 1-350.

SONNTAG, C. F., 1923. On the myology and classification of the wombat, koala and phalangers. *Proc. Zool. Soc. Lond.* **1922:** 863-96.

STRAHAN, R., 1978. What is a Koala? Pp. 3-19 *in* "The Koala" ed by T. J. Bergin. Proceedings of the Taronga Symposium, Sydney, 1976. Zoological Parks Board of NSW: Sydney

WINGE, H. 1893/1941. Originally published in Danish as: Fossil and living marsupials (Marsupialia) from Lagoa Santa, Minas Geraes, Brazil, with a review of the interrelationships of the Marsupials. *E Museo Lundii* **2**(1): 1-132 (1893). English translation (1941): "The interrelationships of the mammalian genera. Vol. I: Monotremata, Marsupialia, Insectivora, Chiroptera, Edentata" ed by S. Jensen, R. Spärck and H. Volsøe. C.A. Reitzels Forlag: Copenhagen.

WOODBURNE, M. O., TEDFORD, R. H. AND CASE, J. A., 1984. Interrelationships of the families of the Diprotodonta. "Possums and opossums: 1984 symposium on evolution". *Abstracts* p. 8. Royal Zoological Society of New South Wales: Sydney.

WOODBURNE, M. O., TEDFORD, R. H., ARCHER, M. AND PLEDGE, N. S., 1987. *Madakoala*, a new genus and two species of Miocene Koala (Marsupialia, Phascolarctidae) from South Australia, and a new species of *Perikoala*. Pp. 293-317 *in* "Possums and opossums: studies in evolution" ed by M. Archer. Surrey Beatty & Sons and the Royal Zoological Society of New South Wales: Sydney.

PLATE 3: The living Common Wombat, *Vombatus ursinus*.
Photograph by E. Beaton courtesy the Australian National Photographic Index.

SPECIMENS OF *Warendja wakefieldi* (VOMBATIDAE: MARSUPIALIA) FROM THE PLEISTOCENE OF SOUTH AUSTRALIA

T. F. FLANNERY[1] and N. S. PLEDGE[2]

A cranial fragment and isolated upper teeth of *Warendja wakefieldi* Hope and Wilkinson, 1982, from Comaum Forest Cave (U118), South Australia, represent a second occurrence of the species. The cranial fragment indicates that the skull of *W. wakefieldi* was much more lightly built, narrower and probably deeper than that of any other vombatid. The palate is dorso-ventrally thin, fragile and relatively broader than that of other vombatids. The palate and the jugal are phenetically closer in structure to those of phascolarctids and plesiomorphic phalangeroids than they are to other vombatids.

Key Words: *Warendja wakefieldi*; Pleistocene; Vombatidae; South Australia.

Pages 365-68 *in* POSSUMS AND OPOSSUMS: STUDIES IN EVOLUTION ed by M. Archer. Surrey Beatty & Sons and the Royal Zoological Society of New South Wales: Sydney, 1987.

INTRODUCTION

WARENDJA WAKEFIELDI Hope and Wilkinson, 1982 is a small, primitive vombatid previously known only from two dentaries and six isolated teeth from Pleistocene sediments in McEachern's Cave, southwestern Victoria (Hope and Wilkinson 1982). This paper documents the occurrence of a right cranial fragment containing P^3, M^{2-5} and left isolated M^2 and M^4s from Pleistocene sediments in Comaum Forest Cave, South Australia. Comaum Forest Cave is situated in the southeastern part of South Australia, approximately 90 km north of McEachern's Cave (Fig. 1). Mammalian fossils associated with the *W. wakefieldi* remains in Comaum Forest Cave include *Zaglossus* sp. cf. *Z. ramsayi, Dasyurus maculatus, Dasyurus* sp. cf. *D. geoffroii, Sarcophilus* sp., *Antechinus/Sminthopsis* indet., *Thylacinus cynocephalus, Perameles* sp., *Pseudocheirus* sp., *Trichosurus* sp., *Phascolarctos* sp., *Thylacoleo carnifex, Vombatus* sp. cf. *V. ursinus, Macropus* sp. cf. *M. giganteus, Macropus rufogriseus, Potorous tridacylus, Bettongia* sp., *Simosthenurus gilli, Simosthenurus maddocki, Simosthenurus* sp. cf. *S. occidentalis, Pseudomys* sp. and *Rattus* sp. (Pledge pers. comm. 1984).

MATERIAL

Warendja wakefieldi is represented in the Comaum Forest Cave Fauna by a fragmentary right jugal and maxilla containing P^3 and M^{2-5} (SAM P24105), a left M^2 and M^4 (SAM P24598), and a left M^4 fragment (SAM P24599).

DESCRIPTION

Maxilla and jugal. The jugal of SAM P24105 lacks only a small posterior segment, being broken away 16 mm posterior to the anterior edge of the jugal-squamosal suture (Fig. 2). It preserves all of its lachrymal and maxillary sutures. The lachrymal-jugal suture is vertically-oriented (parallel with the skull midline) and not near horizontal (nearly parallel with the palate) as in extant vombatids. The maxilla does not contribute to the anteroventral margin of the zygomatic arch as in extant vombatids. The ventral edge of the orbit lies closer to the midline of the skull and does not form a laterally expanded, near-horizontal plate as is seen in other vombatids.

Parts of the palate are preserved up to 11 mm anterior to the P^3 alveolus and the palate extends for a width of at least 8.8 mm medial to the M^2 alveolus (where it is broken away). Neither the maxillary-maxillary nor maxillary-premaxillary sutures are preserved. The palate is strikingly different in morphology from that seen in other vombatids. In adult *Vombatus ursinus* the palate attains a width of only about 3 mm between its midline and M^2, but in species of *Lasiorhinus* this distance can be as great as 6 mm (Dawson 1983). In species of both genera the palatal suture is extremely thick. Because of the molar curvature, and because the palate is so narrow in *V. ursinus*, the left and right molars almost contact at their midpoints. In *W. wakefieldi* the palate is thin and fragile as in phalangerids or macropodoids, and extends for at least 8.8 mm between M^2 and the midline suture of the maxillaries. Thus the left and right molar rows are widely separated. The distance between the anterior end of the cheek-tooth rows is also relatively greater in *W. wakefieldi* than other vombatids because *W. wakefieldi* represents a much smaller animal than any other vombatid. The oral surface of the palate is perforated by several foramina, the larger ones being situated anterior to P^3

[1]Australian Museum, 6-8 College Street, Sydney, New South Wales, Australia 2000.
[2]South Australian Museum, North Terrace, Adelaide, South Australia, Australia 5000.

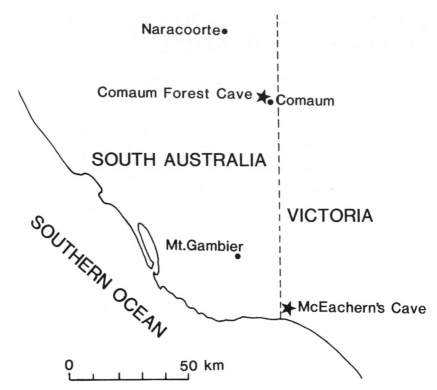

Fig. 1. A map of southeastern South Australia and southwestern Victoria showing the location of McEachern's and Comaum Forest Caves.

and medial to M^2. A portion of the palatine-maxillary suture is preserved opposite the middle portion of M^4. This suture occurs in a similar position in the species of *Lasiorhinus* and *Vombatus*.

The buccal parts of the maxilla and jugal dorsal to the anterior of the cheektooth row are smooth and only slightly concave. This is in strong contrast to extant vombatids, where the area is greatly excavated and a ridge extends above P^3-M^3. There is no masseteric process present. Overall, the cranial fragment suggests an animal with a much narrower, more lightly built and perhaps deeper skull than extant vombatids. Certainly the sites for attachment of the masseters indicate that these muscles were not as powerful as in extant vombatids, as was noted by Hope and Wilkinson (1982) based on the dentary.

Dentition. The P^3 crown does not extend nearly as far ventrally as do those of the molars, which compliments the lower dentition where the anterior part of M_2 and sometimes P_3 jut above the rest of the cheektooth row. Almost certainly, the anterior prominence of the M_2 occluded with the P^3 notch. The P^3 conforms closely in morphology to the isolated tooth tentatively identified as such (NMV P165429) by Hope and Wilkinson (1982). The pulp cavity is shallow compared with extant vombatids, extending only one-quarter to one-fifth the depth of the tooth. M^{2-3} are subequal in height. M^4 is markedly shorter than M^3 and M^5 shorter than M^4. The posterior lobe of M^5 is much smaller than the

anterior lobe. It is apparent from study of this material that NMV P165429 was correctly identified by Wilkinson and Hope (1982) as a left P^3, but that other upper cheekteeth were incorrectly identified by them. NMV P165432 is a right M^2 (not a left M^3), NMV P165431 a left M^2 (not an M^4) and NMV P165430 a left M^3 (not a left M^4). These new identifications are made on the following basis: upper molars of *W. wakefieldi* curve slightly posteriorly, allowing assignation of isolated teeth to either left or right cheektooth rows; M^2 is distinctive because its anterior lobe is much smaller than its posterior lobe and it is much larger than M^5 (the only other molar with greatly unequal lobes); M^4 can be distinguished from M^3 because the shallow buccal groove separating the lobes is more posteriorly placed.

DISCUSSION

The Comaum specimen of *Warendja wakefieldi* is important for several reasons. It represents only the second record of the species and documents a small range extension. It also reveals more of the morphology of this unique vombatid and sheds light on speculation regarding the species put forward in Hope and Wilkinson (1982). It allows for correct identification of the upper molars and also shows that speculation by Hope and Wilkinson that *W. wakefieldi* may have possessed a masseteric process is incorrect.

The Comaum specimens are doubtless conspecfic with the McEachern's Cave material

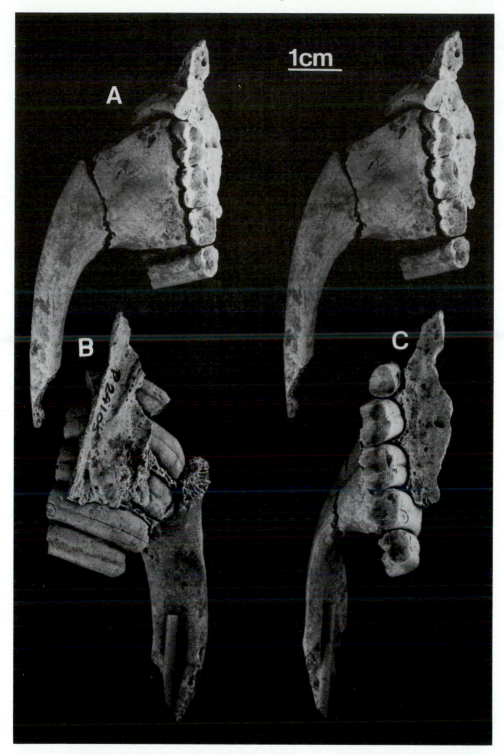

Fig. 2. A, stereopair of buccoventral view; B, lingual view; and C occlusal view of SAM P24105, cranial fragment of *W. wakefieldi* from Comaum Forest Cave, South Australia.

(although they are slightly smaller, see Table 1) because there are very few differences in morphology between this material and the upper cheekteeth from the type locality.

The maxilla fragment of *W. wakefieldi* confirms the idea that the species represents a very primitive vombatid. The structure of the palate and zygomatic arch are more similar to less dentally specialised diprotodontans such as phalangerids and phascolarctids than they are to other vombatids.

Hope and Wilkinson suggested that the palaeoenvironment in the area of McEachern's Cave at the time of the deposition of the *W. wakefieldi* material was wet sclerophyll forest. In support of this hypothesis they cite the absence of

Table 1. Measurements for specimens of *Warendja wakefieldi* from Comaum Forest Cave, South Australia. l = maximum length of wear facet, w = maximum width of wear facet, h = tooth height.

	l	w	h
SAM P24105			
P^3 (right)	5.5	4.3	16.4
M^2 (right)	7.6	5.7	23.5
M^3 (right)	7.8	6.2	22.4
M^4 (right)	6.8	6.0	19.3
M^5 (right)	5.8	4.8	15.2
SAM P24598			
M^2 (left)	7.6	5.7	22.0
M^4 (left)	7.5	5.9	21.3
SAM P24599			
M^4 (left)			15.4

Shortest distance between rim of orbit and M^3 alveolus = 24.9 mm.

heathland species, which are common in other units in the cave, and the existence of species known to inhabit sclerophyll forests such as *Rattus fuscipes, Antechinus stuartii* and *Perameles nasuta* in the unit containing the remains of *W. wakefieldi*. Investigation of the fauna of Comaum Forest Cave is at too early a stage to allow for palaeoenvironmental interpretation. However, the presence of *Dasyurus maculatus* and a koala

at the site lend some support to the idea that sclerophyll forest occurred in the Comaum Forest Cave area when *W. wakefieldi* existed. Further work currently being undertaken by Pledge should add to our knowledge of the palaeoecology of the Comaum Forest Cave Fauna.

CONCLUSIONS

1) A cranial fragment and three isolated molars from Comaum Forest Cave, South Australia represent a second occurrence of *Warendja wakefieldi*.

2) The newly-described maxillary and jugal structure of the species further supports the conclusion that *W. wakefieldi* is a very primitive vombatid.

REFERENCES

HOPE, J. H. AND WILKINSON, H. E., 1982. *Warendja wakefieldi,* a new genus of wombat (Marsupialia: Vombatidae) from Pleistocene sediments in McEachern's Cave, Western Victoria. *Mem. Nat. Mus. Vict.* **43:** 109-20.

DAWSON, L., 1983. The taxonomic status of small fossil wombats (Vombatidae: Marsupialia) from Quaternary deposits, and of related modern wombats. *Proc. Linn. Soc. NSW* **107:** 101-23.

BASICRANIAL ANATOMY OF THE EARLY MIOCENE DIPROTODONTIAN *Wynyardia bassiana* (MARSUPIALIA: WYNYARDIIDAE) AND ITS IMPLICATIONS FOR WYNYARDIID PHYLOGENY AND CLASSIFICATION

K. P. APLIN[1]

The basicranial region of the early Miocene diprotodontian *Wynyardia bassiana* is re-examined, with special reference to the structure of the auditory region and the glenoid fossa. Drawing on broad comparative studies, these regions of the *Wynyardia* basicranium are shown to be distinctly diprotodontian-like rather than didelphoid-like as previously reported. Current inclusion of *Wynyardia* within Diprotodontia, as based on general cranial, postcranial and neuroanatomical features, is thereby reinforced. Compared with other diprotodontians, *Wynyardia* is shown to differ significantly in auditory anatomy from members of each of the three major, infraordinal groups (i.e., Phalangerida, Vombatomorphia and Phascolarctomorphia), including those families with which it has been tied in the past (i.e., Phalangeridae, Macropodidae and Vombatidae). On cladistic grounds, however, it is concluded that *Wynyardia* falls closer to the Vombatomorphia than to either other group. Within this group, *Wynyardia* appears to be the most plesiomorphic and generally least specialised of known forms and as such may represent the sister group of all other vombatomorphians. Formal recognition of such a relationship is deferred pending further studies of the post-cranial skeleton of *Wynyardia* and resolution of its relationship with respect to certain South Australian Miocene age diprotodontians recently referred to Wynyardiidae.

Key Words: Auditory region; Glenoid fossa; *Wynyardia bassiana*; Basicranium; Marsupial; Miocene; Phylogeny.

Pages 369-91 *in* POSSUMS AND OPOSSUMS: STUDIES IN EVOLUTION ed by M. Archer. Surrey Beatty & Sons and the Royal Zoological Society of New South Wales: Sydney, 1987.

INTRODUCTION

IN THE years since Spencer (1901) first described *Wynyardia bassiana,* this phylogenetically enigmatic Miocene diprotodontian has been re-examined by a succession of palaeontologists (e.g., Gill 1957; Ride 1964) and comparative anatomists (e.g., Jones 1931; Haight and Murray 1981). While this has clarified other questions about its age and palaeohabitat (cf. Aplin and Rich 1985), its relationships to other fossil and living marsupials remain in significant doubt. Over the last twenty years, for example, *Wynyardia* has been variously regarded to be: a primitive diprotodontian showing many didelphoid features (Ride 1964; Kirsch 1977); a "vombatoid" (vombatiform of Aplin and Archer 1987) with possible affinities to either Vombatidae (Rich and Archer 1979; Pledge 1987) or Diprotodontoidea (Archer 1984); and a "phalangeroid" (phalangeridan of Aplin and Archer 1987) and possibly close to Phalangeridae (Haight and Murray 1981).

To a large degree this uncertainty over the taxonomic position of *Wynyardia* reflects the inherent limitations of the still unique holotype. In particular, the complete absence of teeth from the broken cranium and jaws has severely restricted comparisons with other fossil taxa, most of which are represented by nothing but teeth. Moreover, the study of skeletal features other than the dentition has been largely neglected within Marsupialia and attempts to determine the phylogenetic position of *Wynyardia* on the basis of cranial and postcranial characters have been hindered by this fact.

One aspect of skeletal structure in *Wynyardia* that has attracted considerable attention is the anatomy of the auditory region. That this region of the cranium is well preserved in *Wynyardia* was noted by both Spencer (1901) and Jones (1931) but neither author gave more than superficial remarks regarding its anatomy or significance. It fell to Ride (1964), therefore, to first adequately prepare this region of the skull and to explore its potential value for phylogenetic interpretation. Based on his analysis and comparisons, Ride (1964:115) concluded that the auditory region in *Wynyardia* is ". . . almost identical with that of modern *Didelphis virginiana* (a didelphid marsupial) . . .". On this basis he characterised *Wynyardia* as a "primitive diprotodont" showing "clear evidence of marsupicarnivoran origins of a didelphid kind . . ." (1964:115). This concept was subsequently formalised by Kirsch (1977) through his recognition of a monotypic superfamily, Wynyardioidea.

[1]School of Zoology, University of New South Wales, P.O. Box 1, Kensington, New South Wales, Australia 2033.

As noted above, a more contemporary view of *Wynyardia* would place it close to either Phalangeridae, Vombatidae or Diprotodontidae, based on aspects of its postcranial anatomy (Tedford *et al.* 1977), other features of the cranium (Rich and Archer 1979; Pledge 1987) and on the gross morphology of its endocranial cast (Haight and Murray 1981). Since in all of these taxa the auditory region is significantly more specialised than that of any didelphid (Kampen 1905; Segall 1969; Stirton 1967), Ride's (1964) observations about *Didelphis*-like basicranium features in *Wynyardia* should probably be reconsidered.

In this article I present the results of such a detailed comparative study of the auditory region and glenoid fossa in *Wynyardia*.

TERMINOLOGY AND ABBREVIATIONS

A comprehensive terminology for the marsupial auditory region was given by Archer (1976), based largely on the earlier works by Kampen (1905), Gregory (1910) and Kampen (1931) but incorporating some new information on homologies with otic structures in the mammal-like reptiles. Although Archer's terminology is unambiguous and has historical continuity with these earlier works, it differs in many respects from the terminology currently in use (and developed largely since Archer's study) for the description of auditory structures in placental mammals. Since many of the structures concerned are undoubtedly homologous between placentals and marsupials, it is in the interests of students of both groups and of basicranial morphology *per se* that a consistent terminology be employed wherever possible. Among recent works on placental auditory structures, that of MacPhee (1981) presents the most comprehensive terminology and contains the best documented glossary; it is also coming into general usage among placental systematists. For these reasons MacPhee's terminology is employed here as it is also in Aplin and Archer (1987). In the case of structures without obvious homologues in placental mammals, I have either followed Archer's terminology or, for previously unnamed structures, coined an appropriate new term. Muscular features not listed by MacPhee (1981) or Archer (1976) are named in accordance with Barbour's (1963) nomenclature for the Brush-tailed Possum, *Trichosurus vulpecula*. Abbreviations of all anatomical terms used in Figs 2-9 are listed below.

All measurements are in millimetres and were taken directly from the specimen using dial calipers.

ABBREVIATIONS

The following list refers to abbreviations used in Figs 2-8 and, in some cases, as a shorthand throughout the text.

ae	articular emminence (of the glenoid fossa)
amr	anterior malleolar ridge
ap	tympanic process of the alisphenoid
ats	sulcus for auditory (eustachian) tube
b	basisphenoid
bf	basicapsular fenestra
bo	basioccipital
cat	auditory tube canal
cp	caudal tympanic process of petrosal
eam	external auditory meatus
em	meatal process of ectotympanic
er	epitympanic recess
fcf	fossula of cochlear fenestra
ff	foramen faciale
fo	foramen ovale
fp	anterior (folian) process of malleus
fr	foramen rotundum
fs	facial sulcus
fvf	fossula of vestibular fenestra
gf	glenoid fossa
hp	hamular process of pterygoid
i	incus
icc	internal carotid canal
icf	internal carotid foramen
if	incudal fossa
ijf	foramen of internal jugular vein
ips	sulcus of inferior petrosal sinus
j	jugal
mes	mastoid epitympanic sinus
mf	mandibular fossa (of glenoid fossa)
mr	mastoid process (nontympanic) of petrosal
oc	occipital condyle
pa	palatine
pf	pterygoid fossa
pgc	postglenoid cavity
pgf	postglenoid foramen
pgp	postglenoid process
plf	posterior lacerate foramen
pp	paroccipital process
pr	promontory of cochlea
pt	pterygoid
ptp	post-tympanic process of squamosal
pw	epitympanic wing of petrosal
pyf	pyriform fenestra
rp	rostral tympanic process of petrosal
sep	entoglenoid process of squamosal
sf	fossa for stapedius m.
sqp	tympanic process of squamosal
sqs	epitympanic sinus of squamosal
sqw	epitympanic wing of squamosal
ssf	subsquamosal foramen
sty	sinus tympani
sw	epitympanic wing of alisphenoid
tc	transverse canal
tf	fossa of tensor tympani

DESCRIPTION

GENERAL OBSERVATIONS

Preservation: General views of the dorsal and ventral surfaces of the damaged cranium are shown in Fig. 1. Comparison with earlier illustrations such as Spencer (1901, plate 49) and Jones

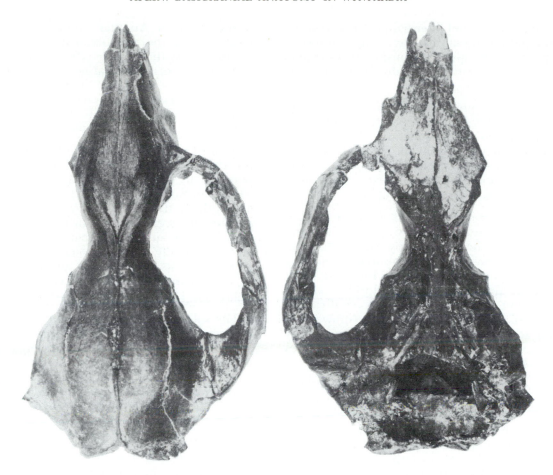

Fig. 1. Wynyardia bassiana Spencer, holotypic cranium in dorsal (left) and ventral (right) views. Scale bar represents 10 mm. Compared with previous illustrations (e.g., Spencer 1901, plate 49; Jones 1931, plate 5), note alteration to shape of anterior part of zygomatic arch, resulting in a more gently sloping contour. Note also that, due to presence of matrix-wedged fracture at posterior end of arch, its anterior part cannot be exactly fitted to its maxillary root.

(1931, plate 5) reveals a slight change in the morphology of the anterior part of the zygomatic arch. This is the result of minor corrective reconstruction of this region by the present author (see caption for Fig. 1). Otherwise preparation of the specimen was limited to some cleaning of the right auditory region. The left tympanic cavity appears to be less complete. It was left unprepared.

Despite its damaged condition, the specimen retains excellent anatomical detail, particularly with regard to muscular and neurovascular impressions, and appears to have suffered little overall distortion of shape.

Sutural relations: Sutural contacts are generally well-defined due to the penetration into even the finest sutures of the contrasting white limestone in which the specimen was preserved. Identification of sutures in some areas proved impossible (e.g., in the sphenoid complex; see below), but this is to be expected given the clearly advanced individual age of the specimen (as indicated by the prominent development of the sagittal and nuchal crests).

As shown in Fig. 2, I regard the alisphenoid to be absent from the middle-ear roof in *Wynyardia*. Because this observation is critical to the present interpretation of the phylogenetic significance of *Wynyardia,* the course of the alisphenoid-squamosal suture, insofar as it pertains to the auditory region, is described in detail.

As noted by Jones (1931), the alisphenoid-squamosal suture is most distinct just behind the complex muscular fossae of the infratemporal fossa. From this point, the suture runs posteromedially, staying well clear of the glenoid fossa, to approach the anterolateral corner of the exposed tympanic cavity. From there it follows the margin of the tympanic cavity, running first across the front and then down the medial margin of this structure. Over the section between the infra-temporal fossa and tympanic cavity, the suture is finely convoluted, forming a low, narrow ridge. By contrast, across the front of the tympanic cavity, the bones interdigitate in complex fashion. Finally, down the medial side of the tympanic cavity, the alisphenoid is represented by a thin, subvertical lamina which overlays but does not interdigitate with the laterally placed squamosal. The ragged appearance of the posterior margin of the alisphenoid suggests the former presence of a substantial tympanic process (see below).

Within the sphenoid complex itself, I am unable to distinguish separate alisphenoid, orbitosphenoid or basisphenoid elements, nor can I accurately determine the former location of the intersphenoidal synchondrosis. This is the usual condition only among very aged individuals of most species of marsupaials.

Non-auditory structures: Most of the important anatomical features outside of the auditory region have been correctly identified and adequately described by one or other of Spencer (1901) or Jones (1931). The following brief comments are thus not intended to be comprehensive but should serve rather to confirm the identity of the major features and to draw attention to some lesser ones that have not previously been noted; a range of non-auditory structures are indicated in Fig. 2.

Basisphenoid: The precise limits of the basisphenoid cannot be determined. In general, however, it is a flat rhomboidal element which lies in the same plane as the (indistinct) presphenoid. Posteriorly the basisphenoid appears to have been incompletely fused to the basioccipital. Approximately one-third to one-half the way along its probable original length, the basisphenoid is perforated by large entocarotid foramina which open directly into the endocranial cavity. Associated with these foramina are well-defined entocarotid sulci which might once have been partially floored (i.e., to form true canals). Large entocarotid processes project posteriorly beyond the main body of the basisphenoid. Other features of note are the strongly developed, paired muscular fossae (for mm. longus capitus) and the presence of well-defined bilateral sulci indicating the line of attachment of the missing pterygoid bones.

Alisphenoid: As indicated above, the alisphenoid does not contribute to the roof of the tympanic cavity but it might have supplied a ventral tympanic process. Other interesting features are the very reduced size of the transverse canal, the lack of a well-defined neurocranial pterygoid fossa, the large and complex pterygoid lamina and the well-defined sulci for the major rami of the mandibular division of n. trigeminus. Whether *Wynyardia* possessed a foramen ovale as against a f. pseudovale (*sensu* Archer 1976) cannot be determined because the relevant area is damaged.

Squamosal: The squamosal is a robust element which is suturally distinct from the petrosal along the entire length of preserved contact. This is of some interest given the advanced individual age of the specimen. Another feature of note is the large size of the neurocranial lamina of the squamosal. With Spencer (1901), I regard this element to meet the frontal along a broad front in the infratemporal fossa. This conflicts with Jones's (1931) interpretation of the squamosal as a more restricted element. All other features relating to this bone are discussed below.

Petrosal: The petrosal is of typical mammalian form (cf. MacIntyre 1972). Three major components are distinguished: a pars cochlearis which formerly contained the membraneous labyrinth of the inner ear; a pars canicularis containing the gyri of the semicircular canals and enclosing the subarcuate (parafloccular) fossa; and a laterally directed "tegmen tympani". Aspects of its endocranial morphology are reported in the accounts of Spencer (1901), Jones (1931) and Haight and Murray (1981).

The sub-vertically oriented medial surface of the petrosal shows several features of interest. Posteriorly, there is a broad, vertically-oriented sulcus marking the position of the posterior lacerate foramen. This lies directly medial to the sinus tympani of the tympanic cavity (see below). An elliptical perilymphatic foramen measuring 2.1 by 1.0 mm is present near the ventral margin of this sulcus.

Anterior to the posterior lacerate sulcus, the medial face of the petrosal displays a ventrally-placed, rugose area and a dorsally-placed sub-horizontal sulcus. Of these the rugose area most likely represents a zone of former bony contact with the now-missing basioccipital bone. The dorsal sulcus, defined endocranially by the inferior petrosal ridge, is clearly of vascular origin. It most likely marks the passage of a large inferior petrosal sinus.

THE MIDDLE-EAR

The anatomy of the right tympanic cavity of *Wynyardia* is illustrated in Fig. 2. Damage to this part of the cranium has resulted in the loss of all tympanic floor elements, including the ectotympanic, and the auditory ossicles. The roof of the middle ear itself is undamaged save for the presence of a clean fracture through the root of the tegmen tympani.

Major roof elements

The roof of the tympanic cavity is comprised of contributions from only two primary basicranial elements: the squamosal and petrosal. The alisphenoid is not represented in the roof of the tympanic cavity other than as part of the thickened base of the now truncated anterior tympanic process.

Squamosal: This element contributes to enclosure of the tympanic cavity in two areas. Anteriorly, it

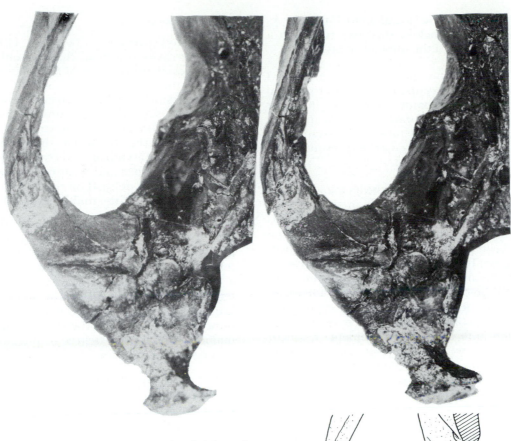

Fig. 2. Wynyardia bassiana Spencer. Stereopair of right auditory region and adjacent structures along with annotated line drawing. Key to most abbreviations used in this and all subsequent figures is given in text. Key to numbered structures: 1, course of alisphenoid-squamosal suture; 2, possible insertion area for m. zygomatico-auricularis; 3, lamina of ?squamosal projecting beneath postglenoid foramen and identified as probable attachment fossa for ectotympanic anterior crus; 4, fissure between tegmen tympani and squamosal identified as probable secondary attachment site for anterior crus; 5, base of truncated alisphenoid tympanic process; 6, small pit on anterior pole of petrosal, possibly for reception of entocarotid process of sphenoid; 7, sulcus indicating former line of attachment of posterior process of pterygoid; 8, pterygoid complex of sphenoid; 9, rugose former contact zone between petrosal and basioccipital; 10, sulci for major branches of mandibular division of 5th cranial nerve; 11, sulcus forming lateral wall of posterior lacerate foramen; 12, rugose "medial crest" of pars cochlearis of petrosal. Broken areas are indicated by cross hatching.

contributes a prominent squamosal epitympanic wing (SQW) which arises medial to the glenoid region and thus forms the greater part of the roof of the anterior tympanic chamber. Anteriorly and medially the SQW is firmly united with the alisphenoid. As noted above, the anterior section of this contact forms a complex, intercalated suture. Laterally, the SQW abuts against the anterolateral surface of the pars cochlearis of the petrosal; the pyriform fenestra is thereby reduced to a narrow slit. Posteriorly, it abuts against the anterior surface of the tegmen tympani (of the petrosal; see below). No septum is developed along this contact.

The tympanic surface of the SQW bears a shallow depression and slopes away ventrally from back to front. Enclosed within the thickened anterolateral margin of this element there is a small cellule which does not appear to communicate with the tympanic cavity (see Fig. 2). This may originally have been marrow-filled.

Posteriorly to the region of the epitympanic wing and medial to the postglenoid cavity, the squamosal supplies the lateral wall of the epitympanic recess. This takes the form of a low, downturned ridge which is continuous posteriorly with the truncated portion of the post-tympanic process of the squamosal. Anteriorly, this flares laterally to shield the postglenoid foramen. As shall be discussed below, this may represent a point of insertion of the missing ectotympanic.

Petrosal: The promontory is a tear-shaped eminence which occupies the postero-medial region of the tympanic cavity. As in other mammals, it is perforated by two primary fenestrae and bears a number of significant neurovascular and muscular impressions (see below).

Medially, the promontory bears a low, rugose crest. This is roughly cresentic in shape, being narrowest anteriorly and expanding to attain its maximum width on the rear of the promontory. The anterior part of the "medial crest" is particularly rugose. Immediately in front of this feature is a small, anteriorly facing depression (also visible on the opposite side of the specimen). As suggested by Ride (1964), this may represent a point of bony contact between the pars cochlearis and some anteriorly-placed element. I concur with this view and suggest the sphenoid entocarotid process as a likely participant. The anterior pole of the petrosal is clearly truncated to some degree but also appears to be naturally foreshortened.

The fenestra vestibuli faces anterolaterally into the epitympanic recess. It measures 1.2×0.7 mm and lies within a shallow fossula.

The fenestra cochleae faces posteriorly into the sinus tympani and lies within a deep cochlear fossula. Externally, the fossula measures 1.7×1.2 mm with its long axis tilted slightly in a clockwise direction (as viewed from behind). Posteroventral to the vestibular fossula, the promontory displays a noticeable bulge.

The tegmen tympani is a relatively small, triangular-shaped structure with its apex directed antero-laterally. This process could well be of compound origin incorporating not only the cartilage preformed body of the true tegmen tympani (strictly an anteriorly-directed extension of the crista parotica) but also any other outgrowths from the main body of the otic capsule. If so, these elements are well-fused in this specimen.

The broken posterior margin of the petrosal represents the basal portion of a truncated caudal tympanic process. Although broken near its base, enough of this process survives to indicate its roughly transverse orientation, its relatively close proximity to the rear face of the promontory and its moderately robust construction. There is nothing about the preserved parts of this region to suggest that it was the site of extensive pneumatisation of the kind seen in many modern diprotodontians (see below).

Tympanic spaces

As noted previously by Ride (1964), the tympanic cavity is a small chamber only slightly larger than that in species of *Didelphis*. Although it lacks the prominent septa, deverticula or accessory sinuses which typically result from significant secondary pneumatisation, it is nonetheless convenient for descriptive purposes to recognise the following three "principal" spaces: an epitympanic recess, an anterior tympanic chamber and a sinus tympani. These are most readily defined on the basis of morphological features of the middle-ear roof.

The epitympanic recess is variously defined as that part of the middle-ear cavity which immediately surrounds the auditory ossicles (e.g., Archer 1976, following Klaauw 1931) or as a distinct pneumatic concavity formed in the tegmen tympani dorsal to the malleo-incudal articulation (e.g., MacPhee 1981). In practice, these amount to much the same thing as most mammals show some degree of pneumatisation of the tegmen tympani (MacPhee 1981).

The epitympanic recess in *Wynyardia* is a shallow, obliquely oriented depression formed in the tegmen tympani of the petrosal. Anteriorly, this is delimited by a low but distinct ridge running obliquely across the tegmen tympani to terminate at its lateral apex. This ridge probably represents the former line of attachment of an anterior malleolar ligament (see below). Laterally, the epitympanic recess is contiguous with a shallow fissure formed between the tegmen tympani and the medial surface of the squamosal. By comparison with the condition in several living diprotodontians (see below), this depression probably represents an insertion point for the anterior crus of the ectotympanic.

The incudal fossa is positioned in the rear of the epitympanic recess on about the same level as the fenestra vestibuli. It is a shallow depression formed primarily within the petrosal but enclosed laterally by the medial rim of the posttympanic process of the squamosal.

Anterior to the epitympanic recess there is a gently concave depression identified herein as the roof of the "anterior tympanic chamber". In Archer's (1976) terminology, this space would be identified as a hypotympanic sinus of the primary enclosing element; e.g., an alisphenoid hypotympanic sinus.

The sinus tympani is that part of the tympanic cavity located between the rear face of the promontory and the caudal tympanic process of the petrosal; in most mammals it surrounds the aperture of the fossula cochleae. This corresponds with the "mastoid epitympanic sinus" of Archer's terminology.

The sinus tympani is a small, mediolaterally oriented space which communicates laterally with the epitympanic recess. Medially it passes out onto the lateral margin of the posterior lacerate foramen. There is no posterior semicircular canal gyrus in the sinus tympani.

Muscular and Neurovascular Impressions

As in other mammals, the root of the tegmen tympani is traversed by a broad sulcus marking the passage of the facial nerve posterior to the geniculate ganglion and of the greater petrosal nerve forward of this point. Although no section of the bony facial nerve canal remains intact, traces of the structure can be detected from the anterior margin of the tegmen tympani, back to a point just anteriad of the vestibular fossula. From here the facial nerve sulcus swings posteroventrally to run down the anterior surface of the caudal tympanic process of the petrosal near its lateral end. Over this section the lateral border of the sulcus is sharply produced.

Immediately behind the posterolateral rim of the facial nerve sulcus there is a distinct pit-like depression which I identify as the stapedial fossa. This is contained medially by a low ridge that crosses the sinus tympani to join the lateral rim of the cochlear fossula.

Anterior to the vestibular fossula the promontory displays a complex pattern of fossae and sulci. The most deeply excavated and otherwise distinct of these is identified as the primary area of origin of a well-developed m.tensor tympani. This is relatively broad anteriorly but becomes progressively narrower and deeper posteriorly. A compact fusiform muscle is indicated, in contrast to the more diffuse spread of fibres found in many diprotodontians (e.g., most phalangeridans).

Dorsal to the tensor tympani fossa but ventral to the facial nerve sulcus there is a complex of minor impressions. While some or all of these features might relate to minor fascicles of the m.tensor tympani, they more likely indicate the submuscular course of various minor neurovascular elements such as the tensor tympani nerve and the anterior continuation of the tympanic nerve. Most of these elements would either enter or exit the tympanic cavity via the slitlike pyriform fenestra.

The surface of the promontory between the fossa tensor tympani and the medial crest has the form of a shallow depression and lacks internal complexity. This is tentatively identified as the site of attachment of the auditory tube.

Several relatively indistinct impressions on the posterior surface of the promontory may relate to the tympanic nerve and associated plexus. The most clearly recognisable of these is a short but distinct sulcus which runs between the anterior rim of the cochlear fossula and the postero-dorsal corner of the vestibular fossula; a similar course is followed by the primary trunk of the tympanic nerve in most marsupials and placentals (MacPhee 1981 and Aplin in prep.).

Located between the cochlear fossula and the posterior end of the medial crest is a second, broader sulcus. Based on position alone, this feature would probably be identified as being for some part of the tympanic plexus. However, for such a minor element, the sulcus appears exceptionally broad. Another possibility is that the sulcus might represent an intrabullar carotid or persistent stapedial ramus analogous to those present in many placental mammals but as yet unreported in any adult marsupial (Tandler 1899; Archer 1976; Archibald 1977, 1979).

The roof of the sinus tympani in marsupials commonly conveys the auricular ramus of the vagus nerve along with associated vessels and these may sometimes leave a distinct sulcus. This is not the case in *Wynyardia*.

Reconstruction of Missing Elements

The primary topic of interest naturally concerns the composition and degree of completeness of the tympanic floor or "bulla". However, some speculative comments may also be made with respect to the morphology of the missing auditory ossicles.

Tympanic floor: Several points have already been made with regard to the mode of construction of the tympanic floor. As noted above, the posterior margin of the alisphenoid is ragged and uneven, as though it formerly gave rise to a prominent tympanic process. As regards the size and complexity of this structure, the lack of bony sculpting along the medial margin of the promontory suggests that there was no direct contact between the alisphenoid and the petrosal in this area. Whether such a process might have reached even further posteriorly to contact the caudal tympanic process of the petrosal cannot be directly commented upon but seems rather unlikely given the lack of evidence for medial contact. Some degree of contact between the alisphenoid process and the anteroventral portion of the ectotympanic can be reasonably inferred, however.

Judging from its surviving basal portion, the caudal tympanic process of the petrosal was a solidly built structure which probably made extensive contact with the posterior surface of the ectotympanic. That this contact did not extend to the rear face of the promontory, however, is indicated by the lack of any appropriate impression in that area.

From the evidence available, *Wynyardia* would thus appear to have had an "incomplete" bulla of the kind seen in several modern diprotodontian families (see below). As in these groups, the medial and posteromedial parts of the bulla would have been formed by a persistent "fibrous membrane" (sensu MacPhee 1981) with attachment to the rugose portion of the promontory.

Some indication as to the size and complexity of the missing ectotympanic bone can be gained through an examination of the probable attachment site of the anterior crus. The first feature of interest is the distinct interosseous "fissure" which lies between the tegmen tympani and the medial end of the postglenoid cavity. In my view, this fissure probably received a bony process developed from the dorsal surface of the anterior crus of the ectotympanic. Certainly this is the case in such taxa as the koala and the wombats (see Figs 4-5).

The second feature which might relate to insertion of the ectotympanic is a small, sub-horizontal lamina of bone which lies anterior of the interosseous fissure noted above and which is continuous with the medial rim of the squamosal. Because this lamina appears to be truncated ventrally in the fossil, I am undecided as to whether it represents an elaboration of the squamosal for reception of the anterior crus, or whether it might actually be a portion of the ectotympanic, fused to the squamosal at this point. Either way, the clear implication from both features is that of fairly rigid attachment of at least the anterior end of the ectotympanic. The question of the lateral extent of the ectotympanic is considered below.

Auditory ossicles: A number of observations may be made with regard to the morphology of the missing auditory ossicles. First, based on the dimensions of the epitympanic recess, the auditory ossicles were clearly not greatly enlarged as in some fossorial and aquatic mammals (cf. Fleischer 1978). Secondly, because the incudal fossa is positioned in the rear of the epitympanic recess, it is clear that the functional axis of the ossicles was subhorizontal (cf. Segall 1969). This in turn would imply some degree of specialisation of the malleus probably including the development of a short and relatively straight neck and reduction of the lamina. Finally, as indicated by the shape of the fenestra vestibuli, the stapedial footplate was moderately elongate (stapedial ratio of 1.71).

The strong development of the fossae for both ossicular muscles suggests that the malleo-incudal complex was rigidly fixed. For the malleus this implies a relatively large anterior or folian process.

OUTER EAR CANAL AND GLENOID FOSSA

Outer ear canal

As in other marsupials, the outer ear canal is enclosed anteriorly, dorsally and posteriorly by portions of the squamosal bone. Ventrally, it would have been at least partially enclosed by the now missing ectotympanic.

The anterior wall of the outer ear canal is formed by a prominent, transversely extensive postglenoid process. This is both thicker and taller laterally than it is at its medial end. A fortuitous fracture through the medial end of this process reveals a small, fully enclosed sinus. This has no obvious connection to the primary tympanic cavity and in life was probably marrow-filled.

Posteriorly, the postglenoid process bears a broad, medially-placed depression and an associated, narrow sulcus. This latter impression runs laterally along the ventral rim of the postglenoid process. One possible interpretation of these features is that they represent (either in whole or in part) a zone of contact between the ectotympanic and the postglenoid process. Ride (1964:115), for example, suggested that the innermost 3 mm of the postglenoid process represented a zone of attachment of the "tympanic wing". With this possibility in mind, I have examined the area in detail, but can find no evidence of any bony contact. My interpretation of this region is that the lateral, more protruberant portion of the postglenoid process represents the origin of a well-developed cranio-mandibular ligament of the kind mentioned by Hofer (1952) and figured by Lightoller (1939) for *Phascolarctos cinereus*. The medially placed depression might in this case be a secondary feature only (i.e., an area lacking ligamentous influence). As regards the associated sulcus, this could conceivably be of vascular origin (perhaps representing a major division of the maxillary artery) but might also represent the line of insertion of a lesser cranio-mandibular ligament.

The postglenoid cavity is transversely broad and deeply U-shaped in longitudinal section. Its lateral rim is undifferentiated from the ventral postzygomatic crest.

Positioned near the lateral margin of the post-glenoid cavity is a low, transversely-oriented ridge of uncertain significance. One possible interpretation of this feature is that it might represent the lateral margin of a shallow epitympanic sinus of the kind that occurs in *Vombatus* (see below). That this is unlikely, however, is indicated by the transverse orientation of the crest, its wide separation from the epitympanic recess and the lack of other evidence of pneumatic activity either within the roof of the postglenoid cavity or in the tympanic cavity proper. An alternative interpretation is that the ridge is related to the insertion of an auriculo-zygomaticus muscle (a derivative of the more commonly developed M. auriculo-mandibularis). A muscular feature of this kind is indeed present in *Vombatus,* although in this taxon it becomes accentuated through pneumatic activity along its medial face.

The postglenoid foramen is located behind the postglenoid process, at the anterolateral corner of the postglenoid cavity. It measures 2.6 × 1.0 mm. As noted above, this foramen is partially hooded medially by a small lamina of bone related to the attachment of the ectotympanic anterior crus.

Because of the presence of matrix within the postglenoid foramen, its route and connections within the squamosal bone could not be determined. Endocranially, however, there is a small aperture positioned immediately anterodorsal to the petrosal which might represent the origin of the postglenoid canal.

The lateral extent of the ectotympanic cannot be confidently determined. As noted above, I regard the ectotympanic to have been firmly anchored at the medial end of the postglenoid cavity but to have remained free of the postero-medial face of the postglenoid process. On the available evidence I suspect that it was probably similar in size and relations to the same element in the diprotodontid *Neohelos* sp., in which the ectotympanic bears a short meatal process (no more than 7 mm in length; Aplin in prep.). Some similarity would probably also be noted to the condition in many macropodoids. In this latter group, as in nearly all phalangeridans, however, the ectotympanic makes direct contact with the posteroventral surface of the promontory (see below). No such contact occurred in *Wynyardia.*

Glenoid Fossa

Since the alisphenoid-squamosal suture in *Wynyardia* passes well medial to the glenoid fossa, the latter structure is formed wholly within the squamosal.

The glenoid fossa in *Wynyardia* is a complex structure which is divisible into distinct morphological and functional units. These are 1, an anteriorly-placed "articular eminence"; and 2, a posteriorly situated "mandibular fossa". In lieu of an appropriate functional label, this morphology shall be referred to herein as the "complex" kind, as against the "rotational" kind seen in didelphids (see below). The specific regional terminology for the glenoid fossa, however, is borrowed from the veterinary literature; as it happens an essentially similar glenoid morphology is found in at least some artiodactyls (e.g., sheep, pig) and some perissodactyls (e.g., horse).

Returning to *Wynyardia,* the articular eminence is a flattened, oval-shaped surface measuring 13.6 mm (transverse width) by 6.4 mm (mesiodistal length). Laterally, it is enclosed by a low preglenoid process formed by the expanded posterior end of the jugal. No such buttressing is present at the medial end of the glenoid fossa. Here the articular surface is smoothly continuous with the neurocranial portion of the squamosal.

The mandibular fossa is a shallow, transversely oriented sulcus which runs behind the articular eminence and thus separates this region from the postglenoid process. It is deepest at its lateral end and becomes shallower and less indistinct toward the medial end of the glenoid fossa.

COMPARISONS

In the following pages, the auditory region of *Wynyardia* is compared firstly with that of a representative didelphid and then with representatives of each of six major groups of diprotodontians as recognised by Aplin and Archer (1987). The basicranial regions of taxa from each of these groups are illustrated in Figs 3-9. In most cases, the middle-ear floor has been removed in order to facilitate comparisons.

COMPARISON WITH DIDELPHIDS

The basicranial anatomy of a species of *Didelphis* is shown in Fig. 3. Of the various didelphids examined, a species of *Didelphis* was selected for detailed illustration because it is of approximately the same body size as *Wynyardia* and because a member of this genus was the subject of former comparisons by Ride (1964). According to Reig *et al.* (1987), *Didelphis* spp. also display relatively few derived basicranial features relative to some other members of this family (e.g., caluromyines). Other accounts of auditory anatomy of didelphids include those by Kampen (1905), Gregory (1910), Segall (1969a,b), Archer (1976) and Reig *et al.* (1987). The latter authors give a comprehensive review of variation in auditory structures in both living and fossil didelphids.

Auditory region: Comparison of the auditory region of *Wynyardia* with that of *Didelphis* reveals many points of resemblance, as emphasised by Ride (1964). In both taxa the principal tympanic cavity is a small space lacking major subdivisions and without obvious inflation of the principal diverticula. As correctly stated by Ride (1964), both taxa also appear to share an absence of epitympanic sinus development either anterior or posterior to the tympanic cavity.

Based on the present reconstruction, the middle-ear floor in *Wynyardia* was probably also similar to that of didelphids in being partially membranous and in being formed of only three elements — an alisphenoid tympanic process (AP), a caudal tympanic process of the petrosal (CTPP) and an ectotympanic. Compared with the majority of didelphids (including *Didelphis*

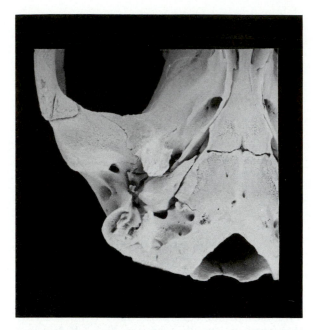
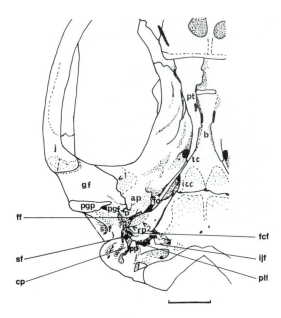

Fig. 3. Didelphis virginiana, right auditory region along with annotated drawing. Ectotympanic and malleus removed. Key to numbered features: 1, course of alisphenoid-squamosal suture; 2, line of medio-dorsal attachment of the "fibrous membrane" of the tympanic floor. Scale bar represents 7.7 mm.

spp.), however, the AP in *Wynyardia* was probably larger and more robust, and may have had more extensive contact with the ectotympanic. It may also have more closely approached the promontory down the medial side of the tympanic cavity.

The CP in *Wynyardia* was probably also larger than it is in *Didelphis,* though perhaps no larger than in certain other polyprotodontans (e.g., some bandicoots). The rostral tympanic process of the petrosal (RP) shown in Fig. 3 of *Didelphis virginiana* is not present in *Wynyardia.*

In contrast to these similarities in general layout and floor construction, the mode of formation of the middle-ear roof differs dramatically between *Wynyardia* and the didelphids. In didelphids the anterior tympanic chamber is roofed either exclusively by an alisphenoid epitympanic wing or else jointly by that process and the anterior part of the tegmen tympani. In these taxa the squamosal lies lateral to the middle-ear and provides only the lateral margin of the epitympanic recess. By contrast, in *Wynyardia* the anterior tympanic chamber is roofed primarily by a large squamosal epitympanic wing with a lesser contribution at the rear from the tegmen tympani. In this species it is therefore the alisphenoid that makes no contribution to the roof of the middle-ear. This difference represents a major point of distinction between *Wynyardia* and didelphids (indeed all polyprotodonts) and, as shall be seen below, a major point of resemblance to other diprotodontians.

Several differences between *Wynyardia* and didelphids appear to relate to the morphology and spatial organisation of the auditory ossicles. As described earlier, the incudal fossa in *Wynyardia*

lies in the rear of the epitympanic fossa, approximately at the level of the fenestra vestibuli. In *Didelphis,* as in the majority of polyprotodonts (Segall (1969b, 1970 notes some exceptions) the incudal fossa lies above the level of the fenestra vestibuli in the roof of the epitympanic recess. The functional ossicular axis in these species therefore deviates markedly from the horizontal. Compared with *Wynyardia,* the fenestra vestibuli in didelphids is more nearly circular (stapedial ratio = 1.4-1.5; Segall 1970, 1971).

Other ways in which *Wynyardia* differs from *Didelphis* include the evidently more rigid attachment of the ectotympanic in *Wynyardia,* its relatively shorter anterior pole of the pars petrosa, the larger diameter of its fenestra cochleae and its more strongly developed cochlear fossula.

Outer ear canal and glenoid fossa: The postglenoid cavity in *Wynyardia* and the didelphids is similar in that both display a broadly U-shaped longitudinal profile and both appear unaffected by secondary pneumatic activity. In both taxa the postglenoid foramen opens at the medial end of the outer ear canal, posterior to the postglenoid process. *Wynyardia* differs from the didelphids, however, in having a transversely broader outer ear canal and in the smaller diameter of the postglenoid canal. In *Didelphis* there is a second vascular foramen located at the end of the postglenoid cavity. This connects via a short conduit with the postglenoid canal, and probably transmits venous return from the external ear. A comparable foramen does not occur in all polyprotodonts.

The ectotympanic in *Didephis* and the majority of didephids is a slender, horseshoe-shaped element which is attached by connective tissue to

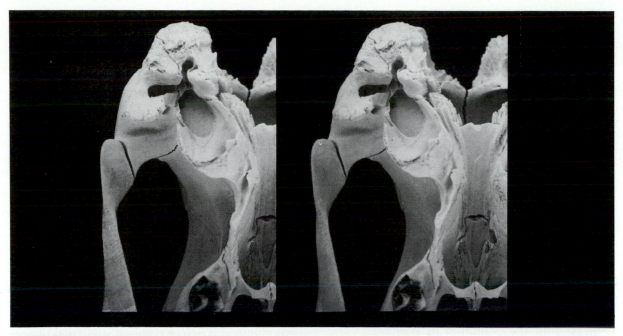

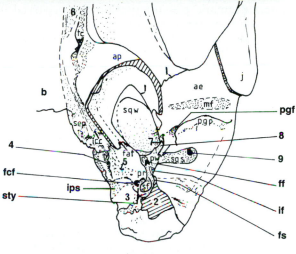

Fig. 4. Phascolarctos cinereus. Stereopair of left auditory region along with annotated drawing. Greater part of alisphenoid tympanic process, ectotympanic and paroccipital process removed to expose tympanic cavity. Key to numbered features: 1, alisphenoid-squamosal suture; 2, base of truncated paroccipital process; 3, sulcus forming lateral wall of posterior lacerate foramen; 4, rugose contact zone between petrosal and basioccipital; 5, "medial crest" of pars cochlearis of petrosal; 6, line of former attachment of pterygoid; 7, entrance to lateral, supraglenoid diverticulum of anterior tympanic cavity; 8, fossa for attachment of ectotympanic anterior crus; 9, entrance to supraglenoid chamber of squamosal epitympanic sinus. Arrow indicates the position of foramen ovale (hidden from view). Scale bar represents 3.8 mm.

a shallow depression along the lateral margin of the tegmen tympani. Other, equally insubstantial connective tissue attachments may occur between the posteromedial rim of the ectotympanic and the RP in many didelphids. In common with nearly all members of its family, the ectotympanic in *Didelphis* has a poorly developed meatal process. Among Australian polyprotodonts, however, the ectotympanic is frequently modified to form an elongate, semitubular structure (cf. Archer 1976). In no case, however, does this element become rigidly attached to the roof of the middle-ear or the postglenoid cavity.

The glenoid fossa in *Didelphis* and the majority of other polyprotodonts is a simple "rotational" joint of the kind found in many different placental groups including insectivores, carnivores and chiropterans. As is typical in this pattern, the articular surface in *Didelphis* has the form of a tranversely elongate and ventrally concave groove and the anterior face of the

postglenoid process participates directly in rotational joint function. This contrasts with the "complex" morphology of the glenoid fossa in *Wynyardia*, wherein the postglenoid process is isolated from the primary articular surface by a distinct mandibular fossa.

In *Didelphis* and most other polyprotodonts the postglenoid process is tranversely less extensive than in *Wynyardia* and is also more smoothly finished posteriorly. Judging from published accounts (e.g., Hiiemae and Jenkins 1969; Turnbull 1970), didelphids probably lack the posterior mandibular ligaments seen in diprotodontians.

COMPARISON WITH DIPROTODONTIANS

The auditory region varies greatly within Diprotodontia, such that six major morphological "types" can be distinguished. These are: 1, a "*Phascolarctos*" type; 2, a vombatomorphian type; 3, a "central" phalangeridan type; 4, a macropodoid type; 5, a *"Tarsipes"* type; and 6, an acrobatid type. These are variably distributed between the three major infraordinal-level taxa recognised by Aplin and Archer (1987) as follows: Phascolarctimorphia (type 1 only); Vombatomorphia (type 2); Phalangerida (types 3-6). In the present context the most important comparisons are with the phascolarctomorphian, vombatomorphian and "central" phalangeridan groups; less extensive comparisons are given for the other groups.

Comparison with the Koala

The anatomy of the auditory region of *Phascolarctos cinereus*, the only living phascolarctomorphian, is illustrated in Fig. 4. Previous accounts of the basicranial region in this species include those of Kampen (1905) and Werner (1964).

Auditory region: The tympanic cavity in the koala is highly inflated and thereby provides a striking contrast with the condition in *Wynyardia*. The most dramatic pneumatisation has occurred in the anterior tympanic cavity which exends far anteriorly, passing beneath the line of passage of the mandibular division of the trigeminal nerve to invade deep into the body of the alisphenoid. Another major result of this pneumatic activity in the koala has been the formation of a deeply invasive squamosal epitympanic sinus (SQS) which connects directly with the epitympanic recess via a patent epitympanic foramen (= supratympanic foramen of Cope 1880). Anteriorly, the SQS in the koala extends well into the body of the postglenoid process.

As in *Wynyardia*, there is a large squamosal epitympanic wing (SQW) in the roof of the tympanic cavity in the koala. Compared with its homologue in *Wynyardia*, however, the SQW in the koala is significantly larger and it supports extensive dorsal and lateral diverticula, the latter penetrating far into the supraglenoid region of the squamosal.

The tegmen tympani in the koala is similar to that of *Wynyardia* in being a simple triangular lamina which is separated from the medial surface of the squamosal by a deep interosseous fossa. In the koala this fossa receives a distinct bony process from the anterior crus of the ectotympanic. Despite the extensive pneumatic activity in the koala middle-ear, the ectotympanic does not fuse with surrounding bones and it is often missing from cleaned skulls.

The middle-ear floor of the koala is comprised of three clearly distinct elements which correspond to those present in didelphids and inferred for *Wynyardia* (i.e., an AP, a CP and an ectotympanic). It differs from the condition in these groups, however, in the anteriorly shifted origin and extreme hypertropy of the AP and in the incorporation of the CP into an elongate (but largely nontympanic) paroccipital process. In keeping with its large size, this latter process is much thickened at the base and highly trabeculated internally.

Despite the considerable size of the AP, the tympanic floor in the koala remains incomplete along its medial margin. As in *Wynyardia*, the medial margin of the pars cochlearis in the koala supports a rugose crest representing the zone of attachment of the fibrous membrane of the middle-ear floor.

Internally, the middle-ear of the koala is similar to that of *Wynyardia* in the lack of inflation of the epitympanic recess and sinus tympani, the near complete closure of the pyriform fenestra and the development of a deep cochlear fossula. It differs from the condition in *Wynyardia*, however, in having less clearly marked fossae for the tensor tympani and stapedius muscles, a more dorsally positioned incudal fossa, a posteriorly less extensive facial nerve canal and a less elongate fenestra vestibuli (stapedial ratio = 1.55; Turnbull and Segall 1984).

Outer ear canal and glenoid fossa: As in *Wynyardia*, the postglenoid cavity in the koala is transversely elongate and is enclosed anteriorly by a prominent postglenoid process. It differs, however, in the presence of a large epitympanic fenestra, opening dorsally into the SQS, in the pneumatic nature of the postglenoid process, and in the positioning of the "postglenoid" foramen medial to, rather than behind, the postglenoid process. This latter feature may in part reflect the development of a prominent process for ligamentous attachment of the complex mandibulo-auricularis fascia of this species (cf. Lightoller 1939).

The glenoid fossa in the koala conforms in general terms to the "complex" morphology described for *Wynyardia*. It differs from the condition in *Wynyardia*, however, in having a better developed mandibular fossa and a ventrally convex articular eminence, and in having the latter surface extend down the front of the auditory bulla to form a distinct "entoglenoid" process.

Comparison with Vombatomorphians

The basicranial region of various taxa now placed within Vombatomorphia have been described and figured by Kampen (1905), Woods (1956), Werner (1964), Segall (1969b), Stirton (1967), Woodburne (1967) and Murray *et al.* (1987). For the present purposes I have illustrated the basicrania of two vombatomorphians: *Vombatus ursinus* (Fig. 5) representing Vombatidae and *Thylacoleo carnifex* (Figs 6 and 7) representing Thylacoleonidae. In general terms, vombatids display perhaps the least degree of auditory specialisation of any vombatomorphan but they have a highly modified glenoid fossa and post-

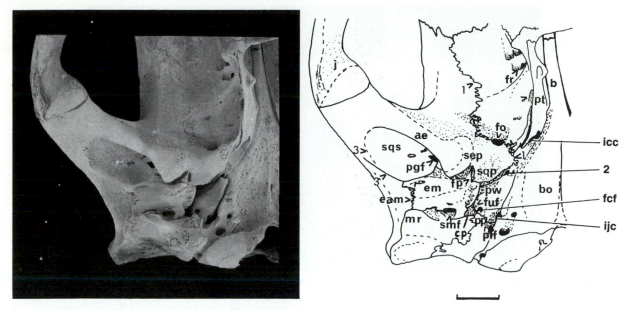

Fig. 5. Vombatus ursinus, right auditory region along with annotated drawing; all elements intact. Key to numbered features: 1, alisphenoid-squamosal suture; 2, remnant of pyriform fenestra; 3, insertion of m. zygomatico-auricularis. Scale bar represents 11.5 mm.

glenoid region. By contrast, thylacoleonids show a less modified glenoid region but have a more specialised auditory region.

Auditory region: The principal tympanic cavity in the vombatomorphians is similar in most respects to that of *Wynyardia.* However, some expansion of the anterior tympanic chamber and of the sinus tympani is evident in the vombatids and in *Thylacoleo.* In all vombatomorphians, the epitympanic recess is a shallow basin.

As in *Wynyardia,* the anterior tympanic chamber in vombatomorphians is roofed primarily by a large SQW which abuts posteriorly against the tegmen tympani. In vombatids and "diprotodontoids" the SQW also closely aproximates the anterolateral face of the petrosal, thereby reducing the pyriform fenestra to a narrow slit. In *Thylacoleo,* the SQW remains widely separated from the petrosal and the pyriform fenestra is corresponding large (Fig. 7).

The tegmen tympani in vombatids and "diprotodontoids" is poorly developed being comparable in shape and relations to that of *Wynyardia* and its lateral margin is modified for reception of the anterior crus of the ectotympanic. In *Thylacoleo,* some lateral expansion of this process has occurred and the anterior crus rests upon the tegmen tympani rather than inserting alongside it.

Enclosure of the anterior tympanic cavity by a squamosal tympanic process (SQP) was long ago noted as a peculiarity of vombatids (e.g., Owen 1841) and more recently recognised as a diagnostic feature for the Vombatomorphia (Archer 1984; Woodburne 1984 and Aplin and Archer 1987). In all members of this group the alisphenoid lies well forward of the tympanic cavity.

Other than for the presence of the SQP, the vombatomorphian tympanic floor differs little from the reconstructed condition in *Wynyardia.* As in the koala, the auditory bulla in the majority of vombatomorphians remains incomplete posteromedially due to a lack of contact between the SQP and the CP, and between these elements and the promontory. The CP in all known vombatomorphians is incorporated into an elongate but largely nontympanic paramastoid complex.

In contrast to *Wynyardia* and didelphids, all vombatomorphians have a squamosal epitympanic sinus. In no two families of vombatomorphians, however, is the SQS identical in form or relation. In *Vombatus* spp., the SQS is a relatively large space but it produces only a shallow, poorly-defined basin in the roof of the already deeply concave outer ear canal (see Fig. 5). Ventrally, the sinus is enclosed by dense connective tissues related to the auriculo-mandibular ligaments. The SQS in *Lasiorhinus* spp. is more deeply excavated but displays the same essential relations as in *Vombatus.*

In both *Thylacoleo* and the "diprotodontoids" the SQS is a large, deeply invasive structure which extends anteriorly into the root of the zygomatic arch and posteriorly into the posttympanic process of the squamosal. Moreover, in at least some "diprotodontoids", the SQS connects dorsally with an even more extensive pneumatic complex formed within the parietal and frontal bones. In diprotodontids, the roof of the outer ear canal is complete, thereby isolating the SQS from the outer ear. By contrast, this lamina is incomplete in both palorchestids and in *Thylacoleo,* such that a large "epitympanic fenestra" is present in the roof of the outer ear

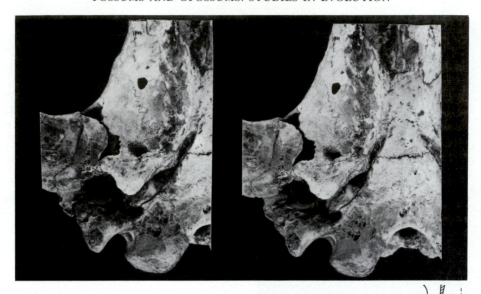

Fig. 6. Thylacoleo carnifex. Stereopair of right auditory region along with annotated drawing. Australian Museum specimen MF9 (holotype of *T. minor*); missing ectotympanic, auditory ossicles, pterygoid and part of roof of postglenoid cavity. Key to numbered features: 1, alisphenoid-squamosal suture; 2, pterygoid complex of sphenoid; 3, zone of former attachment of pterygoid. Scale bar represents 13.6 mm.

canal (Stirton 1967). This was presumably sealed in life by connective tissues and by the meatal epidermis.

Internally, the vombatid middle-ear resembles that of *Wynyardia* in showing an unspecialised epitympanic recess and sinus tympani, a distinct cochlear fossula and a prominent stapedial fossa. This combination of features is also present in *Thylacoleo* and in at least some "diprotodontoids" (e.g., *Ngapakaldia tedfordi, Neohelos* sp.; pers. observ.). The tensor tympani fossa is recognisable in all vombatomorphians but may be less clearly defined than it is in *Wynyardia*. In no vombatomorphian examined is the course of the tympanic nerve so clearly marked as it is in *Wynyardia*. However, a short sulcus on the anterolateral face of the promontory in *Thylacoleo* may relate to this nerve. In no vombatomorphian did I observe any sulcal development ventral to the cochlear fossula.

As in *Wynyardia*, the incudal fossa in all vombatomorphians examined lies in the rear of the epitympanic recess and is formed jointly by the petrosal and the squamosal.

The form of the promontory varies considerably between the various vombatomorphian families. In vombatids the promontory lacks any medial cresting or rugosity. In palorchestids (*Ngapakaldia* species and *Palorchestes painei*) and in *Thylacoleo*, the medial edge of the promontory is produced into a relatively sharp crest. Finally, in the diprotodontids, it supports a variably developed rugose medial crest and is noticeably reduced anteriorly. This latter condition is perhaps most similar to that described for *Wynyardia*, but differs in that the promontory is significantly shorter in the diprotodontids than in *Wynyardia*.

Compared with *Wynyardia*, the fenestra vestibuli in the vombatomorphians is less elongate (stapedial ratio around 1.4-1.5 in all taxa examined).

Outer ear canal and glenoid fossa: With the exception of the vombatids, all known vombatomorphians examined possess "complex" glenoid fossae similar to that of *Wynyardia*. In contrast to the condition in *Wynyardia*, however, the articular eminence in these taxa tends to be more obliquely oriented relative to the long axis of the skull. The glenoid fossa in the vombatids is uniquely specialised, comprising only a ridge-like articular eminence.

In all vombatomorphian families, the articular eminence is expanded medially to form a distinct entoglenoid process. This is continuous with the lateral portion of the SQP

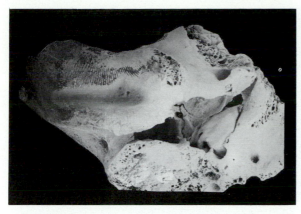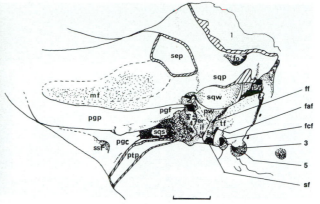

Fig. 7. Thylacoleo carnifex, right auditory region with annotated drawing. National Museum of Victoria specimen P159792; missing ectotympanic, auditory ossicles and greater part of alisphenoid and basioccipital. Key to numbered features: 1, rugose area on squamosal marking former zone of contact with alisphenoid; 2, fossa for ectotympanic anterior crus; 3, foramen for auricular ramus of vagus nerve; 4, lateral margin of epitympanic recess; 5, combined aperture of posterior lacerate and internal jugular foramina. Scale bar represents 10.4 mm.

in vombatids and in *Thylacoleo* but is isolated from this process by a broad groove in the diprotodontoids. Both conditions contrast with an absence of entoglenoid development in *Wynyardia*.

The postglenoid process is also highly variable within the Vombatomorphia. In *Thylacoleo* and diprotodontine diprotodontids (e.g., *Pyramios alcootense* and *Euryzygoma dunense*) the postglenoid process is large and solidly-constructed. It differs from the condition in *Wynyardia*, however, in being more constant in height and thickness along its transverse length and in being recurved posterodorsally so as to partially floor the external acoustic meatus. This tendency is carried to an extreme in *Diprotodon* spp., wherein the postglenoid process meets the post-tympanic region to form a complete, bony outer ear canal. Among zygomaturine diprotodontids, the postglenoid process is a prominent, sinus-bearing structure similar in general terms to that of the koala. Palorchestids, on the other hand, have a small, sinus bearing postglenoid process which is confined to the lateral end of the glenoid fossa (Stirton 1967; Woodburne 1967). In vombatids there is no postglenoid process and they differ in this regard from all other vombatomorphians and from *Wynyardia*.

The ectotympanic shows a variable degree of lateral extension among the vombatomorphians. This is most extreme in the case of the vombatids among which the ectotympanic is an elongate but narrow, tubular structure. In *Thylacoleo* and the diprotodontoids the ectotympanic is moderately elongate laterally; it is complete dorsally in only some taxa (e.g., *Neohelos* spp.).

In no vombatomorphian does the ectotympanic establish intimate contact with the postglenoid process, as it does for example in the phalangeridan possums (see below). On the other hand, the ectotympanic in the vombatids displays a complex pattern of attachments to the anterior and posterior walls of the postglenoid cavity, perhaps

a result in this case of narrowing of the postglenoid cavity in this group. Overall, the ectotympanic in *Wynyardia* was probably most similar in size and relations to that of certain diprotodontids (e.g., *Neohelos* sp.).

The postglenoid foramen opens into the medial end of the postglenoid cavity, behind the postglenoid process (if present), in all vombatomorphans. In all taxa examined in detail (*Thylacoleo carnifex*, *Vombatus ursinus*, *Lasiorhius latifrons*, *Neohelos* sp., various other diprotodontoids) the postglenoid canal fails to make its usual connection with the endocranial dural network. Instead it connects exclusively with a complex of canals which together carry most of the venous return from the posterior temporalis muscle. Dural drainage that previously passed out of the cranium via the postglenoid vein would thus probably be rerouted via the sigmoid sinus and thence into the internal jugular vein and/or the major vertebral sinuses. As noted earlier, the superficies meatus contains a large epitympanic fenestra in *Thylacoleo* and in the palorchestids. This results from incomplete enclosure of the SQS in these groups.

COMPARISON WITH PHALANGERIDANS

With the exception of *Tarsipes rostratus*, the acrobatids and the macropodoids, the various groups of phalangeridans are all essentially alike in auditory anatomy (cf., Kampen 1905; Klaauw 1931; Segall 1969b, 1971 and Aplin and Archer 1987). For the present purposes, it is thus sufficient to illustrate the auditory region for only one of the "central" Phalangerida: *Trichosurus vulpecula* (see Fig. 8). This comparison has added significance inasmuch as it is with phalangerids as a group and species of *Trichosurus* in particular that *Wynyardia* most commonly has been linked.

Auditory region: The auditory region in phalangeridans is characterised most readily by the development of a complete osseous "bulla".

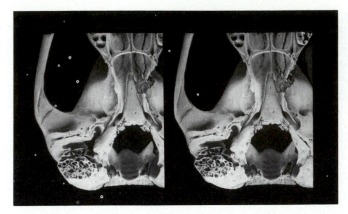
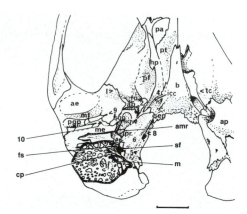

Fig. 8. *Trichosurus vulpecula*. Stereopair of right auditory region with annotated drawing. Whole of tympanic floor, ectoympanic, auditory ossicles and ventral part of mastoid complex removed to expose contents of tympanic cavity. Basioccipital and part of pterygoid of left side also missing. Note that tympanic floor on left side remains intact. Key to numbered features: 1, alisphenoid-squamosal suture; 2, anteromedial lamina of alisphenoid tympanic process (which forms ventral wall of foramen ovale; 3, base of truncated alisphenoid tympanic process; 4, sulcus for auditory tube; 5, sulcus forming lateral wall of posterior lacerate foramen; 6, area of bony contact between alisphenoid tympanic process and pars cochlearis of petrosal; 7, "ectotympanic process" of promontory, indicating area of complex, bony contact with alisphenoid tympanic process, caudal tympanic process of petrosal and ectotympanic; 8, rugose contact zone between petrosal and basioccipital; 9, foramen indicating point of exit of chorda tympani from tympanic cavity; 10, vascular foramina lying on or near to squamoso-ectotympanic suture; 11, non-pneumatic (marrow-filled) cellulae of posttympanic process of squamosal. Arrow indicates position of foramen epitympanicum, leading into squamosal epitympanic sinus. Scale bar represents 7.3 mm.

Although ontogenetic evidence on the formation of this structure is lacking, examination of numerous juvenile crania of various species leads me to believe that the phalangeridan tympanic floor contains only the three same elements that occur in didelphids, the koala and probably *Wynyardia* (i.e., an AP, a CP and an ectotympanic). The tympanic floor in this group thus probably differs from that of *Wynyardia* only in the greater proliferation of the AP and the CP within an already well-established basic pattern.

In all phalangeridans, the middle-ear cavity is roofed in part by a large SQW similar in size and relations to that described for *Wynyardia*. In phalangerids such as *Trichosurus* spp., the anterior tympanic chamber is roofed entirely by this element and the alisphenoid bears the same relation to the tympanic cavity as it does in *Wynyardia* (i.e., present in floor but absent from roof). In many other phalangeridans (e.g., pseudocheirids, petaurids, burramyids), however, expansion of the principal tympanic cavity results in pneumatisation beyond the limits of the SQW. In these taxa, the anterior tympanic chamber is thus roofed jointly by the SQW and an alisphenoid epitympanic wing (SW), an arrangement which occurs also in the koala (see above).

As in *Wynyardia*, the pyriform fenestra in phalangeridans is reduced to a narrow slit lying between the SQW and the anterolateral surface of the pars cochlearis of the petrosal.

In all phalangeridans there is an extensive SQS which is connected with the epitympanic recess via a patent epitympanic foramen and is deeply invasive into the postglenoid region of the squamosal. In some phalangeridans (e.g., *Cercartetus nanus, Petaurus* spp.) the SQS extends anteriorly into the supraglenoid and zygomatic regions, and posteriorly into the posttympanic region of the squamosal. Associated with the development of the SQS in all members of this group is a characteristic dorsal displacement and lateral extension of the tegmen tympani and the resultant elaboration of associated features such as the lateral malleolar ridge.

In addition to the SQS, many phalangeridans (e.g., *T. vulpecula;* see Fig. 8) display a second epitympanic sinus complex at the rear of the tympanic cavity. As with the SQS, this "mastoid epitympanic sinus" (MES) communicates openly with a principal tympanic space, in this case the sinus tympani, via a patent epitympanic foramen. There is considerable variation in the size and internal complexity of this sinus. In the burramyids, there is no MES, despite the fact that the sinus tympani itself is highly inflated in several members of the group (e.g., *Cercartetus concinnus*). At the other extreme, in some phalangerids (e.g., *Phalanger maculatus*) and some pseudocheirids (e.g., *Pseudocheirops cupreus)*, the MES and SQS become so extensive that they meet above the outer ear canal and even come to invade the ventral portion of the ectotympanic (e.g., *Pseudocheirus cupreus*).

In terms of the internal arrangement of the tympanic cavity, phalangeridans are comparable to *Wynyardia* in the positioning of the incudal fossa below the level of the fenestra vestibuli, in the relatively extensive development of a bony facial nerve canal and in the typically elongate form of the fenestra vestibuli. Reported stapedial ratio values for phalangeridans range from around 1.6 in some species of *Pseudocheirus*, to over 2.1 in *Petaurus* spp. (Segall 1970, 1971;

PLATE 4: The cat-sized early Miocene wynyardiid *Wynyardia bassiana* from the Wynyard Local
Fauna, northwestern Tasmania.
Reconstructions by Peter Murray.

Turnbull and Segall 1985). Features in which phalangeridans differ from *Wynyardia,* however, include the absence or weak development of a cochlear fossula in some groups (e.g., burramyids), the development of the incudal fossa wholly within the petrosal and the presence of a marked promontorial bulge anterior to the fenestra vestibuli in most forms (e.g., pseudocheirids, petaurids, phalangerids). At least in the majority of phalangeridans, the primary attachment of the tensor tympani also differs from that inferred for *Wynyardia:* on the anterior part of the expanded "tegmen tympani" rather than the anterolateral face of the promontory.

In many phalangeridans the ectotympanic bears a short posteromedial process which rests directly upon the posteroventral surface of the promontory (see Fig. 8). This results in the formation of a characteristic, knob-like "ectotympanic process of the promontory" which is present in virtually all phalangeridans but absent in all vombatiforms.

Outer ear canal and glenoid fossa: The outer ear canal in phalangeridans is enclosed within a complete bony canal. This is formed ventrally by lateral proliferation of the ectotympanic meatal process, anteriorly by a postglenoid process of variable height, and posteriorly by the combined posterior crus of the ectotympanic and the posttympanic process of the squamosal. Dorsally, the outer ear canal is isolated from the adjacent SQS in most phalangeridans by a thin lamina of bone (squamosal). This lamina is fenestrated in *Burramys parvus.*

In all phalangeridans the posttympanic process of the squamosal is inflated to some degree. In some taxa this area is occupied exclusively by the combined MES and SQS (e.g., *Pseudocheirops cupreus),* while in others it is at least partially filled with numerous marrow-filled cellules (e.g., *Trichosurus vulpecula;* see Fig. 8). One probable side-effect of pneumatisation in this area of the skull might well be the tendency toward fusion (among the phalangeridans) of the posttympanic region of the squamosal and the pars canicularis of the petrosal (Turnbull and Segall 1984). In this regard, the lack of fusion between the squamosal and petrosal in this area in *Wynyardia* can probably be taken as confirmatory evidence as to the absence of a MES in this species.

As noted above, the postglenoid process in the phalangeridans contributes to the anterior wall of the bony outer ear canal. The extent of this contribution varies considerably among the various phalangeridan families. It is largest in certain phalangerids (Flannery *et al.* 1987) wherein the postglenoid process forms the entire anterior wall of the outer ear canal. On the other end of the scale, the postglenoid contribution is negligible in certain burramyids. In many phalangeridans the postglenoid process contains air cellules connected to the SQS.

Incidental to these modifications of the posttympanic region of the skull, the facial nerve in all phalangeridans emerges from the rear of the tympanic cavity through a definitive stylomastoid foramen. This is located between the posterior surface of the ectotympanic and the anteriorly expanded posttympanic process of the squamosal.

The glenoid fossa in all phalangeridans either conforms to the "complex" type described for *Wynyardia* or is a clear derivative of that type. In common with *Wynyardia,* the articular eminence in all phalangeridans is a flattened, transversely-oriented surface. The shape of this area varies considerably within the group, ranging from the nearly round facet of the smaller possums (e.g., *Cercartetus* spp.), to the transversely elongate condition of the larger taxa such as *Phalanger maculatus.* In no phalangeridan is there an entoglenoid process of the kind found in the vombatomorphians. The mandibular fossa is relatively well developed in all phalangeridans, but is particularly deep and well-defined in the larger ringtail possums and phalangerids. In no phalangeridan is the anterior face of the postglenoid process involved in direct rotational articulation with the mandibular condyle.

In all phalangeridans the postglenoid foramen lies anterior to the medial end of the combined postglenoid process-ectotympanic anterior crus, rather than behind the postglenoid process as in *Wynyardia.* Unlike in the vombatomorphians, the postglenoid canal in phalangeridans maintains its primitive connection with the endocranial dural drainage. In terms of absolute size, the postglenoid foramen in the phalangeridans is either roughly equivalent in size to that of *Wynyardia* (e.g., *Pseudocheirops cupreus)* or it is considerably smaller (e.g., many phalangerids).

COMPARISON WITH OTHER "ATYPICAL" PHALANGERIDANS

Less extensive comparisons are required with each of the other major groups of diprotodontians as recognised on basicranial features (see above).

Macropodoidea: Macropodoids are similar to other phalangeridans in all essential features of auditory anatomy (cf. Winge 1941; Kampen 1905; Aplin and Archer 1987; Flannery 1987). Minor differences in auditory anatomy between macropodoids (Fig. 9) and other phalangeridans include: the incomplete ventral and lateral enclosure of the SQS; the antero-medial displacement of the CP to closely approximate the posterolateral face of the promontory; the more rounded stapedial footplate; the lack of inflation of the posttympanic process of the squamosal;

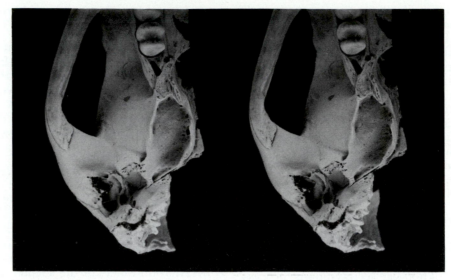

Fig. 9. Dendrolagus lumholtzi. Stereopair of right auditory region. Alisphenoid tympanic process, ectotympanic and auditory ossicles removed; also missing basioccipital and portion of pterygoid.

the location of the tensor tympani on the anterolateral face of the promontory; and the lack of intimate union between the anterior limb of the ectotympanic and the postglenoid process. Other than for the three last-mentioned features, all of which are probably symplesiomorphous, none of the special features of the macropodoid auditory region are paralleled in *Wynyardia.*

As in the phalangeridan possums, the glenoid fossa in macropodoids is either of the "complex" type described for *Wynyardia* or is a clear derivative of that type. Compared with *Wynyardia,* the articulate eminence in the macropodoids is generally more rounded in shape. Otherwise however, it has the same flattened morphology and lack of entoglenoid development.

Tarsipedidae: The auditory region of *Tarsipes rostratus* is highly specialised and differs from that of *Wynyardia* in numerous respects (Aplin and Archer 1987 and Aplin in prep.). Among the more salient distinguishing features are: the small size of the SQW and its apparent replacement by a substantial petrosal epitympanic wing; the presence of a large RP; the presence of a small SQS; and the marked inflation of both the anterior tympanic cavity and the sinus tympani.

The outer ear canal in *Tarsipes* is laterally less extensive than in any other diprotodontian. The glenoid fossa is a simple flattened surface allowing extensive excursions of the much reduced mandibular condyle. There is no mandibular fossa and no postglenoid process.

Acrobatidae: The auditory region in the species of *Acrobates* and *Distoechurus* is likewise highly specialised and thus bears little resemblance to the generalised condition in *Wynyardia* (see Segall 1971; Aplin and Archer 1987). Special features of this group include: the complete withdrawal of the alisphenoid from the auditory region; the presence of a large RP; development of a deeply invasive SQS; completion of the bony facial nerve canal; and the presence of numerous intratympanic septa which extensively subdivide the tympanic cavity proper.

The acrobatid glenoid fossa and outer ear canal are essentially phalangeridan-like in construction but differ from other members of this group, as well as from *Wynyardia,* in showing: a polyprotodont-like, alisphenoid contribution to the anteromedial margin of the glenoid fossa; a separate foramen, located well lateral to the f. ovale, for passage of the masseteric nerve; and a unique ectotympanic "disc" located at the medial end of the outer ear canal.

DISCUSSION

THE ORDINAL AFFINITIES OF *WYNYARDIA*

As documented in the previous section, the basicranial region of *Wynyardia* differs in a number of important respects from that of "typical" didelphids such as the species of *Didelphis.* Of particular significance in this regard are the presence in *Wynyardia* of a squamosal epitympanic wing (SQW) in the tympanic roof, and the unusual "complex" morphology of its glenoid region. In both cases, the condition found in *Wynyardia* is widespread among members of the Diprotodontia.

Within limits of current knowledge, the SQW appears to be a uniquely diprotodontian feature not found in any other marsupial lineage. Within the Diprotodontia, this structure is well-developed in all vombatiforms and in the great majority of phalangeridans. Only in two families of diprotodontians is its status in doubt. In the only known tarsipedid, *Tarsipes rostratus,* the anterior tympanic chamber is roofed jointly by a small lamina of the squamosal (which might be homologous with the SQW), an anterior petrosal epitympanic wing and an alisphenoid epitympanic wing. As argued elsewhere in this volume (Aplin and Archer 1987) this condition is probably best interpreted as an example of secondary reduction of the SQW, rather than of its incipient development. In the second doubtful group, the acrobatid possums, the composition of the anterior portion of the osseous bulla is obscured by extensive fusion of all elements. However, preliminary results from an ongoing ontogenetic study on *Distoechurus pennatus* suggest that this region is formed jointly by squamosal and petrosal derivatives and hence that a SQW is present.

The morphology of the glenoid fossa constitutes a second major point of resemblance between *Wynyardia* and diprotodontian marsupials. In all didelphids and perhaps even all polyprotodonts, the craniomandibular joint is of the simple "condylar" type wherein the centre of rotation remains relatively stationary during jaw function. By contrast, the glenoid fossa in *Wynyardia* displays a specialised condition which is at least structurally analagous to that seen in various ungulate lineages. In terms of function, this appears to reflect a distinctly bimodal pattern of dental use, with the condyle located in the posteriorly-placed "mandibular fossa" for precise cheektooth occlusion, but shifted forward on to the "articular eminence" for incisive function. One important aspect of this pattern in the diprotodontans is that the postglenoid process, if indeed present at all, fails to participate in direct synovial joint operation with the mandibular condyle. Rather it appears to function primarily as an attachment site for complex cranio-mandibular ligaments, as well as possibly acting to restrict posterior excursions of the dentary and to limit mandibular gape.

Based on the preliminary survey of glenoid morphology among diprotodontans undertaken for the present study, I anticipate that this "complex" glenoid type will prove to be an important feature of the dental and masticatory adaptation of many extant diprotodonts and possibly even a key innovation in the origin of Diprotodontia as a whole.

Remarkably, neither the SQW nor the unusual glenoid morphology of diprotodontians have attracted much previous attention. One important exception, however, is provided by the classic work of Kampen (1905) wherein the SQW is described (as a Pars entoglenoidea des Squamosum; e.g., p. 406) for *Phascolarctos*, a wombat, several phalangerids and various macropodoids. These findings were reiterated without further elaboration by Van der Klaauw (1931).

As regards the glenoid fossa, aspects of its functional significance were discussed by Lubosch (1907) and Hofer (1952), but in each case the primary interest appears to have been in other, related structures (i.e., the mandibular meniscus and masticatory musculature respectively) and little attention was given to the glenoid fossa itself.

Other, potentially derived features which are shared between *Wynyardia* and at least some diprotodontians include: the evidently rigid attachment of the ectotympanic anterior crus; emplacement of the incudal fossa in the rear of the epitympanic recess; development of a cochlear fossula; the oval-shaped fenestra vestibuli (i.e. high stapedial ratio); the transversely elongate postglenoid cavity; and the reduced calibre of the postglenoid canal. In no case, however, is any one of these features shared with all other known diprotodontians.

In terms of broad phylogenetic conclusions, the basicranial region of *Wynyardia* thus places it firmly within Diprotodontia. This is consistent with current taxonomic placement of the fossil within this group, as based on other cranial (Jones 1931; Ride 1964), postcranial (Ride 1964) and neuroanatomical (Haight and Murray 1981) evidence.

The present interpretation of the *Wynyardia* auditory region is clearly at odds with that of Ride (1964) over the significance of *Wynyardia's* lacking epitympanic sinuses. As may be gathered from Ride's own comments, he regarded absence of epitympanic sinuses ". . . either anterior or posterior to the epitympanic recess . . ." to be ". . . the most important primitive character of the middle ear of *Wynyardia*" (ibid.:115). A similar sentiment is evident in Ride's (ibid.:115) definition of the familial concept Wynyardiidae as "those diprotodonts which combine diprotodonty with a primitive middle ear which (like that of *Didelphis*) lacks epitympanic sinuses." (brackets original).

The emphasis upon epitympanic sinuses as a taxonomic character in marsupial systematics has a long historical precedent harking back to the writings of Klaauw (1931) and before him, to Kampen (1905). In all of these works, a firm distinction is drawn between the Australian marsupials, which are stated to possess epitympanic sinuses, and the American taxa which reputedly lack them (see also Simpson 1970; Reig and Simpson 1972; Marshall 1977, etc., for recent restatements of this view).

The first potential hurdle to this view arose with Clemens' (1966) description of a squamosal epitympanic sinus in the late Cretaceous American marsupial *Didelphodon vorax*. A more damaging blow, however, was dealt by Archer's (1976, 1982) demonstration that epitympanic sinuses not only occurred amongst some American marsupials (e.g., some borhyaenids) but, more importantly, were decidedly absent in certain structurally generalised peramelids (e.g., *Peroryctes* spp. and *Rhynchomeles prattorum*). The clear message coming from these studies is that epitympanic sinus development is homoplaisious, at least among polyprotodonts.

What then of its status within Diprotodontia? As noted in the previous sections, an epitympanic of some kind or other is indeed present in all diprotodontians — with the probable exception of *Wynyardia*. An important observation, however, is that the epitympanic sinuses vary enormously in degree of invasiveness, ranging from the entirely superficial type of "sinus", such as occurs in the vombatids, through the shallow inpocketing of the roof of the outer ear canal described by Murray *et al.* (1987) for *Wakaleo vanderleuri,* and culminating in the massive pneumatic complexes found in *Thylacoleao carnifex* and, even more dramatically, in certain diprotodontids. Moreover, a variety of different functional factors are almost certainly at play; e.g., adaptations related to auditory function, increase in surface area for muscular insertion, weight reduction in massive skulls, etc. In summary, the likelihood that SQS has arisen multiple times within Diprotodontia alone appears very high. In the case of *Wynayardia*, the absence of epitympanic sinuses is therefore regarded to be less significant than formerly understood by Ride.

AFFINITIES TO OTHER DIPROTODONTIANS

As noted earlier, recent opinion is divided regarding the higher level affinities of *Wynyardia*. According to some who have studied it (e.g., Jones 1931; Haight and Murray 1981), *Wynyardia* is a phalangeridan close to the living phalangerids, yet to others it is a vombatomorphian with special affinities to vombatids (Rich and Archer 1979; Pledge 1987) or diprotodontoids (Archer 1984). Based on the results of the present study, none of these hypotheses appear to be particularly viable.

Wynyardia as a phalangeridan?

The basicranial evidence provides compelling grounds upon which to reject any notion of special affinity between *Wynyardia* and phalangeridan diprotodontians. This applies equally for the group as a whole or with regard to one or other specific family (e.g., Phalangeridae).

As documented in detail within the foregoing sections, the great majority of phalangeridans are remarkably uniform in all major features of auditory and wider basicranial anatomy. Insofar as the auditory region is concerned, the following key, derived features may be regarded as synapomorphic for the Phalangerida: completion of the bony auditory bulla by proliferation of an alisphenoid tympanic process (AP) and a caudal tympanic process of the petrosal (CP); development of a deeply invasive squamosal epitympanic sinus (SQS); pneumatisation of the tegmen tympani resulting in dorsal displacement and lateral expansion of this element; establishment of firm osseous contact between the posteromedial surface of the ectotympanic and the posterior surface of the promontory; formation of a tubular outer ear canal which incorporates the postglenoid process of the squamosal; and, placement of the "postglenoid" foramen anterior of the actual postglenoid process.

In addition, all phalangeridans except the species of *Cercartetus, Burramys* and *Tarsipes* share the further derived condition of pneumatisation within the CP (a mastoid epitympanic sinus). Similarly, all phalangeridans except the macropodoids share the additional features of distension of the posttympanic process (through proliferation of marrow-cellules) with resultant fusion of this element with the pars canicularis of the petrosal and relocation of m. tensor tympani at the root of the tegmen tympani.

Compared with this long list of phalangeridan synapomorphies, *Wynyardia* fails to qualify in respect of every point. For each of the characters listed above, *Wynyardia* displays a more plesiomorphic character state. Unless widespread convergences and/or reversions have occurred, *Wynyardia* cannot be accommodated within Phalangerida. This in turn effectively cancels any possibility of special affinity with the phalangerid possums.

The conclusion that *Wynyardia* is not specially related to phalangerid possums is in conflict with Haight and Murray's (1981) earlier conclusion that ". . . *Wynyardia's* brain is clearly phalangeroid in external morphology, resembling closely that of the extant phalangerid, *Trichosurus vulpecula*". As noted previously by Archer (1984), the basis of Haight and Murray's conclusion is not readily determined from their paper. In general, however, their analysis appears to be entirely phenetic, i.e., affinity judged solely on degrees of overall similarity, with no attempt made to distinguish between ancestral as against derived features. Given this fact, an important question is whether the "close resemblance" between the brains of *Wynyardia* and *Trichosurus vulpecula* might not be due solely to their having both retained a fairly generalised "diprotodontian" type of brain.

Interestingly, where seemingly "special" features of the *Wynyardia* brain are noted by Haight and Murray (e.g., cerebellar shape, parafloccular reduction), the specific similarity is to the vombatomorphians examined (wombats and *Thylacoleo*) rather than to any particular phalangeridan family (Archer 1984).

Wynyardia as a vombatiform?

Somewhat more surprisingly, the basicranial evidence also appears to place *Wynyardia* outside of each of Phascolarctomorphia and Vombatomorphia as currently defined.

In the case of the Phascolarctomorphia, comparisons between the living koala and *Wynyardia* are hindered by the extreme auditory and masticatory specialisations of the living animal. As to whether *Wynyardia* could be regarded as structurally annectant to *Phascolarctos*, the answer would have to be that in terms of its auditory and glenoid structures, it probably is. Taking into account other aspects of its cranial and postcranial anatomy, however, *Wynyardia* may be too specialised in some respects to occupy this position. Moreover, as shall be seen below, many of the features which appear to distinguish *Wynyardia* from the koala also serve to link it (albeit tenuously) with the vombatomorphians.

Turning now to this latter group, vombatomorphians as a group appear to share the following suite of clearly derived auditory features: development of a squamosal tympanic process as the primary bullar element; presence of an entoglenoid extension of the articular eminence; and, elimination of the postglenoid emmissary vein with associated rerouting of dural drainage into the internal jugular vein. Along with a squamosal epitympanic sinus of rather variable relations and likely multiple origins, this suite of features is present in all undoubted vombatomorphians.

In *Wynyardia*, each of these characters appears to be represented by a corresponding plesiomorphic condition: alisphenoid tympanic process probably present; entoglenoid process absent; and patent postglenoid canal present (albeit reduced). On the evidence of the auditory region alone, *Wynyardia* would therefore appear to fall outside of the main vombatomorphian clade. Again, in the absence of considerable homoplaisy, *Wynyardia* cannot be regarded to be specially related to any particular group of vombatomorphians (e.g., vombatids or diprotodontoids).

Despite this formal conclusion, a number of additional observations suggest that *Wynyardia* is probably "closer" to Vombatomorphia than it is to Phalangerida. In the auditory region, I would draw attention to the following features in which *Wynyardia* compares favourably with other vombatomorphians: the mode of attachment of the ectotympanic anterior crus; the well-developed stapedial fossa; the inferred presence of a M. zygomatico-auricularis; and reduction of the postglenoid vein (as a precursor to the vombatomorphian condition). Elsewhere within the cranium, features which appear to link *Wynyardia* specifically to vombatomorphians include: the reduced paraflocculus (Jones 1931; Archer 1984); the elaborate pterygoid complex of the sphenoid and correspondingly reduced neurocranial pterygoid fossa (compare Figs 2, 5, 6; Aplin in prep.); establishment of an extensive squamosal-frontal contact on the sidewall of the braincase (but see comments in Murray *et. al.* 1987); and enlargement of the premaxillae (Jones 1931). Based on comments in Tedford *et al.* (1977), the postcranial skeleton of *Wynyardia* may also display significant similarities to vombatomorphians.

CONCLUSIONS

The basicranial region of *Wynayardia* is decidedly diprotodont-like, displaying the characteristic, derived features of an expanded SQW in the roof of the tympanic cavity and a characteristic "complex" glenoid morphology. These features aside, however, the basicranial region in *Wynyardia* is remarkably plesiomorphic and generally unspecialized, particularly with regard to the complete absence of secondary pneumatic activity in the primary tympanic cavity and the lack of epitympanic sinus development.

In terms of infraordinal affinities, *Wynyaridia* appears to lack the key, derived features which characterise each of the three major clades of diprotodontians (i.e., Phalangerida, Vombatomorphia and Phascolarctomorphia). Various lesser features of the auditory region and of general cranial anatomy, however, indicate that *Wynyardia* may be a vombatiform and perhaps the primitive sister taxon of Vombatomorphia.

ACKNOWLEDGEMENTS

The work reported herein was undertaken in the School of Zoology, University of New South Wales and was supported by a Commonwealth Postgraduate Award. Travel to Hobart to examine the holotype of *Wynyardia bassiana* at the Tasmanian Museum and Art Gallery was made possible by a travel grant from the Australian Museum Trust. For permission to examine the specimen in detail, and for assistance with photography I thank Dr N. Kemp, Curator of Fossils at the Tasmanian Museum. Assoc. Prof. M. Archer and Henk Godthelp provided useful comments on the manuscript. H. Godthelp and L. Worrall also gave considerable assistance with preparation of figures and plates.

REFERENCES

APLIN, K. P. AND ARCHER, M., 1987. Recent advances in marsupial systematics with a new syncretic classification. Pp. xv-lxxii *in* "Possums and opossums: studies in evolution" ed by M. Archer. Surrey Beatty & Sons and the Royal Zoological Society of New South Wales: Sydney.

APLIN, K. P. AND RICH, T. H. V., 1985. *Wynyardia bassiana* Spencer, 1901. The wynyard marsupial. Pp. 219-24 *in* "Kadimakara. Extinct vertebrates of Australia" ed by P. V. Rich and G. H. van Tets. Pioneer Design Studio: Melbourne.

ARCHER, M., 1976. The basicranial region of marsupicarnivores (Marsupialia), interrelationships of carnivorous marsupials, and affinities of the insectivorous peramelids. *Zool. J. Linn. Soc. Lond.* **59:** 217-322.

ARCHER, M., 1982. A review of Miocene thylacinids (Thylacinidae, Marsupialia), the phylogenetic position of the Thylacinidae and the problems of apriorisms in character analysis. Pp. 445-76 *in* "Carnivorous marsupials" ed by M. Archer. Roy. Zool. Soc. NSW: Sydney.

ARCHER, M., 1984. The Australian marsupial radiation. Pp. 633-808 *in* "Vertebrate zoogeography and evolution in Australasia" ed by M. Archer and G. Clayton. Hesperian Press: Perth.

ARCHIBALD, J. D. 1977. Ectotympanic bone and internal carotid circulation of eutherians in reference to anthropoid origins. *J. Human Evol.* **6:** 609-22.

ARCHIBALD, J. D., 1979. Oldest known eutherian stapes and a marsupial petrosal bone from the Late Cretaceous of North America. *Nature, Lond.* **281:** 669-70.

BARBOUR, R. A., 1963. Musculature and limb plexuses of *Trichosurus vulpecula. Aust. J. Zool.* **11:** 488-610.

CLEMENS, W. A., 1966. Fossil mammals of the type Lance Formation, Wyoming. Part II. Marsupialia. *Univ. Calif. Publs. Geol. Sci.,* **62.**

COPE, E. D., 1880. On the foramina perforating the posterior part of the squamosal bone of the Mammalia. *Proc. Amer. Phil. Soc.* **18:** 452-61.

FLANNERY, T. F., 1987. The relationships of the macropodoids (Marsupialia) and the polarity of some morphological features within the Phalangeriformes. Pp. 741-47 *in* "Possums and opossums: studies in evolution" ed by M. Archer. Surrey Beatty & Sons and the Royal Zoological Society of New South Wales: Sydney.

FLANNERY, T. F., ARCHER, M. AND MAYNES, G. M., 1987. The Phylogenetic relationships of living Phalangerids (Phalangeroidea: Marsupialia) with a suggested new taxonomy. Pp. 477-506 *in* "Possums and opossums: studies in evolution" ed by M. Archer. Surrey Beatty & Sons and the Royal Zoological Society of New South Wales: Sydney.

FLEISCHER, G., 1978. Evolutionary principles of the mammalian middle-ear. *Adv. Anat., Embryol., Cell Biol.* **55(5).**

GILL, E. D., 1957. The stratigraphical occurrence and palaeoecology of some Australian Tertiary marsupials. *Mem. Nat. Mus. Vic.* **21:** 135-203.

GREGORY, W. K., 1910. The orders of mammals. *Bull. Amer. Mus. Nat. Hist.* **27:** 3-524.

HAIGHT, J. R. AND MURRAY, P. E., 1981. The cranial endocast of the early Miocene marsupial, *Wynyardia bassiana:* an assessment of taxonomic relationships based upon comparisons with recent forms. *Brain, Behav. Evol.* **19:** 17-36.

HIIEMAE, K. AND JENKINS, F. A., 1969. The anatomy and internal architecture of the muscles of mastication in *Didelphis marsupialis. Postilla* **140.**

HOFER, H., 1952. Uber das gegenwartige Bild der Evolution der Beuteltiere. *Zool. Jahrb. Abt.* **272:** 365-437.

JONES, F. W., 1931. A re-examination of the skeletal characters of *Wynyardia bassiana,* an extinct Tasmanian marsupial. *Pap. Proc. Roy. Soc. Tasm.* **1930:** 96-115.

KAMPEN, P. N. VAN, 1905. Die Tympanalgegend des Saugetierschadels. *Morph. Jahrb.* **34:** 321-722.

KIRSCH, J. A. W., 1977. The comparative serology of Marsupialia and a classification of marsupials. *Aust. J. Zool suppl. Ser.* **No. 52.**

KLAAUW, C. J. VAN DER, 1931. The auditory bulla in some fossil mammals with a general introduction to this region of the skull. *Bull. Amer. Mus. Nat. Hist.* **62:** 1-352.

LIGHTOLLER, G. H. S., 1939. Probable homologues: A study of the comparative anatomy of the mandibular and hyoid arches and their musculature. *Trans. zool. Soc. Lond.* **24:** 349-444.

LUBOSCH, W., 1907. Das Kiefergelenk der Edentaten und marsupialier. *Semon's Zool. Forsch.* **4:** 519-56.

MACINTYRE, G. T., 1972. The trisulcate petrosal pattern of mammals. Pp. 272-303 *in* "Evolutionary biology" vol. 6 ed by T. Dobzhansky, M. K. Hecht and W. C. Steere. Appleton-Century-Crofts: New York.

MACPHEE, R. D. E., 1981. Auditory regions of primates and eutherian insectivores. Morphology, ontogeny and character analysis. *Contrib. Primat.* **18.**

MARSHALL, L. G., 1977. Cladistic analysis of borhyaenoid, dasyuroid and thylacinid (Marsupialia: Mammalia), affinity. *Syst. Zool.* **26:** 410-25.

MURRAY, P., WELLS, R. AND PLANE, M. D., 1987. The cranium of the Miocene thylacoleonid, *Wakaleo vanderleuri:* Click go the shears — a fresh bite at thylacoleonid systematics. Pp. 433-66 *in* "Possums and opossums: studies in evolution" ed by M. Archer. Surrey Beatty & Sons and Royal Zoological Society of New South Wales: Sydney.

OWEN, R., 1841. On the osteology of the Marsupialia. *Trans. zool. Soc. Lond.* **2:** 379-408.

PLEDGE, N. S., 1987. *Muramura williamsi,* a new genus and species of ?Wynyardiid (Marsupialia: Vombatoidea) from the Middle Miocene Etadunna Formation of South Australia. Pp. 393-400 *in* "Possums and opossums: studies in evolution" ed by M. Archer. Surrey Beatty & Sons and the Royal Zoological Society of New South Wales: Sydney.

PRESLEY, R., 1979. The primitive course of the internal carotid artery in mammals. *Acta anat.* **103:** 238-44.

REIG, O. A., KIRSCH, J. A. W. AND MARSHALL, L. G., 1987. Systematic relationships of the Living and Neocenozoic American "Opossum-like" marsupials (Suborder Didelphimorphia), with comments on the classification of these and of the Cretaceous and Paleogene New World and European Metatherians. Pp. 1-89 *in* "Possums and opossums: studies in evolution" ed by M. Archer. Surrey Beatty & Sons and the Royal Zoological Society of New South Wales: Sydney.

REIG, O. A. AND SIMPSON, G. G., 1972. *Sparassocynus* (Marsupialia, Didelphidae), a peculiar mammal from the late Cenozoic of Argentina. *J. Zool., Lond.* **167:** 511-39.

RICH, T. H. V. AND ARCHER, M., 1979. *Namilamadeta snideri,* a new diprotodontan (Marsupialia, Vombatoidea) from the medial Miocene of South Australia. *Alcheringa* **3:** 197-208.

RIDE, W. D. L., 1964. A review of Australian fossil marsupials. *J. Proc. Roy. Soc. West. Aust.* **47:** 97-131.

SEGALL, W., 1969a. The middle ear region of *Dromiciops. Acta anat.* **72:** 489-501.

SEGALL, W., 1969b. The auditory ossicles (malleus, incus) and their relationships to the tympanic: in marsupials. *Acta anat.* **73:** 176-91.

SEGALL, W., 1970. Morphological parallelisms of the bulla and auditory ossicles in some insectivores and marsupials. *Fieldiana, Zool.* **51:** 169-205.

SEGALL, W., 1971. The auditory region (ossicles, sinuses) in gliding mammals and selected representatives of non-gliding genera. *Fieldiana Geol.* **58:** 27-59.

SIMPSON, G. G., 1970. the Argyrolagidae, extinct South American marsupials. *Bull. Mus. Comp. Zool.* **139:** 1-86.

SPENCER, B., 1901. A description of *Wynyardia bassiana*, a fossil marsupial from the Tertiary beds of Table Cape, Tasmania. *Proc. zool. Soc. Lond.* **1900:** 776-94.

STIRTON, R. A., 1967. Diprotodontidae from the Ngapakaldi fauna, South Australia. *Bur. Min. Resour. Aust. Bull.* **85:** 1-44.

TANDLER, J., 1899. Zur vergleichenden Anatomie der Kopfarterien bei den Mammalia. *Denkschr. kais. Akad. Wiss., Wein, Math. nat. Klasse* **67:** 677-784.

TEDFORD, R. H., ARCHER, M., BARTHOLOMAI, A., PLANE, M. D., PLEDGE, N. S., RICH, T. H., RICH, P. V. AND WELLS, R. T., 1977. The discovery of Miocene vertebates, Lake Frome area, South Australia. *Bur. Min. Resour. J. Aust. Geol. Geophys.* **2:** 53-57.

TURNBULL, W. D., 1970. Mammalian masticatory apparatus. *Fieldiana Geol.* **18:** 147-356.

TURNBULL, W. D. AND SEGALL, W., 1984. The ear region of the marsupial sabertooth, *Thylacosmilus*: influence of the sabertooth lifestyle upon it, and convergence with placental sabertooths. *J. Morph.* **181:** 239-70.

WERNER, R. N., 1964. Der Schadel des Beutelbaren. (*Phascolarctos cinereus* Goldfuss 1819) und seine umformung durch lufthaltige Nebenhohlen. *Abhandl. Deutsch. Akad. Wissen., Berlin. Klasse Chem., Geol. Biol.* **T1 4.**

WINGE, H., 1941. "The interrelationships of the mammalian genera. Vol. 1. Monotremata, Marsupialia, Insectivora, Chiroptera, Edentata". (A translation by Deichmann and Allen of "Pattedyrslaegter", 1923.) C. A. Reitzels Forlag: Copenhagen.

WOODBURNE, M. O., 1967. The Alcoota Fauna, Central Australia: an integrated palaeontological and geological study. *Aust. Bur. Min. Resources Bull.* **87.**

WOODBURNE, M. O., 1984. Families of marsupials: relationships, evolution and biogeography. Pp. 48-71 *in* "Mammals: notes for a short course" ed by T. W. Broadhead. *Univ. Tennesee Dept. Geol. Sci. Studies in Geol.* **8.**

WOODS, J. T., 1956. The skull of *Thylacoleo carnifex. Mem. Qd Mus.* **13:** 125-40.

Muramura williamsi, A NEW GENUS AND SPECIES OF ?WYNYARDIID (MARSUPIALIA: VOMBATOIDEA) FROM THE MIDDLE MIOCENE ETADUNNA FORMATION OF SOUTH AUSTRALIA

NEVILLE S. PLEDGE[1]

Two virtually complete skeletons, buried together in the Etadunna Formation at Lake Palankarinna, northern South Australia, represent a new species of marsupial tentatively referred to the extinct family Wynyardiidae. The dentition is described here and the animal named *Muramura williamsi* n. gen, n. sp. Teeth and cranial features support a relationship with *Namilamadeta snideri* and *Wynyardia bassiana*, and a more distant kinship with vombatids.

Key Words: *Muramura williamsi*; Wynyardiidae; Miocene; Etadunna Fm; Palankarinna; South Australia.

Pages 393-400 *in* POSSUMS AND OPOSSUMS: STUDIES IN EVOLUTION ed by M. Archer. Surrey Beatty & Sons and the Royal Zoololgical Society of New South Wales: Sydney, 1987.

INTRODUCTION

*W*YNYARDIA BASSIANA is the name given by Spencer (1901) to a specimen found prior to 1876 in a fallen block of marine Fossil Bluff Sandstone at Table Cape, near Wynyard, Tasmania. It was a well-fossilised partial skeleton in which, unfortunately, erosion had removed the ventral part of the skull including all trace of teeth. Besides early controversy regarding its true age, now regarded as basal Miocene (Quilty 1966), the species has also been placed in several taxonomic positions (Gill 1957; Jones 1930; Osgood 1921; Ride 1964).

In 1973 partial skeletons were found in the Miocene Namba Formation at Lake Pinpa in the Tarkarooloo Basin of northeastern South Australia. One of these (with a skull) was tentatively identified as a wynyardiid (Tedford *et al.* 1977) based on the similarity of its limb bones with *Wynyardia bassiana*. One of the upper molars is figured by Tedford *et al.* (1977, fig. 3A).

The slightly younger Tarkarooloo local fauna, from the Namba Formation in Tom O's Quarry at Lake Tarkarooloo, yielded a further specimen in 1976, named *Namilamadeta snideri* (Rich and Archer 1979). It was related to *Wynyardia* on the basis of a few skull features such as the anterior position of the infraorbital foramen, and to the tooth in figure 3A of Tedford *et al.* (1977) because of some dental characteristics (incorporation of separate stylar cusps into the buccal ends of the lophs).

In 1983, one of my volunteers, David Williams, found a small pile of bone fragments at Lake Palankarinna. On excavation these proved to be from one skull of a double burial of two virtually complete skeletons. The teeth are similar to both those of the Lake Pinpa specimen QMAM 178 and of *Namilamadeta snideri*.

MATERIALS AND METHODS

The following description is based upon the dentary and maxilla found weathering out of the sediment and the skull of the more complete specimen excavated adjacent to them. A fuller description of the skull and postcranial skeleton will be given later after preparation has been completed. Preparation has been hampered by the shattered nature of the bones which were broken by repeated expansion and contraction of the clay matrix just below the weathered zone.

Comparisons have been made with the holotype of *Namilamadeta snideri* Rich and Archer 1979, SAM Pl9951, a cast of an undescribed species provided by Dr R. H. Tedford (American Museum of Natural History) and a cast of the skull of *Wynyardia bassiana*.

Stereoscopic photographs of the teeth were made using the Microflex AFX photomicrographic attachment on a Nikon SMZ-10 stereo microscope. Measurements were made with a calibrated ocular graticule. The specimens are in the palaeontological collections of the South Australian Museum (SAM). NMV refers to specimens in the Museum of Victoria.

SYSTEMATICS

Superorder Marsupialia Illiger, 1811

Order Diprotodonta Owen, 1866

Superfamily Vombatoidea (Iredale and Troughton, 1934)

Family Wynyardiidae Osgood, 1921

Genus *Muramura* nov.

[1]South Australian Museum, Adelaide, South Australia, Australia 5000.

Type species: Muramura williamsi gen. et sp. nov.

Known distribution: Medial Miocene; Lake Eyre Basin (and Tarkarooloo Basin), northern South Australia.

Generic Diagnosis: Distinguished from *Wynyardia bassiana* by its larger size, relatively shorter and deeper snout, slightly flatter cranium, broader frontals, stronger supraorbital crest. Differs from *Namilamadeta snideri* in its smaller size, presence of a small suborbital fossa in the maxilla for insertion of the masseter muscle, enlarged masseteric process on jugal extending below occlusal plane of molars; upper premolar shorter and broader with hypocone less developed, M^2-M^4 smaller and relatively shorter, M^5 larger with better developed posterior moiety, paracone and metacone more medial on the lophs, stylar cusps B and C less markedly separated, canine absolutely larger, I^2 and I^3 relatively smaller. I^1 similar to that of *Namilamadeta snideri* but smaller.

Etymology: In the Diari religion, "Muramura" signifies a "demigod" (Reuther 1981, III, 29. No. 1522) who was one of the primeval ancestors of the Human race. Each "Muramura" originated from a clod of earth and emerged at a particular spot, giving rise to legends and naming features and living creatures in that area. The name is applied here in allusion to the rather central ancestral position of this genus with respect to other vombatoids. The gender is masculine.

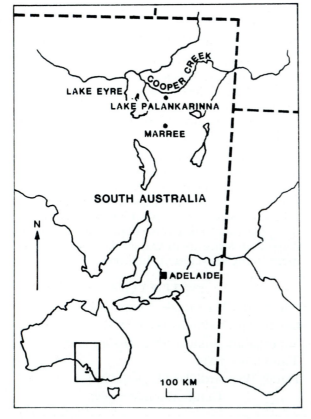

Fig. 1. Locality map.

Muramura williamsi sp. nov.

Holotype: SAM P24525, a complete skeleton with articulated skull and jaws.

Paratype: SAM P24524, an almost complete skeleton with damaged skull (lacking cranium) and jaws.

Type locality: 100 m southeast of Tedford Locality (UCMP loc. V5375), Lake Palankarinna, 100 km north of Marree, South Australia (Fig. 1). The type material comes from an horizon in the Etadunna Formation (Stirton, Tedford and Miller 1981) about 3 m below V5375.

Age: Medial Miocene, about 15 million years (Archer and Bartholomai 1978), Ditjimanka Local Fauna (Woodburne 1986).

Diagnosis: That of the genus until other species are described.

Etymology: Named for the discoverer, David W. Williams, volunteer on several expeditions and camp cook *par excellence*.

Description (Figs 2-4): Until the skulls and skeletons of the holotype and paratype are finally prepared, only a brief description of the holotype skull will be given here. Because the jaws of the holotype skull are clenched, description of the teeth is based upon those of part of the paratype. See also Table 1 for skull measurements.

Skull (Fig. 3): The skull is about the size of that of *Phascolarctos cinereus*, larger than *Wynyardia bassiana* and smaller than *Namilamadeta snideri*. It is relatively high and narrow, with a deep face and rather short muzzle. The highest point is at the anterior end of the sagittal crest. The lambdoidal crest is strong and overhangs the occiput markedly. The sagittal crest profile is convex, the anterior dorsal profile straight. Strong supraorbital crests on the frontals extend back to the sagittal crest. The zygomatic width is slightly more than half the skull length and less than twice the skull height. The zygomatic arch is deep posteriorly and slopes inwards dorsally.

From the anterior part of the zygomatic arch a prominent masseteric process extends downwards, beyond the occlusal plane of the molars, and relatively further than in *Namilamadeta snideri*. Such a process was probably present in *Wynyardia bassiana* (Spencer 1901). Just over half-way down the malar process of the maxilla in *Muramura*, at the root of the zygoma, there is a small but distinct fossa for the insertion of an enlarged masseteric muscle. This feature is totally lacking in *Namilamadeta snideri*. Although there is no indication of its presence in *Wynyardia bassiana*, the bone has been broken off dorsal to that position in that unique specimen.

There is a dorsoventrally elongate infraorbital foramen in *M. williamsi* just above the root of

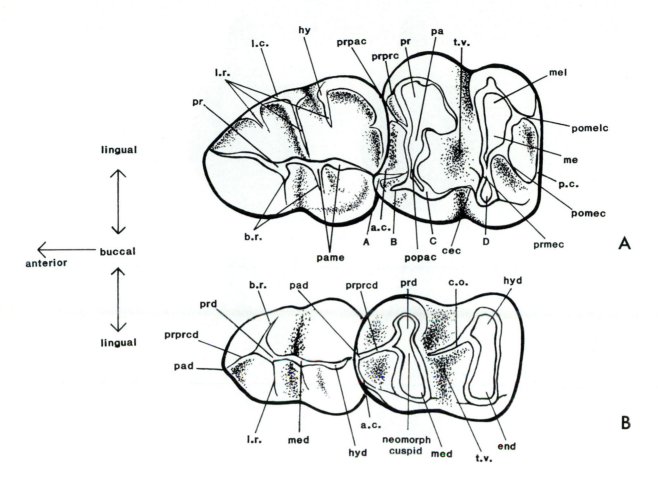

Fig. 2. Muramura williamsi Sketch, illustrating terms used. A, upper right premolar and first molar (M^2). B, lower right premolar and first molar (M_2). Abbreviations: A, "stylar cusp A"; a.c., precingulum; B, stylar cusp B: b.r., buccal rib; C, stylar cusp C; cec, centrocrista; c.o., cristid obliqua; D, stylar cusp D; end, entoconid; hyd, hypoconid; l.c., lingual cingulum; l.r., lingual rib; me, metacone; med metaconid; mel, metaconule; pa, paracone; pad, paraconid; pame, parametacone; p.c., postcingulum; pohyc, posthypocrista; pomec, postmetacrista; pomelc, postmetaconulecrista; popac; postparacrista; pr, protocone; prd, protoconid; prmec, premetacrista; prpac, preparacrista; prprc, preprotocrista; prprcd, preprotocristid; t.v., transverse valley.

P^3 and level with the masseteric fossa. This character seems typical of wynyardiids, seen in *Namilamadeta snideri* and indicated in *Wynyardia bassiana*, as well as vombatids. In contrast, this feature is smaller, round and high up on the malar process of the maxilla in *Phascolarctos*.

Dentary (Fig. 4): The mandibles are rather deep, with anteroposteriorly broad ascending rami, whose leading edges are at about 80° to the occlusal surface and have no masseteric fossae. The dentary has a large, wombat-like angular process rather similar to the preserved fragment of *Wynyardia bassiana* (Spencer 1901), but does not have the marked mid-ventral ridge on the corpus of the dentary that the latter shows. Dentaries are not fused at the symphysis, which is fairly long (about 25 mm).

Dentition (Figs 2-4): Description of the teeth is based primarily upon the right maxilla and dentary of the paratype (P24524). Unfortunately, these teeth are rather heavily worn.

They exhibit a wear pattern similar to *Namilamadeta snideri*. The adult dental formula is: I_1^{1-3}, C_0^1, P_3^3, M_{2-5}^{2-5}. The incisors cannot, at this stage, be seen clearly. They are all smaller than those of *Namilamadeta snideri*.

The I^1 is by far the largest. It is semicircular to elliptical in cross section, with a diameter of 5.5 mm (anteroposteriorly) and is strongly curved with a radius of about 17 mm, passing up and back over the roots of I^2 and I^3. The enamel extends back to the root. It cannot yet be determined whether this tooth tapers towards the root as in *Namilamadeta snideri*.

The I^2 is the smallest tooth, slightly less than I^3. It is peg-like and about 3 mm in diameter.

The I^3 is also peg-like, with a greater diameter of about 3.2 mm.

The C^1 is a small bluntly-rounded subconical tooth, laterally compressed with a somewhat

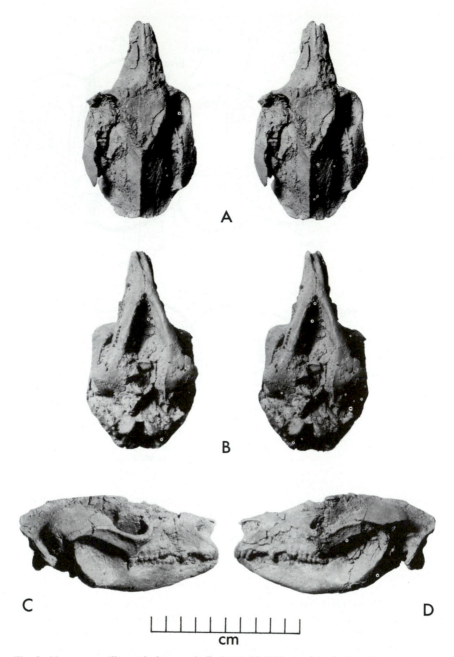

Fig. 3. Muramura williamsi holotype skull, SAM P24525. A, dorsal view; B, ventral view — note atlas vertebra which has slipped forwards into the throat cavity; C, right lateral view; D, left lateral view. Scale in centimetres.

sharper crest on the posterior side than on the anterior. It is slightly larger but similar to that of *N. snideri*.

The P_3 is a relatively large, sub-ovate to subtriangular tooth with buccal and lingual emarginations has a primary crest extending the full length, slightly buccal of the mid-line. A lingual cingulum extends around the tooth from one end of the crest to the other with a notch and discontinuity a little more than halfway along, level with the central cusp. The crest supports three sub-equal cusps. The anterior one is just before the mid-point where there is a deep transverse fissure on the buccal side and a somewhat shallower one directed slightly forwards on the lingual side. The other two cusps — the para-metacone — are closely placed, the third one being at about the three-quarter mark. Strong buccal ribs extend towards the base of the crown from the apices of all cusps. On the lingual side these ribs are somewhat truncated apically by wear but they extend to the cingulum. Each rib terminates in a small cuspule which is least for the anterior rib and largest for the third, where it is interpreted as the hypocone. This tooth is much smaller than that in *N. snideri*.

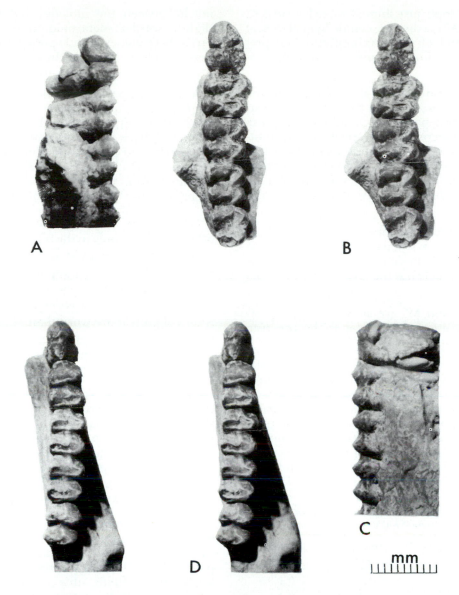

Fig. 4. Muramura williamsi paratype, SAM P24524. A, right maxilla in lateral view; B, right maxilla in occlusal view (stereopair); C, right dentary in lateral view; D, right dentary in occlusal view (stereopair). Scale in millimetres.

Table 1. Measurements (mm) of skull and jaws of *Muramura williamsi* (SAM P24525).

maximum length of skull	141
length, anterior edge of I^1 to occipital condyle	133
length, I^1 to anterior edge of zygomatic arch	52
length, anterior edge of zygomatic arch to lamboidal crest	91
length I^3 to P^3	25
height of premaxilla	c.17
maximum width of zygomatic arches	83
maximum width of nasals	c.27
maximum width of frontals	31
maximum width of occiput	60
maximum height of skull	54
maximum length of dentary to tip of I_1	110
maximum depth of dentary, at M_3	23
length of diastema I_1 to P_3	18
width of angular shelf, normal to sagittal plane	24
length of symphysis	25

The M^2 as with the other molars, is functionally bilophodont at the advanced stage of wear exhibited. There are six major cusps evenly spaced in two transverse lines, with the paracone and metacone just buccal of the midpoint. The anterior row comprises the protocone, paracone and stylar cusp C, while the posterior row contains the metaconule (Tedford and Woodburne 1986) or hypocone of earlier authors, metacone and stylar cusp D. The posterior row is straight, but stylar cusp C is slightly displaced posteriorly from a line extended from protocone and paracone. This displacement is less than in *N. snideri*. Cusps of the anterior row are linked by the protoloph and postparacrista, those of the posterior row by the metaloph and premetacrista. Stylar cusp B at the anterobuccal corner of the tooth is linked by a shorter spur to a point on the anterior loph just lingual to stylar cusp C. "Stylar cusp A" is a discrete conule lingual to B, at the buccal end of the precingulum. The precingulum passes lingually into a weak preprotocrista.

A posterior ridge from stylar cusp D to the posterobuccal corner of the tooth joins the postcingulum which passes lingually into the weak postmetaconulecrista. A relatively strong postmetacrista extends towards but does not reach it. The postprotocrista and premetaconulecrista meet across the transverse valley as a low median swell. A fine centrocrista links stylar cusps C and D.

Centrally, the transverse valley is relatively broad; buccally, it is narrower and deeper but round bottomed; lingually it is a very narrow crevice that initially has the appearance of a fracture.

Molars M^3 and M^4 are similar but the postmetacrista is much weaker, as is the precrista from stylar cusp B.

The M^5 differs in being trapeziform, the metaloph being about ⅔ the length of the protoloph. The metacone is a more discrete cusp on the metaloph; the postcingulum forms a basin between the postmetaconulecrista and the postcrista from stylar cusp D; and the precingulum is well developed between the preprotocrista and stylar cusp A. The "parastyle" (stylar cusp B) has disappeared. Measurements are given in Table 2.

Table 2. Measurements (mm) of upper cheek teeth of *Muramura williamsi* (SAM P24524, paratype).

	Length	Anterior width	Posterior width
P^3	8.8	6.3	7.1
M^2	8.0	8.4	8.1
M^3	7.5	8.0	7.5
M^4	7.5	7.3	6.6
M^5	7.0	6.4	5.2
P^3-M^5	37.9	—	—

The I_1 in the holotype (P24525) is seen to be a large, gently curving tooth, apparently not tapering, narrowly elliptical in cross section and heavily worn on a truncated occlusal surface which is the length of the combined upper incisors. It is not yet possible to determine the distribution of enamel, but it seems to extend from the convex outer surface anteroventrally onto the flatter medial surface.

The P_3 is almost as large as the succeeding molars. The anterior moiety is parallel-sided at the level of the protoconid, with a width equal to its length; the posterior moiety is shorter, wider and inflated. A major longitudinal crest, just lingual of the midline, starts at the protoconid, continues back sinuously across the intervening

notch between the moieties and through the slightly smaller metaconid to the hypoconid. From there it goes slightly posterobuccally to the hypoconulid adjacent to the anterior end of the preprotocristid of the first molar.

A weak preprotocristid descends in a slight lingual curve to the weak paraconid at the anterior end of the tooth. There is a posterolingual rib from the protoconid apex, bordering the notch, and a corresponding buccal expansion. The major cusps are (except the paraconid) subequal, with the protoconid marginally the largest and the hypoconid smallest.

Lower molars. These are very similar to each other and differ only in the relative widths of protolophid and hypolophid. In M_2, the protolophid is shorter than the hypolophid, while in M_5 the reverse situation applies; in M_3 and M_4, the lophids are approximately equal. The first molar is only slightly more worn than M_3 and M_4, with full breaching of the lophid enamel. On M_5, only the enamel of the cusps is breached.

The M_2 is low crowned and bilophodont with a weakly developed mid-link (cristid obliqua) and forelink (preprotocristid). The protolophid is slightly narrower than the hypolophid and convergent with it lingually at about 5°. There is, compared with M_3 and M_4, a lingual curvature of the anterior half of the tooth: whereas the mid- and fore-links of M_3 and M_4 align longitudinally, the fore-link on M_2 is noticeably turned inwards.

The protolophid contains three cusps, the buccal-most being the protoconid. The highest cusp is the least worn and is just buccal to the mid-point of the lophid. Following Tedford and Woodburne (1986), this is interpreted as a neomorph, with the metaconid being at the lingual extremity of the lophid. In its unworn state the metaconid probably would have been the tallest cusp on the lophid. From the buccal side of the neomorph cuspid the preprotocristid extends longitudinally at first, then curves gently to meet the anterior cingulum just lingual to the midline of the tooth. It forms a cutting edge that continues with the sectorial longitudinal cristid of the premolar. The buccal face of the preprotocristid has the same slope as that of the premolar. The buccal precingulum descends with it and fades out at the base of the tooth. The lingual precingulum curves backwards and meets the anterior face of the protolophid just below the metaconid.

The hypolophid is transverse, with the highest point at the entoconid. The advanced degree of occlusal wear is such that there is no indication of a cuspid intermediate between entoconid and hypoconid. (Nor do the more posterior molars suggest such a feature.) The hypoconid was apparently rather conical. A low, broad, but well-developed cristid obliqua extends from it, across the transverse valley and up the posterior face of the protolophid. The transverse valley is thus divided into a rather more open, rounded buccal half and a deeper angular lingual part. Weak longitudinal ribs on the cusps at the lingual ends of the lophids tend to make a *de facto* "entolophid".

As stated above, the more posterior molars are essentially the same.

Differences are: lophids are parallel and normal to tooth length, the "paraconid" is apparently not developed and a weak postcingulum develops as a small pocket on the posterolingual corner.

The measurements for the lower cheekteeth are given in Table 3.

Table 3. Measurements (mm) of lower cheek teeth to *Muramura williamsi* (SAM P24524 paratype).

	Length	Anterior width	Posterior width
P_3	6.9	4.5	5.2
M_2	8.0	5.6	6.0
M_3	7.5	5.8	5.9
M_4	7.6	5.8	5.7
M_5	7.8	5.5	5.3
P_3-M_5	36.2	—	—

DISCUSSION

The full significance of *Muramura williamsi* will not be realised until the delicate and fractured otic region has been cleaned and the postcranial skeleton prepared and described. Suffice to say, it was a dog-sized short-tailed herbivore and, on the basis of its ungual phalanges which are not laterally compressed, probably terrestrial but not fossorial. First indications are that the feet are relatively unspecialised. The pes is most similar to that of a wombat: it is slightly smaller and has similar tarsal bones. The metatarsals are relatively longer than in the wombat studied. Syndactyly, as indicated by the reduction in diameter of the second and third metatarsals, was about as well-developed as it is in *Phascolarctos*. The first digit, however, is strong and sturdy, larger than in the wombat yet shorter than in *Phascolarctos*.

While the feet of *Muramura* may be relatively unspecialised, the reverse is true of the dentition. Nearly all trace of the primitive tritubercular cusp pattern has been lost in the molars and they are convergent on diprotodontids and some macropodids in developing relatively simple transverse lophs. This development is also seen in the somewhat younger (Woodburne 1986) Tarkarooloo Local Fauna species *Namilamadeta snideri*, in which the teeth are slightly larger and show clearer and more pronounced stylar cusps, features considered to be primitive (Rich and

Archer 1979). However, the elongate nature of the upper premolar of *Namilamadeta* is deemed to be a derived character (*ibid*) so *Muramura* is more primitive in this respect while its first upper incisor shows derived characteristics as in *Namilamadeta* and wombats.

A pit for the insertion of the masseteric muscle may be a derived character. It can be seen in vombatids and variously developed on some Koalas and some sthenurines (pers. obs.) but only in the last and in *Muramura williamsi* is it found in combination with a prominent masseteric process. This pit is not seen in *Namilamadeta snideri* and cannot be detected in *Wynyardia bassiana* because the maxilla has been broken away just above that position.

Following the list in Rich and Archer (1979), *Muramura williamsi* shares fourteen character states with *Namilamadeta snideri*, including six derived features: enamel distribution, relative size and cross-sectional shape of I^1, reduced C^1, and position and shape of the infraorbital foramen. It is more derived than *Namilamadeta* in development of stylar cusps B and C, in having a pit for masseteric muscle insertion, and possibly also in the I_1 diameter and upper molar abrasion pattern. *Muramura* is more primitive in proportions and lack of a prominent lingual cusp in P^3 and in the more central position of the paracone and metacone on the lophs of M^3.

The wynyardiid from Lake Pinpa, briefly mentioned and figured by Tedford *et al.* (1977), seems to be a species of *Muramura*. It shares one primitive and two derived characters with *M. williamsi* with respect to *Namilamadeta*, viz. position of the paracone and metacone on M^3, development of stylar cusps B and C and the abrasion pattern on the upper molars. All share the primitive features of stylar cusp D size and incipient transverse lophs on the upper molars.

Muramura williamsi shares with *Wynyardia* and *Namilamadeta* the characters of narrow nasal bones, infraorbital foramen position and shape and well developed masseteric process. It shares with vombatids characters of I^1, a deep premaxilla, features of the infraorbital foramen, development of the masseteric pit and (with *Wynyardia* also) a broad angular process on the dentary. However, the brachydonty of molars in *Muramura* argues against a close relationship with vombatids which were taxonomically distinct by Tarkarooloo Local Fauna times as evidenced by a hypsodont molar, NMV P48996 (Rich and Archer 1979).

The taphonomic significance of these specimens is noteworthy. The Etadunna Formation at Lake Palankarinna has been interpreted as being deposited in a near-lake shore/flood-plain situation where animal remains were subject to considerable disturbance by wading birds, scavengers and crocodiles. Until discovery of the specimens described here, no articulated skeletons had been recorded from the extensive outcrops at Lake Palankarinna, although articulated and partially articulated partial skeletons of *Ngapakaldia* spp. (Stirton 1967) had been found at Lakes Pitikanta and Ngapakaldi to the north. Exceptional conditions of carcase protection and/or burial are therefore indicated here.

No other fossils, except a surface scatter of isolated fish bones, were found at this site. The compact nature of the burial, with the two skeletons curled together and barely disrupted, suggests at least two possible modes of death: a) the secretion of two carcases for later consumption in a crocodile wallow, or b) the catastrophic drowning and burial *in situ* of two animals sleeping in a hollow. More evidence for the mode of preservation may arise as the skeletons are excavated from the enclosing matrix.

CONCLUSIONS

On the basis of dental similarity with *Namilamadeta snideri* (Rich and Archer 1979) and the Lake Pinpa specimen whose limb-bones are *Wynyardia*-like (Tedford *et al.* 1977), *Muramura williamsi* is possibly a wynyardiid diprotodont. It is structurally close to the origin of the vombatids, but too late to be regarded as ancestral to them. There is a close but probably not ancestral relationship to *Namilamadeta snideri*.

ACKNOWLEDGEMENTS

I thank Dave Williams for his enthusiastic help in the field. Fruitful discussion was had with Drs M. Archer and M. O. Woodburne who both encouraged me to write this description. Dr R. H. Tedford provided a cast of the undescribed specimen from the Tarkarooloo Basin and critical comment on an earlier draft. Manuscripts were typed by Mrs Joan Murphy, Janine Flory and Diana Massacci. Figures 1 and 2 were drawn by Kathy Bowshall.

REFERENCES

ARCHER, M. AND BARTHOLOMAI, A., 1978. Tertiary mammals of Australia: a synoptic review. *Alcheringa* **2**: 1-19.

GILL, E. D., 1957. The stratigraphical occurrence and palaeoecology of some Australian Tertiary marsupials. *Mem. Nat. Mus. Vict.* **21**: 135-203.

JONES, F. W., 1930. A re-examination of the skeletal characters of *Wynyardia bassiana*, an extinct Tasmanian marsupial. *Pap. Proc. R. Soc. Tasm.* **1930**: 96-115.

OSGOOD, W. H., 1921. A monographic study of the American marsupial, *Caenolestes. Publ. Field Mus. Nat. Hist., Zool. Ser.* **14**: 1-156.

QUILTY, P. G., 1966. The age of Tasmanian marine Tertiary rocks. *Aust. Jour. Sci.* **29**: 143-44.

REUTHER, J. G., 1981. The Diari. (Trans. P. A. Scherer 1974). *Aust. Inst. Abor. Studies, Microfiche* No. 2.

RICH, T. H. V. AND ARCHER, M., 1979. *Namilamadeta snideri,* a new diprotodontan (Marsupialia: Vombatoidea) from the medial Miocene of South Australia. *Alcheringa* **3**: 197-208.

RIDE, W. D. L., 1964. A review of Australian fossil marsupials. *Jour. R. Soc. W. Aust.* **47**: 97-134.

SPENCER, G., 1901. A description of *Wynyardia bassiana,* a fossil marsupial from the Tertiary beds of Table Cape, Tasmania. *Proc. Zool. Soc. Lond.* **1900**: 776-94.

STIRTON, R. A., 1967. The Diprotodontidae from the Ngapakaldi Fauna, South Australia. *Bur. Min. Resour. Aust. Bull.* **85**: 1-44.

STIRTON, R. A., TEDFORD, R. H. AND MILLER, A. H., 1961. Cenozoic stratigraphy and vertebrate palaeontology of the Tirari Desert, South Australia. *Rec. S. Aust. Mus.* **14**: 19-61.

TEDFORD, R. H., ARCHER, M., BARTHOLOMAI, A., PLANE, M., PLEDGE, N. S., RICH, T., RICH, P. AND WELLS, R. T., 1977. The discovery of Miocene vertebrates, Lake Frome area, South Australia. *B.M.R. Jour. Aust. Geol. and Geophys.* **2**: 53-57.

TEDFORD, R. H. AND WOODBURNE, M. O., 1987. The Ilariidae, a new family of Vombatiform Marsupials from Miocene strata of South Australia and an evaluation of the holomogy of molar cusps. Pp. 401-18 *in* "Possums and opossums: studies in evolution" ed by M. Archer. Surrey Beatty & Sons and the Royal Zoological Society of New South Wales: Sydney.

WOODBURNE, M. O., 1986. Biostratigraphy and biochronology. Ch. 6 *in* "Revision of the Ektopodontidae (Mammalia, Marsuplia, Phalangeroidea) of the Australian Neogene" ed by M. O. Woodburne and W. A. Clemens. *Univ. Calif. Publs. Geol. Sci.* (in press).